LONDON MATHEMATICAL SOCIE

47 Introductory lectures on rings and modules. JOHN A. BEACHY
48 Set theory, ANDRÁS HAJNAL & PETER HAMBURGER. Translated by ATTILA MA
49 An introduction to K-theory for C*-algebras, M. RØRDAM, F. LARSEN & N. J. LAUSTSEN
50 A brief guide to algebraic number theory, H. P. F. SWINNERTON-DYER
51 Steps in commutative algebra: Second edition, R. Y. SHARP
52 Finite Markov chains and algorithmic applications, OLLE HÄGGSTRÖM
53 The prime number theorem, G. J. O. JAMESON
54 Topics in graph automorphisms and reconstruction, JOSEF LAURI & RAFFAELE SCAPELLATO
55 Elementary number theory, group theory and Ramanujan graphs, GIULIANA DAVIDOFF, PETER SARNAK & ALAIN VALETTE
56 Logic, induction and sets, THOMAS FORSTER
57 Introduction to Banach algebras, operators and harmonic analysis, GARTH DALES *et al*
58 Computational algebraic geometry, HAL SCHENCK
59 Frobenius algebras and 2-D topological quantum field theories, JOACHIM KOCK
60 Linear operators and linear systems, JONATHAN R. PARTINGTON
61 An introduction to noncommutative Noetherian rings: Second edition, K. R. GOODEARL & R. B. WARFIELD, JR
62 Topics from one-dimensional dynamics, KAREN M. BRUCKS & HENK BRUIN
63 Singular points of plane curves, C. T. C. WALL
64 A short course on Banach space theory, N. L. CAROTHERS
65 Elements of the representation theory of associative algebras I, IBRAHIM ASSEM, DANIEL SIMSON & ANDRZEJ SKOWROŃSKI
66 An introduction to sieve methods and their applications, ALINA CARMEN COJOCARU & M. RAM MURTY
67 Elliptic functions, J. V. ARMITAGE & W. F. EBERLEIN
68 Hyperbolic geometry from a local viewpoint, LINDA KEEN & NIKOLA LAKIC
69 Lectures on Kähler geometry, ANDREI MOROIANU
70 Dependence logic, JOUKU VÄÄNÄNEN
71 Elements of the representation theory of associative algebras II, DANIEL SIMSON & ANDRZEJ SKOWROŃSKI
72 Elements of the representation theory of associative algebras III, DANIEL SIMSON & ANDRZEJ SKOWROŃSKI
73 Groups, graphs and trees, JOHN MEIER
74 Representation theorems in Hardy spaces, JAVAD MASHREGHI
75 An introduction to the theory of graph spectra, DRAGOŠ CVETKOVIĆ, PETER ROWLINSON & SLOBODAN SIMIĆ
76 Number theory in the spirit of Liouville, KENNETH S. WILLIAMS
77 Lectures on profinite topics in group theory, BENJAMIN KLOPSCH, NIKOLAY NIKOLOV & CHRISTOPHER VOLL
78 Clifford algebras: An introduction, D. J. H. GARLING
79 Introduction to compact Riemann surfaces and dessins d'enfants, ERNESTO GIRONDO & GABINO GONZÁLEZ-DIEZ
80 The Riemann hypothesis for function fields, MACHIEL VAN FRANKENHUIJSEN
81 Number theory, Fourier analysis and geometric discrepancy, GIANCARLO TRAVAGLINI
82 Finite geometry and combinatorial applications, SIMEON BALL
83 The geometry of celestial mechanics, HANSJÖRG GEIGES
84 Random graphs, geometry and asymptotic structure, MICHAEL KRIVELEVICH *et al*
85 Fourier analysis: Part I – Theory, ADRIAN CONSTANTIN
86 Dispersive partial differential equations, M. BURAK ERDOĞAN & NIKOLAOS TZIRAKIS

London Mathematical Society Student Texts 85

Fourier Analysis

Part I – Theory

ADRIAN CONSTANTIN
Universität Wien, Austria

CAMBRIDGE
UNIVERSITY PRESS

University Printing House, Cambridge CB2 8BS, United Kingdom

Cambridge University Press is part of the University of Cambridge.

It furthers the University's mission by disseminating knowledge in the pursuit of education, learning, and research at the highest international levels of excellence.

www.cambridge.org
Information on this title: www.cambridge.org/9781107044104

First published 2016

Printed in the United States of America by Sheridan Books, Inc.

A catalogue record for this publication is available from the British Library.

Library of Congress Cataloguing in Publication data
Names: Constantin, Adrian.
Title: Fourier analysis / Adrian Constantin, Universität Wien, Austria.
Description: Cambridge : Cambridge University Press, 2016- |
Series: London Mathematical Society student texts ; 85 |
Includes bibliographical references and index.
Identifiers: LCCN 2016001236 |
ISBN 9781107044104 (hardback : alk. paper) |
ISBN 9781107620353 (pbk. : alk. paper) Subjects: LCSH: Fourier analysis. |
Mathematical analysis. Classification: LCC QA403.5 .C66 2016 |
DDC 515/.2433–dc23 LC record
available at http://lccn.loc.gov/2016001236

ISBN 978-1-107-04410-4 Hardback
ISBN 978-1-107-62035-3 Paperback

To my family

Contents

Preface

Fourier analysis is a central area of modern mathematics, comprising deep results that rely on advanced principles, as well as numerous aspects that require manipulative ingenuity. The power of the theory is illustrated by its wide applicability. Ideas originating in Fourier analysis permeate many essential developments of modern mathematics, bridging analysis with algebra and providing effective tools for an astonishing variety of applications. We list here an alphabetical sample of subjects, illustrating either areas of mathematics with strong links to Fourier analysis or real world applications of Fourier analysis, that are covered briefly in this textbook:[1] acoustics, complex analysis, functional analysis/operator theory, group theory/representation theory, heat flow, hydrodynamics, image processing, medical imaging, number theory, optics and astronomy, partial differential equations, probability and statistics, quantum mechanics, signal processing.

While formal approaches to Fourier analysis can be informative, to appreciate the subject fully and to strengthen the ability to use it in other contexts, one has to acquire a certain mathematical sophistication that draws on measure theory and functional analysis. Lebesgue's integral and the concepts of Hilbert and Banach spaces are intimately connected to Fourier analysis, providing not only an adequate setting but also being useful in obtaining fundamental results, often with surprisingly little effort. A detailed presentation of measure theory and functional analysis would be out of place in an introductory textbook, but

[1] The list is not exhaustive, being only indicative of the relevance of Fourier analysis to pure and applied mathematics. Part I of the textbook covers the theoretical background of Fourier analysis, while Part II is devoted to applications. Since it is impossible to draw a clear dividing line between Fourier theory and applications – for example, these aspects strongly overlap and intertwine in our discussion (in Part II) of the discrete Fourier transform and of the uncertainty principle – the difference between the material presented in the two volumes is more a question of emphasis.

ignoring these topics would amount to a lamentable attempt to run before we have learned to walk. For this reason we outline in Chapters 2 and 3 the principal facts about the Lebesgue integral and Hilbert/Banach spaces, as needed later, emphasising and illustrating the relevant conceptual ideas. This should provide some essential intuition that must, nevertheless, be adequately backed up by analytic rigour, so that we present at least sketchy proofs, avoiding only the proofs that demand an advanced degree of technical versatility. The reader may take on faith the results stated without proof, but detailed references for further study are provided. This material offers, to the interested reader, a basis for a solid grounding in these aspects and has been "class-tested" to groups of graduate students at Lund University and at the University of Vienna (during the academic years 2002–2004 and 2014–2015, respectively). However, the material in Chapters 2 and 3 is not an integral part of a standard course. Depending on the predilections of the lecturer, an all-encompassing or a minimalist point of view can be adopted. In the latter case, one can dispense altogether with measure-theoretic issues and work with the Lebesgue integral as if it were a Riemann integral, with the added bonus of Fatou's lemma and the monotone and dominated convergence theorems. The completeness, separability and density results for the spaces of integrable and square integrable functions can be taken for granted. Chapters 4–6 represent the core of the theory underpinning Fourier analysis, with various applications presented in Part II. Some applications are aimed at pure mathematicians, while others illustrate the relevance of Fourier analysis to physics and engineering. Each application was selected by virtue of its relevance and interest, but each is self-contained: the formulation of the problem is accessible and a full solution is presented. We avoid topics that can be covered only in part within a first course on Fourier analysis. The even distribution of pure and applied topics aims to cater to both mathematical backgrounds. Realistically, only about a third to a half of the applications presented in Part II can be covered in a lecture course. The available flexibility in the specific choice permits a suitable mix of pure and applied topics – the separation of pure and applied topics being, in the long run, detrimental to both areas. Whether the entire Chapter 6 belongs to the basic material on Fourier analysis is a matter of personal opinion, and thus open to debate. Parts of it could be viewed as optional reading material. Chapter 7 contains various selected advanced topics in Fourier analysis, illustrating some of the main directions in which the subject has developed. The material in Chapters 4–6, with the exception of the aspects related to distribution theory in Chapter 6, has been taught by the author as a one-semester course at King's College London during the academic years 2012–2014, while the distribution-theory aspects are an outgrowth of a lecture course on this topic at Trinity College

Dublin during 2008. In recent decades, Fourier analysis has known a period of intense technical and conceptual development which has led to a bewildering array of related topics. Nevertheless, there are a relatively small number of concepts that are commonly regarded as the bare essentials in the theory of Fourier analysis. A minimal list that could constitute a short, introductory course consists of: Section 1.1, Section 3.2, Sections 4.1–4.3, Sections 5.1–5.3, and Exercises 1.1, 3.3, 3.23, 4.1, 4.2, 4.4, 4.5, 4.11, 4.16, 4.19, 5.2, 5.5, 5.9, 5.10, 5.15, 5.16, 5.18. Each chapter is denoted by a numeral (for example, Chapter 5). The first section of the sixth chapter is denoted Section 6.1, and its second subsection is 6.1.2. Theorem 5.3 refers to the third theorem in Chapter 5 (without specification of the section or subsection), while Exercise 5.4 refers to the fourth exercise in Chapter 5. However, within Chapter 5, the 5 may be dropped and Exercise 4 used instead of Exercise 5.4.

The prerequisites are a thorough knowledge of advanced calculus and linear algebra. A large number of exercises are provided, ranging from easy to very hard, and these are supplied separately with hints and full solutions. The exercises are to be regarded as an integral part of the text, and the provided hints and solutions offer flexibility in calibrating the scale of the undertaking – in a tour it is better to admire the scenery at ease rather than to keep on schedule. Whenever the reader struggles with solving an exercise, it is worthwhile glancing at the hint prior to going through the solution that is available. Even if this proves insufficient, it might offer some valuable insight. We strive throughout for a somewhat detailed presentation, at the risk of boring those able to proceed faster. Such readers have the option of judicious skipping. We have tried to prevent our natural fondness for simplicity turning into an excuse to avoid difficulties at any price; quite often, certain difficulties are apparently circumvented rather than escaped altogether. Eventually, they may be encountered again, when they have multiplied, become more involved and hidden in a confusion of detail, which has been generated by lots of misdirected industry. On the other hand, we have tried to steer away from emphasising matters of pure technique – it is all too common to see failures of insight hidden under a blanket of excess technical detail, and focus on detail often leads to a narrowing of perspective. We hope to have found an acceptable balance between doing too little and attempting to do too much.

An appendix with some brief historical notes illustrates the international character of the underlying research efforts, being also indicative of the time needed for the crystallisation of specific concepts and ideas, as well as of their lasting value. There is a similarity between the struggle of early research mathematicians who developed and formalised a topic and the challenges embraced as one embarks upon the study of the topic.

In writing this textbook the author has acted primarily as a reporter, not a researcher: nearly all the results can be found in earlier books or in research publications. We try to offer a coherent exposition, arranging separate topics into a unified whole, and occasionally incorporating some recent developments. While we attempt to give credit where it is due, we also found that this is sometimes difficult or impossible and, as a result, in some instances, secondary sources have prevailed. The reading of parts of this book would be, we believe, beneficial during, or as a preparation for, a graduate school in mathematics – at least, the author wishes he had this material before beginning his own graduate studies.

I owe a debt of gratitude to Roger Astley of Cambridge University Press who encouraged this project from the beginning, being patient and understanding beyond the call of duty. I would like to thank the reviewers of an early and incomplete draft of the book for their constructive suggestions, which I have attempted to incorporate. I am grateful to several mathematicians for reading and commenting on the manuscript and for trying out parts of it on their classes. I cannot name one without naming them all, so they shall remain unnamed to avoid offence to those whose names have escaped me as I attempted to draw up a tentative list of acknowledgements, but they are all deeply appreciated. Despite their best efforts, there are very likely undetected errors that are my sole responsibility and for which I ask the reader to accept my apologies.

1
Introduction

The aim of this chapter is to introduce the concept of Fourier series in an accessible way. We present the analytic setting in which Fourier series arise as the natural generalisation of trigonometric polynomials. We also describe how the problem of the vibrating string and the investigation of heat flow mark the beginning of the theory of Fourier series as a useful approach for solving differential equations of physical relevance. A link between trigonometric polynomials and number theory is also explored.

1.1 Trigonometric polynomials and series

A *trigonometric polynomial* of degree n is an expression of the form

$$p(t) = \sum_{k=-n}^{n} c_k \, e^{2\pi ikt} \tag{1.1}$$

where the c_ks are complex numbers with $|c_{-n}| + |c_n| \neq 0$. Thus p_n is a continuous periodic function of the real variable t, of period 1, determined by its values on $[0, 1)$, or any other interval of length 1. Since

$$\int_0^1 e^{2\pi ikt} \, dt = \begin{cases} 0 & \text{if} \quad k \neq 0, \\ 1 & \text{if} \quad k = 0, \end{cases} \tag{1.2}$$

the constants c_k in the representation (1.1) of the trigonometric polynomial p can be computed by means of

$$c_k = \int_0^1 p(t) \, e^{-2\pi ikt} \, dt, \qquad |k| \leq n. \tag{1.3}$$

The function $e_k(t) = e^{2\pi ikt}$ is sometimes referred to as the *character with frequency k* or as the kth *pure frequency*.

The trigonometric polynomials (1.1) can also be looked at geometrically. Namely, we can interpret the complex number $p(t)$ in (1.1) as the vector sum of its components, each complex number c_k being modified by a supplementary phase $2\pi kt$. In the case of real positive coefficients the visual approach is particularly simple: $p(t)$ is the extremity of a polygonal contour formed by successive straight segments with respective lengths c_k, each one making the same angle $2\pi t$ with the preceding (and following) one. A simple example is depicted in Figure 1.1; for more elaborate examples we refer to the discussion in Lévy-Leblond (1997).

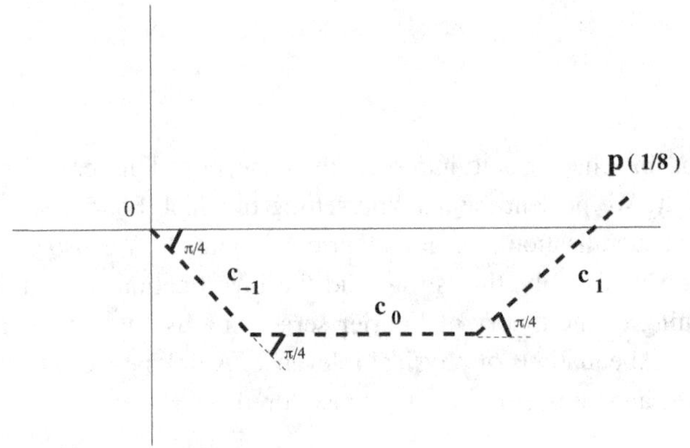

Figure 1.1 The geometric representation of the value at $t = 1/8$ of a trigonometric polynomial $p(t)$ of degree 1 and with positive coefficients.

A fundamental approximation result (to be proved in Chapter 4) is that for any continuous periodic function $f : \mathbb{R} \to \mathbb{C}$ of period 1, given $\varepsilon > 0$, there is a trigonometric polynomial p with

$$|p(t) - f(t)| < \varepsilon, \qquad t \in \mathbb{R}. \tag{1.4}$$

Due to periodicity, it suffices to verify the above inequality for $t \in [0, 1)$.

The role of the multiplicative factor (2π) in the argument of the fundamental trigonometric monomials $e^{2\pi ikt}$ used in (1.1) is to normalise the period to 1. However, given that (1.1) can be expressed as $p(t) = \sum_{k=0}^{n} a_k \cos(2\pi kt) + \sum_{k=0}^{n} b_k \sin(2\pi kt)$, for some $a_k, b_k \in \mathbb{C}$, it is reasonable to wonder why we de not associate the terminology "trigonometric polynomial" with functions of the form

$$q(t) = \sum_{k=0}^{n} \alpha_k \cos^k(2\pi t) + \sum_{k=0}^{n} \beta_k \sin^k(2\pi t) \tag{1.5}$$

for some $a_k, b_k \in \mathbb{C}$. An exercise in trigonometric identities[1] shows that any function of type (1.5)

[1] In this context, it is comforting to know that, see Borzellino and Sherman (2012), polynomial relations between $\cos(2\pi t)$ and $\sin(2\pi t)$ are always consequences of the Pythagorean identity $\cos^2(2\pi t) + \sin^2(2\pi t) = 1$; there are no hidden tricks.

can be written in the form (1.1), with the same value of the degree. However, not all trigonometric polynomials are expressible in the form (1.5): for example, $t \mapsto \sin(4N\pi t)$ with $N \neq 0$ integer are not expressible, see Borzellino and Sherman (2012). For this reason,[2] expressions of the form (1.5) are not enough to approximate well continuous periodic functions of period 1.

The approximation result expressed by means of (1.4) leads us naturally to the concept of a *trigonometric series* or *Fourier series*, defined in analogy to (1.1) as an expression of the form

$$\sum_{k=-\infty}^{\infty} c_k \, e^{2\pi ikt}, \tag{1.6}$$

representing formally a function f of period 1. In light of (1.3), we expect that the constants c_k in (1.6) and the function f are connected by the formula

$$c_k = \int_0^1 f(t) \, e^{-2\pi ikt} \, dt, \qquad k \in \mathbb{Z}. \tag{1.7}$$

More generally, the Fourier series associated to a function $f : \mathbb{R} \to \mathbb{C}$ of period $T > 0$ is

$$\sum_{k=-\infty}^{\infty} c_k \, e^{2\pi ikt/T}, \tag{1.8}$$

where

$$c_k = \frac{1}{T} \int_0^T f(t) \, e^{-2\pi ikt/T} \, dt, \qquad k \in \mathbb{Z}. \tag{1.9}$$

The theory of Fourier series studies the classes of periodic functions (of period $T > 0$) and the notions of convergence appropriate for the correspondence $f(t) \approx \sum_{k=-\infty}^{\infty} c_k \, e^{2\pi ikt/T}$, with the constants c_k given by (1.9), expressing the function f in terms of a superposition of oscillations with frequencies $\nu_k = k/T$ that are integer multiples of the fundamental frequency $\nu = 1/T$. As a glimpse at the intricacy of the subject, notice that above we pointed out that for any continuous periodic function $f : \mathbb{R} \to \mathbb{C}$ of period 1 we can find trigonometric polynomials that approximate it uniformly, that is, in the sense of (1.4). Nevertheless, the specific trigonometric polynomials obtained by means of the symmetric partial sums

$$s_n(f,t) = \sum_{k=-n}^{n} c_k \, e^{2\pi ikt} \tag{1.10}$$

with c_k given by (1.9) are not necessarily good approximations: the sequence

[2] The orthogonality considerations made in Chapter 3 show that if we rely only on functions of the form (1.5), then the approximations miss out an infinite-dimensional subspace of the space of square integrable functions.

$\{s_n(f,t)\}_{n\geq 1}$ might diverge for infinitely many values of $t \in [0,1]$; see the discussion in the introduction to Chapter 4. This shows that continuity coupled with the concept of pointwise or uniform convergence is not adequate. The proper setting turns out to be the class of Lebesgue integrable or square integrable functions, with an associated concept of convergence. The need to go beyond the class of continuous functions and the classical theory of Riemann integrable functions is fully justified by the mathematical power and flexibility of the theory within the new setting, and is further emphasised by its wide range of applicability.

1.2 The dawn of the theory

Fourier analysis dates back to late eighteenth/early nineteenth century studies of the vibrating string and of heat propagation. Two basic partial differential equations of one-dimensional mathematical physics are the wave equation

$$\frac{\partial^2 U}{\partial T^2} = c^2 \frac{\partial^2 U}{\partial X^2} \tag{1.11}$$

and the heat equation

$$\frac{\partial U}{\partial T} = \kappa \frac{\partial^2 U}{\partial X^2}, \tag{1.12}$$

where $c > 0$ and $\kappa > 0$ are physical constants. In (1.11), $U = U(X,T)$ represents, at the location X and at time T, the displacement of a homogeneous string placed in the (X,Y)-plane and stretched along the X-axis between $X = 0$ and $X = L$, where it is tied. The value of the constant c is $\sqrt{\tau/\rho}$, where τ is the tension coefficient of the string and ρ is its mass density. Equation (1.11) is to be solved for $T > 0$ and X between 0 and L, subject to the boundary conditions

$$U(0,T) = U(L,T) = 0, \qquad T \geq 0, \tag{1.13}$$

which express the fact that the endpoints of the string are fixed. The solution U describes the vibrations of a violin string. On the other hand, in (1.12), $U = U(X,T)$ is the temperature in a homogeneous, straight wire of length L, whose endpoints are held at constant temperature zero. The value of the constant κ in (1.12) is specific to the conducting material. The problem is to describe the temperature at time T from its knowledge at time $T = 0$. Consequently, we seek solutions to (1.12) for $T > 0$ and X between 0 and L, subject to the boundary conditions

$$U(0,T) = U(L,T) = 0, \qquad T \geq 0, \tag{1.14}$$

and with the initial temperature specified by

$$U(X,0) = U_0(X), \qquad 0 \le X \le L. \tag{1.15}$$

For the physical derivation of (1.11) and (1.12) we refer to Dym and McKean (1972), Krantz (1999), Stein and Shakarchi (2003) and Strauss (2008). We now discuss some mathematical aspects of historical interest that provided the motivation for the development of the rigorous theory of Fourier series.

The first natural step in the mathematical investigation of (1.11) consists of scaling the equation: a change of units permits us to write the equation in non-dimensional form, thus reducing the number of physical parameters involved. This can be accomplished by means of the change of variables

$$X = Lx, \quad T = \frac{L}{c}\,t, \quad U(X,T) = c_0\, u(x,t), \tag{1.16}$$

where $c_0 = 1$ m is the reference length. The fact that X takes values between 0 and L translates into $x \in [0,1]$, the constant c is absorbed into (1.16), and all variables (the independent variables x and t, as well as the dependent variable u) are now numbers, whereas X and U were expressed initially in m (metres) and T in s (seconds). Clearly $\frac{\partial U}{\partial X} = \frac{c_0}{L}\frac{\partial u}{\partial x}$, $\frac{\partial U}{\partial T} = \frac{c_0 c}{L}\frac{\partial u}{\partial t}$, $\frac{\partial^2 U}{\partial X^2} = \frac{c_0^2}{L^2}\frac{\partial^2 u}{\partial x^2}$, $\frac{\partial^2 U}{\partial T^2} = \frac{c_0^2 c^2}{L^2}\frac{\partial^2 u}{\partial t^2}$, so that (1.11) and (1.13) are transformed into

$$\begin{cases} u_{tt} = u_{xx}, & t > 0, \quad 0 \le x \le 1, \\ u(0,t) = u(1,t) = 0, & t \ge 0, \end{cases} \tag{1.17}$$

where $u_{tt} = \frac{\partial^2 u}{\partial t^2}$ and $u_{xx} = \frac{\partial^2 u}{\partial x^2}$. Once we solve (1.17), we can return to the original physical variables by making the change of variables inverse to (1.16). Similarly, the nondimensional scaled version of (1.12) coupled with (1.14)–(1.15) is

$$\begin{cases} u_t = u_{xx}, & t > 0, \quad 0 \le x \le 1, \\ u(0,t) = u(1,t) = 0, & t \ge 0, \\ u(x,0) = u_0(x), & 0 \le x \le 1, \end{cases} \tag{1.18}$$

with $u_0 : [0,1] \to \mathbb{R}$ a given continuous function satisfying $u_0(0) = u_0(1) = 0$; here $u_t = \frac{\partial u}{\partial t}$. The issue of finding proper initial data (at time $t = 0$) for (1.17), playing the role that u_0 plays for (1.18), is discussed in Section 1.2.1.

1.2.1 The vibrating string controversy

For the sake of simplicity, let us first drop the restrictions $0 \le x \le 1$ and $t \ge 0$, and suppose that u is twice differentiable and solves the partial differential equation in (1.17) for all real x and t. If we change variables $\xi = x-t$, $\eta = x+t$, and set $\gamma(\xi,\eta) = u(x,t)$, in terms of the new variables the partial differential equation in (1.17) becomes $\dfrac{\partial^2\gamma}{\partial\xi\partial\eta} = 0$. Integrating this relation twice gives $\gamma(\xi,\eta) = f(\xi) + g(\eta)$ for some functions f and g, so that

$$u(x,t) = f(x-t) + g(x+t)\,. \tag{1.19}$$

Note that the graph of the function $x \mapsto f(x-t)$ at time $t = 0$ is simply the graph of the function f, while at time t it becomes the graph of f translated by t: $f(x-t)$ represents a *travelling wave* (a pattern that travels without change of form) which propagates to the right with unit speed; see Figure 1.2. Similarly, $g(x+t)$ represents a travelling wave that propagates to the left with unit speed.[3] The partial differential equation in (1.17) being linear, the *superposition principle* holds: if $u_1(x,t)$ and $u_2(x,t)$ are solutions, then so is $a\,u_1(x,t) + b\,u_2(x,t)$ for any constants a and b. In particular, (1.19) shows that the general solution is a superposition of two waves travelling in opposite directions.

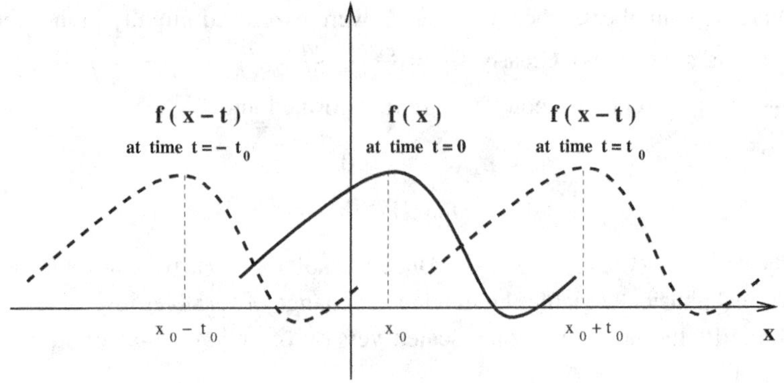

Figure 1.2 A travelling wave propagating to the right: the profile is depicted at three successive times $t = -t_0$, $t = 0$ and $t = t_0$ with $t_0 > 0$.

To connect the obtained result with the original problem (1.17), provided that $u(x,0) = u_0(x)$ for $x \in [0,1]$ specifies the initial shape of the string, extend u_0 to $\mathbb{R}$ as an odd function[4] of period 2. Also, extend the presumed solution

[3] The non-dimensional unit speed corresponds to the speed c in the original physical variables, if we recall the scaling (1.16).

[4] Meaning that $u_0(-x) = -u_0(x)$ and $u_0(x+2) = u_0(x)$ for $x > 0$. Note that if we extend u_0 to the whole real line, relation (1.22) emerges, and this forces oddness. With regard to periodicity, the boundary conditions in (1.17) might seem to indicate the period 1. However, period 2 and oddness combined do not impose any constraint upon $u_0 : [0,1] \to \mathbb{R}$ with $u_0(0) = u_0(1) = 0$, whereas period 1 and oddness require $u_0(1/2) = 0$, for example, due to $u_0(1/2) = u_0(1/2-1) = u_0(-1/2) = -u_0(1/2)$.

$u(x,t)$ from $[0,1]\times[0,\infty)$ to $\mathbb{R}^2$ by requiring that for every fixed $t \geq 0$, the map $x \mapsto u(x,t)$ is odd and periodic of period 2, while for $t < 0$ we simply solve (1.17) backwards in time: we seek a twice differentiable solution $u : \mathbb{R}\times\mathbb{R} \to \mathbb{R}$ to (1.17). Consequently, the solution u must be of the form (1.19) for some functions f and g. We get

$$f(x) + g(x) = u_0(x), \qquad x \in \mathbb{R}, \tag{1.20}$$

by evaluating (1.19) at $t = 0$. To accommodate the boundary conditions in (1.17), (1.19) yields $f(-t) + g(t) = 0$ for all $t \in \mathbb{R}$, so that (1.19) and (1.20) take the form

$$u(x,t) = g(x+t) - g(t-x), \qquad x, t \in \mathbb{R}, \tag{1.21}$$

$$g(x) - g(-x) = u_0(x), \qquad x \in \mathbb{R}, \tag{1.22}$$

respectively. The formula (1.19) was first obtained in 1747 by d'Alembert, who was concerned with finding the general solution of the partial differential equation and ignored the physical context,[5] in particular, the significance of (1.22). A closer look at (1.22) reveals that the form of its left side encodes the fact that the function u_0 is odd, but this relation by itself does not determine the function g. For example, if g is a solution to (1.22), so will be $g + g_0$ for any even function[6] g_0 of period 1. The underlying physics indicates that perhaps the initial velocity[7] $v_0(x) = \frac{\partial u}{\partial t}(x,0)$ for $x \in [0,1]$, might be relevant. Indeed, if v_0 is given on $[0,1]$, we extend it to $\mathbb{R}$ by requiring it to be odd and periodic of period 2. Differentiating (1.21) with respect to the time variable[8] and evaluating the outcome at $t = 0$, we get

$$g'(x) - g'(-x) = v_0(x), \qquad x \in \mathbb{R}. \tag{1.23}$$

Now (1.22) and (1.23) yield $g(x) = \frac{1}{2}\left[u_0(x) + \int_0^x v_0(s)\,ds\right] + \alpha$ for some constant α. Since u_0 and v_0 are odd,[9] using (1.21), we obtain

$$u(x,t) = \frac{u_0(x-t) + u_0(x+t)}{2} + \frac{1}{2}\int_{x-t}^{x+t} v_0(s)\,ds\,, \tag{1.24}$$

[5] It was not unusual for d'Alembert to sacrifice physical reality for a purely philosophical viewpoint, see Wheeler and Crummett (1987).

[6] Meaning that $g_0(-x) = g_0(x)$ for all $x \in \mathbb{R}$.

[7] $u_t = \frac{\partial u}{\partial t}$ is the rate of change of the displacement of a particular point on the string and (generally) differs from the speed c of propagation of the wave along the string. This situation is also encountered for waves in media other than strings.

[8] Since u is twice differentiable, $t \mapsto u(t,t)$ is differentiable. From (1.21) with $x = t$ we then infer that g is differentiable.

[9] So that $-u_0(t-x) = u_0(x-t)$ and $\int_{t-x}^0 v_0(s)\,ds = \int_{x-t}^0 v_0(y)\,dy$, the latter as a consequence of the change of variables $y = -s$.

as shown in 1748 by Euler. One can check directly that for a general twice differentiable function $u_0 : \mathbb{R} \to \mathbb{R}$ and for a general differentiable function $v_0 : \mathbb{R} \to \mathbb{R}$, (1.24) provides us with a classical solution to the wave equation $u_{tt} - u_{xx} = 0$, with $u(x,0) = u_0(x)$ and $u_t(x,0) = v_0(x)$. In our particular setting, observe that the extensions performed for u_0 and v_0 ensure the validity of the boundary conditions in (1.17). The above discussion illustrates the fruitful interplay between abstract mathematics and its relation to nature: physical intuition can provide a feeling for mathematical facts and the other way around.

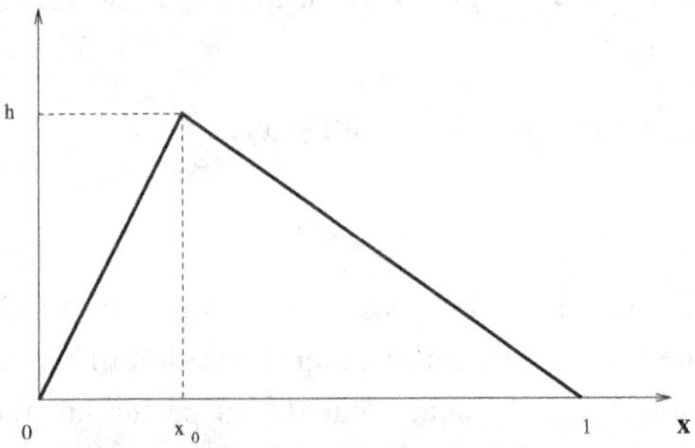

Figure 1.3 Initial position of a plucked string.

Euler's solution (1.24) differed from d'Alembert's (1.21) on the specification[10] of the function g. Furthermore, Euler proclaimed that the function g does not need to be differentiable, but may be any curve drawn by hand.[11] Euler had in mind the plucked string: taking as the initial position of the string the triangular shape given, for some constants $x_0 \in (0,1)$ and $h > 0$, by

$$u_0(x) = \begin{cases} \frac{xh}{x_0} & \text{for} \quad 0 \leq x \leq x_0, \\ \frac{h(1-x)}{1-x_0} & \text{for} \quad x_0 \leq x \leq 1, \end{cases} \tag{1.25}$$

(see Figure 1.3), and choosing zero initial velocity $v_0 \equiv 0$, Euler declared that the subsequent positions of the string are given by

$$u(x,t) = \frac{u_0(x-t) + u_0(x+t)}{2}, \qquad t \geq 0, \quad 0 \leq x \leq 1, \tag{1.26}$$

obtained formally from (1.24). Euler used a physical observation (the fact that the violin string could be released from an initial position with a corner) to impose a mathematical formula. The unsatisfactory aspect of the solution (1.26)

[10] Due to (1.20), the knowledge of g determines f uniquely in terms of the initial position u_0.
[11] The mathematical formalism which we take for granted today was not available at that time: Euler did not perceive a function to be an arbitrary rule that assigns to every point of the domain of definition a single point of the range, see Krantz (1999). This explains the vague formulation.

is that it does not satisfy the partial differential equation we set out to solve: since u_0 is not differentiable at $x = x_0 \in (0, 1)$, the function $u(x, t)$ defined by (1.26) is not differentiable. In light of this, d'Alembert objected to physical arguments for solutions to a partial differential equation and called for the other researchers to engage in mathematics, see Wheeler and Crummett (1987). Euler defended his solutions with corners with mathematically unconvincing arguments. His position was later on vindicated: it turns out that u does solve the equation in an appropriate generalised sense, the understanding of which requires the theory of distributions.

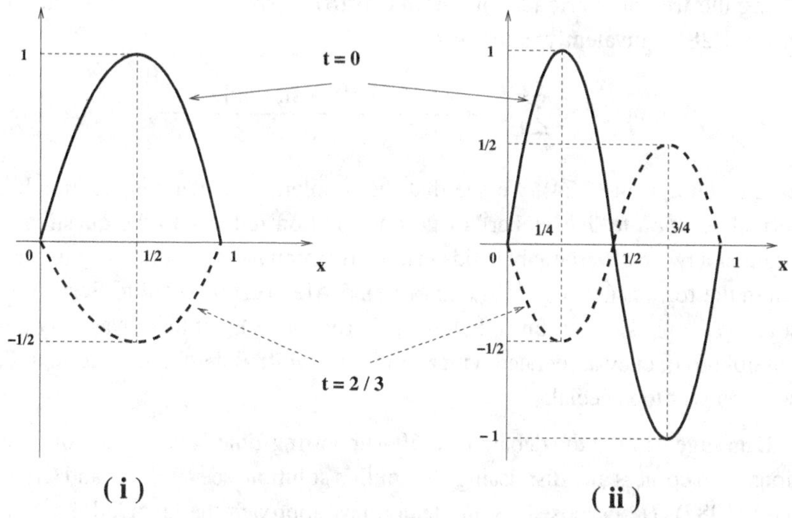

Figure 1.4 Fundamental tone (i) and the first overtone (ii) at two instants in time.

Daniel Bernoulli enters the debate in 1753 in the midst of the d'Alembert–Euler disagreement. His starting point, see Benedetto (1997), is Brook Taylor's observation from 1715 that for any integer $m \geq 1$ the tone[12]

$$u_m(x, t) = \sin(\pi m x) \cos(\pi m t) \tag{1.27}$$

represents a solution to (1.17) with zero initial velocity. In contrast to the travelling waves discussed before, (1.27) represents a *standing wave*. The terminology comes from looking at the graph of $x \mapsto u_m(x, t)$ as t varies (see Figure 1.4): the endpoints $x = 0$ and $x = 1$, as well as all points $x = \frac{k}{m}$ with

[12] The case $m = 1$ is called the *fundamental tone* or *first harmonic* of the vibrating string, and $m \geq 2$ are the *overtones* or *higher harmonics*, $m = 2$ being the *first overtone* or *second harmonic*, see Stein and Shakarchi (2003).

$k \in \{1, \cdots, m-1\}$, remain motionless in time and are called nodes. The points $x = \frac{2k-1}{2m}$ with $k \in \{1, \cdots, m\}$, where the amplitude is maximal, are named anti-nodes. Bernoulli argued formally in terms of the physics of sound and provided no mathematical support for his arguments, see Wheeler and Crummett (1987), claiming that the solution to (1.17) with initial velocity $v_0 \equiv 0$ must be an infinite sum of the tones:

$$u(x,t) = \sum_{m=1}^{\infty} a_m \sin(\pi m x) \cos(\pi m t). \tag{1.28}$$

Using the trigonometric identity $\sin(\alpha)\cos(\beta) = \frac{\sin(\alpha+\beta)+\sin(\alpha-\beta)}{2}$, we can express (1.28) equivalently in the form

$$u(x,t) = \sum_{m=1}^{\infty} a_m \frac{\sin(\pi m[x-t]) + \sin(\pi m[x+t])}{2}. \tag{1.29}$$

Setting $v_0 \equiv 0$ in (1.24), we see that the problem of reconciling Bernoulli's formal solution to d'Alembert's rigorous solution reduces to the question of whether a twice differentiable odd periodic function u_0 of period 2 may be written in the form $u_0(x) = \sum_{m=1}^{\infty} a_m \sin(\pi m x)$. d'Alembert objected to Bernoulli's solution on physical grounds, believing (erroneously) that there was only one possible frequency associated with a vibration, while Euler felt that Bernoulli's solution was too special.

Lagrange entered the debate in 1759, supporting Euler's admission of functions with corners and dismissing Bernoulli's solution, see Wheeler and Crummett (1987). He proposed a completely new approach that avoided the wave equation by viewing the string as a collection of n equally spaced point masses, connected by a light cord. This model leads to a set of n equations of the form $\frac{d^2 y_k}{dt^2} = y_{k-1} - 2y_k + y_{k+1}$. After solving the system for a finite number of masses, Lagrange generated, for the initial position u_0 and the initial velocity v_0 of the string (both odd and of period 2), the solution

$$\begin{aligned} u(x,t) = 2\sum_{m=1}^{\infty}\Big(\int_0^1 \sin(\pi m s)\, u_0(s)\, ds\Big) \cos(\pi m t) \sin(\pi m x) \qquad (1.30)\\ + \frac{2}{\pi}\sum_{m=1}^{\infty}\Big(\int_0^1 \frac{\sin(\pi m s)}{m} v_0(s)\, ds\Big) \sin(\pi m t) \sin(\pi m x). \end{aligned}$$

Note that if we set $t = 0$ in (1.30) and if we interchange summation and integration, then (1.30) gives rise to the Fourier series expansion of the function u_0, while differentiating (1.30) with respect to t and subsequently setting $t = 0$ yields the Fourier series expansion of the function v_0.

To substantiate this claim, note that if u_0 is odd and periodic with period 2, then $\int_{-1}^{1} u_0(s)\,ds = 0$, while $(\frac{1}{2}\int_{-1}^{1} e^{-\pi ims} u_0(s)ds)e^{\pi imx} + (\frac{1}{2}\int_{-1}^{1} e^{\pi ims} u_0(s)ds)e^{-\pi imx} = \int_{-1}^{1} \cos[\pi m(x-s)]\, u_0(s)ds = (\int_{-1}^{1} \cos(\pi ms)u_0(s)\,ds)\cos(\pi mx) + (\int_{-1}^{1} \sin(\pi ms)u_0(s)ds)\sin(\pi mx)$ for $m \geq 1$, and this equals $2(\int_0^1 \sin(\pi ms)u_0(y)dy)\,\sin(\pi mx)$ since the oddness of u_0 ensures the vanishing of the second term, as well as the equality $\int_{-1}^{0} \sin(\pi ms)\,u_0(s)\,ds = \int_0^1 \sin(\pi my)\,u_0(y)\,dy$, by means of the change of variables $y = -s$.

With Lagrange's contribution the intensity of the debate declined quite drastically, in part also because too many mathematical concepts needed clarification. More partial differential equations were derived and when, in 1807, Fourier introduced his approach towards the study of heat flow, Lagrange was the only living participant to the debate. He questioned the convergence of the series arising in Fourier's approach. The issue of convergence became of paramount importance. Riemann introduced his integral in 1854 to investigate Fourier series expansions for a large class of functions, see Benedetto (1997). With the development of a more flexible concept of integral in 1902, Lebesgue opened the modern theory that was continuously improved over the past hundred years.

1.2.2 Fourier's view on heat flow

Let us now describe Fourier's attempt to find the general solution to (1.18). His manuscript was submitted in 1807, when Fourier was almost 40, and had only three unrelated publications, see Benedetto (1997). His point of view proved to be an important breakthrough, containing the germs of an approach with wide applicability to many important equations of mathematical physics.

Fourier first guessed that solutions to the the partial differential equation in (1.18) might be found by separating variables, that is, seeking solutions in the form $u(x,t) = \alpha(x)\beta(t)$. Substituting this into the equation, we get $\alpha(x)\beta'(t) = \alpha''(x)\beta(t)$, and therefore $\frac{\beta'(t)}{\beta(t)} = \frac{\alpha''(x)}{\alpha(x)}$. Since the right-hand side is independent of t and the left-hand side is independent of x, the previous equality is only possible if both sides are equal to a constant, say λ. Consequently $\beta'(t) = \lambda\beta(t)$ and $\alpha''(x) = \lambda\,\alpha(x)$. The first equation yields $\beta(t) = \beta_0\, e^{\lambda t}$. Practical experience tells us that the temperature in the rod dissipates as times goes on, so that the physically relevant case occurs for $\lambda < 0$, say $\lambda = -\theta^2$. The general solution of the differential equation for α is now easily found as being a linear combination of $\cos(\theta x)$ and $\sin(\theta x)$. The boundary conditions imposed in (1.18) eliminate the first of these fundamental solutions and force $\theta = \pi m$ for some integer m.

Thus Fourier found the solutions

$$u_m(x,t) = a_m\, e^{-\pi^2 m^2 t} \sin(\pi m x), \tag{1.31}$$

indexed by the integer $m \geq 1$, where a_m is an arbitrary real constant. Note that in the derivation of (1.31) we divided by the functions α and β, which may vanish (α actually has zeros). However, it is easy to check directly that (1.31) represents a solution to the partial differential equation in (1.18), satisfying the requested boundary conditions.

By linearity, any finite linear combination of solutions of the type (1.31) is also a solution of the differential equation with the boundary conditions imposed in (1.18). Fourier asserted that an arbitrary function u_0 with $u_0(0) = u_0(1) = 0$ can be expanded in a series of the form $u_0(x) = \sum_{m\geq 1} a_m \sin(\pi m x)$ for $x \in [0,1]$, with coefficients[13] $a_m = 2\int_0^1 u_0(x) \sin(\pi m x)\, dx$ for $m \geq 1$, and that the corresponding solution to (1.18) is $u(x,t) = \sum_{m\geq 1} a_m\, e^{-\pi^2 m^2 t} \sin(\pi m x)$ for $t \geq 0$ and $x \in [0,1]$. While Fourier's arguments contained many dubious steps,[14] Fourier's examples and applications are extraordinary, and were influential in establishing the research area of Fourier series. Although the idea of expanding in trigonometric sums or series was implicit (and sometimes even explicit) in earlier works, Fourier had the conviction that his predecessors lacked, and he made good use of it in his studies of heat flow, see Stein and Shakarchi (2003).

The above trigonometric series expansion of the solution is expressed in terms of the exponentials (1.8) with $T = 2$, and not with $T = 1$, the reason being that the problem (1.18) was not formulated for merely spatially periodic functions of period 1. The rationale for the doubling of the apparent period by choosing $T = 2$ and viewing an initial continuous data $u_0 : [0,1] \to \mathbb{R}$ subject to $u_0(0) = u_0(1) = 0$ as the restriction to $[0,1]$ of an odd periodic function of period 2 is explained in Part II, where we provide a rigorous foundation for the previous formal considerations.

1.3 Application: irrationality of π

Following Niven (1947), we present a nice application of trigonometric polynomials to number theory by proving that π is an irrational number.[15]

Assume that $\pi = a/b$, the quotient of positive integers a and b. For $x \in [0,1]$, define the polynomials $f(x) = \frac{x^n(1-x)^n}{n!}$ and $F(x) = \pi^{2n} f(x) - \pi^{2n-2} f^{(2)}(x) + \pi^{2n-4} f^{(4)}(x) + \cdots + (-1)^n f^{(2n)}(x)$, with

13 The series is of type (1.8), with $T = 2$, $c_0 = 0$ and $c_m = -c_{-m} = \frac{a_m}{2i}$ for $m \geq 1$. For u_0 odd and of period 2, we get $c_m - c_{-m} = -i\int_0^2 u_0(x) \sin(\pi m x)\, dx = -2i\int_0^1 u_0(x) \sin(\pi m x)\, dx$ for $m \geq 1$, in view of (1.9)

14 Prompting a critique by Lagrange in 1812, see Benedetto (1997).

15 The first proof, due to Lambert in 1761, exploits a continued-fraction representation of the tangent function. In 1882, von Lindemann proved that π is transcendental, that is, π is not a root of a polynomial equation with rational coefficients.

the positive integer n to be specified later on. Since the polynomial $n!f(x)$ has integer coefficients and all its monomials in x are of degree at least n, every single derivative $f^{(2k)}(0)$ with $0 \le k \le n$ is an integer. Indeed, differentiating an expression of the form $\frac{a_j}{n!}x^{n+j}$, with a_j and $j \ge 0$ integers, $2k$-times with respect to x, the evaluation at $x = 0$ is nonzero only if $2k = n+j$. For $2k = n+j$ we obtain the integer $\frac{a_j(2k)!}{n!}$. Since $f(x) = f(1-x)$, we also have that every single derivative $f^{(2k)}(1)$ with $0 \le k \le n$ is an integer. A straightforward calculation yields $\frac{d}{dx}\left(F'(x)\sin(\pi x) - \pi F(x)\cos(\pi x)\right) = \pi^{2n+2}f(x)\sin(\pi x)$, so that $\pi^{2n+1}\int_0^1 f(x)\sin(\pi x)\,dx = F(1) + F(0)$. Since the previous discussion ensures that $b^{2n+2}\,[F(1) + F(0)]$ is an integer, we deduce that $ba^{2n+1}\int_0^1 f(x)\sin(\pi x)\,dx$ is also an integer. But $0 < f(x) < \frac{1}{n!}$ and $0 < \sin(\pi x) \le 1$ for $0 < x < 1$ yield

$$0 < ba^{2n+1}\int_0^1 f(x)\sin(\pi x)\,dx < \frac{ba^{2n+1}}{n!}. \tag{1.32}$$

However, the power series expansion of the function $a \mapsto ab\,e^{a^2}$ ensures $\lim_{n\to\infty}\frac{ba^{2n+1}}{n!} = 0$. Choosing n large enough we obtain that the right-hand side of (1.32) is less than $1/2$, so that the middle term in (1.32) cannot be an integer. This completes the proof.

To clarify the connection with trigonometric series, note that the above approach involves only values $x \in [0, 1]$. We can think of f as the restriction to $[0, 1]$ of an odd function $f : \mathbb{R} \to \mathbb{R}$ with principal period 2, if we extend f from $[0, 1]$ to $\mathbb{R}$ by requiring[16] $f(x + 1) = -f(x)$ for all $x \in \mathbb{R}$. Since $\int_1^2 f(x)\sin(\pi x)\,dx = \int_0^1 f(y)\sin(\pi y)\,dy$ follows by means of the change of variables $y = x-1$, in terms of the coefficients (1.9) of the trigonometric series associated to the periodic function f, we have $\frac{c_{-1}-c_1}{2i} = \frac{1}{4i}\int_0^2 f(x)\,[e^{i\pi x} - e^{-i\pi x}]\,dx = \frac{1}{2}\int_0^2 f(x)\sin(\pi x)\,dx = \int_0^1 f(x)\sin(\pi x)\,dx$.

1.4 Exercises

The problems provided here illustrate the interplay between trigonometric polynomials and classical algebraic polynomials but also draw attention to a contrasting feature, indicative of structural differences.

1.4.1 Statements

1. Show that a trigonometric polynomial of degree n has at most $2n$ zeros (counted with multiplicity) in $[0, 1)$.

2. Assume that the real Fourier series $\sum_{k\ge1} a_k\cos(2\pi kx)+\sum_{k\ge1} b_k\sin(2\pi kx)$, for suitable $a_k, b_k \in \mathbb{R}$, converges uniformly to a continuous periodic function $f : \mathbb{R} \to \mathbb{R}$. If $n \ge 1$ is the smallest index $k \ge 0$ such that $a_k \ne 0$ or $b_k \ne 0$, show that f has at least $2n$ zeros in $[0, 1)$. [This shows that the number of zeros is at least that of the lowest harmonic:[17] adding higher frequencies, one can only increase the number of zeros.]

3. Show that if the real-valued trigonometric polynomial $f(t) = \sum_{|k|\le n} c_k e^{2\pi ikt}$ is non-negative, then $f = |g|^2$ for some trigonometric polynomial g.

[16] Note that this relation forces $f(x + 2) = -f(x + 1) = f(x)$ for all $x \in \mathbb{R}$.

[17] Note that $a_n\cos(2\pi nx) + b_n\sin(2\pi nx) = \sqrt{a_n^2 + b_n^2}\cos(2\pi n(x - \alpha))$, with $\alpha \in [0, 1/n)$ such that $e^{2\pi in\alpha} = \frac{a_n+ib_n}{\sqrt{a_n^2+b_n^2}}$, has exactly $2n$ zeros $\{\alpha + \frac{2k+1}{4n}\}_{0\le k\le 2n-1}$ in the periodicity window $[\alpha, \alpha + 1)$.

1.4.2 Hints

1. Set $z = e^{2\pi i t}$ to relate to polynomials in $z \in \mathbb{C}$.

2. Assume that f has only $p < 2n$ zeros in $[0, 1)$, and construct a trigonometric polynomial g of degree at most p such that fg does not change sign on $[0, 1]$. Use the orthogonality relations (1.2).

3. Assuming $c_{-n} \neq 0$, show that the roots of the polynomial $P(z) = z^n \sum_{|k|\le n} c_k z^k$ occur in pairs α, $1/\overline{\alpha}$ having equal multiplicity.

1.4.3 Solutions

1. The polynomial $P(z) = \sum_{|k|\le n} c_k z^{n+k}$ has degree at most $2n$, and thus admits at most $2n$ complex roots. Note that $f(t) = e^{-2\pi i n t} P(e^{2\pi i t})$ for $t \in [0, 1)$.

2. Assuming that f has $p < 2n$ zeros on $[0, 1)$, denote them by $t_1 \le t_2 \le \cdots \le t_p$. Let $g(t) = \prod_{k=1}^{p} [\sin(2\pi(t - t_k))]^{m_k}$, with $m_k = 0$ if $f(t)$ does not change sign at $t = t_k$ and $m_k = 1$ if $f(t)$ changes sign at $t = t_k$. Then $\sum_{k=1}^{p} m_k = q \le p$. According to the discussion in Section 1.1, g can be expressed alternatively as a trigonometric polynomial, $g(t) = \sum_{|j|\le q} \beta_j e^{2\pi i j t}$ for some constants $\beta_j \in \mathbb{C}$. By construction, fg does not change sign on $[0, 1)$. Since $fg \not\equiv 0$, we deduce that $\int_0^1 f(t)\, g(t)\, dt = 0$. However, writing $f(t) = \sum_{|k|\ge n} \alpha_k e^{2\pi i k t}$ for some constants $\alpha_k \in \mathbb{C}$, the uniform convergence yields $\int_0^1 f(t)\, g(t)\, dt = \sum_{|k|\ge n} \sum_{|j|\le q} \alpha_k \beta_j \int_0^1 e^{2\pi i (k+j) t} dt = 0$, due to the orthogonality relations (1.2), because $q < n$ prevents $k + j = 0$ if $|k| \ge n$ and $|j| \le q$. The obtained contradiction proves the statement.

3. Since $f(1 - t) = \sum_{|k|\le n} c_{-k} e^{2\pi i k t}$, we can assume without loss of generality that $c_{-n} \neq 0$. Then $P(z) = z^n \sum_{|k|\le n} c_k x^k$ is a polynomial of degree at most $2n$, with $P(0) \neq 0$. Observe that

$$P(z) = z^{2n}\, \overline{P(1/\overline{z})}\,, \qquad z \in \mathbb{C} \setminus \{0\}\,. \tag{$*$}$$

Indeed, $z^{2n}\, \overline{P(1/\overline{z})} = \sum_{|k|\le n} \overline{c}_k z^{n-k}$ is a polynomial in $z \in \mathbb{C}$ that coincides with $P(z)$ on the unit circle $\{z \in \mathbb{C} : |z| = 1\}$ since the fact that

$$f(t) = e^{-2\pi i n t} P(e^{2\pi i t})\,, \qquad t \in [0, 1]\,, \tag{$**$}$$

is real-valued can be expressed as $f(t) = \overline{f(t)}$ for all $t \in [0, 1]$. Factorising $P(z) = \alpha \prod_{k=1}^{p} (z - z_k)^{m_k}$ for $z \in \mathbb{C}$, with $\alpha \in \mathbb{C}$ and the integers $m_k \ge 1$ accounting for the multiplicity, $(*)$ yields that

$$\sum_{k=1}^{p} m_k = 2n\,, \qquad \alpha = \overline{\alpha} \prod_{k=1}^{p} (-1)^{m_k} \overline{z}_k^{m_k}\,, \tag{$***$}$$

and the roots z_k of P with $|z_k| \neq 1$ occur in pairs α_k, $1/\overline{\alpha}_k$ having equal multiplicity. On the other hand, let $\alpha_j = e^{2\pi i t_0}$ with $t_0 \in [0, 1)$ be a root of P with multiplicity $m_j \ge 1$. Recalling $(**)$, the n-times continuously differentiable real-valued function $f(t) = \alpha\, e^{-2\pi i n t} \prod_{k=1}^{p} (e^{2\pi i t} - z_k)^{m_k}$ satisfies $\lim_{t\to t_0} \frac{f(t)}{(t-t_0)^{m_j}} = \frac{f^{(m_j)}(t_0)}{m_j!} \neq 0$. Therefore an odd integer m_j entails a change of sign of $f(t)$ at $t = t_0$. This shows that roots α_j of P with $|\alpha_j| = 1$ have even multiplicity. Thus $p = 2q$ for some integer q and $(***)$ yields $\sum_{k=1}^{q} m_k = n$, $\alpha = \overline{\alpha} \prod_{k=1}^{q} (\frac{\overline{\alpha}_k}{\alpha_k})^{m_k}$, so that $c = (-1)^n \alpha \big/ \prod_{k=1}^{q} \overline{\alpha}_k^{m_k} \in \mathbb{R}$. Consequently, for $z \in \mathbb{C} \setminus \{0\}$ we get $P(z) = \alpha \prod_{k=1}^{q} (z - \alpha_k)^{m_k} (z - 1/\overline{\alpha}_k)^{m_k} = \alpha \prod_{k=1}^{q} [(z - \alpha_k)(\frac{1}{z} - \overline{\alpha}_k) \frac{(-1)z}{\overline{\alpha}_k}]^{m_k} = c\, z^n \prod_{k=1}^{q} [(z - \alpha_k)(\frac{1}{z} - \overline{\alpha}_k)]^{m_k}$. Taking $(**)$ into account, we deduce that $f(t) = c \prod_{k=1}^{q} [(e^{2\pi i t} - \alpha_k)(e^{-2\pi i t} - \overline{\alpha}_k)]^{m_k} = c \left| \prod_{k=1}^{q} (e^{2\pi i t} - \alpha_k)^{m_k} \right|^2$ for $t \in [0, 1]$. In particular, this means that $c > 0$. The desired representation $f(t) = |g(t)|^2$ is obtained by defining $g(t) = \sqrt{c} \prod_{k=1}^{q} (e^{2\pi i t} - \alpha_k)^{m_k}$.

2
The Lebesgue measure and integral

The measurement of lengths and areas has a long history dating back to the ancient cultures of Egypt and Greece. It has led to the development of measure theory as an area of modern mathematics, dealing with systematic approaches for measuring complicated objects based on available measurements for simpler objects. The aim of the present chapter is to provide the basic tools for acquiring working knowledge of the Lebesgue integral and its generalisations to abstract measure and integration theory. Rather than providing in this survey full proofs of the main theorems, we motivate/explain the main building blocks of the theory and we illustrate the flexibility of this powerful concept. For the results that are used but not proved we provide adequate references.

2.1 Historical considerations

A full appreciation of measure theory requires, we believe, some insight into the genesis of the subject. For this we relied on the material provided in Kupka (1986).

2.1.1 Ancient measure theory

In ancient Egypt, several hundreds of years B.C., the relatively flat ground was basically subdivided into rectangular plots whose area could be expressed in whole numbers of square cubits (the unit of length used in the time of the pharaohs). With regard to measure theory, the vastly greater refinement of Greek mathematics over that of the Egyptians is to a large extent attributed to the necessity of the ancient Greeks to ascertain, for agricultural needs, areas in an irregularly shaped hilly/mountaineous terrain. To obtain approximations of areas of more complicated regions, precise enough to satisfy the practical requirements of the time, the Greek mathematicians used the "paving stone technique": the given region was paved as exactly as possible with variously chosen rectangular stones of known area, so that the unknown area is well approximated by the sum of the known areas of the individual nonoverlapping stones. The greater sophistication of Greek mathematicians came about because of their subsequent preference of triangles over rectangles. This is advantageous because any region bounded by straight lines can always be paved exactly by finitely many triangles and such regions provide accurate information about certain areas bounded by curves.

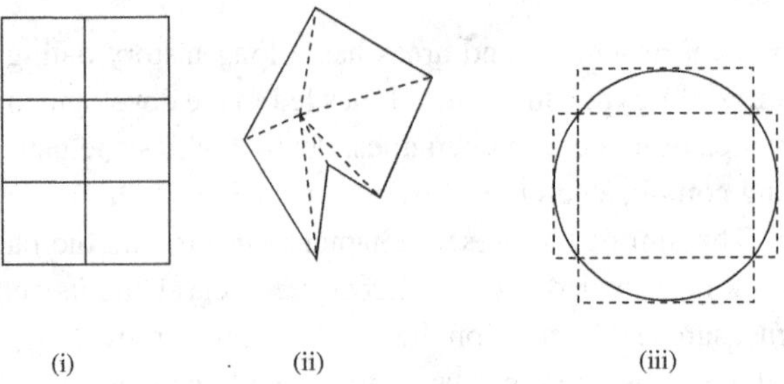

Figure 2.1 (i) Rectangular plot of 6 Egyptian cubits. (ii) Finite triangulation of a region bounded by straight lines using an interior vortex. (iii) "Paving stone" approximation by five rectangles for the area of the disc.

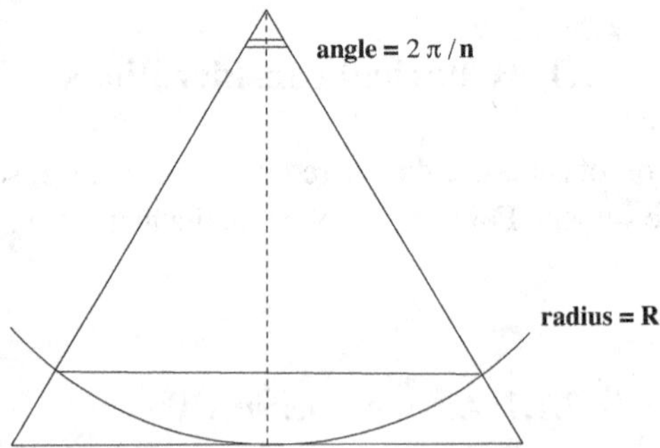

Figure 2.2 The inscribed triangle has base $2R\sin(\pi/n)$ and height $R\cos(\pi/n)$, while the superscribed triangle has height R and base $2R\tan(\pi/n)$.

We illustrate this by deriving the area $A(R)$ of a disc in terms of its radius R, using triangulations in a way that resembles[1] that presented by Euclid in

[1] The ancient version of the modern limiting procedure was more cumbersome and

his mathematical treatise *Elements of Geometry*, written around 300 B.C. The approach consists of subdividing the disc into n equal sectors. With each sector one associates an inscribed and a superscribed triangle, as depicted in Figure 2.2. The sum of the areas of the n inscribed triangles is $nR^2 \sin(\pi/n) \cos(\pi/n)$, while the areas of the n superscribed triangles add up to $nR^2 \tan(\pi/n)$. The first paved area lies entirely within the disc while the second completely covers it. Consequently, $nR^2 \tan(\pi/n) \geq A(R) \geq nR^2 \sin(\pi/n) \cos(\pi/n)$ for any $n \geq 1$. Since $nR^2 \sin(\pi/n)\cos(\pi/n) = \pi R^2 \frac{\sin(\pi/n)}{\pi/n} \cos(\pi/n) \to \pi R^2$ and $nR^2 \tan(\pi/n) = \pi R^2 \frac{\sin(\pi/n)}{\pi/n} \frac{1}{\cos(\pi/n)} \to \pi R^2$ as $n \to \infty$, we deduce that $A(R) = \pi R^2$.

After the spectacular successes of ancient Greek mathematicians, the state-of-the-art in measure theory was stagnant at best for the next two millennia. The Roman conquest of Greek settlements scattered through the Mediterranean started towards the end of the third century B.C. and was completed when the Greek peninsula came under Roman rule towards the middle of the second century B.C. Despite the infatuation of the Roman high society with Greek art and literature, the more practical Roman mindset was less disposed towards philosophy. Greek mathematics was nurtured in philosophical schools and the Romans did not emulate them. No mathematical innovations occurred under Roman rule and the development of Greek mathematics ground to a halt. During the decline and fall of the Western Roman Empire, many of the records of the achievements of ancient Greek mathematics were at least partly lost. The Roman influence permeated the Early Middle Ages, and the use of Roman numerals made calculations painstaking. The High Middle Ages brought a revival of mathematics traced to the introduction of Arabic numerals due to the spread of Islamic culture[2] in Spain and to the rediscovery of Greek antique culture as a byproduct of the Crusades. During this period, universities were founded and ancient Greek works of natural science, mathematics and philosophy were

trigonometric functions had not been invented. Also, the value of π, defined by Euclid as the constant ratio of the area enclosed by a circle to the area of the square on its radius, was estimated later (in the third century B.C.) by Archimedes to be less than 22/7 and greater than 223/71. Chinese mathematicians obtained better values of π in the fifth century A.D. but for the proof that π is irrational (see Section 1.3) more than two millennia were to pass.

[2] In the ninth and tenth centuries A.D., Islamic mathematicians revolutionised algebra, by pioneering its unifying abstract character. They drew upon ideas from various cultures, from Greek to ancient Babylonian, and transmitted ideas related to the Hindu system of numeration (to which they added decimal fractions). The Islamic tradition has discouraged/prohibited the figurative depiction of living creatures, especially human beings. This led to an Islamic art rich in complex abstract patterns and to the development of the abstract idea of an algebraic equation. The Islamic approach to solving equations departed from the strongly geometrical orientation of Greek mathematics in that cases were systematised and formal manipulations were applied to unknown quantities rather than specific numbers (without symbolic representation – all procedures were described verbally). This was a major step towards the concept of an abstract equation.

studied. The subsequent fascination with the ancient Greek culture that was a major characteristic of the Renaissance[3] did resuscitate the ancient Greek findings but impeded the abandonment of the triangular paving stone approach, since this meant psychologically abandoning the ancient Greek scholars. This parting was essential since triangulation has reached its limit. After all, each particular example required an ingenious construction adapted to that specific case and the number of workable settings was practically exhausted.

2.1.2 The Riemann integral

In *Principia Mathematica* (1687) Newton used a systematic method of great power and scope by describing the area of the planar region between the graph of a function $f : [a, b] \to \mathbb{R}$ and the horizontal axis as a limit of the sum of the areas of thin rectangles which cover the given region in the manner of wallpaper or floorboards. This limit, called the integral $\int_a^b f(x)\,dx$ of f over $[a, b]$, can be calculated as being $F(b) - F(a)$, where F is an anti-derivative of f. Approximations by means of rectangular pavings are only required when F is not available. Newton's ideas were clarified[4] in the 1820s in the context of continuous and differentiable functions by Cauchy, who introduced the mathematical notions of derivative and integral based upon limiting procedures.[5] Later on, Riemann's interest in trigonometric series led him to extend Cauchy's work in the 1850s. The concept of a Riemann integral as an averaging process for functions was flexible enough to transform it into an abstract technique of calculation, applicable to a variety of physical problems.[6] While a characterisation of Riemann-integrable functions had to wait until the advent of Lebesgue's theory of integration, it was clear that this class of functions $\mathcal{R}[a, b]$ includes all continuous functions $f : [a, b] \to \mathbb{R}$, as well as some discontinuous bounded functions.

[3] Spurred, after the fall of the Byzantine Empire in the fifteenth century A.D., by the influx of precious ancient manuscripts into the various states of the Italian peninsula as a side-effect of migration.

[4] Newton's brilliant ideas conveyed a useful imagery but had to be put on a firmer foundation. For example, he perceived the derivative as a quotient captured at that precise instant when the numerator and denominator simultaneously vanished, characterising it as "the ratio of the quantities not before they vanish, nor afterwards, but with which they vanish" see Dunham (2005).

[5] Cauchy impressed upon the subject the rigorous character which it bears at the present time, considering the concept of a limit to to be the heart of the matter. However, his methods were subsequently revised since they relied on infinitesimal notions which fell into disrepute – they are legitimised within the framework of "nonstandard analysis".

[6] For example, distance is the integral of velocity with respect to time and work is the integral of force with respect to distance.

The fundamental idea of the Riemann integral is that of averaging. For a bounded function[7] $f : [a,b] \to \mathbb{R}$ and any subdivision $\Delta = \{a = t_0 < t_1 < \cdots < t_n = b\}$ of $[a,b]$, define the lower and upper Darboux sums as

$$\mathcal{L}(f,\Delta) = \sum_{k=1}^{n} (t_k - t_{k-1}) \inf_{t\in[t_{k-1},t_k]} \{f(t)\},$$

$$\mathcal{U}(f,\Delta) = \sum_{k=1}^{n} (t_k - t_{k-1}) \sup_{t\in[t_{k-1},t_k]} \{f(t)\},$$

respectively. If for every $\varepsilon > 0$ there exists a subdivision Δ with $\mathcal{U}(f,\Delta) - \mathcal{L}(f,\Delta) < \varepsilon$, then f is said to be Riemann integrable on $[a,b]$. The notion becomes somewhat clearer by noticing that whenever Δ^* is a finer subdivision than Δ, in the sense that all intermediate points that define Δ are encountered among those defining Δ^*, then $\mathcal{L}(f,\Delta) \le \mathcal{L}(f,\Delta^*) \le \mathcal{U}(f,\Delta^*) \le \mathcal{U}(f,\Delta)$. The previous relation shows that if f is Riemann integrable on $[a,b]$, then there exists a unique real number, denoted by $\int_a^b f(x)\,dx$, such that

$$\mathcal{L}(f,\Delta) \le \int_a^b f(x)\,dx \le \mathcal{U}(f,\Delta)$$

for every subdivision:[8] $\displaystyle\int_a^b f(x)\,dx = \sup_{\Delta}\{\mathcal{L}(f,\Delta)\} = \inf_{\Delta}\{\mathcal{U}(f,\Delta)\}.$

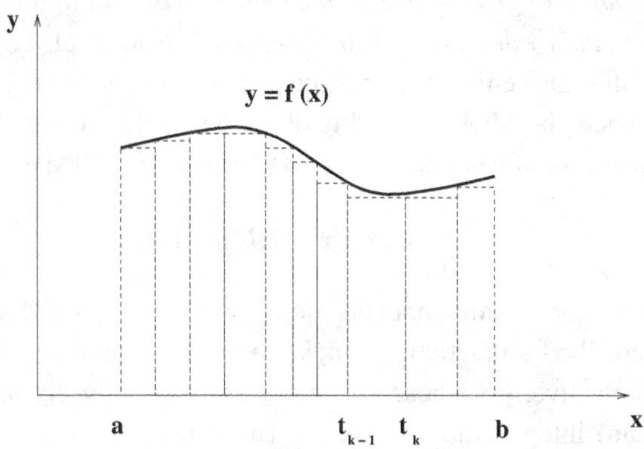

Figure 2.3 The area delimited by the graph of a continuous function $f : [a,b] \to [0,\infty)$ is the limit of the sum of the areas of thin "paving" rectangles. It can often be computed exactly by means of the fundamental theorem of calculus.

[7] Meaning that there are some real numbers $m < M$ with $m < f(x) < M$ for all $x \in [a,b]$.

[8] We thus see that integration is like summation, which is why Leibniz introduced the integral sign $\int$ as an elongated S (sum).

The Riemann integral enjoys properties that are to be expected from an averaging process:

(i) if $f \in \mathcal{R}[a,b]$, then $|f| \in \mathcal{R}[a,b]$ and $\int_a^b |f(x)|\,dx \geq 0$;

(ii) if $f \in \mathcal{R}[a,b]$, then $\alpha f \in \mathcal{R}[a,b]$ for every constant $\alpha \in \mathbb{R}$, and

$$\int_a^b [\alpha\, f(x)]\,dx = \alpha \int_a^b f(x)\,dx.$$

(iii) if $f,\ g \in \mathcal{R}[a,b]$, then $f + g \in \mathcal{R}[a,b]$ and

$$\int_a^b [f(x) + g(x)]\,dx = \int_a^b f(x)\,dx + \int_a^b g(x)\,dx\,;$$

(iv) if $a < c < b$ and $f \in \mathcal{R}[a,b]$, then $f \in \mathcal{R}[a,c]$ and $f \in \mathcal{R}[c,b]$, with

$$\int_a^b f(x)\,dx = \int_a^c f(x)\,dx + \int_c^b f(x)\,dx. \tag{2.1}$$

Moreover, if $f \in \mathcal{R}[a,c] \cap \mathcal{R}[c,b]$, then $f \in \mathcal{R}[a,b]$ and (2.1) holds.

For the proofs of these statements using the definition of the Riemann integral we refer to Hewitt and Stromberg (1965).

While the concept of Riemann integration is useful, the necessity to go beyond it is motivated by the following considerations.

(I) *It is not possible to give an accessible characterisation of the class* $\mathcal{R}[a,b]$ *within Riemann's integration theory*, see Ambrosio et al. (2011). This is possible within the setting of Lebesgue's theory (see Section 2.1.3).

(II) The confidence placed in the validity of the plausible and widely used[9] fundamental theorem of calculus, on the base of which one expected that

$$\int_a^b f'(x)\,dx = f(b) - f(a) \tag{2.2}$$

if f is differentiable with a bounded derivative, was shattered in 1881 when Volterra described a function $f : [a,b] \to \mathbb{R}$ that was differentiable with a bounded derivative, yet whose derivative was so wildly discontinuous that its (Riemann) integral did not exist.[10] The damage inflicted by Volterra's counterexample is fatal since the left side of (2.2) fails even to exist as a Riemann integral. *The Lebesgue integral restores the validity of the fundamental theorem of calculus*, as stated above, so long as we adopt Lebesgue's integral in (2.2); see Section 2.2.3 for an in-depth discussion.

[9] Despite the lack of a rigorous proof.

[10] Exercise 12 presents the alternative example with similar properties due to Goffman (1977).

(III) *The extensions of the Riemann integral to multiple integrals are cumbersome*, often involving technical and basically unnecessary assumptions on the domains of integration, see Ambrosio et al. (2011). Despite the fact that complications must arise in $\mathbb{R}^n$ with $n > 1$, as illustrated by the fact that the simple structure of an open subset[11] of $\mathbb{R}$ has no analogue for $n > 1$ (see Exercise 6), the extension of the Lebesgue integral is quite elegant (see Section 2.3).

(IV) *Many fundamental constructive processes of analysis (e.g. limits, series) are hard to accommodate within the setting of the Riemann integral.*

For example, one can prove that if $f_n : [0,1] \to [0,1]$ are continuous functions converging pointwise to a continuous function $f : [0,1] \to [0,1]$, then

$$\lim_{n\to\infty} \int_0^1 f_n(x)\,dx = \int_0^1 f(x)\,dx$$

as Riemann integrals (see Exercise 4). However, this result is not attainable within the framework of Riemann's theory, see Ambrosio et al. (2011). Moreover, the continuity assumption on the limit function f is not natural since pointwise limits of continuous functions need not be continuous. And if f is not continuous, it might happen that $f \notin \mathcal{R}[0,1]$; see Exercise 9. In contrast to this, the Lebesgue integral behaves quite well (see the discussion in Section 2.3).

(V) Within the class $\mathcal{R}[a,b]$, it is natural[12] to identify two functions for which

$$\int_a^b |f(x) - g(x)|\,dx = 0. \tag{2.3}$$

Within the resulting classes of equivalence, the expression

$$d(f,g) = \int_a^b |f(x) - g(x)|\,dx \tag{2.4}$$

between two representatives of a class of equivalence defines a distance, endowing $\mathcal{R}[a,b]$ with the structure of a metric space. The same construction can be performed within the context of Lebesgue integration, where the integrals in (2.3) and (2.4) are interpreted in the sense of Lebesgue's theory. While the metric space $\mathcal{R}[a,b]$ is not complete (see Exercise 9), this is the case for the Lebesgue integral (see Section 3.2). In this sense, *the construction of the Lebesgue integral parallels the completion of the rational numbers to the real numbers.*

11 The closed subsets of $\mathbb{R}$ do not have such a simple structure (see Exercise 7).

12 If $f \in \mathcal{R}[a,b]$ is modified at a finite number of points, the new function g satisfies (2.3), see Hewitt and Stromberg (1965). However, modifying $f \in \mathcal{R}[a,b]$ at a countable number of points so that the outcome remains bounded might nevertheless result in a function that is not Riemann integrable (see Exercise 1).

2.1.3 Outer measure and measurability

The modern view of measure theory starts with the work of Peano in the 1880s. For a subset S of the plane Peano considered the totality of finite rectangular pavings of S which lay entirely within S, and defined the *inner content* $c_*(S)$ of S to be the supremum of the areas thus covered. He also considered the totality of finite rectangular pavings which completely cover S, and defined the *outer content* $c^*(S)$ of S as the infimum of the areas thus covered. If $c_*(S) = c^*(S)$, the common value must be the area of S, whereas if $c_*(S) < c^*(S)$, then, so far as Peano was concerned, the set S did not have an area. Peano's approach was independently rediscovered and extended[13] by Jordan in the 1890s. The Peano–Jordan notion of area proved to be unsatisfactory. It was felt that a single point in the plane should have zero area, as should a countable union of points. However, if we remove all points with rational coordinates from the unit square $[0, 1] \times [0, 1]$, it is not hard to see that the obtained set S will have inner content 0 and outer content 1, hence no area according to the Peano–Jordan approach. Nevertheless, intuitively, the area of the unit square is 1 and the countably many points that were removed should not affect this. This unfortunate state called out for an extension of the Peano–Jordan definition of area. Borel's spark of inspiration in the 1900s was to allow infinitely many nonoverlapping rectangles instead of just finitely many as in the Peano–Jordan approach. Because of the nonoverlap, this infinity would have to be countable, as one can see by assigning a point with rational coordinates to each single rectangle. To describe Borel's approach we place ourselves in the simplest possible setting by discussing the Borel measure on the real line. The previous discussion and intuition suggest some minimal requirements to be satisfied by the family $\mathcal{B} = \mathcal{B}(\mathbb{R})$ of subsets of $\mathbb{R}$ having a measure (length):

(i) $\emptyset \in \mathcal{B}$;

(ii) if $B \in \mathcal{B}$, then $(\mathbb{R} \setminus B) \in \mathcal{B}$;

(iii) if $B_n \in \mathcal{B}$ for $n \geq 1$, then[14] $\bigcup_{n=1}^{\infty} B_n \in \mathcal{B}$.

These conditions and the general validity of De Morgan's identity

$$\mathbb{R} \setminus \Big(\bigcup_{n\geq 1} A_n\Big) = \bigcap_{n\geq 1} (\mathbb{R} \setminus A_n) \tag{2.5}$$

also ensure stability under countable intersections: if $B_n \in \mathcal{B}$ for $n \geq 1$, then $\bigcap_{n=1}^{\infty} B_n \in \mathcal{B}$. Families satisfying (i)–(iii) are called σ-algebras. If $\mathcal{F}_0$ is a family

[13] To a corresponding definition of volume.

[14] Borel's insight was to impose that the family is stable under countable unions.

of subsets, the minimal σ-algebra including $\mathcal{F}_0$, obtained as the intersection of all σ-algebras including $\mathcal{F}_0$, is called the σ-algebra generated by $\mathcal{F}_0$. The σ-algebra generated by all open sets of $\mathbb{R}^k$ is called the collection of *Borel sets* of $\mathbb{R}^k$ and is denoted by $\mathcal{B}(\mathbb{R}^k)$. As for the concept of a Borel measure, let us first discuss the one-dimensional setting. The Borel measure assigns to every $B \in \mathcal{B}(\mathbb{R})$ a value $\mu(B) \in [0, \infty]$, with $\mu(\emptyset) = 0$, $\mu(\mathbb{R}) = \infty$, and such that the rule is countably additive:[15]

$$\mu(B) = \sum_{n=1}^{\infty} \mu(B_n) \quad \text{if } B = \bigcup_{n=1}^{\infty} B_n \text{ with } B_n \text{ pairwise disjoint} \tag{2.6}$$

(and every B_n a Borel set). For $B \in \mathcal{B}(\mathbb{R})$, its *Borel measure* is defined by[16]

$$\mu(B) = \inf_{I \in \mathfrak{c}} \sum_{I} (b_n - a_n), \tag{2.7}$$

where $\mathfrak{c}$ is the collection of all families of finite or infinite (countable) pairwise disjoint open intervals (a_n, b_n) with $B \subset \bigcup_{n \in I} (a_n, b_n)$. The above construction accommodates exceedingly tiny intervals that spread gradually through the irregularities in the shape of the set B. Indeed, the somewhat abstract definition of Borel sets obscures the fact that these sets can be very complicated. Descriptive set theory provides a constructive characterisation of the Borel sets, revealing their structure by a process giving rise to the Borel hierarchy/stratification; see Srivastava (1998). To generate the Borel sets, one starts with the open sets and taking complements by means of property (ii) above, we obtain the closed sets.[17] This is the first level of complexity. We now perform on these sets the operation (iii) of taking countable unions. The open sets are already closed under this operation, but the closed sets are not; for example, $(0, 1) = \bigcup_{n \geq 1} [1/n, 1 - 1/n]$. Countable unions of closed sets are classically known as F_σ sets, with F and σ standing for the French words "fermé" (closed) and "somme" (sum). The complements of F_σ sets are, in view of De Morgan's identity (2.5), countable intersections of open sets, the so-called G_δ sets, with the notation coming from the German words "Gebiet" (domain) and "Durchschnitt" (intersection). The F_σ and G_δ sets form the second rank in the

15 The reason for requiring countable additivity is that finite additivity is too weak, while uncountable additivity is too strong; for example, since we intend to assign measure zero to the set consisting of a single point, it would imply that the measure of every subset of $\mathbb{R}$ must be zero.

16 In other words, we cover B by pairwise disjoint open intervals, add up their lengths, and define $\mu(B)$ as the infimum of these sums.

17 Exercise 5 provides the elegant structure of open subsets of the real line, while Exercise 7 shows that already on the real line a closed set can exhibit a complicated structure.

Borel stratification, which includes the first level.[18] We can continue this way and form the $F_{\sigma\delta}$ sets – countable intersections of F_σ sets – and the $G_{\delta\sigma}$ sets – countable unions of G_δ sets – and so on. This process can be continued to construct a genuinely new level at step k for every $k \geq 1$. But this does not exhaust the Borel sets $\mathcal{B}(\mathbb{R})$: one has to extend the inductive construction into the transfinite. We therefore see that Borel's approach extended considerably the class of sets to which one can assign a reasonable concept of length. This construction can be performed in any $\mathbb{R}^n$ with $n \geq 1$, the open intervals being replaced by cubes:[19]

$$\{(x_1, \cdots, x_n) : 2^{-m}k_j \leq x_j < 2^{-m}(k_j + 1) \text{ for } j = 1, \cdots, n\}$$

where $k_1, \cdots, k_n$, m are integers with $m \geq 1$. The two alternative constructions of Borel sets carried out in the one-dimensional case can also be performed in dimension n and the result characterising the points of continuity of a function $f : \mathbb{R}^n \to \mathbb{R}$ as a G_δ set remains valid. The only difference is that the complexity of Borel sets increases.[20] In particular, the Borel measure of the set obtained by removing all points with rational coordinates from the unit square is 1: the removed set has Borel measure zero and the full unit square has measure 1.

Despite its resounding success, the Borel measure had one major flaw: intuitively we would expect that all subsets of a set of zero measure should be negligible, and thus asigned zero measure. However, not all subsets of a Borel set of measure zero are Borel sets. For example, the Cantor ternary set C_0 described in Exercise 8 is a Borel set (being a closed set) of measure zero and having uncountably many elements. Consequently the family of subsets of C_0 has the cardinality of the set of subsets of $\mathbb{R}$, which is larger than the cardinality of the Borel sets $\mathcal{B}(\mathbb{R})$, known to be the cardinality of continuum, see Ambrosio et al. (2011). Therefore there exist non-Borel sets[21] among the subsets of C_0,

[18] The set $(0, 1] = \bigcap_{n\geq1} (0, 1 + 1/n) = \bigcup_{n\geq1} [1/n, 1]$ is both F_δ and G_σ, but is not open nor closed. The importance of the G_δ sets is underlined by the fact that the points of continuity of a function $f : \mathbb{R} \to \mathbb{R}$ are always a G_δ set, and, conversely, given a G_δ set $B \subset \mathbb{R}$, there is a function $f : \mathbb{R} \to \mathbb{R}$ continuous at every point $x \in B$ and discontinuous at every other point, see Gelbaum and Olmsted (2003) and Srivastava (1998). In particular, since the set of rational numbers in $[0, 1]$ is not a G_δ set, see Hewitt and Stromberg (1965), it is not possible to construct a function $f : [1/2, 1] \to \mathbb{R}$ that is continuous precisely at the rational points of $[1/2, 1]$. On the other hand, the irrational points of $[1/2, 1]$ form a G_δ set since their complement is easily seen to be an F_σ set. For a function $f : [1/2, 1] \to \mathbb{R}$ that is continuous precisely at the irrational points of $[1/2, 1]$ we refer to Exercise 10.

[19] See Exercise 6, where the reason why one does not rely upon open balls or rectangles is also explained.

[20] For example, already in the planar setting, there are Borel sets of type G_δ whose projection on the horizontal axis is not a Borel set. This subtle issue was discovered by Suslin (1917).

[21] One can also give constructive examples of Lebesgue measurable sets that are not Borel sets, see Federer (1969).

and according to Borel these do not have a measure. This issue was settled by Borel's student Lebesgue. The Lebesgue *outer measure* $m^*(S)$ of a set in $\mathbb{R}^2$ is defined[22] by analogy to the Peano–Jordan outer content of S, except that covers with possibly infinitely many rectangles are used. Lebesgue realised that his outer measure is not countably additive[23] in the sense of (2.6). He created a new notion of *inner measure* $m_*(S)$, defined for a set $S \subset [-n, n] \times [-n, n]$ as

$$m_*(S) = n^2 - m^*\Big(([-n, n] \times [-n, n]) \setminus S\Big),$$

and declared the set S to be measurable if $m^*(S) = m_*(S)$, defining the measure of the set S to be this common value. He also established the countable additivity of the outer measure on the measurable sets: if a measurable set S is subdivided into countably many nonoverlapping measurable subsets, then $m^*(S)$ is the sum of the outer measures of these subsets (the whole is equal to the sum of the measurable parts). Carathéodory later reformulated Lebesgue's notion of measurability as follows: a set S is (Lebesgue) measurable if for every set $A \subset \mathbb{R}^2$ we have

$$m^*(A) = m^*(S \cap A) + m^*(A \cap (\mathbb{R} \setminus S)).$$

In other words, S cuts up A into two pieces: the set $S \cap A$ of common points, and the set $A \cap (\mathbb{R} \setminus S)$ of points in A that do not belong to S. This abstract definition selects the measurable sets as being precisely those which cut up the (outer) measure of any set A into the sum of its parts. Carathéodory's version proved to be more flexible than Lebesgue's approach by means of the inner measure. Useful implications of Carathéodory's definition are, see Hewitt and Stromberg (1965):

- the family $\mathcal{L}$ of (Lebesgue) measurable sets is a σ-algebra and m^* is countably additive on $\mathcal{L}$;
- the restriction of m^* to $\mathcal{L}$, henceforth denoted m and called the Lebesgue measure, is *regular*, in the sense that for every $S \in \mathcal{L}$,

$$m(S) = \inf\{m(O) : \ S \subset O, \ O \text{ open}\} = \sup\{m(K) : \ K \subset S, \ K \text{ compact}\};$$

- all subsets of a set of zero measure are measurable.

The latter point addresses the deficiency of the Borel measure. Actually, the Lebesgue measure is a completion of the Borel measure: every Borel set is Lebesgue measurable with equal Lebesgue and Borel measure, and for every

22 These considerations are valid in any Euclidean space $\mathbb{R}^n$ with $n \geq 1$. We present them in the planar setting (for $n = 2$) to parallel the considerations related to Peano's approach.

23 However, it is countably sub-additive: $m^*(S) \leq \sum_{n=1}^{\infty} m^*(S_n)$ whenever $S \subset \bigcup_{n=1}^{\infty} S_n$. Notice that, unlike the definition of countably additive, the sets S_n need not be disjoint here.

Lebesgue measurable set $S \subset \mathbb{R}^n$ there are Borel sets B, C with $B \subset S \subset C$ and $m(C \setminus B) = 0$ (see Hewitt and Stromberg (1965)). The connection is even more striking: the Lebesgue measure is invariant under isometries[24] and the Lebesgue measure is the only countably additive function defined on the Borel sets $\mathcal{B}(\mathbb{R}^n)$ that is translation invariant and asigns measure 1 to the cube of side 1; see e.g. Elekes and Keleti (2006).

The importance of the sets of Lebesgue measure zero is further emphasised by the following characterisation of Riemann integrable functions.

Theorem 2.1 *A bounded function $f : [a,b] \to \mathbb{R}$ is Riemann integrable if and only if its set of discontinuities has Lebesgue measure zero.*

Proof An important step in the proof consists of quantising the wild behaviour of f at a point of discontinuity $x_0 \in [a,b]$. Since f is bounded, with $M = \sup_{x\in[a,b]} \{|f(x)|\} < \infty$, given $\varepsilon > 0$, we can define the oscillation of f on $[a,b] \cap [x_0 - \varepsilon,\, x_0 + \varepsilon]$ by

$$O(f,\, x_0,\, \varepsilon) = \sup_{x,y\in[a,b]\cap[x_0-\varepsilon,\, x_0+\varepsilon]} \{|f(x) - f(y)|\} \geq 0.$$

Since $\varepsilon \mapsto O(f,\, x_0,\, \varepsilon)$ is clearly a nondecreasing function on $[0,1]$, we can define the oscillation of f at $x_0 \in [a,b]$ by

$$O(f, x_0) = \lim_{\varepsilon\downarrow 0} O(f,\, x_0,\, \varepsilon).$$

Notice that f is continuous at x_0 if and only if $O(f, x_0) = 0$. Since the set $\mathcal{D}_f$ of discontinuities of a function f can be quite complicated, it is convenient to introduce for each $\varepsilon > 0$ the compact set[25]

$$A_\varepsilon = \{x_0 \in [a,b] : O(f, x_0) \geq \varepsilon\}.$$

By means of these nested[26] sets, we can represent the set $\mathcal{D}_f$ as

$$\mathcal{D}_f = \bigcup_{\varepsilon>0} A_\varepsilon = \bigcup_{n\geq 1} A_{1/n}. \tag{2.8}$$

Let us now assume that $m(\mathcal{D}_f) = 0$. Choose some $\varepsilon > 0$. Since $A_\varepsilon \subset \mathcal{D}_f$, we have $m(A_\varepsilon) = 0$. We can therefore cover A_ε by a countable family of open intervals whose lengths add up to less than ε. A_ε being compact, it is covered by finitely many of these intervals, say $I_1, \cdots, I_p$, whose lengths add up to less than ε. The set $J = [a,b] \setminus \bigcup_{k=1}^{p} I_k$ being contained in the complement of A_ε, centred at each point $y \in J$ we can find an open interval $J(y)$ of positive length, with

24 These are distance-preserving maps $F : \mathbb{R}^n \to \mathbb{R}^n$, i.e., $|F(x) - F(y)| = |x - y|$ for $x, y \in \mathbb{R}^n$. Particular examples are translations, defined as $x \mapsto x + x_0$ with $x_0 \in \mathbb{R}^n$ fixed, as well as rotations and reflections in the plane. Note that, see Mazur and Ulam (1932), an isometry of $\mathbb{R}^n$ with $F(0) = 0$ must be a linear map of the form $F(x) = Ax$ for some $n \times n$ matrix A.

25 To see that A_ε is compact, let $x_n \in A_\varepsilon$ be such that $\lim_{n\to\infty} x_n = x^*$. If $x^* \notin A_\varepsilon$, then $O(f, x^*) = \varepsilon - \delta$ for some $\delta \in (0, \varepsilon]$. Choose now $r \in (0,1)$ with $O(f, x^*, r) < \varepsilon - \delta/2$, and subsequently $n_0 \geq 1$ with $|x_{n_0} - x^*| < r/2$. For $x,\, y \in [a,b] \cap [x_{n_0} - r/2,\, x_{n_0} + r/2]$ we have

$$|x - x^*| \leq |x - x_{n_0}| + |x_{n_0} - x^*| \leq r, \quad |y - x^*| \leq |y - x_{n_0}| + |x_{n_0} - x^*| \leq r,$$

so that $O(f, x_{n_0}, r/2) \leq O(f, x^*, r) < \varepsilon$, which contradicts our assumption $x_{n_0} \in A_\varepsilon$. Thus A_ε is closed. Since $A_\varepsilon \subset [a,b]$, A_ε is compact.

26 Clearly $A_{\varepsilon_2} \subset A_{\varepsilon_1}$ if $\varepsilon_2 > \varepsilon_1 > 0$.

$\sup_{y_1,y_2\in J(y)\cap[a,b]}\{|f(y_1)-f(y_2)|\}\le\varepsilon$. Clearly $J\subset\bigcup_{x\in J}J(x)$ so that, J being compact, we can choose a finite subcover, which we denote by $I_{p+1},\cdots,I_{p+q}$. The subdivision Δ of $[a,b]$ defined by all the end points of the intervals $I_1,\cdots,I_{p+q}$ satisfies

$$\mathcal{U}(f,\Delta)-\mathcal{L}(f,\Delta)\le 2M\sum_{k=1}^{p}m(I_k)+\varepsilon\sum_{k=p+1}^{p+q}m(I_k)\le 2M\,\varepsilon+\varepsilon\,(b-a).$$

This shows that $f\in\mathcal{R}[a,b]$.

If $f\in\mathcal{R}[a,b]$, due to (2.8), to prove that $m(\mathcal{D}_f)=0$ it suffices to show that $m(A_{1/n})=0$ for every $n\ge 1$. Given $\varepsilon>0$, choose a subdivision $\Delta=\{a=x_0<x_1<\cdots<x_N=b\}$ of $[a,b]$ with $\mathcal{U}(f,\Delta)-\mathcal{L}(f,\Delta)\le\frac{\varepsilon}{n}$. If $A_{1/n}\cap(x_{k-1},x_k)\ne\emptyset$, we must have

$$\sup_{x\in[x_{k-1},x_k]}\{f(x)\}-\inf_{x\in[x_{k-1},x_k]}\{f(x)\}\ge\frac{1}{n}\,.$$

Thus

$$\frac{1}{n}\sum_{1\le k\le N:\ A_{1/n}\cap(x_{k-1},x_k)\ne\emptyset}(x_k-x_{k-1})\le\mathcal{U}(f,\Delta)-\mathcal{L}(f,\Delta)\le\frac{\varepsilon}{n},$$

so that

$$\sum_{1\le k\le N:\ A_{1/n}\cap(x_{k-1},x_k)\ne\emptyset}\Big(x_k-x_{k-1}+\frac{\varepsilon}{2^k}\Big)<2\varepsilon\,.$$

Since $\bigcup_{1\le k\le N:\ A_{1/n}\cap(x_{k-1},x_k)\ne\emptyset}\Big(x_{k-1}-\frac{\varepsilon}{2^{k+1}},\,x_k+\frac{\varepsilon}{2^{k+1}}\Big)$ is a cover of $A_{1/n}$ with open intervals whose sum of lengths does not exceed 2ε, we have that $m(A_{1/n})\le 2\varepsilon$. The arbitrariness of $\varepsilon>0$ yields $m(A_{1/n})=0$. □

We conclude our discussion by pointing out that to develop a reasonable theory of lengths, areas and volumes, one must limit oneself to some special sets: not every subset of $\mathbb{R}^n$ is measurable. A set that fails to be Lebesgue measurable, found by Vitali in 1905, is presented in Exercise 11. Furthermore, in dimension[27] $n=3$, the unit ball can be subdivided into five nonoverlapping subsets which can be rigidly moved through space (without any deformation, allowing only translations and rotations) and rearranged to form a ball of radius 2; see Wagon (1985). Intuition suggests that the volume of an object remains unchanged when it is rigidly moved through space but the rearranged ball has eight times the volume of the first! This strikingly counterintuitive result is not an inconsistency of mathematics since the "paradox" is resolved by noting that the pieces cannot be Lebesgue measurable (and therefore have no volume). This example explains why the Lebesgue measure cannot measure every set, instead being restricted to measuring sets that are Lebesgue measurable.

[27] A similar construction can be performed in any Euclidean space of dimension $n\ge 3$. However, it is not possible to replicate it in one or two dimensions by splitting into finitely many pieces, although this is possible if, like in Exercise 11, we permit countably many pieces. This difference is caused by the increase in the complexity of sets and the growth of certain special groups of transformations (the isometries of $\mathbb{R}^n$ form a group under the composition of maps) with the dimension: for $n\ge 3$ there are more complicated sets and motions to choose from.

2.2 A brief outline of the Lebesgue integral

Lebesgue was more interested in integration than in measure theory, and he constructed his measure with the aim of extending Riemann's definition of the integral. Intuition suggested that certain specific functions[28] should have an integral, even though they were not Riemann integrable. The Lebesgue integral is a much more flexible tool than Riemann's integral. The wide range of functions becoming integrable and thus falling within the embrace of mathematical analysis explains the unenvisioned wealth of applications made possible by the new concept of integral. The benefit of becoming acquainted with the Lebesgue integral clearly outweighs the transition cost of learning the new formalism. The present section is an outline intended to convey the principal facts about the Lebesgue integral.

2.2.1 Definition and basic properties

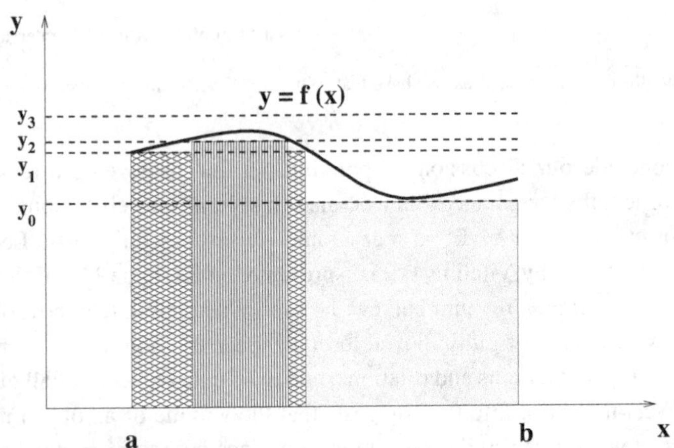

Figure 2.4 An illustration of Lebesgue's approach: lumping together the points where the function takes on approximately the same values, as indicated by the different shadings, the area is approximated by the sum $y_0\, m(S_1) + y_1\, m(S_2) + y_2\, m(S_3)$, where $S_k = \{x \in [0,1] : y_{k-1} \leq f(x) < y_k\}$ for $k = 1, 2, 3$. In this picture the sets S_1 and S_3 consist of one interval each, while S_2 is the union of two disjoint intervals. Contrast this recipe with Riemann's approach, depicted for the same function in Figure 2.3.

Instead of subdividing the domain of definition of a function $f : [a, b] \to \mathbb{R}$, as had traditionally been done for the Riemann integral, Lebesgue defined his integral by subividing the range of the function (see Figure 2.4), comparing his

[28] See Exercise 12.

method with an alternative option of paying off a debt; his exact words were, see Siegmund-Schultze (2008): "I have to pay a certain sum, which I have collected in my pocket. I take the bills and coins out of my pocket and give them to the creditor in the order I find them until I have reached the total sum. This is the Riemann integral. But I can proceed differently. After I have taken all the money out of my pocket I order the bills and coins according to identical values and then I pay the several heaps one after the other to the creditor. This is my integral." This visually perceivable recipe[29] has far-reaching consequences.

Measurable functions

The (Lebesgue) measurable functions $f : [a,b] \to \mathbb{R}$ bear a relationship to the family of all functions defined on $[a,b]$ which resembles the relation between (Lebesgue) measurable subsets of $[a,b]$ and the family of all subsets. We say that a function $f : [a,b] \to \mathbb{R}$ is *Borel measurable* if $f^{-1}[c,d) \in \mathcal{B}(\mathbb{R})$ for every choice of $c < d$, while f is called *Lebesgue measurable* (henceforth referred to simply as measurable) if $f^{-1}[c,d)$ is Lebesgue measurable for every choice of $c < d$. Given a set $A \subset \mathbb{R}$, we denoted by $f^{-1}(A)$ the inverse image under f of the set A: $f^{-1}(A) = \{x \in [a,b] : f(x) \in A\}$. The following flexible properties of inverse images explain why they are better than direct images[30] $f(A) = \{y : y = f(x) \text{ for some } x \in A\}$ with $A \subset [a,b]$:

- $f^{-1}(A \setminus B) = f^{-1}(A) \setminus f^{-1}(B)$ for all subsets $A, B \subset \mathbb{R}$;
- if $\{A_i\}_{i\in I}$ are sets indexed by an arbitrary set I, then

$$f^{-1}\Big(\bigcap_{i\in I} A_i\Big) = \bigcap_{i\in I} f^{-1}(A_i), \quad f^{-1}\Big(\bigcup_{i\in I} A_i\Big) = \bigcup_{i\in I} f^{-1}(A_i).$$

The set-theoretic proofs of the above three relations are easy; see Hewitt and Stromberg (1965). These relations show that instead of using intervals of the form $[c,d)$ in the definition, one might as well use intervals of the form (k,∞) or $[k,\infty)$ with $k \in \mathbb{R}$, since e.g. $[c,d) = \Big(\bigcap_{n\geq 1}(c - \frac{1}{n}, \infty)\Big) \cap \Big(\bigcup_{n\geq 1}(\mathbb{R} \setminus (d - \frac{1}{n}, \infty))\Big)$ and $(k,\infty) = \bigcup_{n\geq 1}[k + \frac{1}{n}, k+n)$. The main point is that the definition uses a minimal set of building blocks that suffice to ensure that the inverse image of a Borel set is measurable, see Exercise 15. While continuous functions are clearly measurable,[31] the concept of measurability is wider than that of con-

[29] Practically inspired by the way shopkeepers counted their daily intake.

[30] The example of $f(x) = x^2$ on $[-1,1]$, with $A_1 = [-1,0]$ and $A_2 = [0,1]$, in which case $f(A_1 \cap A_2) = \{0\}$ but $f(A_1) \cap f(A_2) = [0,1]$, illustrates the difference.

[31] In this context, the definition by means of the open sets (k,∞) proves to be quite economical with regard to the operations that are involved.

tinuity: the characteristic function[32] of the set $[0, 1)$ is Borel measurable but not continuous, while the characteristic function of a set $E \subset [0, 1]$ that is Lebesgue measurable without being a Borel set provides us with an example of a Lebesgue measurable function that is not Borel measurable; clearly any Borel measurable function is Lebesgue measurable. As to why we rely in the definition of Lebesgue measurable functions upon inverse images of Borel sets and not upon inverse images of Lebesgue measurable sets, the reason is that this wider class would restrict the family of measurable functions so much that not even all continuous functions $f : [a, b] \to \mathbb{R}$ would fit in; see Exercise 15.

In dealing with limiting processes, it is convenient to consider functions with values in the extended space $\overline{\mathbb{R}} := \mathbb{R} \cup \{-\infty, \infty\}$, the so-called *extended functions*. We say that $f : [0, 1] \to \overline{\mathbb{R}}$ is (Lebesgue) measurable if $f^{-1}(-\infty)$, $f^{-1}(\infty)$ and $f^{-1}(B)$ are (Lebesgue) measurable whenever $B \in \mathcal{B}(\mathbb{R})$. It suffices to require that $f^{-1}(k, \infty]$ is measurable for all real numbers k, see Hewitt and Stromberg (1965). Linear combinations of extended measurable functions, as well as products of extended measurable functions remain measurable; see Hewitt and Stromberg (1965). While the behaviour under composition of functions is quite tricky (see Exercise 16), extended measurable functions are stable under pointwise limits and countable supremum and infimum: if $f_n : [0, 1] \to \overline{\mathbb{R}}$ are measurable, then the functions $\sup_{n\geq 1} \{f_n\}$, $\inf_{n\geq 1} \{f_n\}$, $\limsup_{n\geq 1} \{f_n\}$, $\liminf_{n\geq 1} \{f_n\}$, are also measurable.[33] In particular, the pointwise limit of measurable functions is measurable. In this context, observe that in the class of (Lebesgue) measurable functions we can define the *almost everywhere convergence*, abbreviated a.e., as pointwise convergence outside a set of measure zero; on this set the limit either fails to exist, or else exists but fails to agree with the prescribed limit.

Let us now elucidate the structure of measurable functions. A function $f : [0, 1] \to \mathbb{R}$ is said to be *simple* if its range is a finite set $\{a_1, \cdots, a_n\}$ with $a_i \neq a_j$ for $i \neq j$. Setting $A_j = f^{-1}(a_j)$, we can represent f as

$$f(x) = \sum_{k=1}^{n} a_k \chi^{A_k}(x), \qquad x \in [0, 1], \tag{2.9}$$

where χ^{A_k} is the characteristic function of the set A_k. In (2.9) we have that $A_1, \cdots, A_n$ is a finite partition[34] of $[0, 1]$, and f is measurable if and only if

32 The characteristic function χ^E of a set E is defined by $\chi^E(x) = \begin{cases} 1 & \text{for} \quad x \in E, \\ 0 & \text{for} \quad x \notin E. \end{cases}$

33 For example, $\{x \in [0, 1] : \sup_{n\geq 1} \{f_n(x)\} > a\} = \bigcup_{n\geq 1} \{x \in [0, 1] : f_n(x) > a\}$, while $\inf_{n\geq 1} \{f_n\} = -\sup_{n\geq 1} \{-f_n(x)\}$ and $\limsup_{n\geq 1} \{f_n(x)\} = \inf_{n\geq 1} \{\sup_{k\geq n} \{f_k(x)\}\}$, $\liminf_{n\geq 1} \{f_n(x)\} = \sup_{n\geq 1} \{\inf_{k\geq n} \{f_k(x)\}\}$.

34 That is, $A_i \cap A_j = \emptyset$ if $i \neq j$ and $\bigcup_{j=1}^{n} A_j = [0, 1]$.

all level sets A_k are measurable.[35] We also consider extended simple functions, by assigning the values $\pm\infty$ to two measurable sets, respectively. The next result shows that any nonnegative extended measurable function can be approximated by simple functions.

Proposition 2.1 *Let* $f : [0,1] \to [0,\infty]$ *be an extended measurable function. For any integer* $n \geq 1$, *define*

$$f_n(x) = \begin{cases} \frac{k-1}{2^n} & \text{if } \frac{k-1}{2^n} \leq f(x) < \frac{k}{2^n}, \ k = 1, 2, \cdots, n2^n, \\ n & \text{if } f(x) \geq n. \end{cases}$$

Then the simple functions f_n *satisfy* $0 \leq f_1 \leq f_2 \leq \cdots$ *and* $\lim_{n\to\infty} f_n(x) = f(x)$ *on* $[0,1]$. *Moreover, if* f *is bounded, the convergence is uniform.*

Proof By construction, the f_n are nonnegative with $f_n \leq f_{n+1}$. Since

$$0 \leq f(x) - f_n(x) \leq \frac{1}{2^n} \quad \text{if} \quad f(x) < n,$$

$$0 \leq f(x) - f_n(x) = f(x) - n \quad \text{if} \quad f(x) \geq n,$$

the conclusion follows. □

The (Lebesgue) integral of a simple nonnegative (Lebesgue) measurable extended function $f : [0,1] \to [0,\infty]$ is defined by

$$\int_0^1 f(x)\,dx = \sum_{k=1}^{n} a_k\, m(A_k) \tag{2.10}$$

(using the measure-theoretic convention that $0 \times \infty = 0$), if f is represented as $f = \sum_{k=1}^{n} a_k \chi^{A_k}$ for some integer $n \geq 0$, some constants $a_1, \cdots, a_n \geq 0$ and some measurable sets $A_1, \cdots, A_n$. One can check that this definition does not depend on the particular representation formula for f; see Ambrosio et al. (2011). Moreover, if $f, g : [0,1] \to [0,\infty]$ are simple nonnegative measurable extended functions, and $\alpha, \beta \geq 0$ are constants, then $\alpha f + \beta g$ is a simple nonnegative measurable extended function with

$$\int_0^1 \big(\alpha f(x) + \beta g(x)\big)\,dx = \alpha \int_0^1 f(x)\,dx + \beta \int_0^1 g(x)\,dx.$$

There is an alternative representation for the integral of a simple nonnegative measurable extended function by means of the repartition function, and its generalisation is useful to define the integral of any nonnegative extended

[35] Note that a simple function f has many representations of the form $f(x) = \sum_{k=1}^{N} a'_k \chi^{A'_k}(x)$ for $x \in [0,1]$, if we do not insist that a'_k are distinct elements in the range of f and $A'_1, \cdots, A'_N$ need not be mutually disjoint. For example, $\chi^{[0,2]} + \chi^{[1,3]} = \chi^{[0,1)} + 2\chi^{[1,2]} + \chi^{(2,3]}$.

function. The *repartition function* $F : \mathbb{R} \to [0,1]$ of a measurable function $f : [0,1] \to \overline{\mathbb{R}}$ is defined by[36]

$$F(t) := m(\{f > t\}), \qquad t \in \mathbb{R}, \tag{2.11}$$

and has the following properties (see Exercise 18):

(i) F is nonincreasing with

$$\lim_{t\to-\infty} F(t) = m(\{f > -\infty\}), \qquad \lim_{t\to\infty} F(t) = m(\{f = \infty\}); \tag{2.12}$$

(ii) for any $t_0 \in \mathbb{R}$ we have

$$\lim_{t\downarrow t_0} F(t) = F(t_0) \quad \text{and} \quad \lim_{t\uparrow t_0} F(t) = m(\{f \geq t_0\}); \tag{2.13}$$

(F is right continuous and has left limits; in particular, F is continuous at t_0 if and only if $m(\{f = t_0\}) = 0$.)

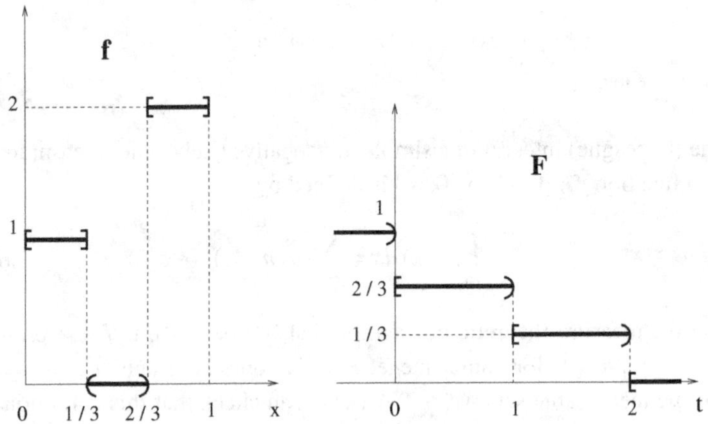

Figure 2.5 The right-continuous repartition function F of the simple function f has jump discontinuities at the points $x = 0$, $x = 1$ and $x = 2$.

Recall the representation (2.9) of a nonnegative simple measurable function $f : [0,1] \to [0,\infty)$, with $n \geq 1$ and $0 < a_1 < \cdots < a_n$. Set $a_0 = 0$. Since $F(t) = 0$ for $t \geq a_n$, we have

$$\int_0^1 f(x)\,dx = \sum_{k=1}^{n} a_k\, m(A_k) = \sum_{k=1}^{n} a_k\,[F(a_{k-1}) - F(a_k)] \tag{2.14}$$
$$= \sum_{k=0}^{n-1} a_{k+1}\, F(a_k) - \sum_{k=1}^{n} a_k\, F(a_k) = \sum_{k=0}^{n-1} (a_{k+1} - a_k)\, F(a_k) = \int_0^{\infty} F(t)\,dt\,.$$

[36] Here $\{f > t\}$ denotes the set $\{x \in [0,1] : f(x) > t\}$.

This observation links the Lebesgue integral to the integral $\int_0^\infty F(t)\,dt$ of a nonincreasing function $F : [0,\infty] \to [0,\infty]$: for functions of this type we define $\int_0^\infty F(t)\,dt$ as the supremum of the partial sums $\sum_{k=0}^{N-1}(t_{k+1}-t_k)\,F(t_{k+1})$ for all possible choices $0 \le t_0 < t_1 < \cdots < t_N < \infty$ with $N \ge 1$. For our purposes the most relevant property of these Riemann-type integrals is their continuity under monotonically nondecreasing sequences.

Proposition 2.2 *If $F_n \uparrow F$ with $F_n : [0,\infty] \to [0,\infty]$ nonincreasing, then $F : [0,\infty] \to [0,\infty]$ is nonincreasing and $\int_0^\infty F_n(t)\,dt \uparrow \int_0^\infty F(t)\,dt$.*

Proof For $0 \le x < y \le \infty$, we have $F(x) \ge F(y)$ since $F_n(x) \ge F_n(y)$ for every $n \ge 1$. Concerning the integrals, clearly $\int_0^\infty F_n(t)\,dt \le \int_0^\infty F(t)\,dt$ for every $n \ge 1$. Let $L = \int_0^\infty F(t)\,dt > 0$, the case $L = 0$ being trivial. For every $\varepsilon > 0$, by definition we can find $N \ge 1$ and $0 \le t_0 < t_1 < \cdots < t_N < \infty$ with $\sum_{k=0}^{N-1}(t_{k+1}-t_k)\,F(t_{k+1}) > L - \frac{\varepsilon}{2}$. For $n \ge 1$ large enough we have $F_n(t_{k+1}) \ge F(t_{k+1}) - \frac{\varepsilon}{2(t_N-t_0)}$ for $k = 0,\cdots,N-1$, so that $\int_0^\infty F_n(t)\,dt \ge \sum_{k=0}^{N-1}(t_{k+1}-t_k)\,F_n(t_{k+1}) \ge \sum_{k=0}^{N-1}(t_{k+1}-t_k)\,F(t_{k+1}) - \frac{\varepsilon}{2} > L - \varepsilon$ and the proof is completed. □

Having the identity (2.14) in mind, we define the Lebesgue integral of an extended nonnegative measurable function $f : [0,1] \to [0,\infty]$ as

$$\int_0^1 f(x)\,dx = \int_0^\infty m(\{f > t\})\,dt\,.$$

If the integral is finite we say that f is integrable on $[0,1]$ and denote $f \in L^1[0,1]$. Note that if $f \in L^1[0,1]$, then the set $\{f = \infty\}$ has measure zero. Indeed, by Markov's inequality (Exercise 25) we have $\lim_{n\to\infty} m(\{f > n\}) = 0$ and $\{f = \infty\} = \bigcap_{n\ge1}\{f > n\}$ yields the assertion in view of Exercise 17 (ii).

The above constructive definition has some important consequences. Firstly, the integral is invariant if we modify f on a set of measure zero: if $f_1 = f_2$ a.e., then for all $t \ge 0$ we have $m(\{f_1 > t\}) = m(\{f_2 > t\})$. Secondly, the additivity property is easily eastablished: if $f_1, f_2 : [0,1] \to [0,\infty]$ are extended measurable functions, we have

$$\int_0^1 \big(f_1(x) + f_2(x)\big)\,dx = \int_0^1 f_1(x)\,dx + \int_0^1 f_2(x)\,dx\,,$$

see Ambrosio et al. (2011). Moreover, the Lebesgue integral behaves nicely under monotone nondecreasing limits.

Theorem 2.2 (Monotone convergence theorem) *Let $f_n : [0,1] \to [0,\infty]$ be a nondecreasing sequence of extended measurable functions. If $f(x) = \lim_{n\to\infty} f_n(x)$ so that $f_n \uparrow f$, then $\int_0^1 f_n(x)\,dx \uparrow \int_0^1 f(x)\,dx$.*

Proof This follows from Proposition 2.2 since the associated repartition functions are such that $F_n \uparrow F$. □

Combining Proposition 2.1 with the monotone convergence theorem we obtain the following approximation property.

Proposition 2.3 *For any extended nonnegative measurable function* $f : [0,1] \to [0,\infty]$ *there exist simple measurable functions* $f_n : [0,1] \to [0,\infty)$ *such that* $f_n \uparrow f$ *and* $\int_0^1 f_n(x)\,dx \uparrow \int_0^1 f(x)\,dx$.

The result in Proposition 2.3 is indicative of an alternative equivalent definition of the Lebesgue integral as the supremum of the integral of minorant simple functions – see Rudin (1974). The present approach is technically less demanding. Either way, Proposition 2.3 resembles the definition of the Riemann integral, the difference being in the available simple functions: in both cases they have finite range, but in Riemann's theory the level sets are finite unions of intervals while for the Lebesgue integral the level sets belong to the richer family of measurable sets.

The following limiting result is of fundamental importance.

Theorem 2.3 (Fatou's lemma) *Let* $f_n : [0,1] \to [0,\infty]$ *be extended measurable functions. Then* $\int_0^1 \liminf_{n\to\infty} f_n(x)\,dx \le \liminf_{n\to\infty} \int_0^1 f_n(x)\,dx$.

Proof Let $g_n(x) = \inf_{k\ge n}\{f_k(x)\}$ and $f(x) = \liminf_{n\to\infty} f_n(x)$, so that $g_n \uparrow f$. By the monotone convergence theorem, $\int_0^1 f(x)\,dx = \lim_{n\to\infty} \int_0^1 g_n(x)\,dx$. On the other hand, clearly $\int_0^1 g_n(x)\,dx \le \int_0^1 f_n(x)\,dx$ for $n \ge 1$, and the assertion follows at once. □

The definition of the integral of functions changing sign is now within grasp. Let $f : [0,1] \to \overline{\mathbb{R}}$ be an extended measurable function. We say that f is (Lebesgue) integrable if both its positive part $f^+(x) := \max\{f(x), 0\}$ and its negative part $f^-(x) := \max\{-f(x), 0\}$ are integrable, in which case the relation $f = f^+ - f^-$ induces the definition

$$\int_0^1 f(x)\,dx = \int_0^1 f^+(x)\,dx - \int_0^1 f^-(x)\,dx.$$

Since $|f| = f^+ + f^-$, as a byproduct of this definition, we see that f is integrable if and only if $\int_0^1 |f(x)|\,dx < \infty$. It is easy to establish the following properties of the integral: if $f, g : [0,1] \to \overline{\mathbb{R}}$ are integrable, then

- $\alpha f + \beta g$ is integrable and $\int_0^1 (\alpha f(x) + \beta g(x))\,dx = \alpha \int_0^1 f(x)\,dx + \beta \int_0^1 g(x)\,dx$ for any constants $\alpha, \beta \in \mathbb{R}$;
- if $f \le g$ a.e., then $\int_0^1 f(x)\,dx \le \int_0^1 g(x)\,dx$;
- $\left|\int_0^1 f(x)\,dx\right| \le \int_0^1 |f(x)|\,dx$,

see Ambrosio et al. (2011). The fundamental result for applications, enabling us to commute limit and integral, is the following.

Theorem 2.4 (Dominated convergence theorem) *Let $f_n : [0,1] \to \mathbb{R}$ be measurable functions converging a.e. to some $f : [0,1] \to \mathbb{R}$. If there exists a nonnegative integrable function $g : [0,1] \to [0,\infty]$ such that for every $n \geq 1$,*

$$|f_n(x)| \leq g(x) \; for \; a.e. \; x \in [0,1],$$

then the functions $\{f_n\}_{n\geq 1}$ and the function f are integrable and

$$\lim_{n\to\infty} \int_0^1 f_n(x)\,dx = \int_0^1 f(x)\,dx\,.$$

Proof The fact that f is measurable with $|f(x)| \leq g(x)$ a.e. on [0,1] is clear. In particular, f is integrable. Since $f+g$ is nonnegative, Fatou's lemma yields $\int_0^1 (f(x)+g(x))\,dx \leq \liminf_{n\to\infty} \int_0^1 (f_n(x)+g(x))\,dx$, so that

$$\int_0^1 f(x)\,dx \leq \liminf_{n\to\infty} \int_0^1 f_n(x)\,dx\,. \tag{2.15}$$

Similary, $\int_0^1 (g(x) - f(x))\,dx \leq \liminf_{n\to\infty} \int_0^1 (g(x) - f_n(x))\,dx$, so that

$$\int_0^1 f(x)\,dx \geq \limsup_{n\to\infty} \int_0^1 f_n(x)\,dx\,. \tag{2.16}$$

The conclusion follows by combining (2.15) and (2.16). □

Remark In our presentation the restriction to $[0,1]$ as the domain of definition was for simplicity. All considerations carry over to any measurable subset $A \subset \mathbb{R}$, even if $m(A) = \infty$. Furthermore, if a measurable function $f : \mathbb{R} \to \mathbb{R}$ is such that $f\chi^A$ is integrable for some measurable set A, then $\int_A f(x)\,dx = \int_{\mathbb{R}} f(x)\chi^A(x)\,dx$. There is also a natural definition for complex valued measurable/integrable functions $f : A \to \mathbb{C}$: its real and imaginary parts, $\Re\mathfrak{e}\{f\}$ and $\Im\mathfrak{m}\{f\}$, must be measurable/integrable, and in the latter case we simply set $\int_A f(x)\,dx = \int_A \Re\mathfrak{e}\{f(x)\}\,dx + i \int_A \Im\mathfrak{m}\{f(x)\}\,dx$, thus reducing the problem to the familiar setting of real-valued functions.

2.2.2 Multiple integrals

The Borel sets of $\mathbb{R}^n$ with $n \geq 2$ are constructed analogously to those of the real line: they form the σ-algebra generated by the open sets. Due to the more complicated structure of open sets in $\mathbb{R}^n$ with $n \geq 2$ (compare the statements of Exercise 5 and Exercise 6) instead of open intervals, the building blocks are boxes of the type $\{x = (x_1, \cdots, x_n) : \; k_i 2^{-N} \leq x_i < (k_i + 1)2^{-N}, \; i = 1, \cdots, n\}$

with $k_1, \cdots, k_n$ and $N \geq 0$ integers. The measure of such a box is 2^{-nN} and the Borel measure of a Borel set $B \subset \mathbb{R}^n$ is defined as the infimum of the sums of the measures of disjoint boxes covering B (see Exercise 6). Just like in the one-dimensional case, we complete the Borel measure to the Lebesgue measure $\mathcal{L}^n$ by means of the sets of zero (Lebesgue) measure, obtained as subsets of the Borel sets of zero measure. This procedure yields the σ-algebra of the Lebesgue measurable sets, with a subset $A \subset \mathbb{R}^n$ being (Lebesgue) measurable if and only if it differs from a Borel set B by a set of Lebesgue measure zero, so that the Lebesgue measures $\mathcal{L}^n$ of A and B are equal. This is the length measure in $\mathbb{R}$, the area measure in $\mathbb{R}^2$ and the volume measure in $\mathbb{R}^3$. The following properties hold:

- (translation invariance) if the set $A \subset \mathbb{R}^n$ is (Lebesgue) measurable, then for any $x \in \mathbb{R}^n$ the set $A + x = \{y + x : y \in A\}$ is measurable with $\mathcal{L}^n(A) = \mathcal{L}^n(A + x)$;
- (regularity) for any (Lebesgue) measurable set $A \subset \mathbb{R}^n$ with $\mathcal{L}^n(A) < \infty$, given $\varepsilon > 0$, we can find a compact set $K_\varepsilon \subset A$ and an open set $O_\varepsilon \supset A$ with $\mathcal{L}^n(O_\varepsilon \setminus K_\varepsilon) < \varepsilon$;
- (transformation under linear maps[37]) for any $n \times n$ matrix $L : \mathbb{R}^n \to \mathbb{R}^n$ and any (Lebesgue) measurable set $A \subset \mathbb{R}^n$, the image set $L(A)$ is measurable with $\mathcal{L}^n(L(A)) = |\det(L)|\, \mathcal{L}^n(A)$;

see Ambrosio et al. (2011). Given a (Lebesgue) measurable set $A \subset \mathbb{R}^n$, the monotone convergence theorem and Fatou's lemma hold for (Lebesgue) measurable functions $f_n : A \to [0, \infty]$, and the dominated convergence theorem holds for sequences of (Lebesgue) measurable functions $f_n : A \to \mathbb{R}$; a glance at their proofs confirms that the same approach can also be implemented in this higher-dimensional context. A fundamental result of Lebesgue integration theory in $\mathbb{R}^n$ concerns the interchange of order of integration in a multiple integral.

Theorem 2.5 (Fubini) *If $f : [0,1]^{n+m} \to \mathbb{R}$ is $\mathcal{L}^{n+m}$-integrable, then*

(i) *for $\mathcal{L}^n$ a.e. $x \in [0,1]^n$ the function $y \mapsto f(x,y)$ is $\mathcal{L}^m$ integrable on $[0,1]^m$, and its integral over $[0,1]^m$ is $\mathcal{L}^n$ integrable on $[0,1]^n$;*

(ii) *for $\mathcal{L}^m$ a.e. $y \in [0,1]^m$ the function $x \mapsto f(x,y)$ is $\mathcal{L}^n$ integrable on $[0,1]^n$, and its integral over $[0,1]^n$ is $\mathcal{L}^m$ integrable on $[0,1]^m$;*

[37] In particular, $\mathcal{L}^n$ is rotationally invariant: if L is orthogonal, then $\mathcal{L}^n(L(A)) = \mathcal{L}^n(A)$. Recall that orthogonal matrix means that its columns and rows are orthogonal unit vectors, or, equivalently, its transpose is equal to its inverse. Orthogonal matrices preserve the inner product.

(iii) *we have*

$$\int_{[0,1]^{n+m}} f(x,y)d\mathcal{L}^{n+m}(x,y) = \int_{[0,1]^n} \Big(\int_{[0,1]^m} f(x,y)\, d\mathcal{L}^m(y) \Big)\, d\mathcal{L}^n(x)$$

$$= \int_{[0,1]^m} \Big(\int_{[0,1]^n} f(x,y)\, d\mathcal{L}^n(x) \Big)\, d\mathcal{L}^m(y)\,. \tag{2.17}$$

The special case when Fubini's theorem is specialised to the characteristic function of a measurable set in $\mathbb{R}^n \times \mathbb{R}^m$ is known as "Cavalieri's principle".

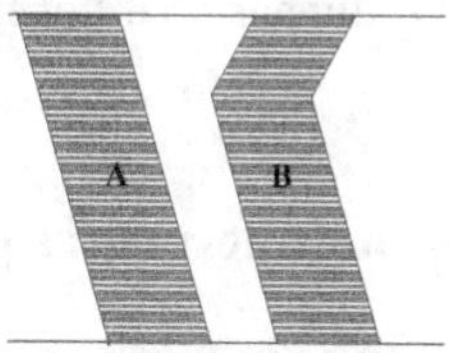

Figure 2.6 Simple illustration of Cavalieri's principle in $\mathbb{R}^2$: if two regions in the plane lie between two parallel lines and every line parallel to them intersects both regions in segments of the same length, then the two regions have equal areas.

Despite being merely a corollary of Fubini's theorem, the next result is often very useful since the finiteness of the iterated integral implies higher-dimensional integrability.

Theorem 2.6 (Tonelli) *Let* $f : [0,1]^{n+m} \to [0,\infty]$ *be* $\mathcal{L}^{n+m}$*-measurable. Then for* $\mathcal{L}^n$ *a.e.* $x \in [0,1]^n$ *the function* $y \mapsto f(x,y)$ *is* $\mathcal{L}^m$*-measurable on* $[0,1]^m$ *with its integral* $\mathcal{L}^n$*-measurable on* $[0,1]^n$*, and for* $\mathcal{L}^m$ *a.e.* $y \in [0,1]^m$ *the function* $x \mapsto f(x,y)$ *is* $\mathcal{L}^n$*-measurable on* $[0,1]^n$ *with its integral* $\mathcal{L}^m$*-measurable on* $[0,1]^m$*. Moreover, the equalities in (2.17) hold.*

The proof of Fubini's theorem is quite tedious; see Hewitt and Stromberg (1965) and Rudin (1974). While Fubini's theorem is very powerful for applications, it is not a steamroller result: the theory of multiple integrals is quite subtle[38] and caution is advisable. For example, the classes of Borel sets and Lebesgue measurable sets are both closed under the operations of complementation, countable unions, and countable intersections. None of them are, however, closed under continuous images.[39] Perhaps the most spectacular fail-

[38] For example, Fubini's original proof was incomplete, see Bruckner et al. (1997).

[39] One can prove that a continuous function $f : [0,1] \to \mathbb{R}$ maps (Lebesgue) measurable sets into (Lebesgue) measurable sets if and only if it maps sets of measure zero into sets of measure zero, and either property is specific to absolutely continuous functions, see Natanson (1955). It actually suffices that Borel sets of Lebesgue measure zero are mapped into sets of Lebesgue measure zero, see Leoni (2009). A brief discussion of absolutely continuous functions is to be found in Section 2.2.3.

ure is that of projections: there exists Borel sets $B \subset \mathbb{R}^2$ whose projection on the real axis are not Borel sets.[40] For Lebesgue measurable sets the situation is worse: there are $\mathcal{L}^2$ measurable sets $A \subset [0,1]^2$ whose sections $y \mapsto A_y = \{x \in [0,1] : (x,y) \in A\}$ are not $\mathcal{L}^1$-measurable subsets of $\mathbb{R}$ for certain values of $y \in [0,1]$, see Exercise 30. This pathology explains the need of the specification "a.e." in items 1 and 2 of Fubini's theorem: one cannot expect measurability or integrability for every single section $x \mapsto f(x,y)$ or $y \mapsto f(x,y)$. To appreciate the importance of the integrability assumption in Fubini's theorem, see Exercises 32–33. For simplicity, from now on we will write dx instead of $d\mathcal{L}^n(x)$ in n-dimensional Lebesgue integrals.

2.2.3 The anti-derivative problem

This section is devoted to a discussion of the relation between integration and differentiation. The existence of continuous and bounded functions $f : \mathbb{R} \to \mathbb{R}$ that are nowhere differentiable (see e.g. Exercise 34) is to be contrasted with the following result.

Theorem 2.7 (Lebesgue) *A nondecreasing function $f : [a,b] \to \mathbb{R}$ has a finite derivative almost everywhere on $[a,b]$.*

For the technically demanding proof of this result we refer to Hewitt and Stromberg (1965). It is remarkable that no continuity assumptions are required and that the result is sharp (see Exercise 35). Throughout our discussion of nondecreasing functions it turns out to be advantageous to set $f'(x) = 0$ at those points x where $f : [a,b] \to \mathbb{R}$ is not differentiable. The function f' thus defined turns out to be integrable on $[a,b]$, with

$$\int_a^b f'(x)\,dx \leq f(b) - f(a). \tag{2.18}$$

Indeed, extend f to $\mathbb{R}$ by setting $f(x) = f(b)$ for $x \geq b$ and $f(x) = f(a)$ for all $x \leq a$.

40 The study of structural properties of Borel sets got an impetus from an error in Lebesgue's treatise (in a lemma, stated without proof, claiming that clearly a set of real numbers that is the projection of a Borel set in the plane is a Borel set), spotted by the undergraduate student Suslin who introduced the family of analytic sets as the class of sets obtainable as continuous images of some Borel set. Analytic subsets of $\mathbb{R}^n$ are closed under countable unions and intersections, continuous images, and inverse images of continuous maps. However, the complement of an analytic set need not be analytic: Suslin proved that if the complement of an analytic set is analytic then the set is Borel, see Suslin (1917). (Conversely, any Borel set is analytic and Borel sets are closed under complements.) There are analytic sets that are not Borel sets but every analytic set is Lebesgue measurable, see Bruckner et al. (1997).

At all points $x \in (a,b)$ where f is differentiable we have $f'(x) = \lim_{n\to\infty} \frac{f(x+1/n)-f(x)}{1/n}$. Consequently f' is a.e. the pointwise limit of measurable functions,[41] therefore measurable. Being nonnegative, the integral on the left-hand side of (2.18) is well-defined. Fatou's lemma yields $\int_a^b f'(x)\,dx \le \liminf_{n\to\infty} \{n \int_a^b [f(x+1/n) - f(x)]\,dx\}$. But monotone functions are Riemann integrable, see Exercise 3. Using a standard change of variables for Riemann integrals we obtain $\int_a^b f(x+1/n)\,dx = \int_{a+1/n}^{b+1/n} f(x)\,dx$, so that $n\int_a^b [f(x+1/n) - f(x)]\,dx = n\int_{a+1/n}^{b+1/n} f(x)\,dx - n\int_a^b f(x)\,dx = n\int_b^{b+1/n} f(x)\,dx - n\int_a^{a+1/n} f(x)\,dx \le f(b) - f(a)$ and the inequality (2.18) emerges.

A striking feature of (2.18) is that even if we require that f is continuous, we cannot expect equality (see Exercise 36). This is in marked contrast to the second fundamental theorem of calculus[42] which states that if $f : [a,b] \to \mathbb{R}$ is differentiable with f' Riemann integrable, then $\int_a^x f'(s)\,ds = f(x) - f(a)$ for all $x \in [a,b]$. Thus the integral of a Riemann-integrable derivative is the function, up to the constant $f(a)$. The fact that f' is Riemann integrable is used in the proof as follows. For any partition $a = x_0 < x_1 < \cdots < x_n = x$ of $[a,x]$ with $x \in (a,b]$, by the mean-value theorem for derivatives we have $f(x) - f(a) = \sum_{k=1}^n [f(x_k) - f(x_{k-1})] = \sum_{k=1}^n f'(c_k)\,[x_k - x_{k-1}]$ for some $c_k \in (x_k, x_{k-1})$. As the partition gets finer, the Riemann sums converge to $\int_a^x f'(s)\,ds$. The second fundamental theorem of calculus is very useful in view of the need to evaluate integrals, such as area and work, in applications. As soon as the integrand can be recognised as the Riemann-integrable derivative of some function, the theorem works. Unfortunately there are differentiable functions f such that f' is bounded but not Riemann integrable (see Exercise 12). The Lebesgue integral is a remedy to this shortcoming.

1. If $f : [a,b] \to \mathbb{R}$ is Lebesgue integrable, then $F(x) = \int_a^x f(x)\,dx$ is a.e. differentiable with $F' = f$. Note that f might have no points of continuity, see Exercise 1!
2. If $f : [a,b] \to \mathbb{R}$ is differentiable with f' Lebesgue integrable, then for all $x \in [a,b]$ we have $\int_a^x f'(s)\,ds = f(x) - f(a)$. In particular, this holds if f' is bounded.

Exercise 36 shows that we cannot expect the validity of the fundamental theorem of calculus[43] for functions $f : [a,b] \to \mathbb{R}$ that are nondecreasing and

[41] Both f and $x \mapsto f(x+1/n)$ are Lebesgue measurable, being nondecreasing: nonempty inverse images of intervals of the form $[c,d)$ are intervals.

[42] The first fundamental theorem of calculus states that if $f : [a,b] \to \mathbb{R}$ is Riemann integrable, then $F(x) = \int_a^x f(x)\,dx$ is differentiable with $F'(x) = f(x)$ at every point $x \in [a,b]$ where f is continuous. The proof consists of noticing that $\frac{F(x+h)-F(x)}{h} = \frac{1}{h}\int_x^{x+h} f(x)\,dx$ and then using the continuity of f at x to get $f(x)$ for the limit as $h \to 0$.

[43] In the sense of equality in (2.18).

continuous. A stronger notion of continuity is needed. A function $f : \mathbb{R} \to \mathbb{R}$ is called **absolutely continuous** if for every $\varepsilon > 0$ there exists $\delta > 0$ such that for any finite[44] family of disjoint intervals $(a_1, b_1), \cdots, (a_n, b_n)$, we have

$$\sum_{k=1}^{n} |f(b_k) - f(a_k)| < \varepsilon \quad \text{if} \quad \sum_{k=1}^{n} |b_k - a_k| < \delta . \tag{2.19}$$

A function $f : [a, b] \to \mathbb{R}$ is said to be absolutely continuous if (2.19) holds whenever the intervals (a_k, b_k) all lie in $[a, b]$. The concept of absolute continuity is stronger than that of uniform continuity. Indeed, the latter corresponds to taking $n = 1$ in (2.19) while Exercise 36 provides an example of a uniformly continuous function (being continuous on a compact set) that is not absolutely continuous.[45] Exercise 27 shows that the indefinite integral of a Lebesgue integrable function is absolutely continuous. The converse is also true.

Theorem 2.8 (Fundamental theorem of the integral calculus for Lebesgue integrals) *If $f : [a, b] \to \mathbb{R}$ is absolutely continuous, then f is differentiable a.e. with f' Lebesgue integrable on $[a, b]$ and for every $x \in [a, b]$ we have*

$$f(x) = f(a) + \int_a^x f'(s)\, ds . \tag{2.20}$$

For the intricate proof of Theorem 2.8 we refer the interested reader to Hewitt and Stromberg (1965). Let us emphasise the following important aspect. It is tempting to think that differentiability a.e. and (Lebesgue) integrability of the derivative could be sufficient to ensure the validity of (2.20). After all, these are the minimal requirements to give a meaning to the right-hand side of the formula. The absolute continuity of the Lebesgue indefinite integrals (Exercise 27) shows that this is not so, and Exercise 36 provides us with a concrete counter-example. In connection with Exercise 36, notice that the nondecreasing continuous function under consideration is not absolutely continuous and maps the Cantor set (of measure zero) into [0,1], a set with strictly positive Lebesgue measure. This is indicative of a general fact: *a monotone function is absolutely continuous if and only if it maps sets of measure zero into sets of zero measure.*[46] Note that Exercise 36 shows that a continuous, a.e. differentiable function need not be the integral of its derivative. It is a highly nontrivial

44 One can replace "finite" by "countable" – see Exercise 37. However, allowing the intervals to overlap leads to a stronger concept of continuity – see Exercise 41.

45 We shall see that the absolute continuity of f ensures equality in (2.18).

46 The result extends to the difference of two monotone functions (these are the functions of bounded variation – see Section 2.2.4). On the other hand, there are continuous functions $f : [a, b] \to \mathbb{R}$ that map sets of measure zero into sets of measure zero but fail on a set of positive measure to have a derivative, thus not being absolutely continuous. For these matters see Hewitt and Stromberg (1965).

result that if $f : [a,b] \to \mathbb{R}$ is continuous and $f'(x)$ exists and is finite for all but a countable set N of points $x \in (a,b)$, with f' integrable[47] on $[a,b]$, then f is absolutely continuous and $f(x) - f(a) = \int_a^x f'(x)\,dx$ for all $x \in [a,b]$. A proof can be found in Hewitt and Stromberg (1965). This result highlights the difference between countable sets and sets of measure zero within the framework of Lebesgue's differentiation theory.

A corollary of Theorem 2.8 is the formula for integration by parts.

Theorem 2.9 *If $f, g : [a,b] \to \mathbb{R}$ are absolutely continuous, then*

$$\int_a^b f(x)\,g'(x)\,dx = f(b)g(b) - f(a)g(a) - \int_a^b f'(x)\,g(x)\,dx\,.$$

Proof From (2.19), using the inequality $|f(x)g(x) - f(y)g(y)| \le \sup_{s\in[a,b]}\{|f(s)|\}\,|g(x) - g(y)| + \sup_{s\in[a,b]}\{|g(s)|\}\,|f(x) - f(y)|$, we see that fg is absolutely continuous. Consequently $f(b)g(b) - f(a)g(a) = \int_a^b (f(x)g(x))'\,dx$ by Theorem 2.8. Since the functions f, g and fg are a.e. differentiable on $[a,b]$, we get that $(fg)'$ is a.e. the sum of the two functions $f'g$ and fg', each of which being integrable as a product of an integrable function with a bounded function. □

Let us now present some considerations about changes of variables in a Lebesgue integral. If the function $F : [c,d] \to \mathbb{R}$ possesses a finite derivative $f(y)$ at every point $y \in [c,d]$, and if $g : [a,b] \to [c,d]$ is differentiable a.e., from the definition of differentiability we see that at every point $x \in [a,b]$ where g has a derivative we have at once that the composite function $F \circ g$ is differentiable and the chain rule $(F \circ g)'(x) = f(g(x)) \cdot g'(x)$ holds. It is of interest to extend this result to more general situations where F does not have a finite derivative everywhere.

Theorem 2.10 *If $F : [c,d] \to \mathbb{R}$ is absolutely continuous and $g : [a,b] \to [c,d]$ as well as $F \circ g : [a,b] \to \mathbb{R}$ have finite derivatives a.e. on $[a,b]$, then the chain rule $(F \circ g)' = (f \circ g) \cdot g'$ holds a.e. on $[a,b]$, where $f = F'$ a.e.*

We refer to Leoni (2009) for a proof of Theorem 2.10. The delicate character of this result is revealed in the following example. Let $g : [0,1] \to [0,1]$ be the increasing function in Exercise 36 and set $F = g^{-1}$. Then $(F \circ g)' = 1$ everywhere, which together with $g' = 0$ a.e. on $[0,1]$ shows that the chain rule does not hold. A further issue relative to Theorem 2.10 is the requirement of a.e. differentiability of the function $F \circ g$. This is especially important in the context of changes of variables for Lebesgue integrals.

[47] f' is a measurable function, defined a.e. Modifying it on the set N of measure zero, e.g. by setting $f'(x) = 0$ for $x \in N$, has no bearing on the integrability issue (or on the value of the integral, if f' turns out to be integrable on $[a,b]$).

Theorem 2.11 *Let $f : [c,d] \to \mathbb{R}$ be (Lebesgue) integrable and $g : [a,b] \to [c,d]$ have a finite derivative a.e. on $[a,b]$. Then $(f \circ g) \cdot g'$ is integrable, and the formula*

$$\int_{g(\alpha)}^{g(\beta)} f(x)\,dx = \int_{\alpha}^{\beta} f(g(s)) \cdot g'(s)\,ds \tag{2.21}$$

holds for all $\alpha, \beta \in [a,b]$ if and only if $F \circ g$ is absolutely continuous, where $F(x) = \int_c^x f(s)\,ds$ for $x \in [c,d]$.

We refrain from presenting the lengthy proof – see Leoni (2009). Instead, we concentrate on exploring the hypotheses and the implications of this result. Note that the absolute continuity of g is not required. For example, the change of variable formula holds on $[a,b] = [c,d] = [-1,1]$ for $f(x) = x^2$, $g(s) = s\sin(s^{-1})$ when $s \neq 0$ and $g(0) = 0$, even though g is not absolutely continuous.[48] Indeed, the hypotheses of Theorem 2.11 are fulfilled since $(F \circ g)(x) = g^3(x)/3$ is continuously differentiable on $[-1,1]$. On the other hand, the integrability of f and the absolute continuity of g would generally not suffice for $F \circ g$ to be absolutely continuous. The reason for this is that the necessary and sufficient condition for $F : \mathbb{R} \to \mathbb{R}$ to be such that $F \circ g : [a,b] \to \mathbb{R}$ is absolutely continuous for all absolutely continuous functions $g : [a,b] \to \mathbb{R}$, is that F is locally Lipschitz,[49] see Leoni (2009), and this condition is stronger than absolute continuity (see Exercise 41).

Let us now discuss an important example. The function $f : [0,1] \to \mathbb{R}$ defined by $f(0) = 0$ and $f(x) = x^2 \sin(\pi/x^2)$ for $x \in (0,1]$ is continuous and has a finite derivative $f'(x) = 2x\sin(\pi/x^2) - 2\pi x^{-1}\cos(\pi/x^2)$ at $x \in (0,1)$. But f' is not integrable[50] on $[0,1]$. This example raises the following question: if the continuous function $f : [0,1] \to \mathbb{R}$ has a finite derivative $f'(x)$ at every point $x \in (0,1)$, how can f be recovered from f'? Lebesgue's theory solves this problem if f' is integrable on $[0,1]$. If f' is not integrable,[51] the Lebesgue integral is insufficient. Other concepts of integration were devised just for this purpose – see the discussion in Henstock (1991). These theories of integration assign a meaning to $\int_a^b f'(x)\,dx$ by exploiting cancellations similar to those encoun-

48 To see this, the explicit form of $g'(s)$ for $s \neq 0$ precludes its integrability on $[-1,1]$.

49 That is, for every $M > 0$ there is a constant $L_M > 0$ such that $|f(x) - f(y)| \leq L_M\,|x-y|$ whenever $|x|, |y| \leq M$.

50 Indeed, for $0 < u \leq s \leq v \leq 1$, we have $|f'(s)| \leq 2v + 2\pi u^{-1}$, so that f' is bounded and therefore integrable on $[u,v]$. Denoting $a_k = (2k+1/2)^{-1/2}$ and $b_k = (2k)^{-1/2}$ for $k \geq 1$, we get $\int_0^1 |f'(s)|\,ds \geq -\sum_{k=1}^{\infty} \int_{a_k}^{b_k} f'(s)\,ds = -\sum_{k=1}^{\infty} f(s)\Big|_{a_k}^{b_k} = \sum_{k=1}^{\infty} \frac{1}{2k+\frac{1}{2}} = \infty$.

51 Luzin showed that the a.e. finite function $f : [a,b] \to [-\infty,\infty]$ is a.e. the derivative of a continuous function F if and only if f is measurable – see Bruckner and Leonard (1966).

tered in conditionally convergent Riemann integrals.[52] However, the utility of such integration theories is not sufficiently broad to make a compelling case for their study outside a relatively narrow circle of experts. Furthermore, the problem of providing a convenient characterisation of functions $f : [a, b] \to \mathbb{R}$ that are derivatives of a function $F : [a, b] \to \mathbb{R}$, in the sense that $F'(x) = f(x)$ for all $x \in [a, b]$, proves to be quite elusive.[53] Some necessary conditions are:

(c1) the existence of F ensures that the function f is a pointwise limit of continuous functions (see Exercise 38), and the points of continuity of such functions must be dense in $[a, b]$, see Natanson (1955).

(c2) any derivative f has the Darboux property (or intermediate value) property[54] (see Exercise 38).

Note that the Darboux property is far weaker than continuity: there are functions with this property that are discontinuous everywhere. However, in view of the regularising property (c1), a derivative must be continuous on a dense set, even if it can be very badly discontinuous – it is possible for it to be discontinuous on a dense set. But the two properties (c1) and (c2) do not characterise derivatives since, unlike the class of derivatives, the class of functions enjoying both properties is not closed under addition.[55] Further restrictions have to be imposed. While characterisations are possible, the available approaches do not supply practical tests.

2.2.4 Length of curves

A curve in $\mathbb{R}^m$ is the image of a continuous function $f : I \to \mathbb{R}^m$, where $I \subset \mathbb{R}$ is an interval. To estimate the length of a curve $f : [a, b] \to \mathbb{R}^m$, we can compute the length of polygonal approximations whose vertices lie on the curve, as shown in Figure 2.7. More precisely, if $x_0 = a < x_1 < \cdots < x_n = b$ is a partition of $[a, b]$, then $\sum_{k=1}^{n} |f(x_k) - f(x_{k-1})|$ is the length of the polygonal approximation of f whose vertices are located at the points $f(x_0), f(x_1), \cdots, f(x_n)$. As we refine the partition, intuitively at least, these polygonal lengths get closer to the true length. We define the length of the curve represented by f as the **total**

52 A simple illustration being the above example: for $x \in (0, 1]$ we set $\int_0^x f'(s)\,ds$ equal to the improper integral $\lim_{\varepsilon \downarrow 0} \int_\varepsilon^x f'(s)\,ds$.

53 See Bruckner and Leonard (1966) for an in-depth discussion.

54 Meaning that if $f(x_1) = y_1$ and $f(x_2) = y_2$, and y is between y_1 and y_2, then there exists an x between x_1 and x_2 with $f(x) = y$.

55 For this reason the Darboux functions are of marginal interest in analysis.

variation $V_f[a,b]$ of f, given by

$$V_f[a,b] = \sup \sum_{k=1}^{n} |f(x_k) - f(x_{k-1})|\,, \tag{2.22}$$

where the supremum is taken over all partitions $\{x_0, \cdots, x_n\}$ of $[a,b]$. If $V_f[a,b]$ is finite we say that the curve is **rectifiable** and f has **bounded variation**. While the term "curve" only refers to a continuous function f, note that (2.22) is meaningful even if we drop the continuity assumption on f. This is the general definition of a function of bounded variation, as introduced by Jordan. It is easy to see that $f = (f_1, \cdots, f_m) : [a,b] \to \mathbb{R}^m$ has finite variation if and only if each component $f_k : [a,b] \to \mathbb{R}$ with $k = 1, \cdots, m$, has bounded variation. Therefore it suffices to study real-valued functions of bounded variation. Note that in the case $m = 1$ this terminology is slightly misleading: the graph $x \mapsto (x, f(x))$ of a function $f : [a,b] \to \mathbb{R}$ has finite length if and only if $V_f[a,b] < \infty$, even if $V_f[a,b]$ is not its length!

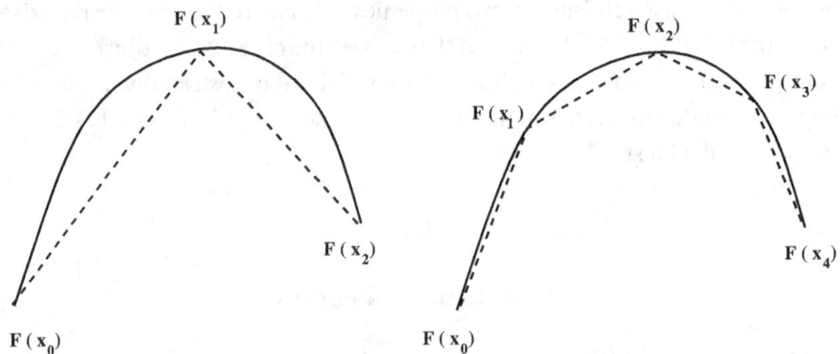

Figure 2.7 Approximating a planar curve by broken line segments. By refining the partition (from three points to five) we increase the polygonal length due to the triangle inequality.

If $f : [a,b] \to \mathbb{R}$ is monotone (either nondecreasing or nonincreasing), then it is obvious that f is of bounded variation, with $V_f[a,b] = |f(b) - f(a)|$, this value being attained for any partition. In general, we have by definition that $V_f[a,b] \geq 0$. Also, it is easy to check that the length is additive on intervals: for any $c \in (a,b)$, we have $V_f[a,b] = V_f[a,c] + V_f[c,b]$. Consequently, the nonnegative extended map[56] $x \mapsto V_f[a,x]$ is nondecreasing on $[a,b]$; we define $V_f[a,a] = 0$. These preliminary considerations lead at once to the fundamental theorem for functions of bounded variation.

[56] ∞ being an allowed value.

Theorem 2.12 (Jordan decomposition theorem) *A function $f : [a,b] \to \mathbb{R}$ is of bounded variation if and only if it is the difference of two nondecreasing functions.*

Proof Write $f(x) = V_f[a,x] - [V_f[a,x] - f(x)]$ for $x \in [a,b]$. The proof is completed by noticing that the function $x \mapsto [V_f[a,x] - f(x)]$ is nondecreasing on $[a,b]$. Indeed, if $a \le x < y \le b$, then $[V_f[a,y] - f(y)] - [V_f[a,x] - f(x)] = V_f[x,y] - [f(y) - f(x)] \ge 0$ since by definition $V_f[x,y] \ge |f(y) - f(x)|$. □

Combining Theorem 2.7 and Theorem 2.12, we deduce that a function $f : [a,b] \to \mathbb{R}$ of bounded variation has a finite derivative a.e. Exercise 36 provides an example of a function of bounded variation that is not absolutely continuous. On the other hand, we have the following.

Theorem 2.13 *An absolutely continuous function $f : [a,b] \to \mathbb{R}$ is of bounded variation*[57] *and*

$$V_f[a,b] = \int_a^b |f'(x)|\,dx\,. \tag{2.23}$$

Proof For $a = x_0 < x_1 < \cdots < x_n = b$ we have $\sum_{k=1}^n |f(x_k) - f(x_{k-1})| = \sum_{k=1}^n \left| \int_{x_{k-1}}^{x_k} f'(x)\,dx \right| \le \sum_{k=1}^n \int_{x_{k-1}}^{x_k} |f'(x)|\,dx = \int_a^b |f'(x)|\,dx = \|f'\|_{L^1[a,b]}$. Therefore $V_f[a,b] \le \|f'\|_{L^1[a,b]}$. To get the opposite inequality, let $\varepsilon > 0$ and choose[58] a continuous function $h : [a,b] \to \mathbb{R}$ such that $\|h - f'\|_{L^1[a,b]} < \varepsilon$. If $g(x) = \int_a^x h(s)\,ds$, then g is continuously differentiable with $g' = h$. For a partition $a = x_0 < x_1 < \cdots < x_n = b$, the mean-value theorem yields $\sum_{k=1}^n |g(x_k) - g(x_{k-1})| = \sum_{k=1}^n \left| \int_{x_{k-1}}^{x_k} h(x)\,dx \right| = \sum_{k=1}^n (x_k - x_{k-1})\,|h(\xi_k)|$ for some $\xi_k \in (x_{k-1}, x_k)$. Refining the partition, the convergence of the Riemann sums yields $V_g[a,b] = \|h\|_{L^1[a,b]}$. The first part of the proof shows that $V_{g-f}[a,b] \le \|h - f'\|_{L^1[a,b]}$ and the definition of the total variation ensures $V_g[a,b] \le V_f[a,b] + V_{g-f}[a,b]$. Thus

$$\begin{aligned}
\|f'\|_{L^1[a,b]} &\le \|f' - h\|_{L^1[a,b]} + \|h\|_{L^1[a,b]} = \|f' - h\|_{L^1[a,b]} + V_g[a,b] \\
&< \varepsilon + V_g[a,b] \le \varepsilon + V_f[a,b] + V_{g-f}[a,b] \le \varepsilon + V_f[a,b] + \|h - f'\|_{L^1[a,b]} < 2\varepsilon + V_f[a,b]\,.
\end{aligned}$$

Since $\varepsilon > 0$ was arbitrary, we have $\|f'\|_{L^1[a,b]} \le V_f[a,b]$ and (2.23) is established. □

The above proof of (2.23) also holds for absolutely continuous functions $f : [a,b] \to \mathbb{C}$, in which case $|x + iy|$ in the definitions (2.22), (2.19) and in (2.23) is to be interpreted as $\sqrt{x^2 + y^2}$ for $x, y \in \mathbb{R}$ and f is absolutely continuous if and only if both its real and imaginary parts are absolutely continuous, so that Theorem 2.8 is also valid for complex-valued functions.

Let us return to the issue of computing the length of a curve. For the curve

[57] To appreciate the leverage offered by the difference of two monotone functions, see Exercise 40 for an absolutely continuous function that fails to be monotone on any subinterval.

[58] For a proof of this approximation result by means of Fourier series, see Section 4.4.

represented by the continuous function $f : [a, b] \to \mathbb{C}$ of bounded variation, its length is $V_f[a, b]$. The example of $f : [0, 1] \to \mathbb{R}$ given by

$$f(0) = 0 \quad \text{and} \quad f(x) = x \sin(\pi/x) \quad \text{for} \quad x \neq 0\,, \tag{2.24}$$

shows that continuous functions might have graphs of infinite length.[59] On the other hand, if the curve is the graph of a continuously differentiable function $f_0 : [a, b] \to \mathbb{R}$, then, using Riemann sums, one can immediately check that the length is given by the formula

$$\int_a^b \sqrt{1 + [f_0'(x)]^2}\, dx\,, \tag{2.25}$$

which corresponds to the complex version of (2.23) applied to $f(x) := x + i f_0(x)$. The previous considerations ensure the validity of the formula (2.25) whenever f_0 is absolutely continuous. This fact coupled with the inequality $\sqrt{1 + [f_0'(x)]^2} > |f_0'(x)|$ ensure that the formula (2.25) holds for f_0 continuous and of bounded variation if and only if f_0 is absolutely continuous.

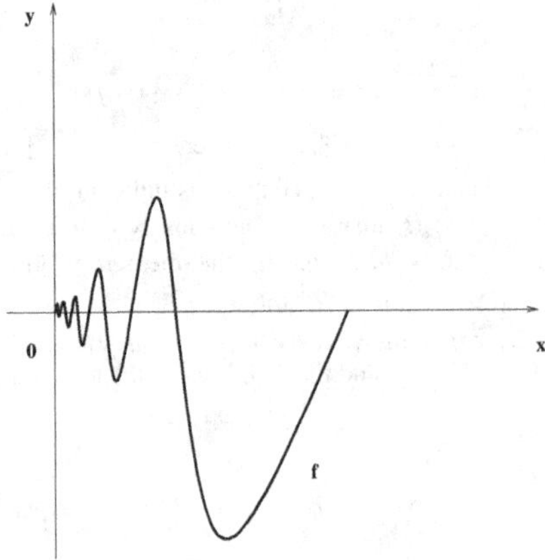

Figure 2.8 The graph of the continuous function $f : [0, 1] \to [-1, 1]$ defined by (2.24) has infinite length.

Let now $f : [a, b] \to \mathbb{C}$ be a rectifiable curve of length (that is, total variation) $L = V_f[a, b]$. The function $s : [a, b] \to [0, L]$ defined by

$$s(x) = V_f[a, x]$$

59 For $n \geq 3$, let $x_0 = 0$ and $x_k = 2/(2n - k)$ for $k = 1, \cdots, 2n - 2$. Then $V_f[0, 1] \geq \sum_{k=0}^{2n-3} |f(x_{k+1}) - f(x_k)| = 2 \sum_{j=0}^{n-2} |f(x_{2j+1}| = 2 \sum_{j=0}^{n-2} \frac{2}{2n-2j-1} > 2 \sum_{j=0}^{n-2} \frac{1}{n-j} \to \infty$. In this context, it is worth pointing out the existence of space-filling curves; these are continuous maps $F : [0, 1] \to \mathbb{R}^2$ whose range cover an entire unit square in $\mathbb{R}^2$, also called "Peano curves" since the first example was provided by Peano (1890). For a simpler construction we refer to Piranian et al. (1952).

is called the **arc-length function**. The definition of the total variation of a complex-valued function shows at once that s is nondecreasing. Moreover, the continuity of f (which is part of the definition of a curve) ensures that s is continuous, see Exercise 39. We define the **arc length parametrisation** of the curve $\tilde{f} : [0, L] \to \mathbb{C}$ as follows. For $t \in [0, L]$, due to the intermediate value theorem for the continuous function s, there is some $x \in [a, b]$ such that $s(x) = t$. There might be more than one such value of x: this can happen only if f is constant on some interval $[x_-, x^+]$, in which case $s(x) = s(x_-) = t$ for all $x \in [x_-, x_+]$. Despite this possible ambiguity, the fact that s is nondecreasing and continuous shows that $\tilde{f}(t) := f(x)$ is a well-defined function, that is, the assigned value $f(x)$ is independent of the choice of $x \in [a, b]$ with $s(x) = t$. We clearly have $\tilde{f}(0) = f(a)$ and $\tilde{f}(L) = f(b)$. If $0 \leq s_1 < s_2 \leq L$ with $s(x_1) = s_1$ and $s(x_2) = s_2$, then $|\tilde{f}(s_1) - \tilde{f}(s_2)| = |f(x_1) - f(x_2)| \leq V_f[x_1, x_2] = V_f[a, x_2] - V_f[a, x_1] = s(x_2) - s(x_1) = s_2 - s_1$. The absolutely continuity of $\tilde{f}$ follows now directly from the definition (2.19). Formula (2.25) yields $V_{\tilde{f}}[0, s] = \int_0^s |\tilde{f}'(r)|\, dr$ for every $s \in [0, L]$. But $V_{\tilde{f}}[0, t] = t$ for all $t \in [0, L]$. Indeed, if $x \in [a, b]$ is such that $s(x) = t$, then, using partitions, we see at once that $V_f[a, x] = V_{\tilde{f}}[0, t]$, while $V_f[a, x] = s(x)$ by definition. Consequently, Theorem 2.8 ensures $|\tilde{f}'| = 1$ a.e. on $[0, L]$. Geometrically, this means that a.e. there is a unit tangent vector to a rectifiable curve. Among the various possible parametrisations of a curve, the arc length parametrisation is to be preferred analytically and geometrically.

We conclude our discussion by pointing out an interesting aspect of continuous functions $f : [a, b] \to \mathbb{R}$. If $m = \min_{s \in [a,b]} \{f(s)\}$ and $M = \max_{s \in [a,b]} \{f(s)\}$, define the Banach indicatrix $N_f : [m, M] \to [0, \infty]$ by setting $N_f(y)$ equal to the number of solutions $x \in [a, b]$ of the equation $f(x) = y$. It turns out that N_f is integrable with $\int_m^M N_f(y)\, dy = V_f[a, b]$, see Natanson (1955). In particular, if the continuous function has bounded variation, then N_f is integrable on $[m, M]$, so that the set of values $y \in [m, M]$ that are attained infinitely often has zero measure.

2.3 Abstract measure theory

Let us discuss some aspects of abstract measure theory. The abstract setting is of fundamental importance in probability theory (in the context of the distribution of a random variable – see Section 7.4), while the generalisation of the notion of measure to encompass signed measures and complex-valued measures has a functional-analytic significance (see Section 3.4).

If X is a set and $\mathcal{A}$ a σ-algebra of subsets[60] of X, a *measure* μ on the measurable space $(X, \mathcal{A})$ is a function $\mu : \mathcal{A} \to [0, \infty]$ such that

(i) $\mu(\emptyset) = 0$;

(ii) if $\{A_j\}_{j\geq 1}$ are disjoint sets in $\mathcal{A}$, then $\mu\big(\bigcup_{j\geq 1} A_j\big) = \sum_{j\geq 1} \mu(A_j)$.

Note that if $Z \in \mathcal{A}$ is such that $\mu(Z) = 0$, then any $A \subset Z$ has $\mu(A) = 0$, provided that $A \in \mathcal{A}$. The case when $\mathcal{A}$ is the family of the Borel subsets of $X = [0, 1]$, with μ the Lebesgue measure, shows that it need not be true that all subsets of null sets belong to $\mathcal{A}$. By the process of completion, the enlargement of the σ-algebra includes all subsets of null sets. In the case of Borel sets one obtains the Lebesgue measurable sets. The procedure can be performed even in the general abstract setting, as follows. If $(X, \mathcal{A}, \mu)$ is a measure space, let $\mathcal{N} = \{N \in \mathcal{A} : \mu(N) = 0\}$ and set $\overline{\mathcal{A}} = \{A \cup B : A \in \mathcal{A} \text{ and } B \subset N \text{ for some } N \in \mathcal{N}\}$. One can show that $\overline{\mathcal{A}}$ is a σ-algebra and there is a unique extension $\overline{\mu}$ of μ to a *complete measure* on $\overline{\mathcal{A}}$, with $\overline{\mu}(A \cup B) = \mu(A)$ if $A \in \mathcal{A}$ and $B \subset N$ for some $N \in \mathcal{N}$; see Hewitt and Stromberg (1965). Since completeness can always be achieved and since it sometimes obviates annoying technical points, throughout this section we tacitly assume that all measures are complete.

In addition to the Lebesgue measure, other examples of measures include:

(a) the *point mass* or *Dirac measure* at $x_0 \in X$, defined by $\mu(A) = 1$ if $x_0 \in A$ and $\mu(A) = 0$ if $x_0 \notin A$, with $\mathcal{A}$ being the family of all subsets of X;
(b) on the σ-algebra of subsets A of an uncountable set X that are either countable or whose complement $X \setminus A$ is countable, set $\mu(A) = 0$ if A is countable and $\mu(A) = 1$ if $X \setminus A$ is countable.

A standard terminology in measure theory is that of a *σ-finite measure*, occurring when $X = \bigcup_{j\geq 1} A_j$ with $A_j \in \mathcal{A}$ such that $\mu(A_j) < \infty$ for all $j \geq 1$. The Lebesgue measure on $\mathbb{R}^n$ and the point mass at $x_0 \in \mathbb{R}^n$ are both σ-finite but the measure defined in item (b) above is not σ-finite.

If $\mathcal{A}$ is a σ-algebra of subsets of the set X, a *signed measure* on $(X, \mathcal{A})$ is a map $\nu : \mathcal{A} \to [-\infty, \infty]$ such that

(i) $\nu(\emptyset) = 0$;
(ii) ν assumes at most one of the values $\pm\infty$;
(iii) if $\{A_j\}_{j\geq 1}$ are disjoint sets in $\mathcal{A}$, then $\nu\big(\bigcup_{j\geq 1} A_j\big) = \sum_{j\geq 1} \nu(A_j)$, with the latter sum converging absolutely if $\nu\big(\bigcup_{j\geq 1} A_j\big)$ is finite.

[60] That is, a nonempty collection of subsets that is closed under complements and countable unions and intersections. In particular, $\emptyset \in \mathcal{A}$ and $X \in \mathcal{A}$, for if $A \in \mathcal{A}$, then $\emptyset = A \cap (X \setminus A)$ and $X = A \cup (X \setminus A)$.

A *complex measure* on $(X, \mathcal{A})$ is a map $\nu : \mathcal{A} \to \mathbb{C}$ such that

(i) $\nu(\emptyset) = 0$;
(ii) if $\{A_j\}_{j\geq 1}$ are disjoint sets in $\mathcal{A}$, then $\nu(\bigcup_{j\geq 1} A_j) = \sum_{j\geq 1} \nu(A_j)$, with the series converging absolutely.

If ν is a complex measure, we write ν_r and ν_i for its real and imaginary parts, respectively. Since infinite values are not allowed, ν_r and ν_i are signed measures that do not assume the values $\pm\infty$: they are both finite, so that the range of the complex measure ν is a bounded subset of $\mathbb{C}$.

Two examples of signed measures come to mind. Firstly, if μ_1 and μ_2 are (positive) measures on the σ-algebra $\mathcal{A}$, at least one being finite, then $\nu = \mu_1 - \mu_2$ is a signed measure. Secondly, if μ is a measure on $\mathcal{A}$ and $f : X \to [-\infty, \infty]$ is an $\mathcal{A}$-measurable function such that at least one of the expressions $\int_X f^+ d\mu$ and $\int_X f^- d\mu$ is finite, then the set function $\nu(A) = \int_A f^+ d\mu - \int_A f^- d\mu$ is a signed measure. It turns out that these are really the only examples – every signed measure originates from either of these construction procedures. Indeed, the *Hahn decomposition* of a signed measure ν consists of the existence of a positive set $P \subset X$ and a negative set $M \subset X$ such that $X = P \cup M$; a set $A \in \mathcal{A}$ being called positive (respectively negative) for ν if $\nu(B) \geq 0$ (respectively $\nu(B) \leq 0$) for all $B \in \mathcal{A}$ such that $B \subset A$. Every signed measure admits a Hahn decomposition (see e.g. Hewitt and Stromberg (1965)) leading to its representation as the difference of two (positive) measures – the *Jordan decomposition*: there exists unique (positive) measures ν^+ and ν^- such that $\nu = \nu^+ - \nu^-$ and $\nu^+ \perp \nu^-$, the latter symbol expressing the fact that ν^+ is *singular with respect to* ν^- (or, equivalently, the other way around), in the sense that there exists $A \in \mathcal{A}$ with $\nu^+(A \cap B) = 0$ and $\nu^-((X \setminus A) \cap B) = 0$ for all $B \in \mathcal{A}$, that is, ν^+ is null on A and ν^- is null on $X \setminus A$, so that ν^+ and ν^- live on disjoint sets. While the Jordan decomposition is unique, the Hahn decomposition need not be so since ν-null sets can be transferred from P to M or from M to P. The measures ν^+ and ν^- are called the *positive and negative variations* of ν, and the (positive) measure $|\nu| = \nu^+ + \nu^-$ is called the *total variation* of ν, by analogy with the representation of a real function of bounded variation as the difference of two nondecreasing functions (see Section 2.2.4). We say that ν is finite, respectively σ-finite, if $|\nu|$ is finite, respectively σ-finite. A measurable function $f : X \to [-\infty, \infty]$ is called integrable with respect to the signed measure ν if it is integrable with respect to both $\nu^=$ and ν^-, in which case we set $\int_X f\, d\nu = \int_X f\, d\nu^+ - \int_X f\, d\nu^-$. These concepts readily generalise to complex measures ν, merely by applying the procedure to the real and imaginary parts ν_r and ν_i separately. The versatility of Lebesgue's integral can be appreciated by noting that the procedure adopted in Sections 2.2.1–2.2.2, comprising monotone convergence, dominated convergence, Fatou's lemma and Fubini's theorem, carry over to (complete) abstract measures.[61]

In contrast to the above, the interplay between integrability and differentiability that was described in Sections 2.2.3–2.2.4, is quite intimately connected to the properties of the Lebesgue integral on finite intervals. Nevertheless, an abstract concept of differentiation exists. The antithesis of a signed measure ν that is singular with respect to the (positive) measure μ, both on $(X, \mathcal{A})$, is *absolute continuity*: $\nu \ll \mu$ (ν is absolutely continuous with respect to μ) if $\nu(A) = 0$ whenever $\mu(A) = 0$; here $A \in \mathcal{A}$. The *Radon–Nikodym theorem* gives for a wide class of measures a complete picture of the notion of absolute continuity: if ν is a σ-finite signed measure and μ is a positive measure on $(X, \mathcal{A})$, then there exists unique σ-finite signed measures λ, ρ on $(X, \mathcal{A})$ such that $\rho \ll \mu$, $\lambda \perp \mu$ and the so-called *Lebesgue decomposition* $\nu = \rho + \lambda$ of ν with respect to μ holds, with $d\rho = f\,d\mu$ for some μ-integrable function $f : X \to [-\infty, \infty]$, determined uniquely with exception of a set of μ-measure zero, and called the *Radon–Nikodym derivative* of ν with respect to μ; we denote $f = d\nu/d\mu$. For a proof we refer to Hewitt and Stromberg (1965). This result extends to the case when ν is a complex measure[62] by applying it to the real and imaginary parts – the only difference being that λ is now a complex measure and $f : X \to \mathbb{C}$. However, the result may fail without the assumption of σ-finiteness (see Exercise 43).

The "derivative" of a signed/complex measure with respect to another measure is a convenient tool. For example, it is useful in defining products of signed/complex measures on Cartesian product spaces: if μ, ν are measures on $\mathbb{R}^n$ and $\mathbb{R}^m$ respectively, we define the product measure $\mu \times \nu$ on $\mathbb{R}^{n+m}$ by $d(\mu \times \nu)(x, y) = \frac{d\mu}{d|\mu|}(x)\,\frac{d\nu}{d|\nu|}(y)\; d(|\mu| \times |\nu|)(x, y)$. The notion of Radon–Nikodym derivative of a measure ν with respect to another measure μ becomes transparent when μ is the Lebesgue measure on $\mathbb{R}^n$, and ν is a regular signed or complex Borel measure[63] on $\mathbb{R}^n$. Indeed, in this context, if $d\nu = d\lambda + f\,d\mu$ is the Lebesgue decomposition, then for almost every $x \in \mathbb{R}^n$ we have $\lim_{r \downarrow 0} \frac{\nu(B_r)}{\mu(B_r)} = f(x)$, see Evans and Gariepy (1992), where $B_r = \{y \in \mathbb{R}^n : |x - y| < r\}$ is the family of open balls shrinking to $x \in \mathbb{R}^n$. In this setting the abstract derivative can be understood as a pointwise derivative.

61 And, by means of decompositions into real/imaginary parts and differences between positive measures, it even extends to signed measures and complex measures.

62 Since all complex measures are finite, the condition of σ-finite becomes redundant.

63 That is, $|\nu|(K) < \infty$ for every compact set $K \subset \mathbb{R}^n$, and $|\nu|(B) = \inf\{|\nu|(O) : O \text{ open}, B \subset O\}$ for every Borel set $B \subset \mathbb{R}^n$.

2.4 Exercises

We recommend attempting to solve the problems devoted to Riemann's integral only after familiarising oneself with the main concepts and tools of Lebesgue integration presented in Sections 2.1 and 2.2 of this chapter. The gained overview permits more accessible and elegant approaches.

2.4.1 Statements

1. Prove that the Dirichlet function

$$f(x) = \begin{cases} 1 & \text{if } x \text{ is rational}, \\ 0 & \text{if } x \text{ is irrational}, \end{cases}$$

is not Riemann integrable on $[0, 1]$.

2. Let $\{x_n\}_{n\geq 1}$ be an enumeration of the rational numbers in $[0, 1)$. Define $f : [0, 1] \to \mathbb{R}$ by the rule

$$f(x) = \sum_{n\in S(x)} \frac{1}{2^n},$$

where $S(x)$ are all integers $n \geq 1$ such that $x_n < x$. Prove that f is strictly monotone and Riemann integrable, but has infinitely many discontinuities.

3. Show that a monotone function $f : [a, b] \to \mathbb{R}$ is Riemann integrable.

4. Let $f_n : [0, 1] \to [0, 1]$ be continuous functions converging pointwise to a continuous function $f : [0, 1] \to [0, 1]$. Show that $\lim_{n\to\infty} \int_0^1 f_n(x)\,dx = \int_0^1 f(x)\,dx$ as Riemann integrals.

5. Prove the following structural characterisation of open sets in $\mathbb{R}$: if $O \subset \mathbb{R}$ is a nonvoid open set, then there exists one and only one family $\mathcal{F}$ of pairwise disjoint, finite or countably infinite open intervals of $\mathbb{R}$, such that $O = \bigcup \mathcal{F}$.

6. (i) It is known that in $\mathbb{R}^2$ the open discs play the analogous role that open intervals play on $\mathbb{R}$: they are the building blocks for the open sets, in the sense that every open set is a union of such sets. Nevertheless, show that the open set $(0, 1) \times (0, 1)$ is not a union of disjoint open discs. The same result holds if instead of open discs we consider open rectangles.

(ii) Consider the family $\mathcal{S}$ of squares of the form

$$\{(x, y) : 2^{-m}k_1 \leq x < 2^{-m}(k_1 + 1),\ 2^{-m}k_2 \leq y < 2^{-m}(k_2 + 1)\},$$

where k_1, k_2, m are integers with $m \geq 1$. Show that any open set $O \subset \mathbb{R}^2$ is the disjoint union of countably many squares in $\mathcal{S}$, but that this decomposition is not unique. (The situation in $\mathbb{R}^n$ with $n \geq 3$ is analogous to that in the planar case.)

7. Let $\lambda \in (0, 1/3]$. Remove the open interval $((1 - \lambda)/2,\ (1 + \lambda)/2)$ of length λ from the middle of $[0, 1]$. This leaves us with the set C_1, consisting of two disjoint closed intervals of equal length. From each of these intervals remove the middle open interval of length $\lambda/3$, so that the remaining set C_2 will consist of the union of four closed disjoint intervals of equal lengths (see Figure 2.9). The lengths of the three disjoint open intervals forming $[0, 1] \setminus C_2$ add up to $\lambda + 2\lambda/3$. Remove now the middle open interval of length $\lambda/3^2$ from each of the four components of C_2 to obtain the set C_3. Continuing this process, in the nth step of the construction we obtain a set C_n consisting of 2^n closed disjoint intervals of equal length, the first containing 0 and the last 1, interlaced by $(2^n - 1)$ open intervals of varying lengths that add up to $\lambda + 2\lambda/3 + \cdots + 2^{n-1}\lambda/3^{n-1}$. Let $C = \bigcap_{n\geq 1} C_n$. Show that C is a compact set, of measure $(1 - 3\lambda)$, containing no interval and with no isolated points.[64]

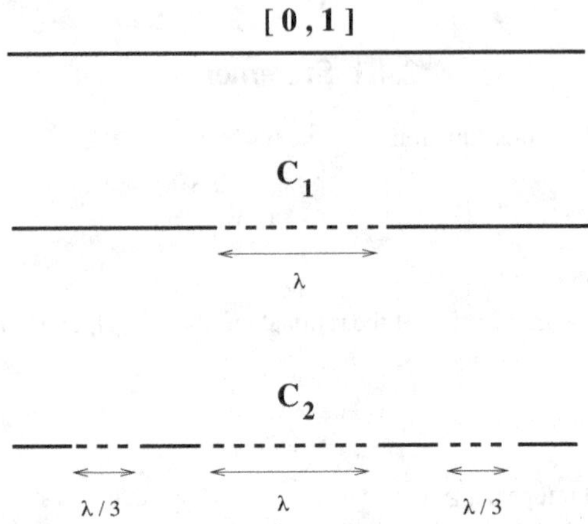

Figure 2.9 The first steps in the construction of the Cantor set of Exercise 7.

8. Setting $\lambda = 1/3$ in Exercise 7 we obtain a set of measure zero, the *Cantor ternary set* C_0. Show that there are uncountably many points in C_0.

9. Use the construction of a Cantor set C of positive measure, described in Exercise 7, to exhibit a sequence of continuous functions $f_n : [0, 1] \to [0, 1]$ that converges pointwise to the characteristic function χ^C of the Cantor set and is a nonconvergent Cauchy sequence with respect to the distance defined by (2.4). Show that χ^C is not Riemann integrable.

10. For integers $m, n \geq 1$, denote by (m, n) their greatest common divisor. Show that the function $f : [1/2, 1] \to \mathbb{R}$, defined by $f(x) = 0$ if x is irrational and $f(x) = 1/p$ if $x = q/p$ with the integers $q,\ p \geq 1$ relatively prime, is continuous at all irrational points and discontinuous at all rational points.

11. Consider the equivalence relation in $\mathbb{R}$ defined by $x \approx y$ if $x - y$ is rational, and pick up a single element $x \in [0, 1]$ as a representative of an equivalence class induced by this relation, thus forming a set $V \subset [0, 1]$. Show that V is not Lebesgue measurable.

12. Let C be the Cantor set constructed for $\lambda = 1/6$ in Exercise 7 and decompose the open set $O = [0, 1] \setminus C$ into $O = \bigcup_{n\geq 1} I_n$, with the open intervals $I_n = (a_n, b_n)$ of length l_n. For each $n \geq 1$, let $J_n = [c_n, d_n]$ be the closed interval centred at the middle of I_n and of length l_n^2. Define the continuous function $f : J_n \to [0, 1]$ by

$$f(x) = \begin{cases} 2(x - c_n)/l_n^2 & \text{for} \quad c_n \leq x \leq (c_n + d_n)/2\,, \\ 2(d_n - x)/l_n^2 & \text{for} \quad (c_n + d_n)/2 \leq x \leq d_n\,, \end{cases}$$

64 That is, for every $x \in C$ we can find a sequence $\{x_k\}_{k\geq 1} \in C \setminus \{x\}$ with $\lim_{k\to\infty} x_k = x$.

(see Figure 2.10) and extend it to a (discontinuous) function $f : [0,1] \to [0,1]$ by setting $f(x) = 0$ if $x \notin J_n$ for some $n \geq 1$. Prove that f is a derivative and that it is not Riemann integrable.

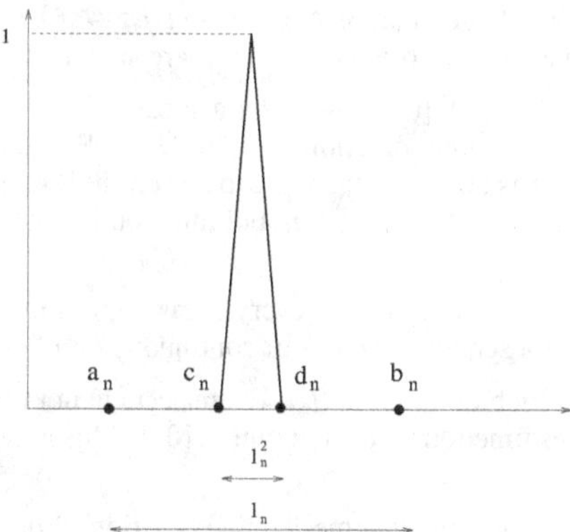

Figure 2.10 A typical "spike" in the graph of the function considered in Exercise 12.

13. Let $f, g : [a,b] \to \mathbb{R}$ be bounded functions. Show that if $f, g \in \mathcal{R}[a,b]$, then $f + g$, $fg \in \mathcal{R}[a,b]$.

14. Let $A \subset [0,1]$ be a Lebesgue measurable set. If $m(A) > 0$, show that A contains a subset $E \subset A$ that is not Lebesgue measurable.

15. (i) Show that $f : [0,1] \to \mathbb{R}$ is Borel/Lebesgue measurable if and only if the inverse image of a Borel set is Borel/Lebesgue measurable.

(ii) Find a continuous function $f : [0,1] \to [0,1]$ such that the inverse image of a Lebesgue measurable set is not a Lebesgue measurable set.

16. Prove that if $f : [a,b] \to [a,b]$ is Lebesgue measurable and $g : [a,b] \to [a,b]$ is continuous, then $g \circ f$ is Lebesgue measurable. In contrast to this, show that if g is Lebesgue measurable and f is continuous, $g \circ f$ might not be Lebesgue measurable.

17. (i) If $A_1 \subseteq A_2 \subseteq \cdots \subseteq A_n \subseteq \cdots$ are (Lebesgue) measurable sets and $A = \bigcup_{n\geq 1} A_n$, prove that $\lim_{n\to\infty} m(A_n) = m(A)$.

(ii) If $B_1 \supseteq B_2 \supseteq \cdots \supseteq B_n \supseteq \cdots$ are (Lebesgue) measurable sets and $B = \bigcap_{n\geq 1} B_n$, prove that $\lim\limits_{n\to\infty} m(B_n) = m(B)$ if $m(B_1) < \infty$. Show that this might fail if $m(B_1) = \infty$.

18. Prove the fundamental properties (2.12) and (2.13) satisfied by the repartition function F of an extended measurable function $f : [0,1] \to \overline{\mathbb{R}}$.

19. (The Borel–Cantelli lemma) If $\{B_n\}_{n\geq 1}$ are measurable sets with $\sum_{n\geq 1}^{\infty} m(B_n) < \infty$, prove that the set of points that belong to an infinite number of the sets B_n has zero Lebesgue measure.

20. By construction, every Lebesgue measurable subset of $[0,1]$ is the union of a Borel set and a set of measure zero. Prove that any measurable function $f : [0,1] \to \mathbb{R}$ becomes Borel measurable after suitable modification on a set of measure zero.

21. Show that for a function $f : [0,1] \to \mathbb{R}$ the fact that all its level sets $f^{-1}(c)$ with

$c \in \mathbb{R}$ are measurable is a necessary condition for measurability. Does it ensure measurability?

22. (Egorov's theorem) For $n \geq 1$, let $f_n : [0,1] \to \overline{\mathbb{R}}$ be extended measurable functions that converge a.e. to a function $f : [0,1] \to \mathbb{R}$. Then for any $\varepsilon > 0$ there exists a measurable set $A_\varepsilon \subset [0,1]$ such that $m(A_\varepsilon) < \varepsilon$ and $f_n \to f$ uniformly on $[0,1] \setminus A_\varepsilon$. (Thus the notions of a.e. and uniform convergence are surprisingly close.)

23. (Luzin's theorem) Let $f : [0,1] \to \mathbb{R}$ be a measurable function. Then for each $\varepsilon > 0$ there exists a continuous function $g_\varepsilon : [0,1] \to \mathbb{R}$ such that $m(\{x : f(x) \neq g(x)\} < \varepsilon$. Moreover, if f is bounded, then g_ε can be selected so that $\sup_{x\in[0,1]} \{|g_\varepsilon(x)|\} \leq \sup_{x\in[0,1]} \{|f(x)|\}$. (This shows that a measurable function is continuous if we discard a "small set".)

24. (i) Using Luzin's theorem, show that every measurable function $f : [0,1] \to \mathbb{R}$ is a.e. a pointwise limit of a sequence $\{f_n\}_{n\geq 1}$ of continuous functions $f_n : [0,1] \to \mathbb{R}$.

(ii) Show that the Dirichlet function (see Exercise 1) is not obtainable[65] as a pointwise limit of continuous functions at every point of $[0,1]$; this underlines the importance of sets of measure zero.

25. Let $f : [0,1] \to \overline{\mathbb{R}}$ be an extended measurable function. Prove that:

(i) (Markov's inequality) for any $a > 0$, we have $m(\{|f| \geq a\}) \leq \frac{1}{a} \int_0^1 |f(x)|\, dx$;

(ii) (Chebyshev's inequality) for any $a > 0$, we have $m(\{|f| \geq a\}) \leq \frac{1}{a^2} \int_0^1 |f(x)|^2\, dx$.

26. Let $f : [0,1] \to [0,\infty]$ be a nonnegative extended measurable function. Show that if $\int_0^1 f(x)\, dx = 0$, then $f = 0$ a.e.

27. Prove the absolute continuity of the Lebesgue integral: if $f : [0,1] \to \mathbb{R}$ is integrable, then for any $\varepsilon > 0$ there exists $\delta > 0$ such that if $A \subset [0,1]$ is a measurable set with $m(A) < \delta$, then $\int_A |f(x)|\, dx < \varepsilon$.

28. (Vitali's theorem) A family $\{f_i\}_{i\in I}$ of extended measurable functions $f_i : [0,1] \to \overline{\mathbb{R}}$ is said to be uniformly integrable[66] if for any $\varepsilon > 0$ there exists $\delta > 0$ such that $\int_A |f_i(x)|\, dx < \varepsilon$ for all $i \in I$ whenever $A \subset [0,1]$ is a measurable set with $m(A) < \delta$. Prove the following extension of the dominated convergence theorem: if $\{f_n\}_{n\geq 1}$ is a sequence of uniformly integrable functions that converge a.e. to some $f : [0,1] \to \mathbb{R}$, then f is integrable and $\lim_{n\to\infty} \int_0^1 f_n(x)\, dx = \int_0^1 f(x)\, dx$.

29. (i) Show that one cannot expect equality in Fatou's lemma.

(ii) Show that the presence of the dominating function in Theorem 2.3 is of the utmost importance.

65 However, it can be obtained as the pointwise iterated limit $\lim_{n\to\infty} \lim_{m\to\infty} [\cos(2\pi n!x)]^m$. Discontinuous functions that are pointwise limits of a sequence of continuous functions are called functions of the first Baire class. The Dirichlet function belongs to the second Baire class: these are functions that do not belong to the class 1 but are limits of a sequence of class 1 functions (the continuous functions form the class zero). One can define inductively the n-th Baire class B_n. Every class can be shown to be nonempty. Clearly, as pointwise limits of Borel measurable functions, all functions in B_n are Borel measurable. However, the (countable) union of the classes B_n does not exhaust the collection of all Borel measurable functions: for this one has to introduce Baire classes B_α with transfinite numbers – see Natanson (1955).

66 This extends from single functions to families of functions the absolute continuity property of the integral, defined in Exercise 27.

30. Give an example of a Lebesgue measurable set $A \subset [0,1]^2$ whose projection on the real axis is not Lebesgue measurable.

31. Let $f : [0,1] \to [0,1]$ be an $\mathcal{L}^1$-measurable function. Prove that its graph $G_f = \{(x, f(x)) : x \in [0,1]\}$ is $\mathcal{L}^2$-measurable and compute its measure.

32. Show that the iterated Lebesgue integrals corresponding to the double integral $\int_0^1 \int_0^1 \frac{x^2-y^2}{(x^2+y^2)^2}\, dxdy$ both exist, but their values differ.

33. Show that the double Lebesgue integral $\int_{-1}^1 \int_{-1}^1 \frac{xy}{(x^2+y^2)^2}\, dxdy$ does not exist, even though both iterated integrals exist and their values coincide.

34. Prove that $\sum_{k=0}^{\infty} 3^{-k} \cos[(21)^k \pi x]$ defines a continuous and bounded function $f : \mathbb{R} \to \mathbb{R}$ that is nowhere differentiable.[67]

35. Show that for every set $E \subset (0,1)$ of measure zero there exists a continuous nondecreasing function $f_E : [0,1] \to \mathbb{R}$ with $f_E'(x) = \infty$ at every $x \in E$. (One can show that for any function $f : [0,1] \to \mathbb{R}$ the set on which f' is infinite must be a set of measure zero; see Bruckner and Leonard (1966).)

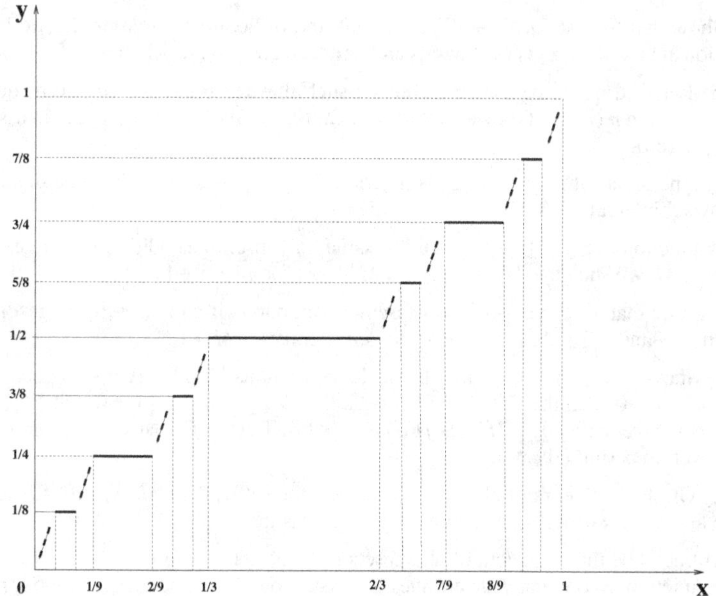

Figure 2.11 Part of the graph of "devil's staircase" constructed in Exercise 36.

36. Let $C_0 \subset [0,1]$ be Cantor's ternary set (see Exercise 8). The intervals in the open set $[0,1] \setminus C_0$ can be divided into classes as follows: the first class consists of the interval $(1/3, 2/3)$, the second of the two intervals $(1/9, 2/9)$ and $(7/9, 8/9)$, the third of the four intervals $(1/27, 2/27)$, $(7/27, 8/27)$, $(19/27, 20/27)$, $(25/27, 26/27)$ etc. In the

[67] Riemann was the first researcher to conjecture the existence of continuous functions that are nowhere differentiable but the first example (very similar to the one presented here) was constructed by Weierstrass in 1875.

nth class we will have 2^{n-1} intervals of equal length 3^{-n}. Let us define a function $f : [0,1] \to [0,1]$ by setting $f(x) = 1/2$ for $x \in (1/3, 2/3)$, $f(x) = 1/4$ for $x \in (1/9, 2/9)$, $f(x) = 3/4$ for $x \in (7/9, 8/9)$. On the four intervals in the third class we set $f(x)$ equal to $1/8, 3/8, 5/8, 7/8$, respectively. In general, on the 2^{n-1} intervals of the nth class, we assign to $f(x)$ successively the values $2^{-n}, 3 \cdot 2^{-n}, 5 \cdot 2^{-n}, \cdots, (2^n - 1) \cdot 2^{-n}$. This procedure defines an nondecreasing function on the open set $[0,1] \backslash C_0$. On C_0 we define f by setting $f(0) = 0$, $f(1) = 1$ and $f(x) = \sup_{s \in [0,1] \backslash C_0,\, s<x} \{f(s)\}$ for $x \in C_0 \cap (0,1)$. Prove that the function thus defined is continuous and nondecreasing, with $f' = 0$ a.e. on $[0,1]$, so that $\int_0^1 f'(x)\,dx = 0 < f(1) - f(0) = 1$. (See Hewitt and Stromberg (1965) for a pathological example of a strictly increasing function $F : [0,1] \to [0,1]$ such that $F(0) = 0$, $F(1) = 1$ and $F' = 0$ a.e.)

37. Show that if $f : [a,b] \to \mathbb{R}$ is absolutely continuous, then for every $\varepsilon > 0$ there is some $\delta > 0$ such that for all countable disjoint intervals $\{(a_n, b_n)\}_{n\geq 1}$ contained in $[a,b]$ with $\sum_{n\geq 1}(b_n - a_n) < \delta$, we have $\sum_{n\geq 1} |f(b_n) - f(a_n)| < \varepsilon$.

38. Let $f : [a,b] \to \mathbb{R}$ be a derivative, that is, there is a function $F : [a,b] \to \mathbb{R}$ with $F'(x) = f(x)$ for all $x \in [a,b]$. Show that f has the Darboux property. Moreover, f is the pointwise limit of a sequence of continuous functions.[68]

39. Show that for $f : [a,b] \to \mathbb{C}$ continuous and of bounded variation the arc length function $s(x) = V_f[a,x]$ is continuous and nondecreasing on $[a,b]$.

40. (i) Let $E \subset [0,1]$ be a measurable set such that for some $\alpha > 0$, the inequality $m(E \cap I) \geq \alpha\, m(I)$ holds for all intervals $I \subset [0,1]$. Show that $[0,1] \setminus E$ is a set of measure zero.

(ii) Find an absolutely continuous function $f : [0,1] \to \mathbb{R}$ that fails to be monotone on any subinterval of $[0,1]$.

41. A function $f : [a,b] \to \mathbb{R}$ is said to satisfy a Lipschitz condition if there exists a constant $M > 0$ such that $|f(x) - f(y)| \leq M\,|x - y|$ for all $x, y \in [a,b]$.

(i) Show that $f : [a,b] \to \mathbb{R}$ is Lipschitz continuous if and only if f is absolutely continuous and $|f'| \leq M$ a.e. on $[a,b]$ for some constant $M > 0$.

(ii) Show that $f : [a,b] \to \mathbb{R}$ is Lipschitz continuous if and only if for every $\varepsilon > 0$ there is a $\delta > 0$ such that for all subintervals $\{[a_k, b_k]\}_{k=1}^n$ of $[a,b]$ for which $\sum_{k=1}^n (b_k - a_k) < \delta$, the inequality $\sum_{k=1}^n |f(b_k) - f(a_k)| < \varepsilon$ holds. That is, f is "absolutely continuous with overlap permitted".

42. (i) On the σ-algebra of all subsets of an infinite set X, define $\mu(A) = 0$ if $A \subset X$ is finite and $\mu(A) = \infty$ if $A \subset X$ is infinite. Is μ a measure?

(ii) Let $\mathcal{A}$ be the σ-algebra of all subsets of some set X, and let $f : X \to [0, \infty]$ be some function. Show that f determines a measure on $(X, \mathcal{A})$ be means of the formula $\mu(A) = \sum_{x \in A} f(x)$, where we identify an uncountable sum of strictly positive numbers with ∞. When is this measure σ-finite? (If $f(x) = 1$ for all $x \in X$, this measure is called the *counting measure*.)

43. Let ν be the counting measure, defined in Exercise 42, and let μ be the Lebesgue measure, both defined on the σ-algebra of Lebesgue measurable subsets of $[0,1]$. Show that $\mu \ll \nu$ but there is no ν-integrable function $f : [0,1] \to [-\infty, \infty]$ with $d\mu = f\,d\nu$, and ν has no Lebesgue decomposition with respect to μ. (This shows the relevance of the σ-finiteness in the Radon–Nikodym theorem.)

[68] That is, f is either continuous or of the first Baire class (see Exercise 24).

2.4.2 Hints

1. Compute the lower and upper Darboux sums for an arbitrary subdivision.

2. Show that f is strictly monotone, continuous at $x = 1$ and at all irrational numbers in $(0, 1)$, and discontinuous at $x = 0$ and at all rational numbers in $(0, 1)$.

3. Show that f has finite right- and left-hand limits at all points of $[a, b]$; and thus only jump discontinuities are possible.

4. Use the dominated convergence theorem.

5. To every $x \in O$ associate $a_x = \inf\{t \in \mathbb{R} : (t, x] \in O\}$, $b_x = \sup\{t \in \mathbb{R} : [x, t) \in O\}$, and show that $\mathcal{F} = \{(a_x, b_x) : x \in O\}$.

6. (i) Assume there is a disjoint family of open discs/rectangles with this property and look at the diagonal of the open square. (ii) For existence, consider for any $x \in O$ the largest square S_x in $\mathcal{S}$ such that $x \in S_x$ and $S_x \subset O$. Concerning uniqueness, think of partitioning a square in $\mathcal{S}$.

7. Take advantage of the detailed structural information available for C_n.

8. C_0 corresponds to the set of numbers $x \in [0, 1]$ that can be expressed by a ternary (base three) expansion not using the digit 1.

9. Use straight line segments in the gaps of the characteristic function of the set C_n to obtain a continuous function f_n that is very close to χ^{C_n}.

10. To prove continuity at an irrational point $x^* \in [1/2, 1]$, show that if $q_k, p_k \geq 1$ are integers with $(q_k, p_k) = 1$ for all $k \geq 1$, and such that $q_k/p_k \to x^*$ as $k \to \infty$, then $p_k \to \infty$ as $k \to \infty$.

11. Consider for all rational numbers r the sets $V_r = \{x + r : x \in V\}$ and take advantage of the translation invariance of the Lebesgue measure.

12. To show that f is not Riemann integrable, prove that $\mathcal{U}(f, \Delta) - \mathcal{L}(f, \Delta) \geq 1/2$ for every subdivision Δ of $[0, 1]$. To show that f is a derivative, prove that the expression $F(x) = \sum_{n=1}^{\infty} \int_{K_n(x)} f(s)\, ds$, where $K_n(x) = J_n \cap [0, x]$ for $n \geq 1$ and each integral in the sum is a Riemann integral, defines an anti-derivative of f.

13. Use Theorem 2.1.

14. Try to adapt the construction in Exercise 11 to this context.

15. (i) Use the properties of inverse images. (ii) Let P be the Cantor set constructed for $\lambda = 1/6$ in Exercise 7 and let C be Cantor's ternary set. Construct a continuous strictly increasing function $f : [0, 1] \to [0, 1]$ with $f(P) = C$ by taking advantage of the structure of the complement in $[0, 1]$ of a Cantor set, obtained as a disjoint countable union of open intervals; the function f should be linear on corresponding intervals. By Exercise 14, there exists a nonmeasurable subset $S \subset P$. Then $A = f(S) \subset C$ is Lebesgue measurable.

16. For the first part, use inverse images. For the second, let $f : [0, 1] \to [0, 1]$, and choose the sets $A,\ S \in [0, 1]$ as in Exercise 15. Set g equal to the characteristic function of the set A.

17. (i) Write $A = A_1 \cup \big(\bigcup_{n\geq 1}(A_{n+1} \setminus A_n)\big)$. (ii) If $m(B_1) < \infty$, let $A = B_1 \setminus B$, set $A_n = B_1 \setminus B_n$ and apply (i). For the second part, generate an example with $B = \emptyset$ and $m(B_n) = \infty$ for all $n \geq 1$.

18. Use the set-theoretic considerations of Exercise 17.

19. The set of points that belong to B_n for infinitely many values of n is $\bigcap_{n\geq 1} \bigcup_{k\geq n} B_k$.

20. Use the definition of measurability by means of inverse images of sets of the form (k, ∞); consider k of the form $k = n/2^m$ with n and $m \geq 0$ integers.

21. The necessity is clear. A counterexample for sufficiency is provided by the function defined by $f(x) = x + 1$ if $x \in A$ and $f(x) = -x - 1$ if $x \notin A$, where $A \subset [0, 1]$ is a nonmeasurable set.

22. For any integer $n \geq 1$, the sets $B_{k,n} = \{x \in [0, 1] : |f_i(x) - f(x)| < 1/n \text{ for } i \geq k\}$ are nested and their union is $[0, 1]$, except perhaps for a set of measure zero. Choose $k(n)$ such that $m(B_{k(n),n}) > 1 - 2^{-n}\varepsilon$ and set $A_\varepsilon = \bigcup_{n\geq 1}\big([0, 1] \setminus B_{n(k),n}\big)$.

23. The presented proof is adapted from Feldman (1981); see Hewitt and Stromberg (1965) for an alternative approach. Choose a countable basis $\{U_n\}_{n\geq 1}$ of the open subsets of $\mathbb{R}$: a family such that every open subset of $\mathbb{R}$ is a countable union of such sets. For each $n \geq 1$, the regularity of the Lebesgue measure ensures the existence of a compact set K_n and of an open set O_n such that $K_n \subset f^{-1}(U_n) \subset O_n$ and $m(O_n \setminus K_n) < \varepsilon 2^{-n}$. Define the open set $F_\varepsilon = \bigcup_{n\geq 1}(O_n \setminus K_n)$ with $m(F_\varepsilon) < \varepsilon$ and infer that the restriction of f to $[0, 1] \setminus F_\varepsilon$ is continuous by showing that inverse images of open subsets of $\mathbb{R}$ are open in $[0, 1] \setminus F_\varepsilon$.

24. (i) First reduce the problem to bounded functions. (ii) Assume the contrary. Infer that the set of rational numbers would be a countable intersection of open sets, since $f^{-1}(-\infty, 1/2] = \bigcup_{n\geq 1}\bigcap_{k\geq n} f_k^{-1}(-\infty, 1/2]$.

25. Use the fact that $a^{-1}\,|f(x)| \geq 1$ on the set $\{|f| \geq a\}$.

26. Apply Exercise 17 (i) in view of $\{f > 0\} = \bigcup_{n\geq 1}\{f > \frac{1}{n}\}$.

27. Assuming the failure of this property, we can find for some $\varepsilon > 0$ measurable sets A_n with $m(A_n) < 2^{-n}$ and $\int_{A_n} |f(x)|\,dx \geq \varepsilon$. By the Borel–Cantelli lemma (Exercise 19), the set B of points that belong to A_n for infinitely many values of n has measure zero. Infer a contradiction from the dominated convergence theorem applied to the functions $f\chi^{B_n}$, where $B_n = \bigcup_{k\geq n} A_k \setminus B$.

28. Given $\varepsilon > 0$, choose $\delta > 0$ so that $m(A) < \delta$ ensures $\int_A |f_n(x)|\,dx < \varepsilon/3$ for every $n \geq 1$. If A_ε is as in Egorov's theorem, evaluate the integral of $|f_n - f|$ over A_ε by means of Fatou's lemma, to deduce that for some $N \geq 1$ we have $\int_0^1 |f_n(x) - f(x)|\,dx \leq \varepsilon$ for $n \geq N$.

29. (i) Let $f_n = \chi^{[0,1/2]}$ if $n \geq 1$ is odd and $f_n = \chi^{(1/2,1]}$ if $n \geq 1$ is even.

(ii) Consider the sequence $f_n = n^2\chi^{(0,1/n)}$ for $n \geq 1$.

30. Let $V \subset [0, 1]$ be (Lebesgue) nonmeasurable and set $A = \{(1, x) : \ x \in V\}$.

31. As in Exercise 23, choose a countable basis $\{U_n\}_{n\geq 1}$ of the open subsets of $\mathbb{R}$. For $y \neq f(x)$ we can find $n \geq 1$ with $f(x) \in U_n$ but $y \notin U_n$. The relation $G_f = ([0, 1]\times\mathbb{R})\setminus \bigcup_{n\geq 1}[f^{-1}(U_n)\times(\mathbb{R}\setminus U_n)]$ shows that G_f is $\mathcal{L}^2$-measurable. Tonelli's theorem applied to the characteristic function of the set $G(f)$ yields $\mathcal{L}^2(G_f) = 0$.

32. Notice that $\frac{x^2-y^2}{(x^2+y^2)^2} = \frac{\partial}{\partial y}\left(\frac{y}{x^2+y^2}\right) = -\frac{\partial}{\partial x}\left(\frac{x}{x^2+y^2}\right)$ for $(x, y) \in [0, 1]^2$ with $x \neq 0$ and $y \neq 0$, respectively.

33. Clearly both iterated integrals are zero. To see that the function is not $\mathcal{L}^2$-integrable, compute $\int_0^1 \frac{xy}{(x^2+y^2)^2}\,dy$.

34. The uniform convergence of the series ensures continuity. To show that f is not differentiable at $x \in \mathbb{R}$, split the series for $\frac{f(x+h)-f(x)}{h}$ into the sum $S_n(h)$ of the first n differences and the rest $R_n(h)$. Subsequently devise a sequence $h_n \to 0$ with $|R_n(h_n)| \geq \frac{2}{3}\,7^n$ and $|S_n(h_n)| \geq \frac{\pi}{6}\,7^n$.

35. Given $n \geq 1$, choose an open set $G_n \supset E$ with $m(G_n) < 2^{-n}$, and set $f_E(x) = \sum_{n=1}^{\infty} m(G_n \cap [0, x])$.

36. Use the fact that the values of f on $[0, 1] \setminus C_0$ are dense in $[0, 1]$ to prove its continuity.

37. Rely on the definition of absolute continuity.

38. Use the mean-value theorem for derivatives[69] to prove the Darboux property. For the Baire class statement explore the definition of the derivative.

39. Take advantage of the fact that by adding points to the partition, the value of the sum on the right-hand side of (2.22) increases.

40. (i) Use the hypothesis to estimate the derivative of the absolutely continuous function $x \mapsto m(E \cap [0, x])$. (ii) Using Cantor-like sets of positive measure, construct a measurable set $A \subset [0, 1]$ such that $m(A \cap I) > 0$ and $m(A \cap ([0, 1] \setminus I)) > 0$ for every open interval $I \subset [0, 1]$. Now consider the indefinite integral of $\chi^A - \chi^{[0,1]\setminus A}$.

41. (i) Take advantage of Theorem 2.8. (ii) One direction is clear. For the other direction, choose $\delta > 0$ corresponding to $\varepsilon = 1$ and set $M = 2/\delta$. Unless $|f'| \le M$ a.e., for some $n \ge 1$ one of the sets $\{x \in (a, b) : \ |f(x + \delta 2^{-n}) - f(x)| > M\delta 2^{-n}\}$ must have positive measure.

42. (i) μ is not countably additive. (ii) μ is σ-finite if and only if $f(x) < \infty$ for all $x \in X$ and the set $\{x \in X : \ f(x) > 0\}$ is countable.

43. Note that $\nu(\{x\}) = 1$ for every $x \in [0, 1]$, while $\nu(A) = \infty$ for every Lebesgue measurable set $A \subset [0, 1]$ with $\mu(A) > 0$.

2.4.3 Solutions

1. Let $\Delta = \{t_0 = 0 < t_1 < \cdots < t_n = 1\}$ be a subdivision. Since both the rational and the irrational numbers are dense in any of the intervals $[t_{j-1}, t_j]$, we have $\mathcal{L}(f, \Delta) = \sum_{k=1}^n (t_k - t_{k-1}) \cdot 0 = 0$, while $\mathcal{U}(f, \Delta) = \sum_{k=1}^n (t_k - t_{k-1}) \cdot 1 = 1$.

2. Since $\sum_{n=1}^\infty \frac{1}{2^n} = 1$, we have that $f(1) = 1$. By convention, the summation over the empty set is zero, so that $f(0) = 0$. Clearly $0 < f(x) < 1$ for $x \in (0, 1)$, due to the existence of rational numbers in $(0, x)$ and in $(x, 1)$. The presence of rational numbers in the interval (x, y) if $0 \le x < y \le 1$ ensures $f(x) < f(y)$. Consequently $f : [0, 1] \to [0, 1]$ is strictly increasing. At any point $y^* \in (0, 1]$, the function f is left continuous since $\lim_{y \uparrow y^*} f(y) = f(y^*)$, all rational numbers strictly less than y^* being exhausted by the limiting process. In particular, f is continuous at $x = 1$. If $q \in [0, 1)$ is a rational number, then $q = x_{n_0}$ for some integer $n_0 \ge 1$, so that $f(x) \ge f(q) + \frac{1}{2^{n_0}}$ if $q < x \le 1$. This yields $\lim_{x \uparrow q} f(x) > f(q)$, so that f is not continuous at q. On the other hand, if $y^* \in (0, 1)$ is an irrational number, then $\lim_{y \downarrow y^*} f(y) = f(y^*)$, so that f is right continuous at $y = y^*$, and thus continuous. Since f is bounded and the set of its discontinuities is countable (and therefore of Lebesgue measure zero), it is Riemann integrable in view of Theorem 2.1.

3. Assume without loss of generality[70] that f is nondecreasing: $f(x) \le f(y)$ if $x \le y$. Let $x_0 \in (a, b)$ and set $f_+(x_0) = \inf\{f(x) : \ x_0 < x \le b\}$. The existence of $f_+(x_0) \in \mathbb{R}$ is ensured since $f(x_0) \le f(x)$ whenever $x_0 < x \le b$. For $\varepsilon > 0$, by definition, $f_+(x_0) + \varepsilon$ is not a lower bound for the set $\{f(x) : \ x_0 < x \le b\}$, so that we can find $\delta > 0$ with $x_0 + \delta \in (a, b)$ and $f(x_0 + \delta) < f_+(x_0) + \varepsilon$. By monotonicity, for $x \in (x_0, x_0 + \delta)$ we have $f_+(x_0) \le f(x) \le f_+(x_0) + \varepsilon$. This shows that $\lim_{x \downarrow x_0} f(x) = f_+(x_0)$. Similarly we show that $\lim_{x \uparrow x_0} f(x) = \sup\{f(x) : a \le x < x_0\} := f_-(x_0)$. Therefore $x_0 \in [a, b]$ is a point of discontinuity of f if and only if $f_+(a) > f(a)$ if $x_0 = a$, or $f_-(b) < f(b)$ if $x_0 = b$, or $f_-(x_0) < f_+(x_0)$ if $x_0 \in (a, b)$; Note that if $f_-(x_0) = f_+(x_0)$, then, due to monotonicity, this common value must be precisely $f(x_0)$. Since by monotonicity $f_+(x) \le f_-(y)$ whenever $a \le x < y \le b$, the family $\mathcal{F}$ of intervals $\{(f_-(x), f_+(x)) :$

[69] Which states that *if $F : [a, b] \to \mathbb{R}$ is differentiable, then for every $x, y \in [a, b]$ with $x < y$ we can find some $z \in (x, y)$ such that $F(y) - F(x) = F'(z)\,[y - x]$*. The proof of this result is immediate if we notice that the differentiable function $F_0 : [x, y] \to \mathbb{R}$ defined by $F_0(s) = \frac{F(y)-F(x)}{y-x}\,(s - x) - F(s) + F(x)$ for $s \in [x, y]$, satisfies $F_0(x) = F_0(y) = 0$. Therefore F_0 must have a local maximum or a local minimum at some point $z \in (x, y)$. Looking at the derivative of F_0 from the left and from the right at the point z, one sees that $F_0'(z) = 0$. This amounts to $F(y) - F(x) = F'(z)\,[y - x]$.

[70] Otherwise, perform the same argument with $(-f)$ instead of f.

$x \in (a, b)$ point of discontinuity of $f\}$ is a pairwise disjoint family of nonvoid open intervals. In each such interval I we can find a rational number $r_I \in I$. The family being pairwise disjoint ensures that the mapping $I \mapsto r_I$ of $\mathcal{F}$ into the set of rational numbers is one-to-one, so that $\mathcal{F}$ is at most countable. Since countable sets have zero Lebesgue measure and f is bounded, as $f(a) \leq f(x) \leq f(b)$ for $x \in [a, b]$ due to monotonicity, Theorem 2.1 is applicable.

4. By the dominated convergence theorem, the stated limit holds with the integrals interpreted as Lebesgue integrals. Since each integrand is a continuous function, the Lebesgue integral coincides with the Riemann integral.

5. For $x \in O$, let $a_x = \inf\{t \in \mathbb{R} : (t, x] \in O\}$ and $b_x = \sup\{t \in \mathbb{R} : [x, t) \in O\}$. Then $(a_x, b_x) \subset O$ with $a_x \notin O$ and $b_x \notin O$. Set $\mathcal{F} = \{(a_x, b_x) : x \in O\}$ and notice that two intervals (a_x, b_x) and (a_y, b_y) with $x, y \in O$ are either disjoint or identical, that is, the elements of $\mathcal{F}$ are pairwise disjoint. For each $I \in \mathcal{F}$ there is a rational number $r_I \in I$ and since the elements of $\mathcal{F}$ are pairwise disjoint, the mapping $I \mapsto r_I$ of $\mathcal{F}$ into the rational numbers is one-to-one. Thus $\mathcal{F}$ is finite or countable. The uniqueness of $\mathcal{F}$ is clear. (If this proof is too sketchy, we refer to Hewitt and Stromberg (1965) for further details.)

6. (i) Assume there is a family of disjoint open discs/rectangles whose union is $(0, 1) \times (0, 1)$. Then the diagonal $\{(x, x) : 0 < x < 1\}$ would be a union of (more than one) disjoint open intervals. But this is prevented by the structural characterisation of one-dimensional open sets provided by Exercise 5. (ii) For $x \in O$, let S_x be the largest square in S such that $x \in S_x$ and $S_x \subset O$. This square is uniquely defined. Indeed, for any integer $m \geq 1$ there is exactly one square $S_{x,m}$ such that $x \in S_{x,m}$. We define $S_x = S_{x,n}$, where $n \geq 1$ is the smallest integer such that $x \in S_{x,n}$. The family $\{S_x\}_{x \in O}$ is a partition of O since if $S_x \cap S_y \neq \emptyset$ for some $x, y \in O$, by the structure of these squares, one of them must be contained in the other, say $S_x \subset S_y$. Then $x \in S_x \subset S_y \subset O$, so $S_x = S_y$ by the maximality property of S_x. The fact that we have at most countably many disjoint sets S_x with $x \in O$ follows, since by the partition property the lower left-hand corners of these sets are all distinct and have both coordinates rational numbers, thus defining an injective map from the family $\{S_x\}_{x \in O}$ to a countable set. Concerning the lack of uniqueness, it suffices to point out that every single square in S can be partitioned into four smaller squares.

7. Each C_n is closed, so that C is closed. Being clearly bounded, C is compact. Since C_n consists of 2^n intervals of equal length $l_n = (\frac{1}{2^n} - \frac{3\lambda}{2^n} + \frac{\lambda}{3^{n-1}})$, and $C \subset C_n$ for each $n \geq 1$, it follows that C contains no interval. The inclusions $C_n \subset C_{n-1}$ ensure that the measure of C is the limit of the sum of the lengths of the 2^n intervals into which C_n is decomposed: $\mu(C) = \lim_{n\to\infty} \mu(C_n) = \lim_{n\to\infty} [1 - 3\lambda(1 - \frac{2^n}{3^n})] = 1 - 3\lambda$. Finally, let $x \in C$. For each $k \geq 1$ we have $x \in C_k$, so that x belongs to one of the 2^k intervals of length l_k each. The endpoints of this interval belong to C and at least one of them is distinct from x and located at a distance at most l_k from x. Define x_k to be this specific endpoint. This yields the requested sequence.

8. Any number $x \in [0, 1]$ can be represented with a ternary (base three), instead of a decimal, expansion: we can write $x = \sum_{k\geq 1} \frac{x_k}{3^k} = 0.x_1x_2x_3\ldots$ with the ternary digits $x_k \in \{0, 1, 2\}$ for $k \geq 1$. As it is the case for the decimal expansion, this representation is not unique; the only cases of nonuniqueness being the points $x = p\,3^{-q}$ for some integers $q \geq 0$ and $0 < p \leq 3^q$, with p not divisible by 3 (for instance, $1/3 = 0.1 = 0.0222\ldots$). One can easily check that C_1 corresponds to the set of numbers that can be expressed by a ternary representation not having 1 as first digit, C_2 corresponds to those numbers that admit a representation not having 1 as a first or second digit etc. Consequently $x \in C$ if and only if x admits a ternary representation with $x_k \in \{0, 2\}$ for each $k \geq 1$. The map $(x_1, x_2, \cdots) \mapsto \sum_{k\geq 1} \frac{x_k}{3^k}$ provides a bijection of $\{0, 2\}^{\mathbb{N}^*}$ with C, where we denoted by $\mathbb{N}^*$ the set of integers $n \geq 1$. C is uncountable since the set $\{0, 2\}^{\mathbb{N}^*}$ is uncountable. The last assertion holds since for each $x \in [0, 1)$ there is a unique representation in the form $x = \sum_{k=1}^{\infty} \frac{a_k}{2^k}$ where each a_k is 0 or 1 and $a_k = 0$ for infinitely many integers $k \geq 1$. Setting $x_k = 2$ if $a_k = 1$ and $x_k = 0$ if $a_k = 0$, we define a one-to-one map from $[0, 1)$ into $\{0, 2\}^{\mathbb{N}^*}$ into $[0, 1)$, so that $\{0, 2\}^{\mathbb{N}^*}$ has to be uncountable.

9. Let $[a_k, b_k]$ and $[a_{k+1}, b_{k+1}]$ be consecutive closed intervals of the set C_n constructed in Exercise 7. Then $b_k - a_k = b_{k+1} - a_{k+1} = \frac{1}{2^n} - \frac{3\lambda}{2^n} + \frac{\lambda}{3^{n-1}}$ and $a_{k+1} - b_k \in \{\lambda, \lambda/3, \cdots, \lambda/3^{n-1}\}$. We define f_n

on the interval $[a_k, b_{k+1}]$ by setting $f_n(x) = \begin{cases} 1 & \text{for} \quad x \in [a_k, b_k] \cup [a_{k+1}, b_{k+1}], \\ 0 & \text{for} \quad x \in [b_k + \lambda/3^n, a_{k+1} - \lambda/3^n], \end{cases}$ with the graph of f_n restricted to $[b_k, b_k + \lambda/3^n]$ and to $[a_{k+1} - \lambda/3^n, a_{k+1}]$ being a straight line segment joining $(b_k, 1)$ to $(b_k + \lambda/3^n, 0)$ and $(a_{k+1} - \lambda/3^n, 0)$ to $(a_{k+1}, 1)$, respectively; see Figure 2.12. Then $\chi^{C_n}(x) \leq f_n(x) \leq \chi^{C_n}(x) + \chi^{D_n}(x)$ for $x \in [0, 1]$, with the set D_n consisting of the union of $2(2^n - 1)$ open intervals of length $\lambda/3^n$ each. Therefore $\lim_{n\to\infty} f_n(x) = \chi^C(x)$ at every fixed $x \in [0, 1]$. The function $f = \chi^C$ is bounded and discontinuous at every single point of C. Indeed, if $x \in C$, then $\chi^C(x) = 1$. However, since no interval lies in C, arbitrarily close to x we can find points in $[0, 1] \setminus C$. Consequently there is a sequence $\{x_k\}_{k\geq 1} \in [0, 1] \setminus C$ with $x_k \to x$. Since $\chi^C(x_k) = 0$ for every $k \geq 1$, the point x is a discontinuity point of χ^C. The fact that for $\lambda < 1/3$ the set C has positive Lebesgue measure ensures that χ^C is not Riemann integrable (see Theorem 2.1). Notice that for $n > m \geq 1$ we have $C_n \subset C_m$, $f_k(x) = 1$ for $x \in C_k$ and $D_k \subset [0, 1] \setminus C_k$ for $k = m, n$, so that $\int_0^1 |f_n(x) - f_m(x)|\, dx \leq \mu(C_n \setminus C_m) + \mu(D_n) + \mu(D_m)$. The previous inequality shows that $\{f_n\}_{n\geq 1}$ is a Cauchy sequence with respect to the distance (2.4). The dominated convergence theorem yields that $f_n \to f$ in $L^1[0, 1]$. Since $\mathcal{R}[0, 1] \subset L^1[0, 1]$, were $\{f_n\}_{n\geq 1}$ convergent in $\mathcal{R}[0, 1]$, the possible limit could only be f. However, we proved that $f \notin \mathcal{R}[0, 1]$.

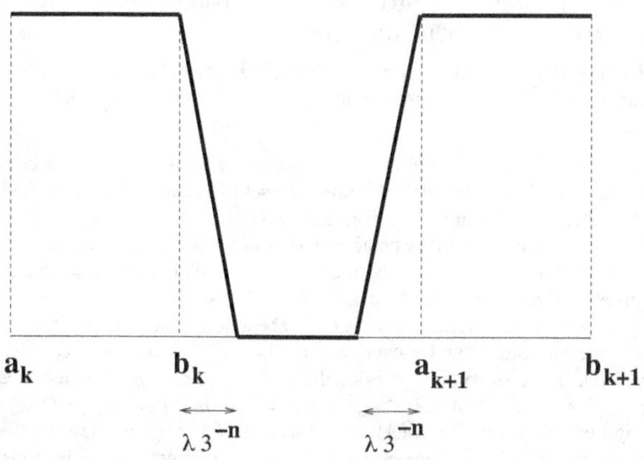

Figure 2.12 In Exercise 9, for a fixed $n \geq 1$, any two consecutive closed intervals of C_n are separated by an open interval of length at least $\lambda/3^{n-1}$. (The vertical and horizontal scales are not the same, to facilitate the drawing.)

10. If $x^* \in [1/2, 1]$ is a rational number, choose a sequence $\{x_n\}_{n\geq 1}$ of irrational numbers in $[1/2, 1]$ converging to x^*. Since $f(x^*) \neq 0$ but $f(x_n) = 0$ for all $n \geq 1$, the function f is discontinuous at x^*. On the other hand, if $x^* \in [1/2, 1]$ is irrational, choose rational numbers $q_k/p_k \in [1/2, 1]$ with $(q_k, p_k) = 1$ for all $k \geq 1$, and converging to x^*. Given $\varepsilon > 0$, let $n_0 \geq \varepsilon^{-1}$ be an integer and notice that the set $\mathcal{F}$ of rational numbers $q/p \in [1/2, 1]$ with $(q, p) = 1$ and $p \leq n_0$ is finite. Since x^* is irrational, we can find an integer $N \geq 1$ such that $q_k/p_k \notin \mathcal{F}$ for $k \geq N$. Consequently $p_k \geq n_0 + 1$ for $k \geq N$, so that $f(\frac{q_k}{p_k}) = \frac{1}{p_k} < \frac{1}{n_0} \leq \varepsilon$ for $k \geq N$. This yields $\lim_{k\to\infty} f(q_k/p_k) = 0$. Since f vanishes for irrational numbers, f must be continuous at x^*.

11. For r rational number, define the sets $V_r = \{x + r : x \in V\}$. Were V Lebesgue measurable with measure v, each of these sets would be measurable with measure v, by translation invariance. This would provide us with a countable partition of $\mathbb{R}$ into sets of equal measure v. If $v = 0$, we reach a contradiction since $m(\mathbb{R}) = \infty$, while if $v > 0$ the contradiction is due to $2 = m([0, 2]) \geq \sum_{r\in[0,1]} m(V_r) = \infty$.

12. The properties of C ensure that O is dense in $[0, 1]$ and $m(O) = 1/2$. Let $\Delta = \{0 = t_0 < t_1 < \cdots < t_m = 1\}$ be a subdivision. On each interval $[t_{k-1}, t_k]$ which intersects one of the J_ns, the oscillation of f, defined by $\sup_{x,y\in[t_{k-1},t_k]} \{|f(x) - f(y)|\}$, is precisely 1. Since the sum of the lengths of these intervals $[t_{k-1}, t_k]$ must exceed the measure of O, we must have $\mathcal{U}(f,\Delta) - \mathcal{L}(f,\Delta) \geq 1/2$. This shows that f, while bounded, is not Riemann integrable. In particular, $f : [0,1] \to [0,1]$ is not continuous, despite the fact that its restriction to every I_n is continuous; the intricate structure of the Cantor set C forces this.To show that f is a derivative, we prove that $F(x) = \sum_{n=1}^{\infty} \int_{K_n(x)} f(s)\,ds$, where the disjoint intervals $K_n(x)$ are defined by $K_n(x) = J_n \cap [0,x]$ for $n \geq 1$, is an anti-derivative. Each integral in the above sum is a Riemann integral (recall that the restriction of f to J_n is continuous). Since for some constant α_N we have $F(x) = \alpha_N + \int_{a_N}^{x} f(s)\,ds$ for $x \in I_N$, clearly $F'(x) = f(x)$ on I_N. Let now $I \subset [0,1]$ be an open interval such that $I \cap C \neq \emptyset$. If $I \cap J_n \neq \emptyset$, then $m(I \cap I_n) \geq c_n - a_n = b_n - d_n = \frac{l_n - l_n^2}{2} \geq \frac{l_n}{4}$, so that

$$m(I \cap J_n) \leq m(J_n) = l_n^2 \leq 16\,[m(I \cap I_n)]^2. \qquad (*)$$

Relation $(*)$ is also valid if $I \cap J_n = \emptyset$. For $x_0 \in C$ and $y \neq x_0$, say, $y > x_0$, let $I = (x_0, y)$. Since $\left| \int_{K_n(x_0)} f(s)\,ds - \int_{K_n(y)} f(s)\,ds \right| \leq m(I \cap J_n)$, we have $|F(y) - F(x_0)| \leq \sum_{n=1}^{\infty} \left| \int_{K_n(x_0)} f(s)\,ds - \int_{K_n(y)} f(s)\,ds \right| \leq \sum_{n=1}^{\infty} m(I \cap J_n)$. Using $(*)$, we deduce that $|F(y) - F(x_0)| \leq 16 \sum_{n=1}^{\infty} [m(I \cap I_n)]^2 \leq 16 \Big(\sum_{n=1}^{\infty} m(I \cap I_n) \Big)^2 = 16 m^2(I) = 16 (y - x_0)^2$. This ensures $F'(x_0) = 0$. Consequently F is differentiable at every $x \in [0,1]$, with $F'(x) = f(x)$.

13. The boundedness of $f + g$ and fg is clear. Also, if $\mathcal{D}_F \subset [a,b]$ stands for the set of discontinuities of the function $F : [a,b] \to \mathbb{R}$, note that $\mathcal{D}_{f+g}, \mathcal{D}_{fg} \subset \mathcal{D}_f \cup \mathcal{D}_g$. The conclusion follows from Theorem 2.1.

14. Since countable sets have zero Lebesgue measure, $m(A) > 0$ ensures that A contains uncountably many elements. Let D be a countably infinite subset of A and define H as the set of all finite sums $\sum_{k=1}^{n} m_k x_k$, where the m_ks are integers and $x_k \in D$: this is the smallest additive subgroup of $\mathbb{R}$ that contains D. By construction it is clear that H is countable. Consider now in $\mathbb{R}$ the equivalence relation $x \approx y$ if[71] $x - y \in H$. Let Γ be the family of equivalence classes having a nonempty intersection with A. For each equivalence class $\gamma \in \Gamma$, choose one representative $x_\gamma \in \gamma \cap A$ and let $E = \{x_\gamma : \gamma \in \Gamma\} \subset A$. We claim that E is not Lebesgue measurable. Indeed, assume that the set E is Lebesgue measurable. Then for every $h \in H$, the set $S_h = \{x_\gamma + h : x_\gamma \in E\}$ is measurable with $m(S_h) = m(E)$. Since every $x \in A$ is of the form $x = x_\gamma + h$ for some $h \in H$ and some $x_\gamma \in \gamma$ with $\gamma \in \Gamma$, we deduce that $A \subset \bigcup_{h \in H} S_h$, so that $m(A) \leq \sum_{h \in H} m(S_h) = \sum_{h \in H} m(E)$. The set H being countably infinite, $m(A) > 0$ forces $m(E) > 0$. In view of the definition of E, two sets in the family $\{S_h\}_{h \in H}$ are either identical or disjoint. Let J be those elements of H that correspond to the distinct members of the family of sets $\{S_h\}_{h \in H}$. Thus $D \subset J \subset H$. For $h \in J$, let $U_h = S_h \cap [-1,2]$. Then $\bigcup_{h \in J} U_h \subset [-1,2]$, and we obtain the contradiction $3 \geq m\big(\bigcup_{h \in J} U_h\big) = \sum_{h \in J} m(U_h) \geq \sum_{h \in D} m(U_h) = \sum_{h \in D} m(S_h) = \sum_{h \in D} m(E) = \infty$. Therefore the assumption that E is measurable must be rejected.

15. (i) Let us deal first with the Borel measurability. If the inverse image of a Borel set is in $\mathcal{B}(\mathbb{R})$, then, in particular $f^{-1}[a,b)$ is in $\mathcal{B}(\mathbb{R})$ for every $a < b$, so that f is Borel measurable. Conversely, let f be Borel measurable. Define $\mathcal{S} = \{S \subset \mathbb{R} : f^{-1}(S) \in \mathcal{B}(\mathbb{R})\}$. One can easily see that $\mathbb{R} \in \mathcal{S}$, that countable unions of sets in $\mathcal{S}$ are again in $\mathcal{S}$, and that $\mathcal{S}$ is closed under complementation: if $S \in \mathcal{S}$, then $\mathbb{R} \setminus S \in \mathcal{S}$. Therefore $\mathcal{S}$ is a σ-algebra of subsets of $\mathbb{R}$. We next show that $\mathcal{S}$ contains every open subset of $\mathbb{R}$ by proving that $(a,b) \in \mathcal{S}$ whenever $a < b$. This is immediate due to $f^{-1}(a,b) = \bigcup_{n \geq 1} f^{-1}[a + \frac{1}{n}, b)$. Consequently all open sets are in $\mathcal{S}$. Thus $\mathcal{S}$ is a σ-algebra containing $\mathcal{B}(\mathbb{R})$. The proof in the case of Lebesgue measurability follows similarly by defining this time $\mathcal{S}$ as the family of sets whose inverse image under f is Lebesgue measurable.

(ii) Let P be the Cantor set constructed for $\lambda = 1/6$ in Exercise 7 and let C be Cantor's ternary set. Let $\{O_k^P\}_{k \geq 1}$ and $\{O_k^C\}_{k \geq 1}$ be the countably many open disjoint intervals whose union is the complement of P and C in $[0,1]$, respectively. We index these intervals in the order they are eliminated in the successive steps of the construction of the respective sets; among the 2^{n-1} open

[71] The construction of H starting from D was necessary to validate the properties required by an equivalence relation: $x \approx x$ (reflexivity), $x \approx y$ implies $y \approx x$ (symmetry), $x \approx y$ together with $y \approx z$ implies $x \approx z$ (transitivity); see Hewitt and Stromberg (1965).

intervals that are eliminated at step n we assign the index in increasing order (left to right). We now define $f : [0,1] \to [0,1]$ as follows: $f(0) = 0$; for $k \geq 1$, define f on O_k^P linearly by mapping the left/right end point of the interval O_k^P to the left/right end point of O_k^C and joining with a line segment; and for $x \in (0,1] \cap P$, set $f(x) = \sup\{f(t) : t < x, t \in \bigcup_{k\geq 1} O_k^P = [0,1] \setminus P\}$. Then $f(P) = C$ and f is a continuous strictly increasing function from $[0,1]$ onto $[0,1]$. Indeed, by the construction of the Cantor set P, between any points $x_1, x_2 \in P$ with $x_1 < x_2$, we can find a point $x_0 \in [0,1] \setminus P$, so that $f(x_1) < f(x_2)$. If $x_1 \in P$ lies to the left of $x_2 \in [0,1] \setminus P$, we can find a point $x_0 < x_2$ in the same open interval O_k^P as x_2, so that again $f(x_1) < f(x_2)$. Also, if $x_1 \in [0,1] \setminus P$ lies to the left of $x_2 \in P$, we can find a point $x_0 > x_1$ in the same open interval O_k^P as x_1, and again $f(x_1) < f(x_2)$. If $x_1, x_2 \in [0,1] \setminus P$ are such that $x_1 < x_2$, clearly $f(x_1) < f(x_2)$ if x_1 and x_2 belong to the same open interval O_k^P. If they belong to two disjoint intervals, $x_1 \in O_m^P = (a_m, b_m)$ and $x_2 \in O_n^P = (a_n, b_n)$, let $O_m^C = (c_m, d_m)$ and $O_n^P = (c_n, d_n)$. While $m \neq n$ is granted, the intricate positioning of the open intervals permit both cases $m < n$ and $m > n$. Nevertheless, $b_m < a_n$ and $d_m < c_n$. Since $f(x_1) < f(b_m) = d_m$ and $f(x_2) > f(a_n) = c_n$, we have $f(x_1) < f(x_2)$. Consequently f is strictly increasing on $[0,1]$. According to the solution of Exercise 3, f can only have jump discontinuities. However, f is clearly continuous on the open set $\bigcup_{k\geq 1} O_k^P = [0,1] \setminus P$. For any $x_0 \in (0,1) \cap P$ we can find a strictly increasing and a strictly decreasing sequence with elements in $[0,1] \setminus P$, both converging to x_0. This prevents a jump discontinuity of f at x_0. Also, 0 and 1 being limit points of left/right end points of intervals O_k^P and O_k^C, we clearly have $\lim_{x\downarrow 0} f(x) = 0$ and $\lim_{x\uparrow 1} f(x) = 1$. Consequently f is continuous on $[0,1]$ and $f(1) = 1$. Since $m(P) = 1/2$, according to Exercise 14, there exists a subset S of P that is not Lebesgue measurable. Let $A = f(S) \subset C$. Since $m(C) = 0$, we have that A is Lebesgue measurable. However $f^{-1}(A) = S$ is not Lebesgue measurable.

16. For the first part, since continuous functions are Borel measurable, by Exercise 15, the inverse image of a Borel set under a continuous function is a Borel set. If B is a Borel set, then $(g \circ f)^{-1}(B) = f^{-1}(g^{-1}(B))$. Since $g^{-1}(B)$ is a Borel set, we can reach our conclusion by the first part of Exercise 15. For the second part, let $f[0,1] \to [0,1]$, and the sets A, $S \subset [0,1]$ be as in Exercise 15. Let g be the characteristic function of the Lebesgue measurable set A. Then f is continuous, g is measurable, while $g \circ f$ is not measurable, being the characteristic function of the Lebesgue nonmeasurable set S.

17. (i) Assume without loss of generality that $m(A_n) < \infty$ for every $n \geq 1$. Then $A = A_1 \cup \bigcup_{n\geq 1}(A_{n+1} \setminus A_n)$, the union being disjoint. By σ-additivity, $m(A) = m(A_1) + \sum_{k=1}^{\infty} (m(A_{n+1}) - m(A_n)) = m(A_1) + \lim_{N\to\infty} \sum_{n=1}^{N} (m(A_{n+1}) - m(A_n)) = \lim_{N\to\infty} m(A_{N+1})$.

(ii) If $m(B_1) < \infty$, let $A = B_1 \setminus B$ and set $A_n = B_1 \setminus B_n$ for $n \geq 1$. Then $A_1 \subseteq A_2 \subseteq \cdots$ and $A = \bigcup_{n\geq 1} A_n$, so that by (i) we have $\lim_{n\to\infty} m(A_n) = m(A)$. Since $m(A_n) = m(B_1) - m(B_n)$ and $m(A) = m(B_1) - m(B)$, the assertion follows. Notice that for $B_n = [n, \infty)$ we have $B_1 \supset B_2 \supset \cdots$ with $m(B_n) = \infty$ for all $n \geq 1$, but $B = \bigcap_{n\geq 1} B_n = \emptyset$.

18. If $t > s$, then $\{f > t\} \subseteq \{f > s\}$ so that $F(t) \leq F(s)$. To prove (2.12), notice that if $x_n \to \infty$, then $\{f > -\infty\} = \bigcup_{n\geq 1}\{f > -x_n\}$, $\{f = \infty\} = \bigcap_{n\geq 1}\{f > x_n\}$, and $\lim_{n\to\infty} F(-x_n) = m(\{f > -\infty\})$ as well as $\lim_{n\to\infty} F(x_n) = m(\{f = \infty\})$ follow in view of Exercise 17, as $\{f > x_1\} \subseteq [0,1]$ has measure at most 1. To prove (2.13), again use Exercise 17. For $t_n \downarrow t_0$, $\{f > t_0\} = \bigcup_{n\geq 1}\{f > t_n\}$. On the other hand, if $t_n \uparrow t_0$, then $\{f \geq t_0\} = \bigcap_{n\geq 1}\{f > t_n\}$.

19. First, note that the set of points that belong to B_n for infinitely many values of n is $\bigcap_{n\geq 1} \bigcup_{k\geq n} B_k$. For any $N \geq 1$ we have $m(\bigcap_{n\geq 1} \bigcup_{k\geq n} B_k) \leq m(\bigcup_{k\geq N} B_k) \leq \sum_{k\geq N} m(B_k)$. As the tail of a convergent series, the right side can be made as small as desired by choosing $N \geq 1$ large enough.

20. Each set of the countable collection of sets $f^{-1}(n2^{-m}, \infty)$, where n and $m \geq 0$ are integers, becomes a Borel set after some set of measure zero is discarded. For any $k \in \mathbb{R}$ we have $f^{-1}(k, \infty) = \bigcup_{[n,m:\, n2^{-m} \geq k]} f_k^{-1}(n2^{-m}, \infty)$. Therefore it suffices to redefine $f = 0$ on the countable union of all the discarded sets (forming a set of measure zero).

21. The necessity follows from Exercise 15 (i) since $\{c\}$ is a Borel set. As a conterexample for sufficiency, let $A \subset [0,1]$ be a nonmeasurable set (see Exercise 11) and let $f(x) = \begin{cases} x+1 & \text{if } x \in A, \\ -x-1 & \text{if } x \notin A. \end{cases}$

As a single-point set, every level set is measurable. However, due to $f^{-1}(0,\infty) = A$, the function f is not measurable.

22. For any integer $n \geq 1$, the sets $B_{k,n} = \{x \in [0,1] : |f_i(x) - f(x)| < \frac{1}{n}$ for $i \geq k\}$ are such that $\bigcup_{k=1}^{\infty} B_{k,n} = [0,1] \setminus B$ for some set B of measure zero, while $B_{k+1,n} \supseteq B_{k,n}$ for all $k \geq 1$. By Exercise 17 (i) we have $\lim_{k\to\infty} m(B_{k,n}) = 1$. Choose $k(n)$ such that $m(B_{k(n),n}) > 1 - 2^{-n}\varepsilon$ and set $A_\varepsilon = \bigcup_{n\geq 1}\big([0,1] \setminus B_{k(n),n}\big)$. Then $m(A_\varepsilon) \leq \sum_{n=1}^{\infty} m([0,1] \setminus B_{k(n),n}) \leq \varepsilon \sum_{n=1}^{\infty} 2^{-n} = \varepsilon$. Now, given any $\delta > 0$, for $n > \delta^{-1}$ we have $|f_i(x) - f(x)| \leq n^{-1} < \delta$ for all $x \in B_{k(n),n}$ and all $i \geq k(n)$. Since $[0,1] \setminus A_\varepsilon \subseteq B_{k(n),n}$, this proves the uniform convergence of f_n to f on $[0,1] \setminus A_\varepsilon$.

23. Consider the countable family $\mathcal{F} = \{U_n\}_{n\geq 1}$ of open subsets of $\mathbb{R}$ of the form (r^-, r^+) with $r^- < r^+$ both rational numbers. For $x < y$, choosing two sequences of rational numbers $\{r_k^-\}_{k\geq 1}$ and $\{r_k^+\}_{k\geq 1}$ with $r_k^- < r_k^+$ for any $k \geq 1$ and such that $r_k^- \downarrow x$, while $r_k^+ \uparrow y$, we write (x, y) as the countable union of the sets $(r_k^-, r_k^+) \in \mathcal{F}$. In view if Exercise 5, every open subset of $\mathbb{R}$ is a countable union of sets in $\mathcal{F}$.

For each $n \geq 1$, the regularity of the Lebesgue measure ensures the existence of a compact set K_n and of an open set O_n such that $K_n \subset f^{-1}(U_n) \subset O_n$ and $m(O_n \setminus K_n) < \varepsilon 2^{-n}$. Define the open set $F_\varepsilon = \bigcup_{n\geq 1}(O_n \setminus K_n)$ with $m(F_\varepsilon) < \sum_{n\geq 1} m(O_n \setminus K_n) \leq \varepsilon$ and set $g = f|_{[0,1]\setminus F_\varepsilon}$. We show that inverse images of open subsets of $\mathbb{R}$ under g are open in $[0,1] \setminus F_\varepsilon$, so that $g : [0,1] \setminus F_\varepsilon \to \mathbb{R}$ is continuous. For this, we first claim that

$$g^{-1}(U_n) = O_n \cap ([0,1] \setminus F_\varepsilon), \qquad n \geq 1. \tag{$**$}$$

Indeed, clearly $g^{-1}(U_n) \subset O_n \cap ([0,1] \setminus F_\varepsilon)$. On the other hand, $O_n \cap ([0,1] \setminus F_\varepsilon) \subset O_n \cap ([0,1] \setminus (O_n \setminus K_n)) = K_n \subset f^{-1}(U_n)$. Intersecting the previous set with $[0,1] \setminus F_\varepsilon$, we obtain $O_n \cap ([0,1] \setminus F_\varepsilon) \subset g^{-1}(U_n)$, which proves ($**$). Now, if $U \subset \mathbb{R}$ is open, there exists a set I of integers $i \geq 1$ such that $U = \bigcup_{i\in I} U_i$. From ($**$) we infer that $g^{-1}(U) = \bigcup_{i\in I} g^{-1}(U_i) = \big(\bigcup_{i\in I} O_i\big) \cap ([0,1] \setminus F_\varepsilon)$ is open in $[0,1] \setminus F_\varepsilon$. Consequently $g : [0,1] \setminus F_\varepsilon \to \mathbb{R}$ is continuous. Extend g to a continuous function $g : [0,1] \to \mathbb{R}$ by defining it to be linear in the disjoint intervals that form the open set F_ε (recall Exercise 5). If $x = 0$ and/or $x = 1$ do not belong to F_ε, set $g = 0$ there. This way, well-determined values are assigned at both endpoints of an interval in F_ε.

24. (i) The bounded function $\hat{f}_n(x) = \begin{cases} f(x) & \text{if } |f(x)| \leq n, \\ n & \text{if } |f(x)| > n, \end{cases}$ is measurable for $n \geq 1$. By Luzin's theorem there is a continuous function $f_n : [0,1] \to \mathbb{R}$ so that the measurable set $A_n = \{x \in [0,1] : f_n(x) \neq \hat{f}_n(x)\}$ satisfies $m(A_n) < 2^{-n}$. Set $A = \bigcap_{n\geq 1} A_n$. Then $m(A) = 0$ by Exercise 17(ii). Now $\lim_{n\to\infty} f_n(x) = f(x)$ for $x \in [0,1] \setminus A$ since $f_n(x) = f(x)$ for all $n \geq |f(x)| + 1$.

(ii) Assume the existence of continuous functions $f_n : [0,1] \to \mathbb{R}$ with $\lim_{n\to\infty} f_n(x) = f(x)$ for every $x \in [0,1]$. Then $\mathbb{R} \setminus \mathbb{Q} = f^{-1}(-\infty, \frac{1}{2}] = \bigcup_{n\geq 1} \bigcap_{k\geq n} f_k^{-1}(-\infty, \frac{1}{2}]$, with the right-hand side being a countable union of closed sets. Consequently $\mathbb{Q}$ would be a countable intersection of open sets. But according to Exercise 5, if $\mathbb{Q}$ is a subset of an open set, that set must be $\mathbb{R}$, which leads to a contradiction.

25. $\int_0^1 |f(x)|\,dx \geq \int_{\{|f|\geq a\}} |f(x)|\,dx \geq \int_{\{|f|\geq a\}} a\,dx = a\,m(\{|f| \geq a\})$ proves Markov's inequality. Chebychev's inequality follows along similar lines.

26. Note that $\{f > 0\} = \bigcup_{n\geq 1}\{f > 1/n\}$. From Markov's inequality, stated in Exercise 25 (i), we infer that $m(\{f > 1/n\}) = 0$ for every $n \geq 1$. This yields $m(\{f > 0\}) \leq \sum_{n=1}^{\infty} m(\{f > 1/n\}) = 0$.

27. Assume that this property does not hold. For some $\varepsilon > 0$ and any $n \geq 1$, we can then find measurable sets A_n with $m(A_n) < 2^{-n}$ and $\int_{A_n} |f(x)|\,dx \geq \varepsilon$. By the Borel–Cantelli lemma (Exercise 19), the set $B = \bigcap_{n\geq 1} \bigcup_{k\geq n} A_k$ of points that belong to A_n for infinitely many values of n has measure zero. The sets $B_n = \bigcup_{k\geq n} A_k \setminus B$ are such that $[0,1] \supseteq B_1 \supseteq B_2 \supseteq \cdots$ with $\bigcap_{n\geq 1} B_n = \emptyset$. The dominated convergence theorem applied to the functions $|f|\chi^{B_n}$ that converge a.e. to zero yields $0 = \lim_{n\to\infty} \int_0^1 |f(x)|\chi^{B_n}(x)\,dx = \lim_{n\to\infty} \int_{B_n} |f(x)|\,dx$. But $A_n \subset B_n \cup B$ with $m(B) = 0$, so that $\int_{B_n} |f(x)|\,dx \geq \int_{A_n} |f(x)|\,dx \geq \varepsilon$ for every $n \geq 1$. The obtained contradiction proves the result.

28. Let $\varepsilon > 0$. Choose $\delta \in (0, \varepsilon)$ so that for all $n \geq 1$ we have $\int_A |f_n(x)|\,dx < \varepsilon/3$ whenever

$m(A) < \delta$. Egorov's theorem ensures the existence of a measurable set A_ε with $m(A_\varepsilon) < \delta$ so that $\lim_{n\to\infty} \{\sup_{x\in[0,1]\setminus A_\varepsilon} |f_n(x) - f(x)|\} = 0$. Choose $N \geq 1$ such that $|f_n(x) - f(x)| < \varepsilon/3$ for all $x \in [0,1] \setminus A_\varepsilon$ and all $n \geq N$. Then for $n \geq N$ we have $\int_0^1 |f_n(x) - f(x)|\, dx = \int_{[0,1]\setminus A_\varepsilon} |f_n(x) - f(x)|\, dx + \int_{A_\varepsilon} |f_n(x) - f(x)|\, dx \leq \int_{[0,1]\setminus A_\varepsilon} |f_n(x) - f(x)|\, dx + \int_{A_\varepsilon} |f_n(x)|\, dx + \int_{A_\varepsilon} |f(x)|\, dx < \frac{2\varepsilon}{3} + \int_{A_\varepsilon} |f(x)|\, dx$. By Fatou's lemma, $\int_{A_\varepsilon} |f(x)|\, dx \leq \liminf_{n\to\infty} \int_{A_\varepsilon} |f_n(x)|\, dx \leq \frac{\varepsilon}{3}$, so that $\left|\int_0^1 [f_n(x) - f(x)]\, dx\right| \leq \int_0^1 |f_n(x) - f(x)|\, dx \leq \varepsilon$ for $n \geq N$. This completes the proof.

29. (i) If we set $f_n = \chi^{[0,1/2]}$ if $n \geq 1$ is odd and $f_n = \chi^{(1/2,1]}$ if $n \geq 1$ is even, then $\liminf_{n\to\infty} f_n(x) = 0$ for all $x \in [0,1]$, while $\int_0^1 f_n(x)\, dx = 1/2$ for all $n \geq 1$. (ii) The sequence $f_n = n^2 \chi^{(0,1/n)}$ for $n \geq 1$ is such that $\liminf_{n\to\infty} f_n(x) = 0$ for all $x \in [0,1]$ but $\int_0^1 f_n(x)\, dx = n$ for all $n \geq 1$.

30. Let $V \subset [0,1]$ be the nonmeasurable set constructed in Exercise 11. Set $A = \{(1,x) :\ x \in V\} \subset \mathbb{R}^2$. Since $A \subset K_n = [1 - \frac{1}{n}, 1] \times [0,1]$ and K_n is compact with $\mathcal{L}^2(K_n) = 1/n \to 0$ as $n \to \infty$, we have $\mathcal{L}^2(A) = 0$. However, the projection of A on the real axis is the set V.

31. As in Exercise 23, choose a countable basis $\{U_n\}_{n\geq 1}$ of the open subsets of $\mathbb{R}$. If $y \neq f(x)$, then there is an open set O_x containing $f(x)$ and not containing y. Choosing $n \geq 1$ with $U_n \subset O_x$, we get that $f(x) \in U_n$ but $y \notin U_n$. Consequently $G_f = ([0,1] \times \mathbb{R}) \setminus \bigcup_{n\geq 1} [f^{-1}(U_n) \times (\mathbb{R} \setminus U_n)]$. Every set $[f^{-1}(U_n) \times (\mathbb{R} \setminus U_n)]$ is $\mathcal{L}^2$-measurable since the set $f^{-1}(U_n)$ differs by a set of zero $\mathcal{L}^1$-measure from a Borel subset of $\mathbb{R}$, so that this product differs by a set of $\mathcal{L}^2$-measure zero from a Borel set of $\mathbb{R}^2$. We deduce that G_f is $\mathcal{L}^2$-measurable. Applying Tonelli's theorem for the characteristic function of the set $G(f)$, we get $\mathcal{L}^2(G_f) = \int_0^1 \left(\int_0^1 \chi^{\{f(x)\}}(y)\, dy\right) dx = \int_0^1 0\, dx = 0$, where $\chi^{\{f(x)\}}(y)$ is the characteristic function of the single-point set $\{f(x)\}$. In this context, note that for a (Lebesgue) measurable set $A \subset [0,1]^2$:

- (i) $\mathcal{L}^2(A) = 0$ if and only if for a.e. $x \in [0,1]$ the x-section of A, defined as $A_x = \{y \in [0,1] : (x,y) \in A\}$, has $\mathcal{L}^1$ measure zero;
- (ii) $\mathcal{L}^2(A) > 0$ if and only if there is some $\delta > 0$ and a measurable set $S \subset [0,1]$ of positive $\mathcal{L}^1$-measure such that $\mathcal{L}^1(A_x) > \delta$ for every $x \in S$;

see van Douwen (1989). Moreover, the measurability is essential – there is a nonmeasurable set $A \subset [0,1]^2$ such that no set A_x contains more than one point, see van Douwen (1989). Assigning to $x \in [0,1]$ the element in A_x, if it exists, and the value -1 otherwise, we obtain a function $f : [0,1] \to [-1,1]$ whose graph $G(f)$ is not $\mathcal{L}^2$-measurable. Indeed, were $G(f)$ $\mathcal{L}^2$-measurable, so would $A = G(f) \cap ([0,1] \times [-1/2, 1])$ be. In particular, $G(f)$ does not have $\mathcal{L}^2$-measure zero!

32. Let $f(x,y) = \frac{x^2-y^2}{(x^2+y^2)^2} = \frac{\partial}{\partial y}\left(\frac{y}{x^2+y^2}\right) = -\frac{\partial}{\partial x}\left(\frac{x}{x^2+y^2}\right)$ for $(x,y) \in [0,1]^2$ with $x \neq 0$ and $y \neq 0$, respectively. Thus $\int_0^1 f(x,y)\, dy = \frac{1}{x^2+1}$ for $x \neq 0$, so that $\int_0^1 \left(\int_0^1 f(x,y)\, dy\right) dx = \frac{\pi}{4}$. Similarly, $\int_0^1 \left(\int_0^1 f(x,y)\, dx\right) dy = -\frac{\pi}{4}$. However, $\int_0^1 \int_0^1 \frac{|x^2-y^2|}{(x^2+y^2)^2}\, dxdy = \int_0^1 \left(\int_0^1 \frac{|x^2-y^2|}{(x^2+y^2)^2}\, dy\right) dx = \int_0^1 \left(\int_0^x \frac{x^2-y^2}{(x^2+y^2)^2}\, dy + \int_x^1 \frac{y^2-x^2}{(x^2+y^2)^2}\, dy\right) dx = \int_0^1 \left(\int_0^x \frac{\partial}{\partial y}\left(\frac{y}{x^2+y^2}\right) dy - \int_x^1 \frac{\partial}{\partial y}\left(\frac{y}{x^2+y^2}\right) dy\right) dx = \int_0^1 \left(\frac{1}{x} - \frac{1}{x^2+1}\right) dx = \infty$, by Tonelli's theorem.

33. For $y \neq 0$, we have $\int_{-1}^1 \frac{xy}{(x^2+y^2)^2}\, dx = -\frac{y}{2} \int_{-1}^1 \frac{\partial}{\partial x}\left(\frac{1}{x^2+y^2}\right) dx = 0$. Similarly, $\int_{-1}^1 \frac{xy}{(x^2+y^2)^2}\, dy = 0$ for $x \neq 0$, so that both iterated integrals are zero. To see that the function is not $\mathcal{L}^2$-integrable, compute $\int_0^1 \frac{xy}{(x^2+y^2)^2}\, dy = \frac{1}{2x} - \frac{x}{2(x^2+1)}$ for $x \neq 0$. Tonelli's theorem yields $\int_0^1 \int_0^1 \frac{xy}{(x^2+y^2)^2}\, dxdy = \int_0^1 \left(\frac{1}{2x} - \frac{x}{2(x^2+1)}\right) dx = \infty$. But the above integral is less than $\int_{-1}^1 \int_{-1}^1 \frac{|xy|}{(x^2+y^2)^2}\, dxdy$.

34. Since $|3^{-n} \cos[(21)^n \pi x]| \leq 3^{-n}$ for $n \geq 1$ and $x \in \mathbb{R}$, we deduce that f is continuous on $\mathbb{R}$. We now show that f is not differentiable at any fixed $x \in \mathbb{R}$. First, for $n \geq 1$ and $h > 0$, write

$$\frac{f(x+h) - f(x)}{h} = S_n(h) + R_n(h), \qquad n \geq 1, \quad h > 0, \qquad (*)$$

where

$$S_n(h) = \sum_{k=0}^{n-1} 3^{-k}\,\frac{\cos[(21)^k\pi(x+h)] - \cos[(21)^k\pi x]}{h},$$

$$R_n(h) = \sum_{k=n}^{\infty} 3^{-k}\,\frac{\cos[(21)^k\pi(x+h)] - \cos[(21)^k\pi x]}{h}.$$

For every $k \geq 1$, $\frac{\cos[(21)^k\pi(x+h)] - \cos[(21)^k\pi x]}{h} = -(21)^k\pi\,\sin[(21)^k\pi(x+\xi_k)]$ for some $\xi_k \in (0,h)$, by the mean-value theorem, so that

$$|S_n(h)| \leq \sum_{k=0}^{n-1} 3^{-k}\,(21)^k\pi = \pi\frac{7^n-1}{6} < \frac{\pi}{6}\,7^n, \quad n \geq 1. \tag{**}$$

On the other hand, for every $n \geq 1$ we can find an integer a_n with

$$-\frac{1}{2} \leq b_n = (21)^n x - a_n \leq \frac{1}{2}, \tag{***}$$

so that

$$\frac{3}{2\,(21)^n} \geq h_n = \frac{1-b_n}{(21)^n} \geq \frac{1}{2\,(21)^n}, \qquad n \geq 1. \tag{****}$$

We also have $(21)^k\pi(x+h_n) = (21)^{k-n}(21)^n\pi(x+h_n) = (21)^{k-n}\pi(1+a_n)$ yields $\cos[(21)^k\pi(x+h_n)] = (-1)^{1+a_n}$, as well as $\cos[(21)^k\pi x] = \cos[(21)^{k-n}(21)^n\pi x] = \cos[(21)^{k-n}\pi(a_n+b_n)] = \cos[(21)^{k-n}\pi a_n]\,\cos[(21)^{k-n}\pi b_n] = (-1)^{a_n}\,\cos[(21)^{k-n}\pi b_n]$. In view of (****) and the fact that (***) ensures $\cos(\pi b_n) \geq 0$, we infer that

$$|R_n(h_n)| = \frac{1}{h_n}\sum_{k=n}^{\infty} 3^{-k}\,(1+\cos[(21)^{k-n}\pi b_n]) \geq \frac{1}{h_n 3^n} \geq \frac{2}{3}\,7^n.$$

Combining this with (**), we get that $\left|\frac{f(x+h_n)-f(x)}{h_n}\right| \geq |R_n(h_n)| - |S_n(h_n)| \geq \frac{4-\pi}{6}\,7^n \to \infty$ for $h_n \to 0$. This means that f is not differentiable at x.

35. For $n \geq 1$ let $G_n \supset E$ be an open set with $m(G_n) < 2^{-n}$ and set $f_n(x) = m(G_n \cap [0,x])$ for $x \in [0,1]$. Since $0 \leq m(G_n \cap [0,x]) - m(G_n \cap [0,y]) \leq m([y,x])$ for $0 \leq y \leq x \leq 1$, each function $f_n : [0,1] \to [0,2^{-n}]$ is continuous and nondecreasing. Consequently $f_E : [0,1] \to [0,1]$ defined by $f_E(x) = \sum_{n=1}^{\infty} f_n(x)$ is nondecreasing and continuous. If $x_0 \in E$ and $n \geq 1$ is fixed, then for $|h|$ sufficiently small, the segment $[x_0, x_0+h]$ lies entirely within G_n. For such values of $h > 0$ (the case $h < 0$ being entirely analogous), we have $G_n \cap [0, x_0+h] = (G_n \cap [0,x_0]) \cup (x_0, x_0+h]$, so that $f_n(x_0+h) = m(G_n \cap [0,x_0+h]) = m(G_n \cap [0,x_0]) + h = f_n(x_0) + h$ and thus $\frac{f_n(x_0+h)-f_n(x_0)}{h} = 1$. Consequently, for any integer $N \geq 1$ and $h > 0$ sufficiently small, we have $\frac{f_E(x_0+h)-f_E(x_0)}{h} \geq \sum_{n=1}^{N} \frac{f_n(x_0+h)-f_n(x_0)}{h} = N$, so that $f_E'(x_0) = \infty$.

36. The monotone character of f is ensured by construction. Moreover, the image of the set $[0,1] \setminus C_0$ under f is dense in $[0,1]$ since the set $\bigcup_{n\geq 1}\{\frac{1}{2^n}, \frac{3}{2^n}, \cdots, \frac{2^n-1}{2^n}\}$ is dense in $[0,1]$. Together with the fact that f is nondecreasing on $[0,1]$, this ensures the continuity of f. Indeed, if x_0 were a point of discontinuity, then either $\liminf_{h\downarrow 0}\{f(x_0+h)\} = \lim_{h\downarrow 0}\{f(x_0+h)\} := f_+(x_0) > f(x_0)$, or $\limsup_{h\uparrow 0}\{f(x_0+h)\} = \lim_{h\uparrow 0}\{f(x_0+h)\} := f_-(x_0) < f(x_0)$. If, say, the latter occurs, then the interval $(f_-(x_0), f(x_0))$ contains no values of f. The obtained contradiction proves the continuity of the function f on $[0,1]$. Finally, clearly $f'(x) = 0$ on the open set $[0,1] \setminus C_0$ of measure 1.

37. For $\varepsilon > 0$ choose $\delta > 0$ such that for any finite collection of disjoint intervals $\{(a_k, b_k)\}_{1\leq k\leq n}$ with $\sum_{k=1}^{n}(b_k - a_k) < \delta$, we have $\sum_{k=1}^{n} |f(b_k) - f(a_k)| < \frac{\varepsilon}{2}$. If now $\{(c_i, d_i)\}_{i\geq 1}$ is a countable family of disjoint intervals with $\sum_{i\geq 1}(d_i - c_i) < \delta$, then for any $n \geq 1$ we have $\sum_{i=1}^{n} |f(d_i) - f(c_i)| < \frac{\varepsilon}{2}$. Consequently $\sum_{i\geq 1} |f(d_i) - f(c_i)| \leq \frac{\varepsilon}{2} < \varepsilon$.

38. Extend F to $\mathbb{R}$ by setting $F(x) = \begin{cases} F(b) + f(b)\,[x-b] \text{ for } x > b\,, \\ F(a) + f(a)\,[x-a] \text{ for } x < a\,. \end{cases}$ This corresponds to extending $f = F'$ to $\mathbb{R}$ by setting $f(x) = f(b)$ for $x > b$ and $f(x) = f(a)$ for $x < a$. Given $a \le x < y \le b$, define the continuous function $\sigma : [0,1] \to \mathbb{R}$ by $\sigma(t) = \frac{F([1-t]b+ty) - F([1-t]a+tx)}{[1-t]b+ty-[1-t]a-tx}$ for $0 \le t \le 1$. The image $\sigma[0,1]$ of $[0,1]$ under σ is an interval. But $\sigma(0) = \frac{F(b)-F(a)}{b-a}$ and $\sigma(1) = \frac{F(y)-F(x)}{y-x}$. Consequently the set $\mathcal{S} = \{\frac{F(y)-F(x)}{y-x} : a \le x < y \le b\}$ of all slopes of chords joining any two points lying on the graph of F is an interval. If $\mathcal{D} = \{f(x) : a \le x \le b\}$, by the mean-value theorem $\mathcal{S} \subset \mathcal{D}$, while the definition of the derivative (as a limit of slopes of chords) ensures $\mathcal{D} \subset \overline{\mathcal{S}}$. It follows at once that $\mathcal{D}$ is an interval, which simply is an alternative way of expressing the Darboux property.[72]

39. If $a \le x < y \le b$, then $V_f[a,x] + V_f[x,y] = V_f[a,y]$ holds. This proves that the function $s : [a,b] \to [0, V_f[a,b]]$ is nondecreasing. To prove the continuity of the function s, let $\varepsilon > 0$ and choose $\delta > 0$ so that $|f(x) - f(y)| < \varepsilon$ whenever $x, y \in [a,b]$ are such that $|x-y| < \delta$. By the definition of $V_f[a,b]$ there is a partition $a = x_0 < x_1 < \cdots < x_n = b$ of $[a,b]$ so that

$$V_f[a,b] - \varepsilon < \sum_{k=1}^{n} |f(x_k) - f(x_{k-1})| \le V_f[a,b]\,. \qquad (*)$$

Due to the triangle inequality, if we refine the partition by adding to it intermediate points that lie in the intervals (x_{k-1}, x_k) for $k = 1, \cdots, n$, the corresponding sum in the centre of relation $(*)$ is not diminished, so that $(*)$ continues to hold. Repeating this, if necessary, we may asssume that $x_1 - x_0 = x_k - x_{k-1} < \delta$ for all $k = 1, \cdots, n$. If $x \in [a,b]$, we can find some $j \in \{1, \cdots, n\}$ with $x \in [x_{j-1}, x_j]$. Using $|f(x_j) - f(x_{j-1})| \le |f(x) - f(x_{j-1})| + |f(x) - f(x_j)|$, we deduce from $(*)$ that $V_f[a,b] - \varepsilon < (\sum_{k=1}^{j-1} |f(x_k) - f(x_{k-1})|) + |f(x) - f(x_{j-1})| + |f(x) - f(x_j)| + (\sum_{k=j+1}^{n} |f(x_k) - f(x_{k-1})|) \le V_f[a,x] + |f(x) - f(x_j)| + V_f[x_j, b] = s(x) + |f(x) - f(x_j)| + V_f[a,b] - s(x_j)$. This yields $0 \le s(x_j) - s(x) < |f(x) - f(x_j)| + \varepsilon < 2\varepsilon$ for all $x \in [x_{j-1}, x_j]$. The partition being equidistant and of mesh $\delta_1 = x_1 - x_0 < \delta$, we infer that if $x, y \in [a,b]$ are such that $|x-y| < \delta_1$, then there is some $j \in \{1, \cdots, n\}$ with $|x - x_j| \le \delta_1$ and $|y - x_j| \le \delta_1$, so that $|s(x) - s(y)| \le |s(x) - s(x_j)| + |s(y) - s(x_j)| < 4\varepsilon$. This proves the continuity of the function s.

40. (i) Let χ^E be the characteristic function of the set E and define the absolutely continuous function $f : [0,1] \to [0,1]$ by $f(x) = m([0,x] \cap E) = \int_0^x \chi^E(s)\,ds$ for $x \in [0,1]$. Since for $x \in [0,1)$ and $h > 0$ sufficiently small we have $\frac{f(x+h)-f(x)}{h} = \frac{1}{h}\int_x^{x+h} \chi^E(s)\,ds = \frac{m([x,x+h]\cap E)}{h} \ge \alpha$, we deduce that at every point $x \in [0,1)$ where f is differentiable, that is, a.e., we have $f'(x) \ge \alpha > 0$. But by Theorem 2.8 we know that $f' = \chi^E$ a.e. on $[0,1]$. Consequently the set E has measure 1.

(ii) The proof relies on the following observation: if $[a,b] \subset [0,1]$ is a nontrivial interval (that is, with $b > a$), then we can find two disjoint Cantor-like[73] sets $C_1, C_2 \subset [a,b]$ of positive measure $\frac{b-a}{2}$ and $\frac{b-a}{36}$, respectively. To see this, first eliminate from $[a,b]$ the open interval $(\frac{7a+5b}{12}, \frac{7b+5a}{12})$ and implement the procedure from Exercise 7 to construct a Cantor-like set $C_1 \subset [a, \frac{7a+5b}{12}] \cup [\frac{7b+5a}{12}, b]$ of measure $\frac{b-a}{2}$. Subsequently construct a Cantor-like set $C_2 \subset [\frac{17b+19a}{36}, \frac{19b+17a}{36}] \subset (\frac{7a+5b}{12}, \frac{7b+5a}{12})$ of measure $\frac{b-a}{36}$.

Let now $\{I_n\}_{n\ge1}$ be an enumeration of all segments $[a,b] \subset [0,1]$ with $b > a$ both rational numbers. We construct inductively sequences $\{C_n\}_{n\ge1}$ and $\{D_n\}_{n\ge1}$ of Cantor-like subsets of $[0,1]$ of positive measure as follows. C_1 and D_1 are disjoint subsets of I_1. Once $C_1, D_1, \cdots, C_{n-1}, D_{n-1}$, are chosen, their finite union is a Cantor-like set U_n of positive measure, so that $I_n \setminus U_n$ contains an open set (c,d). Within $[\frac{2c+d}{3}, \frac{2d+c}{3}] \subset (c,d)$ we can find a disjoint pair C_n, D_n. Note that by

[72] Note that the Darboux property means that the image set is connected. While every derivative has a connected graph, there are Darboux functions whose graphs are not connected – see Bruckner and Leonard (1966).

[73] In analogy to the set constructed in Exercise 7, by a Cantor-like set we understand a compact set C containing no interval and no isolated points.

construction all sets in $\{C_n\}_{n\geq 1}$ and $\{D_n\}_{n\geq 1}$ are disjoint. Let $A = \bigcup_{n=1}^{\infty} C_n$. If $I \subset [0,1]$ is an open interval, for some $n \geq 1$ we have $I_n \subset I$, so that $C_n \subset I$ and $D_n \subset I$. Therefore $0 < m(C_n) \leq m(A \cap I) < m(A \cap I) + m(K_n) \leq m(I)$, since A and $D_n \subset I$ are disjoint sets. The previous inequality grants the desired properties of the set A.

Define now the absolutely continuous function $f : [0,1] \to [-1,1]$ by $f(x) = \int_0^x (\chi^A(s) - \chi^{[0,1]\setminus A}(s))\, ds$ for $x \in [0,1]$. By Theorem 2.8, if the restriction of f to the open interval $I \subset [0,1]$ is monotone, then f' has a.e. the same sign on I: at every point $x \in I$ where $f'(x)$ exists and is finite, the expression $\frac{f(x+h)-f(x)}{h}$ has the same sign for $h > 0$ sufficiently small to ensure that $x + h$ lies within I. But by construction $f' = \pm 1$ on subsets of I of positive measure.

41. (i) Assume the existence of a constant $M > 0$ such that $|f(x) - f(y)| \leq M\,|x - y|$ whenever $x, y \in [a,b]$. Then (2.19) yields the absolute continuity of f. Moreover, from Theorem 2.8 we infer the a.e. validity of $|f'(x)| = \left|\lim_{h\to 0} \frac{f(x+h)-f(x)}{h}\right| \leq \limsup_{h\to 0} \left|\frac{f(x+h)-f(x)}{h}\right| \leq M$. Conversely, if $f : [a,b] \to \mathbb{R}$ is absolutely continuous with $|f'(x)| \leq M$ a.e. for some constant $M > 0$, then $|f(y) - f(x)| = \left|\int_x^y f'(s)\, ds\right| \leq \int_x^y |f'(s)|\, ds \leq \int_x^y M\, ds = M(x - y)$ whenever $a \leq x \leq y \leq b$.

(ii) The existence of a constant $M > 0$ such that $|f(x) - f(y)| \leq M\,|x - y|$ whenever $x, y \in [0,1]$ ensures $\sum_{k=1}^n |f(b_k) - f(a_k)| \leq M \sum_{k=1}^n |b_k - a_k| < \varepsilon$ if $\sum_{k=1}^n |b_k - a_k| < \frac{\varepsilon}{M}$. For the converse, note first that f must be absolutely continuous, so that f' exists a.e. on $[a,b]$. Choose $\delta > 0$ so that $\sum_{k=1}^n |f(b_k) - f(a_k)| < 1$ whenever $\sum_{k=1}^n |b_k - a_k| < \delta$ and set $M = \frac{2}{\delta}$. We claim that $|f'(x)| \leq M$ a.e. on $[a,b]$, in which case we conclude by (i). If the claim does not hold true, then the set $A = \{x \in (a,b) :\ f'(x)$ exists, is finite, and $|f'(x)| > M\}$ has positive measure. But for every $x \in A$ we can find some $h_x > 0$ with $|f(x+h) - f(x)| > M\,|h|$ for all $|h| < h_x$. In particular, $A \subset \bigcup_{k\geq 1} \{x \in (0,1) : |f(x + \frac{\delta}{2^k}) - f(x)| > M\,\delta 2^{-k}\}$ since for every $x \in A$ we can find some integer $k \geq 1$ with $\frac{\delta}{2^k} < h_x$. We deduce that at least one of the sets $\{x \in (0,1) : |f(x + \frac{\delta}{2^k}) - f(x)| > M\,\frac{\delta}{2^k}\}$ has positive measure. In particular, it must contain 2^{k-1} disjoint points $x_1, \cdots, x_{2^{k-1}}$. But then $\sum_{j=1}^{2^{k-1}} |f(x_j + \frac{\delta}{2^k}) - f(x_k)| > \sum_{j=1}^{2^{k-1}} M\,\frac{\delta}{2^k} = \frac{M\delta}{2} = 1$ although $\sum_{j=1}^{2^{k-1}} \frac{\delta}{2^k} = \frac{\delta}{2} < \delta$.

42. (i) Let $\{x_n\}_{n\geq 1}$ be a sequence in X such that $x_n \neq x_m$ for $n \neq m$. Since $\mu\big(\bigcup_{n\geq 1}\{x_n\}\big) = \infty \neq \sum_{n\geq 1} \mu(\{x_n\}) = 0$, μ is not a measure. This example illustrates the difference between finitely additive and countably additive set functions since $\mu\big(\bigcup_{i\in I} A_i\big) = \sum_{i\in I} \mu(A_i)$ if I is finite.

(ii) The fact that μ is a measure is immediate. Also, if $f(x_0) = \infty$ for some $x_0 \in X$, then μ cannot be σ-finite: a decomposition $X = \bigcup_{i\geq 1} A_i$ with $A_i \in \mathcal{A}$ with $\mu(A_i) < \infty$ is impossible since if $x_0 \in A_{i_0}$, then $\mu(A_{i_0}) = \mu(\{x_0\}) + \mu(A \setminus \{x_0\}) \geq \mu(\{x_0\}) = f(x_0) = \infty$. Moreover, if the set $F = \{x \in X :\ f(x) > 0\}$ is countable (or finite), it can be represented as a countable or finite set $\{x_j\}_{j\in J}$. By definition, $\mu(X \setminus F) = 0$, and if $\mu(\{x_j\}) < \infty$ for all $j \in J$, then μ is σ-finite since $X = (X \setminus F) \cup \bigcup_{j\in J}\{x_j\}$. On the other hand, if the set F is uncountable, then a representation $X = \bigcup_{j\in I} A_i$ with I finite or countable is possible only if at least one of the sets A_i contains uncountably many elements of F, in which case this set has infinite measure, by definition.

43. The only set A with $\nu(A) = 0$ is the empty set, so that $\mu \ll \nu$. But $d\mu = f\, d\nu$ would imply $0 = f(x)$ for every $x \in [0,1]$ since $\mu(\{x\}) = 0$, thus $\mu \equiv 0$. On the other hand, if $\nu = \lambda + \rho$ with $\rho \ll \mu$ and $\lambda \perp \mu$, then $\lambda(\{x\}) = 1$ for all $x \in [0,1]$. This leads to $\lambda = \nu$ but $\nu \perp \mu$ cannot hold since for every Lebesgue measurable set $A \subset [0,1]$ with $\mu(A) > 0$, we have $\nu(A) = \infty$.

2.5 Notes to Chapter 2

1. It is possible to develop the Lebesgue integral without relying on measure theory. One such approach is provided by the Daniell integral. Daniell took the axiomatic properties of the integral as the starting point and developed integration for broader classes of functions, see Loomis (1953). More precisely, one starts with an averaging process I defined for a certain set of functions $\mathcal{F}$ (the so-called elementary functions) that is a linear space with the usual operations of addition and scalar multiplication, and such that if $f \in \mathcal{F}$, then $|f| \in \mathcal{F}$. The averaging process is extended to a wider class of functions by requiring

- linearity: $l(\alpha f + \beta g) = \alpha l(f) + \beta l(g)$ for $f, g \in \mathcal{F}$ and $\alpha, \beta \in \mathbb{R}$;
- nonnegativity: if $f \geq 0$ then $l(f) \geq 0$;
- continuity with respect to monotone convergence: if $f_n(x) \downarrow 0$ for every x, then $l(f_n) \to 0$.

The extension to nonnegative functions outside of $\mathcal{F}$ is performed by taking pointwise monotone nondecreasing limits of functions in $\mathcal{F}$. Splitting subsequently a function into its positive and negative part one bypasses the nonnegativity restriction. Starting from the Riemann integral on the space of continuous functions $f : [0, 1] \to \mathbb{R}$, one obtains the Lebesgue integrable functions.

The Daniell integral is closely related to the Riesz approach: the powerful Riesz representation theorem from functional analysis ensures that any linear map from the set of continuous functions $f : [0, 1] \to \mathbb{R}$ to $\mathbb{R}$ that has the nonnegativity property is obtained by integrating f against an (abstract) Borel measure. When applied to the Riemann integral, the resulting measure is the Lebesgue measure, see Rudin (1974). In both approaches, starting from the concept of the integral, one can develop measure theory by specialising the integral to characteristic functions.

In the middle of the twentieth century there was a widespread bias in favour of presenting the theory of integration following abstract approaches, since these avoid the subtleties of measure theory and lead one straight to an extended notion of integration. However, the lack of structural information, due to the economy of these schemes, becomes detrimental when facing fine properties of integrable functions, e.g. in the context of the fundamental theorem of calculus, and one cannot escape measure theory. The advantage of an abstract technique lies in that the filtering down to the bare essence discloses a great generality and structural properties that enable its use in contexts that appear to be totally unrelated. The drawback lies in the lure of art-for-art's sake: while abstractisation and generalisation can be intelectually satisfying, this often leads to a "cancerous overgrowth of abstraction, frequently marked by willful obscurantism", see Kupka (1986). However, usually these destructive facets pass into oblivion within the frame of one generation – eventually the arid labyrinths of pure technique wear out even the creativity of the most gifted researchers.

2. The main purpose of the Lebesgue integral is to provide a flexible concept where limits of integrals hold under mild assumptions. It may happen that even functions that are Riemann integrable are at times not Lebesgue integrable. One example would be $\int_0^\infty \frac{\sin(x)}{x}\,dx$, which exists as the improper Riemann integral $\lim\limits_{x\to\infty} \int_0^x \frac{\sin(s)}{s}\,ds$. However, this function is not Lebesgue integrable.[74] An equivalent concept of improper Lebesgue integral is not that convenient because such a perspective is deficient from the viewpoint of the convergence theorems.

3. The difficulties arising in the characterisation of functions by the knowledge of the derivative a.e. show the limitations of an approach that relies on computing pointwise derivatives. It is desirable to introduce a concept that is less localised, being therefore able to capture global features of the function. The concept of distributional derivative (see Chapter 6) turns out to be helpful. Indeed, the distributional derivative of an absolutely continuous real-valued function defined on $[a, b]$ is an integrable function on $[a, b]$, while the distributional derivative of a function of bounded variation is a signed Radon measure (the difference of two finite regular Borel measures) on $[a, b]$. In particular, the distributional derivative of the function $f : [0, 1] \to \mathbb{R}$ defined in Exercise 36 is a measure μ concentrated on the Cantor ternary set C_0 defined in Exercise 8 (meaning that $\mu(A) = 0$ whenever $A \subset [0, 1] \setminus C_0$). The fact that $f' = 0$ on $[0, 1] \setminus C_0$ (that is, a.e.) is less precise. Moreover, the distributional approach offers an elegant generalisation of the concepts of absolute continuity and bounded variation to functions of several variables: if $\Omega \subset \mathbb{R}^n$ is an open set, an integrable function $f : \Omega \to \mathbb{R}$ is of bounded variation if all its distributional partial derivatives are signed Radon measures, and is absolutely continuous if they are integrable functions; see the in-depth analysis in Evans and Gariepy (1992). The difficulties encountered in attempting to extend the definitions

[74] Indeed, $\int_0^\infty \frac{|\sin(x)|}{x}\,dx \geq \sum_{n=1}^\infty \int_{(n+1/2)\pi}^{(n+2/3)\pi} \frac{|\sin(x)|}{x}\,dx \geq \sum_{n=1}^\infty \frac{1}{12(n+2/3)} = \infty$.

(2.19) and (2.22) are due to the complexity of open sets and their possibly very irregular boundaries in higher dimension. In contrast to this, the case of functions $f = (f_1, \cdots, f_n) : [a, b] \to \mathbb{R}^n$ is plain since the definitions (2.19) and (2.22) can be easily adapted, the problem being reduced to that of dealing with each of the individual components f_k with $k = 1, \cdots, n$.

3

Elements of functional analysis

The Fourier series representation of a function is the function space counterpart of the decomposition of an n-dimensional vector into components with respect to an orthonormal basis for $\mathbb{R}^n$ or $\mathbb{C}^n$. To deal with the underlying infinite-dimensional setting some acquaintance with functional analysis is required. The present chapter aims to present the basic functional analytic framework. We introduce some powerful tools that will be used in Chapter 4 to gain insight into the behaviour of Fourier series.

3.1 An overall perspective

Despite Fourier's optimistic program for representing an arbitrary function by a trigonometric series, the convergence issue for Fourier series is a delicate matter. The challenge is twofold: with regard to the appropriate choice of functions, as well as concerning the suitable notion of convergence (with links between these two issues). Two results illustrate the intricate nature of trigonometric series. In 1872 Weierstrass used a trigonometric series to provide an example of a continuous but nowhere differentiable function (see Exercise 34 of Chapter 2). On the other hand, it was for a long time supposed that every function f which is periodic and continuous possesses a Fourier series which converges at every point to the function. In 1873 du Bois-Reymond exhibited a continuous periodic function with a divergent Fourier series at a point.[1] Nevertheless, the intuition of Fourier, that for a large class of functions one can define the Fourier coefficients and recover the function from the knowledge of its Fourier coefficients, was essentially correct. Riemann's integral, introduced

[1] See Exercise 7 in Chapter 4 for a simpler example due to Fejér.

to deal with trigonometric series, showed great promise at first, but proved in the end to be insufficient to cope with the complexity of the problem. The surprising examples mentioned above shattered the confidence of mathematicians that Fourier series represent a convenient tool, so that towards the end of the nineteenth century, the subject of Fourier series appeared to be intractable and research in this direction reached a standstill. The resurgence of the topic at the beginning of the twentieth century was enabled by Lebesgue's theory of integration. It turns out that square Lebesgue integrable functions represent the class of functions with which Fourier series are most naturally associated. This specific setting was the main source of basic concepts in functional analysis, such as completion, separability and orthogonality.

The class $L^2[0,1]$ of square Lebesque integrable functions consists of the Lebesgue measurable functions $f : [0,1] \to \mathbb{C}$ with $\int_0^1 |f(x)|^2 dx < \infty$, two such functions being identified if they agree a.e. (so that $L^2[0,1]$ is actually an equivalence class of functions). Endowed with the inner product

$$\langle f, g \rangle = \int_0^1 f(x)\,\overline{g(x)}\,dx, \tag{3.1}$$

the space $L^2[0,1]$ turns out to be a separable Hilbert space with orthonormal basis $\{e^{2\pi ikx}\}_{k\in\mathbb{Z}}$ (see Chapter 4). The generalisation of the Fourier series expansion to separable Hilbert spaces is of the same degree of sophistication as that required to deal with the case of $L^2[0,1]$. Moreover, pursuing the approach in a Hilbert space setting has the advantage of providing the rationale for this remarkable fact. For this reason, in Section 3.2 we discuss Fourier series in separable Hilbert spaces.

It is also possible to associate a Fourier series to a function in the class $L^1[0,1]$ of Lebesque integrable functions, consisting of all Lebesgue measurable functions $f : [0,1] \to \mathbb{C}$ with $\int_0^1 |f(x)|dx < \infty$; again, we identify functions in $L^1[0,1]$ if they agree a.e. While $L^1[0,1]$ does not posses the rich structure of a Hilbert space, it is nevertheless a Banach space. Because of the important role played by $L^1[0,1]$ in the theory of Fourier series, we present some background information about Banach spaces in Section 3.3.

The interplay between Fourier analysis and functional analysis is very rich, and cannot be reduced to the considerations we allude to in the present chapter. Other examples of fruitful interaction will appear in subsequent chapters,[2] and in due course we provide references for more advanced topics that are beyond our scope.

[2] For example, the functions of Schwartz class that arise in the context of the Fourier transform do not form a Banach or Hilbert space, nor is this the case for the tempered distributions.

3.2 Hilbert spaces

In calculus one takes advantage of the interconnected structures of the Euclidean space $\mathbb{R}^n$: algebraic (as a linear vector space) and topological (as a metric space), enhanced to geometric by means of the concept of orthogonality. Among the various infinite-dimensional counterparts of Euclidean spaces, the most refined analysis can be performed in the setting of a Hilbert space. In this section we develop the basic Hilbert space theory.

A *real (complex) vector space* is a set X, whose elements are called vectors and in which two operations, called *addition* and *scalar multiplication*, are defined, with the following algebraic properties:

(1) to every pair of vectors $(x, y) \in X \times X$ corresponds a vector $x + y \in X$, in such a way that $x + y = y + x$ (commutativity) and $x + (y + z) = (x + y) + z$ (associativity); X contains a unique vector 0 (called the origin of X) such that $x + 0 = x$ for every $x \in X$; and to each $x \in X$ corresponds a unique vector $-x$ such that $x + (-x) = 0$;

(2) to each pair $(\alpha, x) \in \mathbb{R}\times X$ or $(\alpha, x) \in \mathbb{C}\times X$, α being a real or complex scalar, there is associated a vector $\alpha x \in X$ in such a way that $1x = x$, $\alpha(\beta x) = (\alpha\beta)x$, and the two distributive laws $\alpha(x + y) = \alpha x + \alpha y$, $(\alpha + \beta)x = \alpha x + \beta x$ hold.

The above conditions cover the whole algebraic structure but are not logically independent.[3] Examples of infinite-dimensional vector spaces are spaces of real or complex sequences $x = \{x_n\}_{n\geq 1}$, and function spaces – collections of functions $f : S \to \mathbb{R}$ ($\mathbb{C}$) defined on some set S. In the first case the space is singled out by some specific property of the sequence, a property that must be preserved under addition and scalar multiplication. For function spaces, additional leeway to ensure a variety of examples is provided by the choice of the domain of definition S. Both constructions are illustrated below.

A *pre-Hilbert space* is a real/complex vector space H endowed with an inner product that confers it a geometric structure. An *inner product* (or *scalar product*) is a map $(\cdot, \cdot) : H \times H \to \mathbb{R}$ ($\mathbb{C}$) subject to the rules:

(i) $(x, x) \geq 0$ for all $x \in H$, with equality if and only if $x = 0$;
(ii) $(\alpha x + \beta y, z) = \alpha(x, z) + \beta(y, z)$ for all $\alpha, \beta \in \mathbb{R}$ ($\mathbb{C}$) and $x, y, z \in H$;
(iii) $(x, y) = \overline{(y, x)}$ for all $x, y \in H$.

[3] For example, the existence of $-x$ follows from the other conditions. Indeed, since $1 + 0 = 1$, we have $1x = 1x + 0x$. Knowing that $1x = x$, we get $x = x + 0x$ for all $x \in X$. But $0 \in X$ is the only point $y \in X$ with $x + y = x$ for all $x \in X$, so that $0x = 0 \in X$. Now $(1 - 1)x = 0$ shows that $-x$ could be defined as $(-1)x$. Finding out the minimal set of conditions is possible but pedantic.

Observe that (i) and (ii) imply $(z, \alpha x+\beta y) = \bar{\alpha}(z, x)+\bar{\beta}(z, y)$ for all $\alpha, \beta \in \mathbb{R}\,(\mathbb{C})$ and $x, y, z \in H$. Consequently, scalar products are linear in both variables for real pre-Hilbert spaces, while for complex pre-Hilbert spaces they are linear in the first variable and conjugate-linear in the second. Since complex vector spaces are more relevant for Fourier analysis, throughout the text we assume that the scalars are complex unless otherwise specified.

The scalar product allows us to introduce the concept of orthogonality: $x, y \in H$ are said to be *orthogonal* if $(x, y) = 0$. It also endows H with a metric structure: denoting $\|x\| = \sqrt{(x, x)}$, in view of (i), $\|x - y\|$ defines a distance between two elements $x, y \in H$ if the triangle inequality $\|x + y\| \le \|x\| + \|y\|$ holds for all $x, y \in H$. To verify this, we need the following result.

Proposition 3.1 (Cauchy–Schwarz inequality) *In any pre-Hilbert space H,*

$$|(x, y)| \le \|x\| \cdot \|y\|, \tag{3.2}$$

with equality if and only if $x = \alpha y$ or $y = \alpha x$ for some scalar α.

Proof Define $f : \mathbb{R} \to \mathbb{R}$ by $f(t) = \|x+t\, e^{i\varphi}\, y\|^2$, where $\varphi \in [0, 2\pi)$ is the argument of the complex number (x, y), i.e., $(x, y) = |(x, y)|\, e^{i\varphi}$. Note that $f(t) = (x + t\, e^{i\varphi}\, y,\ x + t\, e^{i\varphi}\, y) = t^2\|y\|^2 + 2t|(x, y)| + \|x\|^2 \ge 0$ for all $t \in \mathbb{R}$. The positivity of this polynomial in t, of degree 2, with real coefficients, requires that $|(x, y)|^2 - \|x\|^2\|y\|^2 \le 0$, which yields (3.2). If $x = \alpha y$ or $y = \alpha x$ for some scalar α, it is clear that equality holds in (3.2). Conversely, if equality holds in (3.2), then $(x, y) = e^{i\theta}\|x\| \cdot \|y\|$ for some $\theta \in [0, 2\pi)$. If $y \ne 0$, for $\alpha = e^{i\theta}\,\|x\|/\|y\|$ we have $\|x - \alpha y\|^2 = (x - \alpha y,\ x - \alpha y) = |\alpha|^2\|y\|^2 - \bar{\alpha}\,(x, y) - \alpha\,\overline{(x, y)} + \|x\|^2 = \|x\|^2 - \|x\|^2 - \|x\|^2 + \|x\|^2 = 0$. This implies $x = \alpha y$. The case $y = 0$ is covered by $y = \alpha x$ with $\alpha = 0$. □

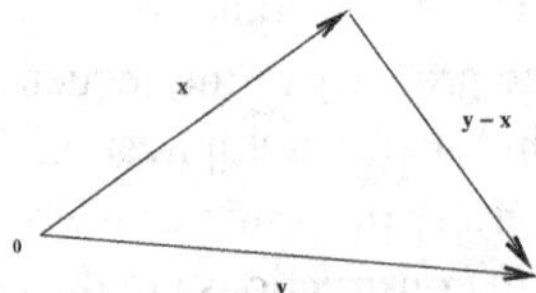

Figure 3.1 A geometric interpretation of the triangle inequality $\|x - y\| \le \|x\| + \|y\|$: the far side of a triangle is shorter than the sum of the lengths of two adjacent sides.

The validity of the triangle inequality is now easily proved. Indeed, $\|x + y\|^2 = (x+y,\ x+y) = \|x\|^2+(x, y)+\overline{(x, y)}+\|y\|^2 \le \|x\|^2+2\|x\|\cdot\|y\|+\|y\|^2 = (\|x\|+\|y\|)^2$, in view of (3.2). Using (3.2), we can define the angle $\theta \in [0, \pi)$ between two non-zero vectors $x, y \in H$ by $\cos\theta = \Re e\{\frac{(x,y)}{\|x\|\cdot\|y\|}\}$, obtaining the "law of cosines": $\|x + y\|^2 = \|x\|^2 + 2\|x\| \cdot \|y\| \cos\theta + \|y\|^2$. In particular, if x and y are orthogonal, this becomes the Pythagorean rule: $\|x + y\|^2 = \|x\|^2 + \|y\|^2$.

In a real Hilbert space two non-zero vectors x and y are orthogonal if and only if the angle between them is $\pi/2$, while in a complex Hilbert space, the angle between two non-zero vectors x and y is $\pi/2$ if and only if $\Re e\{(x, y)\} = 0$. The reason for this discrepancy can be explained as follows.

If we identify $\mathbb{C}^n$ with $\mathbb{R}^{2n}$, then the real part of the scalar product of $x = (a_1 + ib_1, \ldots, a_n + ib_n) \in \mathbb{C}^n$ and $y = (c_1 + id_1, \ldots, c_n + id_n) \in \mathbb{C}^n$ corresponds to the scalar product of the vectors $(a_1, b_1, \ldots, a_n, b_n) \in \mathbb{R}^{2n}$ and $(c_1, d_1, \ldots, c_n, d_n) \in \mathbb{R}^{2n}$. For example, the scalar product of $1, i \in \mathbb{C}$ is $(-i)$, while the scalar product of the corresponding vectors $(1, 0), (0, 1) \in \mathbb{R}^2$ vanishes: the imaginary part is "invisible". In the context of complex Hilbert spaces the angle concept is misleading, the correct notion of orthogonality being $(x, y) = 0$, and not $\Re e\{(x, y)\} = 0$. This is justified geometrically by the projection theorem (Theorem 3.2), applied to the case of the distance to a one-dimensional closed subspace – in analogy to the planar distance from a point to a line.

Since the scalar product induces a distance, pre-Hilbert spaces are metric spaces. The concepts of convergent and Cauchy sequences are therefore meaningful: a sequence $\{x_n\}_{n\geq 1}$ in H converges to $x \in H$ if $\lim\limits_{n\to\infty} \|x_n - x\| \to 0$ (denoted $x_n \to x$), while $\{x_n\}_{n\geq 1}$ is a Cauchy sequence in H if $\lim\limits_{n,m\to\infty} \|x_n - x_m\| = 0$. The validity of the triangle inequality ensures that a convergent sequence is a Cauchy sequence since $\|x_n - x_m\| \leq \|x - x_n\| + \|x - x_m\|$. The converse is not necessarily true, a pre-Hilbert space in which all Cauchy sequences are convergent being called a *Hilbert space*. It turns out that all finite-dimensional pre-Hilbert spaces are Hilbert spaces, but in infinite dimensions there is a distinction between these two concepts, every pre-Hilbert space H being however a dense subset[4] of a Hilbert space by a process of completion (see Theorem 3.5).

The typical example of a finite-dimensional Hilbert space is $\mathbb{C}^n$ with the scalar product $(z, \xi) = \sum_{k=1}^n z_k \overline{\xi_k}$. Examples of pre-Hilbert spaces are the space of continuous functions $x : [0, 1] \to \mathbb{C}$ with the scalar product $(x, y) = \int_0^1 x(t)\overline{y(t)}\,dt$, and the collection of complex sequences $x = \{x(k)\}_{k\geq 1}$ with finitely many non-zero entries, with the scalar product $(x, y) = \sum_{k=1}^\infty x(k)\overline{y(k)}$. The completion of the first space is $L^2[0, 1]$, see Chapter 4. For the latter example, simply note that the Cauchy sequence $x_n = (1, 1/2, \ldots, 1/n, 0, 0, \ldots)$ has no limit in the pre-Hilbert space.[5]

A precise meaning of the term "infinite-dimensional" pre-Hilbert space is provided by the concept of a basis. To explain this we need to use some facts from set theory. Since building a consistent theory of sets from basic axioms is a long, difficult process that is remote from analysis, we will only present a survey, appealing throughout it to intuition and elementary logic.

Definition *A relation R on a set X is a subset R of the Cartesian product $X \times X$: we say that x is R-related to y, denoted xRy, if (x, y) belongs to R. A relation R is called reflexive if xRx for every $x \in X$, symmetric if xRy implies yRx, antisymmetric if xRy and yRx implies $x = y$, transitive if xRy and yRz implies xRz. An equivalence relation is a reflexive, symmetric and transitive relation, and a partial ordering is a reflexive, transitive and antisymmetric relation.*

For example, the relation $x \equiv y \pmod 3$ is an equivalence relation on the integers (with xRy if and only if $x - y$ is a multiple of 3), and on the collection X of all subsets of a given set Ω the relation

[4] That is, every element of the larger space can be arbitrarily well approximated by a sequence from H.

[5] The completion of this space is the space l^2 of complex sequences $x = \{x(k)\}_{k\geq 1}$ with $\sum_{k=1}^\infty |x(k)|^2 < \infty$, endowed with the scalar product $(x, y) = \sum_{k=1}^\infty x(k)\overline{y(k)}$ (see Exercise 8).

$A \prec B$ if $A \subseteq B$ defines a partial ordering. The latter reveals the rationale for the affix "partial": two elements of Ω might be unrelated, as they need not obey $A \prec B$ or $B \prec A$.

Definition *Let $\prec$ be a partial ordering relation on a set X. An element $l \in X$ is called an upper bound for a subset $Y \subset X$ if $y \prec l$ for all $y \in Y$. If $m \in X$ has the property that $m \prec x$ implies $x = m$, we say that m is a maximal element of X, while an element $L \in X$ with the property that $x \prec L$ for every $x \in X$ is called the largest element*[6] *of X. Finally, if for all $x, y \in X$ we have $x \prec y$ or $y \prec x$, then X is totally ordered.*

For example, $\mathbb{R}$ with the familiar order relation $x \leq y$ is totally ordered. On the other hand, the relation $(x_1, y_1) \prec (x_2, y_2)$ if $x_1 \leq x_2$ and $y_1 \leq y_2$ defines a partial ordering on $X = [0, 1] \times [0, 1]$ such that X is not totally ordered,[7] with $(0, 1)$ an upper bound of the subset $Y = \{(0, y) : y \in [0, 1]\} \subset X$, and with the largest element $(1, 1)$ being the only maximal element of $(X, \prec)$. For the collection of closed discs contained in $[0, 1] \times [0, 1]$ there are maximal elements[8] but there is no largest element relative to inclusion.

The tools of elementary set theory are insufficient to permit certain constructions. In particular, the following statement is actually independent of the other axioms of set theory but for our considerations its validity is presumed.

Zorn's lemma *Let X be a nonempty partially ordered set with the property that every totally ordered subset has an upper bound in X. Then each totally ordered set has some upper bound that is also a maximal element of X.*

To take advantage of Zorn's lemma, we first generalise the familiar concept of a coordinate system. A set B in a vector space X is said to be a *Hamel basis* for X if it is linearly independent[9] and if it spans[10] X.

For example, any two noncollinear vectors in $\mathbb{R}^2$ form a Hamel basis of $\mathbb{R}^2$. On the other hand, if X is the vector space of real sequences $x = \{x(k)\}_{k \geq 1}$, then, while the vectors $e_n = \{\delta_{n,k}\}_{k \geq 1}$ for $n \geq 1$ are linearly independent, they do not form a Hamel basis for X. Indeed, since only finite linear combinations of basis elements are allowed, the span of the e_ns is the linear subspace made up of all sequences that have a finite number of nonzero entries.

Theorem *Every vector space X has a Hamel basis.*

Proof Let $\mathcal{X}$ denote the collection of all linearly independent sets in X. We define a partial ordering $\prec$ on $\mathcal{X}$ by saying that $B_1 \prec B_2$ if $B_1 \subseteq B_2$. Then, if $\{B_i\}_{i \in I}$ is a totally ordered subset of $\mathcal{X}$, it is clear that $B = \bigcup_{i \in I} B_i$ is an upper bound for it. Indeed, we only need to prove that B is a linearly independent set. If $\{x_1, \ldots, x_n\}$ are finitely many elements of B, for every $k \in \{1, \ldots, n\}$ we have that $x_k \in B_{i_k}$ for some $i_k \in I$. Since $\{B_i\}_{i \in I}$ is a totally ordered subset of $\mathcal{X}$, among the n sets B_{i_k} there is one, say $B_{i_k^*}$, that contains all the other $(n - 1)$. Consequently $\{x_1, \ldots, x_n\} \subset B_{i_k^*}$ must be linearly independent in X, proving our claim. By Zorn's lemma, $\mathcal{X}$ has a maximal element B_0. To see that B_0 is a Hamel basis, with linear independence granted by $B_0 \in \mathcal{X}$, we only need to prove

[6] Due to antisymmetry there is at most one such element.

[7] For example, $(1/2, 1/3)$ and $(1/3, 1/2)$ are unrelated.

[8] Any disc tangent to two sides of the square is a maximal element.

[9] That is, a finite linear combination $\sum_{k=1}^{n} \alpha_k x_k$ of elements in $x_k \in B$, where α_k are scalars, is zero if and only if all coefficients α_k are zero.

[10] That is, any $x \in X$ has a (unique) representation as a finite linear combination of basis elements.

that every $x \in X$ is a finite linear combination of elements in B_0. This must be the case since if we find some $x \in X$ that does not have this property, then the set $B_0 \cup \{x\}$ would belong to $\mathcal{X}$, contradicting the maximality of B_0. □

The previous result enables us to say that a vector space X is *finite-dimensional* if it admits a Hamel basis with finitely many vectors. One can show (see Exercise 15) that in a finite-dimensional vector space X all Hamel bases have the same number of elements,[11] called the dimension of X. This is the *algebraic dimension* of X. The pursuit of this direction is hindered by the fact that in Hilbert and Banach spaces a Hamel basis is either finite or uncountably infinite (see Exercise 15), and in the latter case explicit examples of a Hamel basis are unfathomable.[12] The limited interest of Hamel bases for infinite-dimensional spaces is also underlined by the fact that not all associated projections are continuous (see the solution to Exercise 19). Fortunately, for Hilbert spaces, a more germane notion of basis is available.

Attaching a meaning in a pre-Hilbert space H to infinite linear combinations of the form $\sum_{k=1}^{\infty} \alpha_k x_k$ leads to the notion of an orthonormal basis, a concept developed from the related concept of an orthonormal set. A collection (finite or infinite, possibly uncountable) of points $\{x_i\}_{i\in I}$ in H is said to be an *orthogonal set* if $(x_i, x_j) = 0$ whenever $i \neq j$. If, in addition, $\|x_i\| = 1$ for all $i \in I$, then the set is called *orthonormal*. An orthogonal set of nonzero vectors $\{x_i\}_{i\in I}$ can be orthonormalised by replacing x_i with $\frac{x_i}{\|x_i\|}$. Note that orthonormal sets are linearly independent (see Exercise 4). Moreover, calling a maximal orthonormal set in H an *orthonormal basis* of H, every orthonormal set in H is contained in an orthonormal basis (see Exercise 4). In particular, any Hilbert space has an orthonormal basis. The fundamental property of orthonormal bases is that they have a natural interpretation as orthogonal coordinate systems.

Theorem 3.1 (Fourier series theorem) *For a finite or countable orthonormal set $\{e_n\}_{n\geq 1}$ in a Hilbert space H, the following statements are equivalent:*

(a) $\{e_n\}_{n\geq 1}$ is an orthonormal basis;

(b) (Fourier series expansion) for any $x \in H$ we have[13] $x = \sum_{n\geq 1} (x, e_n)\, e_n$ *;*

(c) (Parseval identity) we have $(x, y) = \sum_{n\geq 1} (x, e_n)\, \overline{(y, e_n)}$ for all $x, y \in H$;

(d) (Plancherel identity) for any $x \in H$ we have $\|x\|^2 = \sum_{n\geq 1} |(x, e_n)|^2$;

(e) any linear subspace of H that contains $\{e_n\}_{n\geq 1}$ is dense in H.

11 More generally, if B_1 and B_2 are two Hamel bases of a vector space, then B_1 and B_2 have the same cardinality, that is, there is a bijection $b : B_1 \to B_2$; see Hewitt and Stromberg (1965).

12 The fact that the presented existence proof is not constructive being indicative of this feature.

13 The scalars (x, e_n) are called the Fourier coefficients of x.

Theorem 3.1 is formulated for Hilbert spaces since any pre-Hilbert space can be viewed as a subspace of a Hilbert space (see Theorem 3.5). Also, the assumption that the orthonormal basis is finite or countable amounts to H being *separable*, in the sense that it admits a countable dense subset, see Exercise 15. For a discussion of the pathological case of uncountable orthonormal sets in H see also Exercise 15.

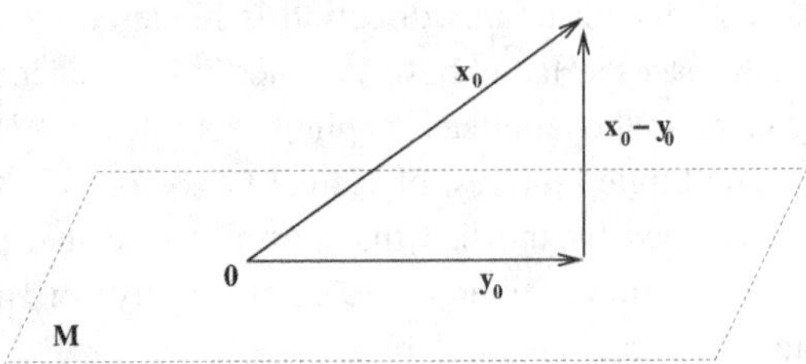

Figure 3.2 Theorem 3.2 captures an intuitive geometric idea: the unique point in M closest to x_0 is found by dropping a perpendicular from x_0 to M.

In order to prove Theorem 3.1 we need some preliminary results that are of interest on their own. They relate to approximation results in a Hilbert space setting. Consider the following problem: let M be a proper subspace of a Hilbert space H and choose a point $x_0 \in H \setminus M$. Does there exist a point $y_0 \in M$ that is closest to x_0, that is, $\|x_0 - y_0\| \leq \|x_0 - y\|$ for all $y \in M$? If it exists, is it unique, and if so, how do we find it? Since infinite-dimensional Hilbert spaces admit nonclosed subspaces (see Exercise 17), and the distance $d(x_0, M) = \inf_{y \in M}\{\|x_0 - y\|\}$ from x_0 to M equals[14] the distance $d(x_0, \overline{M})$ from x_0 to the closure $\overline{M}$ of M, we restrict our attention to closed subspaces M. We will see that in this setting, $d(x_0, M)$ is always attained. Without M being closed, the odd situation $d(x_0, M) = 0$ with $x_0 \notin M$ would occur.

Theorem 3.2 (Projection on closed subspaces) *Let $M \subset H$ be a closed linear subspace of a Hilbert space H. Let $x_0 \in H$ and define $\delta = \inf_{y \in M}\{\|x_0 - y\|\}$. Then there is precisely one $y_0 \in M$ such that $\|x_0 - y_0\| = \delta$. Moreover, $(x_0 - y_0)$ is orthogonal to all elements of M, and y_0 is the only point in M with this orthogonality property.*

[14] By definition, $\overline{M}$ is the smallest closed set containing M, consisting of all points $x \in X$ that are limits of convergent sequences whose elements lie in M. Now $d(x_0, \overline{M}) \leq d(x_0, M)$ since $M \subset \overline{M}$. On the other hand, if $y_n \in \overline{M}$ satisfy $\|x_0 - y_n\| < d(x_0, \overline{M}) + 1/n$, then for $x_n \in M$ with $\|x_n - y_n\| < 1/n$ we have $\|x_0 - x_n\| \leq \|x_0 - y_n\| + \|y_n - x_n\| < d(x_0, \overline{M}) + 2/n$. This yields $d(x_0, M) \leq d(x_0, \overline{M})$.

The comments preceding the statement explain the rationale of assuming M to be closed. For the importance of completeness see Exercise 8.

Proof If $x_0 \in M$, then clearly $y_0 = x_0$. This is the only point $z \in M$ with $(z - x_0)$ orthogonal to all elements of M, since $(z - x_0) \in M$ forces $(z - x_0, z - x_0) = 0$, that is, $z = x_0$.

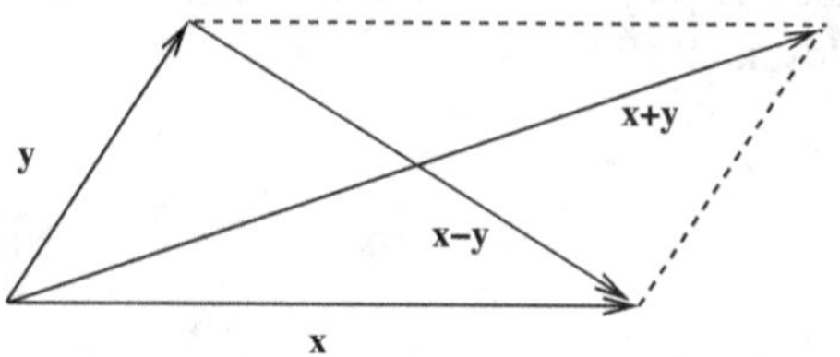

Figure 3.3 A geometric illustration of the parallelogram law.

If $x_0 \notin M$, then, by definition, there exists a sequence $\{y_n\}_{n\geq 1}$ in M with $\lim_{n\to\infty} \|x_0 - y_n\| = \delta$. We will prove that $\{y_n\}_{n\geq 1}$ is a Cauchy sequence in H, and we identify its limit as being the sought-after point $y_0 \in M$. Throughout the proof we will take advantage of the following identity, called the parallelogram law,

$$\|x - y\|^2 + \|x + y\|^2 = 2\|x\|^2 + 2\|y\|^2 , \qquad x, y \in H , \tag{3.3}$$

illustrated in Figure 3.3 and easily checked by expanding the left side using the properties of the scalar product. Using (3.3), we get $\|y_n - y_m\|^2 = \|(y_n - x_0) - (y_m - x_0)\|^2 = 2\|y_n - x_0\|^2 + 2\|y_m - x_0\|^2 - \|y_n + y_m - 2x_0\|^2$. Since $\delta = \inf_{y\in M}\{\|x_0 - y\|$ and $\frac{y_n+y_m}{2} \in M$, we must have $\|\frac{y_n+y_m}{2} - x_0\| \geq \delta$. Consequently, letting $n, m \to \infty$ in the previously displayed inequality, we infer that $\{y_n\}_{n\geq 1}$ is a Cauchy sequence in H. By assumption M is a closed subspace of the Hilbert space H. Therefore there is some $y_0 \in M$ with $y_n \to y_0$ as $n \to \infty$, so that $\|x_0 - y_0\| = \lim_{n\to\infty} \|x_0 - y_n\| = \delta$. The relation $\|x_0 - y_0\| = \delta$ ensures the uniqueness of $y_0 \in M$. Indeed, if $y_0' \in M$ is such that $\|x_0 - y_0'\| = \delta$ and $y_0' \neq y_0$, then $\frac{y_0+y_0'}{2} \in M$ leads us by means of (3.3) to the contradiction $\delta^2 \leq \|x_0 - \frac{y_0+y_0'}{2}\|^2 = 2\|\frac{x_0}{2} - \frac{y_0}{2}\|^2 + 2\|\frac{x_0}{2} - \frac{y_0'}{2}\|^2 - \|\frac{y_0-y_0'}{2}\|^2 < \frac{\|x_0-y_0\|^2}{2} + \frac{\|x_0-y_0'\|^2}{2} = \delta^2$.

Let us now show that y_0 is the only point in M such that $(x_0 - y_0)$ is orthogonal to all elements of M. If $y \in M$, define $f : \mathbb{R} \to \mathbb{R}$ by $f(\lambda) = \|x_0 - (y_0 + \lambda y)\|^2 = (x_0 - y_0 - \lambda y,\ x_0 - y_0 - \lambda y) = \|x_0 - y_0\|^2 - 2\lambda\,\Re e\{(x_0 - y_0, y)\} + \lambda^2\|y\|^2$. Since f attains its minimum δ^2 at $\lambda = 0$, we have $f'(0) = \Re e\{(x_0 - y_0, y)\} = 0$. Similar considerations for $g : \mathbb{R} \to \mathbb{R}$ defined by $g(\lambda) = \|x_0 - (y_0 + i\lambda y)\|^2$ yield $\Im m\{(x_0 - y_0, y)\} = 0$. Thus $(x_0 - y_0, y) = 0$, as claimed. The existence of another $y_0' \in M$ such that $(x_0 - y_0')$ is orthogonal to all vectors in M yields $(y_0 - y_0', y) = (x_0 - y_0', y) - (x_0 - y_0, y) = 0$ for all $y \in M$. The choice $y = y_0 - y_0'$ leads to $y_0 = y_0'$. □

A consequence of Theorem 3.2 is that if $M_1 \subsetneq M_2$ are both closed linear subspaces of a Hilbert space H, then there is a nonzero vector $z \in M_2$ that is orthogonal to every vector in M_1. To see this, it suffices to pick some $x_0 \in M_2 \setminus M_1$ and to define $z = x_0 - y_0$, where $y_0 \in M_1$ is the element of M_1 that is closest to x_0. Let us define the *orthogonal complement*[15] $M^\perp$ of a subset $M \subset H$ in the Hilbert space H as the collection of the vectors in H that are orthogonal to all elements of M: $M^\perp = \{x \in H : \ (x, y) = 0 \text{ for all } y \in M\}$. It is

[15] Theorem 3.3 justifies the terminology.

easy to see that $M^\perp$ is always a closed linear subspace of H, even if M is not closed or is not a linear subspace (see Exercise 12). Given a closed subspace $M \subset \mathbb{R}^n$ and its orthogonal complement $M^\perp$, we can resolve any vector $x \in \mathbb{R}^n$ into a sum $y + z$ of two unique vectors, $y \in M$ being the vector in M that is closest to x (called the projection of x on M), and $z \in M^\perp$ (see Figure 3.4). This property holds in all Hilbert spaces.

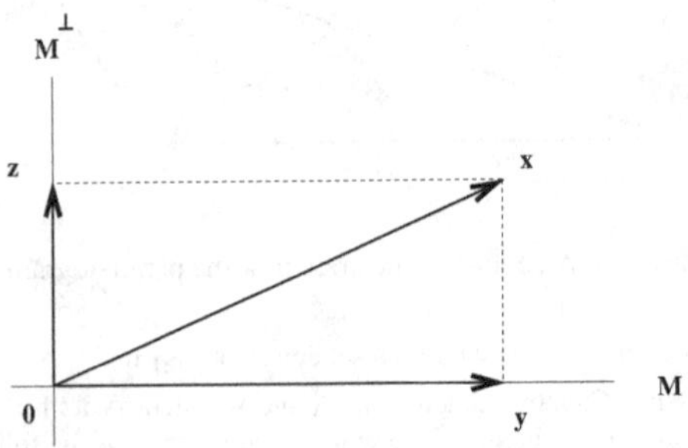

Figure 3.4 Illustration of the projection theorem.

Theorem 3.3 (The projection theorem) *Let M be a closed linear subspace of a Hilbert space H. Then any $x \in H$ can be written uniquely as $x = y + z$ with $y \in M$ and $z \in M^\perp$.*

Proof Let y be the projection of x on M. Set $z = x - y$. By Theorem 3.2, we have that $z \in M^\perp$. To prove uniqueness, let $x = y_1 + z_1 = y_2 + z_2$ with $y_1, y_2 \in M$ and $z_1, z_2 \in M^\perp$. Then $0 = (y_1-y_2)+(z_1-z_2)$, so that $0 = \|(y_1-y_2)+(z_1-z_2)\|^2 = \|y_1-y_2\|^2+2\,\Re e\{(y_1-y_2,\, z_1-z_2)\}+\|z_1-z_2\|^2 = \|y_1-y_2\|^2 + \|z_1-z_2\|^2$ since $(y_1-y_2) \in M$ and $(z_1-z_2) \in M^\perp$. Therefore $y_1 = y_2$ and $z_1 = z_2$. □

Before presenting the proof of Theorem 3.1, we highlight some specific features of orthonormal sets.

Lemma (Bessel's inequality) *Let $\{e_k\}_{k\geq 1}$ be a finite or countable orthonormal set in a pre-Hilbert space H. Then $\sum\limits_{k\geq 1} |(x, e_k)|^2 \leq \|x\|^2$ for all $x \in H$.*

Proof For a finite subset $\{e_1, \ldots, e_n\}$ of $\{e_k\}_{k\geq 1}$, $0 \leq \|x - \sum_{k=1}^n (x, e_k)\, e_k\|^2 = (x - \sum_{k=1}^n (x, e_k)\, e_k\,,\, x - \sum_{j=1}^n (x, e_j)\, e_j) = \|x\|^2 - 2\sum_{k=1}^n (x, e_k)\,(e_k, x) - \sum_{j=1}^n \overline{(x, e_j)}\,(x, e_j) + \sum_{k=1}^n \sum_{j=1}^n (x, e_k)\,\overline{(x, e_j)}\,(e_k, e_j) = \|x\|^2 - \sum_{k=1}^n |(x, e_k)|^2$ since $(e_k, e_j) = \delta_{kj}$. The inequality therefore holds for finite sums. Since the right side is fixed, a limiting process ensures its validity even for an infinite sum. □

Let us now have a closer look at series of the form $\sum_{k=1}^\infty \alpha_n e_n$ in a Hilbert space, where α_n are scalars and $\{e_n\}_{n\geq 1}$ is an orthonormal set.

Lemma *Let $\{e_n\}_{n\geq 1}$ be an orthonormal set in a Hilbert space H. Then*

(a) the series $\sum_{n=1}^\infty \alpha_n e_n$ converges if and only if $\sum_{n=1}^\infty |\alpha_n|^2 < \infty$;

(b) if $\sum_{n=1}^{\infty} \alpha_n e_n = \sum_{n=1}^{\infty} \beta_n e_n$, in the sense that both series converge to the same vector x in H, then $\alpha_n = \beta_n$ for all $n \geq 1$ and $\sum_{n=1}^{\infty} |\alpha_n|^2 = \|x\|^2$.

Proof (a) If $\lim_{N\to\infty} \sum_{n=1}^{N} \alpha_n e_n = x$ in H, then, by the continuity of the scalar product, $(x, e_n) = \big(\sum_{k=1}^{\infty} \alpha_k e_k, e_n\big) = \sum_{k=1}^{\infty} \alpha_k (e_k, e_n) = \alpha_n$ for all $n \geq 1$. Bessel's inequality ensures the convergence of the numerical series $\sum_{n=1}^{\infty} |\alpha_n|^2$. Conversely, if $\sum_{n=1}^{\infty} |\alpha_n|^2 < \infty$, let $S_n = \sum_{k=1}^{n} \alpha_k e_k$. Then for $n > m \geq 1$ we have $\|S_n - S_m\|^2 = (S_n - S_m, S_n - S_m) = \big(\sum_{k=m+1}^{n} \alpha_k e_k, \sum_{j=m+1}^{n} \alpha_j e_j\big) = \sum_{k=m+1}^{n} \sum_{j=m+1}^{n} \alpha_k \overline{\alpha_j} (e_k, e_j) = \sum_{k=m+1}^{n} |\alpha_k|^2$, by orthonormality. Thus $\{S_n\}_{n\geq 1}$ is a Cauchy sequence in H, and must have a limit.

(b) If $x = \sum_{k=1}^{\infty} \alpha_k e_k = \sum_{k=1}^{\infty} \beta_k e_k$, then we have $0 = \lim_{N\to\infty} \big[\sum_{k=1}^{N} \alpha_k e_k - \sum_{k=1}^{N} \beta_k e_k\big] = \lim_{N\to\infty} \sum_{k=1}^{N} (\alpha_k - \beta_k) e_k$, so that $0 = \sum_{k=1}^{\infty} (\alpha_k - \beta_k) e_k$. Bessel's inequality yields $\alpha_k = \beta_k$ for all $k \geq 1$. Moreover, if $x = \sum_{k=1}^{\infty} \alpha_k e_k$, in the sense that the series converges in H to x, then $\|x\|^2 - \sum_{k=1}^{n} |\alpha_k|^2 = (x, x - \sum_{k=1}^{n} \alpha_k e_k) + (x - \sum_{k=1}^{n} \alpha_k e_k, \sum_{k=1}^{n} \alpha_k e_k) \leq \|x - \sum_{k=1}^{n} \alpha_k e_k\| \{\|x\| + \|\sum_{k=1}^{n} \alpha_k e_k\|\}$ by (3.2). Since $\sum_{k=1}^{n} \alpha_k e_k \to x$ as $n \to \infty$, the right-hand side converges to zero. Therefore $\|x\|^2 \leq \sum_{k=1}^{\infty} |\alpha_k|^2$. Combining this with Bessel's inequality yields $\|x\|^2 = \sum_{k=1}^{\infty} |\alpha_k|^2$. □

In view of Theorem 3.2, the next result identifies $\sum_{k=1}^{n} (x, e_k) e_k$ as the orthogonal projection of x on the n-dimensional linear space spanned by the orthonormal vectors $\{e_1, \ldots, e_n\}$. This is a best-approximation result.

Lemma *Let $\{e_1, \ldots, e_n\}$ be an orthonormal set in a Hilbert space H and let $x \in H$. Then for any choice of complex numbers $\alpha_1, \ldots, \alpha_n \in \mathbb{C}$, we have*

$$\|x - \sum_{k=1}^{n} (x, e_k) e_k\| \leq \|x - \sum_{k=1}^{n} \alpha_k e_k\|$$

for all $x \in H$, with equality if and only if $\alpha_k = (x, e_k)$ for all $k = 1, \ldots, n$.

Proof We have $(x - \sum_{k=1}^{n} (x, e_k) e_k, \sum_{j=1}^{n} \alpha_j e_j) = \sum_{k=1}^{n} \overline{\alpha_k} (x, e_k) - \sum_{j=1}^{n} \sum_{k=1}^{n} (x, e_k) \overline{\alpha_j} (e_k, e_j) = \sum_{k=1}^{n} \overline{\alpha_k} (x, e_k) - \sum_{k=1}^{n} \overline{\alpha_k} (x, e_k) = 0$ by orthonormality. The statement follows from Theorem 3.3.□

Every Hilbert space H has an orthonormal basis (see Exercise 15). Moreover, given a finite or countable linearly independent set $\{y_n\}_{n\geq 1}$, the Gram–Schmidt orthonormalisation process defines $x_1 = \frac{y_1}{\|y_1\|}$ and x_{k+1} inductively by[16] $x_{k+1} = \alpha_{k+1} \{y_{k+1} - \sum_{j=1}^{k} (y_{k+1}, x_j) x_j\}$ for $k \geq 1$, with the scalar α_{k+1} chosen so that $\|x_{k+1}\| = 1$. Then $\{x_n\}_{n\geq 1}$ is an orthonormal set with the property that for all $k \geq 1$, the vectors $x_1, ..., x_k$, span the same linear subspace as the vectors $y_1, ..., y_k$ (see Exercise 14). This procedure is very effective in constructing orthonormal sets.

We now prove the fundamental result on Fourier series in Hilbert spaces.

Proof of Theorem 3.1 We first show that (a) implies (b). If $\{e_n\}_{n\geq 1}$ is a maximal orthonormal set, then the only vector orthogonal to all of the e_ns is the zero vector. Denoting by M the linear subspace spanned by the vectors $\{e_n\}_{n\geq 1}$, this means that $M^{\perp} = \{0\}$. But then $(M^{\perp})^{\perp} = H$. On the

[16] The geometrical content of the recipe for x_{k+1} is that you project out all components of y_{k+1} in the directions x_j with $1 \leq j \leq k$, so what is left over is perpendicular to all x_js with $j = 1, \ldots, k$.

other hand, $(M^\perp)^\perp = \overline{M}$ (see Exercise 10), so that $\overline{M} = H$. Consequently, for every $x \in H$ we can find a sequence $\{y_n\}_{n\geq 1}$ of finite linear combinations of vectors from the orthonormal basis, $y_n = \sum_{j=1}^{N(n)} \alpha_{j,n} e_j$, with $y_n \to x$ as $n \to \infty$. Adding, if necessary, a number of zero terms, we may assume that $n \mapsto N(n) \geq n$ is increasing. The previous lemma yields $\|x - \sum_{j=1}^{N(n)} (x, e_j)\, e_j\| \leq \|x - \sum_{j=1}^{N(n)} \alpha_{j,n}\, e_j\| = \|x - y_n\| \to 0$ for $n \to \infty$. Consequently $x = \sum_{k\geq 1} (x, e_k)\, e_k$.

To show that (b) implies (c), for $x = \sum_{n\geq 1} (x, e_n)\, e_n$ and $y = \sum_{n\geq 1} (y, e_n)\, e_n$, observe that $(x, y) = \big(\sum_{k\geq 1} (x, e_k)\, e_k, \sum_{j\geq 1} (y, e_j)\, e_j \big) = \sum_{k\geq 1} \sum_{j\geq 1} (x, e_k)\, \overline{(y, e_j)}\, (e_k, e_j) = \sum_{n\geq 1} (x, e_n)\, \overline{(y, e_n)}$ by orthonormality. (d) is obtained by simply setting $x = y$ in (c).

To show that (d) implies (e), assume the existence of some orthonormal set that spans a subspace M which is not dense in H but satisfies the property specified in (d). As before, $M^\perp = \{0\}$ leads to $\overline{M} = (M^\perp)^\perp = H$, so that $M^\perp$ must contain a vector $e_0 \neq 0$. But $(e_0, e_n) = 0$ for all $n \geq 1$ in combination with (d) yield the contradiction $\|e_0\|^2 = \sum_{n\geq 1} |(e_0, e_n)|^2 = 0$.

To show that (e) implies (a), repeat the arguments used above to show that (a) implies (b). □

We now define the *dimension* of a Hilbert space as the cardinal number of an orthonormal basis. Exercises 4 and 15 show that this concept is well-defined, yielding the same categorisation finite/infinite-dimensional as that obtained using a Hamel basis. While an orthonormal basis in finite dimensions is also a Hamel basis, the two concepts are in marked contrast for an infinite-dimensional separable Hilbert space (see Exercise 15).

A consequence of Theorem 3.1 is that l^2 is the universal model for all separable infinite-dimensional Hilbert spaces.

Theorem 3.4 *An infinite-dimensional separable Hilbert space is isometrically isomorphic*[17] *with* l^2.

Proof Let $\{e_n\}_{n\geq 1}$ be an orthonormal basis of H (see Exercise 17). If to $x \in H$ we associate the Fourier coefficients $\hat{x} = \{(x, e_k)\}_{k\geq 1} \in l^2$, then, by Theorem 3.1, $x \mapsto \hat{x}$ is a linear map from H into l^2. The Plancherel identity ensures that the map is isometric (length-preserving). In particular, the map is one-to-one. The map is also onto. Indeed, if $\hat{x} = \{x_k\}_{k\geq 1} \in l^2$, then $S_n = \sum_{k=1}^n x_k e_k$ for $n \geq 1$ defines a Cauchy sequence in H, since for $n > m$ the Plancherel identity yields $\|S_n - S_m\|^2 = \sum_{k=m+1}^n |x_k|^2$. If $x = \lim_{n\to\infty} S_n \in H$, then $\hat{x}(j) = (x, e_j) = \big(\lim_{n\to\infty} \sum_{k=1}^n x_k e_k,\, e_j\big) = \lim_{n\to\infty} \big(\sum_{k=1}^n x_k e_k, e_j\big) = x_j$ for all $j \geq 1$, in view of orthonormality. □

Remarks (i) The series $\sum_{k=1}^\infty x_k$ in Theorem 3.1 might not be absolutely convergent.[18]

(ii) Despite the spectacular result stated in Theorem 3.4, often one considers specific Hilbert spaces (such as $L^2[0, 1]$) since many natural operators acting on these spaces become appalling when they are expressed in a basis. Especially in mathematical physics, there is more to the problem than a mere formulation in a Hilbert space setting. This additional structure must be retained if we

[17] That is, there is a linear length-preserving map of H onto l^2. Linearity means that multiplication by complex numbers and addition are preserved (see Section 3.3).

[18] We may have that $\sum_{k=1}^\infty |x_k| = \infty$. For example, by Theorem 3.4, the sequence $x(k) = k^{-1}$ for $k \geq 1$ is the Fourier series of some $x \in H$.

are to gain insight. Consequently, Theorem 3.4 is unquestionably a result of fundamental importance but it is not the ultimate result.

(iii) A powerful technique of constructing orthonormal bases for separable Hilbert spaces consists in taking the eigenvectors[19] of a compact self-adjoint operator (see Section 3.4.4).

3.3 Banach spaces

Hilbert spaces turn out to be too restrictive for some very significant purposes. Banach spaces are more general and, while they have less structure than Hilbert spaces, they still have properties that accord with many intuitive notions associated with the geometries of the plane and of three-dimensional space and permit a systematic study of fundamental aspects in great generality. The study of Banach spaces is a subject permeated by elegant arguments leading to powerful results that illuminate many other areas of mathematical analysis, being frequently useful in unanticipated situations. In addition to economy of thought, these general results are often easier than particular versions – the generality makes it easier to focus on the essence of the argument, thus making the reasoning more transparent. In particular, the theory of Banach spaces is very useful in Fourier analysis.

Definition *A function* $\|\cdot\| : X \to [0, \infty)$ *on a linear vector space* X *is said to be a norm on* X *if*

(i) $\|x\| = 0$ *if and only if* $x = 0$;

(ii) $\|x + y\| \leq \|x\| + \|y\|$ *for all* $x, y \in X$;

(iii) $\|\alpha x\| = |\alpha| \cdot \|x\|$ *for all scalars* α *(real or complex) and* $x \in X$.

A normed linear space is a pair $(X, \|\cdot\|)$, *where* X *is a linear space and* $\|\cdot\|$ *is a norm on* X.

The norm $\|\cdot\|$ on a linear vector space X endows it with the structure of a metric space, the distance d being defined by $d(x, y) = \|x - y\|$ for $x, y \in X$. We recall the notion of a metric space.

[19] For example, we define a compact self-adjoint operator $T : L^2[0, 1] \to L^2[0, 1]$, with simple eigenvalue $\lambda_0 = 1$ and double eigenvalues $\lambda_n = \frac{1}{1+4\pi^2 n^2}$ for $n \geq 1$, by associating to $x \in L^2[0, 1]$ the function $(Tx)(t) = \frac{\sinh(1-t)}{\sinh(1)} \int_0^t \sinh(s)\, x(s)\, ds + \frac{\sinh(t)}{\sinh(1)} \int_t^1 \sinh(1 - s)\, x(s)\, ds$; this integral operator is the inverse of the differential operator $(1 - \partial_x^2)$. The corresponding eigenfunctions are $x_0(t) = 1$ and $x_n^{\pm}(t) = e^{\pm 2\pi i n t}$ for $n \geq 1$.

Definition *A metric space (M, d) is a set M and a function $d : M \times M \to [0, \infty)$ wich satisfies*

(i) $d(x, y) = 0$ if and only if $x = y$;

(ii) $d(x, y) = d(y, x)$ for all $x, y \in M$ [symmetry];

(iii) $d(x, y) \le d(x, z) + d(z, y)$ for all $x, y, z \in M$ [triangle inequality].

The function d is called a metric on M.

In a metric space (M, d), the distance induced by d leads to the concept of convergent sequences: $x_n \to x$ in (M, d) if $d(x_n, x) \to 0$ for $n \to \infty$. Moreover, we can introduce the notions of open and closed sets: $O \subset M$ is open if for every $x_0 \in O$ there is some $r > 0$ such that the open ball $B(x_0, r) = \{x \in M : d(x_0, x) < r\}$ is contained in O, while $C \subset M$ is closed if it contains all its limit points, that is, if $x_n \in C$ for all $n \ge 1$ and $x_n \to x$, then $x \in C$. Using the definition, one can easily check that $C \subset M$ is closed if and only if $M \setminus C$ is open. To understand the concepts of open and closed sets it is useful to keep in mind the case of the real line. However, without additional structure, there are pathological cases – see e.g. Exercise 24.

Deciding whether a sequence $\{x_n\}_{n\ge1}$ in a metric space (M, d) is convergent seems to depend heavily on guessing its limit. To overcome this problem, let us introduce the notion of a *Cauchy sequence*: $\{x_n\}_{n\ge1}$ is a Cauchy sequence if for every $\varepsilon > 0$ there is an integer $N(\varepsilon) \ge 1$ such that $d(x_n, x_m) < \varepsilon$ for all $n, m \ge N(\varepsilon)$. Clearly any convergent sequence is a Cauchy sequence.[20] The example of the rational numbers $\mathbb{Q}$ with the usual distance shows that there might exist Cauchy sequences that are not convergent.[21] However, the elements of a Cauchy sequence form a bounded set, being contained in some open ball centred at some point – see Exercise 26.

Definition *A metric space in which all Cauchy sequences converge is called complete. Complete normed linear space are called Banach spaces.*

Important examples of Banach spaces are provided by certain types of sequences and certain functions spaces – see the end of this section. On the other hand, the example of $\mathbb{Q} \subset \mathbb{R}$ is indicative of a possible procedure to follow when dealing with a metric space (M, d) that is not complete: enlarge it by all possible limits of Cauchy sequences. To implement this approach, we have first to make it precise and we have to introduce some new concepts.

20 Indeed, if $d(x_n, x) \to 0$ for $n \to \infty$, given $\varepsilon > 0$, choosing $N(\varepsilon) \ge 1$ such that $d(x_n, x) < \varepsilon/2$ for $n \ge N(\varepsilon)$, we get $d(x_n, x_m) \le d(x_n, x) + d(x_m, x) < \frac{\varepsilon}{2} + \frac{\varepsilon}{2} = \varepsilon$ for $n, m \ge N(\varepsilon)$.

21 Take the rational numbers x_n obtained by truncating the infinite decimal expansion of $\sqrt{2}$ to n decimals, so that $x_1 = 1.4$, $x_2 = 1.41$, $x_3 = 1.414$ etc., with $x_n \to \sqrt{2}$.

Definition *Let (M, d) be a metric space. A set $D \subset M$ in called dense if every $x \in M$ is a limit of a sequence in D. Metric spaces admitting a countable dense subset are called separable.*

In other words, a separable metric space is a space in which one can find a sequence such that every point in the space can be approximated to any level of accuracy with elements in that sequence.[22]

Definition *Let (X, d) and (Y, ρ) be two metric spaces. A function $f : X \to Y$ is called continuous if $\rho(f(x_n), f(x)) \to 0$ whenever $d(x_n, x) \to 0$. A bijection $h : X \to Y$ which preserves the distance, that is, such that $\rho(h(x), h(y)) = d(x, y)$ for all $x, y \in X$, is called an isometry. The metric spaces (X, d) and (Y, ρ) are called isometric if such an isometry exists.*

Isometric spaces are essentially identical as metric spaces: a theorem concerning only the metric structure of (X, d) will hold in all metric spaces isometric to it. It is tempting to believe that asking for an isometry is too much and perhaps a homeomorphism $h : X \to Y$ (a bijection such that both h and its inverse h^{-1} are continuous) should suffice. To see that this is not so, consider the spaces $X = (0, 1]$ and $Y = [1, \infty)$ with the usual distance on $\mathbb{R}$. Then $h(x) = 1/x$ is a homeomorphism between these spaces, but not an isometry, and while $\{n^{-1}\}_{n\geq 1}$ is a Cauchy sequence, $\{h(n^{-1})\}_{n\geq 1}$ is not. This explains the need for isometries.

The next result states precisely in what sense an incomplete metric space can be completed.

Theorem 3.5 *Let (X, d) be a metric space. Then there exists a metric space $(\hat{X}, \hat{d})$ with the following properties:*

(i) $(\hat{X}, \hat{d})$ is complete;

(ii) (X, d) is isometric to a dense subset of $(\hat{X}, \hat{d})$;

(iii) $(\hat{X}, \hat{d})$ is unique, up to isometries.

Moreover, the completion of a linear normed space is a Banach space, while that of a pre-Hilbert space is a Hilbert space.

Proof Consider the family of Cauchy sequences $\{x_n\}_{n\geq 1}$ in X, and call two Cauchy sequences $\{x_n\}_{n\geq 1}$ and $\{y_n\}_{n\geq 1}$ equivalent if $d(x_n, y_n) \to 0$ for $n \to \infty$. Let $\hat{X}$ be the family of equivalence classes of Cauchy sequences in X under this equivalence relation. The metric on $\hat{X}$ is defined as $\lim_{n\to\infty} d(x_n, y_n)$ for any two Cauchy sequences $\{x_n\}_{n\geq 1}$ and $\{y_n\}_{n\geq 1}$ of (X, d), while the isometry from (X, d) to a dense subset of $(\hat{X}, \hat{d})$ assigns to $x \in X$ the constant Cauchy sequence whose elements are all equal to x. After sketching the approach, let us divide the proof into several steps.

(1) The relation introduced above is clearly reflexive, symmetric and transitive. For the equiv-

[22] For example, $\mathbb{Q}$ is separable (being countable) and since $\mathbb{Q}$ is dense in $\mathbb{R}$, $\mathbb{R}$ is also separable. See Exercise 24 for a simple example of a metric space that is not separable.

alence classes $\hat{x}$ and $\hat{y}$, we define

$$\hat{d}(\hat{x}, \hat{y}) = \lim_{n\to\infty} d(x_n, y_n), \tag{3.4}$$

where $\{x_n\}_{n\ge1}$ and $\{y_n\}_{n\ge1}$ are some Cauchy sequences in the respective equivalence classes. To show that the limit exists, note that $d(x_n, y_n) \le d(x_n, x_m)+d(x_m, y_m)+d(y_m, y_n)$ yields $d(x_n, y_n) - d(x_m, y_m) \le d(x_n, x_m) + d(y_n, y_m)$. Interchanging n with m in the previous inequality, we see that $|d(x_n, y_n) - d(x_m, y_m)| \le d(x_n, x_m) + d(y_n, y_m)$ for $n, m \ge 1$. Thus the real sequence $\{d(x_n, y_n)\}_{n\ge1}$ is Cauchy, so that the limit (3.4) exists. We now check that $\hat{d}$ is unambiguously defined, that is, if $\{x'_n\}_{n\ge1}$ and $\{y'_n\}_{n\ge1}$ are other Cauchy sequences in the respective equivalence classes, then $\lim_{n\to\infty} d(x_n, y_n) = \lim_{n\to\infty} d(x'_n, y'_n)$. To see this, note that $d(x_n, y_n) \le d(x_n, x'_n) + d(x'_n, y'_n) + d(y'_n, y_n)$ yields $\lim_{n\to\infty} d(x_n, y_n) \le \lim_{n\to\infty} d(x'_n, y'_n)$. The reverse inequality can be obtained similarly.

(2) We prove now that $\hat{d}$ is a metric on $\hat{X}$. Clearly $\hat{d}(\hat{x}, \hat{y}) \ge 0$, while if $\hat{d}(\hat{x}, \hat{y}) = 0$, there are Cauchy sequences $\{x_n\}_{n\ge1} \in \hat{x}$ and $\{y_n\}_{n\ge1} \in \hat{y}$ with $\lim_{n\to\infty} d(x_n, y_n) = 0$. But this ensures $\hat{x} = \hat{y}$ by definition. Since the symmetry of $\hat{d}$ is obvious, it remains to prove the triangle inequality. This is easy: if $\{x_n\}_{n\ge1} \in \hat{x}$, $\{y_n\}_{n\ge1} \in \hat{y}$ and $\{z_n\}_{n\ge1} \in \hat{z}$, then $\hat{d}(\hat{x}, \hat{y}) = \lim_{n\to\infty} d(x_n, y_n) \le \lim_{n\to\infty} [d(x_n, z_n) + d(z_n, y_n)] \le \limsup_{n\to\infty} d(x_n, z_n) + \limsup_{n\to\infty} d(z_n, y_n) = \hat{d}(\hat{x}, \hat{z}) + \hat{d}(\hat{z}, \hat{y})$.

(3) To establish the completeness of $(\hat{X}, \hat{d})$, let $\{\hat{x}_n\}_{n\ge1}$ be a Cauchy sequence in $(\hat{X}, \hat{d})$. In each equivalence class $\hat{x}_n$, choose a Cauchy sequence $\{x_n(k)\}_{k\ge1}$ in (X, d). Then for each $n \ge 1$ there exists $k_n \ge 1$ such that $d(x_n(m), x_n(k_n)) \le 1/n$ for all $m \ge k_n$. We claim that $\{x_n(k_n)\}_{n\ge1}$ is a Cauchy sequence in (X, d), henceforth denoted by $\hat{x} \in \hat{X}$. To see this, let $\hat{x}_n^* \in \hat{X}$ be the equivalence class of the constant sequence whose elements are all equal to $x_n(k_n) \in X$. Since

$$\hat{d}(\hat{x}_n, \hat{x}_n^*) = \lim_{j\to\infty} d(x_n(j), x_n(k_n)) \le \frac{1}{n}, \tag{3.5}$$

we have that $d(x_n(k_n), x_m(k_m)) = \hat{d}(\hat{x}_n^*, \hat{x}_m^*) \le \hat{d}(\hat{x}_n^*, \hat{x}_n) + \hat{d}(\hat{x}_n, \hat{x}_m) + \hat{d}(\hat{x}_m, \hat{x}_m^*) \le \frac{1}{n} + \hat{d}(\hat{x}_n, \hat{x}_m) + \frac{1}{m}$, so that $\lim_{n,m\to\infty} d(x_n(k_n), x_m(k_m)) = 0$. Thus $\hat{x} \in \hat{X}$. Now $\hat{d}(\hat{x}, \hat{x}_n) \le \hat{d}(\hat{x}, \hat{x}_n^*) + \hat{d}(\hat{x}_n^*, \hat{x}_n) \le \hat{d}(\hat{x}, \hat{x}_n^*) + \frac{1}{n}$, due to (3.5). But, for any $\varepsilon > 0$, $\hat{d}(\hat{x}, \hat{x}_n^*) = \lim_{j\to\infty} d(x_j(k_j), x_n(k_n)) < \varepsilon$ if n is sufficiently large, since $\{x_n(k_n)\}_{n\ge1}$ is a Cauchy sequence in (X, d). From this we get $\lim_{n\to\infty} \hat{d}(\hat{x}, \hat{x}_n^*) = 0$, proving that $\hat{x}_n \to \hat{x}$ in $(\hat{X}, \hat{d})$. Thus $(\hat{X}, \hat{d})$ is complete.

(4) The map $\sigma : x \mapsto \hat{x}^*$, associating to $x \in X$ the constant sequence $\hat{x}^*$ whose elements are all equal to x, is an isometry from (X, d) to $(\sigma(X), \hat{d})$. We now show that its image $\sigma(X)$ is dense in $(\hat{X}, \hat{d})$. Let $\hat{x} \in \hat{X}$ and choose a representative Cauchy sequence $\{x_n\}_{n\ge1}$ in (X, d). The class $\hat{x}_n^* \in \hat{X}$ that contains the constant sequence whose elements are all equal to $x_n \in X$ belongs to $\sigma(X)$ and $\hat{d}(\hat{x}, \hat{x}_n^*) = \lim_{j\to\infty} d(x_j, x_n) \to 0$ for $n \to \infty$.

(5) To prove the uniqueness of $(\hat{X}, \hat{d})$, up to isometries, suppose that (X^*, d^*) is another metric space satisfying (i)–(ii), and let $\sigma^* : X \to \sigma^*(X)$ be the isometry from (X, d) to the dense subset $\sigma^*(X)$ of X^*. Given $x^* \in X^*$, choose a sequence $\{x_n\}_{n\ge1}$ in X such that $\sigma^*(x_n) \to x^*$ in (X^*, d^*). Since σ^* is an isometry and $\{\sigma^*(x_n)\}_{n\ge1}$ is a Cauchy sequence in (X^*, d^*), the sequence $\{x_n\}_{n\ge1}$ must be Cauchy in (X, d). Denote its equivalence class by $\hat{x} \in \hat{X}$. We now claim that the function $f : X^* \to \hat{X}$ defined by $f(x^*) = \hat{x}$ is an isometry. To see this, note that if $y^* \in X^*$ and $\{y_n\}_{n\ge1}$ is a sequence in X with $\sigma^*(y_n) \to y^*$ in (X^*, d^*), then $d^*(x^*, y^*) = \lim_{n\to\infty} d^*(\sigma^*(x_n), y^*) = \lim_{n\to\infty} d^*(\sigma^*(x_n), \sigma^*(y_n))$ since $|d^*(\sigma^*(x_n), \sigma^*(y_n)) - d^*(\sigma^*(x_n), y^*)| \le d^*(\sigma^*(y_n), y^*) \to 0$. But, by the definition of $\sigma^* : X \to X^*$, $d^*(\sigma^*(x_n), \sigma^*(y_n)) = d(x_n, y_n)$, so that $d^*(x^*, y^*) = \lim_{n\to\infty} d(x_n, y_n) = \hat{d}(\hat{x}, \hat{y})$. This confirms that f preserves distances and, in particular, is injective. To show that it is a bijection, given $\hat{x} \in \hat{X}$, choose a Cauchy sequence $\{x_n\}_{n\ge1}$ in (X, d) to represent it. The sequence $\{\sigma^*(x_n)\}_{n\ge1}$ is then a Cauchy sequence in the complete metric space (X^*, d^*). If $x^* \in X^*$ is its limit, then clearly $f(x^*) = \hat{x}$, proving that f is also onto.

(6) Finally, let us show that the completion process preserves the structure: if the metric space in question is a normed linear space this process yields a Banach space, while a pre-Hilbert space is completed to a Hilbert space. For this, note that if X is a linear vector space, so will $\hat{X}$ be: the addition of vectors and the scalar multiplication carry over. For a normed linear space $(X, \|\cdot\|)$,

we define the norm of $\hat{x} \in \hat{X}$ by $\|\hat{x}\| = d(\hat{x}, \hat{0})$, where $\hat{0} \in \hat{X}$ is the equivalence class of the zero sequence in X. For pre-Hilbert spaces, the inner product $(\cdot, \cdot)$ of X extends to an inner product in $\hat{X}$ by setting $(\hat{x}, \hat{y}) = \lim_{n\to\infty} (x_n, y_n)$ for $\{x_n\}_{n\geq 1} \in \hat{x}$ and $\{y_n\}_{n\geq 1} \in \hat{y}$. □

The drawback of the previous theorem is that the elements of the completed space are more complicated entities than those of the original space. When a completion process needs to be carried out, the problem of providing a simple model for the completion arises. In the ideal case, the procedure parallels that for $\mathbb{Q}$ and $\mathbb{R}$: both spaces being familiar, the fact that $\mathbb{Q}$ is dense in $\mathbb{R}$ shows by uniqueness that its completion must be $\mathbb{R}$. For other examples see Exercise 35.

In Section 3.2 we proved that every vector space has a Hamel basis. Since infinite-dimensional linear spaces have uncountable Hamel bases, this concept is unwieldy. In the context of a separable Hilbert space one has the notion of a countable orthonormal basis. Let us discuss briefly the situation in normed linear spaces. For this, we start with some considerations about series in a normed linear space $(X, \|\cdot\|)$. Given a sequence $\{x_k\}_{k\geq 1}$ in X, we say that the series $\sum_{k=1}^{\infty} x_k$ *converges* if there is some $x \in X$ with $\|x - \sum_{k=1}^{n} x_k\| \to 0$ as $n \uparrow \infty$. The series is called *absolutely convergent* if $\sum_{k=1}^{\infty} \|x_k\|$ converges (in $\mathbb{R}$) and is said to be *unconditionally convergent* if all its re-arrangements are convergent and have the same limit $x \in X$, that is, $\sum_{k=1}^{\infty} x_{\sigma(k)}$ converges to x for any permutation σ of the positive integers.

Theorem 3.6 *Absolutely convergent series are unconditionally convergent in Banach spaces.*

Proof Let us first prove that if $\sum_{k=1}^{\infty} \|x_k\| < \infty$, then $\sum_{k=1}^{\infty} x_k$ converges. Since $\|\sum_{k=m}^{n} x_k\| \leq \sum_{k=m}^{n} \|x_k\|$ for $n \geq m \geq 1$, $\{\sum_{k=1}^{n} x_k\}_{n\geq 1}$ is a Cauchy sequence, and thus converges to some limit x. Consider now a re-arrangement $\{x_{\sigma(k)}\}_{k\geq 1}$ of the sequence $\{x_k\}_{k\geq 1}$ and let $\varepsilon > 0$. Choose the integer $N = N(\varepsilon) \geq 1$ such that $\sum_{k=N}^{\infty} \|x_k\| < \varepsilon$. If the integer $p = p(\varepsilon) \geq 1$ is such that $x_1, \dots, x_N$ are all contained in the set $\{x_{\sigma(1)}, x_{\sigma(2)}, \dots, x_{\sigma(p)}\}$, then $\sum_{k\geq p} \|x_{\sigma(k)}\| \leq \varepsilon$, so that the series $\sum_{k\geq p} x_{\sigma(k)}$ is absolutely convergent, therefore also convergent. To see that $x = \sum_{k\geq p} x_{\sigma(k)}$, choose $N_1 = N_1(\varepsilon) \geq N$ so that $\|x - \sum_{k=1}^{n} x_k\| \leq \varepsilon$ for all $n \geq N_1$. Since $N_1 \geq N$, we also have that $\sum_{k=N_1}^{\infty} \|x_k\| \leq \varepsilon$. If $q \geq N_1$ is such that $\{x_1, \dots, x_{N_1}\} \subset \{x_{\sigma(1)}, \dots, x_{\sigma(q)}\}$, then for $n \geq q$ we have $\|x - \sum_{k=1}^{n} x_{\sigma(k)}\| \leq \|x - \sum_{k=1}^{n} x_k\| + \|\sum_{k=1}^{n} x_k - \sum_{k=1}^{n} x_{\sigma(k)}\| \leq \varepsilon + 2\sum_{k\geq N_1}^{n} \|x_k\| \leq 3\varepsilon$. Since $\varepsilon > 0$ was arbitrary, $\sum_{k\geq p} x_{\sigma(k)}$ must converge to x. □

Note that the completeness of the normed linear space is essential[23] in Theorem 3.6. Moreover, in finite-dimensional vector spaces unconditional convergence is equivalent to absolute convergence but in every infinite-dimensional

[23] In the normed linear space of complex sequences $\{x(j)\}_{j\geq 1}$ with at most finitely many nonzero entries, with the norm $\|x\| = \left(\sum_{j\geq 1} |x(j)|^2\right)^{1/2}$, the sequence $\{x_n\}_{n\geq 1}$ with $x_n(j) = 1/j$ for $1 \leq j \leq n$ and $x_n(j) = 0$ for $j \geq n+1$, is absolutely convergent but not convergent.

normed linear space there is a series that converges unconditionally but not absolutely[24], see Dvoretzky and Rogers (1950).

These preliminary considerations permit us to define the *Schauder basis* of a Banach space $(X, \|\cdot\|)$ as a sequence $\{e_n\}_{n\geq 1}$ in X such that every $x \in X$ admits an expansion of the form $x = \sum_{k=1}^{\infty} x_k\, e_k$ for some unique sequence of scalars $\{x_k\}_{k\geq 1}$. One can prove that a Banach space with a Schauder basis is separable.[25] For a sequence $\{e_n\}_{n\geq 1}$ to be a Schauder basis, finite linear combinations of the e_ks should be dense but this is not sufficient.[26] A familiar example of a Schauder basis is provided by an orthonormal basis in a separable Hilbert space. In particular, the trigonometric polynomials $\{e^{2\pi ikt}\}_{k\in\mathbb{Z}}$ are a Schauder basis in $L^2[0, 1]$, as we shall prove in Chapter 4. However, they are not a Schauder basis for $L^1[0, 1]$ or for the Banach space $C_{\text{per}}[0, 1]$ of continuous periodic functions of period 1, endowed with the supremum norm, see Exercise 4.25, although $L^1[0, 1]$ as well as $C_{\text{per}}[0, 1]$ do admit a Schauder basis (see Exercises 34–35). Neither space admits an unconditionally convergent Schauder basis, see James (1982): as opposed to a Hamel basis, the elements of a Schauder basis must be ordered since the series $\sum_{n=1}^{\infty} x_n\, e_n$ may not converge unconditionally. Moreover, while most classical examples of separable Banach spaces have a Schauder basis, there are counterexamples – see Lindenstrauss and Tzafriri (1977). The proofs of these results involve methods and concepts that we do not wish to describe. For our purposes, it suffices to note that in the study of infinite-dimensional separable Banach spaces we cannot rely on Schauder bases. Moreover, in a normed linear space $(X, \|\cdot\|)$, in general, there is no concept of orthogonality because there is no inner product. For this reason, the geometric intuition that, given a closed subspace $M \subsetneq X$, we can always find $x \in X \setminus M$ such that $\|x\| = d(x, M) = \inf_{y\in M} \{\|x - y\|\}$ might fail even in a Banach space setting (see e.g. Exercise 21), despite being valid for Hilbert spaces, see Theorem 3.2. However, the desired result is almost true.

Theorem 3.7 *Let M be a closed proper linear subspace of a normed linear space $(X, \|\cdot\|)$. Then for every $\varepsilon > 0$ there is an $x_\varepsilon \in X$ with $\|x_\varepsilon\| = 1$ and $1 \geq d(x_\varepsilon, M) \geq 1 - \varepsilon$.*

24 For example, if $\{e_n\}_{n\geq 1}$ is an orthonormal basis of the Hilbert space H, then the series $\sum_{n=1}^{\infty} \frac{e_n}{n}$ converges unconditionally but not absolutely.

25 For this, approximate the scalars x_k by $a_k + ib_k$ with $a_k, b_k \in \mathbb{Q}$, anticipating the fact that the considerations in Section 3.4 ensure the continuity of the projection $x \mapsto \sum_{k=1}^{n} x_k\, e_k$ on the finite-dimensional subspace spanned by $\{e_1, \ldots, e_n\}$ for every $n \geq 1$ (see Exercise 45).

26 For example, the poynomials $\{t^{n-1}\}_{n\geq 1}$ are not a Schauder basis of $C[0, 1]$ since a continuous function that is nowhere differentiable (see e.g. Exercise 2.34) cannot be uniformly approximated on $[0, 1]$ by a power series. The additional necessary and sufficient condition is the existence of a constant $K > 0$ with $K\, \|\sum_{k=1}^{n+p} x_k e_k\| \geq \|\sum_{k=1}^{n} x_k e_k\|$ for all integers $n, p \geq 1$; see James (1982).

Proof Let $x \in X \setminus M$. Then $\delta = d(x, M) \in (0, \|x\|]$ since $0 \in M$ yields $\|x\| \geq d(x, M)$, and $d(x, M) = 0$ would ensure the existence of a sequence $x_n \in M$ with $\|x - x_n\| \to 0$, which would force $x \in M$ since M is closed. Choose $y_0 \in M$ with $\|x - y_0\| < \delta(1+\varepsilon)$. If $z = x - y_0$, then for every scalar $\alpha \neq 0$ we have that $\|\alpha z\| = |\alpha| \cdot \|x - y_0\| < |\alpha|\,\delta(1+\varepsilon)$ and $d(\alpha z, M) = \inf_{y \in M} \|\alpha z - \alpha y\| = |\alpha| \inf_{y \in M} \|z - y\| = |\alpha|\delta$ since $d(z, M) = d(x, M) = \delta$. Choose $\alpha = \frac{1}{\|z\|}$ and set $x_\varepsilon = \frac{1}{\|z\|} z$ to get $\|x_\varepsilon\| = 1$ and $d(x_\varepsilon, M) = \delta/\|z\| > 1/(1+\varepsilon) > 1 - \varepsilon$. □

We can now characterise topologically finite-dimensional linear spaces. A subset K of a metric space (X, d) is called *compact* if every sequence $\{x_n\}_{n\geq 1}$ in K has a subsequence x_{n_k} that converges to a limit $x \in K$.

In topology, a set K is called compact if every covering of K by open sets[27] admits a finite subcovering. In metric spaces, this amounts to the definition of sequential compactness, stated above and used most in applications. Indeed, if K is a compact subset of a metric space and $\{x_n\}_{n\geq 1}$ is a sequence in K, then for every $\varepsilon \in (0, 1)$, the open covering of K by open balls $B(x, \varepsilon) = \{y \in K :\ d(x, y) < \varepsilon\}$, centred at $x \in K$, must admit a finite subcovering, so that one of these open balls, say $B(x_1, \varepsilon)$, contains an infinite number of the x_ns. The closed set $\overline{B(x_1, \varepsilon)} = \{y \in K :\ d(x, y) \leq \varepsilon\}$ being compact,[28] we can repeat the procedure with the covering of $\overline{B(x_1, \varepsilon)}$ by open balls centred at its points, and having radius ε^2. Continuing, we obtain a decreasing sequence of nested closed balls $\overline{B}_k$ of radius ε^k (that is, with $\overline{B}_{k+1} \subset \overline{B}_k$). Their intersection must be nonempty[29] and must consist of a single point $x \in K$ as $\varepsilon^k \to 0$ for $k \to \infty$. Selecting at each step the centre x_{n_k} of the ball B_k, we obtain a subsequence $x_{n_k} \to x$. Conversely, let K be a sequentially compact subset in a metric space. Then K is closed and K is *totally bounded*,[30] meaning that for every $\varepsilon > 0$ we can find an ε-net K_ε: a finite subset $K_\varepsilon \subset K$ such that for every $y \in K$ there is some $x \in K_\varepsilon$ with $d(x, y) < \varepsilon$. Indeed, failure of this entails the existence of a sequence of points $\{x_n\}_{n\geq 1}$ in K such that $d(x_n, x_m) \geq \varepsilon$ whenever $n \neq m$, thus preventing the existence of a convergent subsequence. With this claim proved, note that the union of the $\frac{1}{n}$-nets for all $n \geq 1$ yields a countable dense subset $\{\theta_k\}_{k\geq 1}$ of K. Assuming that $\{O_i\}_{i\in I}$ is an open covering of K without a finite subcovering, the total boundedness of K implies that there are a finite number of closed balls $B_{11}, \dots, B_{1n_1}$ of radius 2^{-1} which cover K. At least one of the sets $K \cap B_{1j}$ with $1 \leq j \leq n_1$, say $K_1 = K \cap B_{11}$, cannot be covered by a finite subcollection of O_is. As a subset of K, K_1 will also be totally bounded. Let $B_{21}, \dots, B_{2n_2}$ be a finite set of closed balls of radius 2^{-2} which cover K_1. Again, one of the sets $K \cap B_{2j}$ with $1 \leq j \leq n_2$, say $K_2 = K \cap B_{21}$, cannot be covered by a finite subcollection of O_i's. Continuing in this way we construct a sequence of nested closed sets $K_1 \supset K_2 \supset \dots$, none of which can be covered by a finite subcollection of O_is. K_n being a subset of a closed ball of radius 2^{-n}, centred at some $x \in X$, we have $d(y, z) \leq d(x, y) + d(y, z) \leq 2^{-n} + 2^{-n} = 2^{1-n}$ for all $y, z \in K_n$.

27 That is, a collection of open sets such that K is contained in their union.

28 Just add to any open covering the open complement of $\overline{B(x_1, \varepsilon)}$ to obtain a covering of K, and the resulting finite subcovering of K will also cover $\overline{B(x_1, \varepsilon)}$.

29 Since otherwise their complements form an open covering of K with no finite subcovering.

30 Note that a totally bounded set A in a metric space (X, d) has to be bounded, since if $\{x_1, \dots, x_n\}$ is an ε-net, then for any $x \in A$ there is some x_k with $d(x, x_k) < \varepsilon$, so that $d(x, x_1) \leq d(x, x_k) + d(x_k, x_1) < \varepsilon + \max_{2\leq j\leq n}\{d(x_1, x_j)\}$. The converse is generally false (see e.g. Theorem 3.8). Actually, the considerations showing that a sequentially compact set must be compact actually prove that in a complete metric space a totally bounded set A is relatively compact, meaning that its closure is compact: every sequence admits a convergent subsequence whose limit may, however, not belong to A.

Choosing $x_n \in K_n$, the limit $\xi \in K$ of a convergent subsequence $\{x_{n_k}\}$ belongs to some open set O_j. Thus, for some $\varepsilon > 0$ the open ball $B(\xi, \varepsilon)$, centred at ξ and with radius ε, lies in O_j. But every $y \in K_{n_k}$ satisfies $d(\xi, y) \le d(\xi, x_{n_k}) + d(x_{n_k}, y) \le d(\xi, x_{n_k}) + 2^{-n_k}$ since $\xi \in K_{n_k}$. Thus n_k large enough ensures $K_n \subset B(\xi, \varepsilon)$, which entails $K_n \subset O_j$. This contradiction shows that K is compact.

Theorem 3.8 *The normed linear space $(X, \|\cdot\|)$ is finite-dimensional if and only if the closed unit ball $B = \{x \in X : \|x\| \le 1\}$ is compact.*

Proof If X is finite-dimensional, let $\{e_1, \ldots, e_n\}$ be a Hamel basis of X. The function $f : \mathbb{C}^n \to [0, \infty)$ defined by $f(\alpha_1, \ldots, \alpha_n) = \| \sum_{k=1}^n \alpha_k e_k \|$ is continuous.[31] Let $M \ge m \ge 0$ be the supremum and infimum of f on the compact set $K = \{\alpha = (\alpha_1, \ldots, \alpha_n) \in \mathbb{C}^n : |\alpha_1|^2 + \cdots + |\alpha_n|^2 = 1\}$. Then both m and M are attained, and, in particular, $m > 0$. By scaling, the inequality $m \le f(\alpha) \le M$ for $\alpha \in K$ yields $m^2 \sum_{k=1}^n |\alpha_k|^2 \le \| \sum_{k=1}^n \alpha_k e_k \|^2 \le M^2 \sum_{k=1}^n |\alpha_k|^2$ whenever $\alpha_k \in \mathbb{C}$. Thus, for a sequence $\{x_j\}_{j\ge1}$ in B with $x_j = \sum_{k=1}^n \alpha_{kj} e_k$, we can find a sequence $i_j \to \infty$ such that $\{\alpha_{k i_j}\}$ converges to some α_{k0} for $1 \le k \le n$ and $x_{i_j} \to x_0 = \sum_{k=1}^n \alpha_{k0} e_k$ in X.

Conversely, assume that B is compact. Let $e_1 \in X$ be a unit vector, $\|e_1\| = 1$. Let M_1 be the (closed) linear space generated by e_1. If $X \ne M_1$, by Theorem 3.7 we can find some $e_2 \in X$ with $\|e_2\| = 1$ such that $\|e_1 - e_2\| > \frac{1}{2}$. Inductively, if the unit vectors $e_1, \ldots, e_k$ have been constructed such that $\|e_i - e_j\| > \frac{1}{2}$ whenever $1 \le i < j \le n$, let M_k be the closed linear subspace of all their linear combinations. If $X \ne M_k$, we can find a unit vector e_{k+1} with $\|e_{k+1} - e_j\| > \frac{1}{2}$ for all $j \in \{1, \ldots, k\}$. In an infinite-dimensional space this produces a sequence $\{e_k\}_{k\ge1}$ with no Cauchy subsequence, contradicting the compactness of B. □

A compact set K has to be closed and bounded[32] but the converse is not necessarily true, see Theorem 3.8. Examples of compact sets in Banach spaces are closed and bounded subsets of finite-dimensional subspaces.[33] There are, however, genuine infinite-dimensional compact sets – see Exercise 16.

We now present a list of important examples of Banach and Hilbert spaces.

Spaces of sequences

(i) The linear vector space l^2 of real or complex sequences $x = \{x(k)\}_{k\ge1}$ with $\sum_{k\ge1} |x(k)|^2 < \infty$ is a separable Hilbert space, the norm being given by $\|x\| = \big(\sum_{k\ge1} |x(k)|^2 \big)^{1/2}$. Moreover, $\{e_n\}_{n\ge1}$ defined by $e_n(k) = \delta_{kn}$ for $n, k \ge 1$ is an orthonormal basis.[34] See Exercise 11.

(ii) For $1 \le p < \infty$, the family $l^p(\mathbb{Z})$ of real or complex sequences $x = \{x(k)\}_{k\in\mathbb{Z}}$ satisfying $\sum_{k\in\mathbb{Z}} |x(k)|^p < \infty$ is a separable Banach space with the norm $\|x\| = \big(\sum_{k\in\mathbb{Z}} |x(k)|^p \big)^{1/p}$, and $\{e_k\}_{k\in\mathbb{Z}}$ with $e_k(j) = \delta_{kj}$ for $k, j \in \mathbb{Z}$ as an

[31] Since $|f(\alpha_1, \ldots, \alpha_n) - f(\beta_1, \ldots, \beta_n)| \le \sum_{k=1}^n |\alpha_k - \beta_k| \max_{1\le j\le n} \|e_j\|$.

[32] Otherwise one could find either a sequence $\{x_n\}_{n\ge1}$ in K that converges to a limit that does not belong to K, or with $\|x_{n+1} - x_k\| \ge n$ for $k = 1, \ldots, n$.

[33] In this case, a suitablly chosen Hamel basis shows that these can be identified with closed and bounded (and therefore compact) subsets of $\mathbb{R}^n$ or $\mathbb{C}^n$.

[34] Sometimes it is convenient to consider doubly-infinite sequences with the index $k \in \mathbb{Z}$, in which case these properties persist with obvious modifications, e.g. the index of the members of the orthonormal basis runs over $\mathbb{Z}$.

unconditional Schauder basis.[35] Only $p = 2$ corresponds to a Hilbert space. For $p = 1$ see Exercise 30. For $p \in (1,2) \cup (2,\infty)$ we refer to Yosida (1995).

(iii) The linear vector space $l^\infty(\mathbb{Z})$ of bounded real or complex sequences $x = \{x(k)\}_{k\in\mathbb{Z}}$ is a nonseparable Banach space[36] with the norm $\|x\| = \sup_{k\geq 1} |x(k)|$, see Exercise 30.

(iv) c_0 is the separable Banach space of all real or complex sequences $x = \{x(k)\}_{k\geq 1}$ that converge to zero ($\lim\limits_{k\to\infty} x(k) = 0$), endowed with the norm $\|x\| = \sup_{k\geq 1} |x(k)|$, see Yosida (1995). An unconditional Schauder basis is $\{e_n\}_{n\geq 1}$ with $e_n(k) = \delta_{nk}$ for $n, k \geq 1$, see James (1982).

(v) The family c of convergent real or complex sequences $x = \{x(k)\}_{k\geq 1}$ is a separable Banach space with the norm $\|x\| = \sup_{k\geq 1} |x(k)|$, see Yosida (1995). An unconditional Schauder basis of c is[37] $\{e_n\}_{n\geq 0}$ with $e_n(k) = \delta_{nk}$ for $n, k \geq 1$, and $e_0(k) = 1$ for all $k \geq 0$.

Function spaces

(i) The space $C[0,1]$ of continuous functions[38] $f : [0,1] \to \mathbb{C}$ (or with real values, $f : [0,1] \to \mathbb{R}$) is a separable Banach space with the norm $\|f\|_\infty = \sup_{t\in[0,1]} |f(t)|$. The family $C^m_{\text{per}}[0,1]$ of periodic m-times continuously differentiable functions $f : \mathbb{R} \to \mathbb{C}$ (or with real values), with the norm $\|f\| = \sum\limits_{k=0}^{m} \sup\limits_{t\in[0,1]} |\partial_t^k f(t)|$, is also a separable Banach space. See Exercise 28.

(ii) The space $BV_{\text{per}}[0,1]$ of periodic functions $f : \mathbb{R} \to \mathbb{C}$ with period 1 and such that the restriction of f to $[0,1]$ has bounded variation is a nonseparable Banach space with the norm $\|f\| = |f(0)| + V_f[0,1]$, see Exercise 33.

(iii) For $\alpha \in (0,1]$, the Hölder space h_α of all periodic functions $f : \mathbb{R} \to \mathbb{C}$ with period 1 and that satisfy $|f(t) - f(s)| \leq K|t-s|^\alpha$ for all $0 \leq s,t \leq 1$, for some constant $K > 0$, is a non-separable Banach space with the norm $\|f\|_\alpha = [f]_\alpha + \sup\limits_{t\in[0,1]} |f(t)|$, see Exercise 34. Here $[f]_\alpha = \sup\limits_{0\leq s<t\leq 1} \left\{\frac{|f(t)-f(s)|}{(t-s)^\alpha}\right\}$.

(iv) For $1 \leq p < \infty$ and $(\Omega, \mathcal{A}, \mu)$ a measure space with a positive measure μ, we define $L^p(\Omega,\mu)$ to be the class of μ-measurable functions $f : \Omega \to \mathbb{C}$ (or $\mathbb{R}$) with $\int_\Omega |f(x)|^p \, d\mu(x) < \infty$. With the norm $\|f\| = \left(\int_\Omega |f(x)|^p \, d\mu(x)\right)^{1/p}$, this is a Banach space for $p \neq 2$ and a Hilbert space for $p = 2$, see Yosida (1995).

35 Sometimes sequences indexed by $k \geq 1$ are of interest. They form a closed subspace, l^p, of $l^p(\mathbb{Z})$, and therefore present analogous properties.

36 Sometimes sequences indexed by $k \geq 1$ are considered. They form a Banach space, denoted by l^∞, being a closed subspace of $l^\infty(\mathbb{Z})$. The property of being non-separable persists, as shown by an argument similar to that used in the solution to Exercise 28(ii).

37 To see this, note that if $\lim_{k\to\infty} x(k) = l$, then $(x - le_0)$ belongs to the familiar space c_0.

38 $[0,1]$ can be replaced by any compact set $K \subset \mathbb{R}$.

This Banach space[39] will not always be separable, even if it is separable in two important settings: (1) when $\Omega = \mathbb{Z}$ with the counting measure defined by $\mu(\{k\}) = 1$ for all $k \in \mathbb{Z}$, so that $L^p(\Omega, \nu)$ is precisely $l^p(\mathbb{Z})$; (2) when $\Omega \subset \mathbb{R}^n$ is a Lebesgue measurable set and μ is the Lebesgue measure;[40] see Brézis (2011). See Exercise 4.14 for $p = 1$ and $\Omega = [0, 1]$, with μ the Lebesgue measure.

(v) If $(\Omega, \mathcal{A}, \mu)$ is a measure space with a positive measure μ, we define $L^\infty(\Omega, \mu)$ to be the linear space of μ-measurable functions $f : \Omega \to \mathbb{C}$ (or $\mathbb{R}$) such that there exists a constant $M > 0$ with $|f(x)| \leq M$ for μ-a.e. $x \in \Omega$. Endowed with the norm[41] $\|f\|_\infty = \inf\{M > 0 : \ |f(x)| \leq M$ for μ-almost every $x \in \Omega\}$, $L^\infty(\Omega, \mu)$ is a Banach space, see Brézis (2011), that is not separable except when Ω consists of a finite number of atoms.[42] As in the previous example, for $\Omega = \mathbb{Z}$ with the counting measure we recover $l^\infty(\mathbb{Z})$. The case $\Omega = [0, 1]$ in the setting of the Lebesgue measure is discussed in Exercise 29.

(vi) For $1 \leq p \leq \infty$ and $I \subset \mathbb{R}$ an interval, the Sobolev spaces $W^{1,p}(I)$, consisting of all functions $f : I \to \mathbb{C}$ (or $\mathbb{R}$) that are locally absolutely continuous with $f' \in L^p(I)$, are Banach spaces, separable for $p \neq \infty$, with $W^{1,2}(I)$ the only Hilbert space and $W^{1,\infty}(I)$ being nonseparable, see Brézis (2011).

(vii) For $1 \leq p < \infty$ the Hardy space H^p is defined as the space of all analytic functions in the unit disc $\mathbb{D} = \{z \in \mathbb{C} : \ |z| < 1\}$ for which the norm $\|f\| = \sup_{0<r<1} \left(\int_0^1 |f(re^{2\pi it})|^p \right)^{1/p}$ is finite, while H^∞ is the space of all bounded analytic functions $f : \mathbb{D} \to \mathbb{C}$ with the norm $\|f\| = \sup_{|z|<1} |f(z)|$. These are Banach spaces, separable only for $1 \leq p < \infty$, with H^2 the only Hilbert space, see Duren (2000).

(viii) Given $1 \leq p < \infty$, the Bergman space B^p is the family of all analytic functions $f : \mathbb{D} \to \mathbb{C}$, where $\mathbb{D}$ is the unit disc, for which the norm $\left(\int_{\mathbb{D}} |f(x + iy)|^p dxdy \right)^{1/p}$ is finite.[43] B^2 is a Hilbert space, while all B^p are separable Banach spaces, see Zhu (2007).

39 Let us point out a subtle point here. The norm $\| \cdot \|_p$ does not distinguish between two functions that agree for μ-almost every point $x \in \Omega$. To deal with this nuisance one should redefine $L^p(\Omega, \mu)$ so that its elements are not functions but equivalence classes $\tilde{f}$ of functions, whereby $\tilde{f}$ is the set of all functions that differ from f on a set of μ-measure zero. To avoid this cumbersome notation, we will refer to functions, but one should keep in mind the technical point that $f = g$ in $L^p(\Omega, \mu)$ means agreement μ-almost everywhere.

40 In this case we denote these spaces simply by $L^p(\Omega)$, a convention that we also adopt for the case $p = \infty$ that comes up for discussion.

41 Note that, unlike the usual supremum, this so-called *essential supremum* $\|f\|_\infty$ ignores sets of μ-measure zero, so that we should make the same stipulation (concerning equivalence classes) as in the previous example.

42 That is, there are points $x_1, \ldots, x_m \in \Omega$ with $\{x_k\} \in \mathcal{A}$ and $\mu(\{x_k\}) > 0$ for $k = 1, \ldots, m$, while $\mu(\Omega \setminus \{x_1, \ldots, x_m\}) = 0$.

43 The case $p = \infty$ coincides with the Hardy space H^∞.

3.4 Functionals and operators

Linear algebra could be thought of as the theory of finite-dimensional vector spaces and the linear mappings between these spaces, while operator theory studies the infinite-dimensional version of such transformations. This subject of intrinsic beauty has many useful and significant applications, with Fourier analysis among the beneficiaries. We aim to provide a selection of material that we regard as essential, adding to this utilitarian goal an attempt to convey the aesthetic appeal of the underlying elegant abstract structures.

Definition *A linear operator from a normed linear space* $(X, \|\cdot\|_X)$ *to a normed linear space* $(Y, \|\cdot\|_Y)$ *over the same scalar field is a map* $T : X \to Y$ *satisfying* $T(\alpha x + \beta y) = \alpha Tx + \beta Ty$ *for all* $x, y \in X$ *and all scalars* α, β. *The linear operator* T *is called bounded if there exists a constant* $c > 0$ *with*

$$\|Tx\|_Y \leq c\,\|x\|_X\,, \qquad x \in X, \tag{3.6}$$

the smallest such constant being called the norm $\|T\|$ *of* T*:*

$$\|T\| = \sup_{x\in X,\ x\neq 0} \frac{\|Tx\|_Y}{\|x\|_X}\,. \tag{3.7}$$

Due to homogeneity, we can write (3.7) in the equivalent form

$$\|T\| = \sup_{x\in X,\, \|x\|_X=1} \{\|Tx\|_Y\} \tag{3.8}$$

that is often preferrable to (3.7). In general, we do not expect the norm of T to be attained – see Exercise 21.

Note that the range of a nonzero bounded linear operator is not bounded, rather the image of the open unit ball $B(0,1) = \{x \in X : \|x\|_X < 1\}$ is bounded. The next result explains the rationale for the adjective "bounded".

Theorem 3.9 *Let* T *be a linear operator between two normed linear spaces* $(X, \|\cdot\|_X)$ *and* $(Y, \|\cdot\|_Y)$*, over the same scalar field. The following assertions are equivalent: (i)* T *is continuous at* $0 \in X$*; (ii)* T *is continuous on* X*; (iii)* T *is bounded.*

Proof Let us first show the equivalence of (i) and (ii). Clearly (ii) implies (i). Conversely, assume the validity of (i). Given $x \in X$ and a sequence $\{x_n\}_{n\geq 1}$ that converges to x in X, from $\|x_n - x\|_X \to 0$ we infer that $\|T(x_n - x)\|_Y = \|Tx - Tx_n\|_Y \to 0$. Thus T is continuous at x.

To show that continuity and boundedness are equivalent, note that if T is bounded, then the validity of (3.6) for some $c > 0$ ensures that $Tx_n \to 0$ in Y whenever $x_n \to 0$ in X. Conversely, suppose T is continuous and not bounded. Then for every integer $n \geq 1$ there is some $x_n \in X$ with $\|Tx_n\|_Y > n\,\|x_n\|_X$. Since $T(0) = 0$ ensures $x_n \neq 0$, $x_n' = \frac{x_n}{n\|x_n\|_X} \to 0$ in X (being of norm $1/n$) satisfies $\|Tx_n'\|_Y > 1$ for all $n \geq 1$. But this contradicts the continuity of T at $0 \in X$. □

If the normed linear space $(X, \|\cdot\|_X)$ is finite-dimensional, then any linear operator $T : X \to Y$, where $(Y, \|\cdot\|_Y)$ is a normed linear space over the same scalar field, will be bounded. However, if X is infinite-dimensional and $Y \neq \{0\}$, then there exists an unbounded linear operator – see Exercise 19.

3.4.1 The family of bounded linear operators

We now study the collection $\mathcal{B}(X, Y)$ of all bounded linear operators from the normed linear space $(X, \|\cdot\|_X)$ to the normed linear space $(Y, \|\cdot\|_Y)$.

Theorem 3.10 *If $(X, \|\cdot\|_X)$ is a normed linear space and $(Y, \|\cdot\|_Y)$ is a Banach space, then $\mathcal{B}(X, Y)$, endowed with the operator norm (3.7), is a Banach space.*

Proof Clearly a finite linear combination of bounded linear operators is again a bounded linear operator, so $\mathcal{B}(X, Y)$ is a linear space. Also, it is easy to see that (3.7) is a norm on $\mathcal{B}(X, Y)$. Indeed, if $\|T\| = 0$, then $Tx = 0$ for all $x \in X$, so that $T = 0 \in \mathcal{B}(X, Y)$. Also, it is immediate that $\alpha T \in \mathcal{B}(X, Y)$ with $\|\alpha T\| = |\alpha| \cdot \|T\|$, for all scalars α and all $T \in \mathcal{B}(X, Y)$. Moreover, if $T_1, T_2 \in \mathcal{B}(X, Y)$, then $\|T_1 + T_2\| = \sup\limits_{x\in X, \|x\|_X\neq 0} \frac{\|(T_1+T_2)x\|_Y}{\|x\|_X} \leq \sup\limits_{x\in X, \|x\|_X\neq 0} \frac{\|T_1x\|_Y + \|T_2x\|_Y}{\|x\|_X} \leq \sup\limits_{x\in X, \|x\|_X\neq 0} \frac{\|T_1x\|_Y}{\|x\|_X} + \sup\limits_{x\in X, \|x\|_X\neq 0} \frac{\|T_2x\|_Y}{\|x\|_X} = \|T_1\| + \|T_2\|$. Thus $\mathcal{B}(X, Y)$ is a normed linear space.

To see that $\mathcal{B}(X, Y)$ is complete,[44] let $\{T_n\}_{n\geq 1}$ be a Cauchy sequence. Since $\|T_nx - T_mx\|_Y \leq \|T_n - T_m\| \cdot \|x\|_X$ for all $n, m \geq 1$ and $x \in X$, we have that for every fixed $x \in X$, the sequence $\{T_nx\}_{n\geq 1}$ is a Cauchy sequence in Y. Let $y = \lim_{n\to\infty} T_nx \in Y$ and define $Tx = y$. It is easy to see that this defines a linear operator from X to Y. Since $\{T_n\}_{n\geq 1}$ is a Cauchy sequence in the normed linear space $\mathcal{B}(X, Y)$, $\{\|T_n\|\}_{n\geq 1}$ is a Cauchy sequence in $\mathbb{R}$, with limit $K \geq 0$. Thus $\|Tx\|_Y = \lim_{n\to\infty} \|T_nx\|_Y \leq \lim_{n\to\infty} (\|T_n\| \cdot \|x\|_X) = K\|x\|_X$ for all $x \in X$, proving that $T \in \mathcal{B}(X, Y)$. To complete the proof we have to show that $\|T_n - T\| \to 0$ in $\mathcal{B}(X, Y)$. For this, note that $\|(T_n - T)x\|_Y = \lim_{m\to\infty} \|(T_n - T_m)x\|_Y$ for $x \in X$ yields $\frac{\|(T_n-T)x\|_Y}{\|x\|_X} \leq \limsup_{m\to\infty} \|T_n - T_m\|$ for all $x \in X$, $x \neq 0$. We infer that $\|T_n - T\| = \sup\limits_{x\in X, x\neq 0} \frac{\|(T_n-T)x\|_Y}{\|x\|_X} \leq \limsup\limits_{m\to\infty} \|T_n - T_m\|$. The right side can be made arbitrarily small for n large enough since $\{T_k\}_{k\geq 1}$ was a Cauchy sequence in $\mathcal{B}(X, Y)$. Thus $\|T_n - T\| \to 0$, completing the proof that $\mathcal{B}(X, Y)$ is a Banach space. □

In Theorem 3.10, it is remarkable that only the target space needs to be complete. However, this requirement is essential.[45]

The norm of an operator $T \in \mathcal{B}(X, Y)$ is in general hard to compute, so that we have content ourselves with some crude estimate. To somewhat compensate for this drawback, let us show that the spaces $\mathcal{B}(X, X)$, henceforth denoted by

[44] Note the similarity between this approach and the proof that the continuous periodic functions with the supremum norm form a Banach space – see Exercise 28(i).

[45] For example, if the complex linear normed space $(Y, \|\cdot\|_Y)$ is not complete, then $\mathbb{Y} = \mathcal{B}(\mathbb{C}, Y)$ is not a Banach space. Indeed, if $\{y_n\}_{n\geq 1}$ is a nonconvergent sequence in Y, one can easily check that the sequence $\{T_n\}_{n\geq 1}$ in $\mathbb{Y}$, where $T_nz = zy_n$ for $z \in \mathbb{C}$ and $n \geq 1$, is a Cauchy sequence as $\|T_n - T_m\| = \|y_n - y_m\|_Y$ for all $n, m \geq 1$. Were $\mathbb{Y}$ complete, there would exist $T = \lim_{n\to\infty} T_n$ in $\mathbb{Y}$. But this yields the contradiction $\lim_{n\to\infty} y_n = \lim_{n\to\infty} T_n1 = T1 \in Y$.

$\mathcal{L}(X)$, have an additional structure that sets them apart from most normed linear spaces.

Definition *A Banach space $(A, \|\cdot\|_A)$ together with an associative and distributive multiplication of vectors*[46] *such that $\|ab\|_A \leq \|a\|_A\, \|b\|_A$ for all $a, b \in A$, is called a Banach algebra.*

Note that the multiplication is jointly continuous.[47] A Banach algebra A is said to be *commutative* if $ab = ba$ for all $a, b \in A$, and is said to be *unital* if it possesses a multiplicative unit – also called an identity, that is, a vector $e \in A$ such that $a = ea = ae$ for all $a \in A$. Note that the identity, if it exists, is unique: if e and e' are identities, then $e = ee' = e'$. Also, $\|e\|_A \geq 1$ since $e = e^2$ yields $\|e\|_A \leq \|e\|_A^2$. It may happen that[48] $\|e\|_A > 1$, but this quirk can be easily corrected: if $(A, \|\cdot\|_A)$ is a Banach algebra with identity e, then[49] there is an equivalent norm $\|\cdot\|$ such that $(A, \|\cdot\|)$ is a unital Banach algebra with $\|e\| = 1$. Thus we can assume that the identity of a unital Banach algebra has norm 1.

Examples of infinite-dimensional commutative Banach algebras are the spaces $C[0, 1]$ and $L^\infty[0, 1]$ with the usual multiplication of functions, while the family of compact linear maps on an infinite-dimensional Banach space (to be studied later in Section 3.4.3), with the composition of operators as multiplication, is an example of a nonunital, noncommutative Banach algebra. Note that if the Banach algebra $(A, \|\cdot\|_A)$ does not have an identity, we can "adjoin" one as follows: on the linear space $A' = A \times \mathbb{C}$, define an associative and distributive multiplication by $(x, \alpha) \cdot (y, \beta) = (xy + \beta x + \alpha y, \alpha\beta)$. It is easy to check that A' endowed with the norm $\|(x, \alpha)\|_{A'} = \|x\|_A + |\alpha|$ is a Banach algebra with identity $(0, 1) \in A'$. We can identify A with the co-dimension 1 ideal $\{(x, 0) : x \in A\} \subset A'$ by means of the isometric isomorphism $x \mapsto (x, 0)$.

Theorem 3.11 *If $(X, \|\cdot\|_X)$ is a Banach space, then $\mathcal{L}(X)$, with the composition of operators as multiplication, is a Banach algebra with identity. Unless X is one-dimensional, this Banach algebra is noncommutative.*

Proof We have $\|TS\| = \sup_{x\in X,\, \|x\|=1} \{\|T(Sx)\|_X\} \leq \sup_{x\in X,\, \|x\|=1} \{\|T\| \cdot \|Sx\|_X\} = \|T\| \cdot \|S\|$ for all $T, S \in \mathcal{L}(X)$, due to (3.8) and (3.6). Since the other properties are immediate, we deduce that $\mathcal{L}(X)$ is a Banach algebra with unital element the identity operator $Ix = x$ for $x \in X$. In the special case when $\dim(X) = 1$, this Banach algebra is precisely the space of 1×1 matrices with scalar entries, thus clearly commutative. If X is not one-dimensional, there are at least two different elements $e_1 \neq e_2$ in a Hamel basis $\{e_i\}_{i\in I}$ of X. Motivated by the case of noncommutative 2×2 matrices, we define the operators $T, S \in \mathcal{L}(X)$ by $Te_1 = e_2$, $Te_2 = 0$, $Se_1 = 0$, $Se_2 = e_1$, $Te_i = Se_i = 0$ for $i \in I \setminus \{1, 2\}$, extended by linearity to the entire vector space X. Since $(ST)e_1 = e_1 \neq 0 = (TS)(e_1)$, we see that $\mathcal{L}(X)$ is not commutative. □

46 That is, the multiplication satisfies $(ab)c = a(bc)$ and $(a + b)c = ac + bc$, $a(b + c) = ab + ac$, $\lambda(ab) = (\lambda a)b = a(\lambda b)$ for all $\lambda \in \mathbb{C}$ and $a, b, c \in A$.

47 Indeed, for any a, b, a', b' in a Banach algebra we have $\|ab - a'b'\|_A = \|a(b - b') + (a - a')b'\|_A \leq \|a\|_A\, \|b - b'\|_A + \|a - a'\|_A\, \|b'\|_A$.

48 For example, the linear space $\mathbb{C}$ with the norm $\|z\| = 2|z|$ for $z \in \mathbb{C}$ is a Banach algebra with an identity of norm 2.

49 Indeed, set $\|a\| = \|L_a\|$, where $L_a \in \mathcal{L}(A)$ is the operator $b \mapsto ab$. Using the definition of the operator norm, it is easy to check that $\frac{\|a\|_A}{\|e\|_A} \leq \|a\| \leq \|a\|_A$ for all $a \in A$, so that $\|\cdot\|_A$ and $\|\cdot\|$ are equivalent norms on A. Note that $\|e\| = 1$, while $\|ab\| = \|L_{ab}\| = \|L_a L_b\| \leq \|L_a\| \cdot \|L_b\| = \|a\|\, \|b\|$ for all $a, b \in A$ shows that $(A, \|\cdot\|)$ is a Banach algebra.

The enhanced structure of a Banach algebra leads to a result[50] with no correspondent for plain Banach spaces.

Theorem 3.12 *Let $(A, \|\cdot\|_A)$ be a complex Banach algebra with identity e. If $\|e\|_A = 1$, then any $a \in A$ such that $\|e-a\|_A < 1$ is invertible: there is some $a' \in A$ with $aa' = a'a = e$.*

Proof If $b = e - a$, then $\|b\|_A = r < 1$. Since $\|b^n\|_A \leq \|b\|_A^n = r^n$ for all integers $n \geq 1$, we have that $\sum_{n=1}^{\infty} b^n$ converges in A. Let $a' = e + \sum_{n=1}^{\infty} b^n$. Since $(e-b)(e+b+b^2+\cdots+b^n) = (e-b)(e+b+b^2+\cdots+b^n)(e-b) = (e+b+\cdots+b^n) - (b+b^2+\cdots+b^{n+1}) = e - b^{n+1} \to e$ for $n \uparrow \infty$, we see that $aa' = a'a = e$. □

The previous results illustrate the rich structure of the space $\mathcal{L}(X)$ for a general Banach space $(X, \|\cdot\|_X)$. One might wonder whether there are Banach spaces $(X, \|\cdot\|_X)$ such that $\mathcal{L}(X)$ is a Hilbert algebra with identity, that is, a Hilbert space $(H, \|\cdot\|)$ which is also an associative Banach algebra with identity e satisfying $\|e\| = 1$. It turns out[51] that a complex Hilbert algebra with identity is isomorphic to the complex numbers, while a real Hilbert algebra with identity is isomorphic to the real numbers, the complex numbers or the quaternions.[52] Since most classical Banach spaces X are separable, what about the separability of $\mathcal{L}(X)$? This is clearly the case if X is finite-dimensional, as $\mathcal{L}(X)$ is the family of $n \times n$ matrices with complex entries.[53] If H is an infinite-dimensional separable Hilbert space with orthonormal basis $\{e_k\}_{k\geq 1}$, then to each sequence $\lambda = \{\lambda_k\}_{k\geq 1} \in l^\infty$ we can associate a diagonal operator $T_\lambda : H \to H$ defined by $T_\lambda(e_k) = \lambda_k e_k$ for $k \geq 1$. Clearly $T_{c\lambda+c'\lambda'} = c\,T_\lambda + c'\,T_{\lambda'}$

50 We will take advantage of it in Section 3.4.4.

51 See the discussion in Ingelstam (1963). At first sight, this seems perplexing since, after all, the 2×2-matrices form the Banach algebra $\mathcal{L}(\mathbb{C})$, and they can be identified with the Hilbert space $\mathbb{C}^4$. However, one needs an inner product producing a sub-multiplicative norm such that the identity has norm 1. These simultaneous conditions are very restrictive. Also, the cross product of vectors in $\mathbb{R}^3$ might seem a good candidate, but this "multiplication" is not associative: for the standard basis $e_1 = (1, 0, 0)$, $e_2 = (0, 1, 0)$, $e_3 = (0, 0, 1)$ of $\mathbb{R}^3$ we have $(e_1 \times e_1) \times e_2 = 0$ while $e_1 \times (e_1 \times e_2) = -e_2$.

52 The quaternions $\mathbb{H}$ are members of a noncommutative division algebra, invented in 1843 by Hamilton as a number system that extends the complex numbers. $\mathbb{H}$ is a four-dimensional real vector space endowed with a noncommutative multiplication of vectors, distributive with respect to the addition of vectors and the multiplication by scalars. To define the product of two vectors in $\mathbb{H}$, one specifies a basis $\{\mathbf{1}, \mathbf{i}, \mathbf{j}, \mathbf{k}\}$ such that $\mathbf{1}$ is the multiplicative identity and $\mathbf{i}^2 = \mathbf{j}^2 = \mathbf{k}^2 = -\mathbf{1}$, $\mathbf{ij} = -\mathbf{ji} = \mathbf{k}$, $\mathbf{jk} = -\mathbf{kj} = \mathbf{i}$, $\mathbf{ki} = -\mathbf{ik} = \mathbf{j}$. Defining the norm $\|x\|$ of a quaternion $x = a\mathbf{1} + b\mathbf{i} + c\mathbf{j} + d\mathbf{k} \in \mathbb{H}$ as the Euclidean norm $\sqrt{a^2+b^2+c^2+d^2}$ of the associated vector $(a, b, c, d) \in \mathbb{R}^4$, one has $\|xy\| = \|x\| \cdot \|y\|$. Analogous to complex numbers, there is an operation of conjugation, with $\|x\|^2$ equal to the scalar product of x and its conjugate x^*, and each quaternion $x \neq 0$ has an reciprocal $x^{-1} = x^*/\|x\|^2$. Quaternions are a concise method of representing rotations of $\mathbb{R}^3$ and $\mathbb{R}^4$, see Weiner and Wilkens (2005). For this reason quaternions are used in computer graphics. They are also closely related to the Pauli spin matrices in quantum mechanics.

53 Indeed, the bounded linear operator $T : X \to X$ is completely determined by the way it acts on an orthonormal basis, with $Tx = \sum_{k=1}^{n} \alpha_k\, Te_k$ if $x = \sum_{k=1}^{n} \alpha_k\, e_k$.

for all $c, c' \in \mathbb{C}$ and all $\lambda, \lambda' \in l^\infty$, and it is easy to check that $\|T_\lambda\| = \|\lambda\|_\infty$ for all $\lambda \in l^\infty$. Thus $\lambda \mapsto T_\lambda$ is an isometric inclusion $i : l^\infty \to \mathcal{L}(X)$. This means that we can view l^∞ as a subspace of $\mathcal{L}(X)$. Since l^∞ is not separable, see Exercise 29, neither is $\mathcal{L}(X)$. The same arguments can be adapted to show that for all Banach spaces X with an unconditional Schauder basis, $\mathcal{L}(X)$ is not separable. As to the question whether $\mathcal{L}(X)$ may be separable for some infinite-dimensional Banach space X, this appears to be an open question, see the discussion in Androulakis et al. (2006).

If the vector space Y coincides with the scalar field $\mathbb{R}$ or $\mathbb{C}$ of the normed linear space X, then $\mathcal{B}(X, Y)$ is denoted X^* and is called the *dual space* of X, its elements being the *bounded linear functionals* on X. Dual spaces are interesting in their own right and play an important role in many applications.[54] In a Hilbert space H with the scalar product $(\cdot,\cdot)$, one can verify by invoking Proposition 3.1 that any $x \in H$ yields a bounded linear functional F_x, defined by $F_x(y) = (y, x)$ for $y \in H$. A beautiful and fundamental result states that all bounded linear functionals on a Hilbert space are of this form.

Theorem 3.13 (Riesz representation theorem) *For any functional F in the dual H^* of a Hilbert space $(H, (\cdot,\cdot))$ there is a unique $x_F \in H$ such that $F(y) = (y, x_F)$ for all $y \in H$.*

Proof The idea is to observe that x_F is singled out by being orthogonal to the null space of $F \in H^*$: if $M = \{y \in H : F(y) = 0\}$, then $x_F \in M^\perp$ and F has the desired representation. Indeed, to prove the existence of x_F, define M as above and note that, by the continuity of the linear map F, M is a closed subspace. If $M = H$, then $F \equiv 0$ and we may take $x_F = 0$. If $M \neq H$, then $M^\perp$ is a one-dimensional subspace of H. Indeed, there exists some $x_0 \in M^\perp$ with $F(x_0) \neq 0$; in particular, $x_0 \neq 0$. Since every $y \in H$ can be written in the form $y = (y - \frac{F(y)}{F(x_0)} x_0) + \frac{F(y)}{F(x_0)} x_0$ with $y - \frac{F(y)}{F(x_0)} x_0 \in M$, the projection theorem (Theorem 3.3) ensures that $M^\perp$ is the subspace generated by x_0. If we set $x_F = \frac{\overline{F(x_0)}}{\|x_0\|^2} x_0$, then clearly $0 = (y, x_F) = F(y)$ for all $y \in M$, while for $y = \alpha x_0$ with α a scalar, we also have $F(y) = \alpha F(x_0) = \alpha (x_0, x_F) = (y, x_F)$. Thus $F(y) = (y, x_F)$ for all $y \in H$. To prove that x_F is unique, note that $F(y) = (y, x'_F)$ for all $y \in H$ yields $\|x_F - x'_F\|^2 = (x_F - x'_F, x_F - x'_F) = (x_F - x'_F, x_F) - (x_F - x'_F, x'_F) = F(x_F - x'_F) - F(x_F - x'_F) = 0$, so that $x_F = x'_F$. □

In contrast to the Hilbert space setting, the situation in Banach spaces is more complicated: the dual space procedure yields new Banach spaces from old ones, or even from merely normed linear spaces[55] even if not all Banach spaces arise this way.[56] The rich variety of Banach spaces that arise as dual spaces is illustrated by the following results.

[54] For example, the construction of the Green's function for the Dirichlet problem in planar domains can be performed by using the Hahn–Banach theorem – see e.g. Friedman (1982).

[55] Recall that X^* is a Banach space even if X is not complete, see Theorem 3.10.

[56] For example, $L^1[0, 1]$ as well as $C[0, 1]$ are not the dual of any Banach space, see Albiac and Kalton (2006).

- $c_0^* = c^* = l^1$; see Yosida (1995). This shows that the pre-dual of a Banach space, if it exists, need not be unique.
- For $1 < p < \infty$ and $(\Omega, \mathcal{A}, \mu)$ a measure space with a positive measure μ, the dual of $L^p(\Omega, \mu)$ is $L^q(\Omega, \mu)$, where $q = \frac{p}{p-1} \in (1, \infty]$ is the conjugate exponent, see Brézis (2011), in the sense that every $T \in [L^p(\Omega, \mu)]^*$ has the form $Tf = \int_\Omega f(x)g(x)\,d\mu(x)$ for some unique $g \in L^q(\Omega, \mu)$; this holds also for $p = 1$, provided that the measure space is σ-finite.[57] Incidentally, $(l^p)^* = l^q$ for $1 \le p < \infty$; the case $p = 1$ being discussed in Section 3.4.2, after Theorem 3.16.
- For a measure space $(\Omega, \mathcal{A}, \mu)$ with a positive measure μ, the dual of $L^\infty(\Omega, \mu)$ is not $L^1(\Omega, \mu)$ but the much larger space $ba(\Omega, \mu)$ of finitely additive set functions ψ on $\mathcal{A}$, which are absolutely continuous with respect to the measure μ, equipped with the total variation norm.[58] This result is not that useful since finitely additive measures are not easy to work with. For these matters we refer to Yosida (1995).
- The dual of $C[0, 1]$ is the space $\mathcal{M}[0, 1]$ of all regular Borel measures on $[0, 1]$, equipped with the total variation norm; see Yosida (1995). One can also identify it with the Banach space $NBV[0, 1]$ of functions f of bounded variation on $[0, 1]$, with[59] $f(0) = 0$ and $\lim\limits_{t \downarrow t_0} f(t) = f(t_0)$ for every $t_0 \in (0, 1)$, endowed with the norm $\|f\| = V_f[0, 1]$, see Mukherjea and Pothoven (1986). The dual of $\mathcal{M}[0, 1]$ is an intricate, less useful space, the so-called inverse limit of the spaces $L^\infty(\mu)$ with $\mu \in \mathcal{M}[0, 1]$; see Conway (1990).
- Given $1 \le p < \infty$, the dual space of the Sobolev space $W^{1,p}[0, 1]$ is denoted by $W^{-1,q}[0, 1]$ with $q = \frac{p}{p-1}$, and its elements F can be represented with the help of two functions $f_0, f_1 \in L^q[0, 1]$ by $F(u) = \int_0^1 u(x) f_0(x)\,dx + \int_0^1 u'(x) f_1(x)\,dx$ for $u \in W^{1,p}[0, 1]$; see Brézis (2011).

57 To see that the dual of $L^1(\Omega, \mu)$ might be larger than $L^\infty(\Omega, \mu)$, consider the counting measure ν on the σ-algebra of all sets $S \subset [0, 1]$ such that either S or $[0, 1] \setminus S$ is countable. Note that the measure ν is not σ-finite, while $L^1([0, 1], \nu)$ consists of functions $f : [0, 1] \to \mathbb{C}$ that vanish on all but at most countably many points $\{x_j\}_{j \in J}$, with $\|f\| = \sum_{j \in J} |f(x_j)| < \infty$. However, although the linear functional $F : L^1([0, 1], \nu) \to \mathbb{C}$ defined by $F(f) = \sum_{x \in [0,1]} x f(x)$ lies in the dual since $|F(f)| \le \|f\|$ for all $f \in L^1([0, 1], \nu)$, there is no $g \in L^\infty([0, 1], \nu)$ with $F(f) = \int_0^1 f(x)\overline{g(x)}\,d\nu(x) = \sum_{x \in [0,1]} f(x)\overline{g(x)}$ for all $f \in L^1([0, 1], \nu)$. Indeed, this relation forces $g(x) = x$ for all $x \in [0, 1]$, and this function is not measurable as $g^{-1}[\frac{1}{2}, 1) = [\frac{1}{2}, 1)$ is not measurable.

58 Meaning that $\psi(A_1 \cup A_2) = \psi(A_1) + \psi(A_2)$ for $A_1, A_2 \in \mathcal{A}$ with $A_1 \cap A_2 = \emptyset$, $\mu(A) = 0$ implies $\psi(A) = 0$ if $A \in \mathcal{A}$, and the real part ψ_1 and the imaginary part ψ_2 of ψ satisfy $\sup\limits_{A \in \mathcal{A}} |\psi_j(A)| < \infty$ for $j = 1, 2$.

59 The normalisation requiring $f(0) = 0$ and right-continuity is necessary to provide a one-to-one correspondence between bounded linear functionals and functions of bounded variation (that define a regular Borel measure).

- If $p \in (1, \infty)$ and $q = \frac{p}{p-1}$, then one can identify the dual of the Hardy space H^p with H^q under the integral pairing $(F, G) = \frac{1}{2\pi} \int_0^{2\pi} f(\theta)\, \overline{g(\theta)}\, d\theta$, where[60] f and g are the nontangential limits of $F \in H^p$ and $G \in H^q$. On the other hand, $(H^1)^* \cong BMOA$ under the pairing $(F, g) = \frac{1}{2\pi} \int_0^{2\pi} f(\theta)\, \overline{g(\theta)}\, d\theta$ with $F \in H^1$ and $g \in BMOA$, where f is the nontangential limit of F and $BMOA$ is the nonseparable Banach space ontained as the intersection of the space BMO with the traces of H^2 on the unit circle; see Zhu (2007). Here BMO is the Banach space of functions of bounded mean oscillation on $[0, 2\pi]$, consisting of the functions $f \in L^2[0, 2\pi]$ having a finite norm $\|f\|_{BMO} = \sup_I \{\frac{1}{|I|} \int_I |f(\theta) - f_I|^2 d\theta\}^{1/2}$, where $f_I = \frac{1}{|I|} \int_I f(\theta)\, d\theta$ is the mean of f over the interval $I \subset [0, 2\pi]$ of length $|I|$.
- For $p \in (1, \infty)$, the dual of the Bergman space B^p is the Bergman space B^q under the integral pairing $(f, g) = \int_{\mathbb{D}} f(x+iy)\, \overline{g(x+iy)}\, dxdy$, where $q = \frac{p}{p-1}$. On the other hand, $(B^1)^* \cong \mathfrak{B}$ under the analogous integral pairing, where $\mathfrak{B}$ is the Bloch space of analytic functions $g : \mathbb{D} \to \mathbb{C}$ such that $\|g\|_{\mathfrak{B}} = \sup_{z \in \mathbb{D}} \{(1 - |z|^2)\, |f'(z)|\} < \infty$. Endowed with the norm $\|g\| = |f(0)| + \|g\|_{\mathfrak{B}}$, $\mathfrak{B}$ is a nonseparable Banach space. See Zhu (2007).

3.4.2 The Hahn–Banach theorems

Since the Riesz representation theorem is specific to Hilbert spaces, for a general Banach space it is not *a priori* clear that X^* is larger than $\{0\}$. However, this pathological situation never occurs, unless $X = \{0\}$.

Theorem 3.14 (Hahn–Banach) *Let X be a linear vector space and let $p : X \to [0, \infty)$ be a semi-norm, that is, a function satisfying $p(\alpha x) = |\alpha|\, p(x)$ and $p(x + y) \le p(x) + p(y)$ for all $x, y \in X$ and all scalars α. Suppose f is a linear functional, defined on a linear subspace Y of X and satisfying $|f(y)| \le p(y)$ for all $y \in Y$. Then we can extend[61] f to a linear functional F on X such that $|F(x)| \le p(x)$ for all $x \in X$.*

Before providing the proof of Theorem 3.14, let us point out that the main issue is[62] the fact that $|F(x)| \le p(x)$ for all $x \in X$.

Proof Let us first discuss the case of real vector spaces. The idea of the proof is the following:

[60] One can show that if $F \in H^p$ with $1 \le p \le \infty$, then the limit $f(\theta) = \lim_{r \uparrow 1} F(re^{i\theta})$ exists for almost all $\theta \in [0, 2\pi]$, see Duren (2000).

[61] In the sense that $F(x) = f(x)$ for all $x \in Y$.

[62] Without this restriction, an extension of f could be obtained simply as follows. Completing a Hamel basis $\{e_i\}_{i \in J}$ of Y to a Hamel basis $\{e_i\}_{i \in I}$ of X, we set $F(e_i) = f(e_i)$ for $i \in J$ and $F(e_i) = 0$ for $i \in I \setminus J$, defining F on X by finite linear combinations.

if $z \in X \setminus Y$, we show that we can extend f to a functional having the right properties on the subspace spanned by z and Y. Zorn's lemma will ensure that the process can be continued until the whole space X is exhausted. To provide the details, note first that the extension F of f to the subspace $\tilde{Y}$, spanned by z and Y, is specified as soon as we define $F(z) = c \in \mathbb{R}$, since by linearity $F(\alpha z+y) = \alpha F(z)+f(y)$ for all $\alpha \in \mathbb{R}$ and $y \in Y$. We need to ensure that $|f(y)+\alpha c| \leq p(\alpha z+y)$ for all $\alpha \in \mathbb{R} \setminus \{0\}$ and all $y \in Y$. Writing $y = -\alpha\, y_0$ and dividing both sides of the desired inequality by $|\alpha|$, we obtain the equivalent inequality $|c - f(y_0)| \leq p(z - y_0)$ for all $y_0 \in Y$. In other words, we have to show that there is some $c \in \mathbb{R}$ with $f(y_0) - p(z - y_0) \leq c \leq f(y_0) + p(z - y_0)$ for all $y_0 \in Y$. For this, it suffices that $\sup_{y_0 \in Y} \{f(y_0) - p(z - y_0)\} \leq \inf_{y_0' \in Y} \{f(y_0') + p(z - y_0')\}$ for all $z \in X$. This is ensured since for all $y_0,\, y_0' \in Y$ we have $f(y_0) - f(y_0') = f(y_0 - y_0') \leq p(y_0 - y_0') = p(z - y_0 + y_0' - z) \leq p(z - y_0) + p(z - y_0')$, which yields $f(y_0) - p(z - y_0) \leq f(y_0') + p(z - y_0')$. Having secured one-dimensional extensions, let us conclude the argument. Consider the collection C of all pairs (Y_1, f_1), where Y_1 is a subspace of X that contains Y and f_1 is a linear functional on Y_1, extending f and having the property that $|f_1(y)| \leq p(y)$ for all $y \in Y_1$. The family C is nonempty since $(Y, f) \in C$. We define a partial ordering on C: $(Y_1, f_1) \prec (Y_2, f_2)$ if $Y_1 \subset Y_2$ and if the restriction of f_2 to Y_1 coincides with f_1. Each totally ordered set $\{(Y_i, f_i)\}_{i\in I}$ has an upped bound in C, namely, $(\bigcup_{i\in I} Y_i,\, f_0)$, where f_0 is the functional that coincides with f_i on Y_i. By Zorn's lemma, C contains a maximal element $(\tilde{Y}, \tilde{f})$. Were $\tilde{Y} \neq X$, we could find some $z \in X \setminus \tilde{Y}$ and, by the previous considerations, we could extend, thus contradicting the maximality assumption. Thus $\tilde{Y} = X$ and we may set $F = \tilde{f}$.

Let us now discuss the case of complex vector spaces. Set $\lambda(y) = \Re\{f(y)\}$ for $y \in Y$. One can easily check that λ is a real linear functional on Y: $\lambda(\alpha_1 y_1 + \alpha_2 y_2) = \alpha_1\, \lambda(y_1) + \alpha_2\, \lambda(y_2)$ for all $\alpha_1,\, \alpha_2 \in \mathbb{R}$ and all $y_1,\, y_2 \in Y$. Since $\lambda(iy) = \Re\{f(iy)\} = \Re\{i\, f(y)\} = -\Im\{f(y)\}$ for all $y \in Y$, we deduce that $f(y) = \lambda(y) - i\,\lambda(iy)$ for all $y \in Y$. Thus f and λ are equivalent pieces of information. The procedure used for real vector spaces is applicable to the real functional λ, yielding some $\Lambda : X \to \mathbb{R}$ such that $\Lambda(\alpha_1 x_1 + \alpha_2 x_2) = \alpha_1 \Lambda(x_1) + \alpha_2 \Lambda(x_2)$ for all $\alpha_1,\, \alpha_2 \in \mathbb{R}$ and all $x_1,\, x_2 \in X$, while $|\Lambda(x)| \leq p(x)$ for all $x \in X$. We claim that $F(x) = \Lambda(x) - i\,\Lambda(ix)$ for $x \in X$ completes our quest. Indeed, F clearly extends f. Moreover, $F(x_1 + x_2) = F(x_1) + F(x_2)$ for all $x_1,\, x_2 \in X$, while for all $\alpha,\, \beta \in \mathbb{R}$ and all $x \in X$ we have $F((\alpha + i\beta)x) = F(\alpha x) + F(i\beta x) = \alpha\, F(x) + \beta\, F(ix) = \alpha\, F(x) + \beta\{\Lambda(ix) - i\,\Lambda(-x)\} = \alpha\, F(x) + \beta\{\Lambda(ix) + i\,\Lambda(x)\} = \alpha\, F(x) + i\beta\, F(x)$. Thus $F : X \to \mathbb{C}$ is a linear functional. To complete the proof, we only need to check that $|F(x)| \leq p(x)$ for all $x \in X$. For this, given $x \in X$, since $p(\xi x) = p(x)$ if $|\xi| = 1$, if $\theta \in [0, 2\pi)$ is the argument of the complex number $F(x)$, note that $|F(x)| = e^{-i\theta}\, F(x) = F(e^{-i\theta} x) = \Lambda(e^{-i\theta} x) \leq p(e^{-i\theta} x) = p(x)$. □

For real normed spaces, the following variation of Theorem 3.14 holds.

Theorem 3.15 (Hahn–Banach) *Let X be a real linear vector space and let $p : X \to [0, \infty)$ be such that $p(\lambda x) = \lambda\, p(x)$ and $p(x + y) \leq p(x) + p(y)$ for all $x,\, y \in X$ and all $\lambda > 0$. Then a linear functional f, defined on a linear subspace Y of X and satisfying $f(y) \leq p(y)$ for all $y \in Y$, can be extended to a linear functional F on X satisfying $F(x) \leq p(x)$ for all $x \in X$.*

Proof The approach is entirely analogous to the one pursued in the proof of Theorem 3.14, the only aspect that needs consideration being the existence of suitable one-dimensional extensions. To clarify this point, for $z \in X \setminus Y$ we must ensure the existence of some $c \in \mathbb{R}$ with $f(y)+\alpha c \leq p(y+\alpha z)$ for all $\alpha \in \mathbb{R} \setminus \{0\}$. Setting $y = |\alpha|\, y_0$, a separate glance at the cases $\alpha > 0$ and $\alpha < 0$ confirms that this amounts to checking that $f(y_0) + c \leq p(y_0 + z)$ and $f(y_0) - c \leq p(y_0 - z)$ for all $y_0 \in Y$, that is, we must find some $c \in \mathbb{R}$ with $\sup_{x\in Y} \{f(x) - p(x - z)\} \leq c \leq \inf_{y\in Y} \{p(y + z) - f(y)\}$. But from the hypothesis it follows that $f(y) + f(x) \leq p(y + x) \leq p(y + z) + p(x - z)$ for all $x,\, y \in Y$, and this implies $f(x) - p(x - z) \leq p(y + z) - f(y)$ for all $x,\, y \in Y$. Therefore a suitable $c \in \mathbb{R}$ will exist. □

The Hahn–Banach theorem yields the existence of a wide range of bounded linear functionals, as illustrated by the following results.

Corollary 1 *Let Y be a linear subspace of the normed linear space $(X, \|\cdot\|)$. Then every $f \in Y^*$ admits an extension $F \in X^*$ with $\|F\|_{X^*} = \|f\|_{Y^*}$.*

Proof Consider the semi-norm $p(x) = \|f\|_{Y^*}\|x\|$. By Theorem 3.14 there exists a linear functional F on X that extends f and satisfies $|F(x)| \leq \|f\|_{Y^*}\|x\|$ for all $x \in X$. Thus $\|F\|_{X^*} \leq \|f\|_{Y^*}$. The reverse inequality holds since F extends f. □

To appreciate the strength of Corollary 1, let us discuss briefly the following related issue. If X_0 is a linear subspace of the normed linear space $(X, \|\cdot\|_X)$ and $(Y, \|\cdot\|_Y)$ is a Banach space with the same scalar field, does every $T \in \mathcal{B}(X_0, Y)$ have an extension to some $\tilde{T} \in \mathcal{B}(X, Y)$? Note that an extension to the linear closed space $\overline{X}_0$ is ensured.[63] We may therefore assume that X_0 is a closed subspace of X. If X is a Hilbert space, the desired extension of $T \in \mathcal{B}(X_0, Y)$ is granted by the following construction: let $S : X \to X_1$ be the projection on X_0 (see Section 3.2) and set $\tilde{T} = TS$. What about the case when X is a Banach space? The Hilbert space construction is indicative for the following approach: assume that the closed subpace X_0 of X has a (topological) complement in X, that is, there is a closed subspace X_1 of X such that $X_0 \cap X_1 = \{0\}$ and every $x \in X$ can be written in the form[64] $x = x_1 + x_2$ with $x_1 \in X_1$ and $x_2 \in X_2$. As a consequence of the open mapping theorem (to be discussed later on in this section), the linear operator S defined by $x \mapsto x_1$ belongs to $\mathcal{B}(X, X_0)$; see Exercise 22. Therefore $\tilde{T} = TS \in \mathcal{B}(X, Y)$ is a suitable extension of $T \in \mathcal{B}(X_0, Y)$, since the composition of two continuous operators clearly produces a continuous operator. The problem with this approach is that while in a Banach space closed subspaces admitting a complement abound (see Exercise 22), in every Banach space that is not isomorphic to a Hilbert space, there exist closed subspaces without any complement, see Lindenstrauss and Tzafriri (1977). This failure casts doubt on the possibility of extending bounded operators. It turns out that if X_0 is a closed subspace without a complement, then the identity operator $T : X_0 \to X_0$ does not admit an extension to some operator $\tilde{T} \in \mathcal{B}(X_0, X_0)$ since the existence of such an extension $\tilde{T}$ would produce $\ker(\tilde{T}) = \{x \in X : \tilde{T}x = 0\}$ as a complement[65] of X_0 in X.

Corollary 2 *Let $x \neq 0$ in the normed linear space $(X, \|\cdot\|)$. Then there exists $F \in X^*$ with $\|F\|_{X^*} = 1$ and $F(x) = \|x\|$.*

[63] Indeed, if $x_n \in X_0$ converge to some $x \in X$, then $\|Tx_n - Tx_m\|_Y \leq \|T\|_{\mathcal{B}(X_0,Y)}\|x_n - x_m\|_X$ ensures that $\{Tx_n\}_{n\geq 1}$ is a Cauchy sequence in Y. If we denote its limit by $\tilde{T}x \in Y$, one can easily check that $\tilde{T} \in \mathcal{B}(\overline{X}_0, Y)$ extends T, and $\|\tilde{T}\|_{\mathcal{B}(\overline{X}_0,Y)} = \|T\|_{\mathcal{B}(X_0,Y)}$. This being the only possible extension, and the example of the identity acting on a dense proper subspace X_0 of a Banach space X, with $Y = X_0$, show the necessity of assuming completeness for Y.

[64] Note that this representation is unique.

[65] Indeed, $\ker(\tilde{T})$ is a closed subspace by the linearity and continuity of $\tilde{T}$, and clearly $\ker(\tilde{T}) \cap X_0 = \{0\}$. Moreover, for any $x \in X$ we would have $T(x - Tx) = Tx - Tx = 0$, so $(x - Tx) \in \ker(\tilde{T})$. We are done since $Tx \in X_0$.

Proof Let Y be the one-dimensional subspace of X generated by x. For $\alpha x \in Y$, set $f(\alpha x) = \alpha \,\|x\|$. Since $f \in Y^*$ with $\|f\|_{Y^*} = 1$, we can conclude by Corollary 1. □

Corollary 3 *Let Y be a subspace of the normed linear space $(X, \|\cdot\|)$. If $x_0 \in X$ is at distance $\delta > 0$ from Y, that is,*[66] *if $\inf_{y\in Y}\{\|x_0 - y\|\} = \delta$, then there exists $F \in X^*$ with $\|F\|_{X^*} = 1$, $F(x_0) = \delta$ and $F(y) = 0$ for $y \in Y$.*

Proof Let Y_1 be the subspace of X generated by x_0 and Y. Since $x_0 \notin Y$, we have that every $x \in Y_1$ has a unique representation of the form $x = \alpha x_0 + y$ for some $y \in Y$ and some scalar α. Define on Y_1 the linear functional $f(\alpha x_0 + y) = \alpha\,\delta$ for $y \in Y$ and α a scalar. Then $f \in Y_1^*$ with $\|f\|_{Y_1^*} \le 1$ since for $\alpha \neq 0$ we have $\|\alpha x_0 + y\| = |\alpha| \cdot \|x_0 + \frac{1}{\alpha} y\| \ge |\alpha| \cdot d(x_0, Y) = |\alpha|\,\delta = |f(\alpha x_0 + y)|$. On the other hand, given $\varepsilon > 0$, choosing some $y \in Y$ with $\|x_0 - y\| < \delta + \varepsilon$, we get $\delta = f(x_0) = f(x_0 - y) \le \|f\|_{Y_1^*}\|x_0 - y\| \le (\delta + \varepsilon)\,\|f\|_{Y_1^*}$. Thus $\|f\|_{Y_1^*} \ge \frac{\delta}{\delta+\varepsilon} \to 1$ as $\varepsilon \downarrow 0$, which proves that actually $\|f\|_{Y_1^*} = 1$. Applying Corollary 1 we find some $F \in X^*$ extending f, with $\|F\|_{X^*} = 1$. □

The next consequence of the Hahn–Banach theorem is often used to show that certain subspaces are dense (see e.g. Exercise 20).

Corollary 4 *Let Y be a subspace of the normed linear space $(X, \|\cdot\|)$. If $\overline{Y} \neq X$, then there is some $f \in X^*$, $f \neq 0$, with $f(y) = 0$ for all $y \in Y$.*

Proof Choose some $x_0 \in X \setminus \overline{Y}$ and apply Corollary 3. □

Corollary 5 *In any normed linear space $(X, \|\cdot\|)$ we have*[67]

$$\|x\| = \sup_{f\in X^*,\ \|f\|_{X^*}\le 1} \{|f(x)|\} = \max_{f\in X^*,\ \|f\|_{X^*}\le 1} \{|f(x)|\}\,. \tag{3.9}$$

Proof Clearly $\sup_{f\in X^*,\ \|f\|_{X^*}\le 1} \{|f(x)|\} \le \|x\|_X$ for $x \neq 0$. We conclude by Corollary 2. □

We now exhibit a link between structural properties of X^* and X.

Theorem 3.16 *Let $(X, \|\cdot\|)$ be a Banach space. If X^* is separable, then X is separable.*

Proof To simplify the exposition, assume that the scalars are real. Let $\{f_n\}_{n\ge 1}$ be dense in X^*, and choose $x_n \in X$ with $\|x_n\| = 1$ such that $|f_n(x_n)| \ge \dfrac{\|f_n\|_{X^*}}{2}$ for all $n \ge 1$. We claim that the countable set L of all finite linear combinations of the x_n's, with rational coefficients, is dense in L. Indeed, note that $\overline{L}$ is a linear subspace.[68] If $\overline{L} \neq X$, then Corollary 4 grants the existence of some $f \in X^*$ with $f(y) = 0$ for all $y \in \overline{L}$, while $f(y_0) \neq 0$ for some $y_0 \in X$. However, choosing a subsequence $\{f_{n_k}\}$ with $\|f_{n_k} - f\|_{X^*} \to 0$, from $\|f_{n_k} - f\|_{X^*} \ge |(f_{n_k} - f)(x_{n_k})| = |f_{n_k}(x_{n_k})| \ge \frac{\|f_{n_k}\|_{X^*}}{2}$ we get $\|f_{n_k}\|_{X^*} \to 0$, and therefore $f = 0$. This contradiction completes the proof. □

66 Note that zero distance means that $x_0 \in \overline{Y}$.

67 Exercise 21 shows that, in contrast to (3.9), the supremum in the defining formula $\|f\|_{X^*} = \sup_{x\in X,\ \|x\|\le 1} \{|f(x)|\}$ may not be attained.

68 If $x, y \in \overline{L}$, then $x + y$ is the limit of the sum of the two sequences in L that approximate x and y, respectively, so that $x + y \in \overline{L}$. As for why $\alpha x \in \overline{L}$ for all $\alpha \in \mathbb{R}$ and all $x \in \overline{L}$, it suffices to multiply the sequence in L that converges to x in X with rational numbers that approximate α.

The following example shows that the converse of Theorem 3.16 is not true.

Let us show that $l_\infty = l_1^*$. If $\{\lambda(k)\}_{k\geq 1} \in l_\infty$, then for each $\{a(k)\}_{k\geq 1} \in l_1$ we have that the numerical series $\sum_{k=1}^\infty \lambda(k)a(k)$ converges to a limit $T_\lambda a$, thus defining a linear operator $T_\lambda : l^1 \to \mathbb{C}$. Since $\|T_\lambda\|_{l_1^*} = \sup_{\|a\|_{l_1}\leq 1}\{|T_\lambda a|\} \leq \sup_{\|a\|_{l_1}\leq 1}\{\sum_{k=1}^\infty |\lambda(k)|\cdot|a(k)|\} \leq \sup_{k\geq 1}\{|\lambda(k)|\}\cdot \sup_{\|a\|_{l_1}\leq 1}\{\sum_{k=1}^\infty |a(k)|\} \leq \sup_{k\geq 1}\{|\lambda(k)|\} = \|\lambda\|_{l_\infty}$, we see that $T_\lambda \in l_1^*$, with $\|T_\lambda\|_{l_1^*} \leq \|\lambda\|_{l_\infty}$. Actually, $\|T_\lambda\|_{l_1^*} = \|\lambda\|_{l_\infty}$. Indeed, if $\lambda = 0$, then clearly $T_\lambda = 0$. If $\lambda \in l_\infty$, $\lambda \neq 0$, we argue as follows. For $k \geq 1$, denote by $e_k \in l_1 \cap l_\infty$ the sequence with $e_k(n) = \delta_{kn}$ for $n \geq 1$. Choose now a subsequence $n_k \to \infty$ such that $\lambda(n_k) \neq 0$ and $|\lambda(n_k)| \to \|\lambda\|_{l_\infty}$. Note that $a_{n_k} = \frac{|\lambda(n_k)|}{\lambda(n_k)}\, e_{n_k} \in l_1$ satisfies $\|a_{n_k}\|_{l_1} = 1$ and $T_\lambda a_{n_k} = |\lambda(n_k)| \to \|\lambda\|_{l_\infty}$.

Let us now show that all continuous linear functionals on l_1 arise in the way described above. For $T \in l_1^*$ define the numerical sequence λ by $\lambda(k) = Te_k$ for $k \geq 1$, and let $a_n = \frac{|\lambda(n)|}{\lambda(n)}\, e_n \in l_1$, with the understanding that if some $\lambda(k)$ is zero, that term is omitted and $a_k = 0$. Thus $\|a_n\|_{l_1} \in \{0, 1\}$ for all $n \geq 1$. Since $Ta_n = |\lambda(n)|$ and $|Ta_n| \leq \|T\|_{l_1^*}\|a_n\|_{l_1}$ for all $n \geq 1$, we deduce that $\sup_{n\geq 1}\{|\lambda(n)|\} \leq \|T\|_{l_1^*}$. Consequently $\lambda \in l_\infty$ and $T_\lambda \in l_1^*$. Since T and T_λ agree on the dense set of l_1 formed by the finite linear combinations of the e_k, we must have $T = T_\lambda$. Thus $l_\infty = l_1^*$, since the underlying spaces, as well as the norms, coincide.

We now discuss a geometric form of Theorem 3.16, in the context of the separation of convex sets. If $(X, \|\cdot\|)$ is a real normed linear space, we call a set $C \subset X$ *convex* if $tx + (1-t)y \in C$ whenever $x, y \in C$ and $t \in [0, 1]$. The set C is called *symmetric* if $x \in C$ if and only if $-x \in C$. An *affine hyperplane* $H \subset X$ is a level set of a linear functional that does not vanish identically, that is, $H = \{x \in X : f(x) = \alpha\}$ for some linear map $f : X \to \mathbb{R}$, $f \not\equiv 0$, and some constant $\alpha \in \mathbb{R}$. Note that if $x_0 \in H$, then $H = \{x_0 + y : y \in \ker(f)\}$, where $\ker(f) = \{x \in X : f(x) = 0\}$ is a linear subspace of X. It turns out that the affine hyperplane H is closed if and only if f is continuous (see Exercise 20). We say that the hyperplane $[f = \alpha]$ *separates* two subsets A and B of X if A lies in one of the half-spaces determined by the hyperplane, and B lies in the other, that is, $f(x) \leq \alpha$ for all $x \in A$ and $f(y) \geq \alpha$ for all $y \in B$. We say that the hyperplane separates A and B strictly if there is some $\varepsilon > 0$ such that $f(x) \leq \alpha - \varepsilon$ for all $x \in A$ and $f(x) \geq \alpha + \varepsilon$ for all $x \in B$; see Figure 3.5. The next result is a geometric form of the Hahn–Banach theorem.

Theorem 3.17 *Let A and B be disjoint nonempty convex subsets of a real normed linear space $(X, \|\cdot\|)$. If one of them is open, then there exists a closed hyperplane that separates A and B. If A is closed and B is compact, then there exists a closed hyperplane that separates A and B strictly.*

Proof The approach hinges on the possibility of associating to an open, convex and symmetric set $C \subset X$, with $0 \in C$, a semi-norm p_C, called the gauge of C. This is achieved by defining $p_C(x) = \inf\{t > 0 : x/t \in C\}$. Moreover, p_C also has the following additional properties:

(i) there is a constant $M > 0$ such that[69] $0 \leq p_C(x) \leq M\,\|x\|$ for all $x \in X$;
(ii) $C = \{x \in X : p(x) < 1\}$.

[69] We cannot expect an estimate of the form $m\|x\| \leq p(x)$: for $C = \{(x_1, x_2) \in \mathbb{R}^2 : |x_1| < 1\}$ we have $p_C((0, 1)) = 0$.

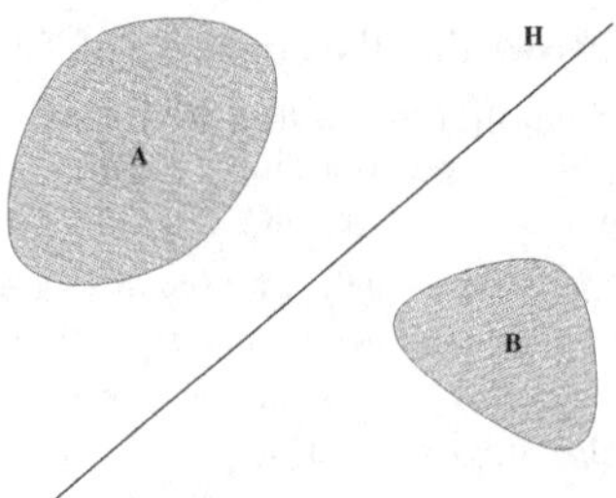

Figure 3.5 An illustration of Theorem 3.15.

Indeed, first note that, since C is convex and symmetric, with $0 \in C$, for $w \in C$ we have[70] $\pm\lambda w \in C$ for all $\lambda \in [0, 1]$. In particular, $p_C(\lambda x) = |\lambda|\, p_C(x)$ for all $\lambda \in \mathbb{R}$ and $x \in X$. On the other hand, given $x, y \in X$ and $\varepsilon > 0$, we have $\dfrac{x}{p_C(x)+\varepsilon}, \dfrac{y}{p_C(y)+\varepsilon} \in C$. For $t = \dfrac{p_C(x)+\varepsilon}{p_C(x)+p_C(y)+2\varepsilon} \in [0,1]$ we have $\dfrac{x+y}{p_C(x)+p_C(y)+2\varepsilon} = t\,\dfrac{x}{p_C(x)+\varepsilon} + (1-t)\,\dfrac{y}{p_C(y)+\varepsilon} \in C$, which yields that $p_C(x+y) \le p_C(x)+p_C(y)+2\varepsilon$. Letting $\varepsilon \downarrow 0$ leads to $p_C(x+y) \le p_C(x)+p_C(y)$. Therefore p_C is a semi-norm. To verify (i), let $r > 0$ be such that $B(0,r) = \{x \in X : \|x\| < r\} \subset C$. Note that for every $\varepsilon \in (0,r)$ we have that $\dfrac{(r-\varepsilon)\,x}{\|x\|} \in B(0,r)$ whenever $x \neq 0$. Thus $p_C(x) \le \dfrac{\|x\|}{r-\varepsilon}$ for all $x \in X$, and we may choose $M = r^{-1}$. Finally, if $x \in C$, then the fact that C is open ensures that $(1+\varepsilon)x \in C$ if $\varepsilon > 0$ is small enough, so that $p_C(x) \le (1+\varepsilon)^{-1} < 1$. To check (ii), it suffices to see that if $p_C(x) < 1$, then the existence of $\lambda \in (0,1)$ with $\dfrac{x}{\lambda} \in C$ yields $x = \lambda\dfrac{x}{\lambda} + (1-\lambda)\,0 \in C$. If we drop the condition of symmetry for C, then by inspection we see that all previous considerations carry on, with exception of the property $p_C(\lambda x) = |\lambda|\, p_C(x)$ for all $\lambda \in \mathbb{R}$ and $x \in X$. Instead, we only get that $p_C(\lambda x) = \lambda\, p_C(x)$ for all $\lambda \ge 0$ and $x \in X$: we obtain a so-called quasi-semi-norm.

Assume now that $A \subset X$ is a nonempty open and convex set, and $B \subset X$ is a nonempty convex set with $A \cap B = \emptyset$. Then the set $K = \{b - a : b \in B, a \in A\}$ is convex and $0 \notin K$. Moreover, $K = \bigcup_{b \in B} \{b - a : a \in A\}$ is open. Choose some $x_0 \in K$ and let p_C be the gauge of the convex open set $C = \{x_0 - x : x \in K\}$, with $0 \in C$ and $x_0 \notin C$. Define on the linear subspace $Y = \{\alpha x_0 : \alpha \in \mathbb{R}\}$ the linear functional $f(\alpha x_0) = \alpha$ for $\alpha \in \mathbb{R}$. Then $f(y) \le p_C(y)$ for all $y \in Y$. Indeed, for $\alpha > 0$ we have $p_C(\alpha x_0) = \alpha\, p_C(x_0) \ge \alpha = f(\alpha x_0)$ since $x_0 \notin C$ ensures $p_C(x_0) \ge 1$ by property (ii), while for $\alpha \le 0$ we have $f(\alpha x_0) = \alpha \le 0 \le p_C(\alpha x_0)$. Theorem 3.14 therefore ensures the existence of an extension of f to a linear functional $F : X \to \mathbb{R}$ such that $F(x) \le p_C(x)$ for all $x \in X$. Property (i) shows that $F \in X^*$. Since $p_C(x) < 1$ for $x \in C$, due to property (ii), we have that $F(x) < 1$ for all $x \in C$. This means that for all $a \in A$ and $b \in B$ we have $F(x_0) - F(b) + F(a) = F(x_0 - b + a) < 1 = F(x_0)$, or $F(a) < F(b)$. Choosing a constant $\alpha \in \mathbb{R}$ such that $\sup_{a \in A}\{f(a)\} \le \alpha \le \inf_{b \in B}\{f(b)\}$, we see that the closed hyperplane $[F = \alpha]$ separates A and B.

To prove the second statement, let now $A \subset X$ be a nonempty closed and convex set, and $B \subset X$ be a nonempty compact and convex set, with $A \cap B = \emptyset$. For $\varepsilon > 0$, let $B(0,\varepsilon) = \{x \in X : \|x\| < \varepsilon\}$, and set $A_\varepsilon = \{a + x : a \in A, x \in B(0,\varepsilon)\}$ and $B_\varepsilon = \{b + x : b \in B, x \in B(0,\varepsilon)\}$. The sets A_ε and B_ε are clearly convex, open, and nonempty. We claim that we can choose $\varepsilon > 0$ sufficiently small so that $A_\varepsilon \cap B_\varepsilon = \emptyset$. Indeed, otherwise there exists a sequence $\varepsilon_n \downarrow 0$ and points $x_n \in A$, $y_n \in B$ with $\|x_n - y_n\| < \varepsilon_n$ for all $n \ge 1$. Since the set B is compact, we can extract

[70] Thus, given $x \in X$, we have $sx \in C$ for $s < \frac{1}{p_C(x)}$. Since by definition $sx \notin C$ for $s > \frac{1}{p_C(x)}$, we can think of $\frac{1}{p_C(x)}$ as the "escape time from C" as we approach x starting from 0, along the ray $\{tx : t \ge 0\}$.

a convergent subsequence $y_{n_k} \to y \in B$, with $\|y_{n_k} - y\| < \varepsilon_{n_k}$. But this yields $x_{n_k} \to y$ since $\|x_{n_k} - y\| \le \|x_{n_k} - y_{n_k}\| + \|y_{n_k} - y\| < 3\,\varepsilon_{n_k}$, and A being closed forces $y \in A$, so that $y \in A \cap B$. This contradicts the assumption $A \cap B = \emptyset$. Choosing now $\varepsilon > 0$ with $A_\varepsilon \cap B_\varepsilon = \emptyset$, in accordance with the first statement, proved above, we can find some $f \in X^* \setminus \{0\}$, and some constant $\alpha \in \mathbb{R}$ with $f(x) \le \alpha \le f(y)$ for all $x \in A_\varepsilon$ and $y \in B_\varepsilon$. That is, $f(a + \varepsilon x) \le \alpha \le f(b + \varepsilon y)$ for all $a \in A$, $b \in B$, and all $x, y \in B(0, \varepsilon)$. But for every $a \in A$ we have $\sup_{\|x\|\le 1} \{f(a + \varepsilon x)\} = \sup_{\|x\|\le 1} \{f(a) + \varepsilon f(x)\} = f(a) + \varepsilon \sup_{\|x\|\le 1} \{f(x)\} = f(a) + \varepsilon \|f\|_{X^*}$, and, similary, for every $b \in B$, we have $\inf_{\|y\|\le 1} \{f(b + \varepsilon y)\} = f(b) - \varepsilon \|f\|_{X^*}$. Thus $f(a) \le \alpha - \varepsilon \|f\|_{X^*} \le \alpha + \varepsilon \|f\|_{X^*} \le f(b)$ for all $a \in A$ and $b \in B$. □

Since the dual X^* of a Banach space $(X, \|\cdot\|)$ is itself a Banach space (Theorem 3.10), it also has a dual space, denoted by X^{**} and called the *second dual* or *bidual* of the space X. There is a canonical injection $J : X \to X^{**}$ defined as follows: given $x \in X$, the map $f \mapsto f(x)$ is a continuous linear functional on X^*, and therefore an element of X^{**}, denoted by Jx. Note that J is a linear isometry.[71] If J turns out to be surjective,[72] we say that X is *reflexive* and we identify X^{**} with X. Many Banach spaces that are important in applied mathematics are reflexive, e.g. finite-dimensional spaces[73] and Hilbert spaces (use Theorem 3.13 twice), while others are not, e.g. $L^1[0, 1]$ and $C[0, 1]$. To prove that a specific Banach space is reflexive, two approaches proved to be quite successful, even if neither is the ultimate tool:

(i) in many situations,[74] one can show that the range of J exhausts X^{**};
(ii) a *uniformly convex*[75] Banach space is reflexive, but the converse is false; see Brézis (2011).

Due to Theorem 3.8, the best one can hope for is that the closed unit ball in an infinite-dimensional Banach space $(X, \|\cdot\|)$ presents some *weak com-*

71 Indeed, linearity is clear and, since by definition $(Jx)(f) = f(x)$ for all $x \in X$ and $f \in X^*$, we get $\|Jx\|_{X^{**}} = \sup_{f \in X^*, \|f\|_{X^*} \le 1} \{|(Jx)(f)|\} = \sup_{f \in X^*, \|f\|_{X^*} \le 1} \{|f(x)|\} = \|x\|$ by Corollary 5.

72 It is essential to use the canonical injection: James (1951) provides a striking example of a nonreflexive Banach space such that there is a surjective isometry from X to X^{**}.

73 Since X, X^* and X^{**} have the same dimension.

74 This is the case for l^p and $L^p[0, 1]$ with $1 < p < \infty$; see Brézis (2011).

75 We say that $(X, \|\cdot\|)$ is uniformly convex if for every $\varepsilon > 0$ there is some $\delta > 0$ such that $\|\frac{x+y}{2}\| < 1 - \delta$ whenever $\|x\| \le 1$, $\|y\| \le 1$ and $\|x - y\| > \varepsilon$. This is a geometric condition: for example, it ensures that the midpoint of a segment of length exceeding ε, determined by two distinct points on the unit sphere $S = \{x \in X : \|x\| = 1\}$, lies within a ball centred at 0 and of radius $(1 - \delta)$ for some $\delta > 0$. It is a striking fact that reflexivity, a topological property (a reflexive space remains so for an equivalent norm), is a consequence of a geometric property of a specific norm. Note that an equivalent norm need not be uniformly convex, as illustrated by the case of $\mathbb{R}^2$ with the Euclidean norm and the equivalent norm $\|(x_1, x_2)\| = \max\{|x_1|, |x_2|\}$: in the latter case, the unit sphere is the square with corners at the points $(\pm 1, 0)$ and $(0, \pm 1)$. Infinite-dimensional examples of uniformly convex Banach spaces are the $L^p(\Omega)$-spaces for $1 < p < \infty$, where Ω is an open subset of $\mathbb{R}^N$ for some integer $N \ge 1$ (with the Lebesgue measure); see Brézis (2011). Also, the Bergman spaces B^p with $1 < p < \infty$ are uniformly convex, see Zhu (2007). Note that there exist reflexive Banach spaces that admit no uniformly convex equivalent norm, see Day (1941).

pactness property. To elaborate this, let us call a sequence $\{x_n\}_{n\geq 1}$ in X *weakly convergent* to $x \in X$, denoted by $x_n \rightharpoonup x$, if $f(x_n) \to f(x)$ for all $f \in X^*$. Corollary 2 shows that the weak limit is uniquely determined. Since $|f(x_n) - f(x)| \leq \|f\|_{X^*} \|x_n - x\|$, we see that a convergent sequence is weakly convergent. The converse, however, is not true in general.[76] However, anticipating the considerations from Section 3.4.3, a weakly convergent sequence $\{x_n\}_{n\geq 1}$ is always bounded[77] (see Exercise 43). In this context, we draw attention to the following useful characterisation of reflexivity.

Theorem 3.18 (Eberlein–Shmulyan) *A Banach space is reflexive if and only if every bounded sequence admits a weakly convergent subsequence.*

We omit the rather involved proof of Theorem 3.18, for which we refer to Yosida (1995). Instead, we present a corollary that explains the reason why reflexive spaces and convex functions play a central role in the calculus of variations and in optimisation.

Theorem 3.19 *Let $K \subset X$ be a nonempty, closed, convex set in a reflexive Banach space $(X, \|\cdot\|)$. Then a convex lower-semicontinuous function*[78] φ : $K \to \mathbb{R}$ *such that* $\lim_{x\in K:\ \|x\|\to\infty} \varphi(x) = \infty$ *attains its infimum on* K.

Proof Choose some $x \in K$ and consider the closed, convex and bounded set $K_0 = \{y \in K : \varphi(y) \leq \varphi(x)\}$. By Theorem 3.18, if $y_n \in K_0$ are such that $\lim_{n\to\infty} \varphi(y_n) = \inf_{y\in K_0} \varphi(y)$, then we can find a subsequence $\{y_{n_k}\}$ that converges weakly to some $x_0 \in X$. We claim that $x_0 \in K_0$. Indeed, if $x_0 \notin K_0$, then by Theorem 3.17 there exists a closed hyperplane that separates K_0 and x_0 strictly, contradicting the weak convergence of $\{y_{n_k}\}$ to x_0. Thus $x_0 \in K_0$ and $\varphi(x_0) = \inf_{y\in K_0} \varphi(y) = \inf_{y\in K} \varphi(y)$. □

For the dual X^* of a Banach space, in addition to convergence and weak convergence, we can also define the weak*-convergence:[79] the sequence $\{f_n\}_{n\geq 1}$ converges weak* to f in X^* if[80] $f_n(x) \to f(x)$ for all $x \in X$. This third type

76 In finite-dimensional spaces weak convergence coincides with convergence: one can model the Banach space by $\mathbb{R}^n$ or $\mathbb{C}^n$, so that convergence is equivalent to convergence on components, that is, under the n functionals given by the projections on the coordinates. Usually, in infinite-dimensional spaces there exist sequences that converge weakly and do not converge. For example, see Theorem 3.1, an orthonormal basis $\{e_n\}_{n\geq 1}$ of a separable Hilbert space is weakly convergent to 0 but fails to be convergent by not being a Cauchy sequence. However, l^1 is a pathological example, having the property that every weakly convergent sequence actually converges, see Brézis (2011).

77 As a consequence of the uniform boundedness principle (Theorem 3.22).

78 This means that $\varphi(tx + (1-t)y) \leq t\varphi(x) + (1-t)\varphi(y)$ for all $x, y \in K$ and all $t \in [0, 1]$, and, if $x_n \to x$ in K, then $\liminf_{n\to\infty} \varphi(x_n) \geq \varphi(x)$.

79 The superscript * emphasises that this type of convergence is specific to dual spaces. As pointed out in Section 3.4.1, not all Banach spaces arise this way.

80 Clearly weak*-limits, if they exist, are unique.

of convergence is even weaker than weak convergence, with the two notions coinciding for reflexive spaces X.

The canonical injection $J : X \to X^{**}$ ensures that weakly convergent sequences in X^* are weak*-convergent. An example of a sequence that is weak*-convergent to 0 but fails to be weakly convergent is the sequence $e_n(k) = \delta_{nk}$ for $n, k \geq 1$ in l^1: since $l_1 = c_0^*$, for every $x \in c_0$ we have $e_n(x) = \sum_{k\geq 1} x(k)\,\overline{e_n(k)} = x(n) \to 0$, while in l^1 weakly convergent sequences are convergent and $\{e_n\}_{n\geq 1}$ fails to be a Cauchy sequence since $\|e_n - e_m\|_{l^1} = 2$ for $n \neq m$.

The rationale for introducing the new concept is the following result (see Yosida (1995) for the proof).

Theorem 3.20 (Banach–Alaoglu) *Any sequence in the closed unit ball $B = \{x \in X^* : \|x\|_{X^*} \leq 1\}$ of the separable*[81] *dual X^* of a Banach space X admits a weak*-convergent subsequence.*

The importance of weak*-convergence lies in the role played by compact sets in existence results for minimisation problems in nonreflexive Banach spaces (like $L^\infty[0,1]$). One can view weak*-convergence in X^* as a "pointwise convergence" of the operators. This is especially useful in function spaces X for which pointwise evaluations $f \mapsto f(t)$ are continuous linear functionals[82] for every t in the domain of definition of $f \in X$. In contrast to this, note that point evaluations are not well-defined on the $L^p(U)$-spaces with $1 < p \leq \infty$, where $U \subset \mathbb{R}^n$ is an open, bounded subset (with the Lebesgue measure). For $1 < p < \infty$ these spaces are reflexive, so that bounded sets are weakly compact, see Theorem 3.18, while for $p = \infty$, Theorem 3.20 guarantees the weak*-compactnes of bounded sets. These facts are quite useful in approximation problems. However, these weak compactness results are false if $p = 1$: since $L^1(U)$ is not the dual of a Banach space, the concept of weak*-convergence is not applicable, and there are bounded sequences that do not admit weakly convergent subsequences.[83] This failure is offset by regarding $L^1(U)$ in the natural way[84] as a subset of the space $M(U)$ of complex-valued regular Borel mea-

[81] Separability is the necessary and sufficient condition for the weak*-topology to be induced by a metric, thus allowing the characterisation of convergence and compactness by means of sequences. If X^* is not separable, then B is compact in the nonmetrisable weak*-topology. For these matters we refer to Brézis (2011). We recall from Theorem 3.16 that if X^* is separable, then X is separable, but the converse is not necessarily the case. On the other hand, one can show that X is reflexive if and only if X^* is reflexive, see Brézis (2011).

[82] This is the case for $C[0,1]$, see Exercise 43, and for the Bergman spaces – see Zhu (2007).

[83] For example, define $f_n : (-1,1) \to \mathbb{C}$ for $n \geq 1$ by $f_n(t) = n$ if $|t| < \frac{1}{2n}$ and $f_n(t) = 0$ otherwise. Then $\|f_n\|_{L^1(-1,1)} = 1$ for all $n \geq 1$, but no subsequence converges weakly. Indeed, $f_{n_k} \rightharpoonup f$ in $L^1(-1,1)$ means $\int_{-1}^1 f_{n_k}(t)g(t)\,dt \to \int_{-1}^1 f(t)g(t)\,dt$ for all $g \in L^\infty(-1,1)$, and taking g to be the characteristic function of open intervals (a,b) with $0 \notin (a,b)$ yields $\int_a^b f(t)\,dt = 0$, so that $f = 0$ a.e. by absolute continuity. But $1 = \int_{-1}^1 f_{n_k}(t)g(t)\,dt \not\to 0$ if $g \equiv 1$.

[84] That is, $f(x)\,dm(x)$ is a measure if $f \in L^1(U)$; here m is the Lebesgue measure.

sures μ on U, with finite mass,[85] which is the dual of $C_0(U)$. Thus Theorem 3.20 takes on the form of a weak compactness result for measures;[86] for further insight see Evans (1990). In particular, Theorem 3.20 yields that if $\{f_n\}_{n\geq 1}$ is a bounded sequence in $L^1(U)$, then there exists a subsequence $\{f_{n_k}\}$ and a regular measure $\mu \in \mathcal{M}(U)$ such that $\lim_{n_k\to\infty} \int_U f_{n_k}(t)g(t)\,dt = \int_U g(t)\,d\mu(t)$ for all $g \in C_0(U)$. For example, for the previously considered bounded sequence $\{f_n\}_{n\geq 1}$ in $L^1(-1,1)$ with no weakly convergent subsequence, regarding $L^1(-1,1)$ as a subset of $\mathcal{M}(-1,1)$, the sequence $\{f_n\}_{n\geq 1}$ is weak*-convergent to the unit mass at the origin.[87]

3.4.3 Baire category and consequences

In this section we will discuss the three basic principles of functional analysis – the principle of uniform boundedness, the open mapping theorem, the closed graph theorem – which rank in importance with the Hahn–Banach Theorem, providing the foundation for many far-reaching results in analysis.

We first show that sets with a nonempty interior are of interest.

Proposition *Let X and Y be normed linear spaces. Then a linear operator*

[85] On a locally compact metric space K, meaning that for every element there is an open set containing it and having compact closure, the Borel σ-algebra $\mathcal{B}_K$ is the smallest σ-algebra that contains all open sets of K, and a positive measure μ_+ on $(K, \mathcal{B}_K)$ is called regular if $\mu_+(K_0) < \infty$ for every compact subset K_0 of K and if $\mu_+(B) = \sup\{\mu_+(K_0) : K_0 \subset B \text{ and } K_0 \text{ compact}\} = \inf\{\mu_+(O) : B \subset O \text{ and } O \text{ open}\}$ for all $B \in \mathcal{B}_K$, while a complex-valued measure $\mu = \mu_1 - \mu_2 + i(\mu_3 - \mu_4)$ on $(K, \mathcal{B}_K)$, with μ_i positive regular measures, is called regular if each μ_i has finite total mass. The variation of μ, $\|\mu\| = \sum_{j=1}^4 \mu_j(K)$, makes the family of all complex-valued regular measures on K a normed linear space $\mathcal{M}(K)$, that can be identified with the dual of the Banach space $C_0(K)$ of all continuous functions $f : K \to \mathbb{C}$ such that for all $\varepsilon > 0$ the set $\{t \in K : |f(t)| \geq \varepsilon\}$ is compact, endowed with the norm $\|f\|_\infty = \sup_{t\in K} |f(t)|$; see the discussion in Conway (1990). $C_0(K)$ can be obtained as the completion of the normed linear space of continuous functions $w : K \to \mathbb{C}$ that vanish outside a compact set $K_0 \subset K$, with the supremum norm; see Problem 33 for the case $K = \mathbb{R}$. Note that if K is not compact, then $\mathcal{M}(K)$ is not the dual of the non-separable Banach space $C_b(K)$ of all bounded continuous functions $f : K \to \mathbb{C}$, with the supremum norm, the dual of $C_b(K)$ being the space $rba(K)$ of all regular bounded finitely additive measures on K, see Aliprantis and Burkinshaw (1999). In this context, regular means that for every $B \in \mathcal{B}_K$ and every $\varepsilon > 0$ there exists a closed set $F \subset B$ and an open set $O \supset B$ such that $|\mu(C)| < \varepsilon$ for every $C \subset O \setminus F$ in $\mathcal{B}_K$. The proof that $C_b(K)$ is a Banach space is analogous to that for $C[0,1]$, discussed in Exercise 31. To see that $C_b(\mathbb{R})$ is not separable, define $f_t(x) = e^{itx}$ for $t \in \mathbb{R}$. Since $f_t(x) - f_s(x) = 2i\, e^{i\frac{t+s}{2}x} \sin(\frac{t-s}{2}x)$ yields $\|f_t - f_s\|_\infty = 2$ for $t \neq s$, we infer non-separability (see the end of the solution to Exercise 28). In contrast to this, if K can be covered by countably many compact sets, then $C_0(K)$ is separable, see Brézis (2011).

[86] The corresponding notion of weak*-convergence is referred to in probability theory as *convergence in distribution (or in law)*.

[87] Since the regular measures $\mu_n \in \mathcal{M}(-1,1)$ corresponding to f_n act on $g \in C_0(-1,1)$ by $\mu_n(g) = n\int_{-1/(2n)}^{1/(2n)} g(t)\,dt = \int_{-1/2}^{1/2} g(\frac{s}{n})\,ds \to g(0)$.

$T : X \to Y$ is bounded if and only if the pre-image of the unit closed ball $B = \{y \in Y : \|y\|_Y \le 1\}$ has nonempty interior.

Proof If T is bounded, let $\|T\| = M < \infty$ and note that $T^{-1}(B)$ contains the open set $\{x \in X : \|x\|_X < \frac{1}{M}\}$ since $\|Tx\|_Y \le \|T\| \cdot \|x\|_X < 1$ if $\|x\|_X < \frac{1}{M}$. Conversely, if $T^{-1}(B)$ contains an open ball $B_\varepsilon(x_0) = \{x \in X : \|x - x_0\|_X < \varepsilon\}$ for some $\varepsilon > 0$, then $\|Tx\|_Y \le \|T(x + x_0)\|_Y + \|Tx_0\|_Y \le 1 + \|Tx_0\|_Y$ whenever $\|x\|_x < \varepsilon$. Since for $x \neq 0$ we have $\frac{\varepsilon}{2\|x\|_X} Tx = T(\frac{\varepsilon x}{2\|x\|_X})$, scaling yields $\|Tx\|_Y \le \frac{2(\|Tx_0\|_Y+1)}{\varepsilon} \|x\|_X$ for all $x \in X$, so that T is bounded. □

A set S in a metric space X is called *nowhere dense* if its closure $\overline{S}$ has an empty interior, that is, contains no open sets.

Theorem 3.21 (Baire category) *A complete metric space (X, d) is never a countable union of nowhere dense sets.*

Proof The idea is to show that if $X = \bigcup_{n\ge1} A_n$ with each A_n nowhere dense, then we can construct a Cauchy sequence $\{x_k\}_{k\ge1}$ such that its limit cannot belong to any A_n, which is impossible. Indeed, assume that $X = \bigcup_{n\ge1} A_n$ with each A_n nowhere dense. Since A_1 is nowhere dense, there is some $x_1 \notin \overline{A_1}$. There is an open ball B_1, centred at x_1, such that $B_1 \cap A_1 = \emptyset$, since otherwise we could find a sequence in A_1 that converges to x_1. Since A_2 is nowhere dense, there is some x_2 in the open set $B_1 \setminus \overline{A_2}$. Let B_2 be an open ball, centred at x_2 and of radius less than $\frac{1}{2}$, such that $\overline{B_2} \subset B_1$ and $B_2 \cap \overline{A_2} = \emptyset$. We construct inductively the open set B_k and choose $x_{k+1} \in B_k \setminus \overline{A_{k+1}}$ and an open ball B_{k+1} of radius less than 2^{-k}, centred at x_{k+1}, with $\overline{B_{k+1}} \subset B_k$ and $B_{k+1} \cap \overline{A_{k+1}} = \emptyset$. Then $\{x_k\}_{k\ge1}$ is a Cauchy sequence since, by construction, $x_n, x_m \in B_N$ if $n, m \ge N$, which entails $d(x_n, x_m) \le d(x_n, x_N) + d(x_n, x_m) \le 2^{1-N} + 2^{1-N} = 2^{2-N} \to 0$ as $N \to \infty$. Let $x = \lim_{n\to\infty} x_n$. But $x_n \in B_N$ for $n \ge N$ yields $x \in \overline{B_N} \subset B_{N-1}$ for all $N \ge 2$, so that $x \notin A_{N-1}$ for any $N \ge 2$, contradicting the fact that $X = \bigcup_{n\ge1} A_n$. □

Theorem 3.21 is often used in the following form: if a nonempty complete metric space is the countable union of closed sets, then at least one of these closed sets has nonempty interior. It is instrumental in proving the following remarkable result that permits one to derive global (uniform) estimates from pointwise estimates.

Theorem 3.22 (Uniform boundedness principle) *Let X be a Banach space and Y be a normed linear space. If $L_i : X \to Y$, $i \in I$, is a family (not necessarily countable) of bounded linear maps such that for each $x \in X$, the set $\{\|L_i x\|_Y\}_{i\in I}$ is bounded, then $\{\|L_i\|\}_{i\in I}$ is bounded.*

Proof Consider the closed sets $A_n = \{x \in X : \|L_i x\|_Y \le n$ for all $i \in I\}$. By hypothesis, $X = \bigcup_{n\ge1} A_n$, so that for some $N \ge 1$, A_N has a nonempty interior. Thus there are $x_0 \in A_N$ and $\varepsilon > 0$ such that $\{x \in X : \|x - x_0\| < \varepsilon\} \subset A_N$. For every $i \in I$, we get that $\|L_i x\|_Y \le \|L_i(x - x_0)\|_Y + \|L_i x_0\|_Y \le 2N$ whenever $\|x\|_X < \varepsilon$. Since for $x \neq 0$ we have $L_i x = \frac{2\|x\|_X}{\varepsilon} L_i \hat{x}$ with $\hat{x} = \frac{\varepsilon x}{2\|x\|_X}$ such that $\|\hat{x}\|_X = \frac{\varepsilon}{2} < \varepsilon$, we deduce that $\|L_i x\|_Y \le \frac{4N}{\varepsilon} \|x\|_X$ for all $x \in X$ and all $i \in I$. □

An immediate consequence of Theorem 3.22 is the following result.

Corollary 1 *Let $T_n : X \to Y$ be bounded linear operators between the Banach* *spaces X and Y. If the pointwise limit* $\lim_{n\to\infty} T_n x \in Y$ *exists for every* $x \in X$, *then* $Tx = \lim_{n\to\infty} T_n x$ *defines a bounded linear operator* $T : X \to Y$, *with* $\|T\| \le \sup_{n\ge 1} \|T_n\| < \infty$.

Linearity is essential in Corollary 1 since a pointwise limit of continuous functions need not be continuous.[88] Note that the general setting of Corollary 1 does not ensure[89] $\lim_{n\to\infty} \|T - T_n\|_{\mathcal{B}(X,Y)} = 0$, even if $\{T_n\}_{n\ge 1}$ converges uniformly to T on compact sets $K \subset X$, that is,[90] $\lim_{n\to\infty} \sup_{x\in K} \|(T - T_n)x\|_Y = 0$.

Another useful direct consequence of Theorem 3.22 is the following characterisation of bounded sets.[91]

Corollary 2 *A set M in a Banach space X is bounded if and only its image under any bounded linear functional* $f \in X^*$ *is bounded (in* $\mathbb{R}$ *or in* $\mathbb{C}$*).*

Proof Since $|f(x)| \le \|f\|_{X^*} \|x\|$, one direction is clear. Conversely, define for every $x \in M$ the bounded linear map $T_x : X^* \to \mathbb{C}$ by $T_x f = f(x)$. By assumption, $\sup_{x\in M} |f(x)| < \infty$ for every fixed $f \in X^*$. It follows from Theorem 3.22 that there exists a constant $c > 0$ such that $|f(x)| \le c\|f\|_{X^*}$ for all $x \in M$ and all $f \in X^*$. Corollary 5 from Section 3.4.2 yields $\|x\|_X \le c$ for all $x \in M$. □

The image of a bounded linear operator between Banach spaces might be a closed set.[92] However, if the image happens to be open,[93] then open sets are mapped into open sets.

Theorem 3.23 (Open mapping theorem) *Let L be a bounded linear map from a Banach space X onto a Banach space Y. Then the image L(O) of an open set* $O \subset X$ *is open in Y.*

Proof We claim that the translation property $L(x + y) = Lx + Ly$ and the fact that unions of open sets are open show that it suffices to prove that $L(O_s)$ has nonempty interior for some $s > 0$; here $O_s = \{x \in X : \|x\|_X < s\}$. Indeed, if $\{y \in Y : \|y - y_0\|_Y < \delta\} \subset L(B_s)$, then for $y \in Y$ with $\|y\|_Y < \delta$ we have $y_0 + y \in L(B_s)$, so that there is some $x \in X$ with $\|x\|_X < s$ and $Lx = y_0 + y$. Applying this

[88] For example, look at $f_n(t) = (1 - t)^n$ for $t \in [0, 1]$.

[89] For example, if $\{e_k\}_{k\ge 1}$ is an orthonormal basis of the Hilbert space H and we define T_n for $n \ge 1$ by $T_n x = \sum_{k=1}^n x_k e_k$ for $x = \sum_{k\ge 1} x_k e_k$, then $\lim_{n\to\infty} T_n x = x$ for every $x \in H$, so that T is the identity. However, $\|T - T_n\|_{\mathcal{L}(H)} = 1$ for all $n \ge 1$.

[90] Let $M = \sup_{n\ge 1} \|T_n\|_{\mathcal{B}(X,Y)} < \infty$. Given $\varepsilon > 0$, we have $K \subset \bigcup_{j=1}^m \{x \in X : \|x - x_j\|_X < \varepsilon\}$ for some $x_1, \dots, x_m \in K$. Since $\lim_{n\to\infty} \|(T - T_n)x_j\|_Y = 0$ for $1 \le j \le m$, there is some $N \ge 1$ with $\|(T - T_n)x_j\|_Y < \varepsilon$ for $1 \le j \le m$ and $n \ge N$. Now, for every $x \in K$, choose x_j with $\|x - x_j\|_X < \varepsilon$ to get $\|(T - T_n)x\|_Y \le \|(T - T_n)x_j\|_Y + \|(T - T_n)(x - x_j)\|_Y \le \varepsilon + (\|T\|_{\mathcal{B}(X,Y)} + \|T_n\|_{\mathcal{B}(X,Y)})\|x - x_j\|_X \le \varepsilon(1 + 2M)$ for $n \ge N$.

[91] The procedure for checking boundedness in finite-dimensional spaces consists of verifying that all projections on the axes are bounded. This amounts to considering the linear functionals that represent the components with respect to some basis. Viewed this way, Corollary 2 replaces the use of components in infinite-dimensional spaces. In this context, Exercise 19 highlights the disadvantage of components for Hamel bases in infinite-dimensional spaces.

[92] For example, this is the case for the projection on a closed subspace of a Hilbert space H.

[93] In this case, since 0 lies in the image, by scaling we see that the operator must be surjective.

to $y = 0$ ensures the existence of some $x_0 \in X$ with $\|x_0\|_X < s$ and $Lx_0 = y_0$. Thus $y = L(x - x_0)$ and $\|x - x_0\|_X < 2s$, which means that $\{y \in Y : \|y\|_Y < \delta\} \subset L(B_{2s})$. Translation and scaling now yield that the image of the open ball of radius ε, centred at some $x^* \in X$, contains the open ball of radius $\frac{\delta\varepsilon}{2s}$, centred at $Lx^* \in Y$. This proves our claim.

Since L is onto, $Y = \bigcup_{n\geq 1} L(O_n)$, and Theorem 3.21 ensures that at least one of the sets $\overline{L(O_n)}$, say $\overline{L(O_N)}$, has nonempty interior. Going through the previous considerations, we see that $O_{\varepsilon N} \subset \overline{L(O_N)}$ for some $\varepsilon > 0$, and scaling yields $O_\varepsilon \subset \overline{L(O_1)}$. We now show that $\overline{L(O_1)} \subset L(O_2)$, which proves the statement. Choose $y \in \overline{L(O_1)}$ and pick $x_1 \in O_1$ with $(y - Lx_1) \in O_{\varepsilon/2} \subset \overline{L(O_{1/2})}$. Now pick $x_2 \in O_{1/2}$ with $(y - Lx_1) - Lx_2 \in O_{\varepsilon/4} \subset \overline{L(O_{1/4})}$ and, inductively, choose $x_n \in O_{2^{1-n}}$ with $y - \sum_{k=0}^{n} Lx_k \in O_{\varepsilon 2^{-n}} \subset \overline{L(O_{2^{-n}})}$. Then $\sum_{k\geq 1} x_k = x \in O_2$ since $\|x\|_X \leq \sum_{k\geq 1} \|x_k\|_X \leq 1 + \frac{1}{2} + \frac{1}{2^2} + \cdots = 2$, and $y = \sum_{k\geq 1} Lx_k = Lx \in L(O_2)$. □

The importance of Theorem 3.23 lies in that for invertible bounded linear maps, being open translates into the continuity of the inverse map:

Theorem 3.24 (Inverse mapping theorem) *A continuous linear bijection L from a Banach space onto another Banach space has a continuous inverse.*

Proof L open means that the inverse L^{-1} is continuous. Indeed, continuity is equivalent to the condition that pre-images of open sets are open.[94] □

The *graph* of a linear map L between two Banach spaces X and Y is the set $\Gamma(L) = \{(x, y) \in X \times Y : y = Lx\}$. The next result gives an elegant characterisation of continuity of L in terms of its graph $\Gamma(L)$.

Theorem 3.25 (Closed graph theorem) *A linear map $L : X \to Y$ between two Banach spaces X and Y is bounded if and only if $\Gamma(L)$ is closed.*

Proof If L is bounded and $(x_n, y_n) \in \Gamma(L)$ converge in the Banach space $X \times Y$, endowed with the norm $|||(x, y)||| = \|x\|_X + \|y\|_Y$, to (x, y), then $(x, y) \in \Gamma(L)$ since by continuity $Lx_n \to Lx$ in Y. Thus $\Gamma(L)$ is closed.

Conversely, assume that $\Gamma(L)$ is closed in $X \times Y$. Since L is linear, $\Gamma(L)$ is a linear space. Endow it with the norm $\|(x, Lx)\| = \|x\|_X + \|Lx\|_Y$ for $(x, Lx) \in \Gamma(L)$. We claim that $(\Gamma(L), \|\cdot\|)$ is a Banach space. Indeed, it is clearly a normed linear space and for a Cauchy sequence $\{(x_n, Lx_n)\}_{n\geq 1}$ in $\Gamma(L)$ we get that $\{x_n\}_{n\geq 1}$ and $\{Lx_n\}_{n\geq 1}$ must be Cauchy sequences in X and Y, respectively. Both must converge. If $x_n \to x$ and $y_n \to y$, then $y = Lx$ since $\Gamma(L)$ is a closet subset of $X \times Y$. Thus $\|(x_n, y_n) - (x, Lx)\| \to 0$, proving that $(\Gamma(L), \|\cdot\|)$ is a Banach space. Define now the continuous maps $\Theta_1 : \Gamma(L) \to X$ and $\Theta_2 : \Gamma(L) \to Y$ by $\Theta_1(x, Lx) = x$ and $\Theta_2(x, Lx) = Lx$ for $(x, Lx) \in \Gamma(L)$. Note that $L = \Theta_2 \circ \Theta_1^{-1}$. Since Θ_1 is a continuous bijection, Θ_1^{-1} will be continuous by the inverse mapping theorem, so L is continuous. □

To appreciate the strength of Theorem 3.25, note that L continuous requires us to show that (i) implies (ii) as well as (iii), the statements being (i) "$x_n \to x$ in X", (ii) "Lx_n converges in Y to some y", (iii) "$Lx = y$". By Theorem 3.25, it suffices to show that (i) and (ii) imply (iii).

[94] This being the $\varepsilon - \delta$ definition of continuity of a map f between two metric spaces (X, d) and (Y, ρ): for any $x_0 \in X$, given $\varepsilon > 0$, there is $\delta > 0$ such that $\rho(f(x), f(x_0)) < \varepsilon$ if $d(x, x_0) < \delta$. The equivalence to the definition $\rho(f(x_n), f(x)) \to 0$ if $d(x_n, x) \to 0$ is clear.

3.4.4 The spectral theorem

The case of operators acting on a finite-dimensional space (matrix theory) provides ample justification for the expectation that the spectrum of an operator is of fundamental importance in uncovering structural properties. In a general setting the task of obtaining information even of a qualitative nature about the spectrum of a bounded linear operator is mostly unassailable but for self-adjoint compact operators acting on a Hilbert space this is often possible. Moreover, according to the spectral theorem, such operators can be diagonalised by choice of a suitable orthonormal basis.[95] The aim of this section is to discuss this major achievement of linear operator theory.[96] Throughout this section H will be a Hilbert space with scalar product $(\cdot,\cdot)$.

Theorem 3.26 *Given a bounded linear operator $T : H \to H$, the relation*

$$(Tx, y) = (x, T^*y) \quad \textit{for all} \quad x, y \in H, \tag{3.10}$$

defines a unique bounded linear operator $T^* : H \to H$, called the adjoint of T, satisfying $\|T^*\| = \|T\|$ and $T^{**} = T$.

Proof The existence of T^* is a consequence of the Riesz representation theorem (Theorem 3.13): given $y \in H$, $x \mapsto \overline{(Tx,y)}$ is a bounded linear functional on H as $|(Tx,y)| \le \|T\| \cdot \|x\| \cdot \|y\|$, so that there is a unique $y^* \in H$ with $\overline{(Tx,y)} = (y^*, x) = \overline{(x, y^*)}$, which defines $T^*y = y^*$. Using (3.10), it is clear that T^* is linear. Moreover, replacing T by T^* in (3.10) allows us to define T^{**}: $(T^*x, y) = (x, T^{**}y)$ for all $x, y \in H$. Note that $T^{**} = T$ since (3.10) yields $(T^*x, y) = \overline{(y, T^*x)} = \overline{(Ty, x)} = (x, Ty)$. We have $\sup_{\|y\|=1}\{\|T^*y\|\} = \|T^*\| \le \|T\|$, since $\|T^*y\|^2 = (T^*y, T^*y) = (TT^*y, y) \le \|T\| \cdot \|T^*y\| \cdot \|y\|$. Replacing T by T^*, we get $\|T\| \le \|T^*\|$, so that $\|T\| = \|T^*\|$. □

Examples of adjoint operators are:

- On $H = \mathbb{C}^n$ with the inner product $(x,y) = \sum_{k=1}^n x_k \overline{y_k}$, a linear operator $T : H \to H$ is represented by a $n \times n$ matrix with complex entries $t_{ij} = (Te_i, e_j)$, while T^* is represented by a $n \times n$ matrix with complex entries $t^*_{ij} = (T^*e_i, e_j)$. Note that $(Tx, y) = \left(\sum_{i=1}^n x_i Te_i, \sum_{j=1}^n y_j e_j\right) = \sum_{i=1}^n \sum_{j=1}^n t_{ij} x_i \overline{y_j}$, while $(x, T^*y) = \left(\sum_{i=1}^n x_i e_i, \sum_{j=1}^n y_j T^* e_j\right) = \sum_{i=1}^n \sum_{j=1}^n \overline{t^*_{ji}}\, x_i \overline{y_j}$. Choosing $x = e_q$ and $y = e_p$ with $1 \le q, p \le n$ yields $t_{qp} = \overline{t^*_{pq}}$. Since these n^2 relations clearly validate (3.10), we see that T^* corresponds to the conjugate transpose matrix of T. On the real Hilbert space $H = \mathbb{R}^n$, these considerations show that T^* corresponds to the transpose matrix of T.
- Let $K : L^2[0,1] \to L^2[0,1]$ be the so-called *Hilbert–Schmidt operator* with integral kernel $k \in L^2([0,1] \times [0,1])$, defined by $(Kf)(t) = \int_0^1 k(t,s) f(s)\, ds$ for $f \in L^2[0,1]$. Then K is bounded, with[97] $(K^*g)(t) = \int_0^1 \overline{k(s,t)}\, g(s)\, ds$ for $g \in L^2[0,1]$.

[95] This is a far-reaching generalisation of the finite-dimensional result that Hermitian matrices can be diagonalised.

[96] For the case of unbounded self-adjoint operators see the discussion in Part II.

[97] Indeed, $\|Kf\|^2_{L^2[0,1]} = \int_0^1 \left| \int_0^1 k(t,s) f(s)\, ds \right|^2 dt \le \int_0^1 \left(\int_0^1 |k(t,s)|^2\, ds \right) \cdot \left(\int_0^1 |f(s)|^2\, ds \right) dt = \|f\|^2_{L^2[0,1]} \int_0^1 \int_0^1 |k(t,s)|^2\, ds dt = \|f\|^2_{L^2[0,1]} \|k\|^2_{L^2([0,1]\times[0,1])}$. By Fubini's theorem, $(Kf, g)_{L^2[0,1]} = \int_0^1 \left(\int_0^1 k(t,s) f(s)\, ds \right) \overline{g(t)}\, dt = \int_0^1 f(s) \left(\overline{\int_0^1 \overline{k(t,s)}\, g(t)\, dt} \right) ds = (f, K^*g)_{L^2[0,1]}$.

- To $f \in L^\infty(\Omega,\mu)$ associate the multiplication operator $M_f : L^2(\Omega,\mu) \to L^2(\Omega,\mu)$, defined by $x(t) \mapsto x(t)f(t)$ for $t \in [0,1]$. M_f is a linear operator with $\|M_f\| \le \|f\|_\infty$ and $(M_f)^* = M_{\overline{f}}$.

The adjoint is useful in providing criteria for the solvability of the equation $Tx = y$ for operators T with a closed range $\mathcal{R}(T)$, in analogy to the finite-dimensional matrix case, as shown by the following result.

Theorem 3.27 *Let $T : H \to H$ be a bounded linear operator. Then*[98] $\overline{\mathcal{R}(T)} = \{ker(T^*)\}^\perp$ *and* $\{ker(T)\}^\perp = \overline{\mathcal{R}(T^*)}$.

Proof Since $T^{**} = T$, it suffices to prove only the first relation. Note that $z \in \{\mathcal{R}(T)\}^\perp$ means $(z, Tx) = 0$ for all $x \in H$, or $(T^*z, x) = 0$ for all $x \in H$, that is, $z \in \ker(T)$. The statement follows since $\{\mathcal{R}(T)\}^{\perp\perp} = \overline{\mathcal{R}(T)}$, see Exercise 10. □

Let us now introduce some important classes of operators.

- A linear operator $U : H \to H$ is called *unitary*[99] if $\mathcal{R}(U) = H$ and U preserves the inner product, that is, $(Ux, Uy) = (x, y)$ for all $x, y \in H$.
- A bounded linear operator $T : H \to H$ is called *self-adjoint* if $T = T^*$.
- A linear operator $K : H \to H$ is called *completely continuous*[100] or *compact* if it maps bounded sets into sets with a compact closure.

Unitary operators preserve the full structure of Hilbert spaces (linear, topological, metric and geometric structure), self-adjoint operators are quite common in mathematical physics, while compact operators present features that resemble to those familiar from matrix theory. We will show that compact self-adjoint operators have a transparent spectral structure. As for unitary operators, we can characterise them easily as follows.

Theorem 3.28 *A bounded linear operator $U : H \to H$ is unitary if and only if*[101] $UU^* = U^*U = I$.

Proof If U is unitary, then $(U^*Ux, y) = (Ux, Uy) = (x, y)$ for all $x, y \in H$ yields $U^*U = I$, and similarly $UU^* = I$. Conversely, $UU^* = I$ ensures $\mathcal{R}(U) = H$, while from $U^*U = I$ we get $(x, y) = (U^*Ux, y) = (Ux, Uy)$ for all $x, y \in H$. □

98 Here $\ker(S)$ and $\mathcal{R}(S)$ denote the null space and the range of $S : H \to H$, respectively, defined by $\ker(S) = \{x \in H : Sx = 0\}$ and $\mathcal{R}(S) = \{y \in H : y = Sx$ for some $x \in H\}$. Note that if the linear operator S is bounded, then $\ker(S)$ is a closed subspace, see Exercise 18(i), but $\mathcal{R}(S)$ is not necessarily closed, see Exercise 58.

99 If H is finite-dimensional, then $\mathcal{R}(U) = H$ follows from the preservation of the inner product but this is not guaranteed in infinite-dimensional settings, see Exercise 52. Note that $(Ux, Ux) = (x, x)$ yields $\|U\| = 1$.

100 This terminology is justified by the fact that such operators are always continuous. Indeed, otherwise there exists a sequence $\{x_n\}_{n\ge1}$ with $\|x_n\| \le 1$ and $\|Kx_n\| \ge n$ for all $n \ge 1$, and the set $\bigcup_{n\ge1}\{Kx_n\}$ fails to have a compact closure, being unbounded.

101 I stands for the identity $Ix = x$.

From linear algebra we know that if $H = \mathbb{C}^n$ or $H = \mathbb{R}^n$, then a linear operator is unitary if and only if the columns (or, alternatively, the rows) of the corresponding $n \times n$ matrix[102] form an orthonormal basis of H. In Chapter 6 we will see that the Fourier transform is a unitary map on $L^2(\mathbb{R})$.

Self-adjoint operators on finite-dimensional Hilbert spaces correspond to Hermitian matrices (square matrices that are equal to their conjugate transpose). On the other hand, on any Hilbert space there exists a special class of self-adjoint operators, called orthogonal projections: given a closed subspace M of H, by Theorem 3.3, any $x \in H$ can be written uniquely as $x = m + n$ with $m \in M$ and $n \in M^\perp$, the operator $P : H \to H$ defined by $P(m + n) = m$ being called *the orthogonal projection* onto M. The name stems from the fact that $\ker(P) = M^\perp$ is orthogonal to the range $\mathcal{R}(P) = M$, while self-adjointness follows at once from the chain of equalities $(Px, y) = (m, m_1 + n_1) = (m, m_1) = (m + n, m_1) = (x, Py)$, valid for $x = m + n$, $y = m_1 + n_1$ with $m, m_1 \in M$ and $n, n_1 \in M^\perp$. In general, a *projection* is a bounded linear operator $P : H \to H$ with $P^2 = P$, with self-adjointness characterising the orthogonal projections. Indeed, while[103] the range $\mathcal{R}(P)$ of every projection P is closed and every $x \in H$ can be written uniquely as $x = u + v$ with $u \in \ker(P)$ and $v \in \mathcal{R}(P)$, $P^* = P$ is equivalent[104] to $\mathcal{R}(P) = \{\ker(P)\}^\perp$. Note that the self-adjointness of a projection operator $P \neq 0$ can also be expressed quantitatively: P is orthogonal if and only if $\|P\|_{\mathcal{L}(H)} = 1$.

Indeed, any projection $P \neq 0$ satisfies $\|P\|_{\mathcal{L}(H)} \geq 1$ since $\|P\|_{\mathcal{L}(H)} = \|P^2\|_{\mathcal{L}(H)} \leq \|P\|^2_{\mathcal{L}(H)}$. If P is orthogonal with $\mathcal{R}(P) = M$, then for $x = m + n$ with $m \in M$ and $n \in M^\perp$ we have $\|Px\|^2 = \|m\|^2 \leq \|m\|^2 + \|n\|^2 = \|x\|^2$. Thus $\|P\|_{\mathcal{L}(H)} \leq 1$ and, by the above, $\|P\|_{\mathcal{L}(H)} = 1$. On the other hand, if P is not orthogonal, then we can find $m_0 \in \mathcal{R}(P)$ and $n_0 \in \ker(P)$ with $(m_0, n_0) \neq 0$. Multipliying m_0 by a suitable scalar, we may assume that $(m_0, n_0) = \|n_0\|^2$. But then for $x = m_0 - n_0$ we have that $\|Px\|^2 = \|m_0\|^2 > \|m_0\|^2 - \|n_0\|^2 = \|x\|^2$, so that $\|P\|_{\mathcal{L}(H)} > 1$. For example, on $\mathbb{C}^2$, the operator P corresponding to the matrix $\begin{pmatrix} 0 & 1 \\ 0 & 1 \end{pmatrix}$ is a nonorthogonal projection with $\|P\|_{\mathcal{L}(H)} = \sqrt{2}$.

Due to Theorem 3.8, on any Hilbert space a bounded linear operator with a finite-dimensional range will be compact. Before providing more intricate examples we will first elucidate some basic properties of the class of compact operators.

[102] In the infinite-dimensional setting one can associate an infinite matrix to a bounded operator acting on a separable Hilbert space. Although this representation is helpful in the study of some specific classes of operators, usually it is more convenient to deal with an operator itself than with its matrix since the cumbersome computations with infinite matrices often overshadow the ongoing process and properties of operator (such as continuity, compactness) are often hard to translate in terms of corresponding matrix – see Exercise 56.

[103] Both statements follow at once by noticing that $\mathcal{R}(P) = \ker(I - P)$, where I is the identity.

[104] We already proved one implication. For the other, if a projection P satisfies $P^* = P$, since $\mathcal{R}(P) = \ker(I - P)$ is closed, Theorem 3.27 yields $\mathcal{R}(P) = \{\ker(P)\}^\perp$.

Theorem 3.29 *If $A \in \mathcal{L}(H)$ is such that A^*A is compact, then A is compact. In particular, A is compact if and only if A^* is compact.*

Proof Assume that A^*A is compact and let $\{x_n\}_{n\geq 1}$ be a bounded sequence. Choose $M > 0$ with $\|x_n\| \leq M$ for all $n \geq 1$. Then $\{A^*Ax_n\}_{n\geq 1}$ has a convergent subsequence $\{A^*Ax_{n_k}\}$. But $\|Ax_{n_i} - Ax_{n_j}\|^2 = (A^*A(x_{n_i} - x_{n_j}), x_{n_i} - x_{n_j}) \leq \|A^*A(x_{n_i} - x_{n_j})\| \cdot \|x_{n_i} - x_{n_j}\| \leq 2M\,\|A^*A(x_{n_i} - x_{n_j})\| \to 0$ as $n_i, n_j \to \infty$, so that $\{Ax_{n_k}\}$ is a Cauchy sequence, thus convergent. Consequently A is compact.

For the second assertion, since $A^{**} = A$, it suffices to prove one implication. Assuming that A^* is compact, the boundedness of A yields that A^*A is compact.[105] By the first part of the statement, A is compact. □

Theorem 3.30 *A bounded linear operator $T : H \to H$ is compact if and only if it maps weakly convergent sequences into convergent sequences.*[106]

Proof If $T \in \mathcal{L}(H)$ and $x_n \rightharpoonup x$, then $Tx_n \rightharpoonup Tx$ since for every $y \in H$ we have that $(Tx_n, y) = (x_n, T^*y) \to (x, T^*y) = (Tx, y)$. If T is compact and $\{Tx_n\}_{n\geq 1}$ does not converge to Tx, then we can find $\varepsilon > 0$ and a subsequence $\{x_{n_k}\}$ with $\|Tx_{n_k} - Tx\| > 0$ for every n_k. But, T being compact, $\{Tx_{n_k}\}$ has a convergent subsequence $Tx_{n'_k} \to z \in H$. As is to be expected, $z = Tx$, and we obtain a contradiction. Indeed, for every $y \in H$ we get $(Tx_{n'_k} - Tx, y) \to (z - Tx, y)$ as $Tx_{n'_k} \to z$ and $(Tx_{n'_k} - Tx, y) \to 0$ as $Tx_{n'_k} \rightharpoonup Tx$, so that $z = Tx$.

Conversely, if $Tx_n \to Tx$ whenever $x_n \rightharpoonup x$, then T is compact since every bounded set in H is weakly compact. □

Let us now prove a compactness criterion for bounded linear operators.

Theorem 3.31 *Let $A : H \to H$ be a bounded linear operator.*

(i) A is compact if for every $\varepsilon > 0$ there exists a compact linear operator $A_\varepsilon : H \to H$ with $\|A - A_\varepsilon\|_{\mathcal{L}(H)} \leq \varepsilon$.

(ii) If A is compact then there are finite rank bounded linear operators[107] *$A_n : H \to H$ with $\|A - A_n\|_{\mathcal{L}(H)} \to 0$.*

Proof (i) Choose compact operators A_{ε_n} corresponding to a decreasing sequence $\varepsilon_n \to 0$. Let S be a bounded set in H and choose $M > 0$ such that $\|x\| \leq M$ for all $x \in S$. If $\{x_n\}_{n\geq 1}$ is a sequence in S, by Theorem 3.18 and Theorem 3.30, it admits a subsequence $\{x_{1n}\}_{n\geq 1}$ mapped by A_{ε_1} into a convergent sequence. Select now a subsequence $\{x_{2n}\}_{n\geq 1}$ of $\{x_{1n}\}_{n\geq 1}$ that is mapped by A_{ε_2} into a convergent sequence. Continuing this process, we obtain for every $k \geq 2$ successive subsequences $\{x_{kn}\}_{n\geq 1}$ that are mapped by A_{ε_2} into a convergent sequence. Then the diagonal sequence $\{x_{kk}\}_{k\geq 1}$ is mapped into a convergent sequence by every A_{ε_n}. But $\|Ax_{nn} - Ax_{mm}\| \leq \|(A - A_{\varepsilon_k})x_{nn}\| + \|A_{\varepsilon_k}(x_{nn} - x_{mm})\| + \|(A - A_{\varepsilon_k})x_{mm}\| \leq 2\varepsilon_k M + \|A_{\varepsilon_k}(x_{nn} - x_{mm})\|$. If k is large enough we can make $2\varepsilon_k M$ as small as we want, while choosing subsequently $m, n \geq N(k)$ large enough makes the other term as small as desired. Consequently $\{Ax_{nn}\}_{n\geq 1}$ is a Cauchy sequence, hence convergent. By Theorem 3.30 we deduce that A is compact.

(ii) Let K be the image of the unit ball $\{x \in H : \|x\| \leq 1\}$ under A. Then K is compact since if $y_n = Ax_n$ with $\|x_n\| \leq 1$ for $n \geq 1$, then, see Theorem 3.30, a weakly convergent subsequence $\{x_{n_k}\}$

105 One can easily see this by taking bounded sequences.

106 That is, $x_n \rightharpoonup x$ implies $Tx_n \to Tx$.

107 A finite-rank operator is a bounded linear operator between Banach spaces whose range is finite-dimensional. The validity of the result in Theorem 3.31(ii) does not extend to all Banach spaces; see Brézis (2011).

of $\{x_n\}_{n\geq 1}$ is mapped by A into a convergent sequence $\{Ax_{n_k}\}$. For $n \geq 1$, we can find finitely many open balls B_i, centred at points $y_i = Ax_i$ with $\|x_i\| \leq 1$ and of radius $\frac{1}{n} > 0$, that cover K. Let Y_n be the finite-dimensional (closed) subspace of H generated by the y_is and define $A_n = P_n \circ A$, where P_n is the orthogonal projection onto Y_n. Then, given $x \in H$ with $\|x\| \leq 1$, by construction we can find i_0 with $\|Ax - y_{i_0}\| < \frac{1}{n}$. Since $\|P_n\|_{\mathcal{L}(H)} \leq 1$, we have that $\|A_n x - P_n y_{i_0}\| \leq \|Ax - y_{i_0}\| < \frac{1}{n}$, that is, $\|A_n x - y_{i_0}\| < \frac{1}{n}$. Thus $\|Ax - A_n x\| \leq \|Ax - y_{i_0}\| + \|A_n x - y_{i_0}\| < \frac{2}{n}$, so that $\|A - A_n\|_{\mathcal{L}(H)} \leq \frac{2}{n}$. Clearly A_n is of finite rank since its range is Y_n. □

Theorem 3.31 can be used to produce certain classes of compact operators:

- $A \in \mathcal{L}(H)$ is called a *Hilbert–Schmidt operator* if $\|A\|_2 = \left(\sum_{k\geq 1} \|Ae_k\|^2\right)^{1/2} < \infty$ for some orthonormal basis $\{e_k\}_{k\geq 1}$ of the separable Hilbert space H. Using Plancherel's identity, see Theorem 3.1(d), one can show that $\|A\|_2$ is independent of the basis chosen,[108] with $\|A\|_2 = \|A^*\|_2$; in particular, it is well-defined. The family $\mathcal{B}_2(H)$ of all Hilbert–Schmidt operators is a subset of the closed subspace $C(H) \subset \mathcal{L}(H)$ of all compact operators.[109] While $\mathcal{B}_2(H)$ is a closed subspace of $\mathcal{L}(H)$ if and only if H is finite-dimensional,[110] $\mathcal{B}_2(H)$ endowed with the norm $\|\cdot\|_2$ is a Hilbert space,[111] see Reed and Simon (1980a). Hilbert–Schmidt operators are important since for $H = L^2(\Omega, \mu)$, $\mathcal{B}_2(H)$ has a concrete realisation: if $K \in L^2(\Omega\times\Omega,\, d\mu\times d\mu)$, then the operator $u \mapsto (Ku)(x) = \int_\Omega K(x,y)u(y)\, d\mu(y)$ belongs to $\mathcal{B}_2(L^2(\Omega,\mu))$ with $\|K\|_2^2 = \int_{\Omega\times\Omega} |K(x,y)|^2\, d\mu(x)\, d\mu(y)$, and conversely, every operator in $\mathcal{B}_2(L^2(\Omega,\mu))$ is of the preceding form for some unique kernel $K \in L^2(\Omega\times\Omega,\, d\mu\times d\mu)$; see Reed and Simon (1980a).
- The *trace class operators*[112] are $\mathcal{B}_1(H) = \{AB :\ A, B \in \mathcal{B}_2(H)\}$, the factorisation being far from unique. If $T = AB \in \mathcal{B}_1(H)$, then[113] $\sum_{k\geq 1} |(Te_k, e_k)| < \infty$ and $tr(T) = \sum_{k\geq 1}(Te_k, e_k)$ is independent of the choice of the orthonormal basis $\{e_k\}_{k\geq 1}$ of H, with $(B, A^*)_2 = tr(T)$. While $\mathcal{B}_1(H)$ is a closed sub-

108 Indeed, if $\{f_j\}_{j\geq 1}$ is another orthonormal basis, then $\sum_{k\geq 1} \|Ae_k\|^2 = \sum_{k\geq 1}\sum_{j\geq 1} |(Ae_k, f_j)|^2 = \sum_{k\geq 1}\sum_{j\geq 1} |(e_k, A^* f_j)|^2 = \sum_{j\geq 1}\sum_{k\geq 1} |(e_k, A^* f_j)|^2 = \sum_{j\geq 1} \|A^* f_j\|^2$.

109 Plancherel's identity ensures that $A_n x = \sum_{k=1}^n (x, e_k)\, Ae_k$ are finite-rank operators that approximate $A \in \mathcal{B}_2(H)$ in $\mathcal{L}(H)$. By Theorem 3.34(i), $C(H)$ is a closed subspace of $\mathcal{L}(H)$.

110 Indeed, if $\{e_k\}_{k\geq 1}$ is an infinite orthonormal basis, then $A_{n,m}e_k = k^{-1/m}\, e_k$ for $1 \leq k \leq n$ and $A_{n,m}e_k = 0$ for $k \geq n+1$ defines for $m \geq 2$ and $n \geq 1$ operators $A_{n,m} \in \mathcal{B}_2(H)$ that converge in $\mathcal{L}(H)$ to the operator $A_m \notin \mathcal{B}_2(H)$ defined by $A_m e_k = k^{-1/m}\, e_k$ for all $k \geq 1$. Note that $\|(A - A_n)x\|^2 \leq (n+1)^{-2/m}\, \|x\|^2$ for every $x \in H$, due to Plancherel's identity.

111 In the Euclidean space $\mathbb{R}^n$ or $\mathbb{C}^n$, for $A = (a_{ij})_{1\leq i,j\leq n}$ we have that $\|A\|_2^2 = \sum_{i=1}^n \sum_{j=1}^n |a_{ij}|^2$.

112 The terminology is elucidated by our discussion of spectral theory.

113 Indeed, $\sum_{i\geq 1}\sum_{j\geq 1} |(Be_i, f_j)(f_j, A^* e_i)| \leq \sum_{i\geq 1} \|Be_i\| \cdot \|A^* e_i\| \leq \|B\|_2 \cdot \|A^*\|_2$ for any orthonormal basis $\{f_k\}_{k\geq 1}$, by Plancherel's identity and the Cauchy–Schwarz inequality (in H and in l^2). Due to Parseval's identity, $\sum_{i\geq 1}(ABe_i, e_i) = \sum_{i\geq 1}(Be_i, A^* e_i) = \sum_{i\geq 1}\sum_{j\geq 1}(Be_i, f_j)(f_j, A^* e_i) = \sum_{j\geq 1}\sum_{i\geq 1}(Af_j, e_i)(e_i, B^* f_j) = \sum_{j\geq 1}(Af_j, B^* f_j) = \sum_{j\geq 1}(BAf_j, f_j)$, since the absolute convergence of the series allows us to switch the order of summation. Similarly, $\sum_{j\geq 1}(ABf_j, f_j)$ has the same value, so that, in addition to the claim, we get that $tr(AB) = tr(BA)$.

space of $\mathcal{L}(H)$ only if H is finite-dimensional,[114] endowed with the norm $\|T\|_1 = \sup_{\phi,\psi} \sum_{k\geq 1} |(T\phi_k, \psi_k)|$, where the supremum is taken over all pairs of orthonormal bases $\{\phi_k\}_{k\geq 1}$ and $\{\psi_k\}_{k\geq 1}$ of the separable Hilbert space H, $\mathcal{B}_1(H)$ is a Banach space, see Lax (2002). If $T \in \mathcal{B}_1(H)$ and $S \in \mathcal{L}(H)$, then[115] $TS \in \mathcal{B}_1(H)$. One can show that $S \mapsto tr(TS)$ is a bounded linear functional on $\mathcal{L}(H)$ and, while these are not all the continuous linear functionals on $\mathcal{L}(H)$, they yield the entire dual of $C(H)$, see Reed and Simon (1980a). Note that $\sum_{k\geq 1} |(Te_k, e_k)| < \infty$ for some orthonormal basis $\{e_k\}_{k\geq 1}$ does not guarantee that $T \in \mathcal{L}(H)$ is of trace class,[116] but if the sum is finite for all orthonormal bases then $T \in \mathcal{B}_1(H)$, see Reed and Simon (1980a).

If $T : \mathbb{C}^n \to \mathbb{C}^n$ is a linear map, the eigenvalues of T are those numbers $\lambda \in \mathbb{C}$ for which $(\lambda I - T)$ is not invertible, the set of all such λ's being called the spectrum of T. Using the Jordan canonical form of a matrix, knowledge of the spectrum (and of the associated eigenspaces) provides insight into the structure of the linear map T. The spectral theory of operators on infinite-dimensional Hilbert spaces is more complicated and richer, representing an important tool for understanding the structure of these operators. Complex eigenvalues are unavoidable even in the spectral theory of real matrices. This fact motivates us to extend real Hilbert spaces to complex Hilbert spaces. Complexification allows us to use the rich fabric of analytic functions[117] to obtain in-depth spectral information.

If $(Z, (\cdot,\cdot))$ is a complex Hilbert space, define the real scalar product $(x,y)_{\mathbb{R}} = \Re e\{(x,y)\} = \frac{(x,y)+(y,x)}{2}$ for $x, y \in Z$. If we forget altogether about the possibility of multiplying vectors by scalars in $\mathbb{C} \setminus \mathbb{R}$ and we consider for $x, y \in Z$ only linear combinations $ax + by$ with $a, b \in \mathbb{R}$, the set Z becomes a real vector space denoted $Z_{\mathbb{R}}$: endowed with $(\cdot,\cdot)_{\mathbb{R}}$, this is actually a real Hilbert space, called the *realification* of the original complex Hilbert space.[118] Defining the operator $J : Z_{\mathbb{R}} \to Z_{\mathbb{R}}$ by $Jx = ix$ for $x \in Z$, we see that $J^2 = -I$ and $(Jx, Jy)_{\mathbb{R}} = (x,y)_{\mathbb{R}}$, $(Jx, y)_{\mathbb{R}} = -(x, Jy)_{\mathbb{R}}$ for all $x, y \in Z_{\mathbb{R}}$, so that J is unitary and skew-adjoint (that is, $J^* = -J$). On the other hand, every separable real Hilbert space $(H, \langle\cdot,\cdot\rangle)$ that is not of odd finite dimension admits[119] a unitary operator J with $J^2 = -I$,

114 The example $A_{n,1} = A_{n,2}^2$ with $A_{n,2}$ constructed above clarifies this.

115 Since clearly $BS \in \mathcal{B}_2(H)$ whenever $B \in \mathcal{B}_2(H)$.

116 As illustrated by the example of T defined by $Te_k = e_{k+1}$ for all $k \geq 1$.

117 One can develop a theory of vector-analytic functions which is almost exactly parallel to the classical theory.

118 The imaginary part $\kappa(x,y) = \frac{(x,y)-(y,x)}{2i} \in \mathbb{R}$ of the complex scalar product $(\cdot,\cdot)$ is the so-called *Kähler form* of the complex Hilbert space Z, and $Z_{\mathbb{R}}$ is a symplectic space equipped with the symplectic form κ (that is, $\kappa : Z_{\mathbb{R}} \times Z_{\mathbb{R}} \to \mathbb{R}$ is an anti-symmetric bilinear map which is non-degenerate, in the sense that the only fixed $x_0 \in Z_{\mathbb{R}}$ such that $\kappa(x_0, y) = 0$ for all $y \in Z_{\mathbb{R}}$ is $x_0 = 0$).

119 For example, one possibility is to define J by $Je_k = \text{sgn}(k)\, e_k$ for an orthonormal basis $\{e_k\}_{k\neq 0}$ indexed over the nonzero integers or over a finite set symmetric about zero. Such an operator does not exist for real Hilbert spaces of finite odd dimension, as one can see by taking the determinants in $J^2 = -I$.

called a *complex structure* on H. This can be used to convert H into a complex vector space $H_\mathbb{C}$: as a set $H_\mathbb{C}$ is identical to H, the addition of vectors in $H_\mathbb{C}$ is as in H and[120] scalar multiplication of vectors by complex numbers is defined by $(a+ib)x = ax + bJx$ for $x \in H$ and $a, b \in \mathbb{R}$. $H_\mathbb{C}$ is a *complexification* of the real Hilbert space H. Equipped with the scalar product $(x, y) = \langle x, y\rangle - i\langle Jx, y\rangle$, $H_\mathbb{C}$ is a complex pre-Hilbert space. Since $\langle Jx, x\rangle = 0$ for all $x \in H$, we see that the complex scalar product induces on $H_\mathbb{C}$ a norm that coincides with the norm of H, so that $H_\mathbb{C}$ is a Hilbert space. In particular, for the realification $H = Z_\mathbb{R}$ of the complex Hilbert space Z, choosing J to be the multiplication by i leads to the complexification $H_\mathbb{C} = Z$. Moreover, every linear operator $T : X \to Y$ between two real Hilbert spaces gives rise to a complex linear operator $T_\mathbb{C} : X_\mathbb{C} \to Y_\mathbb{C}$, defined by $T_\mathbb{C}(x+iy) = Tx + iTy$. It is easy to see that if T is bounded, then $\|T_\mathbb{C}\|_{\mathcal{B}(X_\mathbb{C},Y_\mathbb{C})} = \|T\|_{\mathcal{B}(X,Y)}$. The failure of this elegant complexification approach for real Hilbert spaces of odd finite dimension can be addressed by the following *ad hoc* procedure.[121] A complexification of the real vector space X is the complex vector space $\mathbb{X} = \{x+iy : x, y \in X\} \sim X \times X$ whose vector space operations are defined by $(x_1+iy_1)+(x_2+iy_2) = x_1+x_2+i(y_1+y_2)$ and $(a+ib)(x+iy) = ax-by+i(ay+bx)$ for $a, b \in \mathbb{R}$ and $x, x_1, x_2, y, y_1, y_2 \in X$, so that the real vector space X can be identified with the subspace $X + i0$ of $\mathbb{X}$. If $\langle\cdot,\cdot\rangle$ is a scalar product on the real pre-Hilbert space X, we extend it to a scalar product on the complex pre-Hilbert space $\mathbb{X}$ by $(x+iy,\, u+iv) = \langle x, u\rangle + \langle y, v\rangle + i\{\langle y, u\rangle - \langle x, v\rangle\}$ for $x, y, u, v \in X$. The norm $|||\cdot|||$ induced on $\mathbb{X}$ satisfies $|||x+iy|||^2 = \|x\|^2 + \|y\|^2$ for $x, y \in X$, thus ensuring that $\mathbb{X}$ is a Hilbert space. Moreover, every linear operator $T : X_1 \to X_2$ between two real Hilbert spaces gives rise to a complex linear operator $T_\mathbb{C} : \mathbb{X}_1 \to \mathbb{X}_2$, defined by $T_\mathbb{C}(x+iy) = Tx+iTy$. It is easy to see that if T is bounded, then $\|T_\mathbb{C}\|_{\mathcal{B}(\mathbb{X}_1,\mathbb{X}_2)} = \|T\|_{\mathcal{B}(X_1,X_2)}$, so that with regard to linear operators the complexification is norm-preserving. Denoting vectors $x+iy$ as column vectors $\binom{x}{y}$, every linear operator $\mathbb{T} : \mathbb{X}_1 \to \mathbb{X}_2$ is represented in matrix form as $\mathbb{T} = \left(\begin{smallmatrix} T & -S \\ S & T\end{smallmatrix}\right)\binom{x}{y} = Tx - Sy + i(Ty + Sx)$, where $\mathbb{T}x = Tx + iSx$ for $x \in X_1$, with $T, S : X_1 \to X_2$. In particular, $T_\mathbb{C}$ is represented by the matrix $\left(\begin{smallmatrix} T & 0 \\ 0 & T\end{smallmatrix}\right)$. The connection between $\mathbb{X}$ and the complexification $X_\mathbb{C}$ of the real Hilbert space X by means of a complex structure J (if it exists) is that $\mathbb{X}_+ = \{x + iJx : x \in X\} \subset \mathbb{X}$ is[122] the orthogonal complement of the closed subspace $\mathbb{X}_- = \{x - iJx : x \in X\}$ of $\mathbb{X}$, and $x \mapsto \frac{x-iJx}{\sqrt{2}}$ is a bijection from $X_\mathbb{C}$ to $\mathbb{X}_-$ that preserves the inner product.

Throughout the remainder of this section H stands for a separable complex Hilbert space.

Definition $\lambda \in \mathbb{C}$ *belongs to the "resolvent set"* $\rho(T)$ *of* $T \in \mathcal{L}(H)$ *if* $(\lambda I - T)$ *is a bijection with a bounded inverse* $R_\lambda(T) = (\lambda I - T)^{-1}$*, called the "resolvent" of* T *at* $\lambda \in \rho(T)$*. The set* $\sigma(T) = \mathbb{C} \setminus \rho(T)$ *is called the "spectrum" of* T.

By the inverse mapping theorem, $(\lambda I - T)$ will have a bounded inverse if it is bijective. In the finite-dimensional setting $\sigma(T)$ consists of the eigenvalues of the matrix T. This

120 The vector space operations being thus $(x_1+iy_1)+(x_2+iy_2) = x_1+x_2+i(y_1+y_2)$ and $(a+ib)(x+iy) = ax - by + i(ay+bx)$, due to $J(y_1+y_2) = Jy_1 + Jy_2$ and $aJy + J(bx) + J(bJy) = J(ay+bx) - by$ for $a, b \in \mathbb{R}$ and $x, x_1, x_2, y, y_1, y_2 \in H$.

121 An alternative procedure involves tensor products – see e.g. van Neerven (1997). It also relies on an extension of the set of vectors, in contrast to the complex structure approach that uses only the set of vectors of the original real Hilbert space H.

122 Orthogonality is easily checked, while $x + iy = (u - iJu) + (v + iJv)$ with $u = \frac{x+Jy}{2}$ and $v = \frac{x-Jy}{2}$ for $x, y \in X$.

feature fails in general: there are several ways for $\lambda \in \mathbb{C}$ not to belong to $\rho(T)$, and this leads to a subdivision of the spectrum $\sigma(T)$.

Definition *For a bounded linear operator $T : H \to H$ we distinguish three disjoint subsets of its spectrum:*

(i) the "point spectrum" is the set of all $\lambda \in \mathbb{C}$ for which $(\lambda I - T)$ is not one-to-one. Elements of the point spectrum are called "eigenvalues", the nonzero vectors[123] *$x \in H$ satisfying $Tx = \lambda x$ being the corresponding "eigenvectors";*

(ii) the "continuous spectrum" is the set of all $\lambda \in \mathbb{C}$ such that $(\lambda I - T)$ is one-to-one with a dense range but fails to be surjective;[124]

(iii) the "residual spectrum"[125] *is the set of all $\lambda \in \mathbb{C}$ such that $(\lambda I - T)$ is one-to-one but its range is not dense in H.*

Exercise 59 illustrates the above types of spectrum. For an in-depth analysis we first have to establish some basic properties of the resolvent $R_\lambda(T)$ of a bounded linear operator $T : H \to H$.

Theorem 3.32 *Let $T, S : H \to H$ be bounded linear operators. Then for $\lambda, \mu \in \rho(T)$ and for $z \in \rho(T) \cap \rho(S)$ the following resolvent identities hold:*

(i) $R_\lambda(T) - R_\mu(T) = (\mu - \lambda) R_\lambda(T) R_\mu(T)$;

(ii) $R_z(T) - R_z(S) = R_z(T)(T - S)R_z(S) = R_z(S)(T - S)R_z(T)$.

In particular,[126] *$R_\lambda(T)$ and $R_\mu(T)$ commute.*

Proof We prove (i) by writing the identity $R_\lambda(T) - R_\mu(T) = R_\lambda(T)(\mu I - T)R_\mu(T) - R_\lambda(T)(\lambda I - T)R_\mu(T) = (\mu - \lambda)R_\lambda(T)R_\mu(T)$. As for (ii), rely on $T - S = (zI - S) - (zI - T)$ and expand the second and third items. □

Theorem 3.33 (Neumann expansion) *Let $T : H \to H$ be a bounded linear operator. If $\lambda \in \mathbb{C}$ with $|\lambda| > \|T\|$, then $\lambda \in \rho(T)$ and*

$$(\lambda I - T)^{-1} = \sum_{n=0}^{\infty} \frac{T^n}{\lambda^{n+1}} \quad \text{with} \quad \|(\lambda I - T)^{-1}\| \leq \frac{1}{|\lambda| - \|T\|}.$$

123 By linearity, $(\lambda I - T)$ fails to be injective if and only if $\ker(\lambda I - T) \neq \{0\}$.

124 In contrast to finite-dimensions, the range of a bounded linear operator may fail to be closed; see Exercise 36(iii). The name derives from how this part of the spectrum typically looks, but it is a misconception to always expect a continuous appearance – in Part II we will see that the spectrum $[0, \frac{1}{3}] \cup [\frac{2}{3}, 1]$ of the bounded self-adjoint operator $f(t) \mapsto a(t)f(t)$, acting on $L^2[0, 1]$, is purely continuous; here $a(t) = t$ for $t \in [0, \frac{1}{3}] \cup [\frac{2}{3}, 1]$, $a(t) = 0$ for $t \in (\frac{1}{3}, \frac{2}{3})$. However, an isolated part of the spectrum of a self-adjoint operator must be an eigenvalue, as a consequence of the spectral theorem discussed in Part II.

125 Since $(\lambda I - T)^* = \overline{\lambda} I - T^*$, due to Theorem 3.27, $\overline{\lambda}$ must be an eigenvalue of T^*. We will show that the residual spectrum of a self-adjoint operator is always void.

126 Interchange μ and λ in (i) and compare the result.

Proof This is immediate since $\|T^n\| \le \|T\|^n$ for any $n \ge 1$ ensures the convergence of the series $S = \sum_{n=0}^{\infty} \frac{T^n}{\lambda^{n+1}}$ in the Banach space $\mathcal{L}(H)$, with $(\lambda I - T)S = S(\lambda I - T) = I$. □

By Theorem 3.33, $\rho(T)$ is nonempty while $|\mu| \le \|T\|$ for $\mu \in \sigma(T)$. Actually,

Theorem 3.34 *If $T \in \mathcal{L}(H)$, then the resolvent set, $\rho(T)$, is an open subset of $\mathbb{C}$, while the spectrum, $\sigma(T)$, is a nonempty compact subset of $\mathbb{C}$.*

Proof If $\lambda \in \rho(T)$ and if $\mu \in \mathbb{C}$ is such that $|\mu| < 1/\|(\lambda I - T)^{-1}\|$, then $\lambda + \mu \in \rho(T)$ since for $\mu \neq 0$ we have $(\lambda+\mu)I - T = (\lambda I - T)\mu(\frac{1}{\mu} I + (\lambda I - T)^{-1})$ and the first factor is invertible as $\lambda \in \rho(T)$, while the second factor is invertible by the Neumann expansion; clearly T_1T_2 is invertible with inverse $T_2^{-1}T_1^{-1}$ if $T_1, T_2 \in \mathcal{L}(H)$ are invertible. Thus $\rho(T)$ is an open set in $\mathbb{C}$. The fact that $\sigma(T)$ is a compact is now plain, being a bounded set by Theorem 3.33 and a closed set since its complement $\rho(T)$ is open. It remains to prove that $\sigma(T)$ is nonempty. For $\lambda_0 \in \rho(T)$ we showed above that if $|\lambda - \lambda_0| < 1/\|R_{\lambda_0}(T)\|$, then $\lambda \in \rho(T)$, and we now claim that

$$R_\lambda(T) = R_{\lambda_0}(T) \sum_{n=0}^{\infty} (\lambda - \lambda_0)^n [R_{\lambda_0}(T)]^n . \tag{3.11}$$

Indeed, Theorem 3.32(i) with $\mu = \lambda_0$ yields $R_\lambda(T)[I - (\lambda_0 - \lambda)R_{\lambda_0}(T)] = R_{\lambda_0}(T)$. Multiplying this identity on the right by the convergent series $\sum_{n=0}^{\infty}(\lambda - \lambda_0)^n[R_{\lambda_0}(T)]^n$, which represents the inverse of $[I - (\lambda - \lambda_0)R_{\lambda_0}(T)]$, leads to (3.11). On the other hand, for $|\lambda| > \|T\|$ we have by the Neumann expansion that $R_\lambda(T) = \sum_{n=0}^{\infty} \lambda^{-(n+1)} T^n$. Assuming $\sigma(T) = \emptyset$, for any $x \in H$ and $f \in H^* = H$, these properties would imply that $\lambda \mapsto f(R_\lambda(T)x) \in \mathbb{C}$ is analytic and bounded in $\mathbb{C}$. Therefore, by Liouville's theorem (see Section 7.1), the function is constant. Since the above Neumann expansion for $|\lambda| > \|T\|$ yields $\lim_{|\lambda| \to \infty} \|R_\lambda(T)\| = 0$, we must have $f(R_\lambda(T)x) = 0$ for all $x \in H$ and all $f \in H^* = H$. Thus $R_\lambda(T)x = 0$ for all $x \in H$. But this is impossible since $R_\lambda(T)$ is invertible. Thus the spectrum $\sigma(T)$ must be nonempty. □

Insight into the structure of the spectrum is at hand for self-adjoint operators.

Theorem 3.35 *Let $T \in \mathcal{L}(H)$ be self-adjoint, i.e. $T^* = T$. Then T has no residual spectrum, $\sigma(T) \subset \mathbb{R}$ and eigenvectors corresponding to distinct ein-genvalues are orthogonal.*

Proof For $a, b \in \mathbb{R}$ we compute $\|[T - (a + ib)I]x\|^2 = ((T - aI)x - ibx, (T - aI)x - ibx) = \|(T - aI)x\|^2 + b^2\|x\|^2 \ge b^2\|x\|^2$, so that for $b \neq 0$ the operator $[T - (a + ib)I]$ is injective and has a closed range, with a bounded inverse on its range. If the range is not H, then $(a - ib)$ must be an eigenvalue of T since, by Theorem 3.27, the orthogonal complement of the range of $[T - (a + ib)I]$ is the kernel of $[T - (a + ib)I]^* = [T - (a - ib)I]$. But the inequality shows that only real eigenvalues are possible, so that for $b \neq 0$ we proved that $(a + ib) \in \rho(T)$. Thus $\sigma(T) \subset \mathbb{R}$.

Assume now that $\lambda \in \mathbb{R}$ is in the residual spectrum of T. Then, since $(T - \lambda I)^* = T^* - \overline{\lambda} I = T - \lambda I$, Theorem 3.27 yields $\ker\{(T - \lambda I)^*\} = \overline{\mathcal{R}(T - \lambda I)}^{\perp} \neq \{0\}$, so that λ would be an eigenvalue of T, which is impossible by the definition of the residual spectrum.

Finally, if $Tx = ax$ and $Ty = by$ with $a, b \in \mathbb{R}$ eigenvalues of T, then $a(x, y) = (ax, y) = (Tx, y) = (x, Ty) = (x, by) = (x, y)b$. Thus $(x, y) = 0$ if $a \neq b$. □

At the point where we introduced compact operators we did not explain their importance. The main impetus for the study of compact operators arose from the use of integral equations in attempting to solve boundary value problems in mathematical physics. However, from a purely functional-analytic point of view, the basic principle which makes compact operators important is the *Fredholm alternative*:[127] if $T : H \to H$ is a compact linear operator, then either $Tx = x$ has a solution or $(I - T)^{-1}$ exists in $\mathcal{L}(H)$. This property is very useful in the process of solving equations, as it says that compactness and uniqueness together imply existence: if for any $y \in H$ there is at most one $x \in H$ with $x = y + Tx$, then there is always exactly one such $x \in H$. Indeed, assuming the validity of the Fredholm alternative,[128] the hypothesis means that $\ker(I - T) = \{0\}$, so $(I - T)^{-1}$ exists.

We now discuss this aspect in more detail. Let X and Y be Banach spaces and let $T : X \to Y$ be a linear compact operator. The equation

$$Tx = y \tag{3.12}$$

is very rarely well-posed; by well-posedness we mean that the range $\mathcal{R}(T)$ of T should be closed[129] and for every $y \in \mathcal{R}(T)$ there is a unique solution $x \in X$ of (3.12), and continuous dependence holds, i.e., if $y_n \to y$, then $x_n \to x$ (here $Tx_n = y_n$ and $Tx = y$). In other words, well-posedness means that T is injective with $T^{-1} : \mathcal{R}(T) \to X$ continuous. But if $\mathcal{R}(T)$ is closed,[130] then its dimension is finite.[131] For this reason, instead of (3.12) we are rather more interested in equations of the type

$$Tx - x = y \tag{3.13}$$

for a compact linear operator $T : X \to X$ on a complex Banach space X. Equations of type (3.13) have the following advantages:

(a) $\dim\{\ker(I - T)\} < \infty$;
(b) $\mathcal{R}(I - T)$ is closed;
(c) [*Fredholm alternative*] Consider the four equations

$$(i)\ Tx - x = y\,,\quad (ii)\ Tx - x = 0\,,\quad (iii)\ T^*f - f = g\,,\quad (iv)\ T^*f - f = 0\,,$$

where $x, y \in X$ while $f, g \in X^*$, and $T^* : X^* \to X^*$ is the adjoint[132] of T. Then either

(A) equations (ii) and (iv) have only the trivial solution, and equations (i) and (iii) have a unique solution for any right side, which furthermore depends continuously on it; or
(B) equation (ii) has a finite-dimensional space of solutions $X_k \subset X$, equation (iv) has a finite-dimensional space of solutions $\tilde{X}_k \subset X^*$, and $\dim\{X_k\} = \dim\{\tilde{X}_k\}$, in which case equation (i) is solvable for precisely those $y \in Y$ such that $f(y) = 0$ for all $f \in \tilde{X}_k$, and equation (iii) is solvable for precisely those $g \in X^*$ such that $g(x) = 0$ for all $x \in X_k$.

In a Hilbert space setting the Fredholm alternative simplifies to $\mathcal{R}(I - T) = [\ker\{(I - T^*)\}]^\perp$ and $\mathcal{R}(I - T^*) = [\ker\{(I - T)\}]^\perp$.

127 This is a consequence of the spectral theorem for self-adjoint operators, to be discussed.

128 This property is not typical for general bounded linear operators. For example, for the operator $(A\phi)(t) = 2t\,\phi(t)$ on $L^2[0, 1]$, the equation $A\phi = \phi$ has no solutions but $(I - A)^{-1}$ does not exist (as a bounded linear operator acting on $L^2[0, 1]$).

129 So that continuous dependence is meaningful.

130 As already pointed out, this is desirable for a proper meaning of continuous dependence.

131 Since $T : X \to \mathcal{R}(T)$ is then onto the Banach space $\mathcal{R}(T)$, and, by the open mapping theorem, T is open so that a certain ball of radius $\varepsilon > 0$ in $\mathcal{R}(T)$ is contained in the image of the unit ball of X under T. Since T is compact, the latter has a compact closure, so that the unit ball in $\mathcal{R}(T)$ is relatively compact, meaning that $\dim\{\mathcal{R}(T)\} < \infty$, see Theorem 3.8.

132 If $S : X \to Y$ is a bounded linear operator between the two complex Banach spaces X and Y, then $S^* : Y^* \to X^*$ is the bounded linear operator defined by $S^*f = f(S) \in X^*$ and such that $\|S^*\| = \|S\|$; see Brézis (2011).

The Fredholm alternative is contingent upon the following spectral property of compact operators: if $\lambda \neq 0$ is a point of the spectrum of a compact linear operator $A : X \to X$ on a complex Banach space X, then λ is an eigenvalue of finite multiplicity (that is, $\dim\{\ker(A - \lambda I)\} < \infty$). Moreover, λ is also an eigenvalue of A^* with the same multiplicity. The relevant result is the following theorem; for a proof, see Lax (2002).

Theorem 3.36 (Riesz–Schauder theorem) *Given a complex Banach space X, if $A \in \mathcal{L}(X)$ is compact, then $\sigma(A)$ consists of an at most countable set of points with no possible point of accumulation other than 0, and every $\lambda \in \sigma(A) \setminus \{0\}$ is an eigenvalue of finite multiplicity.*

Rather than discussing the proof of Theorem 3.36, we will present a proof of the corresponding result in separable Hilbert spaces, a context in which much more can be said, as we shall see. After this digression into Banach spaces, from now on we will return to the setting of a separable complex Hilbert space.

Example Let H be a separable complex Hilbert space H with orthonormal basis $\{e_n\}_{n\geq 1}$. Given any real sequence $\{\alpha_n\}_{n\geq 1}$ converging to zero and ordered such that $|\alpha_1| \geq |\alpha_2| \geq \ldots$, the operator $T : H \to H$, defined by $Tx = \sum_{n\geq 1} \alpha_n (x, e_n)\, e_n$ for $x = \sum_{n\geq 1}(x, e_n)\, e_n$, is compact, self-adjoint, with $\sigma(T) = \{0\} \cup \{\alpha_n\}_{n\geq 1}$. Indeed, $\|Tx\|^2 = \sum_{n\geq 1} \alpha_n^2\, |(x, e_n)|^2 \leq \alpha^2 \sum_{n\geq 1} |(x, e_n)|^2 = \alpha^2 \|x\|^2$, where $\alpha = \sup_{n\geq 1}\{|\alpha_n|\}$. Furthermore, the finite-rank operators $T_n x = \sum_{k=1}^{n} \alpha_k\, (x, e_k)\, e_k$ are such that $\|(T - T_n)x\|^2 = \|\sum_{k\geq n+1} \alpha_k\, (x, e_k)\, e_k\|^2 = \sum_{k\geq n+1} \alpha_k^2\, |(x, e_k)|^2 \leq \beta_{n+1} \sum_{k\geq 1} |(x, e_k)|^2 = \beta_{n+1}\|x\|^2$, where $\beta_{n+1} = \sup_{k\geq n+1}\{\alpha_k^2\} \to 0$ as $n \to \infty$. By Theorem 3.31, T is compact. Moreover, any α_n is an eigenvalue of T of finite multiplicity, and for any nonzero $\lambda \notin \{\alpha_n\}_{n\geq 1}$ the operator $(\lambda I - T)$ has the bounded inverse $R_\lambda(T) = \sum_{n\geq 1} \frac{1}{\lambda - \alpha_n}\, (x, e_n)\, e_n$. Since $\sigma(T)$ is closed, we deduce that $\sigma(T) = \{0\} \cup \{\alpha_n\}_{n\geq 1}$. In this example, we see that 0 might be an eigenvalue of T or not, according to whether $0 \in \{\alpha_n\}_{n\geq 1}$ or not. Also, note that in this example, we have $\alpha_1 = \pm \sup_{\|x\|=1}(Tx, x)$ since $(Te_1, e_1) = \alpha_1$ and $-|\alpha_1| \leq (Tx, x) = \sum_{n\geq 1} \alpha_n\, |(x, e_n)|^2 \leq |\alpha_1|$ for $x = \sum_{n\geq 1}(x, e_n)\, e_n$ with $\sum_{n\geq 1} |(x, e_n)|^2 = 1$. □

It is remarkable that the converse of the result presented in the above example also holds: every compact self-adjoint operator on a separable Hilbert space may be diagonalised in some suitable orthonormal basis. To prove this, we first establish a preliminary result that is of independent interest, as it characterises the eigenvalue with the largest absolute value without information about the eigenvectors.

Theorem 3.37 *If $T \in \mathcal{L}(H)$ is a compact and self-adjoint operator on a separable complex Hilbert space H, then at least one of the numbers $\pm\|T\|$ is an eigenvalue of T. In particular, $\|T\| = \max_{\|x\|=1} |(Tx, x)|$.*

Proof Assume $T \neq 0$. Then $\|T\| = \sup_{\|x\|=1}\{|(Tx, x)|\} > 0$, see Exercise 53, while $(Tx, x) \in \mathbb{R}$ for all $x \in H$, see Exercise 54. Consequently, $\|T\| = \sup_{\|x\|=1}\{(Tx, x)\}$ or $\|T\| = -\inf_{\|x\|=1}\{(Tx, x)\}$. Either way, there exists $\lambda \in \mathbb{R}$ with $|\lambda| = \|T\|$ and a sequence $\{x_n\}$ in H with $\|x_n\| = 1$ for all $n \geq 1$ such that $(Tx_n, x_n) \to \lambda$ as $n \to \infty$. Since $0 \leq \|Tx_n - \lambda x_n\|^2 = (Tx_n - \lambda x_n, Tx_n - \lambda x_n) = \|Tx_n\|^2 - 2\lambda(Tx_n, x_n) + \lambda^2 \leq 2\lambda^2 - 2\lambda(Tx_n, x_n) \to 0$ for $n \to \infty$. Thus $(Tx_n - \lambda x_n) \to 0$. Since T is compact, $\{Tx_n\}_{n\geq 1}$ has a convergent subsequence $\{Tx_{n_k}\}$ with limit y. Thus $x_{n_k} \to \frac{1}{\lambda}\, y$. But then

$Tx_{n_k} \to \frac{1}{\lambda} Ty$, so that $Ty = \lambda y$. Note that $y \neq 0$ since $\|y\| = \lim_{n_k \to \infty} \|\lambda x_{n_k}\| = |\lambda| = \|T\|$. Therefore y is an eigenvalue of T. □

Theorem 3.38 (Spectral theorem for compact self-adjoint operators) *Let $T \in \mathcal{L}(H)$ be a compact and self-adjoint operator on a separable complex Hilbert space H. Then there exists an orthonormal system $e_1, e_2, \ldots$ of eigenvectors of T and corresponding nonzero (real) eigenvalues $\lambda_1, \lambda_2, \ldots$, with finite-dimensional eigenspaces* $\ker\{(\lambda_k I - T)\}$, *such that*

$$Tx = \sum_{n \geq 1} \lambda_n (x, e_n) e_n \quad \textit{for all} \quad x \in H. \tag{3.14}$$

Moreover, the spectrum of T consists of $\lambda_1, \lambda_2, \ldots$ and 0, and if $\lambda_1, \lambda_2, \ldots$ is an infinite sequence, then it must converge to zero.

Proof Let us first show that each eigenvalue $\lambda_k \neq 0$ has a finite-dimensional eigenspace, $\ker\{(\lambda_k I - T)\}$. Indeed, by the compactness of T, an infinite set of orthonormal eigenvectors $\{e_{k,j}\}_{j \geq 1}$ would have a convergent subsequence since $e_{k,j} = \frac{1}{\lambda_k} Te_{k,j}$, which is impossible since $\|e_{k,j} - e_{k,i}\|^2 = (e_{k,j} - e_{k,i}, e_{k,j} - e_{k,i}) = 2$ for $i \neq j$ prevents this subsequence from being Cauchy. For each λ_k of T, choose a (finite) orthonormal basis of eigenvectors, and if 0 is an eigenvalue of T, choose a (finite or infinite) orthonormal basis of $\ker(T)$. The collection $\{e_n\}_{n \geq 1}$ of all these vectors is an orthonormal set since, by Theorem 3.35, eigenvectors corresponding to different eigenvalues must be orthogonal. Let M be the closure of the finite linear combinations of $\{e_n\}_{n \geq 1}$. Note that $T : M \to M$ by continuity, and $T : M^\perp \to M^\perp$ by self-adjointness.[133] Let $\tilde{T} : M^\perp \to M^\perp$ be the restriction of T to the Hilbert space $M^\perp$. One can easily check that $\tilde{T}$ is compact and self-adjoint since T is. Moreover, any eigenvalue of $\tilde{T}$ is an eigenvalue of T, and since all eigenvalues of T are in M, we deduce from Theorem 3.37 that $\tilde{T} = 0$. But then $M^\perp = \{0\}$ since if $y \in M^\perp$, then $Ty = 0$, which implies $y \in M$ by the definition of M. Therefore $M = H$ and (3.14) follows. Finally, if T has infinitely many nonzero eigenvalues, since by Theorem 3.35 and Theorem 3.37 these all must lie in the interval $[-\|T\|, \|T\|]$, they will have an accumulation point $\lambda \in [-\|T\|, \|T\|]$. Assume that $\lambda_{n_k} \to \lambda$ and let e_{n_k} be eigenvectors from the collection $\{e_n\}_{n \geq 1}$ such that $Te_{n_k} = \lambda_{n_k} e_{n_k}$. Since T is compact and $\{e_{n_k}\}$ is a bounded sequence, the sequence $\{Te_{n_k}\}$ has a convergent subsequence, so that the sequence with elements $e_{n_k} = \frac{1}{\lambda_{n_k}} Te_{n_k}$ has also a convergent subsequence if $\lambda \neq 0$. But this is impossible since, as in the beginning of the proof, this subsequence is not Cauchy. The only possibility is that $\lambda = 0$. Since this holds for any convergent subsequence, we deduce that $\lambda_n \to 0$. The proof is complete. □

In Theorem 3.38, $\lambda = 0 \in \sigma(T)$ is either an eigenvalue whose eigenspace might have finite or infinite dimension (see Exercise 64), or belongs to the continuous spectrum of T; recall that the residual spectrum of T is empty, see Theorem 3.36.

Example Let H be a separable complex Hilbert space H with orthonormal basis $\{e_n\}_{n \geq 1}$. As a particular case of the operators discussed in the example that precedes Theorem 3.39, the operator $Tx = \sum_{n \geq 1} \frac{1}{n} (x, e_n) e_n$ for $x = \sum_{n \geq 1} (x, e_n) e_n$, is compact and self-adjoint, with $\sigma(T) = \{0\} \cup \{\frac{1}{n}\}_{n \geq 1}$. In this case $\lambda_n = \frac{1}{n}$ (with $n \geq 1$) is an eigenvalue with one-dimensional eigenspace $\{ze_n : z \in \mathbb{C}\}$, while $0 \in \sigma(T)$ belongs to the continuous spectrum of T since T is injective. □

133 Indeed, if $y \in M^\perp$, then $(Tx, y) = 0$ for all $x \in M$, so that $(x, Ty) = 0$ and thus $Ty \in M^\perp$.

As a consequence of Theorem 3.38, we can prove the following result that is often more useful in applications.

Theorem 3.39 (Canonical form for compact operators) *If $A \in \mathcal{L}(H)$ is a compact operator on the separable complex Hilbert space H, then there exist orthonormal sets $\{\phi_n\}_{n\geq 1}$ and $\{\psi_n\}_{n\geq 1}$ (not necessarily complete, finite or infinite) and positive numbers*[134] *λ_n, which converge to zero if their number is infinite, such that*

$$Ax = \sum_{n\geq 1} \lambda_n \, (x, \psi_n) \, \phi_n \quad \textit{for all} \quad x \in H \,, \tag{3.15}$$

with the sum in (3.15), which may be finite[135] *or infinite, convergent in norm.*

Proof The operator A^*A is clearly self-adjoint[136] and, by Theorem 3.29, it is also compact. Therefore Theorem 3.38 ensures the existence of an orthonormal set $\{\psi_n\}_{n\geq 1}$ (possibly finite) such that $A^*A\psi_n = \mu_n\psi_n$ with $\mu_n \neq 0$ and such that $A^*Ax = 0$ for all x in the subspace of H that is orthogonal to $\{\psi_n\}_{n\geq 1}$. Since $\mu_n = (A^*A\psi_n, \psi_n) = (A\psi_n, A\psi_n) = \|A\psi_n\|^2$, each $\mu_n > 0$. Set $\lambda_n = \sqrt{\mu_n}$ and $\phi_n = \frac{1}{\lambda_n} A\psi_n$. It is easy to check that the ϕ_ns are orthonormal and that (3.14) with $T = A^*A$ and $e_n = \psi_n$ yields (3.15). □

3.5 Fréchet spaces

Let F be a linear (real or complex) vector space. A countable collection of semi-norms[137] $\{\|\cdot\|_n\}_{n\geq 0}$ defines a unique topology such that $f_j \to f$ for a sequence $\{f_j\}_{j\geq 0}$ if and only if $\lim_{j\to\infty} \|f_j - f\|_n = 0$ for all $n \geq 0$. The topology is *metrisable*,[138] meaning that its open sets are precisely those given by the metric $d(x, y) = \sum_{n\geq 0} \frac{1}{2^n} \frac{\|x-y\|_n}{1+\|x-y\|_n}$, and *translation-invariant*, in the sense that $d(x + x_0, y + x_0) = d(x, y)$ for all x, y, $x_0 \in F$. The topology is called *Hausdorff* if $f = 0$ when $\|f\|_n = 0$ for all $n \geq 0$. The space F is called a *Fréchet space*[139] if every Cauchy sequence[140] converges in F.

Examples (i) A Banach space $(X, \|\cdot\|)$ is a Fréchet space: set $\|x\|_n = n\|x\|$ for $x \in X$ and $n \geq 0$.

134 Called the *singular values* of A.

135 Note that if H is finite-dimensional, then the sum must be finite.

136 In general, using the definition of the adjoint, one can easily check that $(AB)^* = B^*A^*$ for all A, $B \in \mathcal{L}(H)$.

137 See Section 3.4.2 for the definition of a semi-norm.

138 We consider countably many semi-norms but it is possible to take instead of the natural numbers any index set I, in which case the topology is metrisable if and only if it can be defined by a countable collection of semi-norms; see Köthe (1969).

139 Or a complete metrisable locally convex topological vector space, which can also be defined in purely topological terms (with convex, balanced neighbourhoods leading to the construction of semi-norms).

140 A sequence $\{f_j\}_{j\geq 0}$ is called Cauchy if $\lim_{j,k\to\infty} \|f_j - f_k\|_n = 0$ for all $n \geq 0$.

(ii) $\mathbb{R}^\infty$, the space of all sequences $\{a_j\}_{j\ge 0}$ of real numbers, is a Fréchet space if we define the semi-norms $\|\{a_j\}_{j\ge 0}\|_n = \sum_{j=0}^n |a_j|$ for $n \ge 0$.

(iii) $C^\infty[a,b]$, the space of smooth functions $f : [a,b] \to \mathbb{R}$ is a Fréchet space if we set $\|f\|_n = \sum_{j=0}^n \sup_{t\in[a,b]}\{|f^{(n)}(t)|\}$ for $n \ge 0$.

(iv) $C(\mathbb{R})$, the space of all continuous functions $f : \mathbb{R} \to \mathbb{R}$ is a Fréchet space with the semi-norms $\|f\|_n = \sup_{|t|\le n}\{|f(t)|\}$ for $n \ge 0$.

(v) The *Schwartz class* $\mathcal{S}(\mathbb{R}) = \{f \in C(\mathbb{R}) : \sup_{t\in\mathbb{R}}\{|t|^n|f^{(m)}(t)|\} < \infty$ for all $n, m \ge 0\}$ is a Fréchet space with the countable family of semi-norms $\|f\|_{n,m} = \sup_{t\in\mathbb{R}}\{|t|^n|f^{(m)}(t)|\}$ for $n, m \ge 0$.

(vi) Consider the vector space $C_c(\mathbb{R})$ of all continuous functions $f : \mathbb{R} \to \mathbb{R}$ with compact support. Note that this is a proper subset of $C(\mathbb{R})$, while, for any $n \ge 0$, the vector space $C[-n,n] = \{f : [-n,n] \to \mathbb{R} : f \text{ continuous}\}$ is a Banach space with the norm $\|f\|_n = \sup_{t\in[-n,n]}\{|f(t)|\}$. We define[141] the topology of $C_c(\mathbb{R})$ by saying that $f_j \to f$ if there exists some $n \ge 0$ with $f(t) = f_j(t) = 0$ for $|t| \ge n$ and $\sup_{t\in[-n,n]}\{|f_j(t) - f(t)|\} \to 0$ as $j \to \infty$. $C_c(\mathbb{R})$ is a locally convex topological vector space, with the topology defined by the family of semi-norms $\{\|\cdot\|_n\}_{n\ge 1}$. However, $C_c(\mathbb{R})$ is not a Fréchet space. The reason for this is that if it were, then it would be a complete metric space with the metric $d(f,g) = \sum_{n\ge 0} \frac{1}{2^n} \frac{\|f-g\|_n}{1+\|f-g\|_n}$. But then the Baire category theorem holds and $C_c(\mathbb{R})$ cannot be a countable union of nowhere dense subsets. But, we claim, $C_c(\mathbb{R}) = \bigcup_{n\ge 0} X_n$. Indeed, clearly X_n is a closed subset of $C_c(\mathbb{R})$ and we are left to show that it has empty interior. Assume that for some $f_0 \in X_n$ and some $\varepsilon > 0$, the open set $\{f \in C_c(\mathbb{R}) : d(f, f_0) < \varepsilon\}$ is contained in X_n. But $X_0 = \{0\}$ and for $n \ge 1$, the function $\tau_n : \mathbb{R} \to \mathbb{R}$, given by $\tau_n(s) = \frac{\varepsilon}{3}\cos\left(\frac{\pi s}{4n}\right)$ for $|s| \le 2n$ and $\tau_n(s) = 0$ for $|s| > n$, belongs to $X_{2n} \setminus X_n$ and is such that $d(f_0 + \tau_n, f_0) = d(0,\tau_n) = \sum_{k=0}^\infty \frac{1}{2^k}\frac{\|\tau_n\|_k}{1+\|\tau_n\|_k} \le \frac{\varepsilon}{3}\sum_{k=0}^\infty \frac{1}{2^k} < \varepsilon$. Thus each X_n has empty interior and therefore $C_c(\mathbb{R})$ is not a Fréchet space.

(vii) Considerations similar to those in item (vi) above show that the test functions $\mathcal{D}(\mathbb{R}) = C_c(\mathbb{R}) \cap C^\infty(\mathbb{R})$ form a locally convex topological vector space but not a Fréchet space. □

There are similarities but also important differences between Banach spaces and Fréchet spaces. Some Fréchet spaces are normed[142] linear spaces (in which case all the semi-norms in a collection defining the topology may be taken to be norms by adding the one norm to all of them), e.g. Banach spaces but also $C^\infty[0,1]$, while others admit none, e.g. $\mathbb{R}^\infty$ and $C(\mathbb{R})$. Moreover, a closed subspace Y of a Fréchet space X is also a Fréchet space, as is the quotient X/Y. However, although if a Fréchet space X admits a norm, so does any closed subspace Y, a quotient by a closed subspace may not;[143] for these matters we refer to Hamilton (1982) and Köthe (1969).

Why bother with imposing semi-norms and not simply say that a topological vector space X (that is, a vector space with a topology compatible with the vector space operations, i.e. such that $(x,y) \mapsto x + y$ and $(\lambda, x) \to \lambda x$ are both continuous) is Fréchet if there is a translation-invariant metric d generat-

141 This procedure, called *inductive limit*, is suggested by the fact that $C_c(\mathbb{R})$ is the ascending union of its subspaces $X_n = \{f \in C_c(\mathbb{R}) : |f(t)| = 0 \text{ for } |t| \ge n\}$ with $n \ge 1$, each X_n being closed in its successor X_{n+1}.

142 A semi-norm $\|\cdot\|$ is called a norm if $f = 0$ whenever $\|f\| = 0$.

143 For example, the space $Y = C^\infty_{\text{per}}[0,1]$ of smooth 1-periodic functions $f : \mathbb{R} \to \mathbb{R}$ can be regarded as a closed subspace of $X = C^\infty[0,1]$, and the quotient X/Y can be identified with $\mathbb{R}^\infty$ by setting $a_j = f^{(j)}(1) - f^{(j)}(0)$ for $j \ge 0$.

ing the topology and such that (X, d) is complete? It turns out that, as indicated by the discussion in Section 3.4.2, the semi-norms generate a base of convex neighbourhoods of 0 (and, by translation-invariance, of any point in the vector space), and convexity is intimately connected with the existence of nontrivial continuous linear functionals. Without semi-norms there is the definite possibility that the only such functional is the trivial one.

Example For $0 < p < 1$ let $X = L^p[0,1]$ be the space of measurable functions $f : [0,1] \to \mathbb{R}$ with $\|f\|_p = \int_0^1 |f(s)|^p ds < \infty$. This is a topological vector space with the translation-invariant metric[144] $d(f,g) = \|f - g\|_p$, which makes it complete; see Day (1940). We claim that the only continuous linear functional $L : X \to \mathbb{R}$ is $L \equiv 0$. For this, assume the existence of a nontrivial continuous linear map $L : X \to \mathbb{R}$. Then there is some $f \in X$ such that $L(f) \geq 1$. Using this f, map $[0,1]$ continuously and monotonically to $[0, M]$, where $M = \|f\| > 0$, by means of $t \mapsto \int_0^t |f(s)|^p ds$. There is some $t_0 \in (0,1)$ such that $\int_0^{t_0} |f(s)|^p ds = M/2$. Define $g_1, g_2 : [0,1] \to \mathbb{R}$ by

$$g_1(s) = \begin{cases} f(s), & 0 \leq s \leq t_0, \\ 0, & t_0 < s \leq 1, \end{cases} \qquad g_2(s) = \begin{cases} 0, & 0 \leq s \leq t_0, \\ f(s), & t_0 < s \leq 1, \end{cases}$$

so $f = g_1 + g_2$ and $|f|^p = |g_1|^p + |g_2|^p$, with $\int_0^1 |g_1(s)|^p ds = \int_0^1 |g_2(s)|^p ds = M/2$. Since $L(f) \geq 1$ we must have $L(g_j) \geq 1/2$ for some index $j \in \{1,2\}$. Setting $f_1 = 2g_j$ we get $L(f_1) \geq 1$ and $\int_0^1 |f_1(s)|^p ds = 2^p \int_0^1 |g_j(s)|^p ds = 2^{p-1} M$, with $2^{p-1} < 1$. Iterating this procedure we obtain a sequence $\{f_n\}_{n \geq 1}$ in X with $L(f_n) \geq 1$ and $\|f_n\|_p = \int_0^1 |f_n(s)|^p ds = (2^{p-1})^n M \to 0$ for $n \to \infty$, a contradiction with the continuity of L. □

In contrast to the previous example, the dual of a Fréchet space is large.

Theorem 3.40 (Hahn–Banach for Fréchet spaces) *If F is a Fréchet space and $X \subset F$ is a closed linear subspace, then for every $x_0 \in F \setminus X$ there is a continuous linear functional $L : F \to \mathbb{R}$ (or $\mathbb{C}$) with $L(x) = 0$ for all $x \in X$ and with $L(x_0) = 1$.*

For a proof of Theorem 3.40 we refer to Köthe (1969). Despite the similarity between Theorem 3.40 and the corresponding result for Banach spaces (see Section 3.4.2), there is a significant difference. Namely, the dual of a Fréchet space is too large: the dual of a Fréchet space X which is not itself a Banach space is never a Fréchet space, reflecting the fact that the vector space of continuous linear maps from X to another Fréchet space will not form a Fréchet space. For example, the dual of $\mathbb{R}^\infty$ is the family of sequences with all entries except finitely many equal to zero, the dual of $C^\infty[a,b]$ is the space of distributions (see Chapter 6), the dual of $C(\mathbb{R})$ is the space of measures with compact

144 The reason why we do not use $\left(\int_0^1 |f(s)|^p ds\right)^{1/p}$, as for $p \in (1,\infty)$, thus paying the price of a loss of homogeneity since $\|\lambda f\|_p = \lambda^p \|f\|_p$ for $\lambda > 0$, is that we want $\|f + g\|_p \leq \|f\|_p + \|g\|_p$ for all $f, g \in L^p[0,1]$. This is granted since for $s, t > 0$ we have $(s+t)^p < s^p + t^p$, being equivalent, after dividing by $(s+t)^p$, to $1 < \varepsilon^p + (1-\varepsilon)^p$ for $\varepsilon = s/(s+t) \in (0,1)$, and this last inequality is a consequence of the fact that the right side attains its minimum on $[0,1]$ at the endpoints, being strictly monotone for $\varepsilon \in (0,1/2)$ and for $\varepsilon \in (1/2,1)$. On the other hand, for $f = \chi^{[0,1/2]}$ and $g = \chi^{[1/2,1]}$ we have $1 = \|f+g\|_p^{1/p} > \|f\|_p^{1/p} + \|g\|_p^{1/p} = 2^{1-1/p}$.

support, the dual of $\mathcal{S}(\mathbb{R})$ are the tempered distributions $\mathcal{S}'(\mathbb{R})$, discussed in Chapter 6, and the dual of $C_c(\mathbb{R})$ is the set of regular Borel measures on $\mathbb{R}$; for these matters we refer to the discussion in Hamilton (1982).

We conclude our discussion by pointing out the validity of all three fundamental principles (Hahn–Banach, open mapping, uniform boundedness) in Fréchet spaces: in addition to Theorem 3.40, we also have the following.

Theorem 3.41 (Open mapping theorem) *Let F_1 and F_2 be real or complex Fréchet spaces. If $T : F_1 \to F_2$ is a continuous surjective linear map, then T maps open sets into open sets.*

Theorem 3.42 (Closed graph theorem) *Let F_1 and F_2 be real or complex Fréchet spaces. A linear map $T : F_1 \to F_2$ is continuous if and only if its graph is closed in*[145] *$F_1 \times F_2$.*

Theorem 3.43 (Uniform boundedness principle) *Let F_1 and F_2 be real or complex Fréchet spaces. If $L_a : F_1 \to F_2$ is a family of linear maps, indexed by $a \in A$, such that for every $x_0 \in X$ and every semi-norm $\|\cdot\|_j$, the set $\{\|L_a x_0\|_j : a \in A\} \subset \mathbb{R}$ is bounded, then*[146] *for each $j \geq 0$ there is a semi-norm $\|\cdot\|_{k(j)}$ of F_2 and some constant $C_j > 0$ such that $\|L_a x\|_{k(j)} \leq C_j \|x\|_j$ for all $(a, x) \in A \times X$.*

Recall that in a Banach space $(X, \|\cdot\|)$ a set $\mathcal{M}$ is called bounded if there is some $M > 0$ with $\|x\| \leq M$ for all $x \in \mathcal{M}$. In contrast to this, since the distance d in a Fréchet space F is always bounded, we say that a set $B \subset F$ is *bounded* if for any open set O containing $0 \in F$ there is some $\varepsilon > 0$ such that $\varepsilon B = \{\varepsilon x : x \in B\} \subset O$.

3.6 Exercises

Some exercises illustrate concepts that precede (in our presentation) the development of the tools necessary for their proofs. This explains why their enumeration anticipates that of the bulk of the exercises devoted to a specific section – their statement being relevant for the material in earlier sections (even if by proof they belong to a later section).

[145] The space $F_1 \times F_2$, called the direct sum of the two spaces (sometimes denoted $F_1 \bigoplus F_2$), is a Fréchet space if F_1 and F_2 are so.

[146] If we do not insist on countably many semi-norms in the definition of a Fréchet space – some authors do this – then the uniform boundedness principle is messier since some Baire category assumption is needed (which is automatic in our setting).

3.6.1 Statements

1. Let H be a real pre-Hilbert space. Prove the equivalence of the following statements:

(i) x and y are orthogonal;

(ii) $\|x+y\|^2 = \|x\|^2 + \|y\|^2$;

(iii) $\|x\| \le \|x+\lambda y\|$ for all $\lambda \in \mathbb{R}$;

(iv) $\|x+y\| = \|x-y\|$.

Interpret these equivalences geometrically.

2. Show that the equivalences from Exercise 1 fail for a complex pre-Hilbert space H. In this setting, show that the following statements are equivalent:

(i) $(x,y)=0$;

(ii) $\|x+y\|^2 = \|x\|^2 + \|y\|^2$ and $\|x+iy\|^2 = \|x\|^2 + \|y\|^2$;

(iii) $\|x\| \le \|x+\lambda y\|$ for all $\lambda \in \mathbb{C}$;

(iv) $\|x+y\| = \|x-y\|$ and $\|x+iy\| = \|x-iy\|$.

3. (i) Show that the inner product $(\cdot,\cdot)$ on a pre-Hilbert space H is continuous map from $H \times H$ to $\mathbb{C}$.

(ii) Show that a normed linear space is pre-Hilbert if and only if the parallelogram law (3.3) holds.

4. (i) Prove that an orthonormal set $\{x_i\}_{i\in I}$ in a pre-Hilbert space H must be linearly independent and show that the converse is not necessarily true.

(ii) Prove that any orthonormal set $\{x_i\}_{i\in I}$ in a pre-Hilbert space H can be extended to a maximal orthonormal set.

5. Show that in a real pre-Hilbert space the Pythagorean relation $\|x+y\|^2 = \|x\|^2 + \|y\|^2$ is equivalent to x and y being orthogonal. Is this statement true in complex pre-Hilbert spaces?

6. Show that in a pre-Hilbert space $\|x+y\| = \|x\| + \|y\|$ is only possible if x and y are collinear, that is, if there is a scalar $\lambda \ge 0$ with $x = \lambda y$ or $y = \lambda x$.

7. Let H be a pre-Hilbert space. Assume that for some fixed $x, y \in H$ we have $\|\lambda x + (1-\lambda)y\| = \|x\|$ for all $\lambda \in [0,1]$. Prove that $x = y$ and interpret this result as follows: in a pre-Hilbert space, the sphere $\{x \in H : \|x - x_0\| = R\}$ of radius $R > 0$, centred at some $x_0 \in H$, does not have "flat edges".

8. Let H be the pre-Hilbert space of continuous functions $f : [0,1] \to \mathbb{C}$, endowed with the scalar product $(x,y) = \int_0^1 x(t)\,\overline{y(t)}\,dt$. Show that the subspace $M \subset H$ of functions $x \in H$ with $\int_0^{1/2} x(t)\,dt = \int_{1/2}^1 x(t)\,dt$ is closed in H. If $x_0(t) = -4t$ for $t \in [0,1]$, prove that $d(x_0, M) = 1$ but there is no $y_0 \in M$ with $\|x_0 - y_0\| = 1$.

9. A set K in a (real or complex) vector space is called convex if $tx + (1-t)y \in K$ for all $t \in [0,1]$ and all $x, y \in K$. Show that if K is a closed convex subset of a Hilbert space H, then for every $x \in H$ there is a unique solution $\pi_K(x)$ to the minimisation problem

$\inf_{y \in K}\{\|x - y\|\}$, with $\pi_K(x)$ characterised by the one-sided condition[147] $\Re e\{(x - \pi_K(x), y - \pi_K(x))\} \le 0$ for all $y \in K$.

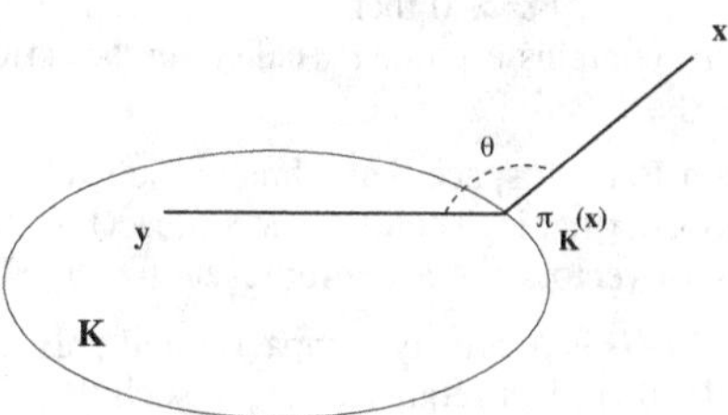

Figure 3.6 The inequality characterising the minimiser in Exercise 9 says that the angle θ determined by the vectors $x - \pi_K(x)$ and $y - \pi_K(x)$ is at least $\pi/2$.

10. Show that $M^\perp$ is a closed linear subspace for every set $M \neq \emptyset$ in a Hilbert space, while $(M^\perp)^\perp = \overline{\text{span}(M)}$, where $\overline{S}$ denotes the closure of the set S and span(M) is the set of finite linear combinations of vectors in M.

11. Prove that l^2 is a separable Hilbert space.

12. (i) Let K be a closed convex set in a Hilbert space H. Show that there is precisely one element of minimal norm in K.

(ii) If $K \subset \mathbb{C}^n$ is closed, convex and bounded, show that there are elements of maximal norm in K. Are they unique?

(iii) Prove that $K = \{x \in l^2 : \sum_{k=1}^{\infty} (1 + \frac{1}{k})^2 |x(k)|^2 \le 1\}$ is a closed, bounded and convex subset of the Hilbert space l^2, with no element of maximal norm.

13. Show that the space H of continuous functions $x : [0, 1] \to \mathbb{C}$ with the scalar product $(x, y) = \int_0^1 x(t)\,\overline{y(t)}\,dt$ is a pre-Hilbert space but not a Hilbert space. (The material in Section 4.3 and Exercise 4.14 show that its completion is $L^2[0, 1]$.)

14. Prove that the Gram–Schmidt process produces an orthonormal set spanning a vector space of the same dimension.

15. (i) Show that a Hilbert space H admits a finite or countably infinite orthonormal basis if and only if it is separable. Prove that there is a bijection between any two orthonormal bases of a Hilbert space H.

(ii) Prove that infinite-dimensional Banach spaces have uncountable Hamel bases.

(iii) Let AP be the closure, with respect to uniform convergence on $\mathbb{R}$, of the linear space generated by complex-valued periodic continuous functions $f : \mathbb{R} \to \mathbb{C}$ (of arbitrary period). Show that, with the scalar product $\lim_{T\to\infty} \frac{1}{2T} \int_{-T}^{T} f(t)\,\overline{g(t)}\,dt$, AP is a pre-Hilbert space. Moreover, the family $\{\phi_a\}_{a\in\mathbb{R}}$, where $\phi_a(t) = e^{iat}$, is an uncountable orthonormal set, with the property that for every $f \in AP$, at most a countable number

[147] The connection between the minimisation problem and a system of inequalities is reminiscent of the situation encountered at points $t_0 \in [0, 1]$ where a differentiable function $F : [0, 1] \to \mathbb{R}$ attains its minimum. Indeed, stating that $F'(t_0)(t_0 - t) \le 0$ for all $t \in [0, 1]$ summarises the fact that either $t_0 \in (0, 1)$ and $F'(t_0) = 0$, or $t_0 = 0$ and $F'(0) \ge 0$, or $t_0 = 1$ and $F'(1) \le 0$.

of the Fourier coefficients $f_a = \lim_{T\to\infty} \frac{1}{2T}\int_{-T}^{T} f(t)e^{-iat}dt$ are nonzero.[148] (One can prove that AP is an example of a nonseparable Hilbert space, $\{\phi_a\}_{a\in\mathbb{R}}$ being an orthonormal basis, see Riesz and Sz.-Nagy (1955). Moreover, $f:\mathbb{R}\to\mathbb{C}$ belongs to AP if and only if it is continuous and for every $\varepsilon > 0$ there is a number $L(f,\varepsilon) > 0$ such that every interval of length $L(f,\varepsilon)$ contains a point τ_ε, called an "ε-period" for f, such that[149] $\sup_{t\in\mathbb{R}} |f(t+\tau_\varepsilon) - f(t)| \le \varepsilon$.)

16. (i) Let $(X, \|\cdot\|)$ be a Banach space that admits a Schauder basis. Prove that a set $K \subset X$ is relatively compact (that is, its closure is compact) if and only if K is bounded and the basis expansion of vectors $x \in K$ converges uniformly.[150]

(ii) Show that a set $K \subset l_2$ is relatively compact if and only if it is bounded and has uniformly decaying tails: there is a sequence $\varepsilon_n \downarrow 0$ such that $\sum_{k\ge n} |x(k)|^2 \le \varepsilon_n^2$ for all $x \in K$. In particular, the Hilbert cube $\{x \in l^2 : |x(k)| \le 1/k\} \subset l^2$ is compact.

(iii) (Ascoli–Arzelà theorem) Prove that a sequence $f_n \in C[0,1]$ is relatively compact if it is equi-continuous and bounded, i.e. $\lim_{\varepsilon\to 0} \sup_{n\ge 1, |t-s|\le\varepsilon} |f_n(t)-f_n(s)| = 0$ and $\sup_{n\ge 1} \|f_n\| < \infty$, respectively. (These conditions are reasonably convenient to check, so that this criterion is often used. Although the Banach space $C[0,1]$ does admit a Schauder basis, the characterisation of relative compactness provided in (i) is impractical in the setting of[151] $C[0,1]$, due to the lack of algorithmic simplicity.)

17. (i) Prove that all finite-dimensional subspaces of a Banach space X are closed.

(ii) Show that in a infinite-dimensional Banach space there exists a nested sequence of infinite-dimensional closed subspaces X_n, with $X \supsetneq X_1 \supsetneq X_2 \supsetneq \dots$

(iii) Prove that infinite-dimensional Banach spaces have uncountable Hamel bases.

148 This property holds for any orthonormal set $\{e_\alpha\}_{\alpha\in I}$ of a Hilbert space $(H, (\cdot,\cdot))$, and enables us to define $x = \sum_{\alpha\in I}(x, e_\alpha)e_\alpha$ in the sense of unconditional convergence by requiring that for every $\varepsilon > 0$ there is a finite set $E \subset I$ such that $\|x - \sum_{\alpha\in F}(x, e_\alpha)e_\alpha\| < \varepsilon$ for all finite sets $F \subset I$ that contain E. The definition of the convergence of $\sum_{\alpha\in I} |(x, e_\alpha)|^2$ is similar. Going through the considerations made to prove Theorem 3.1, one can verify that the proof carries over in exactly the same way, with Bessel's inequality and the Parseval and Plancherel identities still valid, provided the meaning of convergence is modified in the same spirit.

149 Periodic continuous functions (of any period) belong to AP, while $f(t) = e^{2\pi it} + e^{2\pi it\sqrt{2}}$ is an example of an almost-periodic function that is not periodic. The importance of almost-periodic functions is illustrated by their occurence in the theory of ordinary differential equations. Indeed, the general solution to the homogeneous constant-coefficient differential equation $\frac{d^n x}{dt^n} + c_1 \frac{d^{n-1}x}{dt^{n-1}} + \dots + c_n x = 0$ is of the form $x(t) = \sum_{k=1}^m p_k(t)e^{\theta_k t}$, where $\theta_1,\dots,\theta_m$ are the distinct roots of the algebraic equation $z^n + c_1 z^{n-1} + \dots + c_n = 0$, and $p_k(t)$ are polynomials of degree less than the multiplicity of the corresponding exponents θ_k. A solution of this type is bounded if and only if each polynomial $p_k(t)$ which is not identically zero is a constant and corresponds to a purely imaginary root, $x(t) = \sum_{k=1}^m a_k e^{\lambda_k it}$ with $\lambda_1,\dots,\lambda_m \in \mathbb{R}$ and $a_1,\dots,a_m \in \mathbb{C}$, so that the bounded solutions are precisely the almost-periodic functions.

150 That is, there is a sequence of positive real numbers $\{\varepsilon_n\}_{n\ge 0}$ converging to zero such that $\|\sum_{k\ge n} x_k e_k\| = \|x - \sum_{k=1}^n x_k e_k\| \le \varepsilon_n$ for all $x \in K$. Here $x = \sum_{k\ge 1} x_k e_k$ is the expansion of $x \in X$ in the Schauder basis $\{e_k\}_{k\ge 1}$ of X.

151 This is also the case for the Banach space $L^1[0,1]$, where instead of using (i) in practical applications, one relies on the following alternative criterion: a subset K of $L^1(\mathbb{R})$ is relatively compact if and only if it is bounded, $\lim_{t\to 0} \int_{\mathbb{R}} |f(t+s) - f(s)|\,ds = 0$ uniformly in $f \in K$ and $\lim_{\alpha\to\infty} \int_{|s|\ge\alpha} |f(s)|\,ds = 0$ uniformly in $f \in K$; see Yosida (1995) for the proof. For the case of $L^1[0,1]$, this criterion simplifies since the third condition is automatically satisfied: think of $f \in L^1[0,1]$ as a particular type of function $f \in L^1(\mathbb{R})$ that happens to vanish on $\mathbb{R} \setminus [0,1]$.

(iv) Prove that infinite-dimensional Banach spaces admit nonclosed subspaces.

18. Let $(X, \|\cdot\|_X)$ and $(Y, \|\cdot\|_Y)$ be Banach spaces.

(i) Prove that if a linear operator $T : X \to Y$ is bounded, then its kernel, $\ker(T) = \{x \in X : Tx = 0\}$, is a closed subspace of X.

(ii) Is the condition "$\ker(T)$ is a closed linear subspace of X" sufficient for the linear operator $T : X \to Y$ to be continuous?

19. Let $(X, \|\cdot\|_X)$ and $(Y, \|\cdot\|_Y)$ be normed linear spaces with the same scalar field. If X is finite-dimensional, prove that any linear operator $T : X \to Y$ must be bounded. On the other hand, if X is infinite-dimensional and $Y \neq \{0\}$, show that there are unbounded linear operators $T : X \to Y$.

20. Let $(X, \|\cdot\|)$ be a Banach space and let f be a linear functional on X.

(i) Show that f is bounded if and only if its kernel $\ker(f) = \{x \in X : f(x) = 0\}$ is a closed subspace of X. (Compare this result with Exercise 18(ii).)

(ii) Show that if f fails to be continuous, then $\ker(f)$ is a dense proper subset of X. (For examples of discontinuous functionals see Exercise 19.)

21. Let X be the Banach space of all continuous functions $u : [0, 1] \to \mathbb{R}$ with $u(0) = 0$, endowed with the supremum norm $\|\cdot\|_X$. Show that $F(u) = \int_0^1 u(t)\,dt$ defines $F \in X^*$ with $\|F\| = 1$, but there is no $u \in X$ with $\|u\| = 1$ and $F(u) = 1$.

22. (i) Let X_1 be a closed subspace of a Banach space $(X, \|\cdot\|)$ and suppose that it admits a (topological) complement X_2. Show that the projections of X onto X_1 and X_2 are both continuous.

(ii) Show that, in a Banach space, finite-dimensional subspaces and closed subspaces of finite co-dimension[152] admit a complement.

23. (i) To appreciate the simple structure of orthonormal sets in Hilbert spaces, show that if $\{x_n\}_{n\geq 1}$ is an orthonormal set and M is the linear subspace spanned by $\{x_n\}_{n\geq 1}$, then any x in the closure $\overline{M}$ of M is of the form $x = \sum_{n=1}^{\infty} \alpha_n x_n$.

(ii) The property in (i) may be lost if $\{x_n\}_{n\geq 1}$ is just a linearly independent set. Indeed, if $\{e_n\}_{n\geq 0}$ is an orthonormal basis, show that $\{x_n\}_{n\geq 1}$, with $x_n = (\cos \frac{1}{n})\, e_0 + (\sin \frac{1}{n})\, e_n$, is a linearly independent set, e_0 belongs to the closure of the subspace spanned by the x_ns but e_0 cannot be written in the form $\sum_{n=1}^{\infty} \alpha_n x_n$.

24. (i) Show that the Euclidean distance $d(x, y) = \sqrt{(x_1 - y_1)^2 + (x_2 - y_2)^2}$ and the "rough" distance, defined by $d'(x, y) = 1$ if $x \neq y$ and $d'(x, y) = 0$ if $x = y$, both endow the plane $\mathbb{R}^2$ with a metric structure; here $x = (x_1, x_2) \in \mathbb{R}^2$ and $y = (y_1, y_2) \in \mathbb{R}^2$. Prove that the two structures are not identical by characterising convergence and open/closed sets in $(\mathbb{R}^2, d')$. Is $(\mathbb{R}^2, d')$ separable?

(ii) Two norms $\|\cdot\|_1$ and $\|\cdot\|_2$ on a linear space X are said to be equivalent[153] if there are constants $K_1, K_2 > 0$ with $K_1\|x\|_1 \leq \|x\|_2 \leq K_2\|x\|_1$ for all $x \in X$. Show that all norms on finite-dimensional spaces are equivalent, while any infinite-dimensional space admits two norms that are not equivalent.

[152] In the sense that finitely many vectors suffice to complete a Hamel basis of the subspace to a Hamel basis of the whole space. The fact that subspaces of finite co-dimension need not be closed is illustrated by Exercise 20(ii), since $\ker(f)$ for a linear functional f has co-dimension 1, irrespectively whether f is continuous or not.

[153] This means that the metrics generated by these norms are equivalent.

25. Banach spaces $(X, \|\cdot\|)$ with the property that $\|x+y\| = \|x\| + \|y\|$ only holds if x and y are collinear[154] are called strictly convex.

(i) Show that in Banach spaces that are not strictly convex, the unit sphere $S = \{x \in X : \|x\| = 1\}$ contains segments. (Examples of such spaces are $\mathbb{R}^2$ with the norm $\|(x_1, x_2)\| = |x_1| + |x_2|$ and the space $C[0,1]$ of continuous functions $f : [0,1] \to \mathbb{C}$ endowed with the supremum norm $\|f\|_\infty = \sup_{t\in[0,1]} |f(t)|$.)

(ii) Show that the space of continuous functions $f : [0,1] \to \mathbb{C}$ endowed with the norm $\|f\| = \|f\|_\infty + \left(\int_0^1 |f(t)|^2 dt\right)^{1/2}$ is a strictly convex Banach space.

26. Show that a Cauchy sequence $\{x_n\}_{n\geq 1}$ in a metric space (X, d) is bounded.

27. Let X be the linear space of all continuous functions $f : [0,1] \to \mathbb{R}$ with $f(0) = 0$, endowed with the supremum norm $\|f\| = \sup_{t\in[0,1]} |f(t)|$. Let M be the linear subspace of X consisting of all $f \in X$ with $\int_0^1 f(t)\, dt = 0$.

(i) Show that X is a Banach space and M is a closed proper subspace of X.

(ii) If $f_0 \in X \setminus M$ is such that $\|f_0\| = 1$, show that $0 < d(f_0, M) < 1$.

28. (i) Show that $C[0,1]$ is a Banach space.

(ii) (Bernstein polynomials) Associate to $f \in C[0,1]$ the polynomials $(B_n f)(t) = \sum_{k=0}^n f(\frac{k}{n})\, C_n^k t^k (1-t)^{n-k}$, where $C_n^k = \frac{n!}{k!(n-k)!}$. Prove that $B_n f \to f$ in $C[0,1]$. (This is a constructive proof of the Weierstrass theorem: any continuous function on $[0,1]$ is uniformly approximated by polynomials.)

(iii) Show that $C[0,1]$ is separable.

(iv) Show that for every integer $n \geq 0$, the space $C^n_{\mathrm{per}}[0,1]$ is a separable Banach space, with a compact injection $f \mapsto f$ from $C^m_{\mathrm{per}}[0,1]$ to $C^k_{\mathrm{per}}[0,1]$ whenever $m > k$.

29. Show that $L^\infty[0,1]$ is a nonseparable Banach space.

30. (i) Prove that $l^1(\mathbb{Z})$ is a separable Banach space.

(ii) Prove that $l^\infty(\mathbb{Z})$ is a nonseparable Banach space.

31. (i) For $\alpha \in \mathbb{C}$ with $|\alpha| \in (0,1)$, show that the finite linear combinations of $f_0 = (1, 0, 0, \dots)$ and $f_n = (1, \alpha^n, \alpha^{2n}, \alpha^{3n}, \dots)$ with $n \geq 1$ are dense in l^2.

(ii) Show that if $\{e_k\}_{k\geq 1}$ is an orthonormal basis of a Hilbert space H and if $\{f_k\}_{k\geq 1}$ is an orthonormal set with $\sum_{k\geq 1} \|e_k - f_k\|^2 < \infty$, then $\{f_k\}_{k\geq 1}$ is an orthonormal basis of H.

32. (i) Let $f : [0,1] \to \mathbb{R}$ be a continuous functions with $\int_0^1 f(t) t^k\, dt = \frac{1}{k+1}$ for all integers $k \geq 1$. Prove that $f \equiv 1$.

(ii) The Laguerre functions $\{\phi_n\}_{n\geq 0}$ are defined for $t \geq 0$ by $\phi_0(t) = e^{-t/2}$ and $\phi_n(t) = \frac{1}{n!} e^{t/2} \frac{d^n}{dt^n}(t^n e^{-t})$ if $n \geq 1$. Using the identity $\int_0^\infty t^k \frac{d^n}{dt^n}(t^n e^{-t})\, dt = \begin{cases} (-1)^n (n!)^2 \text{ for } k = n, \\ 0 \text{ for } 0 \leq k < n, \end{cases}$ prove that $\{\phi_k\}_{k\geq 0}$ is an orthonormal basis of $L^2[0,\infty)$.

(iii) Prove that the Hermite functions, $\psi_k(x) = \frac{(-1)^k}{\sqrt{2^k k!\, \sqrt{\pi}}}\, e^{x^2/2} \frac{d^k}{dx^k}(e^{-x^2})$ for $k \geq 0$, form an orthonormal basis of $L^2(\mathbb{R})$.

33. Prove that $BV_{\mathrm{per}}[0,1]$ is a nonseparable Banach space.

---

[154] That is, $x = \lambda y$ or $y = \lambda x$ for some scalar $\lambda \geq 0$.

34. For $\alpha \in (0, 1]$, show that the Hölder space h_α is a nonseparable Banach space. (In contrast to this, the closure h_α^0 of $C^1_{\text{per}}[0, 1]$ in h_α, called the *little Hölder space*, is a separable Banach space[155], $C^1_{\text{per}}[0, 1]$ being separable (see Exercise 28).)

35. Prove that the following metric/normed linear spaces are not complete and find their completions: (i) the real line $\mathbb{R}$ with the distance $d(x, y) = |\arctan(x) - \arctan(y)|$; (ii) $\mathbb{R}$ with the distance $d(x, y) = |e^x - e^y|$; (iii) the complex-valued polynomials with the norm $\|P\| = \max_{t\in[0,1]} |P(t)|$; (iv) the linear space of continuous functions $f : \mathbb{R} \to \mathbb{C}$ with compact support,[156] endowed with the norm $\|f\| = \sup_{x\in\mathbb{R}} |f(x)|$; (v) the bounded Riemann-integrable functions $f : [0, 1] \to \mathbb{C}$ with the norm $\|f\| = \int_0^1 |f(t)|\, dt$. (We anticipate from Exercises 4.14 and 4.16 that the space $L^1[0, 1]$ of measurable functions $f : [0, 1] \to \mathbb{C}$ with $\int_0^1 |f(t)|\, dt < \infty$ is a separable Banach space,[157] with $C[0, 1]$ as a dense subset.)

36. (i) Let the operator $T : l^2 \to l^2$ be defined by $Tx = (x(1), \frac{x(2)}{2}, \ldots, \frac{x(n)}{n}, \ldots)$ for $x = (x(1), x(2), \ldots, x(n), \ldots) \in l^2$. Compute its norm.

(ii) Let $T : X \to Y$ be a bounded linear operator between the Banach spaces $(X, \|\cdot\|_X)$ and $(Y, \|\cdot\|_Y)$. Show that if there is a constant $M > 0$ such that $\|Tx\|_Y \geq M\,\|x\|_X$ for all $x \in X$, then the range of T, $T(X) \subset Y$, is a closed linear subspace of Y.

(iii) Give an example of a bounded linear operator on a Hilbert space whose range is not closed.

37. (i) The right-shift $T : l^2 \to l^2$ maps $x = (x(1), x(2), \ldots)$ to $Tx = \big(x(2), x(3), \ldots\big)$. Show that T is a linear bounded operator that is surjective (onto) but not injective (one-to-one).

(ii) The left-shift $S : l^2 \to l^2$ maps $x = (x(1), x(2), \ldots)$ to $Sx = \big(0, x(1), x(2), \ldots\big)$. Show that S is a linear bounded operator that is injective but not surjective.

(iii) Give an example of two bounded linear operators $T, S : H \to H$, with H a

155 One can show that $f \in h_\alpha^0$ if and only if $f \in h_\alpha$ and $\sup\limits_{0<|t-s|<\delta} \frac{|f(t)-f(s)|}{(t-s)^\alpha} \to 0$ for $\delta \downarrow 0$, see Lunardi (2013), while for all $f \in h_\alpha^0$, $\|f - (B_n f)\|_\alpha \to 0$ as $n \to \infty$, see Bustamante and Jiménez (2000), where $B_n f$ are the Bernstein polynomials from Exercise 28. Without imposing periodicity, denote by H_α^0 the Hölder space of all functions $f : [0, 1] \to \mathbb{R}$ with $f(0) = 0$ and $\sup\limits_{0\leq|t-s|<\delta} \frac{|f(t)-f(s)|}{|t-s|^\alpha} \to 0$ as $\delta \to 0$. Define the functions $\Delta_k \in C[0, 1]$, where $\Delta_{-1}(t) = 1$ and $\Delta_0(t) = t$ for $t \in [0, 1]$, while $\Delta_n(t) = \Delta(2^j t - k)$, for $n = 2^j + k$ with $j \geq 0$ and $0 \leq k < 2^j$, is the restriction to $[0, 1]$ of the function obtained by translations and dyadic changes of scales from the triangular function $\Delta : \mathbb{R} \to [0, 1]$ given by $\Delta(t) = 2t$ if $0 \leq t \leq \frac{1}{2}$, $\Delta(t) = 2(1 - t)$ for $\frac{1}{2} \leq t \leq 1$, and $\Delta(t) = 0$ for $t < 0$ and $t > 1$. Then $\{\Delta_k\}_{k\geq -1}$ is a Schauder basis for $C[0, 1]$, $\{\Delta_k\}_{k\geq 0}$ is a Schauder basis of the closed subspace $C_0[0, 1]$ of functions $f \in C[0, 1]$ with $f(0) = 0$, and $\{\Delta_k\}_{k\geq 1}$ is a Schauder basis for the separable Banach space H_α^0, see Ciesielski (1960). Moreover, F belongs to the dual of H_α^0 if and only if $F(f) = \int_0^1 f(t)\, d\mu(t) + \iint_{[0,1]\times[0,1]} \frac{2f(t)-f(t+s)-f(t-s)}{s^2}\, d\nu(t, s)$ for all $f \in H_\alpha^0$, where μ and ν are signed measures on $[0, 1]$ and $[0, 1] \times [0, 1]$, respectively, see Hamadouche (2000).

156 Meaning that f vanishes outside a compact set.

157 A Schauder basis is the *Haar system* $\{\phi_k\}_{k\geq 1}$, defined as follows, see James (1982). ϕ_1 equals 1 on $[0, 1]$; ϕ_2 is 1 on $[0, \frac{1}{2})$ and -1 on $[\frac{1}{2}, 1]$; ϕ_3 is 1 on $[0, \frac{1}{4})$, -1 on $[\frac{1}{4}, \frac{1}{2})$, and 0 on $[\frac{1}{2}, 1]$; ϕ_4 is 1 on $[\frac{1}{2}, \frac{3}{4})$, -1 on $[\frac{3}{4}, 1]$, and 0 on $[0, \frac{1}{2})$. For integers $m \geq 1$ and $0 \leq k < 2^{m-1}$, ϕ_{2^m+k} is 1 on I^m_{2k-1}, -1 on I^m_{2k}, and 0 on $[0, 1] \setminus (I^m_{2k-1} \cup I^m_{2k})$, where $[0, 1]$ is partitioned into 2^m consecutive intervals $\{I^m_j\}_{1\leq j\leq 2^m}$ of equal lengths.

Hilbert space, such that $TS = I$ (identity), but such that neither T nor S is invertible. (Note: If $TS = ST = I$, then both operators are invertible, by definition.)

38. Let $X \neq \{0\}$ be a normed linear space. Show that there are no bounded linear operators $T, S : X \to X$ with $TS - ST = I$, where I is the identity.

39. Let X be a Banach space with a separable dual space X^*. If $\{f_n\}_{n\geq 1}$ is weak*-convergent to f in X^* and if $\{x_n\}_{n\geq 1}$ is weakly convergent to x in X, can we conclude that $\lim_{n\to\infty} f_n(x_n) = f(x)$?

40. (i) Prove the following variant of the closed graph theorem: If X, Y, Z are Banach spaces over the same scalar field, with a continuous inclusion $Y \subset Z$, then for a linear map $T : X \to Z$ the following statements are equivalent:

(a) $Tx \in Y$ for all $x \in X$ (qualitative regularity);

(b) $T : X \to Y$ is a bounded linear operator (quantitative regularity);

(c) $Tx \in Y$ for all x in a dense subset D of X, and for some constant $C > 0$ we have $\|Tx\|_Y \leq C\|x\|_X$ for all $x \in D$ (quantitative regularity on a dense subset).

(ii) Show that $(Tx)(t) = (1-t)^{1/2} \int_0^t x(s)s^{-1/3}ds$ defines a bounded linear map from $C_{\text{per}}[0,1]$ to $C^1_{\text{per}}[0,1]$. (This illustrates the method of *a priori* estimates to establish regularity of a linear map T on a Banach space X by using (i), once the *a priori* estimate $\|Tx\|_Y \leq C\|x\|_X$ for x in a dense subset of X is established.)

41. Let K be a closed convex subset of a reflexive Banach space X. Prove that $\inf\limits_{x\in K} \|x_0 - x\|$ is attained for any $x_0 \in X$. (This result extends Theorem 3.2.)

42. Let $x_n \rightharpoonup x$ be a weakly convergent sequence in a Banach space X.

(i) Prove that $\{x_n\}_{n\geq 1}$ is bounded, with $\|x\| \leq \liminf_{n\to\infty} \|x_n\|$. (Weak limits do not increase the norm.)

(ii) If X is a Hilbert space, show that $x_n \to x$ if and only if $\lim_{n\to\infty} \|x_n\| = \|x\|$.

(iii) Prove that both statements in (i) remain valid if, instead of being weakly convergent, $\{x_n\}_{n\geq 1}$ is weak*-convergent in the dual Y^* of the Banach space Y.

43. (i) Prove that $\{f_n\}_{n\geq 1}$ converges weakly to f in $C[0,1]$ if and only if the functions f_n are uniformly bounded and pointwise convergent.[158]

(ii) Prove that $C[0,1]$ is not a reflexive Banach space.

44. The algebraic dual of a vector space X is the vector space X' of all linear functionals on X. Let J be the natural map from X into X'' (the algebraic dual of X'), defined by $[J(x)](x') = x'(x)$ for $x \in X$ and $x' \in X'$. We say that X is *algebraically reflexive* if J is onto. Prove that X is algebraically reflexive if and only if X is finite-dimensional.

45. Let X be a Banach space with Schauder basis $\{e_k\}_{k\geq 1}$. The unique representation $x = \sum_{k\geq 1} x_k e_k$ defines the coordinate map $e_k^* : X \to \mathbb{C}$ by $e_k^*(x) = x_k$ for all $k \geq 1$. Prove that $e_k^* \in X^*$ and that for all $n \geq 1$, the canonical projections $P_n : X \to X_n$, defined by $P_n x = \sum_{k=1}^n x_k e_k$, are continuous onto the n-dimensional space X_n spanned by $\{e_1, \ldots, e_n\}$, with $\sup_{n\geq 1} \|P_n\|_{\mathcal{L}(X)} < \infty$.

46. (Mean-value theorem) A map u from $(0, 1)$ to a Banach space X is called differen-

158 That is, $\lim_{n\to\infty} f_n(t) = f(t)$ for all $t \in [0,1]$ and there exists some constant $M > 0$ with $|f_n(t)| \leq M$ for all $n \geq 1$ and all $t \in [0,1]$.

tiable at $t_0 \in (0,1)$ if $u'(t_0) = \lim\limits_{t \to t_0} \frac{u(t)-u(t_0)}{t-t_0} \in X$ exists. Show that[159] if u is differentiable on $(0,1)$, then $\|u(t) - u(s)\| \le (t-s) \sup_{\xi \in [s,t]} \|u'(\xi)\|$ for $0 < s < t < 1$.

47. Let X be a complex Banach space and G be an open domain of $\mathbb{C}$. A function $f : G \to X$ is called analytic if $\lim\limits_{h \to 0} \frac{f(z+h)-f(z)}{h}$ exists (in X) at every $z \in G$, while weakly analyticity means that the scalar function $z \mapsto l(f(z))$ is analytic in G (in the classical sense) for every $l \in X^*$. Prove the surprising result that a weakly analytic function must be analytic.

48. (i) Show that if X is a reflexive and separable Banach space, then X^* is separable. (This is a partial converse of Theorem 3.16.)

(ii) Show that a closed subspace Y of a reflexive Banach space X is also a reflexive Banach space.

49. By means of examples, show that completeness is essential for the validity of the uniform boundedness principle (Theorem 3.22), and of the inverse mapping and closed graph theorems (Theorems 3.24–3.25).

50. (i) Let X and Y be real Banach spaces and let $B : X \times Y \to \mathbb{R}$ be a separately continuous bilinear mapping.[160] Show that B is jointly continuous, that is, if $x_n \to 0$ in X and $y_n \to 0$ in Y, then $B(x_n, y_n) \to 0$.

(ii) Show by means of an example that for nonlinear functions $f : \mathbb{R}^2 \to \mathbb{R}$ separate continuity does not imply joint continuity.

51. Let $(H, (\cdot,\cdot))$ be a Hilbert space and $A, B : H \to H$ be linear operators such that $(Ax, y) = (x, By)$ for all $x, y \in H$. Show that A and B are bounded.

52. Give an example of a linear operator $U : H \to H$ that preserves[161] the inner product of the infinite-dimensional Hilbert space H without being onto. Is this possible in finite dimensions?

53. Let $(H, < \cdot, \cdot >)$ be a Hilbert space and $T : H \to H$ be a self-adjoint bounded linear operator. Prove that $\|T\| = \sup_{\|x\|=1} \{|(Tx, x)|\}$ and $\|T^2\| = \|T\|^2$. Give an example of a merely bounded linear operator for which the latter relation fails.

54. Let $T : H \to H$ be a bounded linear operator on a complex Hilbert space H. Show that T is self-adjoint if and only if $(Tx, x) \in \mathbb{R}$ for all $x \in H$.

55. The *Laplace transform* of a function $f : (0, \infty) \to \mathbb{C}$ is defined by $(\mathcal{L}f)(s) = \int_0^\infty f(t)\, e^{-st}\, dt$ for $s > 0$, provided that the integral exists.

(i) Given $1 \le p \le \infty$, show that $\mathcal{L}$ is not bounded as an operator from $L^p(0, \infty) \to L^p(0, \infty)$, unless $p = 2$, in which case $\|\mathcal{L}\| = \sqrt{\pi}$.

(ii) By computing $\mathcal{L}^2$, show that the linear integral operator $(Tf)(s) = \int_0^\infty \frac{f(t)}{t+s}\, dt$ is bounded from $L^2(0, \infty)$ to $L^2(0, \infty)$, and its norm is equal to π.

56. Let H be a separable Hilbert space with orthonormal basis $\{e_k\}_{k \ge 1}$. A bounded linear

[159] If $X = \mathbb{R}$ and if $u : (0,1) \to X$ is differentiable, then for every $0 < s < t < 1$ there exists some $\xi(s,t) \in (s,t)$ with $u(t) - u(s) = u'(\xi)\,(t-s)$. The example $u(t) = e^{6\pi i t}$ with $u(\frac{1}{3}) = u(\frac{2}{3})$ shows that this property fails already for $X = \mathbb{C}$.

[160] That is, for each fixed $x \in X$, $B(x, \cdot) : Y \to \mathbb{R}$ is a bounded linear functional, and for each fixed $y \in Y$, $B(\cdot, y) : X \to \mathbb{R}$ is a bounded linear functional.

[161] That is, $(Ux, Uy) = (x, y)$ for all $x, y \in H$.

operator $A : H \to H$ can be represented as an infinite matrix $\mathfrak{a}$ with respect to this basis, the mnth element of this matrix being $a_{nm} = (Ae_n, e_m)$.

(i) Prove that $\sum_{n\geq 1} |a_{nm}|^2 < \infty$ for every $m \geq 1$ and $\sum_{m\geq 1} |a_{nm}|^2 < \infty$ for every $n \geq 1$ both hold for such a matrix representation. Take $a_{nm} = (n+m)^{-2/3}$ for $n, m \geq 1$ to show that these conditions are not sufficient for the infinite matrix $\{a_{nm}\}_{n,m\geq 1}$ to correspond to a bounded linear operator.

(ii) For an infinite scalar matrix $\{a_{nm}\}_{n,m\geq 1}$, show that $\sum_{n\geq 1} \sum_{m\geq 1} |a_{nm}|^2 < \infty$ ensures the existence of a bounded linear operator $A : H \to H$ having this matrix representation with respect to the given basis, but this condition is not necessary. (The necessary and sufficient conditions[162] have limited practical applicability.)

57. To $f \in L^\infty[0,1]$ associate the multiplication operator $M_f : L^2[0,1] \to L^2[0,1]$, defined by $(M_f g)(t) = f(t)g(t)$ for $t \in [0,1]$.

(i) Taking for granted[163] that $e_k(t) = e^{2\pi i k t}$ with $k \in \mathbb{Z}$ form an orthonormal basis of $L^2[0,1]$, show that the entries $a_{km} = (M_f e_k, e_m)_{L^2[0,1]}$ of the doubly-infinite matrix $(k, m \in \mathbb{Z})$ corresponding to M_f depend on the difference $(k-m)$ only.[164]

(ii) Prove that M_f is compact if and only if $f = 0$ a.e.

58. Find a compact linear operator $T : l^2 \to l^2$ with nonclosed range.

59. Find the spectrum of the following operators:

(i) the identity operator $T : l^2 \to l^2$, defined by $Tx = x$ for $x \in l^2$;

(ii) the right-shift operator $T : l^2 \to l^2$, defined by $Tx = (0, x(1), x(2), \ldots)$;

(iii) the Volterra operator $T : L^2[0,1] \to L^2[0,1]$, defined by $(Tf)(t) = \int_0^t f(s)\,ds$ for $f \in L^2[0,1]$ and $t \in [0,1]$.

60. Let H be a separable complex Hilbert space. Show that $\sigma(T)$ is a subset of the unit circle $|\lambda| = 1$ if $T \in \mathcal{L}(H)$ is unitary.

61. Let H be a separable complex Hilbert space. Show that $T \in \mathcal{L}(H)$ is a *normal operator* (that is, $TT^* = T^*T$) if and only if $\|Tx\| = \|T^*x\|$ for all $x \in H$.

62. Let $T \in \mathcal{L}(X)$ for a Banach space X.

(i) Show that the *spectral radius* $r(T) = \lim\limits_{n\to\infty} \|T^n\|^{1/n}$ exists and satisfies $r(T) \leq \|T\|$. For $X = \mathbb{C}^2$ construct $T \in \mathcal{L}(X)$ with $r(T) = 0$ and $\|T\| = 1$.

(ii) Prove that $r(T) = \|T\|$ for a normal operator T in a complex Hilbert space.

63. For which $\lambda \in \mathbb{R}$ does the integral equation $f(x) - \lambda \int_0^1 e^{x+y} f(y)\,dy = 1$ have a solution $f \in L^2[0,1]$?

64. Find compact self-adjoint operators T on separable Hilbert spaces such that $\sigma(T) = \{0\}$, with 0 being an eigenvalue of given (finite or infinite) multiplicity.

[162] These are the necessary conditions from (i) supplemented by the requirement that $(\mathfrak{a}^*\mathfrak{a})^n$ is defined for all $n \geq 1$ and $\sup_{n\geq 1} \sup_{k\geq 1} |[(\mathfrak{a}^*\mathfrak{a})^n]_{kk}|^{1/n} < \infty$, see Crone (1971). Here $\mathfrak{a}^*$ is the conjugate transpose infinite matrix.

[163] We will prove this in Chapter 4.

[164] Such operators are called *Laurent operators*. One can show that if $T : L^2[0,1] \to L^2[0,1]$ is a bounded linear operator whose associated matrix has this structural property, then $T = M_f$ for some $f \in L^\infty[0,1]$, see Gohberg et al. (2003).

3.6.2 Hints

1. Square the expressions and use the properties of the scalar product. For the interpretation, draw the vectors in $\mathbb{R}^2$.

2. Square the expressions and use the properties of the scalar product.

3. (i) Use the Cauchy–Schwarz inequality. (ii) To define the scalar product starting from the norm, use the so-called polarisation identity $(x,y) = \frac{\|x+y\|^2-\|x-y\|^2}{4}$ in the real case and $(x,y) = \frac{1}{4}\sum_{k=0}^{3} i^k\|x+i^k y\|^2$ in the complex case.

4. (i) Assume linear dependence and take the scalar product with each of the vectors that are involved. For the counterexample, try out the setting of $\mathbb{R}^2$.

(ii) Adapt the approach pursued in proving the existence of a Hamel basis.

5. Develop the Pythagorean relation using the properties of the scalar product.

6. and **7.** Square the relation and use the properties of the scalar product.

8. Define $x_1(t) = 1$ for $t \in [0,1/2]$ and $x_1(t) = -1$ for $t \in (1/2,1]$, so that the constraint characterising those $y \in H$ that lie in M is $(y,x_1) = 0$. Using the relations $\|x_1\|^2 = (x_0,x_1) = 1$, show that $\|y-x_0\|^2 = \|y-(x_0-x_1)\|^2 + 1 \geq 1$ for all $y \in M$. Note that $x_0 - x_1 \notin M$, although $(x_0-x_1,x_1) = 0$.

9. For the existence and uniqueness of $\pi_K(x)$, inspect the proof of Theorem 3.2. For the characterisation, express the minimisation problem in term of square norms and recast it as an inequality for a polynomial expression of degree 2 in t.

10. Combine linearity with the properties of the scalar product.

11. For the Cauchy–Schwarz inequality, pass to the limit in finite sums. For completeness, identify first the scalar limit of the nth term.

12. In (i) use Exercise 9 and in (ii) use a compactness argument. For (iii), use estimates to show that the set is closed, convex and bounded. If x were an element of maximal norm with $x(k) \neq 0$, show that by replacing $x(k)$ by 0 and increasing appropriately the absolute value of $x(2k)$, without changing the other entries, the norm increases but the weighted sum remains the same.

13. Have a close look at Exercise 9 from Chapter 2.

14. Use induction.

15. In (i), take advantage of Theorem 3.1 and the Gram–Schmidt procedure. Also, if E and F are two orthonormal bases, for $e \in E$, set $F_e = \{f \in F : (e,f) \neq 0\}$ and use Bessel's inequality to show that each set F_e is finite or countably infinite. Baire's category theorem is relevant for (ii). In (iii), start from periodic functions to show that the scalar product is well-defined. As for the Fourier coefficients, rely on Bessel's inequality.

16. (i) To prove the necessity, use the uniform boundedness principle for the truncations $S_n(x) = \sum_{k=1}^n \alpha_k e_k$ of the expansion of x in the Schauder basis. For sufficiency rely on total boundedness.

(ii) Exhibit an orthonormal basis of l^2 and use (i). (iii) Prove total boundedness by producing for every $\varepsilon > 0$ a finite ε-net.

17. For (i)–(ii) rely on Hamel bases and the corollaries to the Hahn–Banach theorem. For (iii)–(iv) use Hamel bases and Theorem 3.21.

18. Part (i) follows from the definition and the continuity. In part (ii), try to exhibit an an explicit one-to-one unbounded linear operator as a diagonal operator acting on a Hamel basis of an infinite-dimensional normed linear space.

19. Consider in both settings a Hamel basis $\{e_i\}_{i\in I}$ of X and take advantage of the corresponding coordinates.

20. (i) One direction is easy. For the other, use the Hahn–Banach theorem.

(ii) Assuming otherwise, use the Hahn–Banach theorem to infer that $f \in X^*$.

21. Compute $F(t^{1/n})$ and use the continuity of $u \in X$.

22. Use the closed graph theorem in (i). In (ii), use Corollary 1 to Theorem 3.15 and Hamel bases, respectively.

23. In (i), look at the coefficients as elements of l^2. Use orthonormality in (ii).

24. For (i), show that $d'(x_n, x) \to 0$ if and only if the sequence $\{x_n\}_{n\geq 1}$ is eventually constant, and every subset of $\mathbb{R}^2$ is both open and closed with respect to d'. In (ii) rely on Hamel bases.

25. In (i), for the segment property, use the triangle inequality for $\|x + \beta y\|$ and $\|\alpha x + y\|$ with $\alpha, \beta \in (0, 1)$ if $\|x + y\| = \|x\| + \|y\|$. For (ii), note that $\|\cdot\|$ and $\|\cdot\|_\infty$ are equivalent norms, and use Exercise 6.

26. Rely on the definition.

27. Item (i) is clear. In (ii), show that $\|g - f_0\| \geq 1$ for all $g \in M$ would ensure $\left|\int_0^1 h(t)\,dt\right| \leq \left|\int_0^1 f_0(t)\,dt\right|$ for all $h \in X$ with $\|h\| = 1$. Construct a sequence $h_n \in X$ for which this inequality must fail from an index n onwards.

28. (i) For a Cauchy sequence $\{f_n\}_{n\geq 1}$, fix $t \in [0, 1]$ to identify the limit. (ii) To estimate $|f(t) - (B_n f)(t)|$, rely on the identities $\sum_{k=0}^n C_n^k t^k (1-t)^{n-k} = 1$ and $\sum_{k=0}^n (t - \frac{k}{n})^2 C_n^k t^k (1-t)^{n-k} = \frac{t(1-t)}{n}$ for $n \geq 1$. (iii) Rely on (ii) and the density of $\mathbb{Q}$ in $\mathbb{R}$. (iv) Use Exercise 16(iii).

29. Completeness follows analogous to the solution to Exercise 28(i), after eliminating the problematic sets of measure zero. Concerning separability, look at the family of characteristic functions of the sets $[0, t]$ with $t \in (0, 1)$.

30. Fix $k \in \mathbb{Z}$ to identify the limit. For the separability of $l^1(\mathbb{Z})$, consider sequences with finitely many nonzero entries, all of the form $a + ib$ with a and b rational numbers. As for $l^\infty(\mathbb{Z})$, consider sequences with all entries either 0 or 1.

31. In (i), formulate the problem in terms of power series. For (ii), rely on the Plancherel identity and on Bessel's inequaliy.

32. Use Exercise 28(ii) in (i). For (ii) and (iii), after checking inductively the orthonormality, introduce the analytic functions $z \mapsto \int_0^\infty f(t) e^{-t} e^{-2\pi i z t} dt$ and $z \mapsto \int_{\mathbb{R}} f(x) e^{-x^2/2} e^{-2\pi i x z} dx$, respectively, to show that only $f \equiv 0$ is orthogonal to the given orthonormal set.

33. To prove completeness, use partitions and look at the values at a fixed $x \in [0, 1]$. Note that for every $s \in (0, \frac{1}{2})$, the function f_s, defined by $f_s(x) = 0$ for $x \in [0, s] \cup [1 - s, 1]$ and $f_s(x) = 1$ for $x \in (s, 1 - s)$, belongs to $BV_{\text{per}}[0, 1]$.

34. To prove that h_α is a Banach space, relate it to the space $C_{\text{per}}[0, 1]$. For the separability issue, construct an uncountable set of functions at distance larger than 1 from one another.

35. In (i)–(ii) find suitable isometries. The Banach spaces $C[0, 1]$ and $L^1[0, 1]$ are relevant for (iii) and (v), respectively. In (iv), pay attention to $x \mapsto e^{-x^2}$, which suggests the class of continuous functions that vanish at infinity.

36. For (i), show first that $\|T\| \leq 1$, and find $x \neq 0$ with $Tx = x$ to conclude that $\|T\| = 1$. For (ii), associate to a convergent sequence in the range a Cauchy sequence in the pre-image, and rely on continuity. In (iii) use the operator from (i).

37. Parts (i)–(ii) are easy, and provide operators suitable for part (iii).

38. Assuming $TS - ST = I$, show that $TS^{n+1} - S^{n+1}T = (n + 1)S^n$ for all integers $n \geq 0$. Take now advantage of the fact that $\mathcal{L}(X)$ is a Banach algebra.

39. Consider an orthonormal basis of a separable Hilbert space.

40. For (i) use Theorem 3.25 to show that (a) implies (b), and rely on Y being complete to show that (c) implies (b). For (ii), choose $X = Z = C_{\text{per}}[0, 1]$ and $Y = C^1_{\text{per}}[0, 1]$ and verify (a) in item (i).

41. Use Theorem 3.19.

42. Use the Hahn–Banach theorem and the properties of the scalar product.

43. In (i) use the identification of the dual of $C[0,1]$ with $\mathcal{M}[0,1]$. For (ii), show that the sequence $t \mapsto (1-t)^n$ has no weakly convergent subsequence.

44. Construct dual Hamel bases in X and X'.

45. Use the open mapping theorem to prove that $\|x\|_0 = \sup_{n\geq 1} \|\sum_{k=1}^n x_k e_k\|$ is an equivalent norm on X.

46. Use Corollary 5 in Section 3.4.2.

47. Use the Cauchy integral formula and the uniform boundedness principle.

48. In (i), use the definition of reflexivity and Theorem 3.16. For (ii), combine Theorem 3.18 with Corollary 3 in Section 3.4.2.

49. Let X_0 be the subset of l^1 consisting of sequences with at most finitely many nonzero entries. For $n \geq 1$, define $T_n : X_0 \to X_0$ by $(T_n x)(k) = 0$ if $k \neq n$ and $(T_n x)(n) = nx(n)$. On the normed linear space $X = l^1$ with the norm $\|x\|_X = \sum_{i\in I} |\alpha_i|$, where $x = \sum_{i\in I} \alpha_i e_i$ is the representation of $x \in X$ with respect to a Hamel basis $\{e_i\}_{i\in I}$ of unit vectors of l^1, let $T : X \to l^1$ be given by $T(x) = x$.

50. Apply the uniform boundedness principle in (i). For (ii), define f by $f(0,0) = 0$ and $f(x,y) = \frac{xy}{x^2+y^2}$ if $(x,y) \neq (0,0)$.

51. Apply the closed graph theorem.

52. Use orthonormal bases.

53. Use the parallelogram law. For the example, take a nilpotent square matrix.

54. Rely on the scalar product.

55. For (i), to deal with $p \neq 2$, compute $\mathcal{L}(f_n)$ for $f_n(t) = e^{-\alpha_n t}$, where $\{\alpha_n\}_{n\geq 1}$ is a sequence of positive numbers. For $p = 2$, note that the kernel e^{-st} is not in $L^2((0,\infty) \times (0,\infty))$. Write first $f(t)e^{-st}$ as the product of $f(t)e^{-st/2}t^{1/4}$ and $e^{-st/2}t^{-1/4}$ to take advantage of the Cauchy–Schwarz inequality, and subsequently compute $\|\mathcal{L}(f_n)\|^2$ for $f_n(t) = t^{-1/2}\chi^{[1/n,n]}(t)$. In (ii), deal first with the case $f \geq 0$ and rely on Exercise 53.

56. In (i), express Ae_n and A^*e_m by means of Theorem 3.1, and evaluate (Ax, x) for $x = \sum_{n\geq 1} n^{-2/3} e_n$ for the second part. For the first part in (ii) rely on the Cauchy–Schwarz inequality, and note that the condition fails for the identity operator.

57. For (i), note that $fg \in L^2[0,1]$, while for (ii), assuming compactness, show that M_f is self-adjoint and use Theorem 3.37.

58. Take Exercise 36 into account.

59. For (ii), note that $\|T\| = 1$ and show that any $\lambda \in \mathbb{C}$ with $|\lambda| < 1$ belongs to the residual spectrum. For (iii), show that any $\lambda \neq 0$ belongs to $\rho(T)$ by relying on the Neumann expansion.

60. Show that $(\lambda I - T)$ is bounded from below if $|\lambda| \neq 1$.

61. Use the definition of a normal operator and Exercise 54.

62. For (i), set $a_n = \ln(\|T^n\|)$ for $n \geq 1$ and show that $a_{n+m} \leq a_n + a_m$ for $n, m \geq 1$, which ensures that $\lim_{n\to\infty} \{\frac{a_n}{n}\}$ exists and equals to $\inf_{n\geq 1} \{\frac{a_n}{n}\}$. For (ii), show inductively that $\|T^{2^n}\| = \|T\|^{2^n}$ for all $n \geq 1$, if T is a normal operator.

63. Use the Fredholm alternative.

64. Consider the operator $(T, 0)$ on $L^2[0,1] \bigoplus \mathbb{C}^n$, or on $L^2[0,1] \bigoplus l^2$, where T is the Volterra operator from Exercise 59(iii).

3.6.3 Solutions

1. We have $\|x+y\|^2 - \|x\|^2 - \|y\|^2 = (x+y, x+y) - (x, x) - (y, y) = 2(x, y)$, while $\|x+\lambda y\|^2 - \|x\|^2 = (x+\lambda y,\ x+\lambda y) - (x, x) = \lambda^2\|y\|^2 + 2\lambda(x, y)$ and $\|x+y\|^2 - \|x-y\|^2 = (x+y, x+y) - (x-y, x-y) = 4(x, y)$. The equivalence of (i), (ii) and (iv) and the fact that (i) implies (iii) are plain. If (iii) holds for all $\lambda \in \mathbb{R}$, the second expression above being a polynomial of degree 2 in λ forces $(x, y)^2 \leq 0$, that is, $(x, y) = 0$. As for the geometric interpretation, (ii) is Pythagora's theorem, (iii) means that the perpendicular is shorter than an oblique segment, while (iv) expresses the fact that a parallelogram is a rectangle if and only if its diagonals are equal (see Figure 3.7).

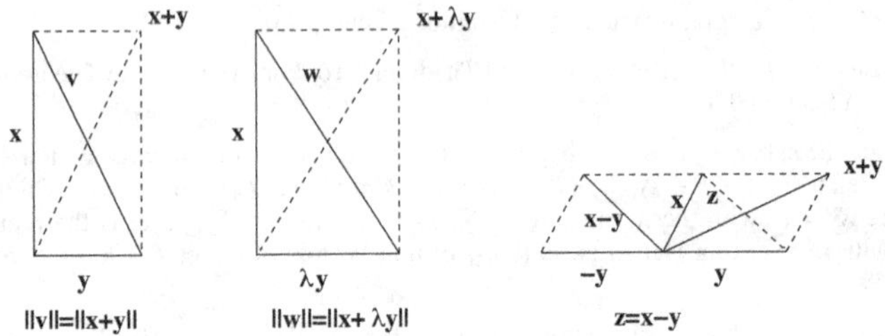

Figure 3.7 The geometric interpretation of the equivalences stated in Exercise 1.

2. We have $\|x+y\|^2 - \|x\|^2 - \|y\|^2 = (x+y, x+y) - (x, x) - (y, y) = 2\,\Re e\{(x, y)\}$, while $\|x+iy\|^2 - \|x\|^2 - \|y\|^2 = (x+iy, x+iy) - (x, x) - (y, y) = 2\,\Im m\{(x, y)\}$ and $\|x+\lambda y\|^2 - \|x\|^2 = (x+\lambda y,\ x+\lambda y) - (x, x) = |\lambda|^2\|y\|^2 + 2\Re e\{\overline{\lambda}(x, y)\}$. Also, $\|x+y\|^2 - \|x-y\|^2 = (x+y, x+y) - (x-y, x-y) = 4\Re e\{(x, y)\}$ and $\|x+iy\|^2 - \|x-iy\|^2 = (x+iy, x+iy) - (x-iy, x-iy) = 4\Im m\{(x, y)\}$. Each of the statements (ii) and (iv) of Exercise 1 amounts to $\Re e\{(x, y)\} = 0$, and the new additions amount to $\Im m\{(x, y)\} = 0$. Clearly (i) implies (iii). On the other hand, if (iii) would hold in a case when $(x, y) \neq 0$, then $y \neq 0$ and for $\lambda = -(x, y)/\|y\|^2$ the third identity above becomes $-|(x, y)|^2/\|y\|^2 < 0$.

3. (i) Let $x_n \to x$ and $y_n \to y$ in H. Then $|(x_n, y_n) - (x, y)| = |(x_n, y_n) - (x_n, y) + (x_n, y) - (x, y)| = |(x_n, y_n - y) + (x_n - x, y)| \leq |(x_n, y_n - y)| + |(x_n - x, y)| \leq \|x_n\| \cdot \|y_n - y\| + \|y\| \cdot \|x_n - x\|$. The boundedness of the real sequence $\{\|x_n\|\}_{n\geq 1}$, ensured by $x_n \to x$, yields $(x_n, y_n) \to (x, y)$.

(ii) In the real case we define $(x, y) = \frac{\|x+y\|^2 - \|x-y\|^2}{4}$. Properties (i) and (iii) of the scalar product are immediate. To show that (ii) holds, note that (3.3) yields $4(x+z, y) + 4(x-z, y) = \|x+z+y\|^2 - \|x+z-y\|^2 + \|x-z+y\|^2 - \|x-z-y\|^2 = (\|x+z+y\|^2 + \|x-z+y\|^2) - (\|x+z-y\|^2 + \|x-z-y\|^2) = 2(\|x+y\|^2 + \|z\|^2) - 2(\|x-y\|^2 + \|z\|^2) = 2(\|x+y\|^2 - \|x-y\|^2) = 8(x, y)$. For $z = x$ we get $(2x, y) = 2(x, y)$ and the previous identity becomes $(x+z, y) + (x-z, y) = (2x, y)$. Choosing $x = \frac{u+v}{2}$ and $y = \frac{u-v}{2}$, this yields $(u, y) + (v, y) = (u+v, y)$ for all $x, u, v \in H$. It remains to show that $(\lambda x, y) = \lambda(x, y)$ for all $x, y \in H$ and all $\lambda \in \mathbb{R}$. Linearity in the first component shows that the desired relation will hold for all integers λ. This, in turn, yields that it will hold for all rational numbers. Since the latter are dense in $\mathbb{R}$, we conclude by the continuity of $x \mapsto (x, y)$.

In the case of complex scalars, we define $(x, y) = \frac{\|x+y\|^2 + i\|x+iy\|^2 - \|x-y\|^2 - i\|x-iy\|^2}{4}$. Properties (i) and (iii) of the scalar product are easily checked. We can take advantage of the above considerations if we observe that $(x, y)_{\mathbb{R}} = \frac{\|x+y\|^2 - \|x-y\|^2}{4}$ satisfies all the properties of a real scalar product and $(x, y) = (x, y)_{\mathbb{R}} + i(x, iy)_{\mathbb{R}}$. Therefore, $(x+z, y) = (x+z, y)_{\mathbb{R}} + i(x+z, iy)_{\mathbb{R}} = (x, y)_{\mathbb{R}} + (z, y)_{\mathbb{R}} + i\{(x, iy)_{\mathbb{R}} + (z, iy)_{\mathbb{R}}\} = (x, y) + (z, y)$. Also, for $\lambda \in \mathbb{R}$ we get that $(\lambda x, y) = (\lambda x, y)_{\mathbb{R}} + i(\lambda x, iy)_{\mathbb{R}} = \lambda(x, y)_{\mathbb{R}} + i\lambda(x, iy)_{\mathbb{R}} = \lambda(x, y)$. It remains to check that $(ix, y) = i(x, y)$ for all $x, y \in H$. For this, note that $(ix, y) = (ix, y)_{\mathbb{R}} + i(ix, iy)_{\mathbb{R}} = -(x, iy)_{\mathbb{R}} + i(x, y)_{\mathbb{R}} = i\{(x, y)_{\mathbb{R}} + i(x, iy)_{\mathbb{R}}\} = i(x, y)$, using the definition of $(\cdot, \cdot)$ in the second step.

4. (i) If $\alpha_1 x_1 + \cdots + \alpha_n x_n = 0$, taking the scalar product of both sides with x_k $(1 \leq k \leq n)$ yields $\alpha_k = 0$ by orthonormality, thus proving the linear independence of the orthonormal set of vectors

$\{x_k\}_{1\le k\le n}$. Note that the unit vectors $(3/5, 4/5)$ and $(1, 0)$ in $\mathbb{R}^2$ are linearly independent without being orthogonal.

(ii) Let $\mathcal{B}$ denote the collection of all orthonormal sets in X that contain the given orthonormal set $\{x_i\}_{i\in I}$. Inclusion defines a partial ordering $\prec$ on $\mathcal{B}$, with the property that if $\{B_i\}_{i\in I}$ is a totally ordered subset of $\mathcal{B}$, then $B = \bigcup_{i\in I} B_i$ is an upper bound for it. By Zorn's lemma, $\mathcal{B}$ has a maximal element B_0.

5. Note that $\|x+y\|^2 - \|x\|^2 - \|y\|^2 = (x+y, x+y) - (x, x) - (y, y) = 2\Re e\{(x, y)\}$. Consequently, in a real/complex pre-Hilbert space the Pythagorean relation is equivalent to $(x, y) = 0$ and $\Re e\{(x, y)\} = 0$, respectively. In particular, in $\mathbb{C}^2$, $x = (i, i-1)$ and $y = (1-i, i)$ satisfy the Pythagorean relation although $(x, y) = 2i \neq 0$.

6. Assume that $x \neq 0$ and $y \neq 0$ are such that $\|x+y\| = \|x\| + \|y\|$. Then $0 = \|x+y\|^2 - (\|x\| + \|y\|)^2 = (x+y, x+y) - (\|x\|+\|y\|)^2 = 2\Re e\{(x, y)\} - 2\|x\|\cdot\|y\|$. For $\lambda = \|x\|/\|y\|$ we get $\|x-\lambda y\|^2 = (x-\lambda y, x-\lambda y) = \|x\|^2 - 2\lambda\,\Re e\{(x, y)\} + \lambda^2\|y\|^2 = 0$, so that $x = \lambda y$. The choice of λ is suggested by re-casting (3.2) in the form $\|x\|^2\|y\|^2 \geq |(x, y)|^2 = (\Re e\{(x, y)\})^2 + (\Im m\{(x, y)\})^2$ since $\Re e\{(x, y)\} = \|x\| \cdot \|y\|$ is seen to force equality and $\Im m\{(x, y)\} = 0$.

7. For $\lambda = 0$ we get $\|x\| = \|y\|$. Since $\|\lambda x + (1-\lambda)y\|^2 = (\lambda x + (1-\lambda)y, \lambda x + (1-\lambda)y) = \lambda^2\|x\|^2 + 2\lambda(1-\lambda)\,\Re e\{(x, y)\} + (1-\lambda)^2\|y\|^2 = \|x\|^2$, we deduce that $\|x\|^2 = \|y\|^2 = \Re e\{(x, y)\}$, setting $\lambda = 0$ and $\lambda = 1/2$. But then $\|x-y\|^2 = \|x\|^2 - 2\Re e\{(x, y)\} + \|y\|^2 = 0$, so that $x = y$. In particular, if $\|x_1 - x_0\| = \|x_2 - x_0\| = R$, then the previous considerations with $x = x_1 - x_0$ and $y = x_2 - x_0$ ensure that all points on the segment $\{\lambda x_1 + (1-\lambda)x_2 : \lambda \in [0, 1]\}$ joining x_1 to x_2 are at distance R from x_0 only if $x_1 = x_2$.

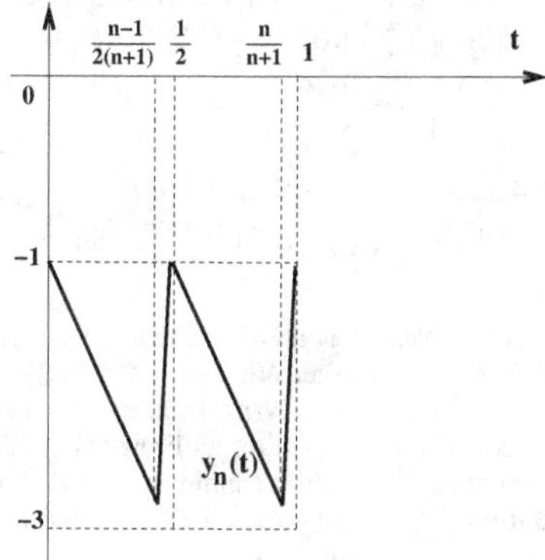

Figure 3.8 The graph of the functions constructed in the solution to Exercise 8.

8. With the notation introduced in the hint, if $z_n \to z$ in H, then by (3.2) we have $|(z_n - z, x_1)| \leq \|z_n - z\| \cdot \|x_1\| = \|z_n - z\| \to 0$, so that $(z, x_1) = 0$ if $(z_n, x_1) = 0$ for all $n \geq 1$. Thus M is closed in H. For $y \in M$ we have $\|y - x_0\|^2 = (y - x_0, y - x_0) = (y - (x_0 - x_1) - x_1, y - (x_0 - x_1) - x_1) = \|y - (x_0 - x_1)\|^2 - 2\Re e\{(y - (x_0 - x_1), x_1)\} + \|x_1\|^2 = \|y - (x_0 - x_1)\|^2 + 1$ since $(y, x_1) = 0$ and $(x_0, x_1) = \|x_1\|^2 = 1$. Thus $d(x_0, M) \geq 1$. Since $(x_0 - x_1)(t) = -4t - 1$ for $t \in [0, 1/2]$ and $(x_0 - x_1)(t) = -4t + 1$ for $t \in (1/2, 1]$, we have $(x_0 - x_1) \notin M$, so that there is no $y_0 \in M$ with $\|x_0 - y_0\| = 1$, since the only candidate is $y_0 = x_0 - x_1$. However, for $n \geq 2$ the functions

$$y_n(t) = \begin{cases} -4t - 1 & \text{for } 0 \leq t \leq \frac{1}{2} - \frac{1}{n+1}, \\ (2n-2)t - n & \text{for } \frac{1}{2} - \frac{1}{n+1} \leq t \leq \frac{1}{2}, \\ -4t + 1 & \text{for } \frac{1}{2} \leq t \leq \frac{n}{n+1}, \\ (2n-2)t - 2n + 1 & \text{for } \frac{n}{n+1} \leq t \leq 1, \end{cases}$$

belong to M and $\|y_n - x_0\|^2 = 1 + \frac{7}{3(n+1)} \to 1$ for $n \uparrow \infty$. This yields $d(x_0, M) = 1$. Theorem 3.2 is not applicable since H is not a Hilbert space: its completion is $L^2[0,1]$; see Chapter 4.

9. The first part of the proof of Theorem 3.2 applies without modifications, yielding the existence and uniqueness of $\pi_K(x)$. As for the characterisation, denote $z = \pi_K(x) \in K$. For all $y \in K$ and $t \in [0,1]$ we have $z + t(y-z) \in K$, so that $\|z + t(y-z) - x\|^2 \geq \|z - x\|^2$. Expanding the left side, we get $t^2\|y-z\|^2 + 2t\,\Re\mathrm{e}\{(z-x,\, y-z)\} \geq 0$ for all $t \in [0,1]$. Dividing by $t > 0$ and subsequently passing to the limit $t \downarrow 0$ yields $\Re\mathrm{e}\{(z-x,\, z-y)\} \leq 0$ for all $y \in K$. Conversely, if there exists some $z \in K$ with $\Re\mathrm{e}\{(z-x,\, z-y)\} \leq 0$ for all $y \in K$, the argument can be reversed. Indeed, we first infer that $t^2\|y-z\|^2 + 2t\,\Re\mathrm{e}\{(z-x,\, y-z)\} \geq 0$ for all $y \in K$ and $t \in [0,1]$. Thus $\|z + t(y-z) - x\|^2 \geq \|z-x\|^2$ for all $y \in K$ and $t \in [0,1]$. The choice $t = 1$ leads us to $\|y - x\| \geq \|z - x\|$ for all $y \in K$.

10. If $x, y \in M^\perp$, then all scalars α and β and every $z \in M$, we have $(\alpha x+\beta y, z) = \alpha(x,z)+\beta(y,z) = 0$. Therefore $M^\perp$ is a linear subspace even if M did not have a linear structure: for example, in the real Hilbert space $\mathbb{R}^3$, for $M = \{(0,1,0),\,(0,0,1)\}$ we have $M^\perp = \{(\alpha,0,0) : \alpha \in \mathbb{R}\}$. To see that $M^\perp$ has to be closed, let $x_n \to x$ in H with $x_n \in M^\perp$ for every $n \geq 1$. Then, for any fixed $y \in M$, we have $(x,y) = \lim_{n\to\infty}(x_n, y) = 0$, so that $x \in M^\perp$, proving that $M^\perp$ is closed. To prove the second statement, notice that $M \subset (M^\perp)^\perp$ since for every $x \in M$ and $y \in M^\perp$, the definition of $M^\perp$ requires $(x,y) = 0$. Since by the first statement $(M^\perp)^\perp$ is a closed linear subspace, we successively infer that $\mathrm{span}(M) \subset (M^\perp)^\perp$, and $\overline{\mathrm{span}(M)} \subset (M^\perp)^\perp$. In the case of a strict inclusion, we could find some $z \in (M^\perp)^\perp \setminus \overline{\mathrm{span}(M)}$. By Theorem 3.2, its projection ξ onto $\overline{\mathrm{span}(M)}$ is such that $(z-\xi) \perp \overline{\mathrm{span}(M)}$. In particular, $(z-\xi) \in M^\perp$, so that $(z, z-\xi) = 0$. Note that $(z-\xi, \xi) = 0$ by the definition of ξ. But then $\|z-\xi\|^2 = (z-\xi, z-\xi) = (z, z-\xi) - (\xi, z-\xi) = 0$ and $z = \xi$ contradicts the choice of z.

11. Checking that $x + y \in l^2$ if $x, y \in l^2$ is the only nontrivial aspect of the linear vector space structure of l^2. Since $\left(\sum_{k=1}^\infty |x(k)|^2\right)\left(\sum_{k=1}^\infty |y(k)|^2\right) - \left(\sum_{k=1}^\infty |x(k)|\cdot|y(k)|\right)^2 = \sum_{k=1}^\infty \sum_{j=1}^\infty \left(|x(k)|^2 \cdot |y(j)|^2 - |x(k)|\cdot|y(k)|\cdot|x(j)|\cdot|y(j)|\right) = \sum_{k\neq j}\left(|x(k)|\cdot|y(j)| - |y(k)|\cdot|x(j)|\right)^2 \geq 0$, we get $\sum_{k=1}^\infty |x(k) + y(k)|^2 = \sum_{k=1}^\infty [x(k)+y(k)]\cdot\overline{[x(k)+y(k)]} = \sum_{k=1}^\infty |x(k)|^2 + \sum_{k=1}^\infty |y(k)|^2 + \sum_{k=1}^\infty [x(k)\,\overline{y(k)} + \overline{x(k)}\,y(k)] \leq \sum_{k=1}^\infty |x(k)|^2 + \sum_{k=1}^\infty |y(k)|^2 + 2\sum_{k=1}^\infty |x(k)|\,|y(k)| \leq \sum_{k=1}^\infty |x(k)|^2 + \sum_{k=1}^\infty |y(k)|^2 + 2\left(\sum_{k=1}^\infty |x(k)|^2\right)^{1/2}\left(\sum_{k=1}^\infty |y(k)|^2\right)^{1/2} = \left[\left(\sum_{k=1}^\infty |x(k)|^2\right)^{1/2} + \left(\sum_{k=1}^\infty |y(k)|^2\right)^{1/2}\right]^2$. Thus $x + y \in l^2$ if $x, y \in l^2$ since[165] $\|x + y\| \leq \|x\| + \|y\|$.

To prove that the linear vector space l^2 is a Hilbert space, let $\{x_n\}_{n\geq 1}$ be a Cauchy sequence in l^2. Then for every $\varepsilon > 0$ there is an integer $N(\varepsilon)$ such that $\|x_n - x_m\| \leq \varepsilon$ for $n, m \geq N(\varepsilon)$, that is, $\sum_{k=1}^\infty |x_n(k) - x_m(k)|^2 \leq \varepsilon^2$ for $n, m \geq N(\varepsilon)$. In particular, for each fixed integer $k \geq 1$, each $\{x_n(k)\}_{n\geq 1}$ is a Cauchy sequence in $\mathbb{C}$. Calling its limit $x(k) \in \mathbb{C}$, we identify the sequence $x = \{x(k)\}_{k\geq 1}$ of complex numbers as a potential candidate for the limit in l^2. For a fixed integer $N \geq 1$, let $m \to \infty$ in the relation $\sum_{k=1}^N |x_n(k) - x_m(k)|^2 \leq \varepsilon^2$, valid for $n, m \geq N(\varepsilon)$, to infer that $\sum_{k=1}^N |x_n(k) - x(k)|^2 \leq \varepsilon^2$ for $n \geq N(\varepsilon)$. Since $N \geq 1$ was arbitrary, we deduce that $\sum_{k=1}^\infty |x_n(k) - x(k)|^2 \leq \varepsilon^2$ for all $n \geq N(\varepsilon)$. Thus $x_n \to x$ in l^2 and, as a by-product, we get $x \in l^2$.

To prove that l^2 is separable, consider the countable set $\mathfrak{S}$ of sequences $x \in l^2$ such that there is an integer $N \geq 1$ with $x(k) = 0$ if $k \geq N$, and all $x(k)$ rational numbers if $k \leq N$. Given $y \in l^2$ and $\varepsilon > 0$, choose the integer $N \geq 1$ such that $\sum_{k=N}^\infty |y(k)|^2 < \varepsilon^2/2$. Define now a sequence $x \in \mathfrak{S}$ by setting $x(k) = 0$ for $k \geq N$, while for $k < N$ set $x(k)$ equal to a rational number such that $|x(k) - y(k)| < \frac{\varepsilon}{2N}$. Then $\|x - y\| = \left(\sum_{k=1}^\infty |x(k) - y(k)|^2\right)^{1/2} < \varepsilon$. Thus $\mathfrak{S}$ is dense in l^2.

12. (i) This is a particular case of Exercise 9, with $x = 0$.

(ii) The norm is a continuous map from $K \subset \mathbb{C}^n$ to $[0, \infty)$ and must attain its maximum on the compact set K. The example of $K = \{x \in \mathbb{C} : \|x\| \leq 1\}$ shows that there might be infinitely many elements of maximal norm.

(iii) Since $(1 + \frac{1}{k})^2 \geq 1$ for $k \geq 1$, by Plancherel's identity we must have $\|x\| \leq 1$ for all $x \in K$,

[165] We saw that (3.2) is valid in a pre-Hilbert space. Here we checked it in order to show that the constraint $x \in l^2$ is compatible with the addition of vectors.

showing that K is bounded. Plancherel's identity together with the inequalities $(1+\frac{1}{k})^2 \le 4$ for $k \ge 1$ yield $|f(x)-f(y)| \le 4\,\|x-y\|^2$ for all $x, y \in l^2$, where $f : l^2 \to [0,\infty)$ is defined by $f(x) = \sum_{k=1}^{\infty} (1+\frac{1}{k})^2\,|x(k)|^2$. If $x_n \to x$ in l^2 and $x_n \in K$ for all $n \ge 1$, then $f(x) = \lim_{n\to\infty} f(x_n) \le 1$, so that K is closed. To prove that K is convex, let $x, y \in K$ and $\lambda \in [0,1]$. Since $|\lambda x(k) + (1-\lambda)y(k)|^2 = \lambda^2|x(k)|^2 + 2\lambda(1-\lambda)\,\Re e\{(x(k), y(k))\} + (1-\lambda)^2|y(k)|^2 \le \lambda\,|x(k)|^2 + (1-\lambda)\,|y(k)|^2$ for all $k \ge 1$ is ensured by the elementary inequality $2\,\Re e\{(x(k), y(k))\} \le 2\,|x(k)|\,|y(k)| \le |x(k)|^2 + |y(k)|^2$, by summing over all $k \ge 1$, we infer that $\sum_{k=1}^{\infty} (1+\frac{1}{k})^2\,|\lambda x(k) + (1-\lambda)y(k)|^2 \le \lambda \sum_{k=1}^{\infty} (1+\frac{1}{k})^2\,|x(k)|^2 + (1-\lambda)\sum_{k=1}^{\infty} (1+\frac{1}{k})^2\,|y(k)|^2 \le \lambda + (1-\lambda) = 1$. This shows that K is convex. Assuming that $x \in K$ is an element of maximal norm, choose an integer $n \ge 1$ with $x(n) \ne 0$. Alter now the sequence $x \in l^2$ by replacing[166] $x(n)$ with 0 and $x(2n)$ with $\sqrt{|x(2n)|^2 + |x(n)|^2\,\frac{4(n+1)^2}{(2n+1)^2}}$. The sequence $\hat{x}$ thus constructed lies in K since $\sum_{k=1}^{\infty} (1+\frac{1}{k})^2\,|\hat{x}(k)|^2 = \sum_{k=1}^{\infty} (1+\frac{1}{k})^2\,|x(k)|^2$. However, $\|\hat{x}\|^2 - \|x\|^2 = |x(2n)|^2 + |x(n)|^2\,\frac{4(n+1)^2}{(2n+1)^2} - |x(n)|^2 - |x(2n)|^2 = |x(n)|^2\,\frac{4n+3}{(2n+1)^2} > 0$ shows that $\|\hat{x}\|$ exceeds the presumed maximum.

13. It is easy to check that this is a pre-Hilbert space. To see that it is not a Hilbert space, recall that in the solution to Exercise 9 from Chapter 2 we constructed continuous functions $f_n : [0,1] \to [0,1]$ with $\lim_{n\to\infty} f_n(t) = \chi(t)$ for every $t \in [0,1]$, where χ is the characteristic function of a Cantor set $C \subset [0,1]$ of positive measure. The dominated convergence theorem yields $\lim_{n\to\infty} \int_0^1 |f_n(t) - \chi(t)|^2\,dt = 0$. Therefore $\{f_n\}_{n\ge1}$ is a Cauchy sequence in H since $\|f_n - f_m\|^2 = \int_0^1 |f_n(t) - f_m(t)|^2\,dt \le \int_0^1 \big[|f_n(t) - \chi(t)| + |f_m(t) - \chi(t)|\big]^2\,dt \le 2\int_0^1 |f_n(t) - \chi(t)|^2\,dt + 2\int_0^1 |f_m(t) - \chi(t)|^2\,dt \to 0$ for $n, m \to \infty$. We claim that $\{f_n\}_{n\ge1}$ does not converge in H. Indeed, if $f_n \to f$ in H, then (f being continuous) we can apply the dominated convergence theorem to the sequence $(f_n - f)$ to infer that $\|\chi - f\|^2 = \int_0^1 |\chi(t) - f(t)|^2\,dt = \lim_{n\to\infty} \int_0^1 |\chi(t) - f_n(t)|^2\,dt$. But the right-hand side vanishes. Exercise 26 of Chapter 2 yields $f(t) = \chi(t)$ a.e. on $[0,1]$. Since χ is discontinuous on the set C of positive measure, f cannot be continuous. The obtained contradiction shows that H is not a Hilbert space.

14. The proof is by induction on $k \ge 1$. For $k = 1$ all is clear. Assume now that $\{x_1, \dots, x_k\}$ form an orthonormal set that spans the same linear subspace as the vectors $y_1, \dots, y_k$. Then the vector $y_{k+1} - \sum_{j=1}^{k} (y_{k+1}, x_j)x_j$ is nonzero since by linear independence the vector y_{k+1} does not belong to the subspace spanned by the vectors $x_1, \dots, x_k$. Choose α_{k+1} equal to the reciprocal of the norm of this vector, thus defining $x_{k+1} = \alpha_{k+1}\big\{y_{k+1} - \sum_{j=1}^{k} (y_{k+1}, x_j)x_j\big\}$ of unit norm. To check that the set $\{x_1, \dots, x_{k+1}\}$ is orthonormal it suffices to verify that $(x_{k+1}, x_j) = 0$ for all $j = 1, \dots, k$. This is plain: $(x_{k+1}, x_j) = \alpha_{k+1}\,(y_{k+1}, x_j) - \alpha_{k+1}\sum_{l=1}^{k} (y_{k+1}, x_l)\,(x_l, x_j) = \alpha_{k+1}\,(y_{k+1}, x_j) - \alpha_{k+1}\,(y_{k+1}, x_j) = 0$ since $(x_l, x_j) = \delta_{lj}$ by the orthonormality of the set $\{x_1, \dots, x_k\}$.

15. (i) If H admits a finite or countable orthonormal basis $\{e_n\}_{n\ge1}$, then H is separable since the collection C of finite sums of vectors e_n multiplied by scalars having both the real and imaginary part rational numbers provides a countable dense subset. Indeed, the closure of C contains the finite linear combinations of the vectors e_n, being, due to part (e) of Theorem 3.1, the entire space H. Conversely, assume H is separable and let $\{f_n\}_{n\ge1}$ be a dense sequence. Use the Gram–Schmidt procedure to construct the corresponding orthonormal set. Part (e) of Theorem 3.1 ensures that this is a basis.

By Exercise 4(ii), in any Hilbert space there is at least one orthonormal basis. Let E and F be two orthonormal bases. If E is a finite set, then combining Theorem 3.1 with the fact that the vectors in E are linearly independent (see Exercise 4) yields that E is a Hamel basis. Let $E = \{x_1, \dots, x_n\}$ and choose some $y_1 \in F$. We write $y_1 = \alpha_1 x_1 + \cdots + \alpha_n x_n$. At least one $a_i \ne 0$, say $a_1 \ne 0$. Then $x_1 = \frac{1}{a_1}\,y_1 - \frac{a_2}{a_1}\,x_2 - \cdots - \frac{a_n}{a_1}\,x_n$, so that $\{y_1, x_2, \dots, x_n\}$ span H. If $n > 1$, then there

166 This procedure is suggested by realising that the maximum of $\sum_{k=1}^{n} |\hat{x}(k)|^2$ under the constraint $\sum_{k=1}^{n} (1+\frac{1}{k})^2\,|\hat{x}(k)|^2 \le 1$ is attained by choosing $\hat{x}(1) = \cdots = \hat{x}(n-1) = 0$ and $\hat{x}(n) = (1+\frac{1}{n})^{-1}$, thus shifting all the "weight" towards the highest index.

must be some $y_2 \neq y_1 \in F$ with $y_2 = b_2x_2 + \cdots + b_nx_n + b_1y_1$ with at least one of the coefficients $b_2, \ldots, b_n$ nonzero, say $b_2 \neq 0$. This yields $x_2 = -\frac{b_1}{b_2} y_1 - \frac{1}{b_2} y_2 - \frac{b_3}{b_2} x_3 - \cdots - \frac{b_n}{b_2} x_n$, so that $\{y_1, y_2, x_3, \ldots, x_n\}$ span H. Continuing along this line of argument we find that we can replace each $x_i \in E$ by some $y_j \in F$. Thus F has at least n elements. Fixing $m \geq n$ elements of F and reversing the roles of the x_is and y_js yields $n \geq m$. Therefore $n = m$. We deduce that E and F have the same number of elements.

Assume now that both E and F have infinitely many elements. For each $e \in E$, define the set $F_e = \{f \in F : (e, f) \neq 0\}$. Since F is a maximal orthonormal set, F_e is nonempty. Furthermore, Bessel's inequality for $x = e$ ensures that for every integer $n \geq 1$ there are not more than n^2 elements $f \in F_e$ with $|(e, f)| \geq \frac{1}{n}$. Consequently, each set F_e is either finite or countably infinite. Because the maximality of the orthonormal set E ensures that each $f \in F$ belongs to some F_e, we have $F = \bigcup_{e \in E} F_e$. Since the cardinality $\aleph_0$ of the natural numbers is the smallest infinite number, with the property that $\aleph_0 \cdot \aleph = \aleph$ for any other cardinality $\aleph$ of an infinite set, we deduce that the cardinality of F does not exceed that of E. Reversing the role of E and F we infer the converse. Thus the only possibility is that E and F are sets of the same cardinality. This means that there is a bijection from E to F. (For an accessible discussion of cardinal and ordinal numbers we refer to Hewitt and Stromberg (1965).)

(ii) If X admits a countable Hamel basis $\{e_k\}_{k \geq 1}$, we may assume that $\|e_k\| = 1$ for all $k \geq 1$. Note that, by Exercise 17(i), the subspace A_n spanned by $e_1, \ldots, e_n$ is closed, while by hypothesis $X = \bigcup_{n \geq 1} A_n$. By Theorem 3.21, one of the sets A_n, say A_N, will contain an open ball, centred at some $x_0 = \sum_{k=1}^N \alpha_k e_k$ and of radius $\varepsilon > 0$. However, $\|x_0 - (x_0 + \frac{\varepsilon}{2} e_{N+1})\| = \frac{\varepsilon}{2}$ but $x_0 + \frac{\varepsilon}{2} e_{N+1} \notin A_N$.

(iii) Note that the scalar product (f, g) is well-defined if f and g are both periodic. To prove this for $f, g \in AP$, let f_n and g_n be finite linear combinations of continuous periodic functions that converge uniformly on $\mathbb{R}$ to f and g, respectively. Then for every $\varepsilon > 0$ there is an integer $N \geq 1$ such that $|f_n(t) - f(t)| \leq \varepsilon$ and $|g_n(t) - g(t)| \leq \varepsilon$ for all $t \in \mathbb{R}$, whenever $n \geq N$. For $T > 0$, thinking of $(f, g)_T = \frac{1}{2T} \int_{-T}^T f(t) \overline{g(t)}\, dt$ as a scalar product for continuous functions defined on $[-T, T]$, with the corresponding norm $\| \cdot \|_T$, for $n \geq N$ we get $\left|(f_n, g_n)_T - (f, g)_T\right| = \left|(f_n - f, g_n)_T + (f, g_n - g)_T\right| \leq \|f - f_n\|_T \|g_n\|_T + \|f\|_T \|g - g_n\|_T \leq \varepsilon(\|g_n\|_T + \|f\|_T) \leq \varepsilon(\|g_N\|_T + \|f_N\|_T + 3\varepsilon)$. Uniform convergence ensures that $\|g_N\|_T + \|f_N\|_T$ is uniformly bounded (for $T > 0$ and $N \geq 1$) by some constant $M > 0$, so that the existence of $(f_n, g_n) = \lim_{T \to \infty} (f_n, g_n)_T$ entails $\left| \limsup_{T \to \infty} \frac{1}{2T} \int_{-T}^T f(t) \overline{g(t)}\, dt - \liminf_{T \to \infty} \frac{1}{2T} \int_{-T}^T f(t) \overline{g(t)}\, dt \right| \leq 2\varepsilon(M + 3\varepsilon)$. Since $\varepsilon > 0$ was arbitrary, this proves the existence of $(f, g) = \lim_{T \to \infty} \frac{1}{2T} \int_{-T}^T f(t) \overline{g(t)}\, dt$. In particular, $(\phi_a, \phi_b) = \lim_{T \to \infty} \frac{e^{iT(a-b)} - e^{-i(a-b)}}{iT(a-b)} = 0$ for all real numbers $a \neq b$. Finally, if $f \in AP$, let A be the set of all ϕ_a such that $|(f, \phi_a)| > 0$. Bessel's inequality for countable orthonormal sets shows that for each $n \geq 2$, the set $A_n = \{\phi_a \in A : |(f, \phi_a)| > \frac{\|f\|}{n}\}$ contains at most $(n^2 - 1)$ functions. Since $A = \bigcup_{n \geq 2} A_n$, we see that A is at most countably infinite.

16. (i) Assume that K is relatively compact and let $S_n x = \sum_{k=1}^n \alpha_k e_k$ be the truncation of the expansion $x = \sum_{k \geq 1} \alpha_k e_k$ in the Schauder basis $\{e_k\}_{k \geq 1}$. By the uniform boundedness principle, there is a constant $M > 0$ with $\|S_n\| \leq M$ for all $n \geq 1$. Let us prove that $S_n \to I$ uniformly on K. Indeed, given $\varepsilon > 0$, choose a cover of the closure $\overline{K}$ of K by a finite set of open balls of radius $\varepsilon > 0$, centred at the points $x_1, \ldots, x_N$. Choose also the integer $p \geq 1$ such that $\|S_n x_k - x_k\| < \varepsilon$ for all $k = 1, \ldots, N$ and all $n \geq p$. Then for every $x \in K$ there is some $k \in \{1, \ldots, N\}$ such that $\|x - x_k\| < \varepsilon$, so that $\|S_n x - x\| \leq \|S_n x - S_n x_k\| + \|S_n x_k - x_k\| + \|x_k - x\| \leq \|S_n\| \cdot \|x - x_k\| + 2\varepsilon \leq \varepsilon(M + 2)$ for all $n \geq p$.

For sufficiency, note that if $\{x_j\}_{j \geq 1}$ is a sequence in K, then by assumption there is some $N \geq 1$ such that $\sum_{k \geq N} \|\alpha_{jk} e_k\| < 1/n$ for all $j \geq 1$, where $x_j = \sum_{k \geq 1} \alpha_{jk} e_k$. But this means that K is totally bounded and therefore relatively compact, see the considerations that precede Theorem 3.8.

(ii) Note that $e_k(j) = \delta_{kj}$ for $k, j \geq 1$ is a Schauder basis for l^2, and apply (i).

(iii) Given $\varepsilon > 0$, choose for each $x \in [0, 1]$ some $\delta_x > 0$ such that $|f_n(x) - f_n(y)| < \frac{\varepsilon}{4}$ for all $n \geq 1$, whenever $|x - y| < \delta_x$. The balls $B(x, \delta_x)$ cover $[0, 1]$, extract a finite subcover which we denote by $\{B(x_j, \delta_j)\}_{1\le j\le N}$. For a fixed $x_j \in [0,1]$, denote by A_j the bounded set of points $\{f_n(x_j)\}_{n\ge1}$. Since A_j is relatively compact, there is a finite $\frac{\varepsilon}{4}$-net $\{z_k(x_j)\}_{1\le k\le m(j)}$ for the set A_j, with z_k a function in the sequence $\{f_n\}_{n\ge1}$. Let $\mu = (\mu_1, \ldots, \mu_N)$ denote an Nth-uple of integers such that $1 \le \mu_j \le m(j)$ for $1 \le j \le N$, and denote by $\mathcal{F}$ their finite union. For each $\mu \in \mathcal{F}$, let f_μ be a function[167] in the sequence $\{f_n\}_{n\ge1}$ such that $|f_\mu(x_j) - z_{\mu_j}(x_j)| < \frac{\varepsilon}{4}$ for all $j = 1, \ldots, N$. We claim that these functions form a finite ε-net for $\{f_n\}_{n\ge1}$. Indeed, given $n \ge 1$, for each $j = 1, \ldots, N$ there is an integer $\mu_j \in [1, m(j)]$ such that $|f_n(x_j) - z_{\mu_j}(x_j)| < \frac{\varepsilon}{4}$. By the defining property of the corresponding[168] f_μ, and since any $x \in [0,1]$ lies in some $B(x_j, \delta_j)$, we get $|f_n(x) - f_\mu(x)| \le |f_n(x) - f_n(x_j)| + |f_n(x_j) - z_{\mu_j}(x_j)| + |z_{\mu_j}(x_j) - f_\mu(x_j)| + |f_\mu(x_j) - f_\mu(x)| < \varepsilon$. The compactness follows now from the considerations that precede Theorem 3.8.

17. (i) If $Y \subset X$ is a finite-dimensional subspace, the considerations made in the proof of Theorem 3.8 show that Y is isomorphic to $\mathbb{C}^n$ or $\mathbb{R}^n$. Thus Y is closed. (ii) Let $\{e_i\}_{i\in I}$ be a Hamel basis of X. Choosing some $i_0 \in I$, by Corollary 2 of Theorem 3. 15, there is some $F \in X^*$ with $F(e_{i_0}) = \|e_{i_0}\| \neq 0$. Note that $X_1 = \ker(F)$ is a proper closed infinite-dimensional subspace of X. Indeed, all statements but the one about the dimension are clear. For the latter, since $e_i \notin X_1$ forces $\frac{1}{F(e_i)}\, e_i - \frac{1}{F(e_{i_0})}\, e_{i_0} \in X_1$, we see that $\dim(X_1) = n < \infty$ would yield that I has at most $(n+1)$ elements. X_1 being infinite-dimensional, we can repeat the procedure to construct X_2, etc. (iii) Assume that the Banach space X admits a countable Hamel basis $\{e_k\}_{k\ge1}$. Then $X = \bigcup_{n\ge1} Y_n$, where Y_n is the finite-dimensional subspace spanned by $\{e_1, \ldots, e_n\}$. By (i), Y_n is closed for any $n \ge 1$. Since for every $y_n = \sum_{k=1}^n \alpha_k e_k \in Y_n$ and every $\varepsilon > 0$, $x_\varepsilon = y_n + \frac{\varepsilon}{2\|e_{n+1}\|}\, e_{n+1} \notin Y_n$ satisfies $\|y_n - x_\varepsilon\| < \varepsilon$, Y_n is nowhere dense. But this contradicts Theorem 3.21, so that out starting hypothesis must be wrong. (iv) Let $\{e_i\}_{i\in I}$ be a Hamel basis. Due to item (iii), we may assume that I contains the positive integers. Consider now the linear subspace Y of all vectors of the form $\sum_{k\ge1} \alpha_k e_k$, with at most finitely many scalars α_k nonzero. The reasoning used in (iii) above shows that Y cannot be a closed subspace.

18. (i) If $Tx_n = 0$ and $x_n \to x$, then by continuity $Tx = \lim_{n\to\infty} Tx_n = 0$.

(ii) Let X be an infinite-dimensional normed linear space. Let $\{e_i\}_{i\in I}$ be a Hamel basis, with the (uncountable) index set I containing the set $\mathbb{N}$ of all integers $k \ge 1$. Define the linear operator $T : X \to X$ by $Tx = \sum_{k=1}^\infty k x_k\, e_k + \sum_{i\in I\setminus\mathbb{N}} x_i\, e_i$ for $x = \sum_{i\in I} x_i\, e_i$; all but finitely many of the scalars x_i are zero. The kernel of T is closed, being equal to $\{0\}$. However, T is not bounded since $\|Te_k\| = k\,\|e_k\|$ for all $k \ge 1$.

19. Let $\{e_i\}_{1\le i\le n}$ be a Hamel basis of the finite-dimensional complex[169] normed space X. The set $K = \{(x_1, \ldots, x_n) \in \mathbb{C}^n : \|\sum_{i=1}^n x_i\, e_i\|_X = 1\}$ is clearly closed and bounded, therefore compact. Consequently, the continuous function $(x_1, \ldots, x_n) \mapsto \sum_{i=1}^n |x_i|$ attains its maximum $m > 0$ on K. If $t = \max_{1\le i\le n}\{\|Te_i\|_Y\}$, we obtain that $\|Tx\|_Y \le \sum_{i=1}^n |x_i| \cdot \|Te_i\|_Y \le tm$ for all $x \in X$ with $\|x\|_X = 1$. Thus $\|T\| \le tm$.

Let now $\{e_i\}_{i\in I}$ be a Hamel basis of the infinite-dimensional complex normed space X, with $\mathbb{N} \subset I$. Choose a countable subset $\{e_k\}_{k\ge1}$ and note that $\sum_{k\ge1} \frac{1}{k^2}\, e_k$ defines an element $e = \sum_{j\in J} x_j\, e_j \in X$; here J is a finite subset of I. Choose $y \in Y$ with $\|y\|_Y = 1$ and define for $m \ge 1$ the linear operators $T_m\big(\sum_{i\in I} x_i\, e_i\big) = x_m\, y$. Note that $T_m(e) = \delta_{jm}\, x_m\, y$. Since $T_m(e) = \lim_{n\to\infty} T_m\big(\sum_{k=1}^{n+m} \frac{1}{k^2}\, e_k\big) = \frac{1}{m^2}\, y$ if $T_m \in \mathcal{B}(X, Y)$, T_m is not continuous if $m \notin J$.

20. $\ker(f)$ is clearly a linear subspace.

(i) If f is continuous, then clearly $\ker(f)$ will be closed since $x_n \to x$ implies $f(x_n) \to f(x)$. Conversely, assume that $\ker(f)$ is closed. If $\ker(f) = X$, then $f \equiv 0$. If $\ker(f) \neq X$, choose $x_0 \in X$ with $f(x_0) \neq 0$. Then $e = \frac{1}{f(x_0)}\, x_0$ satisfies $f(e) = 1$. Note that $x - f(x)\, e \in \ker(f)$ for all $x \in X$.

[167] It may be that for a specific μ there is no such function with this property. In that case there is no corresponding f_μ defined.

[168] Existence is guaranteed since f_n itself satisfies all the required properties, even though it is not likely for f_n to be the f_μ we chose.

[169] The real case is entirely analogous.

Moreover, if $x = \alpha_1\, e+y_1 = \alpha_2\, e+y_2$ with α_1, α_2 scalars and $y_1, y_2 \in \ker(f)$, then $\alpha_1 = \alpha_2 = f(x)$. Thus every $x \in X$ has a unique representation $x = \alpha\, e+y$ with $y \in \ker(f)$ and α scalar. In particular, e is at unit distance from $\ker(f)$. Corollary 3 to Theorem 3.15 yields the existence of $F \in X^*$ with $f(e) = 1$ and $f(y) = 0$ for $y \in \ker(f)$. But for $x = \alpha\, e + y$ we then have $F(x) = \alpha = f(x)$. Thus $f = F$ is continuous.

(ii) If f is not continuous, then $\ker(f)$ must be a proper subspace as otherwise $f \equiv 0$. We show that the assumption $\overline{\ker(f)} \neq X$ leads to a contradiction. For this, if $x_0 \notin \overline{\ker(f)}$, then clearly $f(x_0) = \alpha \neq 0$, while $d(x_0, \overline{\ker(f)}) = \delta > 0$. Since $\overline{\ker(f)}$ is clearly a linear subspace, Corollary 3 to Theorem 3.15 yields the existence of $F \in X^*$ with $F(x_0) = \delta$ and $F(y) = 0$ for all $y \in \overline{\ker(f)}$. In particular, $F(x) = 0$ for $x \in \ker(f)$. On the other hand, if $f(x) = \beta \neq 0$, then $f(x) = \frac{\beta}{\alpha}\, f(x_0)$, so that $x - \frac{\beta}{\alpha}\, x_0 \in \ker(f)$ and $F(x - \frac{\beta}{\alpha}\, x_0) = 0$, that is, $F(x) = \frac{\beta}{\alpha}\, F(x_0) = \frac{\delta}{\alpha}\, f(x)$. We obtained that $F = \frac{\delta}{\alpha}\, f$ on X, which contradicts the fact that $f \notin X^*$.

21. As a closed linear subspace of $C[0,1]$, X is a Banach space. Since $|F(u)| \leq \int_0^1 |u(t)|\, dt \leq \int_0^1 \|u\|_X\, dt = \|u\|_X$, we see that $\|F\| \leq 1$. On the other hand, $F(t^{1/n}) = 1 - 1/n \uparrow 1$ for $n \uparrow \infty$, while $\|t^{1/n}\|_X = 1$. Thus $\|F\| = 1$. But if $\|u\|_X = 1$, then $u(t) \leq 1$ on $[0,1]$, and $u(0) = 0 < 1$ yields $F(u) < \int_0^1 1\, dt = 1$.

22. (i) If $x = x^1 + x^2$ with $x^i \in X_i$, define $\pi_i(x) = x^i$ for $i \in \{1,2\}$. If now $x_n \to x$ and $\pi_1(x_n) \to y$ in X_1, then $\pi_2(x_n) \to x - y$. But $\pi_2(x_n) \in X_2$ and X_2 is closed, so that $x - y \in X_2$. Since $y \in X_1$, we deduce that $\pi_1(x) = y$. Similarly we prove that π_2 is bounded. (ii) Let Y be a finite-dimensional subspace of the Banach space X. If $\{e_k\}_{1\leq k\leq n}$ is a Hamel basis of Y, then the projection $\varphi_k : Y \to \mathbb{C}$ on the kth basis element, defined by $\varphi_k(\sum_{j=1}^n x_j e_j) = x_k$, clearly belongs to Y^*. By Corollary 1 to Theorem 3.15 we can extend φ_k to some $\phi_k \in X^*$. Since $\ker(\phi_k)$ is closed, see Exercise 20(i), so will be $X_1 = \bigcap_{1\leq k\leq n} \ker(\phi_k)$. This is the complement of Y in X since $\phi_k(x - \sum_{j=1}^n x_j e_j) = 0$ for $1 \leq k \leq n$. Finally, if $Y \subset X$ is a closed subspace such that $\{f_1, \dots, f_n\} \subset X \setminus Y$ complete the Hamel basis $\{e_i\}_{i\in I}$ of Y to a Hamel basis of X, then the finite-dimensional subspace of X that is spanned by $\{f_1, \dots, f_n\}$, closed by Exercise 17(i), is a complement of Y in X.

23. (i) Every element $x \in M$ is a finite linear combination of vectors in the orthonormal set $\{x_n\}_{n\geq 1}$. An element in the closure $\overline{M}$ is a limit of elements $x_n \in M$, that is, $\|x - x_n\| \to 0$ as $n \to \infty$, with $x_n = \sum_{k=1}^\infty c_{k,n}\, x_k$ and only finitely many of the coefficients $c_{k,n}$ nonzero. Orthonormality ensures $\|x_n - x_m\|^2 = \sum_{k=1}^\infty |c_{k,n} - c_{k,m}|^2$, where again only finitely many of the coefficients $c_{k,n}$ and $c_{k,m}$ are nonzero. The fact that the sequence $\{x_n\}_{n\geq 1}$ is Cauchy (being convergent) can therefore be expressed by saying that the sequence $\{\mathfrak{c}_n\}_{n\geq 1}$, defined by $\mathfrak{c}_n(k) = c_{k,n}$ for $n, k \geq 1$, is Cauchy in l^2. Since l^2 is complete, the sequence $\{\mathfrak{c}_n\}_{n\geq 1}$ converges to some $\mathfrak{c}$ in l^2. Note that although by construction each sequence $\mathfrak{c}_n$ has only finitely many nonzero terms, the sequence $\mathfrak{c}$ might have infinitely many nonzero elements.[170] The partial sums $s_n = \sum_{k=1}^n \mathfrak{c}(k)\, x_k$ are such that $\|s_n - s_m\|^2 = \sum_{k=m+1}^n |\mathfrak{c}(k)|^2$ for $n > m$. Therefore the limit $s = \sum_{k=1}^\infty \mathfrak{c}(k)\, x_k$ exists. We claim that $s = x$. Indeed, by orthonormality, $\|s_n - x_n\|^2 = \sum_{k=1}^\infty |\mathfrak{c}(k) - c_{k,n}|^2$, and the right side converges to zero as $n \to \infty$ since it is precisely $\|\mathfrak{c} - \mathfrak{c}_n\|_{l^2}^2$. By construction $s_n \to s$. Since $x_n \to x$, we deduce that $x = s$, so that $x = \sum_{k=1}^\infty \mathfrak{c}(k)\, x_k$.

(ii) Let $n \geq 1$. To see that the vectors $\{x_1, \dots, x_n\}$ are linearly independent, assume that there are complex numbers $c_1, \dots, c_n$ such that $c_1 x_1 + \cdots + c_n x_n = 0$. The expression on the left is a linear combination of the vectors $\{e_0, e_1, \dots, e_n\}$, the coefficient of e_0 being $\sum_{k=1}^n c_k \cos\left(\frac{1}{k}\right)$, while for $1 \leq k \leq n$ the coefficient of e_k is $c_k \sin\left(\frac{1}{k}\right)$. Since $\sin\left(\frac{1}{k}\right) \neq 0$, we get $c_1 = \cdots = c_n = 0$. On the other hand, the relation $\lim_{n\to\infty} x_n = e_0$ shows that e_0 belongs to the closure of the subspace spanned by the x_ns. Assume now that[171] $e_0 = \sum_{k=1}^\infty \alpha_k\, x_k$ for some complex numbers α_k. Then

170 For example, $\mathfrak{c}_n$ of the form $(1, \frac{1}{2}, \frac{1}{3}, \dots, \frac{1}{n}, 0, 0, \dots)$ converges in l^2 to $\mathfrak{c}$ given by $\mathfrak{c}(k) = \frac{1}{k}$ that has all terms nonzero.

171 This means that the partial sums $\sum_{k=1}^n \alpha_k\, x_k$ converge in the norm to some e_0, identified with the infinite sum.

for any $n \geq 1$ we would have $0 = (e_0, e_n) = \sum_{k=1}^{\infty} \alpha_k (x_k, e_n) = \sum_{k=1}^{\infty} \alpha_k \sin(\frac{1}{k}) \delta_{kn} = \alpha_n \sin(\frac{1}{n})$, yielding $\alpha_n = 0$. We obtain the contradiction $e_0 = 0$.

24. (i) From calculus we know that $(\mathbb{R}^2, d)$ is a metric space, and it is easy to check that $(\mathbb{R}^2, d')$ is also a metric space. If $d'(x_n, x) \to 0$ as $n \to \infty$, choosing the integer $N \geq 1$ such that $d'(x_n, x) < \frac{1}{2}$ for $n \geq N$ ensures $x_n = x$ for all $n \geq N$. Every subset of $\mathbb{R}^2$ is open with respect to the metric d', since the open ball centred at some $x \in \mathbb{R}^2$ and of radius $\frac{1}{2}$ is precisely the set $\{x\}$. All closed sets arise as complements of open sets, so that every subset of $\mathbb{R}^2$ will also be closed. In particular, a sequence cannot be dense in $(\mathbb{R}^2, d')$ since $\mathbb{R}^2$ is uncountable.

(ii) If X is finite-dimensional, the first part of the proof of Theorem 3.8 shows that any norm $\|\cdot\|$ on X is equivalent to the Euclidean norm on $\mathbb{C}^n$. Consequently, from the point of view of the metric structure,[172] the Euclidean norm is representative. On the other hand, if X is infinite-dimensional, let us partition the index set I of a Hamel basis $\{e_i\}_{i\in I}$ into two nonempty sets I_1 and I_2 with I_1 countable, $I_1 = \{e_k\}_{k\geq 1}$. For $x = \sum_{i\in I} x_i e_i$, define the norms $\|x\|_1 = \sum_{k\geq 1} k\,|x_k| + \sum_{j\in I_2} |x_j|$ and $\|x\|_2 = \sum_{i\in I} |x_i|$. There is no constant $K > 0$ with $\|x\|_1 \leq K\|x\|_2$ for all $x \in X$ since the choice $x = e_k$ forces $K \geq k$ for all $k \geq 1$.

25. (i) If $\|x + y\| = \|x\| + \|y\|$, then for $\beta \in (0, 1)$ we have $\|x + \beta y\| = \|x + y - (1 - \beta)y\| \geq \|x + y\| - (1 - \beta)\|y\| = \|x\| + \beta\|y\|$, and similarly $\|\alpha x + y\| = \alpha\|x\| + \|y\|$ for all $\alpha \in (0, 1)$. Scaling by a scalar $\mu > 0$ yields $\|ax + by\| = a\|x\| + b\|y\|$ for all $a, b > 0$. In particular, if $x \neq 0$ and $y \neq 0$ are not collinear, then $e_1 = \frac{x}{\|x\|}$ and $e_2 = \frac{y}{\|y\|}$ are distinct points in S, with $\|se_1 + (1 - s)e_2\| = 1$ for all $s \in [0, 1]$. As for examples, the case of $\mathbb{R}^2$ with the norm $\|(x_1, x_2)\| = |x_1| + |x_2|$ is settled by drawing the unit sphere, see the solution to Exercise 23(ii). For $C[0, 1]$ with the supremum norm $\|\cdot\|_\infty$, pick two positive functions that are not linearly dependent but attain their maximum at the same point, e.g., $x(t) = 1 - (t - \frac{1}{2})^2$ and $y(t) = 1 - (t - \frac{1}{2})^4$ with $\|x + y\|_\infty = \|x\|_\infty + \|y\|_\infty = 2$.

(ii) As a sum of two standard norms, $\|\cdot\|$ is a norm. Note that $\|f\|_\infty \leq \|f\| \leq 2\|f\|_\infty$ for $f \in C[0, 1]$, so that $\|\cdot\|$ is equivalent to the supremum norm, and endows $C[0, 1]$ with a Banach space structure. To prove strict convexity, note that $\|f + g\| = \|f\| + \|g\|$ forces $\big(\int_0^1 |f(t) + g(t)|^2 dt\big)^{1/2} \geq \big(\int_0^1 |f(t)|^2 dt\big)^{1/2} + \big(\int_0^1 |g(t)|^2 dt\big)^{1/2}$. But $C[0, 1]$ with the $L^2[0, 1]$-norm is a pre-Hilbert space, so that equality must hold, and we can conclude by Exercise 8.

26. By definition there is an integer $N \geq 1$ with $d(x_n, x_N) \leq 1$ if $n \geq N$. Consequently $d(x_1, x_n) \leq d(x_1, x_N) + d(x_N, x_n) \leq d(x_1, x_N) + 1$ for $n \geq N$. This shows that $d(x_1, x_k) \leq 1 + \max_{1\leq j\leq N}\{d(x_1, x_j)\}$ for all $k \geq 1$.

27. (i) X is a Banach space, as a closed subspace of $C[0, 1]$, and M is clearly closed. Note that for every $n \geq 1$, the function h_n defined by $h_n(t) = nt$ for $0 \leq t \leq \frac{1}{n}$ and $h_n(t) = 1$ for $\frac{1}{n} \leq t \leq 1$, belongs to M and satisfies $\|h_n\| = 1$.

(ii) Clearly $1 \geq d(f_0, M) > 0$ whenever $f_0 \in X \setminus M$ satisfies $\|f_0\| = 1$. Now $d(f_0, M) \geq 1$ for such an f_0 would mean that $\|f_0 - g\| \geq 1$ for all $g \in M$. We claim that this forces $\big|\int_0^1 h(t)\,dt\big| \leq \big|\int_0^1 f_0(t)\,dt\big|$ for all $h \in X$ with $\|h\| = 1$. Indeed, this is trivially true if $h \in M$, while if $h \in X \setminus M$, for $g = f_0 - \frac{\int_0^1 f_0(t)\,dt}{\int_0^1 h(t)\,dt} h \in M$ we get $1 \leq \|f_0 - g\| = \frac{|\int_0^1 f_0(t)\,dt|}{|\int_0^1 h(t)\,dt|}\|h\| = \frac{|\int_0^1 f_0(t)\,dt|}{|\int_0^1 h(t)\,dt|}$. This proves our claim. However, the inequality fails for the functions $h_n \in X$ defined by $h_n(t) = nt$ for $0 \leq t \leq \frac{1}{n}$ and by $h_n(t) = 1$ for $\frac{1}{n} \leq t \leq 1$, if $n \geq 1$ is large enough, since $\int_0^1 h_n(t)\,dt = 1 - \frac{1}{2n}$. Indeed, if a continuous function $f_0 : [0, 1] \to \mathbb{R}$ is such that $-1 \leq f_0(t) \leq 1$ for all $t \in [0, 1]$, then $\int_0^1 f_0(t)\,dt = \pm 1$ if and only if $f(t) = \pm 1$ for all $t \in [0, 1]$.

28. (i) If $\{f_n\}_{n\geq 1}$ is a Cauchy sequence in $C[0, 1]$, then, given $\varepsilon > 0$, there is $N = N(\varepsilon) \geq 1$ such

172 However, there are geometric differences: for example, in the complex plane $\mathbb{C} \approx \mathbb{R}^2$, the abstract unit sphere $\{x \in \mathbb{C} : \|x\| = 1\}$ is a circle for the Euclidean norm, while for the equivalent norms $\|x\|_1 = |\alpha| + |\beta|$ and $\|x\| = \max\{|\alpha|, |\beta|\}$, where $x = \alpha + i\beta$ with $\alpha, \beta \in \mathbb{R}$, the shape is that of a square with corners at ± 1, $\pm i$, and $\pm 1 \pm i$, respectively.

that $|f_n(t) - f_m(t)| < \varepsilon$ for all $t \in [0, 1]$, whenever $n, m \geq N$. Consequently, for any fixed $t \in [0, 1]$, the numerical sequence $\{f_n(t)\}_{n\geq 1}$ will be a Cauchy sequence, and must converge to some limit $f(t)$. Letting $m \to \infty$ in the previous inequality yields $|f_n(t) - f(t)| \leq \varepsilon$ for all $t \in [0, 1]$ and all $n \geq N$. Thus f_n converges uniformly to f on $[0, 1]$. Choosing $\delta > 0$ such that $|f_N(t) - f_N(s)| \leq \varepsilon$ for $|t - s| < \delta$, we get $|f(t) - f(s)| \leq |f(t) - f_N(t)| + |f_N(t) - f_N(s)| + |f_N(s) - f(s)| \leq 3\varepsilon$ whenever $|t - s| < \delta$, so that $f \in C[0, 1]$.

(ii) Note the identities $\sum_{k=0}^{n} C_n^k t^k (1-t)^{n-k} = [t + (1-t)]^n = 1$ and $\sum_{k=0}^{n} \frac{k}{n} C_n^k t^k (1-t)^{n-k} = \sum_{k=1}^{n} \frac{(n-1)!}{(k-1)!(n-k)!} t^k (1-t)^{n-k} = t \sum_{j=0}^{n-1} \frac{(n-1)!}{j!(n-1-j)!} t^j (1-t)^{n-1-j} = t$, by the first identity. We also have that $\sum_{k=0}^{n} \left(\frac{k}{n}\right)^2 C_n^k t^k (1-t)^{n-k} = \sum_{k=1}^{n} \frac{k}{n} \frac{(n-1)!}{(k-1)!(n-k)!} t^k (1-t)^{n-k} = \sum_{k=1}^{n} \frac{k}{n} C_{n-1}^{k-1} t^k (1-t)^{n-k} = t \sum_{j=0}^{n-1} \frac{j+1}{n} C_{n-1}^{j} t^j (1-t)^{n-1-j} = \frac{t}{n} \sum_{j=0}^{n-1} C_{n-1}^{j} t^j (1-t)^{n-1-j} + t \sum_{j=0}^{n-1} \frac{j}{n} C_{n-1}^{j} t^j (1-t)^{n-1-j} = \frac{t}{n} + t \sum_{j=0}^{n-1} \frac{j}{n-1} C_{n-1}^{j} t^j (1-t)^{n-1-j} - \frac{t}{n} \sum_{j=0}^{n-1} \frac{j}{n-1} C_{n-1}^{j} t^j (1-t)^{n-1-j}$ using the first identity and the fact that $\frac{j}{n} = \frac{j}{n-1} - \frac{j}{n(n-1)}$. Invoking the second identity leads us to $\sum_{k=0}^{n} \left(\frac{k}{n}\right)^2 C_n^k t^k (1-t)^{n-k} = \frac{t}{n} + t^2 - \frac{t^2}{n}$. The three identities that we verfied above yield $\sum_{k=0}^{n} \left(t - \frac{k}{n}\right)^2 C_n^k t^k (1-t)^{n-k} = t^2 \sum_{k=0}^{n} C_n^k t^k (1-t)^{n-k} - 2t \sum_{k=0}^{n} \frac{k}{n} C_n^k t^k (1-t)^{n-k} + \sum_{k=0}^{n} \left(\frac{k}{n}\right)^2 C_n^k t^k (1-t)^{n-k} = \frac{t(1-t)}{n}$. Now, given $\varepsilon > 0$, choose $M > 0$ and $\delta = \delta(\varepsilon) > 0$ such that $|f(t)| \leq M$ for all $t \in [0, 1]$ and $|f(t) - f(s)| < \varepsilon$ whenever $|t - s| < \delta$, by the boundedness and uniform continuity of $f \in C[0, 1]$. For $n \geq \max\left\{\frac{1}{\delta^4}, \frac{M^2}{\varepsilon^2}\right\}$, partition the set of integers $\{0, 1, \ldots, n\}$ into the two sets $S(t) = \{k \in \{0, 1, \ldots, n\} : |t - \frac{k}{n}| < n^{-1/4}\}$ and $L(t) = \{k \in \{0, 1, \ldots, n\} : |t - \frac{k}{n}| \geq n^{-1/4}\}$ whenever $t \in [0, 1]$. Then $\sum_{k\in S(t)} |f(t) - f(\frac{k}{n})| C_n^k t^k (1-t)^{n-k} \leq \varepsilon \sum_{k\in S(t)} C_n^k t^k (1-t)^{n-k} \leq \varepsilon \sum_{k=0}^{n} C_n^k t^k (1-t)^{n-k} = \varepsilon$ by the first identity, while $\sum_{k\in L(t)} |f(t) - f(\frac{k}{n})| C_n^k t^k (1-t)^{n-k} \leq 2M \sum_{k\in L(t)} \frac{(t-\frac{k}{n})^2}{(t-\frac{k}{n})^2} C_n^k t^k (1-t)^{n-k} \leq 2Mn^{1/2} \sum_{k=0}^{n} (t - \frac{k}{n})^2 C_n^k t^k (1-t)^{n-k} = 2Mn^{1/2} \frac{t(1-t)}{n} \leq Mn^{-1/2} < \varepsilon$, using the fourth identity and the fact that $t(1-t) \leq \frac{1}{2}$ for $t \in [0, 1]$. Adding up these two estimates, we get $|f(t) - (B_n f)(t)| \leq 2\varepsilon$ for all $t \in [0, 1]$, if $n \geq \max\left\{\frac{1}{\delta^4}, \frac{M^2}{\varepsilon^2}\right\}$.

(iii) Since the countable union of countable sets is countable, the set of all polynomials in $t \in [0, 1]$ with rational coefficents, obained as the union for $k \geq 0$ of all such polynomials of degree at most k, is countable. Given $f \in C[0, 1]$ and $\varepsilon > 0$, choose the integer $n \geq 1$ such that $|f(t) - (B_n f)(t)| \leq \frac{\varepsilon}{2}$ for all $t \in [0, 1]$, see (ii). If the rational numbers α_{kn} are chosen so that $|f(\frac{k}{n}) - \alpha_{kn}| \leq \frac{\varepsilon}{2}$ for all $k = 0, 1, \ldots, n$, then $|f(t) - \sum_{k=0}^{n} \alpha_{kn} C_n^k t^k (1-t)^{n-k}| \leq \sum_{k=0}^{n} |f(\frac{k}{n}) - \alpha_{kn}| C_n^k t^k (1-t)^{n-k} + |f(t) - (B_n f)(t)| \leq \frac{\varepsilon}{2} \sum_{k=0}^{n} C_n^k t^k (1-t)^{n-k} + \frac{\varepsilon}{2} = \varepsilon$, by the first identity in the above solution to (ii).

(iv) The closed subspace $\{f \in C[0, 1] : f(1) = f(0)\}$ of $C[0, 1]$ can be identified with $C_{\text{per}}[0, 1]$, by extending[173] such functions to continuous periodic functions $f : \mathbb{R} \to \mathbb{C}$ with period 1. This proves that $C_{\text{per}}[0, 1]$ is a separable Banach space. Note that $F \in C^1_{\text{per}}[0, 1]$ if and only if $F(t) = a + \int_0^t f(s)\,ds$, with $a \in \mathbb{C}$ and $f \in C_{\text{per}}[0, 1]$ subject to the constraint $\int_0^1 f(s)\,ds = 0$. If $\{F_n\}_{n\geq 1}$ is a Cauchy sequence in $C^1_{\text{per}}[0, 1]$, then $\{F_n'\}_{n\geq 1}$ is a Cauchy sequence in $C_{\text{per}}[0, 1]$ and $\{F_n(0)\}_{n\geq 1}$ is a Cauchy sequence in $\mathbb{C}$. If $F_n' \to f \in C_{\text{per}}[0, 1]$ and $F_n(0) \to a$ in $\mathbb{C}$, one can see that $F_n(t) \to a + \int_0^t f(s)\,ds$ in $C^1_{\text{per}}[0, 1]$, proving that $C^1_{\text{per}}[0, 1]$ is a Banach space. It is separable since the above considerations show that the polynomials $(a_0 + ib_0) + \sum_{k=0^n} (a_{k+1} + ib_{k+1}) \int_0^t s^k (1-s)^{n-k}\,ds$, with $n \geq 0$ an integer and a_k, b_k rational numbers for $0 \leq k \leq n+1$, are dense. Moreover, for a bounded sequence $\{F_n\}_{n\geq 1}$ in $C^1_{\text{per}}[0, 1]$ there exists $M > 0$ such that $|F_n(t)| + |F_n'(t)| \leq M$ for all $t \in [0, 1]$ and $n \geq 1$. Since $|F_n(t) - F_n(s)| = |\int_s^t F_n'(s)\,ds| \leq M|t - s|$, by Exercise 16(iii) we see that the set $\{F_n\}_{n\geq 1}$ is compact in $C_{\text{per}}[0, 1]$. We now repeat these arguments to deal with the injection $C^1_{\text{per}}[0, 1] \subset C^2_{\text{per}}[0, 1]$ etc.

29. The fact that $L^\infty[0, 1]$ is a normed linear space is plain. For completeness, if $\{f_n\}_{n\geq 1}$ is a Cauchy sequence, given $n \geq 1$, choose $N(n) \geq 1$ such that $\|f_m - f_k\|_\infty \leq \frac{1}{n}$ and define for $m, k \geq N(n)$ the negligible set A_{nmk} of points $t \in [0, 1]$ where $|f_m(t) - f_k(t)| \leq \varepsilon$ fails. On the complement of

[173] Set $f(t + k) = f(t)$ for $t \in [0, 1]$ and $k \in \mathbb{Z}$.

the set $A = \bigcup_{n\geq 1;\, m,k\geq N(n)} A_{nmk}$ of measure zero, the approach used in Exercise 28(i) shows that f_n converge uniformly to some limit $f \in L^\infty[0,1]$. To prove that $L^\infty[0,1]$ is not separable, define for $t \in (0,1)$ the function f_t by $f_t(s) = 1$ if $s \in [0,t)$ and $f_t(s) = 0$ if $s \in [t,1]$. Since $\|f_t - f_T\|_\infty = 1$ for $t \neq T$, the open sets $O_t = \{f \in L^\infty[0,1] : \|f - f_t\|_\infty < \frac{1}{2}\}$, with $t \in (0,1)$, are disjoint. A set that in dense in $L^\infty[0,1]$ has to contain at least one element from each O_t, and can therefore not be finite or countable.

30. (i) One can easily see that $l^1(\mathbb{Z})$ is a normed linear space. To prove completeness, let $\{x_n\}_{n\geq 1}$ be a Cauchy sequence of elements in $l^1(\mathbb{Z})$. Then, see Exercise 26, there is some $M > 0$ such that $\|x_n\| = \sum_{k\in\mathbb{Z}} |x_n(k)| \leq M$ for all $n \geq 1$. Moreover, for every $\varepsilon > 0$ there is an integer $N(\varepsilon) \geq 1$ such that $\|x_n - x_m\| = \sum_{k\in\mathbb{Z}} |x_n(k) - x_m(k)| < \varepsilon$ for $n, m \geq N(\varepsilon)$. In particular, for a fixed $k \in \mathbb{Z}$, $\{x_n(k)\}_{n\geq 1}$ is a numerical Cauchy sequence. Let $x(k) \in \mathbb{C}$ be its limit. We claim that this defines a sequence $x \in l^1(\mathbb{Z})$ and that $x_n \to x$ in $l^1(\mathbb{Z})$. Since for every fixed $m \geq 1$ we have $\sum_{k=-m}^m |x_n(k)| \leq \|x_n\| \leq M$ for all $n \geq 1$, we get $\sum_{k=-m}^m |x(k)| \leq M$ in the limit $n \to \infty$. Thus $x \in l^1(\mathbb{Z})$, with $\|x\| \leq M$. Similarly, given $\varepsilon > 0$, for every fixed integer $m \geq 1$ we have $\sum_{k=-m}^m |x_i(k) - x_j(k)| \leq \|x_i - x_j\| < \varepsilon$ for all $i, j \geq N(\varepsilon)$. Letting $i \to \infty$ leads to $\sum_{k=-m}^m |x(k) - x_j(k)| \leq \varepsilon$ for all $j \geq N(\varepsilon)$. The subsequent limiting process $m \to \infty$ yields $\|x - x_j\| \leq \varepsilon$ for all $j \geq N(\varepsilon)$. Thus $x_n \to x$ in $l^1(\mathbb{Z})$, proving that $l^1(\mathbb{Z})$ is a Banach space. The set of doubly-infinite complex sequences with finitely many nonzero entries is clearly dense in $l^1(\mathbb{Z})$. Each such sequence can be approximated by a similar type of sequence whose entries are of the form $a + ib$ with a and b rational numbers. Since countable unions of countable sets are countable, this proves the separability of $l^1(\mathbb{Z})$.

(ii) The proof that $l^\infty(\mathbb{Z})$ is a Banach space goes along the lines of that for $l^1(\mathbb{Z})$. To prove nonseparability, take the set S of sequences $x = \{x(k)\}_{k\in\mathbb{Z}} \in l^1(\mathbb{Z})$ with $x(k) = 0$ for $k \leq 0$ and $x(k) \in \{0,1\}$ for $k \geq 1$. The cardinality of the set S equals to that of $[0,1]$, since any real number $s \in [0,1]$ has a representation $s = \sum_{k\geq 1} c_k 2^{-k}$ with $c_k \in \{0,1\}$ for all $k \geq 1$. Two different elements of S are at distance 1, and this proves that $l^\infty(\mathbb{Z})$ is not separable by repeating the reasoning from the end of the solution to Exercise 29.

31. (i) We have to show that if $a = (a_0, a_1, \ldots) \in l^2$ is orthogonal to every f_k with $k \geq 0$, then $a = 0$. Clearly $a \perp f_0$ amounts to $a_0 = 0$, and the other relations yield $\sum_{j=1}^\infty a_j (\alpha^k)^j = 0$ for all $k \geq 1$. This means that the power series $F(z) = \sum_{j=1}^\infty a_j z^j$ converges to 0 for every $z = \alpha^k$ with $k \geq 1$. Consequently, the radius of convergence of the series is at least $|\alpha|$. Actually $a_j = 0$ for all $j \geq 1$. Indeed, if $j_0 \geq 1$ is the smallest index $j \geq 1$ with $a_j \neq 0$, then $F(z) = z^{j_0}(a_{j_0} + \sum_{i\geq 1} a_{j_0+i} z^i) = z^{j_0} g(z)$ and $g(0) \neq 0$ ensures $g(z) \neq 0$ whenever $|z|$ is small enough, which is not possible since $g(\alpha^k) = 0$ for all $k \geq 1$. Thus $a_j = 0$ for all $j \geq 0$.

(ii) Since $(e_j, e_i) - (f_j, e_i) = \delta_{ij} - (f_j, e_i)$ while $(e_i, f_j) - (f_i, f_j) = \overline{(f_j, e_i)} - \delta_{ij}$, we deduce that $|(e_j - f_j, e_i)| = |(e_i - f_i, f_j)|$ for all $i, j \geq 1$. Using the Parseval identity and Bessel's inequality yields $\sum_{j\geq 1} \|e_j - f_j\|^2 = \sum_{j\geq 1}\sum_{i\geq 1} |(e_j - f_j, e_i)|^2 = \sum_{i\geq 1}\sum_{j\geq 1} |(e_i - f_i, f_j)|^2 \leq \sum_{i\geq 1} \|e_i - f_i\|^2$. Having equal finite ends forces equality in Bessel's inequality for every single index: $\sum_{j\geq 1} |(e_i - f_i, f_j)|^2 = \|e_i - f_i\|^2$ for all $i \geq 1$. The Fourier series theorem yields $\sum_{j\geq 1} (e_i - f_i, f_j) f_j = e_i - f_i$ for every $i \geq 1$, that is, $\sum_{j\geq 1} (e_i, f_j) = e_i$ for all $i \geq 1$. Therefore every e_i belongs to the closure of the linear subspace spanned by the f_js, proving that this subspace is the entire space.

32. (i) Since $t \mapsto [f(t) - 1]$ is $L^2[0,1]$-orthogonal to all polynomials, and the polynomials are dense in $C_{\text{per}}[0,1]$, see Exercise 28(ii), we deduce that $\int_0^1 [f(t) - 1]\, \overline{g(t)}\, dt = 0$ for all $g \in C_{\text{per}}[0,1]$. In particular, $\int_0^1 [f(t) - 1]\, e^{-2\pi i k t}\, dt = 0$ for all $k \in \mathbb{Z}$. Since the latter form an orthonormal basis of $L^2[0,1]$, we must have $f = 1$ a.e. By continuity $f(t) = 1$ for all $t \in [0,1]$.

(ii) Repeated integration by parts yields $\int_0^\infty t^k \frac{d^n}{dt^n}(t^n e^{-t})\, dt = (-1)^n [n!]^2$ if $k = n \geq 0$, the value being 0 if $0 \leq k < n$. To prove that $\{\phi_k\}_{k\geq 0}$ is an orthonormal set, first observe that $\int_0^\infty \phi_0(t) \overline{\phi_k(t)}\, dt = \frac{1}{n!} \int_0^\infty \frac{d^n}{dt^n}(t^n e^{-t})\, dt = \delta_{0k}$. Since for $n \geq 1$ we can expand the given expression to[174] $\phi_n(t) = \sum_{k=0}^{n-2}(-1)^k C_n^k \frac{n(n-1)...(k+1)}{n!} t^k e^{-t/2} + \frac{(-1)^{n-1}}{(n-1)!} t^{n-1}e^{-t/2} + \frac{(-1)^n}{n!} t^n e^{-t/2}$, we find that for $0 \le m < n$, $\int_0^\infty \phi_m(t)\,\overline{\phi_k(t)}\,dt = \sum_{k=0}^m c_k \int_0^\infty t^k \frac{d^n}{dt^n}(t^n e^{-t})\,dt$ for some constants c_k. This expression vanishes by the first identity. Assume now that some $f \in L^2[0,\infty)$ is orthogonal on all ϕ_k with $k \ge 0$. Since $\{\phi_k\}_{k\ge0}$ span the same subspace of $L^2[0,\infty)$ as $\{t^k e^{-t/2}\}_{k\ge0}$, we get that $\int_0^\infty f(t)e^{-t}t^k\,dt = 0$ for all $k \ge 0$. The Lebesgue dominated convergence theorem enables us to define in the half-plane $\{z \in \mathbb{C} : \mathcal{I}(z) < \frac{1}{2\pi}\}$ the analytic function $F(z) = \int_0^\infty f(t)e^{-t}e^{-2\pi i t z}dt$. Since $(\partial_z^k F)(0) = (-2\pi i t)^k \int_0^\infty f(t)e^{-t}dt = 0$ for all $k \ge 0$, we see that $F \equiv 0$. Thus $\int_0^\infty f(t)e^{-t}e^{-2\pi i t\xi}dt = 0$ for all $\xi \in \mathbb{R}$. If we set $f(t) = 0$ for $t < 0$, then the function $t \mapsto f(t)e^{-t}$ belongs to $L^2(\mathbb{R})$ and its Fourier transform is zero. Thus $f = 0$ a.e., see Chapter 5. This proves that $\{\phi_k\}_{k\ge0}$ is an orthonormal basis of $L^2[0,\infty)$.

(iii) Introduce the so-called *Hermite polynomials* $H_n(x) = (-1)^n e^{x^2}\,\partial_x^n(e^{-x^2})$ for $n \ge 0$; H_n is a polynomial of degree n, with the coefficient of x^n equal to 2^n. The orthonormality of $\{\psi_k\}_{k\ge0}$ amounts to $\int_{\mathbb{R}} H_n(x)H_m(x)e^{-x^2}\,dx = 2^n n!\,\sqrt{\pi}\,\delta_{nm}$ for $n, m \ge 0$. Since $e^{-x^2}P(x) \to 0$ for $|x| \to \infty$ holds for any polynomial P, for $0 \le m < n$ integration by parts yields $\int_{\mathbb{R}} H_m(x)H_n(x)e^{-x^2}\,dx = (-1)^n \int_{\mathbb{R}} H_m(x)\,\frac{d^n}{dx^n}(e^{-x^2})\,dx = \int_{\mathbb{R}} e^{-x^2}\,\frac{d^n}{dx^n}\,H_m(x)\,dx = 0$ since $\frac{d^n}{dx^n}\,H_m(x) = 0$, while if $m = n$ we get $\int_{\mathbb{R}} H_n^2(x)e^{-x^2}\,dx = 2^n\,n!\int_{\mathbb{R}} e^{-x^2}\,dx$. Since[175] $\int_{\mathbb{R}} e^{-x^2}dx = \sqrt{\pi}$, we checked the orthonormality of the family $\{\psi_k\}_{k\ge0}$ in $L^2(\mathbb{R})$. To prove that $\{\psi_k\}_{k\ge0}$ is an orthonormal basis, assume that $f \in L^2(\mathbb{R})$ satisfies $\int_{\mathbb{R}} f(x)\psi_k(x)\,dx = 0$ for all $k \ge 0$. Since $\{\psi_k\}_{k\ge0}$ span the same subspace as $\{x^k e^{-x^2/2}\}_{k\ge0}$, we get $\int_{\mathbb{R}} f(x)x^k e^{-x^2/2}\,dx = 0$ for all $k \ge 0$. By the Lebesgue dominated convergence theorem, the function $F(z) = \int_{\mathbb{R}} f(x)e^{-x^2/2}e^{-2\pi i x z}\,dx$ is analytic in $\mathbb{C}$. Our hypothesis yields $(\partial_z^k F)(0) = (-2\pi i)^k \int_{\mathbb{R}} f(x)e^{-x^2/2}x^k\,dx = 0$ for all $k \ge 0$, so that $F \equiv 0$. In particular, we get $\int_{\mathbb{R}} f(x)e^{-x^2/2}e^{-2\pi i x t}\,dx = 0$ for all $t \in \mathbb{R}$. Consequently the Fourier transform of the $L^2(\mathbb{R})$-function $x \mapsto f(x)e^{-x^2/2}$ vanishes and therefore $f = 0$ a.e., see Chapter 5.

33. Clearly $BV_{\text{per}}[0,1]$ is a normed linear space. If $\{f_n\}_{n\ge1}$ is a Cauchy sequence, then $\{f_n(0)\}_{n\ge1}$ will be a numerical Cauchy sequence having a limit $f(0)$. Considering the partition $x_0 = 0 < x_1 = 1$ of $[0,1]$, we get $\|f_n - f_k\| \ge V_{(f_n-f_k)}[0,1] \ge |(f_n - f_k)(1) - (f_n - f_k)(0)| \ge |f_n(1) - f_k(1)| - |f_n(0) - f_k(0)|$, which shows that $\{f_n(1)\}_{n\ge1}$ is a numerical Cauchy sequence, and therefore converges to some $f(1)$. Similarly, considering for a fixed $x \in (0,1)$ the partition $x_0 = 0 < x_1 = x < x_2 = 1$, we get that $\{f_n(x)\}_{n\ge1}$ is a numerical Cauchy sequence converging to some limit $f(x)$. This defines a periodic function $f : \mathbb{R} \to \mathbb{C}$. To see that the restriction of f to $[0,1]$ has bounded variation, let $x_0 = 0 < x_1 < \cdots < x_m = 1$ be a partition of $[0,1]$. Then $\sum_{j=0}^{m-1} |f(x_{j+1}) - f(x_j)| \le \sum_{j=0}^{m-1} |f(x_{j+1}) - f_n(x_{j+1})| + \sum_{j=0}^{m-1} |f(x_j) - f_n(x_j)| + V_{f_n}[0,1]$. If $N \ge 1$ is such that $\|f_n - f_k\| \le 1$ for $n, k \ge N$, then $V_{(f_n-f_k)} \le \|f_n - f_k\| \le 1$ for $n, k \ge N$, so that $V_{f_n}[0,1] \le M = 1 + \max\{V_{f_1}[0,1], \ldots, V_{f_N}[0,1]\}$. These considerations yield $\sum_{j=0}^{m-1} |f(x_{j+1}) - f(x_j)| \le \limsup_{n\to\infty}\{\sum_{j=0}^{m-1} |f(x_{j+1}) - f_n(x_{j+1})| + \sum_{j=0}^{m-1} |f(x_j) - f_n(x_j)| + V_{f_n}[0,1]\} \le M$, so that $f \in BV_{\text{per}}[0,1]$. To prove that $\|f - f_n\| \to 0$ for $n \to \infty$ amounts to $V_{(f-f_n)}[0,1] \to 0$ for $n \to \infty$ since $\lim_{n\to\infty} f_n(0) = f(0)$. Given $\varepsilon > 0$, there is some $N(\varepsilon) \ge 1$ with $\|f_n - f_k\| \le \varepsilon$ if $n, k \ge N(\varepsilon)$. Thus $\sum_{j=0}^{m-1} |(f_k - f_n)(x_{j+1}) - (f_k - f_n)(x_j)| \le \varepsilon$ for all partitions $x_0 = 0 < x_1 < \cdots < x_m = 1$ of $[0,1]$. Taking $\limsup_{k\to\infty}$ yields $V_{(f-f_n)}[0,1] \le \varepsilon$ if $n \ge N(\varepsilon)$. Thus $BV_{\text{per}}[0,1]$ is a Banach space. Note that for $s \in (0, \frac{1}{2})$, we have $f_s \in BV_{\text{per}}[0,1]$, where $f_s(x) = 0$ for $x \in [0,s] \cup [1-s,1]$ and $f_s(x) = 1$ for $x \in (s, 1-s)$. Since $\|f_s - f_r\| = 2$ for $0 < r < s < \frac{1}{2}$, the reasoning used in the last part of the solution to Exercise 29 shows that the space $BV_{\text{per}}[0,1]$ is not separable.

34. h_α is clearly a linear normed space. Let now $\{f_n\}_{n\ge1}$ be a Cauchy sequence. Since $h_\alpha \subset C_{\text{per}}[0,1]$, there is $f \in C_{\text{per}}[0,1]$ with $f_n \to f$ uniformly on $[0,1]$. Moreover, by Exercise 26,

[174] A sum extending over an empty range is understood to have the value zero.

[175] Changing to polar coordinates yields
$$\left(\int_{\mathbb{R}} e^{-x^2}dx\right)^2 = \iint_{\mathbb{R}^2} e^{-x^2-y^2}\,dxdy = \int_0^\infty \int_0^{2\pi} e^{-r^2}\,r\,drd\theta = 2\pi \int_0^\infty re^{-r^2}\,dr = -\pi e^{-r^2}\Big|_{r=0}^{r=\infty} = \pi.$$

there is $M > 0$ with $\|f_n\|_\alpha \le M$ for all $n \ge 1$. In particular, $[f_n]_\alpha$ is bounded by M, so that, letting $n \to \infty$ in the inequality $|f_n(t) - f_n(s)| \le M|t-s|^\alpha$, we get $|f(t)-f(s)| \le M|t-s|^\alpha$. Thus $f \in h_\alpha$. On the other hand, given $\varepsilon > 0$, there is some $N(\varepsilon) \ge 1$ such that $\|f_n - f_m\|_\alpha < \varepsilon$ for $n, m \ge N(\varepsilon)$. Therefore $|f_n(t) - f_m(t) - f_n(s) + f_m(s)| \le \varepsilon|t-s|^\alpha$ for all $0 \le s \le t \le 1$ and $n, m \ge N(\varepsilon)$. Letting $m \to \infty$, we get $[f - f_n]_\alpha \le \varepsilon$ if $n, m \ge N(\varepsilon)$. Since $f_n \to f$ uniformly on $[0,1]$, we see that $f_n \to f$ in h_α. Thus h_α is a Banach space.

To prove that h_α is not separable, define for $s \in (0, \frac{1}{2})$ the function $f_s \in C_{\text{per}}[0,1]$ by $f_s(t) = 0$ if $t \in [0,s] \cup [1-s, 1]$, $f_s(t) = (t-s)^\alpha$ if $s \le t \le \frac{1}{2}$, and $f_s(t) = (1-t-s)^\alpha$ if $\frac{1}{2} \le t \le 1-s$. We claim that $f_s \in h_\alpha$ with $\|f_s - f_r\|_\alpha \ge 1$ whenever $0 < r < s < \frac{1}{2}$, which ensures that h_α is not separable, see the argument at the end of the solution to Exercise 29. To prove the claim, note that if $0 \le s \le x < y \le \frac{1}{2}$, then $\frac{(y-s)^\alpha - (x-s)^\alpha}{(y-x)^\alpha} = (1+\lambda)^\alpha - \lambda^\alpha \le 1$, where $\lambda = \frac{x-s}{y-x} \ge 0$, since the function $\lambda \mapsto (1+\lambda)^\alpha - \lambda^\alpha$ is decreasing on $(0,\infty)$ as its derivative is negative. Also, for $\frac{1}{2} \le x < y \le 1$ we have $\frac{(1-x-s)^\alpha - (1-y-s)^\alpha}{(y-x)^\alpha} = (1+\lambda)^\alpha - \lambda^\alpha \le 1$, with $\lambda = \frac{1-y-s}{y-x} \ge 0$, while if $s \le x \le \frac{1}{2} \le y \le 1-s$ with $y > x$, then $\frac{|(1-y-s)^\alpha - (x-s)^\alpha|}{(y-x)^\alpha} \le \frac{|(1-y-s)^\alpha - (1-\theta-s)^\alpha|}{|y-\theta|^\alpha} \le 1$ with $\theta = 1-x$. These inequalities show that $[f_s]_\alpha \le 1$. Finally, if $0 < r < s < \frac{1}{2}$, then for $w = f_s - f_r$ we have that $\|w\|_\alpha \ge \frac{|w(s)-w(r)|}{(s-r)^\alpha} = 1$.

35. (i) $x \mapsto \arctan(x)$ maps $\mathbb{R}$ isometrically onto $(-\frac{\pi}{2}, \frac{\pi}{2})$ with the usual distance. The completion is therefore $[-\frac{\pi}{2}, \frac{\pi}{2}]$. (ii) $x \mapsto e^x$ maps $\mathbb{R}$ isometrically onto $(0,\infty)$, so that the completion is $[0,\infty)$. (iii) Note that $\sum_{k=0}^n (\frac{x}{2})^k$ is a Cauchy sequence. For fixed $x \in [0,1]$, $\lim_{n\to\infty} \sum_{k=0}^{n} (\frac{x}{2})^k = \frac{2}{2-x}$ is not a polynomial as all its derivatives at $x = 0$ are nonzero. In view of Exercise 28, the completion is $C[0,1]$. (iv) It is easy to see that we deal with a normed linear space. Let us now introduce the space $C_0(\mathbb{R})$ of continous functions $f : \mathbb{R} \to \mathbb{C}$ that vanish at infinity, i.e., for every $\varepsilon > 0$, the set $\{x \in \mathbb{R} : |f(x)| \ge \varepsilon\}$ is a compact set. We endow $C_0(\mathbb{R})$ with the supremum norm $\|f\|_\infty = \sup_{x\in\mathbb{R}} |f(x)|$ to obtain a Banach space. Indeed, if $\{f_k\}_{k\ge1}$ is a Cauchy sequence in $C_0(\mathbb{R})$, then applying the approach adpoted in the solution of Exercise 28(i) to the restrictions of the functions f_k to a given interval $[-n,n]$, we see that f_k converge uniformly on $[-n,n]$ to a continuous function f. Repeating this procedure for all integers $n \ge 1$, we construct a continuous function $f : \mathbb{R} \to \mathbb{C}$ to which f_n converge uniformly on any compact subset of $\mathbb{R}$. Moreover, given $\varepsilon > 0$, if $N \ge 1$ is such that $\|f_n - f_m\|_\infty < \varepsilon$ for $n, m \ge N$, and if $K \subset \mathbb{R}$ is a compact set such that $|f_N(x)| < \varepsilon$ for $x \notin K$, then $|f_n(x)| \le |f_n(x) - f_N(x)| + |f_N(x)| < 2\varepsilon$ for all $x \notin K$ and all $n \ge N$. Since $\lim_{n\to\infty} f_n(y) = f(y)$ for all $y \in \mathbb{R}$, we deduce that $|f(x)| \le 2\varepsilon$ for all $x \notin K$. Thus $f \in C_0(\mathbb{R})$. Moreover, the uniform convergence of $\{f_n\}_{n\ge1}$ to f on K yields $\limsup_{n\to\infty} |f_n(y) - f(y)| = 0$ if $y \in K$ and $\limsup_{n\to\infty} |f_n(x) - f(x)| \le 4\varepsilon$ for all $x \notin K$, so that $\|f_n - f\|_\infty \to 0$ as $n \to \infty$. This proves that $C_0(\mathbb{R})$ is a Banach space. Our original normed linear space is dense in $C_0(\mathbb{R})$. Indeed, given $f \in C_0(\mathbb{R})$, choose an integer $k_n \ge 1$ such that $|f(x)| < \frac{1}{n}$ for $|x| \ge k_n$. Define the functions $\phi_n : \mathbb{R} \to \mathbb{C}$ by $\phi_n(x) = f(x)$ for $|x| \le k_n$, $\phi_n(x) = [k_n + 1 - x]f(k_n)$ for $x \in [k_n, k_n+1]$, $\phi_n(x) = [k_n + 1 + x]f(-k_n)$ for $x \in [-k_n - 1, -k_n]$, and $\phi_n(x) = 0$ for $|x| \ge k_n + 1$. Then ϕ is continuous with its support contained in $[-k_n - 1, k_n + 1]$, and $\|f - \phi_n\|_\infty < \frac{1}{n}$. Thus $\{\phi_n\}_{n\ge1}$ is a Cauchy sequence in the original normed linear space, and converges to f in $C_0(\mathbb{R})$. To see that the original normed linear space is not complete, it suffices to choose $f(x) = e^{-x^2}$. The completion is $C_0(\mathbb{R})$. Note that $C_0(\mathbb{R})$ is a separable Banach space. To see this, by the density of the original normed space, it suffices to prove the existence of a sequence $\{f_n\}_{n\ge1}$ of compactly supported functions $f_n \in C_0(\mathbb{R})$ such that arbitrarily close to any compactly supported function in $C_0(\mathbb{R})$ we can find some f_n. For this, note first that a simple change of variables, in combination with Exercise 28, shows that for every $n \ge 1$ the Banach space $C[-n,n]$ is separable.[176] Therefore the closed subspace of $C[-2n, 2n]$ of all functions that are supported in $[-n,n]$ is also separable.[177] Let $\{\xi_{kn}\}_{k\ge1}$ be a dense sequence. The desired sequence $\{f_n\}_{n\ge1}$ is a re-labelling of $\bigcup_{n,k\ge1}\{\xi_{kn}\}$.

176 The function $t \mapsto f(t)$ belongs to $C[0,1]$ if and only if $s \mapsto f(\frac{s+n}{2n})$ belongs to $C[-n,n]$.

177 In general, if X is a separable Banach space, then any closed subspace $Y \subset X$ is a separable Banach space. To prove separability, let $\{x_k\}_{k\ge1}$ be dense in X. For any integer $n \ge 1$, select for every x_k some $y_{kn} \in Y$ with $\|x_k - y_{kn}\| < 1/n$, if such and y_{kn} exists, and collect all y_{kn}s in an at most countable set A_n. Then $\bigcup_{n\ge1} A_n$ is dense in Y.

(v) Recall from Chapter 2 that for bounded Riemann-integrable functions $f : [0,1] \to \mathbb{C}$, the Riemann and Lebesgue integrals coincide. The restriction of polynomials to $[0,1]$ are bounded Riemann-integrable functions that are dense in $L^1[0,1]$, see Exercise 4.16, so that the completion is $L^1[0,1]$. Exercise 2.9 provides a concrete example of a Cauchy sequence of bounded Riemann-integrable functions $f_n : [0,1] \to [0,1]$ that fails to converge (within the setting of the Riemann integral).

36. (i) Linearity is clear. The inequality $\|Tx\|^2 = \sum_{k=1}^{\infty} \frac{|x(k)|^2}{k^2} \le \sum_{k=1}^{\infty} |x(k)|^2 = \|x\|^2$ for $x \in l^2$ shows that $\|T\| \le 1$. Since T maps the sequence having the first entry 1 and all others zero into itself, $\|T\| = 1$.

(ii) Clearly $T(X)$ is a linear subspace of Y. To show that it is closed, let $y_n \to y$ with $y_n \in T(X)$. Then for every $n \ge 1$, $y_n = Tx_n$ for some $x_n \in X$. The inequality $\|x_n - x_m\|_X \le \frac{1}{M}\|Tx_n - Tx_m\|_Y = \frac{1}{M}\|y_n - y_m\|_Y$ shows that $\{x_n\}_{n\ge1}$ is a Cauchy sequence in X. Let $x = \lim_{n\to\infty} x_n$. The continuity of T ensures that $Tx_n \to Tx$ in Y, so that $y = Tx$.

(iii) The operator T defined in (i) is an example. Indeed, the sequence $y \in l^2$ with entries $y(k) = k^{-3/2}$ for $k \ge 1$, clearly does not lie in the range of T. Define for every $r \in (0,1)$, the sequence x_r with entries $x_r(k) = r^k\, k^{-1/2}$ for $k \ge 1$. Since $\sum_{k=1}^{\infty} r^{2k}\, k^{-1} < \sum_{k=1}^{\infty} r^{2k} = \frac{1}{1-r^2}$, we see that $x_r \in l^2$ for every $r \in (0,1)$. The operator T maps $x_r \in l^2$ into the sequence $y_r \in l^2$ with entries $y_r(k) = r^k\, k^{-3/2}$ for $k \ge 1$. Since $\|y_r - y\|_{l^2}^2 = \sum_{k=1}^{\infty} \frac{(1-r^k)^2}{k^3} \to 0$ as $r \uparrow 1$, y lies in the closure of the range of T.

37. Item (i) and (ii) are easily verified. As for item (iii), choose the operators defined in (i) and (ii).

38. If X has finite dimension $n \ge 1$, by choosing a basis, we can identify T and S with some $n \times n$-matrices $(a_{ij})_{1\le i,j\le n}$ and $(b_{ij})_{1\le i,j\le n}$, respectively. If $(c_{ij})_{1\le i,j\le n}$ is the matrix corresponding to $R = TS - ST$, its trace equals $\sum_{k=1}^{n} c_{kk} = \sum_{i=1}^{n}\big(\sum_{j=1}^{n} a_{ij}b_{ji} - \sum_{l=1}^{n} b_{il}a_{li}\big) = 0$, while the identity matrix, corresponding to I, has trace n. Thus $TS - ST \ne I$.

The previous argument does not apply to infinite-dimensional spaces – there are the issues of the basis and also of the trace. Instead, let us prove by induction that if $TS - ST = I$, then

$$TS^{n+1} - S^{n+1}T = (n+1)S^n\,, \qquad (*)$$

for all integers $n \ge 0$. Indeed, by assumption $(*)$ holds for $n = 0$. If it holds for some integer $n = k \ge 0$, then $TS^{k+2} - S^{k+1}TS = (k+1)S^{k+1}$. Since $TS = ST + I$ yields $S^{k+1}TS = S^{k+2}T + S^{k+1}$, we get that $(*)$ holds for $n = k+1$. Thus the validity of $(*)$ is established for all integers $n \ge 0$. From $(*)$ and the fact that $\mathcal{L}(X)$ is a Banach algebra we infer that

$$(n+1)\|S^n\| \le \|T\|\cdot\|S\|\cdot\|S^n\| + \|S\|\cdot\|S^n\|\cdot\|T\| = 2\,\|S\|\cdot\|T\|\cdot\|S^n\| \qquad (**)$$

for all integers $n \ge 0$. Now, if for some $k \ge 1$ we have $S^k = 0$, then $(*)$ with $n = k-1$ yields $S^{k-1} = 0$, so that in a few steps we infer that $S = 0$, which would force $I = 0$. The remaining option is that $\|S^k\| \ne 0$ for all $k \ge 1$. But then $(**)$ yields $(n+1) \le 2\,\|S\|\cdot\|T\|$ for all $n \ge 0$, another contradiction.

39. Let $\{e_n\}_{n\ge1}$ be an orthonormal basis of the Hilbert space H. Set $f_{2k} = e_{2k}$ and $f_{2k-1} = 0$ for $k \ge 1$, while $x_n = e_n$ for all $n \ge 1$. Since H is reflexive, weak* convergence and weak convergence coincide in H^*. Note that $f_n \rightharpoonup 0$ in H^* and $x_n \rightharpoonup 0$ in H but $\{f_n(x_n)\}_{n\ge1}$ is not a convergent numerical sequence since $f_{2k}(x_{2k}) = 1$ while $f_{2k-1}(x_{2k-1}) = 0$ for all $k \ge 1$.

40. (i) Assuming that (a) holds, let $x_n \to x$ in X be such that $Tx_n \to y$ in Y. Then $Tx_n \to y$ in Z and therefore $y = Tx$, $T : X \to Z$ being bounded. Theorem 3.25 yields (b). Assuming now that (c) holds, for an arbitrary $x \in X$, choose $x_n \in D$ with $x_n \to x$. The inequality ensures that $\{Tx_n\}_{n\ge1}$ is a Cauchy sequence in Y, so that $\lim_{n\to\infty} Tx_n = y \in Y$ exists. But $Tx_n \to Tx$ in Z, so that $y = Tx$ and (a) holds. By the above, (b) holds too. The other implications are immediate. (ii) Choose $X = Z = C_{\text{per}}[0,1]$ and $D = Y = C^1_{\text{per}}[0,1]$ in (i). Exercise 28 shows that D is dense in X. Clearly $T : X \to Z$ is linear and bounded. For $x \in Y$, integration by parts yields

$(Tx)(t) = \frac{3}{2}(1-t)^{1/2}t^{2/3}\,x(t) - \frac{3}{2}(1-t)^{1/2}\int_0^1 s^{2/3}x'(s)\,ds$ for all $t \in [0,1]$, so that $\|Tx\|_X \le \frac{3}{2}\|x\|_X + \frac{3}{2}\|x'\|_X = \frac{3}{2}\|x\|_Y$.

41. Apply Theorem 3.19 for the continuous function $\varphi(x) = \|x - x_0\|$.

42. (i) Define $T_n \in X^{**}$ by $T_n(f) = f(x_n)$ for $n \ge 1$. Since convergent scalar sequences are bounded, we have that $\sup_{n\ge1}|T_n(f)| < \infty$ for every $f \in X^*$. The uniform boundedness principle yields $M = \sup_{n\ge1}\|T_n\|_{X^{**}} < \infty$. But, by Corollary 5 from Section 3.4.2, $\|x_n\| = \|T_n\|$ for every $n \ge 1$. For the second part, if $x \neq 0$, choose $f \in X^*$ with $\|f\|_{X^*} = 1$ and $f(x) = x$, see Corollary 2 from Section 3.4.2, to get $\|x\| = f(x) = \liminf_{n\to\infty} f_n(x) \le \liminf_{n\to\infty}\{\|f\|_{X^*}\|x_n\|\} \le \liminf_{n\to\infty}\|x_n\|$. (ii) For the non-trivial direction, note that $\limsup_{n\to\infty}\|x - x_n\|^2 = \limsup_{n\to\infty}(x - x_n, x - x_n) = \limsup_{n\to\infty}\{\|x\|^2 + \|x_n\|^2 - (x, x_n) - (x_n, x)\} = \limsup_{n\to\infty}\|x_n\|^2 - \|x\|^2 = 0$. (iii) Define $T_n \in Y^*$ by $T_n(y) = x_n(y)$ for $n \ge 1$. Then $\|T_n\|_{Y^*} = \|x_n\|_X$, and the statement follows from the uniform boundedness principle.

43. (i) Assume that $f_n \rightharpoonup f$ in $C[0,1]$. Since the Dirac mass at $t \in [0,1]$, defined by $\delta_t(B) = 1$ if $t \in B$ and $\delta_t(B) = 0$ if $t \notin B$, belongs to the dual $\mathcal{M}[0,1]$, we deduce that $\lim_{n\to\infty} f_n(t) = f(t)$ for all $t \in [0,1]$. Uniform boundedness is ensured by Exercise 42(i). Conversely, if $\{f_n\}_{n\ge1}$ is uniformly bounded and converges pointwise to some $f \in C[0,1]$, then the Lebesgue dominated convergence theorem ensures that $\lim_{n\to\infty}\mu(f_n) = \mu(f)$ for every $\mu \in \mathcal{M}[0,1]$, the dual of $C[0,1]$.

(ii) Note that $f_n(t) = (1-t)^n$ are of unit norm in $C[0,1]$ for every $n \ge 1$. By Theorem 3.18, reflexivity would ensure the existence of a weakly convergent subsequence $\{f_{n_k}\}$. But the pointwise limit of $\{f_{n_k}\}$ is the discontinuous function f given by $f(t) = 0$ if $t \in (0,1]$ and $f(0) = 1$. This contradicts item (i).

44. Since the finite-dimensional case is clarified by counting dimensions, suppose that X is infinite-dimensional. Let $\{e_i\}_{i\in I}$ be an infinite Hamel basis B for X. For $i \in I$, define $e_i' \in X'$ by $e_i'(e_j) = \delta_{ij}$ and note that they are linearly independent in X'. Consider the so-called dual Hamel basis B' of X', obtained by completing the family $\{e_i'\}_{i\in I}$. We now specify $x_0'' \in X''$ by requiring that $x_0''(e_i') = c_i$ for $i \in I$ and $x_0''(x') = 0$ for $x' \in B' \setminus \bigcup_{i\in I}\{e_i'\}$, with infinitely many of the scalars c_i nonzero. If J is onto, then we can find $x \in X$ such that $[J(x)](x') = x'(x) = x_0''(x')$ for all $x' \in X'$. However, if $i \in I$ is such that the term e_i is missing in the unique representation of x as a (finite) linear combination of elements from B, and $x''(e_i') = c_i \neq 0$, then $e_i'(x) = 0 \neq c_i = x''(e_i')$, which is a contradiction.

45. Without loss of generality we may assume that the basis is normalised: $\|e_k\| = 1$ for all $k \ge 1$. Since $\lim_{n\to\infty}\sum_{k=1}^n x_k e_k = x$, we see that the norm $\|x\|_0 = \sup_{n\ge1}\|\sum_{k=1}^n x_k e_k\|$ is well-defined, with $\|x\| \le \|x\|_0$ for all $x \in X$. We claim that X endowed with the norm $\|\cdot\|_0$ is not just a normed linear space X_0, but a Banach space. Then the open mapping applied to the identity acting between X_0 and X implies that the inverse is continuous, proving that $\|\cdot\|$ and $\|\cdot\|_0$ are equivalent norms, which yields the statement. To prove that X_0 is complete, note first that if $x = \sum_{k\ge1} x_k e_k$, then $|x_k| = \|x_k e_k\| = \|\sum_{j=1}^k x_j e_j - \sum_{j=1}^{k-1} x_j e_j\| \le 2\|x\|_0$. Let now $\{\xi_n\}_{n\ge1}$ be a Cauchy sequence in X_0, with $\xi_n = \sum_{k\ge1} c_{nk} e_k$ in X. Then, given $\varepsilon > 0$, there exists $N_\varepsilon \ge 1$ such that $\|\xi_n - \xi_m\|_0 = \sup_{k\ge1}\|\sum_{j=1}^k (c_{nj} - c_{mj})e_j\| < \varepsilon$. The previous two estimates ensure that for every $j \ge 1$ the scalar sequence $\{c_{nj}\}_{n\ge1}$ is a Cauchy sequence. Let $c_j \in \mathbb{C}$ be its limit. Note that $\|\sum_{j=1}^k (c_j - c_{mj})e_j\| = \|\sum_{j=1}^k \lim_{n\to\infty}(c_{nj} - c_{mj})e_j\| = \lim_{n\to\infty}\|\sum_{j=1}^k (c_{nj} - c_{mj})e_j\| \le \varepsilon$ for each $k \ge 1$, provided that $m \ge N_\varepsilon$. Since $\|x\|_0 \ge \|x\|$ ensures that $\{\xi_n\}_{n\ge1}$ is a Cauchy sequence in X, we get that $\sum_{j=1}^k c_j e_j$ converges to a limit $\xi \in X$, and $\lim_{n\to\infty}\|\xi_n - \xi\| = 0$. Moreover, for $n \ge N_\varepsilon$ we have $\|\xi - \xi_n\|_0 = \sup_{k\ge1}\|\sum_{j=1}^k (c_j - c_{nj})e_j\| \le \varepsilon$. Thus $\xi_n \to \xi$ in X_0 and X_0 is complete.

46. We have that $\|u(t) - u(s)\| = \max\limits_{f\in X^*,\,\|f\|_{X^*}\le1}\{|f(u(t)) - f(u(s))|\}$, by Corollary 5 in Section 3.4.2. The scalar-valued map $t \mapsto f(u(t))$ is differentiable on $(0,1)$ since $\frac{f(u(t_0+h)) - f(u(t_0))}{h} - f(u'(t_0)) = f\big(\frac{u(t_0+h) - u(t_0)}{h} - u'(t_0)\big) \to 0$ as $h \to 0$. The scalar version of the mean-value theorem from calculus yields $|f(u(t)) - f(u(s))| \le (t-s)\sup\limits_{\xi\in[s,t]}|f(u'(\xi))|$. Since $|f(u'(\xi))| \le \|f\|_{X^*}\|u'(\xi)\|$, by taking the supremum over $f \in X^*$ with $\|f\|_{X^*} \le 1$, we complete the proof of the theorem.

47. Since $z \mapsto l(f(z))$ is analytic in G, we may represent it by the Cauchy integral formula

$l(f(z_0)) = \frac{1}{2\pi i}\int\limits_C \frac{l(f(z))}{z-z-0}\,dz$, where $C \subset G$ is a circle around $z_0 \in G$. Similar formulas (with the same circle C) hold when z_0 is replaced by z_0+h and z_0+k, provided $|h|+|k|$ is small enough. If $h \neq k$, $h \neq 0$ and $k \neq 0$, a straighforward calculation starting from these three formulas yields $\frac{1}{h-k}\{\frac{l(f(z_0+h))-l(f(z_0))}{h} - \frac{l(f(z_0+k))-l(f(z_0))}{k}\} = \frac{1}{2\pi i}\int\limits_C \frac{l(f(z))}{(z-z_0-h)(z-z_0-k)(z-z_0)}$. For a fixed $l \in X^*$ and $|h|+|k|$ small enough, the absolute value of the right side is bounded by a constant $M(l)$ that is independent of h and k. Denoting $x_{hk} = \frac{1}{h-k}\{\frac{f(z_0+h)-f(z_0)}{h} - \frac{f(z_0+k)-f(z_0)}{k}\}$, we can re-write the left side as $l(x_{hk})$, with $|l(x_{hk})| \le M(l)$. The uniform boundedness principle applied to the family of operators $l \mapsto l(x_{hk})$ in X^{ast*}, in combination with Corollary 5 from Section 3.4.2, yield the existence of a constant $C>0$ such that $\|x_{hk}\| \le C$ whenever $|h|+|k|$ is small enough. Recalling the definition of x_{hk}, we write this conclusion in the form $\|\frac{f(z_0+h)-f(z_0)}{h} - \frac{f(z_0+k)-f(z_0)}{k}\| \le C|h-k|$. Since X is a Banach space, the difference quotients tend to a limit as $h,k \to 0$.

48. In (i) reflexivity ensures that X^{**} is separable, and we can conclude by Theorem 3.16. Y is clearly a Banach space, so that (ii) follows from Theorem 3.18 and by Corollary 3 in Section 3.4.2. Indeed, if $\{x_n\}_{n\ge1}$ is a sequence in Y with $x_n \rightharpoonup x$ in X, assuming that $x \notin Y$ would entail the existence of some $F \in X^*$ with $F(x) > 0$ and $F(y) = 0$ for all $y \in Y$. This leads, however, to the contradiction $0 = F(x_n) \not\to F(x)$.

49. Let X_0 be the subset of l^1 of all sequences with at most finitely many nonzero entries. Define $T_n : X_0 \to X_0$ by $(T_n x)(k) = 0$ if $k \neq n$ and $(T_n x)(n) = nx(n)$, with $n \ge 1$. Then $\sup_{n\ge1}\|T_n x\| < \infty$ for every $x \in X_0$. However, $\|T_n\| = n$ since clearly $\|T_n\| \le n$ and $T_n e_n = ne_n$ for $e_n(k) = \delta_{nk}$. This does not contradict Theorem 3.22 since X_0 is an incomplete normed linear space.

Let us now observe that the linear space $X = l^1$ has a Hamel basis $\{e_i\}_{i\in I}$ of vectors with $\|e_i\|_{l^1} = 1$ for every $i \in I$. Endowed with the norm $\|x\|_X = \sum_{i\in I}|\alpha_i|$, where $x = \sum_{i\in I}\alpha_i e_i$, X is clearly a normed linear space. Since $\|e_i\|_{l^1} = 1$, we get $\|x\|_{l^1} \le \|x\|_X$ for all $x \in l^1$, so that $Tx = x$ defines a bounded linear operator from X to l^1. The inverse is, however, not continuous, despite having a closed graph. Indeed, a continuous inverse would ensure that X is separable, l^1 being so by virtue of having a Schauder basis. But I is uncountable, see Exercise 15(ii), so that $\|e_i - e_j\|_X = 2$ for $i \neq j$ ensures that X is not separable, see the argumentation in the end of the solution to Exercise 29. Finally, to verify that $T^{-1} : l^1 \to X$ has a closed graph, note that if $x_n \to x$ in l^1 and $T^{-1}x_n = x_n \to y$ in X, then clearly $x = y$ since $\|x_n - y\|_X \ge \|x_n - y\|_{l^1}$. Comparing these considerations with the statements of Theorems 3.24–3.25 shows that X is not complete.

Having established above the importance of the completeness of the range for the validity of the open mapping and closed graph theorem, let us also show that the domain of definition must be complete. For this, define the bounded linear operator $S : X_0 \to X_0$ by $(Sx)(k) = \frac{x(k)}{k}$ for $k \ge 1$. Its inverse S^{-1} is clearly unbounded since $\|e_n\| = 1$ and $\|S^{-1}e_n\| = n$ for all $n \ge 1$. Moreover, S^{-1} has a closed graph. Indeed, since $(S^{-1}x_n)(k) = kx_n(k)$, we see that $x_n \to x$ in X_0 and $S^{-1}x_n \to y$ in X_0 mean $\sum_{k\ge1}|x_n(k) - x(k)| \to 0$ and $\sum_{k\ge1} k\,|x_n(k) - y(k)| \to 0$ as $n \to \infty$, respectively. In particular, $\lim_{n\to\infty}\{k\,|x_n(k) - y(k)|\} = 0$ and $\lim_{n\to\infty}|x_n(k) - x(k)| \to 0$ for every $k \ge 1$, and thus $x = y$.

50. (i) For $n \ge 1$, define the linear operator $T_n : Y \to \mathbb{R}$ by $T_n y = B(x_n, y)$. By hypothesis, each T_n is bounded. Since $x_n \to 0$ in X and $B(\cdot, y)$ is bounded, $\{|T_n y|\}_{n\ge1}$ is bounded for each fixed $y \in Y$. By the uniform boundedness principle, there is some $C > 0$ with $|T_n y| \le C\|y\|$ for all $n \ge 1$. Thus $|B(x_n, y_n)| = |T_n y| \le C\|y_n\| \to 0$ as $n \to \infty$. (ii) If $f(0,0) = 0$ and $f(x,y) = xy/(x^2+y^2)$ if $(x,y) \neq (0,0)$, then f is separately continuous despite being discontinuous at $(0,0)$.

51. Let us show that the graph $\Gamma(B) = \{(x,y) \in H \times H : y = Bx\}$ is closed. Indeed, let $(x_n, Bx_n) \to (x,y)$. For every $z \in H$ we have that $(z,y) = \lim_{n\to\infty}(z, Bx_n) = \lim_{n\to\infty}(Az, x_n) = (Az, x) = (z, Bx)$, so that $y = Bx$. By the closed graph theorem, B must be bounded. The proof that A is bounded is similar.

52. On $H = l^2$, let $U(x_1, x_2, x_3, \dots) = (0, x_1, x_2, x_3, \dots)$. Then U preserves the inner product but $(1, 0, 0, \dots)$ is not in its range. If H is finite-dimensional, let $\{e_1, \dots, e_n\}$ be an orthonormal basis. Then $(Ue_i, Ue_j) = (e_i, e_j) = \delta_{ij}$ shows that the set $\{Ue_k\}_{1\le k\le n}$ is orthonormal. It must be a basis due to the dimension count, and this proves that U is onto.

53. Let $\alpha = \sup_{\|x\|=1} \{|(Tx, x)|\}$. Since $|(Tx, x)| \le \|Tx\| \cdot \|x\| \le \|T\| \cdot \|x\|^2$, we have that $\alpha \le \|T\|$. To prove that $\|T\| \le \alpha$, let $\beta > 0$ and use the fact that $(Tx, Tx) = (T^2x, x)$ to get $4\|Tx\|^2 = 2\|Tx\|^2 + 2(T^2x, x) = (T(\beta x + \frac{1}{\beta}Tx), \beta x + \frac{1}{\beta}Tx) - (T(\beta x - \frac{1}{\beta}Tx), \beta x - \frac{1}{\beta}Tx)$. The definition of α yields $4\|Tx\|^2 \le \alpha\|\beta x + \frac{1}{\beta}Tx\|^2 + \alpha\|\beta x - \frac{1}{\beta}Tx\|^2 = 2\alpha\{\beta^2\|x\|^2 + \frac{1}{\beta^2}\|Tx\|^2\}$, due to the parallelogram law. If $\|Tx\| \ne 0$, choose $\beta^2 = \frac{\|Tx\|}{\|x\|}$ to get $4\|Tx\|^2 \le 4\alpha\|x\| \cdot \|Tx\|$. Thus $\|Tx\| \le \alpha\|x\|$, proving that $\|T\| \le \alpha$. Thus $\|T\| = \alpha$. Applied to the self-adjoint operator T^2, this yields $\|T^2\| = \sup_{\|x\|=1} \{|(T^2x, x)|\} = \sup_{\|x\|=1} \{|(Tx, Tx)|\} = \|T\|^2$. Note that in $\mathbb{C}^2$, the matrix $T = \begin{pmatrix} 0 & 1 \\ 0 & 0 \end{pmatrix}$ is nil-potent, with $T^2 = 0$, while $\|T\| = 1$.

54. If T is self-adjoint, then $(Tx, x) = (x, Tx) = \overline{(Tx, x)}$ shows that $(Tx, x) \in \mathbb{R}$. Conversely, if $(Tx, x) \in \mathbb{R}$, then clearly $(Tx, x) = (x, Tx)$. Therefore, for any $x, y \in H$ we have $4(x, Ty) = (x+y, T(x+y)) - (x-y, T(x-y)) + i(x+iy, T(x+iy)) - i(x-iy, T(x-iy)) = (T(x+y), x+y) - (T(x-y), x-y) + i(T(x+iy), x+iy) - i(T(x-iy), x-iy)) = 4(Tx, y)$, so that $T = T^*$.

55. (i) For the constant function $f \equiv 1$ we have $(\mathcal{L}f)(s) = 1/s$, which settles the case $p = \infty$. Let now $\{\alpha_n\}_{n\ge1}$ be a sequence of positive numbers, set $f_n(t) = e^{-\alpha_n t}$ for $t > 0$ with $\|f_n\|^p_{L^p(0,\infty)} = 1/(p\alpha_n)$ and compute $(\mathcal{L}f_n)(s) = 1/(s+\alpha_n)$ for $s > 0$, with $\|\mathcal{L}f_n\|^p_{L^p(0,\infty)} = \alpha_n^{1-p}/(p-1)$ for $p > 1$. Thus $\mathcal{L}f_n \notin L^1(0,\infty)$ although $f_n \in L^1(0,\infty)$, while for $\infty > p > 1$ the boundedness of $\mathcal{L}$ from $L^p(0,\infty)$ to $L^p(0,\infty)$ would ensure the existence of some constant $M_p > 0$ with $\alpha_n^{1-p} \le M_p\alpha_n^{-1}$. The choices $\alpha_n = n^{\pm1}$ for $n \ge 1$ force $p = 2$. If $p = 2$, then

$$|(\mathcal{L}f)(s)|^2 = \Big(\int_0^\infty (f(t)e^{-st/2}t^{1/4})(e^{-st/2}t^{-1/4})dt\Big)^2 \le C(s)\int_0^\infty |f(t)|^2 e^{-st}t^{1/2}dt$$

by Cauchy–Schwarz, where $C(s) = \int_0^\infty e^{-st}t^{-1/2}dt$. Since $C(s) = 2\sqrt{s}\int_0^\infty e^{-x^2}dx = \sqrt{s}\int_{\mathbb{R}} e^{-x^2}dx = \sqrt{\pi s}$ by the change of variable $x = \sqrt{st}$, integration yields

$$\|\mathcal{L}f\|^2_{L^2(0,\infty)} \le \sqrt{\pi}\int_0^\infty\int_0^\infty |f(t)|^2 e^{-st}t^{1/2}s^{-1/2}dtds\,.$$

Interchanging the order of integration, the explicit formula for $C(s)$ applies and the upper bound is seen to be $\pi\int_0^\infty |f(t)|^2dt$. Thus $\|\mathcal{L}f\|_{L^2(0,\infty)} \le \sqrt{\pi}\|f\|_{L^2(0,\infty)}$. To show that equality holds, note that $\|f_n\|^2_{L^2(0,\infty)} = 2\ln(n)$ for $f_n(t) = t^{-1/2}\chi^{[1/n,n]}(t)$ with $n \ge 1$, while

$$\|\mathcal{L}f\|^2_{L^2(0,\infty)} = \int_0^\infty \Big(\int_{1/n}^n e^{-st}t^{-1/2}dt\Big)\Big(\int_{1/n}^n e^{-s\tau}\tau^{-1/2}d\tau\Big)ds = \int_0^\infty\int_{1/n}^n\int_{1/n}^n \frac{e^{-s(t+\tau)}}{\sqrt{t\tau}}\,dt\,d\tau\,ds\,.$$

An interchange of integration order shows that this equals to $\int_{1/n}^n\int_{1/n}^n \frac{1}{(t+\tau)\sqrt{t\tau}}\,dtd\tau$. Since the change of variables $y = \sqrt{t/\tau}$ yields

$$\int_{1/n}^n \frac{1}{(t+\tau)\sqrt{t}}\,dt = \frac{2}{\sqrt{\tau}}\int_{1/\sqrt{n\tau}}^{\sqrt{n/\tau}} \frac{dy}{1+y^2} = \frac{2\,[\arctan(\sqrt{n\tau}) - \arctan(1/\sqrt{n\tau})]}{\sqrt{\tau}}$$

and thus

$$\int_{1/n}^n\int_{1/n}^n \frac{1}{(t+\tau)\sqrt{t\tau}}\,dtd\tau = 2\int_{1/n}^n \frac{\arctan(\sqrt{n\tau})}{\tau}\,d\tau - 2\int_{1/n}^n \frac{\arctan(1/\sqrt{n\tau})}{\tau}\,d\tau\,,$$

performing the changes of variables $r = \sqrt{n\tau}$ in the first term and $r = 1/\sqrt{n\tau}$ in the second yields

$$\|\mathcal{L}f_n\|^2_{L^2(0,\infty)} = 4\Big\{\int_1^n \frac{\arctan(r)}{r}\,dr - \int_{1/n}^1 \frac{\arctan(r)}{r}\,dr\Big\}.$$

Using integration by parts, we obtain

$$\|\mathcal{L}f_n\|^2_{L^2(0,\infty)} = 4\Big\{\ln(n)\Big[\arctan(n) - \arctan\Big(\frac{1}{n}\Big)\Big] - \int_1^n \frac{\ln(r)}{1+r^2}\,dr + \int_{1/n}^1 \frac{\ln(r)}{1+r^2}\,dr\Big\}$$
$$= 4\ln(n)\Big\{\arctan(n) - \arctan\Big(\frac{1}{n}\Big)\Big\}$$

since the remaing terms cancel after changing variables $r \mapsto 1/r$. Thus

$$\frac{\|\mathcal{L}f\|^2_{L^2(0,\infty)}}{\|f_n\|^2_{L^2(0,\infty)}} = 2\arctan(n) - 2\arctan\Big(\frac{1}{n}\Big) \to \pi \quad \text{for} \quad n \to \infty\,.$$

This proves that $\mathcal{L}$ has norm $\sqrt{\pi}$.

(ii) Setting $a_\pm = (|a| \pm a)/2 \geq 0$ and observing that $|a+ib|^2 = |(a_+ - a_-) + i(b_+ - b_-)|^2 \leq a_+^2 + a_-^2 + b_+^2 + b\_2$ for real numbers a and b shows that it suffices to prove that $\mathcal{L}^2 f = Tf$ for $f : (0,\infty) \to [0,\infty)$ of class $L^2(0,\infty)$, since $\mathcal{L}^2$ is a bounded operator on $L^2(0,\infty)$. But for $f \geq 0$ in $L^2(0,\infty)$ it is easy to compute $(\mathcal{L}^2 f)(s) = \int_0^\infty (\mathcal{L}f)(t)\,e^{-st}dt = \int_0^\infty \big(\int_0^\infty f(\tau)e^{-t\tau}\,d\tau\big)e^{-st}dt = \int_0^\infty f(\tau)\int_0^\infty e^{-(\tau+s)t}dtd\tau = \int_0^\infty \frac{f(\tau)}{\tau+s}\,d\tau$. Finally, it is easy to see that $\mathcal{L}$ is self-adjoint on $L^2(0,\infty)$, and therefore the norm of $\mathcal{L}^2$ equals to the square of the norm of $\mathcal{L}$, see Exercise 53.

56. (i) By Theorem 3.1, $Ae_n = \sum_{m\geq1}(Ae_n, e_m)e_m = \sum_{m\geq1} a_{nm}e_m$, so that $\sum_{m\geq1}|a_{nm}|^2 < \infty$ for every $n \geq 1$. Similary, $A^*e_m = \sum_{n\geq1}(A^*e_m, e_n)e_n = \sum_{n\geq1}(e_m, Ae_n)e_n = \sum_{n\geq1}\overline{a_{nm}}\,e_n$ yields $\sum_{n\geq1}|a_{nm}|^2 < \infty$ for every $m \geq 1$. Assuming that the specified matrix corresponds to a bounded linear operator A, for $x = \sum_{n\geq1} n^{-2/3}e_n$ we would get $(Ax, x) = \big(\sum_{n\geq1} n^{-2/3}Ae_n, \sum_{m\geq1} m^{-2/3}e_m\big) = \sum_{n\geq1}\sum_{m\geq1}\frac{1}{(m+n)^{2/3}n^{2/3}m^{2/3}} \geq \sum_{n\geq1}\sum_{m\geq1}\frac{1}{(m+n)^2} = \sum_{N\geq1}\frac{N}{(N+1)^2}$ since there are precisely N pairs of integers $n, m \geq 1$ with $n + m = N + 1$ as choosing $1 \leq n \leq N$ determines uniquely $m \geq 1$. But $|(Ax, x)| < \infty$ if A is bounded, while $\sum_{N\geq1}\frac{N}{(N+1)^2} \geq \frac{1}{2}\sum_{N\geq1}\frac{1}{N+1} = \infty$.

(ii) If $M = \sum_{m\geq1}\sum_{k\geq1}|a_{km}|^2 < \infty$, then $\sum_{k\geq1}|x_k|^2 < \infty$ yields

$$\sum_{m\geq1}\Big|\sum_{k\geq1} a_{km}x_k\Big|^2 \leq \sum_{m\geq1}\Big(\sum_{k\geq1}|a_{km}|^2\cdot\sum_{k\geq1}|x_k|^2\Big) \leq M\sum_{k\geq1}|x_k|^2\,,$$

so that $Ax = \sum_{m\geq1}\big(\sum_{k\geq1} a_{km}x_k\big)e_m$ defines a bounded linear operator, due to Theorem 3.1. Note that $(Ae_n, e_m) = a_{nm}$ for $n, m \geq 1$. The condition fails blatantly for the identity operator.

57. (i) Note that $fg \in L^2[0,1]$ with $(M_f e_k, e_m) = \int_0^1 f(t)g(t)e^{2\pi i t(k-m)}dt = \widehat{(fg)}(k-m)$.

(ii) If M_f is compact, being clearly self-adjoint, Theorem 3.37 shows that if f does not vanish a.e., then M_f must have an eigenvalue $\lambda \neq 0$. But then, if $g \in L^2[0,1]$ is an eigenfunction, the equality $(\lambda - f)g = 0$ in $L^2[0,1]$ ensures that $f(t) = \lambda$ for all t in a set of positive measure $A \subset [0,1]$. The restriction of $\frac{1}{\lambda}M_f$ to $L^2(A)$ must then be compact, which is impossible by Theorem 3.8 since it is the identity operator on an infinite-dimensional Hilbert space.

58. Let T be the bounded linear operator with nonclosed range from Exercise 36(i). For $n \geq 1$, let $T_n : l^2 \to l^2$ be the finite-rank operator obtained by retaining the first n entries of $Tx \in l^2$ and setting all others zero. Since $\|Tx - T_n x\|^2_{l^2} = \sum_{k\geq n+1}^\infty \frac{|x(k)|^2}{k^2} \leq \frac{1}{(n+1)^2}\|x\|^2_{l^2}$, Theorem 3.31(ii) yields that T is compact.

59. (i) Clearly $1 \in \sigma(T)$ is an eigenvalue (any nonzero $x \in l^2$ being an eigenvector), while for $\lambda \neq 1$, $\frac{1}{\lambda-1}I$ is the bounded inverse of $(\lambda I - T)$. Thus $\sigma(T) = \{1\}$.

(ii) Let us first show that T has no eigenvalues. If $\lambda \in \mathbb{C}$ and $x \in l^2$ are such that $Tx = \lambda x$, then $0 = \lambda x(1)$, $x(1) = \lambda x(2)$, etc. yields $x = 0$.

Since clearly $\|T\| = 1$, all $\lambda \in \mathbb{C}$ with $|\lambda| > 1$ are in the resolvent set $\rho(T)$.

Let us show that $\lambda = 1$ is in the residual spectrum: $(I - T)$ is injective but its range is not dense.

Indeed, $y = (I - T)x$ means $y(1) = x(1)$, $y(2) = x(2) - x(1)$, $y(3) = x(3) - x(2)$ etc. Therefore $y(k) + \cdots + y(1) = x(k)$ for all $k \geq 1$. Since $x \in l^2$ ensures $\lim_{k\to\infty} x(k) = 0$, we should have $|y(k) + \cdots + y(1)| \to 0$ for $k \to \infty$. Consequently $y = (\alpha, 0, 0, \ldots)$ is not in the range of $(I - T)$ if $\alpha \neq 0$. This means that the range is in the orthogonal complement of the vector $(1, 0, 0, \ldots)$. Similarly, if $|\lambda| < 1$, then $y = (\lambda I - T)x$ means $y(1) = \lambda x(1)$, $y(2) = \lambda x(2) - x(1)$, $y(3) = \lambda x(3) - x(2)$ etc. Therefore $\lambda^{k-1} y(k) + \lambda^{k-2} y(k_1) + \cdots + \lambda y(2) + y(1) = \lambda^k x(k) \to 0$ for $k \to \infty$. Consequently, the range of $(\lambda I - T)$ is orthogonal to the vector $(1, \lambda, \lambda^2, \ldots) \in l^2$, and λ belongs to the residual spectrum.

The previous considerations and the fact that $\sigma(T)$ is closed show that $\sigma(T)$ must coincide with the closed unit disc $\{\lambda \in \mathbb{C} : |\lambda| \leq 1\}$.

(iii) We first show that $(\lambda I - T)$ has a bounded inverse for every complex number $\lambda \neq 0$. For this, we will prove that the Neumann expansion $\sum_{n\geq 0} \frac{T^n}{\lambda^{n+1}}$ converges, by showing inductively that[178] $|(T^n x)(t)|^2 \leq \frac{t^n \|x\|^2}{n!}$ for $t \in [0, 1]$ and $n \geq 1$. Thus $\|T^n x\|^2 \leq \frac{\|x\|^2}{n!} \int_0^1 t^n dt = \frac{\|x\|^2}{(n+1)!}$, so that $\|\frac{T^n}{\lambda^{n+1}}\| \leq \frac{1}{|\lambda|^{n+1}\sqrt{(n+1)!}}$ and $\sum_{k\geq 0} \frac{1}{|\lambda|^{n+1}\sqrt{(n+1)!}}$ converges because of the ratio test. Therefore the limit of the Neumann expansion is a bounded linear operator A with $(\lambda I - T)A = (I - \frac{T}{\lambda}) \sum_{n\geq 0} \frac{T^n}{\lambda^n} = I = \sum_{n\geq 0} \frac{T^n}{\lambda^n}(I - \frac{T}{\lambda}) = A(I - \frac{T}{\lambda})$. This means that $(\lambda I - T)^{-1} = A$. We conclude that $\sigma(T) = \{0\}$, since the spectrum is nonempty. Finally, $\sigma(T)$ is purely continuous. Indeed, clearly 0 is not an eigenvalue of T, and if it were in the residual spectrum, then 0 must be an eigenvalue of T^*. But this is impossible since $(T^* f)(t) = \int_t^1 f(s)\,ds$ for $t \in [0, 1]$. This last claim follows since for $f, g \in L^2[0, 1]$ the absolute continuity of the maps $t \mapsto \int_0^t f(s)\,ds$ and $t \mapsto \int_t^1 g(s)\,ds$ yields $\partial_t\big(\int_0^t f(s)\,ds \int_t^1 \overline{g(s)}\,ds\big) = f(t) \int_t^1 \overline{g(s)}\,ds - \overline{g(t)} \int_0^t f(s)\,ds$ for a.e. $t \in (0, 1)$, so that $(Tf, g)_{L^2[0,1]} = (f, T^* g)_{L^2[0,1]}$ follows after integration on $[0, 1]$.

60. For $|\lambda| \neq 1$ we have $\|Tx - \lambda x\| \geq |\,\|Tx\| - \|\lambda x\|\,| = |(1 - |\lambda|)\|x\||$, so that $R_\lambda(T) = (\lambda I - T)^{-1}$ exists as a bounded operator with norm less than $1/|1 - |\lambda||$.

61. If T is normal, then $(TT^* x, x) = (T^* T x, x) = (T^* x, T^* x) = (Tx, Tx)$, so that $\|Tx\| = \|T^* x\|$ for every $x \in H$. Conversely, if $\|Tx\| = \|T^* x\|$ for all $x \in H$, reversing the order in the previous chain of equalities yields $((TT^* - T^* T)x, x) = 0$ for all $x \in H$. Since $(TT^* - T^* T)$ is self-adjoint due to the general rule $(AB)^* = B^* A^*$ [easy to verify using the definition], we conclude by Exercise 53.

62. (i) If $a_0 = 0$ and $a_n = \ln \|T^n\|$ for $n \geq 1$, then $\|T^{n+m}\| \leq \|T^n\|\,\|T^m\|$ yields $a_{n+m} \leq a_n + a_m$ for all $n, m \geq 1$. For integers $n > m \geq 1$, write $n = mq + r$ with integers $q \geq 1$ and $r \in [0, m - 1]$. Then $a_n = a_{mq+r} \leq q a_m + a_r$, so that $\frac{a_n}{n} \leq \frac{a_m}{m} + \frac{a_r}{n} \leq \frac{a_m}{m} + \frac{\max\{a_0,\ldots,a_{m-1}\}}{n}$ and thus $\limsup_{n\to\infty}\{\frac{a_n}{n}\} \leq \frac{a_m}{m}$ for all $m \geq 1$. But then $\limsup_{n\to\infty}\{\frac{a_n}{n}\} \leq \liminf_{m\to\infty}\{\frac{a_m}{m}\}$, so that $\lim_{n\to\infty}\{\frac{a_n}{n}\}$ exists. Clearly $\lim_{n\to\infty}\{\frac{a_n}{n}\} \geq \inf_{n\geq 1}\{\frac{a_n}{n}\}$. An inequality would mean $\frac{a_N}{N} < \lim_{n\to\infty}\{\frac{a_n}{n}\}$ for some $N \geq 1$, which is impossible as $\limsup_{n\to\infty}\{\frac{a_n}{n}\} \leq \frac{a_N}{N}$. Thus $\lim_{n\to\infty}\{\frac{a_n}{n}\} = \inf_{n\geq 1}\{\frac{a_n}{n}\}$. In particular, this means that $\ln r(T) \leq a_1 = \ln \|T\|$. Note that $T : \mathbb{C}^2 \to \mathbb{C}^2$ given by $T(z_1, z_2) = (z_2, 0)$ is such that $\|T\| = 1$ while $r(T) = 0$ since $T^2 = 0$.

(ii) Let us first show that $\|T^2\| = \|T\|^2$. For this, note that $\|T^* T\| = \|T^2\|$ since for any $x \in H$ we have $\|T^* T x\|^2 = (T^* T x, T^* T x) = (Tx, TT^* T x) = (Tx, T^* T^2 x) = (T^2 x, T^2 x) = \|T^2 x\|^2$. On the other hand, since $T^* T$ is clearly self-adjoint, by Theorem 3.37, $\|T^* T\| = \sup_{\|x\|=1}\{(T^* T x, x)\} = \sup_{\|x\|=1}\{(Tx, Tx)\} = \sup_{\|x\|=1}\{\|Tx\|^2\} = \|T\|^2$. We can now prove inductively that $\|T^{2^n}\| = \|T\|^{2^n}$ for all $n \geq 1$. Indeed, clearly T^{2^n} is a normal operator, and if the claim holds for $n \geq 1$, then,

[178] The simple estimate $\|T^n\| \leq \|T\|^n$ is not good enough for this purpose. As for the claimed estimate, we have $|(Tx)(t)|^2 \leq \big(\int_0^t |x(s)|\,ds\big)^2 \leq t \int_0^t |x(s)|^2\,ds \leq t \int_0^1 |x(s)|^2\,ds = t\,\|x\|^2$, using the Cauchy–Schwarz inequality. The assumed validity of the estimate for $n \geq 1$ yields $|(T^{n+1}x)(t)| = |(T(T^n x))(t)| \leq \int_0^t |(T^n x)(s)|\,ds \leq \|x\| \int_0^t \sqrt{\frac{s^n}{n!}}\,ds$ by the induction hypothesis. The upper bound equals $\frac{t^{1+n/2}\|x\|}{(1+n/2)\sqrt{n!}} \leq \frac{t^{(n+1)/2}\|x\|}{\sqrt{(n+1)!}}$, and the desired inequality follows.

applying the property derived at the outset, $\|T^{2^{n+1}}\| = \|(T^{2^n})^2\| = \|T^{2^n}\|^2 = (\|T\|^{2^n})^2 = \|T\|^{2^{n+1}}$. Thus $\|T^{2^k}\| = \|T\|^{2^k}$ for all $k \geq 1$, so that $r(T) = \lim_{n\to\infty} \|T^n\|^{1/n} = \lim_{k\to\infty} \|T^{2^k}\|^{2^{-k}} = \|T\|$.

63. For $\lambda \in \mathbb{C}$, consider the Hilbert–Schmidt operator $(Tf)(x) = \lambda \int_0^1 e^{x+y} f(y)\, dy$ on $L^2[0,1]$, with adjoint $(T^*f)(x) = \overline{\lambda} \int_0^1 e^{x+y} f(y)\, dy$, and re-write the equation as $f - Tf = 1$. It is easy to check that $\ker\{(I - T^*)\} \neq \{0\}$ if and only if $\lambda = \frac{2}{e^2-1}$, in which case the kernel is $\{\alpha e^x : \alpha \in \mathbb{C}\}$ with $1 \notin [\ker\{(I - T^*)\}]^\perp$. By the Fredholm alternative the equation has no solution for $\lambda = \frac{2}{e^2-1}$, while $\ker\{(I - T)\} = \{0\}$ for $\lambda \neq \frac{2}{e^2-1}$ ensures the existence of a unique solution in these cases.

64. Given two separable Hilbert spaces H_1 and H_2, with orthonormal bases $\{e_n\}_{n\geq 1}$ and $\{f_n\}_{n\geq 1}$, the pairs (x, y) with $x \in H_1$ and $y \in H_2$ form a separable Hilbert space, called the *direct sum* $H = H_1 \bigoplus H_2$; the scalar product being given by $((x_1, y_1), (x_2, y_2))_H = (x_1, x_2)_{H_1} + (y_1, y_2)_{H_2}$, so that $\{(e_n, f_m)\}_{n,m\geq 1}$ is an orthonormal basis of H. Let T be the Volterra operator from Exercise 59(iii), acting on $L^2[0,1]$. Given the separable Hilbert space $H_2 \neq \{0\}$, $T_0(f, g) = Tf$ defines a compact self-adjoint linear operator on $H = L^2[0,1] \bigoplus H_2$ with purely point spectrum $\sigma(T_0) = \{0\}$. For $H_2 = \mathbb{C}^n$ the zero eigenvalue has multiplicity n, with eigenspace $\{(0, y) : y \in \mathbb{C}^n\}$, while for $H_2 = l^2$ its multiplicity is infinite, the eigenspace being $\{(0, y) : y \in l^2\}$.

3.7 Notes to Chapter 3

- If $\|\cdot\|$ is a norm on the real vector space X, we extend it[179] to a norm on the complexification[180] $X_\mathbb{C} = \{x+iy : x, y \in X\}$ of X, by $\|x+iy\| = \frac{1}{\sqrt{2}} \sup_{\theta\in[0,2\pi)} \sqrt{\|x\cos\theta - y\sin\theta\|^2 + \|x\cos\theta + y\sin\theta\|^2}$. Since $\frac{\|x\|+\|y\|}{2} \leq \|x+iy\| \leq \|x\| + \|y\|$ as $\|x+iy\| \geq \max\{\|x\|, \|y\|\}$, we see that $x_n + iy_n \to x + iy$ in $X_\mathbb{C}$ if and only if $x_n \to x$ and $y_n \to y$ in X. Also, $X_\mathbb{C}$ is a Banach space if X is a Banach space. Every linear operator $T : X \to Y$ between two real normed spaces gives rise to a complex linear operator $T_\mathbb{C} : X_\mathbb{C} \to Y_\mathbb{C}$, defined by $T_\mathbb{C}(x+iy) = Tx + iTy$. It is easy to see that if T is bounded, then $\|T_\mathbb{C}\|_{\mathcal{B}(X_\mathbb{C},Y_\mathbb{C})} = \|T\|_{\mathcal{B}(X,Y)}$. In contrast to the setting of Hilbert spaces, discussed in Section 3.4.4, there exist complex Banach spaces which cannot be obtained as the complexification of a real Banach space, see the discussion in van Neerven (1997).
- Any separable infinite-dimensional Hilbert space has genuinly infinite-dimensional compact subsets (that is, not all its compact sets are homeomorphic – in the sense of being images under a continuous bijection – to compact subsets of $\mathbb{R}^n$ or $\mathbb{C}^n$): if $\{e_n\}_{n\geq 1}$ is an orthonormal basis, then the *Hilbert cube* $\{x = \sum_{n\geq 1}(x, e_n)\, e_n : |(x, e_n)| \leq \frac{1}{n}\}$ is of this type; see Exercise 16.
- One of the impediments to the development of a rich parallel spectral theory for operators on Banach spaces, similar to that for Hilbert spaces, is the lack of a notion of an adjoint operator that is as versatile as that on Hilbert spaces.
- The Riesz representation theorem (Theorem 3.13) permits us to identify the dual H^* of the Hilbert space H with H, but caution is sometimes advisable with these identifications. For example, the dense subspace $X = \{u \in l^2 : \sum_{k\geq 1} k^2 |u(k)|^2 < \infty\}$ of $H = l^2$ is a Hilbert space when equipped with the scalar product $(x, y) = \sum_{k\geq 1} k^2\, x(k)\, \overline{y(k)}$; we can identify H^* with H but we cannot simultaneously identify X with $X^* = \{w = \{w(k)\}_{k\geq 1} : \sum_{k\geq 1} \frac{1}{k^2} |w(k)|^2 < \infty\}$, an isometry from X onto X^* being $u \mapsto \{k^2\, u(k)\}_{k\geq 1}$ (see also the discussion in Brézis (2011)).

[179] If X is a pre-Hilbert space, then we get $\|x + iy\| = \sqrt{\|x\|^2 + \|y\|^2}$, which also arises from the complexification $(x + iy, u + iv) = (x, u) + (y, v) + i\{(y, u) - (x, v)\}$ of the scalar product on X. Thus $X_\mathbb{C}$ is also a pre-Hilbert space. It is tempting to use $\|x + iy\| = \sqrt{\|x\|^2 + \|y\|^2}$ in general normed spaces, but this attempt is doomed because it leads by means of $2(\|x\|^2 + \|y\|^2) = |1 + i|^2 \cdot \|x + iy\|^2 = \|(1 + i)(x + iy)\|^2 = \|x - y + i(x + y)\|^2 = \|x - y\|^2 + \|x + y\|^2$ to the parallelogram law, being thus specific to pre-Hilbert spaces, see Exercise 3.

[180] The sum $x + iy$ stands formally for the pair (x, y).

4

Convergence results for Fourier series

To any function $f : [0, 1] \to \mathbb{C}$ that is integrable (in the sense of Lebesgue) we assign its Fourier coefficients $\{\hat{f}(k)\}_{k\in\mathbb{Z}}$ by the prescription

$$\hat{f}(k) = \int_0^1 f(x)\, e^{-2\pi ikx}\, dx\,, \qquad k \in \mathbb{Z}\,. \tag{4.1}$$

The series

$$\sum_{k\in\mathbb{Z}} \hat{f}(k)\, e^{2\pi ikx}$$

is termed the Fourier series of f, and the convention – always in force throughout the book – is to define it as the limit, in $\mathbb{C}$, of its symmetric partial sums

$$(S_n f)(x) = \sum_{k=-n}^{n} \hat{f}(k)\, e^{2\pi ikx}\,, \quad n \geq 0\,. \tag{4.2}$$

Note that the general partial sums of the two-sided infinite Fourier series are of the form

$$(S_{m,n} f)(x) = \sum_{k=m}^{n} \hat{f}(k)\, e^{2\pi ikx}\,,$$

with $m < n$ two integers, and allowing for $m \to -\infty$ and $n \to \infty$ independently might seem appropriate. A compelling reason to rely on symmetric sums is the following: for $k \neq 0$, the kth Fourier coefficient $\hat{f}(k)$ and the kth term of the (formal) Fourier series of a real-valued function f both have generally nonzero imaginary parts[1] but grouping together the terms $\hat{f}(-|k|)\, e^{-2\pi i|k|x}$

[1] Unless $\int_0^1 f(x)\, \sin(2\pi kx)\, dx = 0$ and $\int_0^1 f(x)\, \sin(2\pi kx)\, dx = \int_0^1 f(x)\, \cos(2\pi kx)\, dx = 0$, respectively.

and $\hat{f}(|k|)\,e^{2\pi i|k|x}$ produces[2] a real term, $a_k\cos(2\pi|k|x) + b_k\sin(2\pi|k|x)$. If $f \in L^1[0,1]$ is such that $\{\hat{f}(k)\}_{k\in\mathbb{Z}} \in l^1(\mathbb{Z})$, then the convergence of the symmetric partial sums (4.2) at some $x = x_0 \in [0,1]$ ensures[3] that the general partial sums $(S_{m,n}f)(x)$ will also converge at $x = x_0$. Neverteless, even if $\{\hat{f}(k)\}_{k\in\mathbb{Z}} \in l^1(\mathbb{Z})$, reliance on the more demanding interpretation of the Fourier series is more than we bargained for. To see this, first note that it is equivalent to requiring that each of the series $\sum\limits_{k\geq 0}\hat{f}(k)\,e^{2\pi ikx}$ and $\sum\limits_{k\leq 0}\hat{f}(k)\,e^{2\pi ikx}$ converges. Since knowing $a,\,b \in \mathbb{C}$ amounts to knowing $a \pm b$, the more demanding interpretation asks, in addition to the convergence of (4.2), that

$$(S_n^*f)(x) = \sum_{k=1}^{n}\Big(\hat{f}(k)\,e^{2\pi ikx} - \hat{f}(-k)\,e^{-2\pi ikx}\Big), \quad n \geq 1, \tag{4.3}$$

also converge. The operation of passing from the Fourier series $\sum\limits_{k\in\mathbb{Z}}\hat{f}(k)\,e^{2\pi ikx}$ to the so-called *conjugate* series

$$\sum_{k\in\mathbb{Z}}\{-i\,\mathrm{sgn}(k)\}\,\hat{f}(k)\,e^{2\pi ikx}$$

is a subtle direction of research in harmonic analysis, and (4.3) are, up to a multiplicative factor of i, precisely the symmetric partial sums of the conjugate series. To illustrate the difficulties, note that even if the conjugate series converges (in the conventional sense for Fourier series) at every $x \in [0,1]$, it might not[4] originate as the Fourier series of a function $f^* \in L^1[0,1]$. On the other hand, if the conjugate series is the Fourier series of some $f^* \in L^1[0,1]$, then the more demanding definition of convergence actually conceals two functions, f and f^*. These considerations motivate the conventional wisdom to interpret a Fourier series as the limit of its symmetric partial sums (4.2).

A basic problem is the representation of a more or less arbitrary given function f by its Fourier series. The various meanings of "representation" are tantamount to ways in which the Fourier series may be said to converge to f.

[2] With $a_k = 2\int_0^1 f(s)\cos(2\pi|k|s)\,ds \in \mathbb{R}$ and $b_k = 2\int_0^1 f(s)\sin(2\pi|k|s)\,ds \in \mathbb{R}$.

[3] Since $|(S_{m,n}f)(x_0) - (S_nf)(x_0)| \leq \sum\limits_{-n\leq k<m}|\hat{f}(k)|$ if $|m| < n$, while for $|m| \geq n$ we have $|(S_{m,n}f)(x_0) - (S_{|m|}f)(x_0)| \leq \sum\limits_{n\leq k<|m|}|\hat{f}(k)|$.

[4] For example, it is true (but far from straightforward) that there exists a function $f \in L^1[0,1]$ with $\hat{f}(0) = \hat{f}(\pm 1) = 0$ and $\hat{f}(k) = \frac{1}{\ln|k|}$ for $|k| \geq 2$. Moreover, for all $x \in (0,1)$, the limit $f(x) = \sum\limits_{k\in\mathbb{Z},\,|k|\geq 2}\frac{1}{\ln|k|}\,e^{2\pi ikx} \in \mathbb{R}$ exists, with the conventional interpretation based on (4.2). In this case $(S_n^*f)(x) = 2i\sum\limits_{k\geq 2}^{n}\frac{\sin(2\pi kx)}{\ln(k)}$ for $n \geq 2$. While the sequence $\{(S_n^*f)(x)\}_{n\geq 2}$ converges for all $x \in [0,1]$, it does not correspond to the symmetric partial sums of some $L^1[0,1]$-function (see Exercise 21).

Many areas of mathematics owe a debt of gratitude to the long struggle to come to terms with this issue, e.g., calculus (the definition of a function and the concept of a limit), integration/measure theory, the theory of distributions, but even modern set theory and mathematical logic.[5] In the nineteenth century, the representation was taken to mean the pointwise convergence of the Fourier series at all points $x \in [0, 1]$ to the limit $f(x)$. With the passage of time it was realised that this interpretation leaves much to be desired. For example, the Fourier series of a periodic continuous function, with period 1, may diverge at some points (see Exercise 7). Actually, Carleson (1966) proved that the Fourier series of a periodic continuous function converges almost everywhere to the function, while Kahane and Katznelson (1966) showed that for any set $E \subset [0, 1]$ of measure zero, there is a periodic continuous function (with period 1) whose Fourier series diverges on E. To proceed, one option is to identify tangible classes of functions where pointwise convergence holds. Alternatively, relaxing the demand for convergence of the Fourier series to f at all points $x \in [0, 1]$ to convergence of some trigonometric series

$$\sum_{k \in \mathbb{Z}} c_k \, e^{2\pi i k x}$$

at almost all points of $[0, 1]$ in the sense that corresponds to the meaning assigned for the Fourier series of some $f \in L^1[0, 1]$, that is, $\lim_{n \to \infty} \sum_{k=-n}^{n} c_k \, e^{2\pi i k x} \in \mathbb{C}$ exists, presents the advantage that such a representation exists for every measurable function $f : [0, 1] \to \mathbb{C}$, but the enthusiasm wanes in view of the inherent defect[6] of lack of uniqueness: there are trigonometric series which converge to zero almost everywhere and which nevertheless have at least one nonvanishing coefficient $c_k \in \mathbb{C}$. This surprising result means that a pointwise representation of a trigonometric series at almost all points is never unique. On the other hand, a convergent trigonometric series at all points is necessarily unique, if it exists at all. But unfortunately there are continuous periodic functions that do not admit such a representation.[7] Consequently, using trigonometric series in general does not resolve the limitations of Fourier series, since the alternative of coming to terms with representations that are by no means unique has to be rejected on grounds of utility. These considerations motivate the pref-

[5] The dawn of which can be traced back to Cantor's interest in the convergence of Fourier series.

[6] This pheonomenon cannot occur for the Fourier series associated to some $f \in L^1[0, 1]$ (see Section 4.2).

[7] For an in-depth discussion and suitable references we refer to Edwards (1967). In general, the techniques that were developed to deal with these issues are largely irrelevant for the core theory of Fourier series.

erence[8] for Fourier series, as being more effective operationally. New types of convergence actually justify the installing of Fourier series at the forefront. For example, if pointwise convergence is replaced by convergence in $L^2[0,1]$), the situation improves: the Fourier series of a function $f \in L^2[0,1]$ converges in $L^2[0,1]$ to that function, as a consequence of the Fourier series theorem applied to the orthonormal basis $\{e^{2\pi ikx}\}_{k\in\mathbb{Z}}$ of the Hilbert space $L^2[0,1]$, see Theorem 3.1(b). Other natural questions arise. For example, what happens if $f \in L^1[0,1]$? Fourier series are an important tool in the study of the differential equations of mathematical physics. When is it legitimate to differentiate term-wise a Fourier series representation? This problem leads to the study of a new class of function spaces – Sobolev spaces. By construction, within these spaces one can express differentiation in a transparent way as an operation on the Fourier coefficients. This compatibility and the fact that certain Sobolev spaces have a Hilbert space structure begin to explain why these spaces are prominent in the study of partial differential equations – more than spaces of periodic, continuously differentiable functions.[9] Furthermore, one can resort to a new interpretation that leads to the concept of distributions – a major driving force behind the development of modern analysis. While the level at which this book is written precludes an in-depth study of these topics, we will nevertheless present some basic results that offer a perspective, and we will suggest material for readers willing to embark on a more challenging program. Rather than attempt an inventory of sharp results, we try to convey the flavour of the theory of Fourier series and to impart working knowledge. With this objective in mind, we aim to balance the need to develop some intuitive grasp with the requisite of some essential techniques. While the subtleties can only be illustrated by addressing some rather technical points and by providing a solid base of examples, we will try not to get carried away and lose sight of the core. This means that some interesting aspects will not be pursued at all.

Throughout this chapter it is convenient to think of functions in $L^1[0,1]$ or in $L^2[0,1]$ as restrictions to $[0,1]$ of periodic functions $f : [0,1] \to \mathbb{C}$. Alternatively, we can extend $f : [0,1] \to \mathbb{C}$ to a periodic function $f : \mathbb{R} \to \mathbb{C}$ by setting $f(x+k) = f(x)$ for all $k \in \mathbb{Z}$. This procedure might require us to alter the value of $f(1)$ to $f(0)$, but modifications on sets of measure zero are irrelevant for Lebesgue integrals. When considering smoother classes of functions (for instance, functions that are Hölder continuous or differentiable), we tacitly assume that they are periodic.

[8] The general theory of trigonometric series is mostly dealing with questions that simply do not arise for Fourier series, so that we will not make further references to it.

[9] Which are Banach spaces deprived of a scalar product, and where complications arise in the convergence of the Fourier series – as illustrated by the case of periodic continuous functions.

4.1 Basic properties of Fourier coefficients

Let us denote by $\hat{f}$ the function $k \mapsto \hat{f}(k)$, defined on $\mathbb{Z}$, which encodes the two-way infinite sequence of Fourier coefficients $\{\hat{f}(k)\}_{k\in\mathbb{Z}}$ of $f \in L^1[0,1]$.

(P1) *The mapping $f \mapsto \hat{f}$ is linear on $L^1[0,1]$.*

This is clear from (4.1) since integration is a linear operation on the integrand.

(P2) *Given $f \in L^1[0,1]$, we have $|\hat{f}(k)| \le \|f\|_{L^1}$ for all $k \in \mathbb{Z}$.*

Indeed, $|\hat{f}(k)| = \left|\int_0^1 f(x)\, e^{-2\pi ikx}\, dx\right| \le \int_0^1 |f(x)|\, dx = \|f\|_{L^1}$. While $\hat{f}(0) = \|f\|_{L^1}$ if f is real and nonnegative, the estimate in (P2) is far from being sharp.[10]

(P3) *Given $a \in \mathbb{R}$, denote by T_a the translation operator, acting on the linear space of the functions $f : \mathbb{R} \to \mathbb{C}$ by $(T_a f)(x) = f(x-a)$ for $x \in \mathbb{R}$. For any $f \in L^1[0,1]$ we have*

$$\widehat{(T_a f)}(k) = e^{-2\pi ika}\hat{f}(k) \quad for\ all \quad k \in \mathbb{Z}.$$

The change of variables $y = x+a$ yields $\hat{f}(k) = \int_0^1 f(x)\, e^{-2\pi ikx} dx = \int_a^{1+a} f(y-a)\, e^{-2\pi ik(y-a)} dy = e^{2\pi ika} \int_a^{1+a} f(y-a)\, e^{-2\pi iky} dy = e^{2\pi ika} \int_0^1 f(y-a)\, e^{-2\pi iky} dy$, since the function $y \mapsto f(y-a)\, e^{-2\pi iky}$ is periodic of period 1, so that its average over a period is independent of the particular realisation of the period. The right side of the displayed relation is exactly $e^{2\pi ika}\, \widehat{(T_a f)}(k)$.

(P4) *For $f : \mathbb{R} \to \mathbb{C}$ locally absolutely continuous and periodic of period 1, let $f' \in L^1[0,1]$ be equal almost everywhere to the derivative of f. Then*

$$\widehat{(f')}(k) = 2\pi ik \cdot \hat{f}(k) \quad for\ all \quad k \in \mathbb{Z}.$$

Indeed, integration by parts (see Theorem 2.9) yields the relations $\widehat{(f')}(k) = \int_0^1 f'(x)\, e^{-2\pi ikx} dx = \{f(x)\, e^{-2\pi ikx}\}\big|_{x=0}^{x=1} + 2\pi ik \int_0^1 f(x)\, e^{-2\pi ikx} dx = 2\pi ik \cdot \hat{f}(k)$, with the integrated part vanishing because of periodicity. Note that (P3) is not granted if one merely assumes that f possesses almost everywhere an integrable derivative – see the discussion of absolute continuity in Chapter 2. No such technicalities arise if the function f is continuously differentiable.

(P5) *(Riemann–Lebesgue lemma)* $\lim_{|k|\to\infty} \hat{f}(k) = 0$ *for any $f \in L^1[0,1]$.*

Let E be the set of functions $f \in L^1[0,1]$ for which the statement holds. Clearly E is a linear subspace of $L^1[0,1]$. Moreover, (P2) ensures that E is a closed subset of $L^1[0,1]$. Indeed, if $f_n \in E$ converge to some f in $L^1[0,1]$, given $\varepsilon > 0$, we can choose an integer $n_1 \ge 1$ such that $\|f_{n_1} - f\|_{L^1} < \varepsilon/2$. Since $f_{n_1} \in E$, we find an integer $n_2 \ge 1$ such that $|\hat{f}_{n_1}(k)| < \varepsilon/2$ whenever $|k| \ge n_2$. But $|\hat{f}(k) - \hat{f}_{n_1}(k)| \le \|f - f_{n_1}\|_{L^1}$ due to (P2), so that $|\hat{f}(k)| \le |\hat{f}(k) - \hat{f}_{n_1}(k)| + |\hat{f}_{n_1}(k)| < \varepsilon/2 + \varepsilon/2 = \varepsilon$ for $|k| \ge n_2$. Since $\varepsilon > 0$ was arbitrary, this shows that $f \in E$, so that E is a closed subspace. To prove the lemma it therefore suffices to show that E contains a set of

[10] Indeed, there is an absolute constant $c > 0$ such that for each positive integer $N \ge 1$ there exists a trigonometric polynomial $f_N(x) = \sum_{k=1}^{N-1} c_k\, e^{2\pi ikx}$ with $|c_k| = 1$ for $k = 0, \ldots, N-1$ (so $|\hat{f}(k)| = 1$ for $0 \le k \le N-1$) and $\|f_N\|_{L^1} \ge \sqrt{N} - c$, see Edwards (1967). On the other hand, using the Cauchy–Schwarz inequality (3.2) for $L^2[0,1]$ and the orthogonality relations (1.2), we get $\|f_N\|_{L^1}^2 = \left(\int_0^1 |f_N(x)|\, dx\right)^2 \le \int_0^1 |f_N(x)|^2\, dx = \sum_{k=0}^{N-1} |c_k|^2 = N$.

functions S, the finite linear combinations of which are dense in $L^1[0,1]$. Examples[11] are the set of characteristic functions of intervals (a,b) with $0 \le a < b \le 1$, or the set of smooth periodic functions of period 1.

The basic properties (P1)-(P5) show that $f \mapsto \hat{f}$ is a continuous linear map of $L^1[0,1]$ into the linear space[12] $c_0(\mathbb{Z})$ of two-way infinite sequences $\alpha = \{\alpha_k\}_{k\in\mathbb{Z}}$ for which $\lim_{|k|\to\infty} \alpha_k = 0$. If we denote by $\mathbb{A}(\mathbb{Z})$ the range of this mapping, can we determine whether a given $\alpha \in c_0(\mathbb{Z})$ lies in $\mathbb{A}(\mathbb{Z})$? Were an effective general method known, the theory of Fourier series would be much simpler and much less intriguing than it in fact is. To illustrate the difficulties, note that while the sequence $\alpha^* \in c_0(\mathbb{Z})$ with $\alpha^*(k) = \frac{\mathrm{sgn}(k)}{\ln|k|}$ for $|k| \ge 2$ and $\alpha^*(0) = \alpha^*(\pm 1) = 0$, does not belong to $\mathbb{A}(\mathbb{Z})$, the sequence $\{|\alpha^*(k)|\}_{k\in\mathbb{Z}}$ does belong to $\mathbb{A}(\mathbb{Z})$, see Exercise 21. This shows not only that $\mathbb{A}(\mathbb{Z})$ is a proper subset[13] of $c_0(\mathbb{Z})$, but also that a seemingly harmless orderly change of sign can terminate the membership to $\mathbb{A}(\mathbb{Z})$. As a positive result, let us now discuss a striking property of $\mathbb{A}(\mathbb{Z})$: pointwise multiplication of elements of $\mathbb{A}(\mathbb{Z})$ yields an element of $\mathbb{A}(\mathbb{Z})$. Due to a lack of available characterisations of membership to $\mathbb{A}(\mathbb{Z})$, we cannot address this issue directly.[14] Instead, we try to gain insight by investigating the functions $e_k(x) = e^{2\pi i k x}$, for which

$$\widehat{(e_k)}(q) = \delta_{kq} \quad \text{for all} \quad k, q \in \mathbb{Z}. \tag{4.4}$$

The orthogonality relations (1.2) and the structure of the functions e_k yield $\int_0^1 e_q(x-y)\, e_p(y)\, dy = \delta_{qp}\, e_p(x)$, so that if we define the convolution $f * g$ by

$$(f * g)(x) = \int_0^1 f(x-y)\, g(y)\, dy \tag{4.5}$$

11 Indeed, by the construction of the Lebesgue integral, the set of finite linear combinations of characteristic functions χ^I of measurable subsets $I \subset [0,1]$ is dense in $L^1[0,1]$. Moreover, the regularity of the Lebesgue measure ensures that for every $\varepsilon > 0$ we can find an open set $O \subset \mathbb{R}$ such that $I \subset O$ and $m(O \setminus I) < \varepsilon/2$. By Exercise 2.5, we may write O as a union of pairwise disjoint, finite or countably infinite open intervals. Choosing a sufficiently large finite collection of such intervals and denoting their union by O', we make sure that the measure of the complement of $(I \cap O')$ in $[0,1]$ does not exceed ε. This shows that in $L^1[0,1]$ we can approximate the characteristic function of a measurable set by a finite sum of characteristic functions of open intervals $(a,b) \subset (0,1)$. The fact that $\chi^{(a,b)} \in E$ is easily verified since $|\widehat{(\chi^{(a,b)})}(k)| = \left|\int_a^b e^{-2\pi i k x} dx\right| = \left|\frac{e^{-2\pi i b} - e^{-2\pi i a}}{-2\pi k}\right| \le \frac{1}{\pi|k|}$ for $k \ne 0$. As for the smooth periodic functions of period 1, they are dense in $L^1[0,1]$, see Section 4.3. For such a function f, set $g = f'$. By (P3), $\hat{g}(k) = 2\pi i k \hat{f}(k)$ for all $k \in \mathbb{Z}$. Since g is continuous, (P2) ensures the boundedness of the sequence $\{\hat{g}(k)\}_{k\in\mathbb{Z}}$. Thus $\lim_{|k|\to\infty} \hat{f}(k) = 0$.

12 Equipped with the supremum norm $\|\alpha\| = \sup_{k\in\mathbb{Z}}\{|\alpha_k|\}$, $c_0(\mathbb{Z})$ is a Banach space, being a closed linear subspace of the Banach space $l^\infty(\mathbb{Z})$ of bounded two-way infinite sequences (endowed with the supremum norm).

13 One can actually show that $\mathbb{A}(\mathbb{Z})$ is a meagre, that is, first category (see Section 3.4.2), subset of $c_0(\mathbb{Z})$; see the discussion in Edwards (1967).

14 By trying to find, given $\hat{f}, \hat{g} \in \mathbb{A}(\mathbb{Z})$, an element of $L^1[0,1]$, denoted by $f * g$, such that $\widehat{(f * g)}(k) = \hat{f}(k) \cdot \hat{g}(k)$ for all $k \in \mathbb{Z}$.

then $(\widehat{e_q * e_p})(k) = \delta_{qp}\,\hat{e}_p(k) = \hat{e}_q(k)\cdot\hat{e}_p(k)$ for all $k \in \mathbb{Z}$. Since each of the expressions $f * g$ and $\hat{f}\cdot\hat{g}$ is bilinear in the pair (f, g), the desired relation extends to finite linear combinations of the e_k. Thus (4.5) is a promising starting point. Let now $f, g \in L^1[0, 1]$. Fubini's theorem (see Chapter 2) shows that for a.e. $x \in [0, 1]$, the integrand on the right of (4.5) is an integrable function of y, so that (4.5) defines $(f * g)(x)$ for a.e. $x \in [0, 1]$ as a measurable function with

$$\|f * g\|_{L^1} \le \|f\|_{L^1}\cdot\|g\|_{L^1}\,. \tag{4.6}$$

Moreover, as a consequence of the invariance of the integral in (4.5) under the change of variables $y \mapsto x - y$, we see that at every point $x \in [0, 1]$ for which $(f * g)(x)$ exists, $(g * f)(x)$ also exists and has the same value. Thus

$$f * g = g * f\,. \tag{4.7}$$

Using Fubini's theorem, a computation[15] shows that

$$(\widehat{f * g})(k) = \hat{f}(k)\cdot\hat{g}(k)\,, \qquad k \in \mathbb{Z}\,. \tag{4.8}$$

Relying on invariance and on Fubini's theorem, one can also see that the convolution is associative on $L^1[0, 1]$: $(f * g) * h = f * (g * h)$ for $f, g, h \in L^1[0, 1]$. These properties show that $L^1[0, 1]$ is a commutative Banach algebra under convolution. However, there is no identity, that is, no element $e \in L^1[0, 1]$ such that $e * f = f$ for all $f \in L^1[0, 1]$. Indeed, were such an e to exist, (4.4) and (4.8) would yield $\hat{e}(k) = 1$ for all $k \in \mathbb{Z}$, a possibility ruled out by (P4).

4.2 Pointwise convergence

Addressing the issue of the recovery of a function from its Fourier series, it is natural to start with the most basic notion of convergence, namely pointwise convergence. Exercise 7 shows that the Fourier series of a continuous periodic function $f : \mathbb{R} \to \mathbb{C}$ might diverge at some points. Since mere continuity does not suffice to ensure pointwise convergence, differentiability comes to mind. We first present an elegant, yet very powerful result due to Chernoff (1980).

[15] We have the chain of equalities $\int_0^1 (f * g)(x)\, e^{-2\pi ikx}\, dx = \int_0^1 e^{-2\pi ikx}\big(\int_0^1 f(x-y)g(y)\, dy\big)\, dx = \iint_{[0,1]\times[0,1]} e^{-2\pi ikx} f(x-y)g(y)\, dxdy = \iint_{[0,1]\times[0,1]} e^{-2\pi ik(x-y)} f(x-y)\, e^{-2\pi iky} g(y)\, dxdy = \int_0^1 \int_{-y}^{1-y} e^{-2\pi iks} f(s)\, e^{-2\pi iky} g(y)\, dsdy = \int_0^1 \int_0^1 e^{-2\pi iks} f(s)\, e^{-2\pi iky} g(y)\, dsdy = \int_0^1 \big(\int_0^1 e^{-2\pi iks} f(s)\, ds\big) e^{-2\pi iky} g(y)\, dy = \hat{f}(k) \int_0^1 e^{-2\pi iky}\, g(y)\, dy$, with one intermediate step validated on account of the change of variables $s = x - y$ and another by the periodicity of the function $s \mapsto e^{-2\pi iks} f(s)$. Periodicity makes the particular representation of the period irrelevant in the computation of its integral over a period, so that integrating over $[0, 1]$ or over some interval of the form $[-y, 1 - y]$ yields the same outcome.

Theorem 4.1 *Let $f \in L^1[0,1]$ and suppose that f is differentiable at the point x_0. Then $\sum_{k=-m}^{n} \hat{f}(k)\, e^{2\pi i k x_0} \to f(x_0)$ as $m, n \to \infty$.*

Proof By subtracting a constant from f and shifting the origin, if need be, we may suppose[16] that $x_0 = 0$ and $f(x_0) = 0$. Set $g(0) = g(1) = f'(0)$ and $g(x) = \frac{f(x)}{e^{2\pi i x}-1}$ for $x \in (0,1)$. The differentiability of the periodic function $f(x)$ at $x = 0$ ensures the existence of some $m, \varepsilon > 0$ such that $\left|\frac{f(x)}{e^{2\pi i x}-1}\right| \le m$ for $x \in (0,\varepsilon) \cup (1-\varepsilon, 1)$. Since $\frac{1}{|e^{2\pi i x}-1|} = \frac{1}{2\sin(\pi x)} \le \frac{1}{\sin(\pi\varepsilon)}$ for $x \in (\varepsilon, 1-\varepsilon)$, setting $M = \max\{m, \frac{1}{\sin(\pi\varepsilon)}\}$, we get that the measurable function g satisfies $|g(x)| \le M(1 + |f(x)|)$ for $x \in (0,1)$. Thus $g \in L^1[0,1]$. Since $f(x) = (e^{2\pi i x} - 1)\, g(x)$ for all $x \in [0,1]$, from (4.1) we infer that $\hat{f}(k) = \hat{g}(k-1) - \hat{g}(k)$ for all $k \in \mathbb{Z}$. This telescoping effect yields $\sum_{k=-m}^{n} \hat{f}(k) = \hat{g}(-m-1) - \hat{g}(n) \to 0$ for $n, m \to \infty$, due to (P5). □

A glance at the proof of Theorem 4.1 confirms that differentiability at $x = x_0$ is much stronger than necessary, and we can replace it by the weaker local Hölder condition $|f(x) - f(x_0)| \le K\,|x - x_0|^{\alpha}$ for $|x - x_0| < \varepsilon$, where $K, \varepsilon > 0$ and $\alpha \in (0,1]$ are some constants. This remark leads us to investigate, given $\alpha \in (0,1]$, the class h_α of periodic functions $f : \mathbb{R} \to \mathbb{C}$ of period 1 that are Hölder continuous with exponent α: functions with $[f]_\alpha = \sup_{x\in[0,1],\, y\in(0,1)} \{\frac{|f(x+y)-f(x)|}{|y|^{\alpha}}\} < \infty$, see Chapter 3; equivalently, this can be expressed as requiring the existence of some $M > 0$ with $|f(x+y) - f(x)| \le M\,|y|^{\alpha}$ for all $x, y \in [0,1]$. Note that h_1 are the Lipschitz continuous functions (see Chapter 2). The classes h_α provide a scale of smoothness between continuity and continuous differentiability:[17] $h_\beta \subsetneq h_\alpha$ if $0 < \alpha < \beta \le 1$, and, while all continuously differentiable functions belong to h_1, $f(x) = |\sin(2\pi x)|$ is in h_1 without being differentiable at $x = 0$. By Theorem 4.1, if $f \in h_\alpha$ for some $\alpha \in (0,1]$, then the Fourier series of f converges to f on $[0,1]$. The next result shows that this convergence is actually uniform.

Theorem 4.2 *Given $\alpha \in (0,1]$, there exists a constant $C = C(\alpha) > 0$ with*

$$\sup_{x\in[0,1]} |(S_n f)(x) - f(x)| \le \frac{C}{n^{\alpha^2/(\alpha+2)}} \left([f]_\alpha + \|f\|_\infty\right), \tag{4.9}$$

for every $f \in h_\alpha$; here $\|f\|_\infty = \sup_{t\in[0,1]}\{|f(t)|\}$.

Proof Given $n \ge 1$, a key role is played by the so-called *Dirichlet kernel*

$$D_n(x) = \sum_{k=-n}^{n} e^{2\pi i k x} = e^{-2\pi i n x} \sum_{k=0}^{2n} e^{2\pi i k x} = e^{-2\pi i n x} \frac{e^{2\pi i(2n+1)x} - 1}{e^{2\pi i x} - 1} = e^{-2\pi i n x} \frac{e^{2\pi i(2n+1/2)x} - e^{-\pi i x}}{e^{-\pi i x}\,(e^{2\pi i x} - 1)}$$
$$= \frac{e^{2\pi i(n+1/2)x} - e^{-2\pi i(n+1/2)x}}{e^{\pi i x} - e^{-\pi i x}} = \frac{\sin(\pi(2n+1)x)}{\sin(\pi x)}, \tag{4.10}$$

16 This preliminary reduction makes the procedure more transparent. It may be performed since, in view of (4.1) and (P3), for $F(x) = f(x + x_0) - f(x_0)$ we have $\hat{F}(k) = e^{2\pi i k x_0} \hat{f}(k)$ for all $k \in \mathbb{Z} \setminus \{0\}$, while $\hat{F}(0) = \hat{f}(0) - f(x_0)$.

17 Clearly $h_\beta \subset h_\alpha$, while $f(x) = |\sin(2\pi x)|^{\alpha}$ belongs to $h_\alpha \setminus h_\beta$, see Exercise 8.

for $x \in (0,1)$, while clearly $D_n(0) = D_n(1) = 2n+1$. The rationale behind introducing D_n stems from the realisation that it permits us to express $S_n f$ in a more transparent way. Indeed,

$$(S_n f)(x) = \sum_{k=-n}^{n} e^{2\pi ikx} \int_0^1 f(y)\, e^{-2\pi iky}\, dy = \int_0^1 \sum_{k=-n}^{n} e^{2\pi ik(x-y)} f(y)\, dy$$
$$= \int_0^1 D_n(x-y)\, f(y)\, dy = \int_{|s|\le 1/2} f(x-s)\, D_n(s)\, ds\,, \qquad n \ge 1. \qquad (4.11)$$

The final expression is obtained[18] by making the substitution $s = x - y$, and taking advantage of the fact that for every fixed x, $s \mapsto f(x-s)\, D_n(s)$ is periodic of period 1, which permits us to replace the integral over $[-x, 1-x]$ by the integral over $[-\frac{1}{2}, \frac{1}{2}]$. While the fact that $S_n f = f * D_n$ is essential, note that, see Exercise 9, there is an absolute constant $K > 1$ with

$$|D_n(x)| \le K \min\Big\{n, \frac{1}{|x|}\Big\} \quad \text{for all} \quad |x| \le 1/2 \quad \text{and all} \quad n \ge 2, \qquad (4.12)$$

$$\frac{1}{K}\ \ln(n) \le \|D_n\|_{L^1[0,1]} \le K\, \ln(n) \quad \text{for all} \quad n \ge 2\,. \qquad (4.13)$$

The growth of the bound in (4.13) prevents us from relying solely on general properties of convolutions. To proceed, it is instructive to draw the graph of D_n for different values of n; see Figure 4.1. This visual aid should convey three important features[19]: the graph depicts an even function, its peak at $x = 0$ grows indefinitely as n increases, and the oscillations to either side of the peak become increasingly rapid but, rather than dying away, their amplitude stabilises for $|x| \ge \frac{1}{3}$. The intuition behind the convergence of (4.12) to $f(x)$ is that the sharp central spike of $D_n(s)$ at $s = 0$ picks out the value $f(x)$, and the rapid oscillations away from $s = 0$ obliterate the rest of the integral because of cancellations, on average, between positive and negative values.

[18] The chain of equalities $\int_0^1 D_n(x-y)\, f(y)\, dy = \int_{x-1}^{x} D_n(s)\, f(x-s)\, ds = \int_{-1/2}^{1/2} D_n(s)\, f(x-s)\, ds$ is justified by these features.

[19] Inspecting (4.10), we see that D_n vanishes exactly $2n$ times in $[-\frac{1}{2}, \frac{1}{2}]$, at $x_k = \frac{k}{2n+1}$ with $1 \le |k| \le n$. Between consecutive zeros the smooth real function $D_n(x)$ must have at least one extremum, adding up to at least $2n$ points in $[x_n - 1,\, x_n]$ where $D_n'(x)$ will vanish. But $D_n'(x)$, as a trigonometric polynomial of degree n, has at most $2n$ roots in $[x_n - 1,\, x_n)$, see Exercise 1.1. Thus there is exactly one local extremum between each pair of consecutive zeros of D_n. Also, as a by-product, since $D_n'(\pm\frac{1}{2}) = D_n'(0) = 0$ by a direct computation, $\pm\frac{1}{2}$ and 0 are local extremum points, with further $(n-1)$ points $x_k' \in (x_k,\, x_{k+1})$ and their mirror-images $-x_k'$, with $1 \le k \le n-1$, exhausting the set of local extremum points in $[-\frac{1}{2}, \frac{1}{2}]$. On the other hand, $D_n(x) = 1 + \sum_{k=1}^{n} (e^{2\pi ikx} + e^{-2\pi ikx}) = 1 + 2\sum_{k=1}^{n} \cos(2\pi kx)$ yields $1 - 2n \le D_n(x) \le 2n+1$ for all $x \in \mathbb{R}$. For $0 < |x| < \frac{1}{2}$, due to $|\cos(2\pi x)| < 1$, we have $|D_n(x)| < 2n+1$, while $D_n(\pm\frac{1}{2}) = (-1)^n$. Thus $x = 0$ is the only point with $|x| \le \frac{1}{2}$ where $D_n(x)$ attains its maximum value $(2n+1)$. Combining the strict monotonicity of $x \mapsto \frac{1}{\sin(\pi x)}$ on $(0, \frac{1}{2}]$ with the identity $\sin((2n+1)\pi s) = -\sin\big((2n+1)\pi(s - \frac{1}{2n+1})\big)$, by means of (4.10) we get $|D_n(x_k')| < |D_n(x_k' - \frac{1}{2n+1})|$. But x_{k-1}' being the location of the maximum of $|D_n(x)|$ for $x \in [x_{k-1},\, x_k]$ yields $|D_n(x_{k-1}')| \ge |D_n(x_k' - \frac{1}{2n+1})|$ since $x_k' - \frac{1}{2n+1} \in (x_{k-1},\, x_k)$. Thus $|D_n(x_k')| < |D_n(x_{k-1}')|$, meaning that the amplitude of the local extrema decreases as x runs from 0 to $\frac{1}{2}$, stabilising at 1 for $x = \frac{1}{2}$. Now the fact that in (4.10) the denominator is positive for $x \in (0, \frac{1}{2}]$, combined with the oscillatory behaviour of the numerator, shows that local maxima and minima alternate. In particular, $D_n(x)$ attains its minimum $m_n \ge 1 - 2n$ at $x_1' \in (\frac{1}{2n+1}, \frac{2}{2n+1})$. Note that the value of $D_n(x)$ at the midpoint $x = \frac{3}{2(2n+1)}$ of this interval is $\frac{-1}{\sin\left(\frac{3\pi}{2(2n+1)}\right)} < -\frac{2(2n+1)}{3\pi}$ since $\sin(s) < s$ for $s > 0$. Since $\frac{2}{3\pi} \approx 0.2122$, we have $m_n < -0.21\,(2n+1)$. This analysis confirms the general validity of the oscilatory features depicted in Figure 4.1.

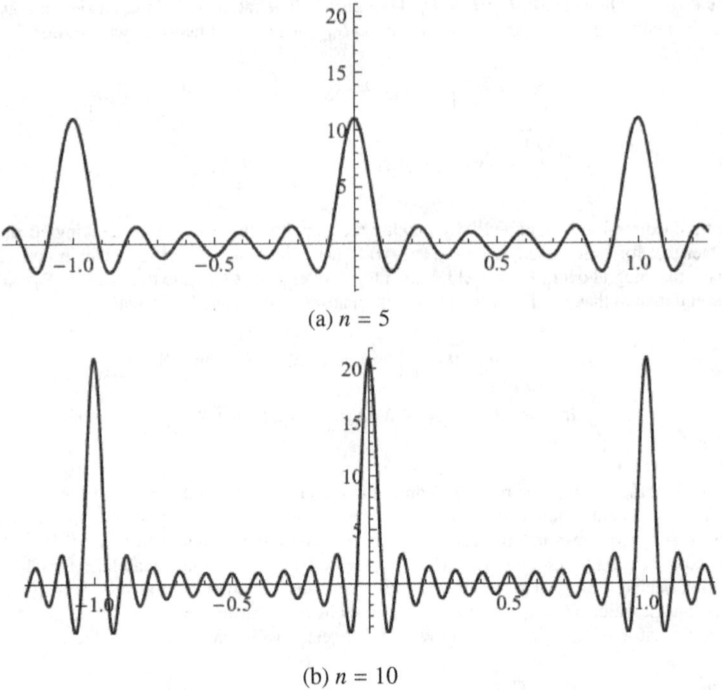

(a) $n = 5$

(b) $n = 10$

Figure 4.1 The graph of the Dirichlet kernel D_n.

To begin with, a fringe benefit of the periodicity and of the orthogonality relations (1.1.2) is

$$\int_{-1/2}^{1/2} D_n(x)\,dx = \int_0^1 D_n(x)\,dx = \sum_{k=-n}^{n} \int_0^1 e^{2\pi ikx}\,dx = 1\,, \quad n \geq 1\,. \tag{4.14}$$

For $\delta \in (0, 1/2)$ to be determined, (4.11) and (4.14) yield

$$(S_n f)(x) - f(x) = \int_{|s|\leq 1/2} \big(f(x-s) - f(x)\big)\, D_n(s)\,ds = A_n(x,\delta) + B_n(x,\delta)\,, \tag{4.15}$$

for all $|x| \leq 1/2$ and all $n \geq 1$, where

$$A_n(x,\delta) = \int_{|s|\leq\delta} \big(f(x-s) - f(x)\big)\, D_n(s)\,ds\,, \qquad B_n(x,\delta) = \int_{\delta\leq|s|\leq 1/2} \big(f(x-s) - f(x)\big)\, D_n(s)\,ds\,.$$

Throughout the rest of the proof, $C = C(\alpha) > 0$ stands for a numerical constant that can change from line to line, but remains independent of $f \in h_\alpha$, $x \in [-1/2, 1/2]$, $n \geq 1$, and $\delta \in (0, 1/2)$. Using (4.12), we now estimate

$$|A_n(x,\delta)| \leq C \int_{|s|\leq\delta} \frac{|f(x-s) - f(x)|}{|s|}\,ds \leq C\,[f]_\alpha \int_{|s|\leq\delta} |s|^{\alpha-1}\,ds \leq C\,[f]_\alpha\,\delta^\alpha\,. \tag{4.16}$$

To estimate the second term in (4.15), we will take advantage of the oscillation of $D_n(s)$. For this, setting $h(x,s) = \frac{f(x-s)-f(x)}{\sin(\pi s)}$, we have

$$
\begin{aligned}
B_n(x,\delta) &= \int\limits_{\delta<|s|<\frac12} \frac{f(x-s)-f(x)}{\sin(\pi s)} \sin((2n+1)\pi s)\,ds = \int\limits_{\delta<|s|<\frac12} h(x,s)\sin((2n+1)\pi s)\,ds \\
&= -\int\limits_{\delta<|s|<\frac12} h(x,s)\sin\Big((2n+1)\pi\Big(s+\frac{1}{2n+1}\Big)\Big)\,ds \\
&= -\int\limits_{\delta<|y-\frac{1}{2n+1}|<\frac12} h\Big(x, y-\frac{1}{2n+1}\Big)\sin((2n+1)\pi y)\,dy\,,
\end{aligned}
$$

using (4.10) in the first line and the change of variables $y = s + \frac{1}{2n+1}$ in the last line. Adding up the integrals in the first and third line above, we can write $2B_n(x,\delta)$ in the form

$$
\begin{aligned}
&\int\limits_{\delta<|y|<\frac12} h(x,y)\sin((2n+1)\pi y)\,dy - \int\limits_{\delta<|y-\frac{1}{2n+1}|<\frac12} h\Big(x,y-\tfrac{1}{2n+1}\Big)\sin((2n+1)\pi y)\,dy \\
&= \int\limits_{\delta<|y|<\frac12} \Big(h(x,y)-h\Big(x,y-\tfrac{1}{2n+1}\Big)\Big)\sin((2n+1)\pi y)\,dy \\
&\quad + \int_{-\frac12}^{-\frac12+\frac{1}{2n+1}} h\Big(x,y-\tfrac{1}{2n+1}\Big)\sin((2n+1)\pi y)\,dy - \int_{-\delta}^{-\delta+\frac{1}{2n+1}} h\Big(x,y-\tfrac{1}{2n+1}\Big)\sin((2n+1)\pi y)\,dy \\
&\quad + \int_{\delta}^{\delta+\frac{1}{2n+1}} h\Big(x,y-\tfrac{1}{2n+1}\Big)\sin((2n+1)\pi y)\,dy - \int_{\frac12}^{\frac12+\frac{1}{2n+1}} h\Big(x,y-\tfrac{1}{2n+1}\Big)\sin((2n+1)\pi y)\,dy\,,
\end{aligned}
$$

by keeping track of the overlap of the sets $\{\delta < |y| < 1/2\}$ and $\{\delta < |y - \frac{1}{2n+1}| < \frac12\}$. We estimate the absolute values of the previous five integrals by bringing the absolute value into the integrand, and taking advantage of the bounds

$$
\begin{cases}
|h(x,y)| \le C\,\dfrac{\|f\|_\infty}{\delta}\,, \\[2mm]
|h(x,y)-h(x,y-\tau)| \le C\Big(\dfrac{|\tau|^\alpha [f]_\alpha}{\delta} + \dfrac{\|f\|_\infty |\tau|}{\delta^2}\Big),
\end{cases}
\qquad \text{for } 0 < 2\tau \le \delta \le |y| \le \frac12\,.
$$

The first bound follows at once from the definition of h since $|\sin(\pi y)| \ge \sin(\pi\delta)$ for $1/2 \ge |y| \ge \delta$, and $\dfrac{\sin(\pi\delta)}{\delta}$ is bounded on $(0,1/2]$. To handle $|h(x,y)-h(x,y-\tau)|$, note that we can rewrite and estimate it as follows:

$$
\begin{aligned}
&\Big|\frac{f(x-y)-f(x-y+\tau)}{\sin(\pi y)} + \big(f(x-y+\tau)-f(x)\big)\cdot\Big(\frac{1}{\sin(\pi y)} - \frac{1}{\sin(\pi(y-\tau))}\Big)\Big| \\
&\le \Big|\frac{f(x-y)-f(x-y+\tau)}{\sin(\pi y)}\Big| + |f(x-y+\tau)-f(x)|\cdot\frac{|\sin(\pi y)-\sin(\pi(y-\tau))|}{|\sin(\pi y)|\cdot|\sin(\pi(y-\tau))|} \\
&\le C\Big(\frac{\tau^\alpha [f]_\alpha}{\sin(\pi\delta)} + \frac{\|f\|_\infty \tau}{\sin(\pi\delta)\cdot\sin(\pi\delta/2)}\Big) \le C\Big(\frac{\tau^\alpha [f]_\alpha}{\delta} + \frac{\|f\|_\infty \tau}{\delta^2}\Big).
\end{aligned}
$$

Returning to the sum of five integrals and recalling (4.16), from (4.15) we obtain that

$$
|(S_n f)(x) - f(x)| \le C\Big([f]_\alpha\,\delta^\alpha + \frac{\tau\,[f]_\alpha}{\delta} + \frac{\tau^\alpha [f]_\alpha}{\delta} + \frac{\|f\|_\infty\,\tau}{\delta^2}\Big).
$$

The choice $\tau = (2n+1)^{-1}$ and $\delta = 2\,(2n+1)^{-\alpha/(\alpha+2)} \ge 2\tau$ yields (4.9). □

While the bound (4.9) is not sharp,[20] it reflects an important feature of Fourier series, namely that their rate of convergence improves with the smoothness of f. A further illustration of this fact is provided in the next result.

Theorem 4.3 *If $f : \mathbb{R} \to \mathbb{C}$ is periodic with period 1 and $p \geq 1$ times continuously differentiable, then, for all $m \geq 1$ and all $x \in \mathbb{R}$,*

$$|(S_m f)(x) - f(x)| \leq \frac{\sqrt{2}}{(2\pi)^p}\, m^{-p+1/2}\, \|f^{(p)}\|_{L^2[0,1]}\,. \tag{4.17}$$

Proof If $1 \leq m < n$, then

$$|(S_n f)(x) - (S_m f)(x)| = \Big|\sum_{m<|k|\leq n} \hat{f}(k)\, e^{2\pi i k x}\Big| = \Big|\sum_{m<|k|\leq n} \widehat{(f^{(p)})}(k)\,\frac{e^{2\pi i k x}}{(2\pi i k)^p}\Big| \leq \sum_{m<|k|\leq n} \frac{|\widehat{(f^{(p)})}(k)|}{|2\pi k|^p}$$

$$\leq \Big(\sum_{m<|k|\leq n} \frac{1}{(2\pi\, k)^{2p}}\Big)^{1/2} \Big(\sum_{m<|k|\leq n} |\widehat{(f^{(p)})}(k)|^2\Big)^{1/2} \leq \frac{\sqrt{2}}{(2\pi)^p}\, m^{-p+1/2} \Big(\int_0^1 |f^{(p)}(s)|^2\, ds\Big)^{1/2} \tag{4.18}$$

using successively (P4), the Cauchy–Schwarz inequality in $l^2(\mathbb{Z})$ and in $L^2[0,1]$, and Bessel's inequality for the orthonormal set $\{e^{2\pi i k x}\}_{k\in\mathbb{Z}}$, in combination with the chain of estimates

$$\sum_{m<|k|\leq n} \frac{1}{(2\pi\, k)^{2p}} \leq \frac{2}{(2\pi)^{2p} m^{2p-2}} \sum_{k=m+1}^{n} \frac{1}{k^2} = \frac{2}{(2\pi)^{2p} m^{2p-1}} \sum_{j=1}^{n-m} \frac{1}{m}\, \frac{1}{(1+\frac{j}{m})^2}$$

$$\leq \frac{2}{(2\pi)^{2p} m^{2p-1}} \sum_{j=1}^{n-m} \int_{\frac{j-1}{m}}^{\frac{j}{m}} \frac{1}{(s+1)^2}\, ds \leq \frac{2}{(2\pi)^{2p} m^{2p-1}} \int_0^{\frac{n-m}{m}} \frac{1}{(s+1)^2}\, ds$$

$$< \frac{2}{(2\pi)^{2p} m^{2p-1}} \int_0^{\infty} \frac{1}{(s+1)^2}\, ds = \frac{2}{(2\pi)^{2p} m^{2p-1}}\,.$$

Since Theorem 4.2 ensures $\lim_{n\to\infty} (S_n f)(x) = f(x)$ for every $x \in [0,1]$, letting $n \to \infty$ in (4.18) yields (4.17). □

After discusing the effect of smoothness on the convergence speed of the Fourier series, let us address the case of merely continuous functions. The discussion in the preamble of this chapter highlights some pathological situations, discouraging expectations for the recovery of a function $f \in C_{\text{per}}[0,1]$, the class of periodic continuous functions of period 1, from its Fourier series. This somewhat depressing[21] state-of-the-art was completely changed with Fejér's brillant idea of considering the averages of the symmetric partial sums instead of the sums themselves. Fejér's insight relied on the fact that the arithmetical means $y_k = \frac{x_0+\cdots+x_{k-1}}{k}$ $(k \geq 1)$ of a sequence $\{x_j\}_{j\geq 0}$ of complex numbers

20 For example, for $\alpha = 2/3$, the choice $\tau = (2n+1)^{-1}$ and $\delta = 2\,(2n+1)^{-2/5}$ in the last line of the proof of Theorem 4.2 yields the bound $Cn^{-1/5}$ for any $f \in h_{1/2}$ with $\|f\|_\infty + [f]_{1/2} \leq 1$.

21 The first example of a continuous function with a divergent Fourier series at a point, produced by du Bois-Reymond (1876), did throw the subject into disarray.

might converge even though the original numerical sequence fails to do so.[22] The complications arising in the context of the pointwise convergence of the Fourier series of functions $f \in C_{\rm per}[0,1]$ accentuate the attractive simplicity of the following result of Fejér (1904).

Theorem 4.4 *The arithmetical means $\sigma_n f = \frac{S_0 f + \cdots + S_{n-1} f}{n}$ of the symmetric partial sums (4.2) of $f \in C_{per}[0,1]$ converge uniformly to f.*

To clarify the statement, note that since $\hat{f}(k)$ appears only in $S_j f$ for $j \geq |k|$,

$$(\sigma_n f)(x) = \sum_{0 \leq |k| \leq n-1} \hat{f}(k)\Big(1 - \frac{|k|}{n}\Big) e^{2\pi i k x}, \qquad x \in \mathbb{R}. \tag{4.19}$$

In (4.19), the factors $(1 - \frac{|k|}{n})$ modify the Fourier coefficients $\hat{f}(k)$ by damping down the amplitude of the high-frequency terms (for $|k|$ near n), producing an improvement of the convergence properties, see Theorems 4.4–4.5.

Proof The proof relies on the properties of the arithmetical means of the Dirichlet kernel (the so-called Fejér kernel): for $0 < x < 1/2$ we have

$$\begin{aligned} F_n(x) &= \frac{D_0(x) + \cdots + D_{n-1}(x)}{n} = \frac{1}{n}\sum_{k=0}^{n-1} \frac{\sin((2k+1)\pi x)}{\sin(\pi x)} = \frac{1}{n}\sum_{k=0}^{n-1} \frac{\Im m\{e^{(2k+1)\pi i x}\}}{\sin(\pi x)} \\ &= \frac{\Im m\Big\{\sum\limits_{k=0}^{n-1} e^{(2k+1)\pi i x}\Big\}}{n\sin(\pi x)} = \frac{\Im m\Big\{e^{\pi i x}\sum\limits_{k=0}^{n-1} e^{2k\pi i x}\Big\}}{n\sin(\pi x)} = \frac{\Im m\Big\{e^{\pi i x}\frac{1 - e^{2\pi i n x}}{1 - e^{2\pi i x}}\Big\}}{n\sin(\pi x)} \\ &= \frac{\Im m\Big\{\frac{e^{\pi i x}(1 - e^{2\pi i n x})(1 - e^{-2\pi i x})}{(1 - e^{2\pi i x})(1 - e^{-2\pi i x})}\Big\}}{n\sin(\pi x)} = \frac{\Im m\{e^{\pi i x}(1 - e^{2\pi i n x})(1 - e^{-2\pi i x})\}}{n\sin(\pi x)\,[2 - 2\cos(2\pi x)]} \\ &= \frac{2\sin(\pi x) - \sin((2n+1)\pi x) + \sin((2n-1)\pi x)}{n\sin(\pi x)\,[2 - 2\cos(2\pi x)]} = \frac{1 - \cos(2\pi n x)}{n\,[1 - \cos(2\pi x)]} \\ &= \frac{1}{n}\Big[\frac{\sin(\pi n x)}{\sin(\pi x)}\Big]^2, \end{aligned} \tag{4.20}$$

using (4.10), trigonometric formulas and sums of geometrical series. Moreover, $F_n(0) = n$ and $F_{2k+1}(\frac{1}{2}) = \frac{1}{2k+1}$ while $F_{2k}(\frac{1}{2}) = 0$ for $k \geq 0$, since $D_k(0) = 2k+1$ and $D_k(\frac{1}{2}) = (-1)^k$. The function F_n is even and periodic of period 1, both properties being inherited from the D_ks. The family $\{F_n\}_{n \geq 1}$ of periodic functions forms what is called an *approximate identity*,[23] due to the following properties:[24]

[22] For example, $y_{2k+1} = -1/(2k+1)$ and $y_{2k+2} = 0$ if $x_k = (-1)^k$ for $k \geq 1$. On the other hand, $x_j \to x$ ensures $y_k \to x$. Indeed, given $\varepsilon > 0$, if $|x_j - x| < \varepsilon$ for $j \geq n$, then for $k \geq n$ we have $|y_k - x| = |\frac{x_0 + \cdots + x_{k-1}}{k} - x| = |\frac{(x_0 - x) + \cdots + (x_{k-1} - x)}{k}| \leq \frac{|x_0 - x|}{k} + \cdots + \frac{|x_{k-1} - x|}{k} \leq \sum_{j=0}^{n-1} \frac{|x_j - x|}{k} + \varepsilon\,\frac{k-n}{k}$, so that $\limsup_{k\to\infty} |y_k - x| \leq \varepsilon$. Thus $y_k \to x$, $\varepsilon > 0$ being arbitrary.

[23] This is not true of the D_n, for which (A2) fails, as we saw in (4.13).

[24] Their validity is easily established. Indeed, (A1) follows at once from (4.14). Since F_n are nonnegative, in view of (4.20), (A2) holds with $K = 1$. As for (A3), note that by (4.20), $\int_{\delta < |x| < 1/2} |F_n(x)|\,dx = \frac{2}{n}\int_\delta^{1/2} \Big[\frac{\sin(\pi n x)}{\sin(\pi x)}\Big]^2 dx \leq \frac{1}{n\sin^2(\pi\delta)}$, since $\sin^2(\pi n x) \leq 1$ and $\sin(\pi x) \geq \sin(\pi\delta)$ for $\delta \leq x \leq \frac{1}{2}$.

(A1) $\int_{-1/2}^{1/2} F_n(x)\,dx = 1$ for all $n \geq 1$;

(A2) $\sup_{n\geq 1} \int_{-1/2}^{1/2} |F_n(x)|\,dx = K < \infty$;

(A3) for any fixed $\delta \in (0, 1/2)$ we have $\int_{\delta<|x|<1/2} |F_n(x)|\,dx \to 0$ as $n \to \infty$.

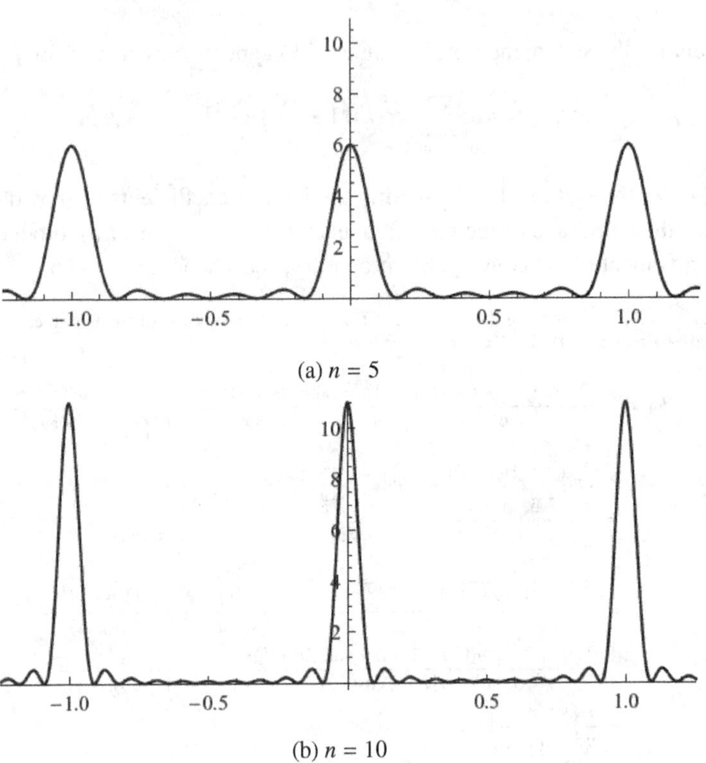

Figure 4.2 The graph of the Fejér kernel F_n.

After all these preparations, we proceed as follows. Using (4.15), we get

$$(\sigma_n f)(x) - f(x) = \frac{1}{n}\sum_{k=0}^{n-1}(S_k f)(x) - f(x) = \frac{1}{n}\sum_{k=0}^{n-1}\int_{-1/2}^{1/2}(f(x-s) - f(x))\,D_k(s)\,ds$$
$$= \int_{-1/2}^{1/2}(f(x-s) - f(x))\,F_n(s)\,ds\,. \tag{4.21}$$

Our aim is to show that this discrepancy is uniformly small. The argumentation is much easier than that in the proof of Theorem 4.2 since, in contrast to the Dirichlet kernels D_n, the tails of the

Fejér kernels F_n are small for large n; see Figure 4.2. Indeed, with $\delta \in (0, 1/2)$ fixed, we have

$$\Big|\int_{-1/2}^{1/2}\big(f(x-s)-f(x)\big)\,F_n(s)\,ds\Big| \le \int_{-1/2}^{1/2}|f(x-s)-f(x)|\cdot|F_n(s)|\,ds$$

$$= \int_{-\delta}^{\delta}|f(x-s)-f(x)|\cdot|F_n(s)|\,ds + \int_{\delta\le|s|\le 1/2}|f(x-s)-f(x)|\cdot|F_n(s)|\,ds$$

$$\le \sup_{x\in[0,1],\,|s|\le\delta}\{|f(x-s)-f(x)|\}\int_{-\delta}^{\delta}|F_n(\tau)|\,d\tau + 2\,\|f\|_\infty\int_{\delta\le|s|\le 1/2}|F_n(s)|\,ds$$

$$\le K \sup_{x\in[0,1],\,|s|\le\delta}\{|f(x-s)-f(x)|\} + 2\,\|f\|_\infty\int_{\delta\le|s|\le 1/2}|F_n(s)|\,ds\,,$$

using (A2) in the last line. The first term can be made as small as we please for $\delta > 0$ sufficiently small, since f is uniformly continuous, while the second term can be made small for fixed $\delta \in (0, 1/2)$ if n is large enough, due to (A3). □

The fact that the Fourier series of $f \in C_{\text{per}}[0, 1]$ might fail to converge to $f(x)$ on a set of measure zero lowers the expectations for the pointwise convergence if $f \in L^1[0, 1]$. The worst-case scenario may actually occur: there is $f \in L^1[0, 1]$ such that its Fourier series diverges at every single point $x \in [0, 1]$, see Kolmogorov (1926). Moreover, see Exercise 24, there is $f \in L^1[0, 1]$ with $\sup_{n\ge1}\{\|S_nf\|_{L^1[0,1]}\} = \infty$, which prevents the approximation of f by S_nf in the norm of $L^1[0, 1]$. In light of this situation,[25] it is surprising that the Fourier coefficients $\{\hat{f}(k)\}_{k\in\mathbb{Z}}$ of a function $f \in L^1[0, 1]$ determine the function. Given that the symmetric partial sums $\{S_nf\}_{n\ge1}$ may diverge, our best option is to rely on the arithmetical means $\{\sigma_nf\}_{n\ge1}$.

Theorem 4.5 *For any $f \in L^1[0, 1]$ we have $\lim_{n\to\infty}\sigma_nf = f$ in $L^1[0, 1]$ and $\lim_{n\to\infty}\sigma_nf = f$ a.e. on $[0, 1]$.*

The proof of Theorem 4.5 requires two auxiliary results from integration theory, regarding the continuity of translation (Lemma 4.6) and differentiation (Lemma 4.7), respectively.

Lemma 4.6 $\lim_{h\to0}\|T_hf - f\|_{L^1[0,1]} = 0$ *if $f \in L^1[0, 1]$.*

Proof In proving (P4) we have already noted that the set of finite linear combinations of characteristic functions $\chi^{(a,b)}$ of open intervals $(a, b) \subset [0, 1]$ is dense in $L^1[0, 1]$. Let f_n be a sequence of functions of this type that approximate f in $L^1[0, 1]$. For each open interval $(a, b) \subset [0, 1]$ we have $\|T_h\chi^{(a,b)} - \chi^{(a,b)}\|_{L^1[0,1]} \le 2|h| \to 0$ for $h \to 0$, so that the same is true for every single f_n. But then $\|T_hf - f\|_{L^1[0,1]} \le \|T_hf - T_hf_n\|_{L^1[0,1]} + \|T_hf_n - f_n\|_{L^1[0,1]} + \|f - f_n\|_{L^1[0,1]} = 2\,\|f - f_n\|_{L^1[0,1]} + \|T_hf_n - f_n\|_{L^1[0,1]}$ yields $\limsup_{h\to0}\|T_hf - f\|_{L^1[0,1]} \le 2\,\|f - f_n\|_{L^1[0,1]}$ for every $n \ge 1$. Letting $n \to \infty$ yields the statement. □

Lemma 4.7 $\lim_{h\to0}\Big(\frac{1}{h}\int_0^h|f(x+s) - f(x)|\,ds\Big) = 0$ *a.e. if $f \in L^1[0, 1]$.*

[25] For the intricate relationship between the a.e. convergence and the convergence in $L^1[0, 1]$ of a sequence of functions, see Exercise 13.

Proof Let $\{t_k\}_{k\geq 1}$ be a countable dense subset of $\mathbb{R}$, e.g. the rational numbers. For $k \geq 1$, define the absolutely continuous function $F_k(x) = \int_0^x |f(y) - t_k|\, dy$ with $F_k'(x) = |f(x) - t_k|$ a.e., see Theorem 2.8. Since the countable union of sets of measure zero remains a set of measure zero, we find some set $A \subset [0, 1]$ of measure zero, such that $\lim\limits_{h\to 0} \frac{1}{h} \int_x^{x+h} |f(y) - t_k|\, dy = |f(x) - t_k|$ for all $x \in [0, 1] \setminus A$ and all $k \geq 1$. Fix $x \in [0, 1] \setminus A$ and let $\varepsilon > 0$. If t_k is such that $|f(x) - t_k| < \frac{\varepsilon}{2}$, then $\limsup\limits_{h\to 0} \left|\frac{1}{h} \int_x^{x+h} |f(y) - f(x)|\, dy\right| \leq \limsup\limits_{h\to 0} \left|\frac{1}{h} \int_x^{x+h} |f(y) - t_k|\, dy\right| + \left|\frac{1}{h} \int_x^{x+h} |f(x) - t_k|\, dy\right| = 2\,|f(x) - t_k| < \varepsilon$. Thus $\lim\limits_{h\to 0} \frac{1}{h} \int_x^{x+h} |f(y) - f(x)|\, dy = 0$ for all $x \in [0, 1] \setminus A$. The change of variables $y = x + s$ yields the statement. □

Proof of Theorem 4.5 Using (4.21), for any $\delta \in (0, \frac{1}{2})$ we get

$$\begin{aligned}
\|\sigma_n f - f\|_{L^1[0,1]} &= \int_{-1/2}^{1/2} \Big| \int_{-1/2}^{1/2} [(T_s f)(x) - f(x)]\, F_n(s)\, ds \Big|\, dx \\
&\leq \int_{-1/2}^{1/2} \Big(\int_{-1/2}^{1/2} |(T_s f)(x) - f(x)|\, dx \Big) F_n(s)\, ds \\
&= \int_{|s|\leq\delta} \|T_s f - f\|_{L^1[0,1]}\, F_n(s)\, ds + \int_{\delta<|s|\leq\frac{1}{2}} \|T_s f - f\|_{L^1[0,1]}\, F_n(s)\, ds \\
&\leq \sup_{|s|\leq\delta} \{\|T_s f - f\|_{L^1[0,1]}\} + 2\,\|f\|_{L^1[0,1]} \int_{\delta<|s|\leq\frac{1}{2}} F_n(s)\, ds
\end{aligned}$$

in view of (A2), since $\|T_s f - f\|_{L^1[0,1]} \leq 2\,\|f\|_{L^1[0,1]}$. Given $\varepsilon > 0$, by (A3) there is $n(\varepsilon) \geq 1$ so that $\|\sigma_n f - f\|_{L^1[0,1]} \leq \sup\limits_{|s|\leq\delta} \{\|T_s f - f\|_{L^1[0,1]}\} + \varepsilon$ for all $n \geq n(\varepsilon)$. Letting $\delta \to 0$, Lemma 4.6 yields $\limsup\limits_{n\to\infty} \|\sigma_n f - f\|_{L^1[0,1]} \leq \varepsilon$. Since $\varepsilon > 0$ was arbitrary, we deduce that $\sigma_n f \to f$ in $L^1[0, 1]$.

Let us now prove that for $x \in [-\frac{1}{2}, \frac{1}{2}]$ with the property that $|f(x)| < \infty$ and $\lim\limits_{h\to 0} \frac{1}{h} \int_0^h |f(x + s) - f(x)|\, ds = 0$, see Lemma 4.7, we have $(\sigma_n f)(x) \to f(x)$ as $n \to \infty$. Given $\varepsilon > 0$, choose $\delta \in (0, \frac{1}{2})$ such that $I(s) = \int_0^s |f(x - y) - f(x)|\, dy$ satisfies $|I(s)| \leq \varepsilon |s|$ for $|s| \leq \delta$. Starting again with (4.21), we get

$$\begin{aligned}
|(\sigma_n f)(x) - f(x)| &= \Big| \int_{-1/2}^{1/2} [f(x - s) - f(x)]\, F_n(s)\, ds \Big| \\
&\leq \int_{|s|\leq\delta} |f(x - s) - f(x)|\, F_n(s)\, ds + \int_{\delta<|s|\leq 1/2} |f(x - s) - f(x)|\, F_n(s)\, ds \\
&\leq \int_{-\delta}^{\delta} |f(x - s) - f(x)|\, \frac{2n}{1+n^2 s^2}\, ds + \int_{\delta<|s|\leq 1/2} \frac{|f(x-s) - f(x)|}{n \sin^2(n\delta)}\, ds
\end{aligned}$$

due to (4.20) and the fact that[26] $0 \leq F_n(s) \leq \frac{2n}{1+n^2 s^2}$ for all $n \geq 1$ and all $|s| \leq \frac{1}{2}$. The second term above tends to zero as $n \to \infty$, by the dominated convergence theorem. As for the first term, using integration by parts, it can be majorised by

$$\begin{aligned}
&\frac{2n\,[I(\delta) - I(-\delta)]}{1+n^2\delta^2} + \int_{-\delta}^{\delta} I(s)\, \frac{4n^3 s}{(1+n^2 s^2)^2}\, ds \leq [I(\delta) - I(-\delta)]\, \frac{2n}{1+n^2\delta^2} + \int_0^{\delta} I(s)\, \frac{4n^3 s}{(1+n^2 s^2)^2}\, ds \\
&\leq \frac{4n\varepsilon\delta}{1+n^2\delta^2} + \int_0^{\delta} \frac{4\varepsilon n^3 s^2}{(1+n^2 s^2)^2}\, ds < 2\varepsilon + 4\varepsilon \int_0^{\infty} \frac{t^2}{(1+t^2)^2}\, dt \leq 2\varepsilon\Big(1 + 2\int_0^{\infty} \frac{dt}{1+t^2}\Big) = 2\varepsilon(1 + \pi)\,.
\end{aligned}$$

Since $\varepsilon > 0$ was arbitrary, the desired conclusion now follows. □

[26] Recall that F_n is even and nonnegative. Given $n \geq 2$, for $0 \leq x \leq \frac{1}{n}$ we have that $F_n(x) \leq n \leq \frac{2n}{1+n^2x^2}$, while $F_n(x) \leq \frac{1}{n \sin^2(\pi x)} \leq \frac{1}{4nx^2} \leq \frac{2n}{1+n^2x^2}$ for $\frac{1}{n} \leq x \leq \frac{1}{2}$ since $\sin(s) \geq \frac{2s}{\pi}$ for $s \in [0, \frac{\pi}{2}]$, see the solution to Exercise 20(ii).

We shall present some deductions from Theorem 4.4 and Theorem 4.5, each of which is of considerable importance. In view of (P1), the first result asserts that a function $f \in L^1[0,1]$ is determined by its Fourier coefficients.

Corollary (Uniqueness of Fourier coefficients) *If* $f \in L^1[0,1]$ *and* $\hat{f} = 0$*, then* $f = 0$ *a.e.*

Proof Since $\sigma_n f = 0$ for all $n \geq 1$, this follows from Theorem 4.5. Alternatively, set $F(x) = c + \int_0^x f(y)\,dy$, where $c \in \mathbb{C}$ is chosen so as to make $\int_0^1 F(x)\,dx = 0$. Since $\hat{f}(0) = \int_0^1 f(x)\,dx = 0$, F is periodic.[27] Moreover, by construction, F is absolutely continuous with $F' = f$ a.e. Due to (P4) and the choice of c, $\hat{F} = 0$. Theorem 4.4 yields $F = 0$, so that $f = 0$ a.e., as alleged. □

We now show that the convergence at a point x_0 of the Fourier series of a function $f \in L^1[0,1]$ depends solely on the behaviour of f near x_0, even though the Fourier coefficients (4.1) depend on the values of f throughout $[0,1]$. Note, however, that the extent of the localisation is limited: one of the less attractive features of Fourier series is that bad behaviour of f near a point affects the global behaviour of the series. For example, the presence of a single jump discontinuity of f prevents the Fourier series from being absolutely convergent (see Exercise 22).

Corollary (The localisation principle) *If* $f, g \in L^1[0,1]$ *coincide throughout some neighbourhood of* x_0*, then* $\lim_{n\to\infty}[(S_n f)(x_0) - (S_n g)(x_0)] = 0$.

Proof Since $f \mapsto S_n f$ are linear operators, we may assume without loss of generality that $g \equiv 0$. Moreover, since $f(x_0) = (T_{-x_0} f)(0)$ by (P3) and $(S_n f)(x_0) = (S_n(T_{-x_0} f))(0)$ by (4.11), we may take $x_0 = 0$. Now, in view of (4.11) and (4.10),

$$
\begin{aligned}
(S_n f)(0) &= \int_{-1/2}^{1/2} f(-s) D_n(s)\,ds = \int_{-1/2}^{1/2} \frac{f(-s)}{\sin(\pi s)} \sin((2n+1)\pi s)\,ds \\
&= \int_{-1/2}^{1/2} \Big\{ f(-s)\cos(2\pi n s) + \frac{f(-s)\cos(\pi s)}{\sin(\pi s)} \sin(2\pi n s) \Big\}\,ds . \qquad (4.22)
\end{aligned}
$$

By assumption there is $\delta \in (0, \frac{1}{2})$ such that $f(s) = 0$ for $|s| < \delta$. Since $|\sin(\pi s)| \geq \sin(\pi\delta)$ for $\delta \leq |s| \leq \frac{1}{2}$, we see that the function $s \mapsto \frac{f(-s)\cos(\pi s)}{\sin(\pi s)}$ belongs to $L^1\big[-\frac{1}{2}, \frac{1}{2}\big]$. Clearly $s \mapsto f(-s)$ also belongs to $L^1\big[-\frac{1}{2}, \frac{1}{2}\big]$. Using the identities $\sin(2\pi n s) = \frac{e^{2\pi i n s} - e^{-2\pi i n s}}{2i}$ and $\cos(2\pi n s) = \frac{e^{2\pi i n s} + e^{-2\pi i n s}}{2}$ in (4.22), we conclude by (P5). □

The next result shows that, irrespective how badly the Fourier series of some function $f \in L^1[0,1]$ might behave in respect to pointwise convergence, we may integrate it termwise.

Corollary (Termwise integration of Fourier series) *If* $f \in L^1[0,1]$ *then for all*

[27] We extend f to $f : \mathbb{R} \to \mathbb{C}$ by setting $f(x+k) = f(x)$ for $k \in \mathbb{Z}$ and $x \in [0,1)$, the possible modification of the value of $f(1)$ being irrelevant for the integrations that we perform.

$x \in [0,1]$ *we have*

$$\int_0^x f(y)\,dy = \lim_{n\to\infty} \int_0^x (S_n f)(y)\,dy = x\hat{f}(0) + \sum_{k\neq 0} \hat{f}(k)\,\frac{e^{2\pi ikx}-1}{2\pi ik}. \tag{4.23}$$

Proof It suffices to establish the formula for $x \in (0,1]$. Let $g : [0,1] \to \mathbb{C}$ be the characteristic function of the interval $(0,x)$, extended to $\mathbb{R}$ so as to have period 1. A direct computation shows that $\hat{g}(0) = x$, while $\hat{g}(k) = \frac{e^{-2\pi ikx}-1}{-2\pi ik}$ for $k \neq 0$, so that some algebraic manipulations yield $(S_n g)(y) = x + \sum_{k=1}^n \frac{\sin(\pi k(x-y))+\sin(\pi ky)}{\pi k}$ for all $y \in [0,1]$. The inequality $\left|\sum_{k=1}^n \frac{\sin(\pi ky)}{k\pi}\right| \leq 3$ for all $y \in \mathbb{R}$ and all $n \geq 1$, see Exercise 20(iii), ensures that the family of functions $\{S_n g\}_{n\geq 1}$ is uniformly bounded on $[0,1]$. On the other hand, due to the localisation principle, $\lim_{n\to\infty}(S_n g)(y) = g(y)$ for $y \in (0,x)\cup(x,1)$, since g is constant in some small neighborhood of such a point. Using the dominated convergence theorem, we get $\int_0^x f(y)\,dy = \int_0^1 f(y)g(y)\,dy = \lim_{n\to\infty}\int_0^1 f(y)\,(S_n g)(y)\,dy = \lim_{n\to\infty}\sum_{k=-n}^n \hat{g}(k)\int_0^1 f(y)\,e^{2\pi iky}\,dy = \lim_{n\to\infty}\sum_{k=-n}^n \hat{g}(k)\,\hat{f}(-k) = \lim_{n\to\infty}\sum_{k=-n}^n \hat{f}(k)\,\hat{g}(-k)$. With $\hat{g}(-k)$ computed above, (4.23) emerges. □

A by-product of the integration identity (4.23) is the following necessary condition on the Fourier coefficients of $f \in L^1[0,1]$: the series $\sum_{k\in\mathbb{Z}} \frac{\hat{f}(2k+1)}{2k+1}$ converges. This fact is justified by setting $x = \frac{1}{2}$ in (4.23). It can be used to exhibit trigonometric series that are not Fourier series – see Exercise 21(i).

We conclude this section by pointing out that while the means $\sigma_n f$ are a very useful substitute when the convergence of the symmetric partial sums $S_n f$ might fail, see Theorem 4.4 and Theorem 4.5, they also present an inherent limitation in their ability to provide fast convergence. For example, the only functions $f \in C_{\text{per}}[0,1]$ with $\lim_{n\to\infty}\{n\|f-\sigma_n f\|_\infty\} = 0$ are constants, see Exercise 17. This is in stark contrast to Theorem 4.3, showing conclusively that $S_n f$ are better approximations to f than $\sigma_n f$ for periodic functions f that are $p \geq 2$ times continuously differentiable.

4.3 Mean-square convergence

Mathematically the most forthright notion of convergence for Fourier series is convergence in the $L^2[0,1]$ norm: as we prove in this section, necessary and sufficient conditions for convergence are readily available, the abstract foundation being the Hilbert space theory developed in Section 3.2. Note that $L^2[0,1]$ is a separable Hilbert space, see Exercise 14. The inequality $|f(x)| \leq \frac{1+|f(x)|^2}{2}$ shows that[28] $L^2[0,1] \subset L^1[0,1]$, thus allowing us to infer facts relative to the convergence of the Fourier series of $f \in L^2[0,1]$ from the results valid within

[28] The inclusion is strict, since the function $f \in L^1[0,1]$, given by $f(0) = 0$ and $f(x) = x^{-1/2}$ for $x \in (0,1]$, does not belong to $L^2[0,1]$.

the more general setting of $L^1[0,1]$-theory. Nevertheless, the $L^2[0,1]$-theory is much richer. As an indication of the improvement with respect to the $L^1[0,1]$-setting, note that, by a difficult result due to Carleson (1966), the Fourier series of $f \in L^2[0,1]$ converges a.e. to f.

Concerning the convergence of the Fourier series in $L^2[0,1]$, note that the relations (1.1.2) mean that $\{e_k\}_{k\in\mathbb{Z}}$, where $e_k(x) = e^{2\pi ikx}$ for $x \in \mathbb{R}$ and $k \in \mathbb{Z}$, is an orthonormal set in $L^2[0,1]$. The uniqueness of the Fourier coefficients of $f \in L^1[0,1]$ (see Section 4.2) and the inclusion $L^2[0,1] \subset L^1[0,1]$ ensure that the orthogonal complement of $\{e_k\}_{k\in\mathbb{Z}}$ is $\{0\}$, so that this family is actually an orthonormal basis, and we can take full advantage of Theorem 3.1 in the setting of $L^2[0,1]$. In particular, we can identify,[29] by means of Theorem 3.4., the Fourier coefficients of $f \in L^2[0,1]$ with the family of sequences in $l^2(\mathbb{Z})$. The validity of the Parseval and Plancherel identities is also granted.

4.4 Convergence at a jump discontinuity

Let us now discuss the possibility of the convergence of the Fourier series of some function $f \in L^1[0,1]$ at every $x \in [0,1]$, without assuming a Hölder-type condition (or differentiability), as in Section 4.2. Of the multitude of results of this type, perhaps the most useful regards the class $BV_{\text{per}}[0,1]$ of periodic functions f whose restriction to $[0,1]$ has bounded variation. As shown in Chapter 2, these functions admit[30] at most countably many discontinuities, all of which must be jump discontinuities, in the sense that the limits $f(x+) = \lim_{y\downarrow x} f(y)$ and $f(x-) = \lim_{y\uparrow x} f(y)$ exist and are unequal. The convergence theorem that we now present was first proved by Jordan (1881), who extended an earlier less general result of Dirichlet (1829).

Theorem 4.8 *If* $f \in BV_{per}[0,1]$, *then* $\lim_{n\to\infty}(S_n f)(x) = \frac{f(x+)+f(x-)}{2}$ *for every* $x \in [0,1]$, *with* $|(S_n f)(x)| \le \|f\|_\infty + 2\,V_f[0,1]$. *Moreover, the convergence is uniform if* f *is also continuous.*

29 This result is known to be the best possible: given a sequence $\{c_k\}_{k\in\mathbb{Z}}$ with $\sum_{k\in\mathbb{Z}} |c_k|^2 = \infty$, one can choose the $\pm$ signs in such a way that the series $\sum_{k\in\mathbb{Z}} \pm c_k\, e^{2\pi ikx}$ is not the Fourier series of any $f \in L^1[0,1]$. Moreover, the inequality $\sum_{k\in\mathbb{Z}} |\hat{f}(k)|^2 < \infty$, valid for $f \in L^2[0,1]$, leaves little place for improvement even for $f \in C_{\text{per}}[0,1]$: it can be shown, see Zygmund (1959), that there are constants $\beta > 1$ and $c > 0$ such that $\sum_{n=2}^{\infty} \frac{e^{2\pi icn\ln(n)}}{n^{1/2}[\ln(n)]^\beta}\, e^{2\pi inx}$ defines $f \in C_{\text{per}}[0,1]$ with the property that $\sum_{k\in\mathbb{Z}} |\hat{f}(k)|^{2-\varepsilon} = \infty$ for all $\varepsilon > 0$.

30 Note that while a function $f : \mathbb{R} \to \mathbb{C}$ having only jump discontinuities has at most countably many, see Bruckner et al. (1997), this feature does not characterise functions of bounded variation. For example, the function in (2.24), extended by periodicity to $\mathbb{R}$, is continuous but its restriction to $[0,1]$ is not of bounded variation.

The proof of Theorem 4.8 relies on the so-called Tauberian result proved in Lemma 4.9. This is a subtle partial converse relating the convergence of the arithmetical means to that of the original sequence.[31]

Lemma 4.9 *Let $\{a_j\}_{j\geq 0}$ be a sequence of complex numbers. If $s_n = \sum_{j=0}^n a_j$ are such that their arithmetical means $\sigma_n = \frac{s_0+\cdots+s_{n-1}}{n}$, $n \geq 1$, converge to A, and if the sequence $\{ja_j\}_{j\geq 0}$ is bounded, then $s_n \to A$ as $n \to \infty$.*

Proof For $\varepsilon > 0$, let $N \geq 1$ be such that $|A - \sigma_n| \leq \varepsilon$ for $n \geq N$. For $p \geq 1$ the identity $(n+p)\sigma_{n+p} - n\sigma_n = s_n + \cdots + s_{n+p-1}$ yields $(n+p)(\sigma_{n+p} - A) - n(\sigma_n - A) = p(s_n - A) + \sum_{k=0}^{p-1}(s_{n+k} - s_n)$. If $M = \sup_{j\geq 0}\{|ja_j|\}$, then $|s_{n+k} - s_n| = \left|\sum_{j=1}^k a_{n+j}\right| \leq \sum_{j=1}^k |a_{n+j}| \leq \frac{Mk}{n}$, so that $\sum_{k=0}^{p-1} k = \frac{p(p-1)}{2}$ leads to $p|s_n - A| \leq (n+p)|\sigma_{n+p} - A| + n|\sigma_n - A| + \frac{Mp(p-1)}{2n}$. Thus $|s_n - A| \leq (\frac{2n}{p} + 1)\varepsilon + \frac{M(p-1)}{n}$ for $n \geq N$ and $p \geq 1$. The upper bound being increasing in $p > n\sqrt{\frac{2\varepsilon}{M}}$, setting p equal to the smallest integer that exceeds $n\sqrt{\frac{2\varepsilon}{M}}$, the upper bound is less than $\varepsilon + 2\sqrt{2\varepsilon M} \to 0$ for $\varepsilon \to 0$. □

Proof of Theorem 4.8 Fix $x \in [-\frac{1}{2}, \frac{1}{2}]$ and let $A = \frac{f(x+)+f(x-)}{2}$. Given $\varepsilon > 0$, choose $\delta \in (0, \frac{1}{2})$ such that $|f(x \pm s) - f(x\pm)| \leq \varepsilon$ for $|s| \leq \delta$. Using (4.15) as we did in (4.21), we get

$$(\sigma_n f)(x) = \int_{-1/2}^{1/2} f(x-s)F_n(s)\,ds = \int_{-1/2}^{1/2} f(x+s)F_n(s)\,ds\,, \tag{4.24}$$

after the change of variables $s \mapsto (-s)$, since F_n is even. Recalling (A1), we get

$$\begin{aligned}
|(\sigma_n f)(x) - A| &= \tfrac{1}{2}\Big|\int_{-1/2}^{1/2} [f(x-s) + f(x+s) - 2A]\,F_n(s)\,ds\Big| \\
&\leq \int_{|s|\leq\delta} |f(x-s) + f(x+s) - 2A|\,F_n(s)\,ds + \int_{|\delta|\leq|s|\leq\frac{1}{2}} |f(x-s) + f(x+s) - 2A|\,F_n(s)\,ds \\
&\leq 2\varepsilon \int_{|s|\leq\delta} F_n(s)\,ds + 4\|f\|_\infty \int_{|\delta|\leq|s|\leq\frac{1}{2}} F_n(s)\,ds\,.
\end{aligned}$$

Since functions of bounded variation are bounded, we have $\|f\|_\infty < \infty$. Thus, the properties (A1) and (A3) of the Fejér kernel F_n ensure that $\limsup_{n\to\infty} |(\sigma_n f)(x) - A| \leq 2\varepsilon$. Since $\varepsilon > 0$ was arbitrary, this proves that $(\sigma_n f)(x) \to \frac{f(x+)+f(x-)}{2}$ as $n \to \infty$. Now $(S_n f)(x)$ converges to the same limit due to Lemma 4.9 since

$$|k\hat{f}(k)| \leq V_f[0,1]\,, \qquad k \in \mathbb{Z}. \tag{4.25}$$

Indeed, for $k \neq 0$, set $h = \frac{1}{2|k|}$ and note that $\hat{f}(k) = \sum_{j=0}^{2|k|-1} \int_{jh}^{(j+1)h} f(x)\,e^{-2\pi ikx}dx = \int_0^h \sum_{j=0}^{2|k|-1} f(y + jh)e^{-2\pi ik(y+jh)}dy$ after performing in the $(j+1)$th integral the change of variables $x = y + jh$. Consequently $|\hat{f}(k)| \leq \left|\int_0^h \sum_{j=0}^{2|k|-1}(-1)^j f(y+jh)\,e^{-2\pi iny}dy\right| \leq \int_0^h \left|\sum_{j=0}^{2|k|-1}(-1)^j f(y+jh)\,e^{-2\pi iny}\right|dy \leq \int_0^h \sum_{j=0}^{|k|-1} |f(x+2jh) - f(y+(2j+1)h)|\,dy \leq \int_0^h V_f[0,1]\,dy = h\,V_f[0,1]$. This validates (4.25). Moreover, (4.25) with (4.19) and (4.11) yield $|(\sigma_n f)(x) - (S_{n+1}f)(x)| \leq \left|\sum_{|k|\leq n} \hat{f}(k)\,\frac{|k|}{n}\,e^{2\pi ikx}\right| \leq \sum_{|k|\leq n} \hat{f}(k)\,\frac{|k|}{n} \leq 2\,V_f[0,1]$. Since $|(\sigma_n f)(x)| \leq \|f\|_\infty \int_{-1/2}^{1/2} F_n(s)\,ds = \|f\|_\infty$ by (4.24) and (A1), we obtain the stated bound of $|(\sigma_n f)(x)|$.

Finally, if f is also continuous on $\mathbb{R}$, then $\sigma_n f \to f$, see Theorem 4.4. By inspection we see that the above considerations hold uniformly for $x \in \mathbb{R}$. □

[31] Recall from the preamble of Theorem 4.4 that the unrestricted converse might fail.

A single jump discontinuity of $f \in BV_{\rm per}[0,1]$ at some point $x_0 \in [0,1]$ prevents the uniform convergence of $S_n f$ on $[0,1]$. The specific form of non-uniform convergence is illustrated by the example of the "reverse sawtooth" function $\phi(x) = \frac{1}{2} - x + [x]$, where $[x]$ is the largest integer $k \leq x$. Note that $\phi \in BV_{\rm per}[0,1]$ is smooth except for jump discontinuities at the integers k, where it presents jumps of unit size due to $\phi(k\pm) = \pm\frac{1}{2}$; see Figure 4.3.

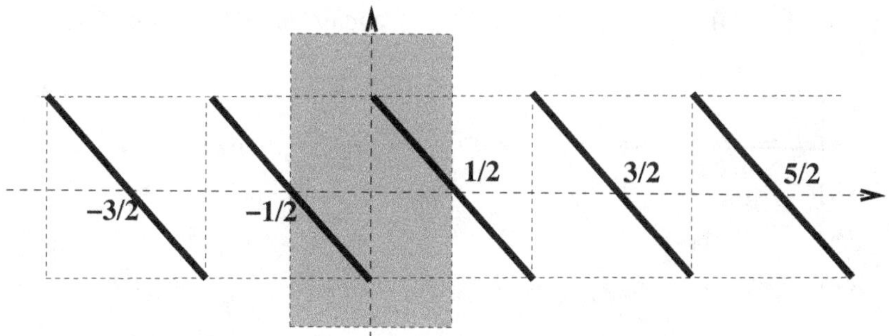

Figure 4.3 Graph of the reverse sawtooth function $\phi(x) = \frac{1}{2} - x + [x]$, with the periodicity window $[-\frac{1}{2}, \frac{1}{2}]$ highlighted.

Note that ϕ is an odd function, with $\hat{\phi}(0) = 0$ and $\hat{\phi}(k) = \frac{1}{2\pi i k}$ for $k \neq 0$, see Exercise 1. Thus $(S_n\phi)(x) = \sum_{k=1}^n \frac{\sin(2\pi k x)}{\pi k}$ for $n \geq 1$. Since $(S_n\phi)'(x) = 2\sum_{k=1}^n \cos(2\pi k x) = D_n(x) - 1$, due to (4.10) we have that $(S_n\phi)(x) = -x + \int_0^x \frac{\sin((2n+1)\pi s)}{\sin(\pi s)}\,ds$ for $n \geq 1$ and $x \in (0, \frac{1}{2})$. The function $g(s) = \frac{1}{\sin(\pi s)} - \frac{1}{\pi s}$ being continuously differentiable[32] on $[-\frac{1}{2}, \frac{1}{2}]$, integration by parts provides us with a constant $M > 0$ such that $\left|\int_0^x g(s)\sin((2n+1)\pi s)\,ds\right| = \left|\int_0^x \frac{g(s)}{(2n+1)\pi}\,\partial_s\{\sin((2n+1)\pi s)\}\,ds\right| \leq \frac{M}{n}$ for all $x \in [0, \frac{1}{2}]$ and all $n \geq 1$. Thus $\left|(S_n\phi)(x) + x - \int_0^x \frac{\sin((2n+1)\pi s)}{\pi s}\,ds\right| \leq \frac{M}{n}$ for all $x \in [0, \frac{1}{2}]$ and all $n \geq 1$. The change of variables $t = (2n+1)\pi s$ enables us to write this as $\left|(S_n\phi)(x) + x - \frac{1}{\pi}\int_0^{(2n+1)\pi x} \frac{\sin(t)}{t}\,ds\right| \leq \frac{M}{n}$ for all $x \in [0, \frac{1}{2}]$ and all $n \geq 1$. For $x \in (0, \frac{1}{2}]$ we infer that $\lim_{n\to\infty}(S_n\phi)(x) = -x + \frac{1}{\pi}\int_0^\infty \frac{\sin(t)}{t}\,dt = -x + \frac{1}{2} = \phi(x)$, see Exercise 3. To study the fluctutations of $(S_n\phi)(x)$ as $n \to \infty$ and $x \to 0$ simultaneously, note that $\sup_{0\leq x\leq \frac{1}{2}}\{[(S_n\phi)(x) + x]\} \leq \frac{1}{2}\sup_{s\geq 0}\{\mathfrak{s}(s)\} + \frac{M}{n} = \frac{\mathfrak{s}(\pi)}{2} + \frac{M}{n}$, while if $nx_n \to \frac{1}{2}$ as $n \to \infty$, then $\lim_{n\to\infty}(S_n\phi)(x_n) = \lim_{n\to\infty}\int_0^{(2n+1)\pi x_n} \frac{\sin(t)}{t}\,dt = \frac{\mathfrak{s}(\pi)}{2}$; here $\mathfrak{s}$ is the function defined in Exercise 3. Keeping in mind that each single function $S_n\phi$ is continuous, with $(S_n\phi)(0) = 0$ and $(S_n\phi)(x) = \sum_{k=1}^n \frac{\sin(2\pi k x)}{\pi k} \geq 0$ for all $x \in [0, \frac{1}{2n}]$, these considerations show that the set of accumulation points of the sequences $\{(S_n\phi)(x_n)\}_{n\geq 1}$, with $x_n \in (0, \frac{1}{2})$ converging to zero, is precisely $[0, \frac{\mathfrak{s}(\pi)}{2}]$. The number[33] $\frac{\mathfrak{s}(\pi)-1}{2} \approx 0.089$ is called the *overshoot* of the symmetric partial sums. Since ϕ is odd, as we approach $x = 0$ from the left, the set of accumulation points is $[-\frac{\mathfrak{s}(\pi)}{2}, 0]$, signalling another overshoot.

[32] The only contentious point is zero, where the issue is resolved by performing a power series expansion of both the numerator and the denominator in the expression $g(x) = \frac{\pi s - \sin(\pi s)}{\pi s\,\sin(\pi s)}$.

[33] The numerical value can be computed by expanding $\frac{\sin(t)}{t}$ in a Taylor series and integrating termwise – see Pinsky (2009). Note that many textbooks provide wrong values for this constant, see the list provided in Hewitt and Hewitt (1979).

We now show that for any function $f \in BV_{\text{per}}[0,1]$, as x approches a discontinuity point a, the discrepancy between $f(x)$ and $(S_n f)(x)$ manifests itself as a sequence of spikes that overshoot the lateral limits of f at a. The peculiar feature, termed *Gibbs' phenomenon*, is that, as n increases, the oscillations get compressed in width but not in height, with the magnitude of the overshoot about 8.9 % of the size of the jump, independent of the function f. This is illustrated in Figure 4.4 for the so-called "square wave" defined by $f(x) = -1$ for $x \in [-\frac{1}{2}, 0)$ and $f(x) = 1$ for $x \in [0, \frac{1}{2})$, and extended by periodicity to $\mathbb{R}$.

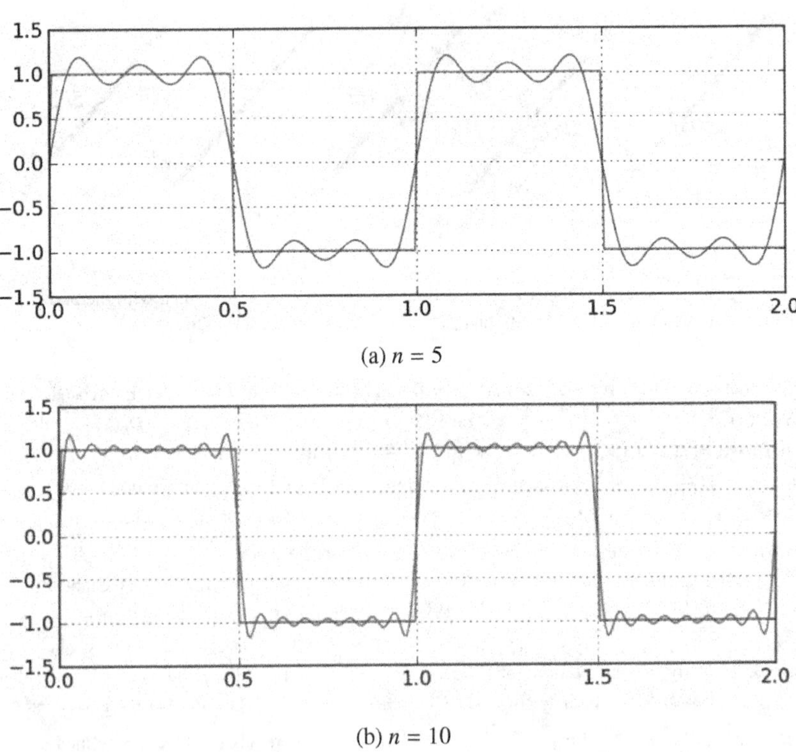

(a) $n = 5$

(b) $n = 10$

Figure 4.4 The graph of the square wave f as broken line segments, and of $S_n f$ as an oscillatory curve, with $n = 3/n = 8$ in the top/bottom figure, respectively (the vertical line segments are not part of the graph of f, but they convey the visual perception of a wave). The overshoot at a discontinuity of f does not die out as n increases. Rather, on each side of the discontinuity, it approaches the constant $g = \frac{s(\pi)-1}{2} \approx 0.089$ times the size of the jump.

To prove the above statement for a general $f \in BV_{\text{per}}[0,1]$, let us label the discontinuities[34] of f in $[-\frac{1}{2}, \frac{1}{2})$ by a_k with $k \geq 1$. We denote by $\delta_k = f(a_k+) - f(a_k-)$ the corresponding jump and by

34 There are finitely or at most countably many, see the discussion in Section 2.2.4.

$m_k = \frac{f(a_k+)+f(a_k-)}{2}$ the average of the lateral limits of f at a_k. Clearly $\sum_{k\geq 1} |\delta_k| \leq V_f[0,1]$. Since $\|T_k\|_{BV_{\mathrm{per}}[0,1]} = \|T_k\phi\|_{L^1[0,1]} + V_{T_k}[0,1] = 2$ for $T_k = T_{a_k}\phi$ with ϕ the reverse sawtooth function defined above, the series[35] $\sum_{k\geq 1} \delta_k\, T_k$ converges[36] in $BV_{\mathrm{per}}[0,1]$ to a function ψ. We claim that the discontinuities of ψ occur in $[-\frac{1}{2}, \frac{1}{2})$ precisely at the points a_k, with the corresponding jump being equal to δ_k. Indeed, given $\varepsilon > 0$, there is $N \geq 1$ such that $\sum_{k\geq N} |\delta_k| < \varepsilon$. Consequently $|\psi(x) - \sum_{k=1}^{n} \delta_k \phi(x-a_k)| \leq \varepsilon$ for all $n \geq N$ and all $x \in [-\frac{1}{2}, \frac{1}{2}]$, which yields $|\psi(a\pm) - \sum_{k=1}^{n} \delta_k \phi((a-a_k)\pm)| \leq \varepsilon$ for any $n \geq N$ and every $a \in [-\frac{1}{2}, \frac{1}{2})$. Since $\varepsilon > 0$ was arbitrary, we obtain that $\psi(a\pm) = \sum_{k\geq 1} \delta_k \psi(a\pm)$ for all $a \in [-\frac{1}{2}, \frac{1}{2})$, proving the claim. We now define the function $F = f - \psi$. Clearly $F \in BV_{\mathrm{per}}[0,1]$ but F is also continuous, as the jumps in f and ψ cancel out. In particular, $\lim_{x\to a_k} F(x) = m_k$. Theorem 4.8 ensures that $S_n F \to F$ as $n \to \infty$, uniformly on $\mathbb{R}$, while the functions $S_n T_k$ are uniformly bounded on $\mathbb{R}$ by some constant $M > 0$, for all $n, k \geq 1$. Given $\varepsilon > 0$, let $N \geq 1$ be so large that $\sum_{k\geq N+1} |\delta_k| \cdot \|S_n T_k\|_\infty < \varepsilon$ for all $n \geq 1$. Then, for all $n \geq 1$ and all $x \in [-\frac{1}{2}, \frac{1}{2}]$, we have that $\left|(S_n\psi)(x) - \sum_{k=1}^{N} \delta_k\, (S_n T_k)(x)\right| = \left|\sum_{k\geq N+1} \delta_k\, (S_n T_k)(x)\right| \leq \varepsilon$. This ensures that the accumulation points of the sequences $\{(S_n\psi)(x_n)\}_{n\geq 1}$, with $x_n \in [-1,1]$ converging to some $a \in [-\frac{1}{2}, \frac{1}{2}]$, are ε-close to those of the sequences $\{\sum_{k=1}^{N} \delta_k\, (S_n T_k)(x_n)\}_{n\geq 1}$. For $a \notin \{a_1, \cdots, a_N\}$ the latter type of sequences all converge to $\sum_{k=1}^{N} \delta_k\, (S_n T_k)(a)$. On the other hand, if $a = a_k$ for some $k \in \{1, \cdots, N\}$, the case study of ϕ shows that if $\delta_k > 0$, then the set of accumulation points is $[0, \delta_k \frac{s(\pi)}{2}]$ if $x_n \downarrow a$ and $[-\delta_k \frac{s(\pi)}{2}, 0]$ if $x_n \uparrow a$, while for $\delta_k < 0$ the two sets are $[\delta_k \frac{s(\pi)}{2}, 0]$ and $[0, -\delta_k \frac{s(\pi)}{2}]$, respectively. Since $\varepsilon > 0$ was arbitrary, this shows that the set of accumulation points of the partial sums $\{(S_n f)(x_n)\}_{n\geq 1}$ is $f(a)$ if $x_n \to a \neq a_k$, while if $\delta_k > 0$, the set is $[m_k, m_k + \delta_k \frac{s(\pi)}{2}]$ if $x_n \downarrow a_k$ and $[m_k - \delta_k \frac{s(\pi)}{2}, m_k]$ if $x_n \uparrow a_k$; for $\delta_k < 0$ the two sets are $[m_k + \delta_k \frac{s(\pi)}{2}, m_k]$ and $[m_k, m_k - \delta_k \frac{s(\pi)}{2}]$, respectively. At the discontinuity point a_k, the lateral limits of f are $f(a_k\pm)$. Since $m_k + \delta_k \frac{s(\pi)}{2} - f(a_k+) = \delta_k(\frac{s(\pi)}{2} - 1) = f(a_k-) - [m - \delta_k \frac{s(\pi)}{2}]$, while $[m_k - \delta_k \frac{s(\pi)}{2}] - f(a_k-) = f(a_k+) - [m_k + \delta_k \frac{s(\pi)}{2}] = \delta_k\, (1 - \frac{s(\pi)}{2})$, the size of the overshoot on either side of a_k stabilises in the limit $n \to \infty$ to $|\delta_k|\, (\frac{s(\pi)}{2} - 1)$. This proves the general validity of Gibbs' phenomenon.

The discovery of the Gibbs phenomenon has a quite interesting history.

In 1898 the experimentalist Michelson constructed a harmonic analyser (one of the first mechanical analogue computers) by which up to the first 80 symmetric partial sums $S_n f$ of a given continuous periodic function f could be graphed with extraordinary precision. But when he looked at the discontinuous sawtooth function[37] $f(x) = x$ for $x \in [-\pi, \pi)$, extended to be 2π-periodic on $\mathbb{R}$, the graphs produced by the machine presented a peculiar protuberance near the points of discontinuity of f. Michelson tried to fine tune his machine as he associated the presence of these wiggles to mechanical defect, but his efforts were fruitless. In a letter to *Nature* he then complained about this undesired overshoot effect that seemed to be contrary to what he believed the textbooks were claiming – see Michelson (1898). Shortly thereafter, Gibbs pointed out that, due to the lack of uniform convergence, there is a difference between the limit of the graphs and the graph of the limit, giving no hint of a proof but calling attention to the fact that on the interval $(-\pi, \pi)$, for large n, the graphs of the curves $y = (S_n f)(x) = 2\sum_{k=1}^{n}(-1)^{k-1}\frac{\sin(kx)}{k}$ fall from the point $(\pi, 0)$ at a steep gradient to a point very nearly at distance $2\int_0^\pi \frac{\sin(t)}{t}$ below the x-axis, then oscillate above and below $y = x$ close to this line until x approaches π, when they rise at a distance $2\int_0^\pi \frac{\sin(t)}{t}$ above the x-axis, and then fall rapidly to $(\pi, 0)$; see Gibbs (1898). Gibbs' paper at first attracted no attention since it was devoted to a particular series. Bôcher (1906) subsequently showed that the same phenomenon occured more generally at a discontinuity of any periodic function f such that f and its derivative f' are both continuous except for a finite number of jump discontinuities in

35 Or sum, if only finitely many discontinuities are present in a periodicity window.

36 Endowed with the norm $\|f\|_{BV} = \|f\|_{L^1[0,1]} + V_f[0,1]$, $BV_{\mathrm{per}}[0,1]$ is a Banach space; see Exercise 3.33.

37 Note that $f(x) = -2\pi\,\phi(\frac{x}{2\pi} + \frac{1}{2})$, where ϕ is the reverse sawtooth function, analysed in detail before. We preferred ϕ to f to locate the discontinuity at zero.

a period window. He called this property Gibbs' phenomenon as he believed Gibbs to be the first person noticing it. However, it is now known that Wilbraham (1848) presented diagrams showing this behaviour for the series[38] $\sum_{k\geq 1}(-1)^{k-1}\frac{\cos((2k-1)x)}{2k-1}$ for $x \in (-\pi, \pi)$. He also remarked that "a similar investigation of the equation $y = \sin(x) - \frac{\sin(2x)}{2} + \frac{\sin(3x)}{3} - \cdots$ would lead to an analogous result". This second series is, up to a multiplicative factor, the same as that which Gibbs used. As a further twist to this story, du Bois-Reymond (1874) investigated the behaviour of Fourier series near points of discontinuity of the function and came across the issue of estimating the values of the integral $\int_0^{nx} \frac{\sin(t)}{t}\,dt$ when $n \to \infty$ and $x \to 0$ simultaneously. Knowing that $\int_0^{\pm\infty} \frac{\sin(t)}{t}\,dt = \pm\frac{\pi}{2}$, he jumped to the conclusion that at a point of discontinuity a the limit points of the Fourier partial sums represent all the values between $f(a-)$ and $f(a+)$, instead of extending beyond this interval by an amount proportional to the jump of f at a. These considerations represent a good illustration of the importance of mathematical rigour in the attempt to unravel a counter-intuitive phenomenon.

4.5 Exercises

Problems 1-3 are related to the material in Section 4.1, Problems 4–22, 24–27 and 33 to that in Section 4.2, Problems 23 and 28 to that in Section 4.3, and Problems 29–32 to that in Section 4.4. Problems 1, 2, 4, 5, 10, 11, 16, 19, 20(i), 22(i), 28, 31, 32(ii), are for routine training, while the other problems present an increased level of difficulty, many being quite advanced (e.g., aspects of Problems 21 and 22 require intricate considerations). Some problems represent amplifications of presented results, while others offer examples/counterexamples. Working them all out is possibly too arduous but we recommend at least a careful examination of the statements.

4.5.1 Statements

1. Let $f \in L^1[0,1]$ be given by $f(x) = \frac{1}{2} - x$ for $x \in [0,1]$. Show that $\hat{f}(0) = 0$ and $\hat{f}(k) = \frac{1}{2\pi i k}$ for $k \neq 0$.

2. Show that $(S_n f)(x) = \frac{4}{\pi}\sum_{\{k:\,1\leq 2k-1\leq n\}} \frac{\sin((2k-1)2\pi x)}{2k-1}$ for $f \in L^1[0,1]$ defined by $f(x) = 1$ for $0 \leq x < 1/2$ and $f(x) = -1$ for $1/2 \leq x < 1$.

3. Show that the function $\mathfrak{s}(x) = \frac{2}{\pi}\int_0^x \frac{\sin(t)}{t}\,dt$, defined for $x \geq 0$, is such that $\mathfrak{s}(0) = 0$, $\lim_{x\to\infty}\mathfrak{s}(x) = 1$ and $0 \leq \mathfrak{s}(x) \leq \mathfrak{s}(\pi)$ for all $x \geq 0$. (Numerically $\mathfrak{s}(\pi) \approx 1.18$.)

4. If $f \in L^1[0,1]$, show that $\lim_{n\to\infty}\left(\frac{1}{n}\sum_{k=0}^{n-1} f(x+\frac{k}{n})\right) = \int_0^1 f(s)\,ds$ in $L^1[0,1]$.

5. For $f \in C_{\text{per}}[0,1]$ continuously differentiable define the *averaging operator* depending on the parameter $h > 0$ by $(\mathcal{A}_h f)(x) = \frac{1}{2h}\int_{x-h}^{x+h} f(t)\,dt$ for $x \in \mathbb{R}$. Thus $\mathcal{A}_h f$ is a new function F_h. Show that F_h is periodic of period 1, as a function of x, and find the Fourier series of F_h in terms of the Fourier series of f.

6. Prove that for every integer $n \geq 1$, the formula $f_n(x) = \sum_{k=1}^{n} \frac{\sin(2\pi k x)}{k}$ defines a continuous function $f_n : \mathbb{R} \to \mathbb{R}$ of period 1, with $|f_n(x)| \leq 2\pi$ for all $x \in \mathbb{R}$.

[38] Note that if w is the 2π-periodic function whose restriction to $[-\pi, \pi)$ is given by $w(x) = \pi/4$ for $-\pi/2 \leq x < \pi/2$ and $w(x) = -\pi/4$ for $x \in [-\pi, -\pi/2) \cup [\pi/2, \pi)$, then for $-1/2 \leq x < 1/2$ we have that $\frac{4}{\pi}\,w(2\pi x - \frac{\pi}{2}) = f(x)$ is the square wave f depicted in Figure 4.4. Therefore Exercise 1 shows that the Fourier series of w is $\sum_{k\geq 1}(-1)^{k-1}\frac{\cos((2k-1)x)}{2k-1}$.

7. Let $Q_n(x) = \sum_{k=1}^{2^{n^2}} \frac{\cos[2\pi(2^{n^2+1}-k)x] - \cos[2\pi(2^{n^2+1}+k)x]}{k}$ for $x \in \mathbb{R}$ and the integer $n \geq 1$. Show that $f(x) = \sum_{n=1}^{\infty} \frac{Q_n(x)}{n^2}$ defines a continuous function of period 1, whose Fourier series diverges at $x = 0$.

8. Given $\alpha \in (0, 1)$, show that $f(x) = |\sin(2\pi x)|^{\alpha}$ belongs to Hölder space h_{α} but $f \notin h_{\beta}$ for $\beta \in (\alpha, 1]$.

9. For $n \geq 2$, let D_n be the Dirichlet kernel. Show that:

(i) $|D_n(x)| \leq 3 \min\{n, \frac{1}{|x|}\}$ for all $|x| \leq \frac{1}{2}$;

(ii) $\frac{4}{\pi^2} \ln(n) \leq \int_{-1/2}^{1/2} |D_n(x)|\, dx \leq 1 + \frac{2}{\pi} \int_0^{\pi} \frac{\sin(s)}{s}\, ds + \frac{4}{\pi} \ln(n)$.

10. Show that the symmetric partial sums of the Fourier series of $f \in L^1[0, 1]$ and their arithmetical means can be expressed in convolution form $S_n f = f * D_n$ and $\sigma_n f = f * F_n$, respectively, for all $n \geq 1$.

11. (i) Show that a continuous function $f : \mathbb{R} \to \mathbb{C}$, periodic with period 1, is even[39] if and only if its Fourier coefficients $\{\hat{f}(k)\}_{k\in\mathbb{Z}}$, given by (4.1), are such that $\hat{f}(k) = \hat{f}(-k)$ for all $k \in \mathbb{Z}$.

(i) Characterise odd[40] continuous functions $f : \mathbb{R} \to \mathbb{C}$, periodic with period 1, by means of a necessary and sufficient condition on their Fourier coefficients.

12. (i) If $f : \mathbb{R} \to \mathbb{C}$ is continuous and periodic with period 1, show the validity of Poisson's formula

$$\sum_{k\in\mathbb{Z}} \hat{f}(k)\, r^{|k|}\, e_k(x) = \int_0^1 \frac{1 - r^2}{1 - 2r\cos(2\pi(x - y)) + r^2}\, f(y)\, dy$$

for every $x \in [0, 1]$ and every $r \in [0, 1)$. Here $e_k(x) = e^{2\pi i k x}$ for $k \in \mathbb{Z}$ and $\{\hat{f}(k)\}_{k\in\mathbb{Z}}$ are the Fourier coefficients of f, given by (4.1).

(ii) Show that

$$\lim_{r\uparrow 1} \{ \sup_{x\in[0,1]} \big|f(x) - \sum_{k\in\mathbb{Z}} \hat{f}(k)\, r^{|k|}\, e_k(x)\big| \} = 0$$

whenever $f : \mathbb{R} \to \mathbb{C}$ is continuous and periodic with period 1. (This is an alternative to Fejér's approach – Theorem 4.4 – for the recovery of f from its Fourier coefficients.)

13. Show that a sequence $\{f_n\}_{n\geq 1}$ that converges in $L^1[0, 1]$ towards a limit $f \in L^1[0, 1]$ is not necessarily convergent a.e. on $[0, 1]$ but has a subsequence $\{f_{n_k}\}$ that converges a.e. on $[0, 1]$. On the other hand, exhibit a sequence $\{f_n\}_{n\geq 1}$ in $L^1[0, 1]$ such that $f_n \to 0$ a.e. on $[0, 1]$ but no subsequence $\{f_{n_k}\}$ converges in $L^1[0, 1]$.

14. Prove that $L^1[0, 1]$ and $L^2[0, 1]$ are separable Banach spaces.

15. Show that there is some $f \in L^1[0, 1]$ with $\sup_{n\geq 1} \{\|S_n f\|_{L^1[0,1]}\} = \infty$.

16. Prove that the trigonometric polynomials are dense in $C_{\text{per}}[0, 1]$ and in $L^1[0, 1]$.

17. (i) If $f \in C_{\text{per}}[0, 1]$ is such that $\sum_{k\in\mathbb{Z}} |k\, \hat{f}(k)| < \infty$, show that there is some constant $c > 0$ such that $\|\sigma_n f - f\|_{\infty} \leq \frac{c}{n}$ for all $n \geq 1$.

(ii) Show that $\lim_{n\to\infty} \{n\, \|f - \sigma_n f\|_{\infty}\} = 0$ forces $f \in C_{\text{per}}[0, 1]$ to be constant.

[39] That is, $f(x) = f(-x)$ for all $x \in \mathbb{R}$.

[40] In the sense that $f(x) = -f(-x)$ for all $x \in \mathbb{R}$.

18. If $f \in C_{\text{per}}[0,1]$ is such that $\hat{f}(k) \geq 0$ for all $k \in \mathbb{Z}$, show that $\sum_{k\in\mathbb{Z}} \hat{f}(k) < \infty$.

19. Show that a trigonometric series $\sum_{k\in\mathbb{Z}} c_k\, e^{2\pi ikx}$ is the Fourier series of some $f \in L^1[0,1]$ if and only if its arithmetical means $\sigma_n = \frac{S_0+\cdots+S_{n-1}}{n}$ converge in $L^1[0,1]$ to some limit; here $S_n(x) = \sum_{|k|\leq n} c_k\, e^{2\pi ikx}$. (This characterisation of Fourier series is of limited practical interest – in specific cases it is advisable to decide on other grounds; see e.g. Exercise 21.)

20. For $n \geq 1$, $\tilde{D}_n(x) = \sum_{k=1}^n \sin(2\pi kx)$ is the so-called *conjugate Dirichlet kernel.* Clearly $\tilde{D}_n$ is smooth, odd and periodic of period 1.

(i) Show that $\tilde{D}_n(x) = \frac{\cos(\pi x)-\cos((2n+1)\pi x)}{2\sin(\pi x)}$ for $x \in (0, \frac{1}{2})$.

(ii) Prove the estimate $|\tilde{D}_n(x)| \leq \frac{1}{2x}$ for all $x \in (0, \frac{1}{2})$ and $n \geq 1$.

(iii) Show that $\left|\sum_{k=1}^n \frac{\sin(k\pi y)}{k\pi}\right| < 3$ for all $y \in [0,1]$ and all $n \geq 1$.

21. (i) Show that $\sum_{n=2}^\infty \frac{\sin(2\pi nx)}{\ln(n)}$ converges to a finite limit at every point without being the Fourier series of some function in $L^1[0,1]$.

(ii) Show that the series $\sum_{n=2}^\infty \frac{\sin(2\pi nx)}{n\ln(n)}$ converges uniformly to some $f \in C_{\text{per}}[0,1]$.

(iii) Show that $\sum_{n=2}^\infty \frac{\cos(2\pi nx)}{\ln(n)}$ converges at every x with $0 < |x| \leq \frac{1}{2}$ to some $F(x)$. Moreover, $F \in L^1[0,1]$ and $\sum_{n=2}^\infty \frac{\cos(2\pi nx)}{\ln(n)}$ is the Fourier series of F, with $2\pi \int_0^x F(s)\,ds = f(x)$ for $x \in [0,1]$. In particular, f is absolutely continuous.

22. (i) Let $f \in L^1[0,1]$. Prove that $\sum_{k\in\mathbb{Z}} |\hat{f}(k)| < \infty$ ensures that, after correction on a set of measure zero, $f \in C_{\text{per}}[0,1]$. Moreover, $S_n f \to f$ uniformly on $[0,1]$.

(ii) Show that the family $\mathcal{A}$ of all $f \in C_{\text{per}}[0,1]$ with $\hat{f} \in l^1(\mathbb{Z})$, endowed with the norm $\|f\|_{\mathcal{A}} = \sum_{k\in\mathbb{Z}} |\hat{f}(k)|$, is a unital commutative Banach algebra under pointwise multiplication.

(iii) Show that there exists $f \in \mathcal{A}$ that is not absolutely continuous, and that there are functions in $C_{\text{per}}[0,1] \setminus \mathcal{A}$ that are absolutely continuous on $[0,1]$. (This should be contrasted to the elegance of $L^2[0,1]$-theory; to appreciate the leverage, see also Exercise 23.)

(iv) If $\{c_k\}_{k\in\mathbb{Z}}$ is a numerical sequence such that $\sum_{k\in\mathbb{Z}} |c_k\, \hat{g}(k)| < \infty$ for each $g \in L^1[0,1]$, prove that $\sum_{k\in\mathbb{Z}} |c_k| < \infty$. Use this to show that $f \in L^1[0,1]$ belongs to $\mathcal{A}$ if and only if $\sum_{k\in\mathbb{Z}} |\hat{f}(k)\,\hat{g}(k)| < \infty$ for each $g \in L^1[0,1]$. (The condition should not be confused with demanding mere convergence of $\sum_{k\in\mathbb{Z}} c_k\hat{g}(k)$ for every $g \in L^1[0,1]$; see Exercise 30.)

23. Let $f \in L^2[0,1]$. Show that, after correction on a set of measure zero, f is absolutely continuous with $f' \in L^2[0,1]$ if and only if $\sum_{k\in\mathbb{Z}} k^2\, |\hat{f}(k)|^2 < \infty$.

24. Show that there exists $f \in L^1[0,1]$ with $\sup_{n\geq 1}\{\|S_n f\|_{L^1[0,1]}\} = \infty$. (Kolmogorov (1926) constructed a function $f \in L^1[0,1]$ for which $\sup_{n\geq 1}\{|(S_n f)(x)|\} = \infty$ at every $x \in [0,1]$. This divergence is generic: there exists a dense G_δ-set G in $L^1[0,1]$ such that $\sup_{n\to\infty}\{|(S_n f)(x)|\} = \infty$ for all $x \in [0,1]$ whenever $f \in G$, see Kahane (2000). In contrast to the abstract result, providing an explicit example is an elaborate undertaking; see e.g. Pinsky (2009).)

25. Show that $\{e^{2\pi ikx}\}_{k\in\mathbb{Z}}$ is not a Schauder basis for any one of the Banach spaces $L^1[0,1]$ and $C_{\text{per}}[0,1]$. (In contrast to this, note that this is an orthonormal basis – and therefore a Schauder basis – for the Hilbert space $L^2[0,1]$.)

26. Show that there is a dense G_δ-set G in the Banach space $C_{\text{per}}[0,1] \times \mathbb{R}$ such that $\limsup_{n\to\infty}\{|(S_n f)(x)|\} = \infty$ for all $(f,x) \in G$. (This shows that for continuous periodic functions the divergence of the Fourier series is generic.)

27. (i) Show that if $f \in h_\alpha$ for some $\alpha \in (0,1)$, then $\sup_{k\in\mathbb{Z}}\{|k|^\alpha |\hat{f}(k)|\} < \infty$. Prove that this result is sharp by showing that[41] $f_\alpha(x) = \sum_{k=0}^\infty 2^{-k\alpha} e^{2\pi i 2^k x}$ belongs to h_α, and $\widehat{(f_\alpha)}(n) = n^{-\alpha}$ whenever $n = 2^k$. For $\alpha \in (N, N+1)$ for some integer $N \geq 1$, define the space[42] h_α as the family of all functions $f : \mathbb{R} \to \mathbb{C}$ that are periodic with period 1, and N-times differentiable with $f^{(N)} \in h_{\alpha-N}$. Is $f_\alpha \in h_\alpha$ for $\alpha \in [1, \frac{3}{2})$?

(ii) Show that $f \in L^1[0,1]$ is a.e. equal to an infinitely differentiable periodic function if and only if its Fourier coefficients are rapidly decreasing, that is, for every integer $n \geq 1$ there is a constant $c_n > 0$ such that $|\hat{f}(k)| \leq \frac{c_n}{|k|^n}$ for all $k \neq 0$. (These results quantify the speed of convergence to zero in (P5) for some classes of functions. However, results of this type are somewhat limited. In particular, a practical characterisation of differentiability of a fixed degree in terms of the behaviour of the Fourier coefficients remains elusive.)

28. In contrast to the $L^2[0,1]$-setting, by considering $S_1 f$ for $f \in C_{\text{per}}[0,1]$ given by $f(x) = x(1-x)$ for $x \in [0,1]$, show that in $C_{\text{per}}[0,1]$ the nth symmetric partial sum of the Fourier series is not the best approximation of f by a trigonometric polynomial of degree n.

29. (i) (Dini's criterion) Show that if $f \in L^1[0,1]$ satisfies a Dini condition at $x_0 \in [0,1]$, meaning that there exists some $\theta \in \mathbb{C}$ and some $\delta > 0$ such that $\int_0^\delta \frac{|f(x_0+s)+f(x_0-s)-2\theta|}{s}\,ds < \infty$, then $\lim_{n\to\infty}(S_n f)(x_0) = \theta$.

(ii) By means of the examples $f_1, f_2 : [-\frac{1}{2}, \frac{1}{2}] \to \mathbb{R}$ given by $f_1(0) = f_2(0) = 0$ and $f_1(x) = x\sin(\frac{\pi}{x})$, $f_2(x) = \frac{1}{-\ln|x|}$ for $0 < |x| \leq \frac{1}{2}$, show that Dini's condition at $x_0 = 0$ and Theorem 4.8 are not comparable.

30. If $f \in L^1[0,1]$, show that $\sum_{k\neq 0} \frac{\hat{f}(k)}{k}$ is convergent for all $x \in [0,1]$.

31. For the "reverse sawtooth" function $\phi(x) = \frac{1}{2} - x + [x]$, prove the following estimate of the arc length of the curves $\{(x, (S_n\phi)(x)) : -\frac{1}{2} \leq x \leq \frac{1}{2}\}$: there is a constant $C > 1$ with $\frac{\ln(n)}{C} \leq \int_{-1/2}^{1/2} \sqrt{1 + [(S_n\phi)'(x)]^2} \leq C\ln(n)$ for all $n \geq 2$. (This behaviour is of interest in the context of Gibbs' phenomenon.)

32. (i) Show that $2\pi|k\hat{f}(k)| \leq V_f[0,1]$ for $f \in BV_{\text{per}}[0,1]$.

(ii) If $f \in BV_{\text{per}}[0,1]$ and $g \in L^1[0,1]$, show that $\sum_{k\in\mathbb{Z}} \hat{f}(-k)\,\hat{g}(k) = \int_0^1 f(x)g(x)\,dx$.

33. Let $f \in L^1[0,1]$. Show that $\hat{f}(k) \geq 0$ for all $k \in \mathbb{Z}$ if and only if f is positive-definite, meaning that $\int_0^1\int_0^1 f(x-y)\,\overline{u(x)}\,u(y)\,dxdy \geq 0$ for each $u \in C_{\text{per}}[0,1]$.

[41] Trigonometric sums of order 2^k are known as *dyadic sums*. They are often used because of the following identities that pertain to finite sums and tail sums: if $N \geq 1$ is an integer and $\alpha, \beta > 0$, then $\sum_{k=N}^\infty 2^{-k\alpha} = \frac{2^{-N\alpha}}{1-2^{-\alpha}}$ and $\sum_{k=0}^{N-1} 2^{k\beta} = \frac{2^{N\beta}-1}{2^\beta - 1}$.

[42] Note that for $\alpha > 1$ the inequality $|f(x+h) - f(x)| \leq M|h|^\alpha$ is not viable since it forces f to be constant by means of $f'(x) = 0$. By combining the considerations (see Chapter 3) for $C_{\text{per}}[0,1]$ with those for h_β when $\beta \in (0,1)$, it is not difficult to show that h_α is a Banach space with respect to the norm $\|f\|_\alpha = \sum_{k=0}^N \sup_{x\in[0,1]}\{|f^{(k)}(x)|\} + [f^{(N)}]_\beta$, where $\beta = \alpha - N$.

4.5.2 Hints

1. For $k \neq 0$, use (4.1) and integration by parts.

2. Use (4.1) and take into account the cancellations in $(S_n f)(x)$.

3. Investigate $\mathfrak{s}((n+1)\pi) - \mathfrak{s}(n\pi)$ to prove that the improper Riemann integral exists. Use (P5) to show that $\lim_{n\to\infty} \mathfrak{s}(n\pi + \frac{\pi}{2}) \to 1$ as $n \to \infty$.

4. Compute the Fourier coefficients of the sum.

5. Integrate the Fourier series of f term by term.

6. Note that $|f_n'(x)| \leq 2\pi n$ for all $x \in \mathbb{R}$. By passing to complex exponentials, show that $f_n'(x) = \pi \frac{\sin(\pi(2n+1)x)}{\sin(\pi x)} - \pi$ for $0 < |x| \leq 1/2$. If $x_k = k/(2n+1)$ for $k = 0, \cdots, n$, show that as x increases from 0 towards $\frac{1}{2}$, the function $g_n(x) = \pi \int_0^x \frac{\sin(\pi(2n+1)s)}{\sin(\pi s)}\, ds$ starts from $g_n(0) = 0$ by increasing on $[0, x_1]$, then it decreases on $[x_1, x_2]$, and subsequently it alternates between intervals of increase and decrease delimited by the points x_k. Moreover, the minimum of $g_n(x)$ on an interval of decrease exceeds the minimum of $g_n(x)$ on the previous interval (of increase), and the maximum of $g_n(x)$ on an interval of increase that lies to the right of x_2 does not exceed the maximum of $g_n(x)$ on the previous interval (of decrease). Infer that $x = x_1$ is the location of the maximum of $g_n(x)$ on $[0, \frac{1}{2}]$.

7. Since $Q_n(x) = 2\sin(2^{n^2+2}\pi x) \sum_{k=1}^{2^{n^2}} \frac{\sin(2\pi k x)}{k}$ for $x \in \mathbb{R}$, the previous exercise yields the uniform convergence of the series defining f. Show that $\sum_{k=2^{n^2}}^{2^{n^2+1}} \hat{f}(k) \not\to 0$.

8. Use the inequality $(a+b)^\alpha \leq a^\alpha + b^\alpha$ for $a, b > 0$.

9. Take advantage of the inequalities $\tan(\pi x) \geq \pi x \geq \sin(\pi x)$ for $0 \leq x \leq \frac{1}{2}$, and $\sin(\pi x) \geq \frac{\pi x}{2}$ for $0 \leq x \leq \frac{1}{3}$. The trigonometric identity $\frac{\sin((2n+1)\pi x)}{\sin(\pi x)} = \cos(2\pi n x) + \frac{\sin(2\pi n x)}{\tan(\pi x)}$ for $0 < x < \frac{1}{2}$ is also useful.

10. Rely on the definition of D_n and F_n.

11. Use Fejér's result (Theorem 4.4).

12. For (i), express $\hat{f}(k)$ in $\sum_{k\in\mathbb{Z}} \hat{f}(k)\, r^{|k|}\, e_k(x)$ as an integral. For (ii), show that the function $Q(r,x) = \frac{1-r^2}{1-2r\cos(2\pi x)+r^2}$ satisfies properties analogous to a continuous version, dependent on the parameter $r \in [0,1)$ rather than on the integer $n \geq 1$, of the key properties (A1)–(A3) of the Fejér kernel F_n.

13. Note that for $f_n(x) = \chi^{[\frac{j}{2^k}, \frac{j+1}{2^k}]}(x)$, where $n = 2^k + j$ with $0 \leq j < 2^k$, we have $f_n \to 0$ in $L^1[0,1]$, but $\{f_n(x)\}_{n\geq 1}$ does not converge for any $x \in [0,1]$. For the second part, pick integers $n_1 < n_2 < \cdots$ such that $\|f_{n_k} - f\|_{L^1[0,1]} < \frac{1}{2^k}$ for $k \geq 1$ and show that $f_{n_k}(x) \to f(x)$ for a.e. $x \in \mathbb{R}$. For the last part, consider $f_n = n\chi^{[0,\frac{1}{n}]}$ for $n \geq 1$.

14. For completeness, if $\{f_n\}_{n\geq 1}$ is a Cauchy sequence in, say, $L^2[0,1]$, produce first a plausible candidate for the limit function f by selecting a sequence $n_1 < n_2 < \cdots$ such that $\|f_{n_k} - f_n\|_{L^2[0,1]} < \frac{1}{2^k}$ for all $n \geq n_k$, and showing that the sequence $\{f_{n_k}\}$ converges a.e. on $[0,1]$ to some measurable function $f : [0,1] \to \mathbb{C}$. Then show that $f \in L^2[0,1]$ and $f_n \to f$ in $L^2[0,1]$.

For the separability of, say, $L^2[0,1]$, it suffices to show that real-valued functions $f \in L^2[0,1]$ can be well-approximated in $L^2[0,1]$ by step functions taking on rational values on a finite number of disjoint intervals with rational endpoints.

15. Show that the operator $S_n : f \mapsto S_n f$, which maps $L^1[0,1]$ to $L^1[0,1]$, has norm $\|D_n\|_{L^1[0,1]}$ and use the uniform boundedness principle. Recall (4.13).

16. Inspect the statements of Theorem 4.4 and Theorem 4.5.

17. For (i), if $M = \sum_{|k|\geq 1} |k\hat{f}(k)|$, using (4.19), show that $\|S_n f - \sigma_n f\|_\infty \leq \frac{M}{n}$ and $\|S_n f - f\|_\infty \leq \frac{M}{n}$ for all $n \geq 1$. For (ii), if $n \geq 1$, using (4.19), show that $|k\hat{f}(k)| \leq n\|f - \sigma_n f\|_\infty$ if $|k| \leq n-1$.

18. Use (4.19) and Theorem 4.4.

19. For the nontrivial implication, rely on (P2) and (4.19).

20. For (i), adapt the approach used in (4.10). For (ii), use the inequalities $\frac{2}{\pi s} \leq \sin(s) \leq s$ for $s \in [0, \frac{\pi}{2}]$. As for (iii), rely on introducing the sequence of differences $\{a_k - a_{k+1}\}_{k\geq 1}$ to estimate partial numerical sums of the type $\sum_{k=m}^n a_k b_k$ by means of the identity $\sum_{k=m}^n a_k b_k = a_n \sum_{k=m}^n b_k + \sum_{k=m}^{n-1} \{(a_k - a_{k+1}) \sum_{j=m}^k b_j\}$, valid for $n > m$.

21. To prove the convergence in (i), use the identity from the hint to Exercise 20(iii). If the trigonometric series were the Fourier series of some $g \in L^1[0,1]$, then $\hat{g}(k) = \frac{k}{2i|k|\ln|k|}$ for $|k| \geq 2$ would yield $\sum_{k\in\mathbb{Z}} \left|\frac{\hat{g}(2k+1)}{2k+1}\right| = \infty$. The convergence of the trigonometric series in (ii) also relies on the identity used in Exercise 20(iii). For (iii), apply twice the identity stated in the hint to Exercise 20(iii) to obtain an alternative form for $\sum_{k=2}^n \frac{\cos(2\pi kx)}{\ln(k)}$. To show that the trigonometric series is the Fourier series of F, look at the anti-derivatives of the symmetric partial sums.

22. For (i), show that $f(x) = \sum_{k\in\mathbb{Z}} \hat{f}(k)\, e^{2\pi ikx}$ a.e. As for (ii), use item (i) and the fact that $l^1(\mathbb{Z})$ is a Banach space. In (iii), try a lacunary series and recall Exercise 21(iii). For (iv), rely on Baire's theorem from Chapter 3.

23. For the direction that is not obvious, show that if $g \in L^2[0,1]$ is such that $\hat{g}(k) = 2\pi ik\,\hat{f}(k)$ for all $k \in \mathbb{Z}$, then $f(x) = f(0) + \int_0^x g(y)\,dy$ a.e. on $[0,1]$.

24. Take advantage of the uniform boundedness principle.

25. Assuming the contrary, determine the coefficients c_k in the associated representation $f(x) = \sum_{k\in\mathbb{Z}} c_k\, e^{2\pi ikx}$.

26. For integers $M, N \geq 1$, consider the open set $G(M,N) = \bigcup_{n\geq N} G_0(M,n)$, where $(S_n^* f)(x) = \int_{-1/2}^{1/2} f(x-s)\, \frac{\sin(2\pi ns)}{\pi s}\, ds$ defines $G_0(M,n) = \{(f,x) \in C_{\text{per}}[0,1] \times \mathbb{R} : \ (S_n^* f)(x)| > M\}$. Show that $G = \bigcap_{M,N\geq 1} G(M,N)$ is dense in $C_{\text{per}}[0,1] \times \mathbb{R}$.

27. For (i), show that $\hat{f}(k) = \frac{1}{2}\int_0^1 [f(x) - f(x+\frac{1}{k})]\, e^{-2\pi ikx} dx$. To estimate $|f_\alpha(x+h) - f_\alpha(x)|$ for $\alpha \in (0,1]$, take advantage of the fact that dyadic sums are involved, while for $\alpha \in (1, \frac{3}{2})$ rely on Exercise 23. In (ii) use (P5) and Exercise 23.

28. Compare $S_1 f$ with the approximation provided by the constant function $\frac{1}{8}$.

29. For (i) express $(S_n f)(x_0) - \theta$ as as integral and use (P5). For (ii), note that f_1 is not of bounded variation (see Chapter 2).

30. Investigate the function $g(x) = \int_0^x f(s)\,ds - \hat{f}(0)\,x$.

31. Use the fact that $(S_n\phi)'(x) = \pi[D_n(x) - 1]$.

32. In (i), rely on Riemann sums to prove the inequality for continuous functions, then argue by approximation. For (ii), use Theorem 4.8 and dominated convergence.

33. Show that $f \in L^1[0,1]$ is positive definite if and only if the defining inequality holds for trigonometric polynomials.

4.5.3 Solutions

1. Clearly $\hat{f}(0) = 0$. For integers $k \neq 0$, $\hat{f}(k) = \int_0^1 (\frac{1}{2} - x)\, e^{-2\pi ikx} dx = -\int_0^1 x\, e^{-2\pi ikx} dx = \frac{x}{2\pi ik}\, e^{-2\pi ikx}\Big|_{x=0}^{x=1} + \frac{1}{2\pi ik}\int_0^1 e^{-2\pi ikx} dx = \frac{1}{2\pi ik}$.

2. Since $\hat{f}(0) = 0$ and $\hat{f}(k) = -\int_{-1/2}^0 e^{-2\pi ikx} dx + \int_0^{1/2} e^{-2\pi ikx} dx = \frac{e^{-2\pi ikx}}{2\pi ik}\Big|_{x=-1/2}^{x=0} - \frac{e^{-2\pi ikx}}{2\pi ik}\Big|_{x=0}^{x=1/2} = \frac{1-(-1)^k}{\pi ik}$ for $k \neq 0$, $(S_n f)(x) = \sum_{1\leq 2k-1\leq n} \frac{2[e^{2\pi i(2k-1)x} - e^{-2\pi i(2k-1)x}]}{(2k-1)\pi i} = \frac{4}{\pi}\sum_{1\leq 2k-1\leq n} \frac{\sin(2\pi(2k-1)x)}{2k-1}$.

3. From the sign of the integrand and the monotonicity of its denumerator we see that the numbers $\mathfrak{s}((n+1)\pi) - \mathfrak{s}(n\pi) = \frac{2}{\pi}\int_{n\pi}^{(n+1)\pi} \frac{\sin(t)}{t}\,dt$ alternate in sign and decrease to zero in absolute value as n increases. This proves the existence of the improper Riemann integral $\lim_{x\to\infty}\mathfrak{s}(x)$ and identifies the location of the local extrema of $\mathfrak{s}(x) \geq 0$: local maxima/minima at the points $(2k-1)\pi$ and $2k\pi$ respectively ($k \geq 1$ integer), with the global maximum at $x = \pi$. Now, the continuity of $g(s) = \frac{1}{\sin(\pi s)} - \frac{1}{\pi s}$ on $[-\frac{1}{2}, \frac{1}{2}]$ yields by (P5) that $\lim_{n\to\infty}\int_{-1/2}^{1/2} g(s)\sin((2n+1)\pi s)\,ds = 0$. Using (4.10) and the fact that $\int_{-1/2}^{1/2} D_n(s)\,ds = 1$, we get $\lim_{n\to\infty}\int_{-1/2}^{1/2}\frac{\sin((2n+1)\pi s)}{\pi s}\,ds = 1$. The change of variables $t = (2n+1)\pi s$ and the evenness of the integrand transforms this into $\lim_{n\to\infty}\mathfrak{s}(n\pi+\pi/2) = 1$. This suffices to conclude.

4. Let $f_n(x) = \frac{1}{n}\sum_{k=0}^{n-1} f(x+\frac{k}{n})$. Then, unless $j < n$, (P3) yields $\hat{f}_n(j) = \frac{1}{n}\sum_{k=0}^{n-1} e^{2\pi i\frac{k}{n}j}\hat{f}(j) = 0$ for $j \neq 0$, while $\hat{f}_n(0) = \hat{f}(0) = \int_0^1 f(s)\,ds$. By the uniqueness of Fourier coefficients, $f_n(x) = \hat{f}(0)$ a.e. on $[0,1]$.

5. Note that $(\mathcal{A}_h f)(x+1) = \frac{1}{2h}\int_{x+1-h}^{x+1+h} f(t)\,dt$. Changing variables $t = s+1$, we get $(\mathcal{A}_h f)(x+1) = \frac{1}{2h}\int_{x-h}^{x+h} f(s+1)\,ds = \frac{1}{2h}\int_{x-h}^{x+h} f(s)\,ds = (\mathcal{A}_h f)(x)$ due to the periodicity of f. Since f is continuously differentiable, its Fourier series converges uniformly to f on $[0,1]$, see Theorem 4.3, so that $f(t) = \sum_{k\in\mathbb{Z}}\hat{f}(k)\,e^{2\pi ikt}$ for all $t \in \mathbb{R}$. We apply $\mathcal{A}_h$ to f by integrating the series term by term: $(\mathcal{A}_h f)(x) = \sum_{k\in\mathbb{Z}}\frac{\hat{f}(k)}{2h}\int_{x-h}^{x+h} e^{2\pi ikt}\,dt$ for $x \in [0,1]$. For $k = 0$ we have $\int_{x-h}^{x+h} e^{2\pi ikt}\,dt = 2h$, while for $k \neq 0$ we have $\int_{x-h}^{x+h} e^{2\pi ikt}\,dt = \frac{1}{2\pi ik}e^{2\pi ikt}\Big|_{t=x-h}^{t=x+h} = \frac{e^{2\pi ik(x+h)} - e^{2\pi ik(x-h)}}{2\pi ik} = e^{2\pi ikx}\frac{e^{2\pi ikh} - e^{-2\pi ikh}}{2\pi ik} = \frac{\sin(2\pi kh)}{\pi k}e^{2\pi ikx}$. Thus $\hat{F}_h(0) = \hat{f}(0)$, while $\hat{F}_h(k) = \frac{\sin(2\pi kh)}{2\pi kh}\hat{f}(k)$ for $k \neq 0$.

6. Note that each f_n is odd and with period 1, so that each f_n' is even and with period 1. Since $e^{\pm 2\pi ix} \neq 1$ for $x \in (0, \frac{1}{2}]$, we can expand $f_n'(x) = 2\pi\sum_{k=1}^n \cos(2\pi kx) = \pi\sum_{k=1}^n [e^{2\pi ikx} + e^{-2\pi ikx}]$ as $f_n'(x) = \pi\Big\{e^{2\pi ix}\sum_{j=0}^{n-1} e^{2\pi ijx} + e^{-2\pi ix}\sum_{j=0}^{n-1} e^{-2\pi ijx}\Big\} = \pi\Big\{e^{2\pi ix}\frac{1-e^{2\pi inx}}{1-e^{2\pi ix}} + e^{-2\pi ix}\frac{1-e^{-2\pi inx}}{1-e^{-2\pi ix}}\Big\}$, or $f_n'(x) = \pi\frac{e^{2\pi ix}(1-e^{2\pi inx})(1-e^{-2\pi ix}) + e^{-2\pi ix}(1-e^{-2\pi inx})(1-e^{2\pi ix})}{(1-e^{-2\pi ix})(1-e^{2\pi ix})}$, which we can further develop to obtain $f_n'(x) = \pi\frac{2\cos(2\pi x) - 2 + 2\cos(2\pi nx) - 2\cos(2\pi(n+1)x)}{2-2\cos(2\pi x)} = \pi\frac{\sin((2n+1)\pi x) - \sin(\pi x)}{\sin(\pi x)} = \pi\frac{\sin((2n+1)\pi x)}{\sin(\pi x)} - \pi$. The first equality in the first chain holds for all $x \in [0, \frac{1}{2}]$, yielding $|f_n'(x)| \leq 2\pi n$ for all $x \in [0, \frac{1}{2}]$, with $f_n'(0) = 2\pi n$.

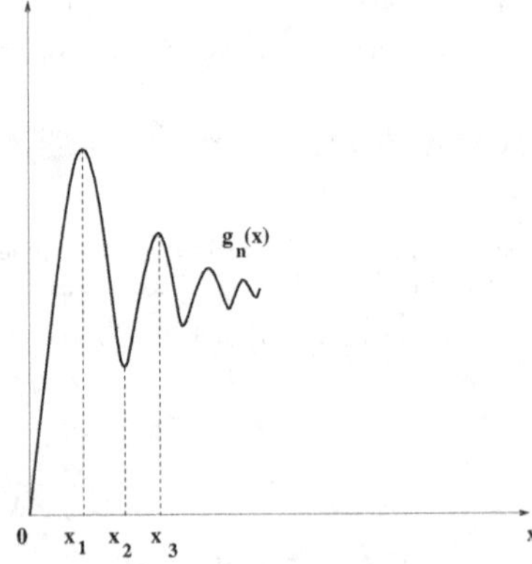

Figure 4.5 The oscillatory behaviour of the function g_n from Exercise 6.

For $x \in [0, \frac{1}{2}]$, set $g_n(x) = \pi\int_0^x \frac{\sin(\pi(2n+1)s)}{\sin(\pi s)}\,ds$. Clearly $g_n(0) = 0$ and the points $x_k = \frac{k}{2n+1}$ with $k = 0, \ldots, n$, are the zeros of the function $x \mapsto \sin(\pi(2n+1)x))$ on $[0, \frac{1}{2}]$. Between two zeros, this function has a constant sign, alternating between $\pm$. Therefore, as x increases from 0 to $\frac{1}{2}$,

$g_n(x)$ alternates between intervals of increase and decrease delimited by the points x_k. Moreover, if $k \in \{0, 1, \dots, n\}$ is an odd integer, the function $g_n(x)$ increases on $[x_{k-1}, x_k]$ and decreases on $[x_k, x_{k+1}]$ or on $[x_k, \frac{1}{2}]$ if $k = n$ with n odd. If $s \in (0, \frac{1}{2n+1})$ is such that $x_k + s \le \frac{1}{2}$, then $g_n(x_k+s) - g_n(x_k - s) = \int_{x_k-s}^{x_k+s} g_n'(x)\,dx = \int_{x_k-s}^{x_k} g_n'(x)\,dx + \int_{x_k}^{x_k+s} g_n'(x)\,dx = \int_0^s \{g_n'(x_k+y) + g_n'(x_k-y)\}\,dy = \pi \int_0^s \Big\{\frac{\sin((2n+1)\pi(x_k+y))}{\sin(\pi(x_k+y))} + \frac{\sin((2n+1)\pi(x_k-y))}{\sin(\pi(x_k-y))}\Big\}\,dy = \pi \int_0^s \sin((2n+1)\pi y)\Big\{\frac{1}{\sin(\pi(x_k-y))} - \frac{1}{\sin(\pi(x_k+y))}\Big\}\,dy > 0$ since $\sin(\pi(x_k+y)) > \sin(\pi(x_k-y)) > 0$ and $\sin((2n+1)\pi y) > 0$ for $y \in (0, s)$. This shows that the minimum of $g_n(x)$ on an interval of decrease exceeds the minimum of $g_n(x)$ on the previous interval (of increase). A similar consideration shows that, to the right of x_2, the maximum of $g_n(x)$ on an interval of increase does not exceed the maximum of $g_n(x)$ on the preceding interval (of decrease). In particular, $g_n(x)$ is nonnegative on $[0, \frac{1}{2}]$, attaining its maximum at $x = x_1$. These considerations, in combination with the fact that $g_n'(x) = f_n'(x) + \pi$, ensure that for all $x \in [0, \frac{1}{2}]$, we have $0 \le g_n(x) \le g_n(x_1) = \int_0^{x_1} g_n'(s)\,ds = \int_0^{x_1} [f_n'(s) + \pi]\,ds \le \int_0^{x_1} \pi(2n+1)\,ds = \pi$. Since $f_n(x) = g_n(x) - \pi x$, we deduce that $-\frac{\pi}{2} \le f_n(x) \le \pi$ for $x \in [0, \frac{1}{2}]$. The claim follows since the function f_n is odd and of period 1.

7. Clearly $Q_n(x) = 2\sin(2^{n^2+2}\pi x)\sum_{k=1}^{2^{n^2}} \frac{\sin(2\pi k x)}{k}$ for $x \in \mathbb{R}$. Exercise 6 therefore ensures the continuity and periodicity of f by uniform convergence. The defining formula shows that the terms in the original sums that define Q_n and Q_m are nonoverlapping for $n \neq m$. Consequently, in the representation $\hat{f}(k) = \sum_{n=1}^{\infty} \frac{1}{n^2} \int_0^1 \Big(\sum_{p=1}^{2^{n^2}} \frac{\cos[2\pi(2^{n^2+1}-p)x] - \cos[2\pi(2^{n^2+1}+p)x]}{p}\Big) e^{-2\pi i k x}\,dx$, with $k \in \mathbb{Z}$, at most one of the terms is nonzero. More precisely, for an integer $k \ge 1$, $\hat{f}(k) \neq 0$ only if $k = 2^{n^2+1} \pm p$ with $p \in \{1, \cdots, 2^{n^2}\}$, in which case $\hat{f}(k) = \mp\frac{1}{2pn^2}$. We get $\Big|\sum_{k=2^{n^2+1}+1}^{2^{n^2+1}+2^{n^2}} \hat{f}(k)\Big| = \frac{1}{n^2}\sum_{p=1}^{2^{n^2}} \frac{1}{2p} > \frac{1}{2n^2}\sum_{p=1}^{2^{n^2}} \int_p^{p+1} \frac{1}{t}\,dt = \frac{1}{2n^2}\int_1^{2^{n^2}+1} \frac{1}{t}\,dt = \frac{\ln(2^{n^2}+1)}{2n^2} > \frac{\ln(2)}{2}$. Thus the Fourier series at $x = 0$, namely $\sum_{k\in\mathbb{Z}} \hat{f}(k)$, diverges since its partial sums fail to be a Cauchy sequence.

8. $\lim_{y\downarrow 0} \frac{|f(y)-f(0)|}{y^\beta} = \infty$ shows that $f \notin h_\beta$, while $f \in h_\alpha$ is ensured by the inequality $|f(x+y) - f(x)| \le (2\pi)^\alpha |y|^\alpha$. This inequality is proved as follows. Differentiation shows that $u(t) = (1+t)^\alpha - t^\alpha - 1$ is decreasing on $(0, \infty)$. Since $u(0) = 0$, we get $u(t) \le 0$ for all $t \ge 0$. If $b > a \ge 0$, the choice $t = a/b$ yields $(a+b)^\alpha \le a^\alpha + b^\alpha$, so that for all $r, s \in \mathbb{R}$ we have $|s|^\alpha \le |s-r|^\alpha + |r|^\alpha$. Thus $|f(x+y) - f(x)| \le |\sin(2\pi(x+y)) - \sin(2\pi x)|^\alpha$ and $|\sin(2\pi(x+y)) - \sin(2\pi x)| \le 2\pi|y|$ is granted by the mean-value theorem for the function $s \mapsto \sin(2\pi s)$.

9. We first prove the following inequalities: $\sin(\pi x) \le \pi x \le \tan(\pi x)$ for $0 \le x \le 1/2$, and $\pi x/2 \le \sin(\pi x)$ for $0 \le x \le \frac{1}{3}$. For the first inequality, note by differentiation that the functions $g_1, g_2 : [0, 1/2] \to \mathbb{R}$ defined by $g_1(x) = \tan(\pi x) - \pi x$ and $g_2(x) = \pi x - \sin(\pi x)$ are both nondecreasing with $g_1(0) = g_2(0)$. For the second inequality, note by differentiation that $g_3 : [0, 1/3] \to \mathbb{R}$ given by $g_3(x) = \sin(\pi x) - \pi x/2$ is also nondecreasing, with $g_3(0) = 0$.

(i) We know that $(2n+1)$ is the maximum of $|D_n|$. On the other hand, using (4.10) and the second inequality from the prelude, $|D_n(x)| \le \frac{2|\sin((2n+1)\pi x)|}{\pi|x|} \le \frac{2}{\pi|x|}$ for $|x| \le \frac{1}{3}$, while for $\frac{1}{3} \le |x| \le \frac{1}{2}$, (4.10) yields $|D_n(x)| \le \frac{2}{\sqrt{3}} \le \frac{1}{\sqrt{3}|x|}$.

(ii) The first inequality from the prelude yields $\int_{-1/2}^{1/2} |D_n(x)|\,dx = 2\int_0^{1/2} \Big|\frac{\sin((2n+1)\pi x)}{\sin(\pi x)}\Big|\,dx \ge 2\int_0^{1/2} \frac{|\sin((2n+1)\pi x)|}{\pi x}\,dx$. Changing variables $s = (2n+1)\pi x$, the last term equals $\frac{2}{\pi}\int_0^{\frac{(2n+1)\pi}{2}} \frac{|\sin(s)|}{s}\,ds > \frac{2}{\pi}\sum_{k=0}^{n-1}\int_{k\pi}^{(k+1)\pi} \frac{|\sin(s)|}{s}\,ds = \frac{2}{\pi}\sum_{k=0}^{n-1}\int_0^{\pi} \frac{|\sin(u)|}{u+k\pi}\,du > \frac{2}{\pi}\sum_{k=0}^{n-1} \frac{1}{(k+1)\pi}\int_0^{\pi} \sin(u)\,du = \frac{4}{\pi^2}\sum_{k=0}^{n-1} \frac{1}{k+1} > \frac{4}{\pi^2}\sum_{k=0}^{n-1}\int_k^{k+1} \frac{1}{s+1}\,ds = \frac{4}{\pi^2}\ln(n+1) > \frac{4}{\pi^2}\ln(n)$, changing variables $s = u + k\pi$ in an intermediate step. For the upper estimate, note that inspection validates the identity $\frac{\sin((2n+1)\pi x)}{\sin(\pi x)} = \cos(2\pi n x) + \frac{\sin(2\pi n x)}{\tan(\pi x)}$ for $0 < x < \frac{1}{2}$, so that $\Big|\frac{\sin((2n+1)\pi x)}{\sin(\pi x)} - \frac{\sin(2\pi n x)}{\tan(\pi x)}\Big| \le 1$. We obtain that $\int_{-1/2}^{1/2} |D_n(x)|\,dx = 2\int_0^{1/2} \Big|\frac{\sin((2n+1)\pi x)}{\sin(\pi x)}\Big|\,dx \le 1 + 2\int_0^{1/2} \Big|\frac{\sin(2n\pi x)}{\tan(\pi x)}\Big|\,dx \le 1 + 2\int_0^{1/2} \frac{|\sin(2n\pi x)|}{\pi x}\,dx$. By means of the change of variables $s = 2n\pi x$, the last term equals $1 + \frac{2}{\pi}\int_0^{\pi n} \frac{|\sin(s)|}{s}\,ds = 1 + \frac{2}{\pi}\sum_{k=0}^{n-1}\int_{k\pi}^{(k+1)\pi} \frac{|\sin(s)|}{s}\,ds =$

$1 + \frac{2}{\pi}\sum_{k=0}^{n-1}\int_0^{\pi}\frac{\sin(t)}{t+k\pi}\,dt \le 1 + \frac{2}{\pi}\int_0^{\pi}\frac{\sin(t)}{t}\,dt + \frac{2}{\pi}\sum_{k=1}^{n-1}\int_0^{\pi}\frac{1}{k\pi}\,dt = 1 + \frac{2}{\pi}\int_0^{\pi}\frac{\sin(t)}{t}\,dt + \frac{2}{\pi}\sum_{k=1}^{n-1}\frac{1}{k} < 1 + \frac{2}{\pi}\int_0^{\pi}\frac{\sin(t)}{t}\,dt + \frac{2}{\pi}\sum_{k=1}^{n-1}\int_k^{k+1}\frac{2}{t}\,dt = 1 + \frac{2}{\pi}\int_0^{\pi}\frac{\sin(t)}{t}\,dt + \frac{4}{\pi}\ln(n)$, relying on the substitution $s = t + k\pi$ in some intermediate step.

10. We have the chain of equalities $(S_n f)(x) = \sum_{|k|\le n}\hat{f}(k)\,e^{2\pi ikx} = \sum_{|k|\le n}e^{2\pi ikx}\int_0^1 f(y)\,e^{-2\pi iky}\,dy = \sum_{|k|\le n}\int_0^1 f(y)\,e^{-2\pi ik(y-x)}\,dy = \int_0^1 f(y)\sum_{|k|\le n}e^{-2\pi ik(y-x)}\,dy = \int_0^1 f(y)\,D_n(x-y)\,dy = (f * D_n)(x)$. Thus $(\sigma_n f)(x) = \frac{(S_0 f)(x)+\cdots+(S_{n-1}f)(x)}{n} = \int_0^1 f(y)\,\frac{D_0(x-y)+\cdots+D_{n-1}(x-y)}{n}\,dy = \int_0^1 f(y)\,F_n(x-y)\,dy = (f * F_n)(x)$.

11. (i) For $k \in \mathbb{Z}$, we have that $\hat{f}(-k) = \int_0^1 f(x)\,e^{2\pi ikx}dx = \int_0^1 f(-y+1)\,e^{2\pi ik(1-y)}dy = \int_0^1 f(-y+1)\,e^{-2\pi iky}dy$, using the change of variables $y = -x + 1$. If f is periodic with period 1, the right-hand side can be expressed as $\int_0^1 f(-y)\,e^{-2\pi iky}dy$. If f is also even, the last expression equals to $\int_0^1 f(y)\,e^{-2\pi iky}dy = \hat{f}(k)$.

Conversely, if $\hat{f}(-k) = \hat{f}(k)$ for every $k \in \mathbb{Z}$, then $(\sigma_n f)(x) = (\sigma_n f)(-x)$ for all $n \ge 1$ and $x \in \mathbb{R}$. Indeed, for all integers $k \ge 1$, we have $\hat{f}(-k)\,e^{-2\pi ikx} + \hat{f}(k)\,e^{2\pi ikx} = \hat{f}(k)\,(e^{-2\pi ikx} + e^{2\pi ikx}) = 2\hat{f}(k)\,\cos(2\pi kx)$, so that each function $x \mapsto (S_n f)(x)$ is even, as will be their arithmetical means $\sigma_n f$. By Theorem 4.4, $\sigma_n f$ converge uniformly to f on $\mathbb{R}$. Therefore f must be even.

(ii) The necessary and sufficient condition is that $\hat{f}(-k) = -\hat{f}(k)$ for every $k \in \mathbb{Z}$. The proof is almost identical to that of (i). Indeed, we showed above that for a continuous and periodic function f, with period 1, $\hat{f}(-k) = \int_0^1 f(-y)\,e^{-2\pi iky}dy$ for all $k \in \mathbb{Z}$. Consequently, if $f(-y) = f(y)$ for all $y \in \mathbb{R}$, then $\hat{f}(-k) = -\hat{f}(k)$. Conversely, if $\hat{f}(-k) = -\hat{f}(k)$ for every $k \in \mathbb{Z}$, then $\hat{f}(-k)\,e^{-2\pi ikx} + \hat{f}(k)\,e^{2\pi ikx} = \hat{f}(k)\,(e^{2\pi ikx} - e^{-2\pi ikx}) = 2i\hat{f}(k)\,\sin(2\pi kx)$ is an odd function of x, so that $S_n f$ as well as their arithmetical means $\sigma_n f$ will be odd functions. By Theorem 4.4, $\sigma_n f$ converge uniformly to f on $\mathbb{R}$. Therefore f must be odd.

12. (i) We have $\sum_{k\in\mathbb{Z}}\hat{f}(k)\,r^{|k|}\,e_k(x) = \sum_{k\in\mathbb{Z}}r^{|k|}\int_0^1 e^{2\pi ik(x-y)}f(y)\,dy = \int_0^1\sum_{k\in\mathbb{Z}}r^{|k|}\,e^{2\pi ik(x-y)}f(y)\,dy = \int_0^1\sum_{k\ge0}r^k\,e^{2\pi ik(x-y)}f(y)\,dy + \int_0^1 r\,e^{-2\pi i(x-y)}\sum_{k\ge0}r^k e^{-2\pi ik(x-y)}f(y)\,dy$. Identifying the geometric series, we write this as $\int_0^1\sum_{k\ge0}(r\,e^{2\pi i(x-y)})^k f(y)\,dy + \int_0^1 r\,e^{-2\pi i(x-y)}\sum_{k\ge0}(re^{-2\pi i(x-y)})^k f(y)\,dy = \int_0^1\Big(\frac{1}{1-re^{2\pi i(x-y)}} + \frac{r\,e^{-2\pi i(x-y)}}{1-re^{-2\pi i(x-y)}}\Big)f(y)\,dy = \int_0^1\frac{1-r^2}{1-2r\cos(2\pi(x-y))+r^2}\,f(y)\,dy$, by the Lebesgue dominated convergence theorem.

(ii) For $r \in [0,1)$ and $x \in \mathbb{R}$, set $Q(r,x) = \frac{1-r^2}{1-2r\cos(2\pi x)+r^2}$ and note that for every fixed $r \in [0,1)$ the function $x \mapsto Q(r,x)$ is even, positive and periodic with period 1. Moreover, evaluating item (i) at $x = 0$ for the constant function 1, we obtain $\int_0^1 Q(r,y)\,dy = \int_0^1 Q(r,-y)\,dy = 1$ since the Fourier coefficients of the constant function 1 are $\hat{1}(0) = 1$ and $\hat{1}(k) = 0$ for integers $k \ne 0$. Combining these facts, we get $\big|\sum_{k\in\mathbb{Z}}\hat{f}(k)\,r^{|k|}\,e_k(x) - f(x)\big| = \big|\int_0^1 Q(r,x-y)\,f(y)\,dy - \int_{-1/2}^{1/2}Q(r,y)\,f(x)\,dy\big| = \big|\int_{-1/2}^{1/2}Q(r,y-x)\,f(y)\,dy - f(x)\int_{-1/2}^{1/2}Q(r,s)\,ds\big|$, and, performing the change of variables $y = x+s$, this equals $\big|\int_{-1/2}^{1/2}Q(r,s)\,f(x+s)\,ds - f(x)\int_{-1/2}^{1/2}Q(r,s)\,ds\big| = \big|\int_{-1/2}^{1/2}Q(r,s)\,[f(x+s)-f(x)]\,ds\big| \le \int_{-1/2}^{1/2}Q(r,s)\,|f(x+s)-f(x)|\,ds$ for $0 \le r < 1$ and $x \in \mathbb{R}$. Given $\delta \in (0,1)$ and $r \in [0,1)$, since $0 \le Q(r,s) \le \frac{1-r^2}{1-2r\cos(2\pi\delta)+r^2}$ for all $s \in [-\frac{1}{2},\frac{1}{2}]$ with $|s| \ge \delta$, we have that $\int_{s\in[-\frac{1}{2},\frac{1}{2}],\,|s|\ge\delta}Q(r,s)\,|f(x+s)-f(x)|\,ds \le \frac{2(1-r^2)\,\sup_{x\in[-1,1]}\{|f(x)|\}}{1-2r\cos(2\pi\delta)+r^2}$, while $\int_{-\delta}^{\delta}Q(r,s)\,|f(x+s)-f(x)|\,ds \le \sup_{|x-y|\le\delta}\{|f(x)-f(y)|\}\int_{-1/2}^{1/2}Q(r,s)\,ds = \sup_{|x-y|\le\delta}\{|f(x)-f(y)|\}$. Adding up the last two estimates and letting subsequently $r \downarrow 0$ yields the inequality $\limsup_{r\uparrow0}\big|\sum_{k\in\mathbb{Z}}\hat{f}(k)\,r^{|k|}\,e_k(x) - f(x)\big| \le \sup_{|x-y|\le\delta}\{|f(x)-f(y)|\}$. Since $\delta > 0$ was arbitrary, we conclude using the uniform continuity of f on $[-1,1]$.

13. For $n = 2^k + j \ge 1$ with $0 \le j < 2^k$, consider the function $f_n = \chi^{[\frac{j}{2^k},\frac{j+1}{2^k}]}$ on $[0,1]$. Since

$\int_0^1 |f_n(x)|\,dx = \frac{1}{2^k}$ for $2^k \le n < 2^{k+1}$, we see that $f_n \to 0$ in $L^1[0,1]$. However, the sequence $\{f_n(x)\}_{n\ge1}$ does not converge for any $x \in [0,1]$, since it attains the values 0 and 1 infinitely often.

Let now $f_n \to f$ in $L^1[0,1]$. Then there is a sequence $n_1 < n_2 < \cdots$ with $\|f_{n_k} - f\|_{L^1[0,1]} < \frac{1}{2^k}$ for $k \ge 1$. By monotone convergence $\int_0^1 \sum_{k=1}^\infty |f_{n_k}(x) - f(x)|\,dx = \sum_{k=1}^\infty \int_0^1 |f_{n_k}(x) - f(x)|\,dx \le \sum_{k=1}^\infty \frac{1}{2^k} = 1$, so that $\sum_{k=1}^\infty |f_{n_k}(x) - f(x)| < \infty$ a.e. on $[0,1]$. At every point $x \in [0,1]$ where the series converges, the general term approaches 0, so that $f_{n_k}(x) \to f(x)$ for $n_k \to \infty$.

Finally, the dominated convergence theorem indicates that we should seek functions that are not uniformly bounded. Let $f_n = n\chi^{[0,\frac{1}{n}]}$ for $n \ge 1$. Clearly $f_n(x) \to 0$ for every $x \in (0,1]$. By the previous result, a subsequence $\{f_{n_k}\}$ that converges in $L^1[0,1]$ to some function f would contain a subsequence $\{f_{n'_k}\}$ converging a.e. to f on $[0,1]$. Thus $f(x) = 0$ a.e. on $[0,1]$. This is incompatible with the relation $\|f\|_{L^1[0,1]} = 1$, ensured by $\|f_{n_k}\|_{L^1[0,1]} = 1$ for all n_k and the presumed convergence $f_{n_k} \to f$ in $L^1[0,1]$.

14. Since the proofs for $L^1[0,1]$ and $L^2[0,1]$ are similar, it suffices to discuss the solution for $L^2[0,1]$. The fact that $L^2[0,1]$ is a pre-Hilbert space is immediate. Let now $\{f_n\}_{n\ge1}$ be a Cauchy sequence in $L^2[0,1]$. Select a subsequence $\{n_k\}_{k\ge1}$ such that $\|f_{n_k} - f_j\|_{L^2[0,1]} < 2^{-k}$ for all $j \ge n_k$. Consider now for $j \ge 1$ the sets $A_j = \{x \in [0,1] : |f_{n_{j+1}}(x) - f_{n_j}(x)| \ge 2^{-j/3}\}$, and let $A = \bigcap_{k\ge1}\bigcap_{j\ge k} A_j$ be the set of points that belong to infinitely many of the sets A_j. Chebyshev's inequality (Exercise 2.25) yields $m(A_j) \le 2^{2j/3}\,\|f_{n_{j+1}} - f_{n_j}\|^2_{L^2[0,1]} \le 2^{-j/3}$, so that $\sum_{j\ge1} m(A_j) < \infty$, which forces $m(A) = 0$ by the Borel–Cantelli lemma (Exercise 2.19). But the set $C = \{x \in [0,1] : \{f_{n_k}\}_{k\ge1}$ not a Cauchy sequence in $\mathbb{C}\} = \{x \in [0,1] : \{f_{n_k}\}_{k\ge1}$ fails to converge$\}$ is a subset of A. Indeed, on the complement of A we have $|f_{n_{k+1}}(x) - f_{n_k}(x)| < 2^{-k/3}$ with exception of at most finitely many $k \ge 1$, so that $\sum_{k\ge1} |f_{n_{k+1}}(x) - f_{n_k}(x)| < \infty$, which ensures that $\{f_{n_k}\}_{k\ge1}$ is a Cauchy sequence in $\mathbb{C}$ since $|f_{n_{k+j}}(x) - f_{n_k}(x)| \le |f_{n_{k+1}}(x) - f_{n_k}(x)| + |f_{n_{k+2}}(x) - f_{n_{k+1}}(x)| + \cdots + |f_{n_{k+j}}(x) - f_{n_{k+j-1}}(x)| \to 0$ for $j \ge 1$ and $k \to \infty$. These considerations show that $m(C) = 0$ and enable us to define the measurable function $f : [0,1] \to \mathbb{C}$ by setting $f(x) = 0$ if $x \in C$ and $f(x) = \lim_{k\to\infty} f_{n_k}(x)$ for $x \notin C$. From Fatou's lemma (Theorem 2.3) applied to $\int_0^1 |f_{n_k}(x) - f_j(x)|^2 dx = \|f_{n_k} - f_j\|^2_{L^2[0,1]} \le 2^{-2n_k}$ with $j \ge n_k$, we deduce that $\|f - f_j\|_{L^2[0,1]} \le 2^{-n_k}$ for $j \ge n_k$. Thus $f \in L^2[0,1]$ and $f_n \to f$ in $L^2[0,1]$. This proves that $L^2[0,1]$ is a Hilbert space.

For the separability issue, it suffices to show that real-valued functions $f \in L^2[0,1]$ can be well-approximated in $L^2[0,1]$ by step functions taking on rational values on a finite number of disjoint intervals with rational endpoints. For this, note that if $(a_n, b_n) \subset [0,1]$ and $c_n \in \mathbb{R}$, choosing rational numbers q_n, p_n, r_n with $a_n < q_n < p_n < b_n$ and $q_n - a_n \le b_n - p_n \le |c_n - r_n| < \frac{1}{3n(c_n^2+1)}$, yields $\|c_n\chi^{(a_n,b_n)} - r_n\chi^{(q_n,p_n)}\|^2_{L^2[0,1]} \le \int_{a_n}^{q_n} c_n^2\,dx + \int_{q_n}^{p_n} |c_n - r_n|^2\,dx + \int_{p_n}^{b_n} c_n^2\,dx \le \frac{1}{n}$. Since in the proof of (P5) from Section 4.1 we showed that finite linear combinations of characteristic sets of open subintervals of $[0,1]$ are dense in $L^1[0,1]$, a space larger than $L^2[0,1]$, we are done.

15. For $f \in L^1[0,1]$ we get $\|S_nf\|_{L^1[0,1]} = \|D_n * f\|_{L^1[0,1]} \le \|D_n\|_{L^1[0,1]}\|f\|_{L^1[0,1]}$ due to (4.6). On the other hand, if $f = F_k$ is the Fejér kernel, then $S_nf = D_n * F_k = F_k * D_n = \sigma_k D_n \to D_n$ uniformly on $[0,1]$, by Theorem 4.4. Thus $\|S_n\|_{\mathcal{L}(L^1[0,1])} = \|D_n\|_{L^1[0,1]}$. The estimate (4.13) and the uniform boundedness principle now show that there exists $f \in L^1[0,1]$ such that $\|S_nf\|_{L^1[0,1]}$ is unbounded when $n \to \infty$.

16. Since σ_nf are trigonometric polynomials, Theorem 4.4 and Theorem 4.5 settle the claim.

17. (i) Let $M = \sum_{|k|\ge1} |k\,\hat{f}(k)| < \infty$. Then $\sum_{k\in\mathbb{Z}} \hat{f}(k)\,e^{2\pi ikx}$ will converge in $L^1[0,1]$ to some limit that by the uniqueness of the Fourier coefficients can only be $f(x)$ a.e. Thus for a countable dense set of $x \in [0,1]$ we will have that $|f(x) - (S_nf)(x)| = \left|\sum_{|k|\ge n+1} \hat{f}(k)\,e^{2\pi ikx}\right| \le \sum_{|k|\ge n+1} |\hat{f}(k)| \le \sum_{|k|\ge n+1} \frac{|k\,\hat{f}(k)|}{n} \le \frac{M}{n}$ for all $n \ge 1$. Since both f and S_nf are continuous, the previous inequality must hold throughout $[0,1]$, so that $\|f - S_nf\|_\infty \le \frac{M}{n}$ for all $n \ge 1$. But (4.19) yields $|(\sigma_nf)(x) -$

$(S_nf)(x)| = \left|\sum_{|k|\le n-1}\frac{k\hat{f}(k)}{n}\right| \le \sum_{|k|\le n-1}\frac{|k\hat{f}(k)|}{n} \le \frac{M}{n}$ for all $n \ge 1$ and all $x \in [0,1]$, so that $\|\sigma_nf - S_nf\|_\infty \le \frac{M}{n}$ for all $n \ge 1$. Thus $\|\sigma_nf - f\|_\infty \le \frac{2M}{n}$ for all $n \ge 1$.

(ii) From (4.19) and (1.1.2) we infer that $\frac{|k|\cdot|\hat{f}(k)|}{n} = \int_0^1 [f(x) - (\sigma_nf)(x)]\,e^{-2\pi ikx}dx$ for all $|k| \le n-1$, so that $|k\hat{f}(k)| \le n\int_0^1 |f(x) - (\sigma_nf)(x)|\,dx \le n\|f - \sigma_nf\|_\infty$. For k fixed, the hypothesis ensures that the right side of the previous inequality vanishes in the limit $n \to \infty$. Thus $\hat{f}(k) = 0$ for all $k \ne 0$, and Theorem 4.4 yields $f = \hat{f}(0)$ since $(\sigma_nf)(x) = \hat{f}(0)$ for all $n \ge 1$.

18. Theorem 4.4. and (4.19) ensure that $(\sigma_nf)(0) = \sum_{|k|\le n-1}\hat{f}(k)\,(1 - \frac{|k|}{n}) \to f(0)$ as $n \to \infty$. Looking at every single term in the sum $(\sigma_nf)(0)$ confirms that the sequence $\{(\sigma_nf(0)\}_{n\ge1}$ is non-decreasing. Thus, for $n \ge N \ge 1$ we get $\sum_{|k|\le N-1}\hat{f}(k)\,(1-\frac{|k|}{n}) \le \sum_{|k|\le n-1}\hat{f}(k)\,(1-\frac{|k|}{n}) = (\sigma_nf)(0) \le f(0)$. Letting $n \to \infty$ yields $\sum_{|k|\le N-1}\hat{f}(k) \le f(0)$ for any $N \ge 1$.

19. One implication is granted by Theorem 4.5. Conversely, assume that $\sigma_n(x) = \sum_{|k|\le n-1} c_k\,(1 - \frac{|k|}{n})\,e^{2\pi ikx}$ converge in $L^1[0,1]$ to some function f. Then (P2) yields $\lim_{n\to\infty}\widehat{(\sigma_n)}(k) = \hat{f}(k)$ for every fixed $k \in \mathbb{Z}$. On the other hand, the explicit representation of $\sigma_n(x)$ yields $\widehat{(\sigma_n)}(k) = c_k\,(1 - \frac{|k|}{n})$ for $|k| \le n-1$ (and zero otherwise), so that $\lim_{n\to\infty}\widehat{(\sigma_n)}(k) = c_k$. Thus $c_k = \hat{f}(k)$ for all $k \in \mathbb{Z}$.

20. (i) We have $\sum_{k=1}^n \sin(2\pi kx) = \sum_{k=1}^n \frac{e^{2\pi ikx}-e^{-2\pi ikx}}{2i} = \frac{e^{2\pi ix}}{2i}\sum_{k=1}^{n-1} e^{2\pi iky} - \frac{e^{-2\pi ix}}{2i}\sum_{k=1}^{n-1} e^{-2\pi iky}$ $= \frac{e^{2\pi ix}(e^{2\pi inx}-1)}{2i(e^{2\pi ix}-1)} - \frac{e^{-2\pi ix}(e^{-2\pi inx}-1)}{2i(e^{-2\pi ix}-1)} = \frac{(e^{2\pi inx}-1)(1-e^{2\pi ix})-(e^{-2\pi inx}-1)(1-e^{-2\pi ix})}{2i(e^{2\pi ix}-1)(e^{-2\pi ix}-1)} =$ $\frac{\sin(2\pi nx)+\sin(2\pi x)-\sin(2\pi(n+1)x)}{2(1-\cos(2\pi x))} = \frac{\sin(2\pi x)-2\sin(\pi x)\cos((2n+1)\pi x)}{4\sin^2(\pi x)} = \frac{\cos(\pi x)-\cos((2n+1)\pi x)}{2\sin(\pi x)}$ for all $x \in (0,\frac{1}{2})$.

(ii) The function $f_1(s) = s - \sin(s)$ has a nonnegative derivative on $(0,\infty)$. Since $f_1(0) = 0$, we deduce that $f_1(s) \ge 0$ for all $s \ge 0$, so that $\sin(s) \le s$ for $s \ge 0$. On the other hand, the function $f_2(s) = \sin(s) - \frac{2s}{\pi}$ satisfies $f_2(0) = f_2(\frac{\pi}{2}) = 0$. There is exactly one point $\xi \in (0,\frac{\pi}{2})$ where $f_2'(s) = 0$, and $f_2''(\xi) = -\sin(\xi) < 0$. Thus ξ is the point where f_2 attains its maximum on $[0,\frac{1}{2}]$, and the only possibility is that the minimum is attained at both endpoints. In particular, this means that $f_2(s) \ge 0$ for all $s \in [0,\frac{1}{2}]$. We proved the validity of the inequalities $\frac{2s}{\pi} \le \sin(s) \le s$ for $s \in [0,\frac{1}{2}]$. Bounding from above the absolute value of the numerator of $\tilde{D}_n(x)$ by 2, the desired estimate follows at once from the previous lower bound on $\sin(\pi x)$.

(iii) Fix $y \in (0,1]$ and let $m \ge 1$ be that integer for which $\frac{1}{m+1} < y \le \frac{1}{m}$. For $n \le m$, using the fact that $0 \le \sin(s) \le s$ for all $s \in [0,\pi]$, we have that $0 \le \sum_{k=1}^n \frac{\sin(\pi ky)}{k\pi} \le \sum_{k=1}^n y = ny \le my \le 1$. If $n = m+1$, then $-1 \le \sin((m+1)\pi y) \le 0$ and the previous estimate yield $\left|\sum_{k=1}^{m+1}\frac{\sin(\pi ky)}{k\pi}\right| \le 1$. On the other hand, if $n = m+N$ for some integer $N \ge 2$, we first verify by inspection the following identity: $\sum_{k=m+1}^{m+N}\frac{\sin(\pi ky)}{k\pi} = \frac{1}{(m+N)\pi}\sum_{k=m+1}^{m+N}\sin(\pi ky) + \sum_{k=m+1}^{m+N-1}\{(\frac{1}{k\pi} - \frac{1}{(k+1)\pi})\sum_{j=m+1}^k \sin(\pi jy)\}$. Thus $\left|\sum_{k=m+1}^{m+N}\frac{\sin(\pi ky)}{k\pi}\right| \le \frac{1}{(m+N)\pi}\left|\tilde{D}_{m+N}(\frac{y}{2}) - \tilde{D}_m(\frac{y}{2})\right| + \frac{1}{\pi}\sum_{k=m+1}^{m+N-1}\frac{1}{k(k+1)}\left|\tilde{D}_k(\frac{y}{2}) - \tilde{D}_m(\frac{y}{2})\right|$. Using (ii), we get $\left|\sum_{k=m+1}^{m+N}\frac{\sin(\pi ky)}{k\pi}\right| \le \frac{4}{(m+N)\pi y} + \frac{4}{\pi y}\sum_{k=m+1}^{m+N-1}(\frac{1}{k} - \frac{1}{k+1}) = \frac{4}{(m+1)\pi y} < \frac{4}{\pi}$ by our choice of m. We get that $\left|\sum_{k=1}^{m+N}\frac{\sin(\pi ky)}{k\pi}\right| \le \left|\sum_{k=1}^{m+1}\frac{\sin(\pi ky)}{k\pi}\right| + \left|\sum_{k=m+1}^{m+N}\frac{\sin(\pi ky)}{k\pi}\right| \le 1 + \frac{4}{\pi}$. By continuity the inequality will hold for all $y \in [0,1]$.

21. (i) To show that $\sum_{k\ge2}\frac{\sin(2\pi kx)}{\ln(k)}$ converges at every $x \in [0,1]$, each term being periodic and odd, it suffices to show that its partial sums form a Cauchy sequence at every fixed $x \in (0,\frac{1}{2}]$. Given $x \in (0,\frac{1}{2}]$, choose the integer $N \ge 1$ with $\frac{1}{N+1} < x \le \frac{1}{N}$. If $n \ge m \ge N+1$, then, relying on the identity used in the solution to Exercise 20(iii), $\left|\sum_{k=m}^n\frac{\sin(2\pi kx)}{\ln(k)}\right| = \left|\frac{1}{\ln(n)}\sum_{k=m}^n \sin(2\pi kx) + \sum_{k=m}^{n-1}\{(\frac{1}{\ln(k)} - \frac{1}{\ln(k+1)})\sum_{j=m}^k \sin(2\pi jx)\right| \le \frac{1}{\ln(n)}|\tilde{D}_n(x) - \tilde{D}_{m-1}(x)| + \sum_{k=m}^{n-1}(\frac{1}{\ln(k)} - \frac{1}{\ln(k+1)})|\tilde{D}_k(x) - \tilde{D}_{m-1}(x)| \le \frac{2}{x}\frac{1}{\ln(m)} \le \frac{2(N+1)}{\ln(m)} \to 0$ if $m \to \infty$. In the previous estimates we relied on the inequality established in Exercise 20(ii).

Assume now that $\sum_{k\ge2}\frac{\sin(2\pi kx)}{\ln(k)}$ is the Fourier series of some $g \in L^1[0,1]$. Then $(S_ng)(x) = \sum_{k=2}^n\frac{\sin(2\pi kx)}{\ln(k)} = \sum_{k=2}^n\frac{e^{2\pi kx}-e^{-2\pi ikx}}{2i\ln(k)}$ yields $\hat{g}(k) = \int_0^1 (S_{|k|}g)(x)\,e^{-2\pi ikx}dx = \frac{k}{2\pi i|k|\ln(|k|)}$ for $|k| \ge 2$, while $\hat{g}(0) = \hat{g}(\pm 1) = 0$. Thus $\sum_{k\in\mathbb{Z}} \Big|\frac{\hat{g}(2k+1)}{2k+1}\Big| = \frac{1}{2}\sum_{k\geq 1}\frac{1}{(2k+1)\ln(2k+1)} \geq \frac{1}{8}\sum_{k\geq 1}\frac{1}{(k+1)\ln(k+1)} \geq \frac{1}{8}\sum_{k\geq 1}\int_k^{k+1}\frac{ds}{(s+1)\ln(s+1)} = \frac{1}{8}\int_1^{\infty}\frac{ds}{(s+1)\ln(s+1)} = \frac{1}{8}\ln(\ln(s+1))\Big|_{s=1}^{\infty} = \infty$, which is impossible in view of the considerations made in the end of Section 4.2.

(ii) Each term being odd, it suffices to address the convergence issue on $[0, \frac{1}{2}]$. Fix $x \in (0, \frac{1}{2}]$ and choose an integer $N \geq 2$ with $\frac{1}{N+1} < x \leq \frac{1}{N}$. For $n \geq N+1$, using the identity from the solution to Exercise 20(iii), we get $\Big|\sum_{k=N+1}^{n}\frac{\sin(2\pi kx)}{k\ln k}\Big| = \Big|\frac{1}{n\ln n}\sum_{k=N+1}^{n}\sin(2\pi kx) + \sum_{k=N+1}^{n-1}\Big\{\Big(\frac{1}{k\ln k} - \frac{1}{(k+1)\ln(k+1)}\Big)\sum_{j=N+1}^{k}\sin(2\pi jx)\Big\} \leq \frac{1}{n\ln n}|\tilde{D}_n(x) - \tilde{D}_{N-1}(x)| + \sum_{k=N+1}^{n-1}\Big(\frac{1}{k\ln k} - \frac{1}{(k+1)\ln(k+1)}\Big)|\tilde{D}_k(x) - \tilde{D}_{N-1}(x)| \leq \frac{2}{x}\frac{1}{(N+1)\ln(N+1)} \leq \frac{2}{\ln(N+1)}$, while for $m \leq n \leq N$ we have that $\Big|\sum_{k=m}^{n}\frac{\sin(2\pi kx)}{k\ln k}\Big| \leq \sum_{k=m}^{n}\frac{2\pi x}{\ln k} \leq \frac{2\pi Nx}{\ln m} \leq \frac{2\pi}{\ln m}$. Thus $\Big|\sum_{k=m}^{n}\frac{\sin(2\pi kx)}{k\ln k}\Big| \leq \frac{2\pi}{\ln m}$ for all $x \in [0, \frac{1}{2}]$ and all $n, m \geq 2$, proving the uniform convergence.

(iii) For $n \geq 4$, using twice the identity from the hint to Exercise 20(iii) in combination with (4.10), we get $\sum_{k=2}^{n}\frac{\cos(2\pi kx)}{\ln k} = \frac{1}{\ln n}\sum_{k=2}^{n}\cos(2\pi kx) + \sum_{k=2}^{n-1}\Big(\frac{1}{\ln k} - \frac{1}{\ln(k+1)}\Big)\sum_{j=2}^{k}\cos(2\pi jx) = \frac{1}{\ln n}\Big[\frac{D_n(x)-1}{2} - \cos(2\pi x)\Big] + \sum_{k=2}^{n-1}\Big(\frac{1}{\ln k} - \frac{1}{\ln(k+1)}\Big)\Big[\frac{D_k(x)-1}{2} - \cos(2\pi x)\Big] = -\Big[\frac{1}{\ln n} + \sum_{k=2}^{n-1}\Big(\frac{1}{\ln k} - \frac{1}{\ln(k+1)}\Big)\Big]\Big(\frac{1}{2} + \cos(2\pi x)\Big) + \frac{D_n(x)}{2\ln n} + \sum_{k=2}^{n-1}\Big(\frac{1}{\ln k} - \frac{1}{\ln(k+1)}\Big)\frac{D_k(x)}{2} = -\frac{1}{\ln 2}\Big(\frac{1}{2} + \cos(2\pi x)\Big) + \frac{D_n(x)}{2\ln n} + \Big(\frac{1}{\ln(n-1)} - \frac{1}{\ln n}\Big)\sum_{k=2}^{n-1}\frac{D_k(x)}{2} + \sum_{k=2}^{n-2}\Big(\frac{1}{\ln k} - \frac{2}{\ln(k+1)} + \frac{1}{\ln(k+2)}\Big)\sum_{j=2}^{k}\frac{D_j(x)}{2}$. Since (4.10) and (4.20) yield $\sum_{j=2}^{k}\frac{D_j(x)}{2} = \frac{(k+1)F_{k+1}(x)}{2} - 1 - \cos(2\pi x)$, the last expression equals $\frac{D_n(x)}{2\ln n} + \frac{1}{2\ln 2} - [1 + \cos(2\pi x)]\Big\{\frac{1}{\ln 2} + \frac{1}{\ln(n-1)} - \frac{1}{\ln n} + \sum_{k=2}^{n-2}\Big(\frac{1}{\ln k} - \frac{2}{\ln(k+1)} + \frac{1}{\ln(k+2)}\Big)\Big\} + \Big(\frac{1}{\ln(n-1)} - \frac{1}{\ln n}\Big)\frac{nF_n(x)}{2} + \sum_{k=2}^{n-2}\Big(\frac{1}{\ln k} - \frac{2}{\ln(k+1)} + \frac{1}{\ln(k+2)}\Big)\frac{(k+1)F_{k+1}(x)}{2} = \frac{D_n(x)}{2\ln n} + \frac{1}{2\ln 2} - \big(1 + \cos(2\pi x)\big)\Big\{\frac{2}{\ln 2} - \frac{1}{\ln 3}\Big\} + \Big(\frac{1}{\ln(n-1)} - \frac{1}{\ln n}\Big)\frac{nF_n(x)}{2} + \sum_{k=2}^{n-2}\Big(\frac{1}{\ln k} - \frac{2}{\ln(k+1)} + \frac{1}{\ln(k+2)}\Big)\frac{(k+1)F_{k+1}(x)}{2}$.
Denoting $a_n = \frac{1}{2\ln n}$, $b = \frac{1}{\ln 3} - \frac{2}{\ln 2}$, $c_n = \frac{n}{2\ln(n-1)} - \frac{n}{2\ln n}$, $d_{k+1} = \frac{k+1}{2\ln k} - \frac{k+1}{\ln(k+1)} + \frac{k+1}{2\ln(k+2)}$, we write the expression in the form $a_n D_n(x) + \frac{1}{2\ln 2} + b(1 + \cos(2\pi x)) + c_n F_n(x) + \sum_{k=2}^{n-2} d_{k+1}F_{k+1}(x)$.
Note that $0 \leq c_n = \frac{n[\ln n - \ln(n-1)]}{2\ln n\cdot\ln(n-1)} = \frac{n}{n-1}\frac{\ln(1+\frac{1}{n-1})^{n-1}}{2\ln n\cdot\ln(n-1)} \leq \frac{1}{2\ln n\cdot\ln(n-1)}$, while $d_{k+1} = \frac{k+1}{2}\{(\frac{1}{\ln k} - \frac{1}{\ln(k+1)}) - (\frac{1}{\ln(k+1)} - \frac{1}{\ln(k+2)})\} = \frac{k+1}{2}\{\int_k^{k+1}\frac{ds}{s\ln^2(s)} - \int_{k+1}^{k+2}\frac{ds}{s\ln^2(s)}\} \geq 0$ since $s \mapsto \frac{1}{s\ln^2(s)}$ is decreasing on $[2, \infty)$. Moreover, $\sum_{k=2}^{n-2} d_{k+1} = \sum_{k=2}^{n-2}\frac{k+1}{2}\{\int_k^{k+1}\frac{ds}{s\ln^2(s)} - \int_{k+1}^{k+2}\frac{ds}{s\ln^2(s)}\} = \frac{3}{2}\int_2^3\frac{ds}{s\ln^2(s)} + \frac{1}{2}\sum_{k=3}^{n-2}\int_k^{k+1}\frac{ds}{s\ln^2(s)} - \frac{n-1}{2}\int_{n-1}^{n}\frac{ds}{s\ln^2(s)} \leq \frac{3}{2}\int_2^3\frac{ds}{s\ln^2(s)} + \frac{1}{2}\int_3^{\infty}\frac{ds}{s\ln^2(s)} = \frac{3}{2}\int_2^3\frac{ds}{s\ln^2(s)} + \frac{1}{2\ln 3} < \infty$.
Since $F_{k+1}(x) \geq 0$ with $\int_0^1 F_{k+1}(x)\,dx = 1$, the monotone convergence theorem ensures that $F(x) = \frac{1}{2\ln 2} + b[1 + \cos(2\pi x)] + \sum_{k=2}^{\infty} d_{k+1}F_{k+1}(x)$ defines a function in $L^1[0, 1]$. Since (4.12) and (4.20) ensure the existence of a constant $K > 1$ with $|D_n(x)| \leq \frac{K}{|x|}$ and $F_n(x) \leq \frac{K}{|x|}$ for all $0 < |x| \leq \frac{1}{2}$ and all $n \geq 1$, the fact that $\lim_{n\to\infty} a_n = \lim_{n\to\infty} c_n = 0$ yields $\sum_{k=2}^{n}\frac{\cos(2\pi kx)}{\ln k} \to F(x)$ whenever $0 < |x| \leq \frac{1}{2}$, even if the convergence fails at $x = 0$. Note that $\sum_{k=2}^{n}\frac{\cos(2\pi kx)}{\ln k}$ does not converge to $F(x)$ in $L^1[0, 1]$ since, while $\frac{1}{2\ln 2} + b[1 + \cos(2\pi x)] + \sum_{k=2}^{n-2} d_{k+1}F_{k+1}(x) \to F(x)$ in $L^1[0, 1]$ by monotone convergence and $\|c_nF_n\|_{L^1[0,1]} = c_n \to 0$ as $n \to \infty$, due to (4.13) there exists a constant $K > 1$ such that $\|a_nD_n\|_{L^1[0,1]} \geq a_n\frac{\ln n}{K} = \frac{1}{2K}$ for all $n \geq 4$. Thus proving that $\sum_{k=2}^{\infty}\frac{\cos(2\pi kx)}{\ln k}$ is the Fourier series of $F(x)$ is not straightforward. For this, recall from (ii) that $\sum_{k=2}^{\infty}\frac{\sin(2\pi kx)}{k\ln k}$ converges uniformly to some $f \in C_{\text{per}}[0, 1]$. If we set $f_n(x) = \sum_{k=2}^{n}\frac{\sin(2\pi kx)}{2\pi k\ln k}$ for $n \geq 2$, then $f_n'(x) = \sum_{k=2}^{n}\frac{\cos(2\pi kx)}{\ln k}$ with $f_n(0) = 0$, so that $f_n(x) = \int_0^x\sum_{k=2}^{n}\frac{\cos(2\pi ks)}{\ln k}\,ds = \frac{x}{2\ln 2} + a_n\int_0^x D_n(s)\,ds + b\int_0^x[1 + \cos(2\pi s)]\,ds + c_n\int_0^x F_n(s)\,ds + \sum_{k=2}^{n-2} d_{k+1}\int_0^x F_{k+1}(s)\,ds$ for all $x \in [0, 1]$. Since $0 \leq c_n\int_0^x F_n(s)\,ds \leq c_n \to 0$ and $a_n \to 0$, using Exercise 20(iii) in the alternative form $a_nx + a_n\sum_{k=1}^{n}\frac{\sin(2\pi kx)}{\pi k}$ of the first term, and the monotone convergence theorem for the last term, we get $\lim_{n\to\infty}\int_0^x\sum_{k=2}^{n}\frac{\cos(2\pi ks)}{\ln k}\,ds = \frac{x}{2\ln 2} + b\int_0^x[1 + \cos(2\pi s)]\,ds + \int_0^x\sum_{k=2}^{\infty} d_{k+1}F_{k+1}(s)\,ds$ for all $x \in [0, 1]$. Thus $\frac{f(x)}{2\pi} = \int_0^x F(s)\,ds$ for all $x \in [0, 1]$ since $f_n(x) \to \frac{f(x)}{2\pi}$ uniformly on $[0, 1]$. f being absolutely continuous on $[0, 1]$, (P4) yields $\hat{F}(k) = ki\hat{f}(k)$ for all $k \in \mathbb{Z}$. By (ii),

$f_n(x) = \sum_{|k|=2}^{n} \frac{e^{2\pi ikx}}{2ik \ln|k|}$ converges uniformly to $\frac{f(x)}{2\pi}$ on $[0,1]$. Thus $\hat{f}(k) = \frac{1}{2ik\ln|k|}$ and $\hat{F}(k) = \frac{1}{2\ln|k|}$ for $|k| \geq 2$, while $\hat{f}(k) = \hat{F}(k) = 0$ for $|k| \leq 1$. Thus $\sum_{k=2}^{\infty} \frac{\cos(2\pi kx)}{\ln k}$ is the Fourier series of F.

22. (i) By hypothesis $\sum_{k\in\mathbb{Z}} \hat{f}(k)\, e^{2\pi ikx}$ converges in $L^1[0,1]$ to some limit, that has to be equal to $f(x)$ a.e. by the uniqueness of Fourier coefficients. Thus, after correction on a set of measure zero, $|(S_nf)(x) - f(x)| = \left|\sum_{|k|\geq n+1} \hat{f}(k)\, e^{2\pi ikx}\right| \leq \sum_{|k|\geq n+1} |\hat{f}(k)|$.

(ii) Clearly $\mathcal{A}$ is a normed linear space. If $\{f_n\}_{n\geq 1} \in \mathcal{A}$ is a Cauchy sequence, then $\{\hat{f}_n\}_{n\geq 1} \in l^1(\mathbb{Z})$ will have a limit $\hat{f} \in l^1(\mathbb{Z})$. Then, in view of (i), after correction on a set of measure zero, $f(x) = \sum_{k\in\mathbb{Z}} \hat{f}(k)\, e^{2\pi ikx}$ will be the limit of $\{f_n\}_{n\geq 1}$ in $\mathcal{A}$: $\mathcal{A}$ is a Banach space. Moreover, if $f,\, g \in \mathcal{A}$, then (i) ensures $\widehat{(fg)}(k) = \int_0^1 f(x)g(x)\, e^{-2\pi ikx}dx = \sum_{q\in\mathbb{Z}}\sum_{p\in\mathbb{Z}} \int_0^1 \hat{f}(q)\,\hat{g}(p)\, e^{2\pi i(q+p-k)x}dx = \sum_{q\in\mathbb{Z}} \hat{f}(q)\,\hat{g}(k-q)$, due to (1.1.2). Thus $\sum_{k\in\mathbb{Z}} |\widehat{(fg)}(k)| \leq \sum_{k\in\mathbb{Z}} |\hat{f}(q)| \cdot |\hat{g}(k-q)| \leq \left(\sum_{q\in\mathbb{Z}} |\hat{f}(q)|\right) \cdot \left(\sum_{p\in\mathbb{Z}} |\hat{g}(p)|\right) = \|f\|_{\mathcal{A}}\|g\|_{\mathcal{A}}$. This proves that $\mathcal{A}$ is a Banach algebra under pointwise multiplication. The constant function $1 \in \mathcal{A}$ is the unital element.

(iii) Let $c_k = \frac{1}{k}$ if $|k| = 2^{-n}$ for some integer $n \geq 1$, and let $c_k = 0$ otherwise. By (i), the lacunary series $\sum_{k\in\mathbb{Z}} c_k\, e^{2\pi ikx}$ defines, after possible correction on a set of measure zero, a function $f \in \mathcal{A}$. However, since $|2\pi ik\,\hat{f}(k)| = 2\pi$ for $k = 2^n$, (P4)-(P5) prevent f from being absolutely continuous. On the other hand, an absolutely continuous function $f \notin \mathcal{A}$ is provided in Exercise 21(ii). Indeed, for this choice of f we have that $\hat{f}(0) = \hat{f}(\pm 1) = 0$ and $\hat{f}(k) = \frac{1}{2ik\ln|k|}$ for $|k| \geq 2$, see the solution to Exercise 21(iii). Thus $\sum_{|k|\geq 2} |\hat{f}(k)| = \sum_{k\geq 2} \frac{1}{k\ln k} \geq \sum_{k\geq 2} \int_k^{k+1} \frac{ds}{s\ln s} = \ln(\ln s)\Big|_{s=2}^{s=\infty} = \infty$.

(iv) If $f \in \mathcal{A}$, then (P2) yields $\sum_{k\in\mathbb{Z}} |\hat{f}(k)\hat{g}(k)| \leq \|g\|_{L^1[0,1]} \sum_{k\in\mathbb{Z}} |\hat{f}(k)| < \infty$ for all $g \in L^1[0,1]$. For the converse, by (i), it suffices to show that if $c_k \in \mathbb{C}$ are such that $\sum_{k\in\mathbb{Z}} |c_k\hat{g}(k)| < \infty$ for all $g \in L^1[0,1]$, then $\sum_{k\in\mathbb{Z}} |c_k| < \infty$. To prove this, note first that the sets $C_n = \{g \in L^1[0,1] : \sum_{k\in\mathbb{Z}} |c_k\hat{g}(k)| \leq n\}$ are closed in $L^1[0,1]$ for all $n \geq 1$, since their complements O_n are open. Indeed, if $g_0 \in O_n$, then there is some $\varepsilon > 0$ and some integer $p \geq 1$ with $\sum_{|k|\leq p} |c_k\widehat{(g_0)}(k)| > n + \varepsilon$. If $0 < \eta < \frac{\varepsilon}{\sum_{|k|\leq N}|c_k|+1}$, (P2) ensures that if $g \in L^1[0,1]$ is such that $\|g - g_0\|_{L^1[0,1]} < \eta$, then $\sum_{|k|\leq p} |c_k\widehat{(g-g_0)}(k)| \leq \sum_{|k|\leq p} |c_k| \cdot \|g - g_0\|_{L^1[0,1]} < \varepsilon$, so that $\sum_{|k|\leq p} |c_k\hat{g}(k)| \geq \sum_{|k|\leq p} |c_k\widehat{(g_0)}(k)| - \sum_{|k|\leq p} |c_k\widehat{(g-g_0)}(k)| > n$ and $g \in O_n$. By assumption, $L^1[0,1] = \bigcup_{n\geq 1} C_n$. Baire's theorem ensures that at least one of the sets C_n, say C_N, has nonempty interior. Thus we find $f_0 \in L^1[0,1]$ and $\delta > 0$ such that $f \in C_N$ whenever $\|f - f_0\|_{L^1[0,1]} < \delta$. Set $g = f - f_0$ to get $\sum_{k\in\mathbb{Z}} |c_k\hat{g}(k)| \leq \sum_{k\in\mathbb{Z}} |c_k\hat{f}(k)| + \sum_{k\in\mathbb{Z}} |c_k\widehat{(f_0)}(k)| \leq 2N$ whenever $\|g\|_{L^1[0,1]} < \delta$. Scaling yields $\sum_{k\in\mathbb{Z}} |c_k\hat{g}(k)| \leq \frac{2N}{\delta}\|g\|_{L^1[0,1]}$ for all $g \in L^1[0,1]$. Defining for $h \in (0,1)$ the function $H(x) = \frac{1}{h}$ if $x \in [0,h]$ and $H(x) = 0$ if $x \in (h,1]$, we see that $\|H\|_{L^1[0,1]} = 1$ while $\hat{H}(0) = 1$ and $\hat{H}(k) = \frac{e^{2\pi ikh}-1}{2\pi ikh}$ for $k \neq 0$. The previous inequality yields $1 + \sum_{|k|\leq n} \left|c_k\, \frac{e^{2\pi ikh}-1}{2\pi ikh}\right| \leq \frac{2N}{\delta}$ for all $h \in (0,1)$ and all $n \geq 1$. Let $h \to 0$ to infer that $1 + \sum_{|k|\leq n} |c_k| \leq \frac{2N}{\delta}$ for all $n \geq 1$. Thus $\sum_{k\in\mathbb{Z}} |c_k| < \infty$.

23. If f is absolutely continuous with $f' \in L^2[0,1]$, then (P4) and the Plancherel identity (Theorem 3.1) yield $\|f\|^2_{L^2[0,1]} + \|f'\|^2_{L^2[0,1]} = \sum_{k\in\mathbb{Z}} |\hat{f}(k)|^2[1 + 4\pi^2k^2]$. Conversely, if $f \in L^2[0,1]$ is such that $\sum_{k\in\mathbb{Z}} k^2\,|\hat{f}(k)|^2 < \infty$, then Theorem 3.1 ensures that there exists $g \in L^2[0,1]$ such that $\hat{g}(k) = 2\pi ik\,\hat{f}(k)$ for all $k \in \mathbb{Z}$. Then (4.23) yields $\int_0^x g(y)\,dy = \sum_{k\in\mathbb{Z}}(e^{2\pi ikx} - 1)\,\hat{f}(k)$ for all $x \in [0,1]$. On the other hand, since $\sum_{k\in\mathbb{Z},\,k\neq 0} |\hat{f}(k)| \leq \left(\sum_{k\in\mathbb{Z},\,k\neq 0} k^2\,|\hat{f}(k)|^2\right)^{1/2}\left(\sum_{k\in\mathbb{Z},\,k\neq 0} \frac{1}{k^2}\right)^{1/2} < \infty$, defining $h(x)$ to be the sum $\sum_{k\in\mathbb{Z}} \hat{f}(k)\, e^{2\pi ikx}$, by dominated convergence $h \in L^1[0,1]$. By the same token, $\hat{h}(j) = \int_0^1 h(x)\, e^{-2\pi ijx}dx = \sum_{k\in\mathbb{Z}} \hat{f}(k) \int_0^1 e^{2\pi ikx}\, e^{-2\pi ijx}dx = \hat{f}(j)$ for all $j \in \mathbb{Z}$, using (1.1.2) in the last step. The uniqueness theorem (see Section 4.2) forces $h = f$ a.e. Consequently, after correcting f on a set of measure zero, we get $f(x) = \sum_{k\in\mathbb{Z}} \hat{f}(k)\, e^{2\pi ikx}$ for all $x \in [0,1]$, so that $f(x) - f(0) = \sum_{k\in\mathbb{Z}} \hat{f}(k)\,(e^{2\pi ikx} - 1) = \int_0^x g(y)\,dy$ for all $x \in [0,1]$. Thus f is absolutely continuous with $f' = g$ a.e.

24. For an integer $N \geq 1$, consider the operator $S_N : L^1[0,1] \to L^1[0,1]$ that associates to f its symmetric partial sum $S_N f$. Using Exercise 10 and (4.6), we get $\|S_N f\|_{L^1[0,1]} = \|D_N * f\|_{L^1[0,1]} \leq \|D_N\|_{L^1[0,1]}\|f\|_{L^1[0,1]}$, so that $\|D_N\|_{L^1[0,1]}$ is an upper bound for the operator norm of S_N. On the other hand, for $f = F_n$ with $n \geq N$, we have $\|S_N f\|_{L^1[0,1]} = \|D_N * F_n\|_{L^1[0,1]} = \|\sigma_n(D_N)\|_{L^1[0,1]}$. Since $D_N \in C_{\text{per}}[0,1]$, from Theorem 4.4 we infer that $\lim_{n\to\infty} \|\sigma_n(D_N)\|_{L^1[0,1]} = \|D_N\|_{L^1[0,1]}$. Consequently the operator norm of S_N is given by $\|D_N\|_{L^1[0,1]} \geq \frac{\ln(N)}{K}$, see (4.13). By the uniform boundedness principle, there must exist some $f \in L^1[0,1]$ with $\{\|S_N f\|_{L^1[0,1]}\}_{N\geq 1}$ unbounded.

25. If it were a Schauder basis, then to each element f of the space would correspond precisely one sequence of scalars c_k such that $f = \sum_{k\in\mathbb{Z}} c_k e_k$, with the limit in the norm of the space; here $e_k(x) = e^{2\pi i k x}$. In either space we get $c_k = \hat{f}(k)$ for all $k \in \mathbb{Z}$ since for all $n \geq |k|$ we have that $|c_k - \hat{f}(k)| = \|(\sum_{|j|\leq n} c_j e_j - f)\overline{e_k}\|_{L^1[0,1]} = \|\sum_{|j|\leq n} c_j e_j - f\|_{L^1[0,1]} \to 0$ as $n \to \infty$. For $C_{\text{per}}[0,1]$ we obtain a contradiction with Exercise 7, while for $L^1[0,1]$ the contradiction is with Exercise 24.

26. Let $X = C_{\text{per}}[0,1] \times \mathbb{R}$. For $f \in C_{\text{per}}[0,1]$ and $n \geq 1$, define $S_n^* f \in C_{\text{per}}[0,1]$ by $(S_n^* f)(x) = \int_{-1/2}^{1/2} f(x-s)\,\frac{\sin(2\pi n s)}{\pi s}\,ds$. In view of the obvious identities $\frac{\sin(\pi(2n+1)s)}{\sin(\pi s)} - \frac{\sin(2\pi n s)}{\pi s} = \frac{e^{2\pi i n s}+e^{-2\pi i n s}}{2} + \frac{e^{2\pi i n s}-e^{-2\pi i n s}}{2i}\,\frac{\pi s\cos(\pi s)-\sin(\pi s)}{\pi s\sin(\pi s)}$, the fact that $s \mapsto \frac{\pi s\cos(\pi s)-\sin(\pi s)}{\pi s\sin(\pi s)}$ is continuous on $[-\frac{1}{2},\frac{1}{2}]$ yields $\lim_{n\to\infty}[(S_n f)(x) - (S_n^* f)(x)] = 0$ for every $x \in [-\frac{1}{2},\frac{1}{2}]$, due to (P5) and (4.10)–(4.11). For integers $M, N \geq 1$, we define the open sets $G_0(M,n) = \{(f,x) \in X : \ (S_n^* f)(x)| > M\}$ and $G(M,N) = \bigcup_{n\geq N} G_0(M,n)$. By construction we have that $\limsup_{n\to\infty}\{(S_n f)(x)\} = \infty$ on $G = \bigcap_{M,N\geq 1} G(M,N)$. By Baire's category theorem, it suffices to show that each $G(M,N)$ is dense in X. For $g_n : [-\frac{1}{2},\frac{1}{2}] \to \mathbb{R}$ given by $g_n(x) = \chi^{[0,\frac{1}{2}]}(x)\,\sin(2\pi n x)$, and extended by continuity to yield a function $g_n \in C_{\text{per}}[0,1]$, we have $(S_n^* g_n)(0) = \int_0^{1/2} \frac{\sin^2(2\pi n s)}{\pi s}\,ds = \int_0^{n/2} \frac{\sin^2(2\pi t)}{\pi t}\,dt = c_n > 0$. For $n \geq 2$ we have $c_n \geq \sum_{k=1}^n \int_{(4k-3)/8}^{(2k-1)/4} \frac{\sin^2(2\pi t)}{\pi t}\,dt \geq \frac{1}{2\pi}\sum_{k=1}^n \int_{(4k-3)/8}^{(2k-1)/4} \frac{1}{t}\,dt = \frac{1}{2\pi}\sum_{k=1}^n \ln(\frac{4k-2}{4k-3}) = \frac{1}{2\pi}\sum_{k=1}^n \int_{4k-3}^{4k-2} \frac{1}{t}\,dt \geq \frac{1}{2\pi}\sum_{k=1}^n (\frac{1}{4}\int_{4k}^{4k+4} \frac{1}{t}\,dt)$ since $t \mapsto \frac{1}{t}$ is decreasing for $t > 0$, and the latter expression equals $\frac{\ln(n+1)}{8\pi} \to \infty$ as $n \to \infty$. Since by construction $(\frac{f_1 \pm f_2}{2}, 0) \notin G_0(M,n)$ whenever $(f_1,0), (f_2,0) \notin G_0(M,n)$, given $(f_0,x_0) \in X$ and $h > \frac{M}{c_n}$, at least one of the pairs $(f_0(x+x_0) \pm hg_n(x), 0)$ belongs to $G_0(M,n)$. By a simple change of variables, this means that one of the pairs $(f_0(x) \pm hg_n(x-x_0), x_0)$ will belong to $G_0(M,n)$, and thus also to $G(M,N)$ as soon as $n \geq N$. But these pairs are arbitrarily close to (f_0,x_0) in X when h is small enough, and this is the case if n is large enough. Therefore $G(M,N)$ is dense in X, as we had to show.

27. (i) For $k \neq 0$ we have that $\hat{f}(k) = \int_0^1 f(x)e^{-2\pi i k x}dx = -\int_0^1 f(x)e^{-2\pi i k(x-\frac{1}{2k})}dx = -\int_{-\frac{1}{2k}}^{1-\frac{1}{2k}} f(y+\frac{1}{2k})e^{-2\pi i k y}dy = -\int_0^1 f(y+\frac{1}{2k})e^{-2\pi i k y}dy$ after the change of variable $y = x - \frac{1}{2k}$, taking into account the fact that by periodicity allows a free choice of the periodicity window. Thus $|\hat{f}(k)| = \frac{1}{2}\left|\int_0^1 [f(x) - f(x+\frac{1}{2k})]\,e^{-2\pi i k x}dx\right| \leq \frac{1}{2^{\alpha+1}}\,[f]_\alpha\,\frac{1}{|k|^\alpha}$.

On the other hand, for f_α it is clear that $\widehat{(f_\alpha)}(k) = 0$ unless $k = 2^j$ for some $j \geq 0$, in which case it equals to $2^{-j\alpha} = k^{-\alpha}$. If $\alpha \in (0,1)$, given $h \in (0,\frac{1}{2})$, we have that $|f_\alpha(x+h) - f_\alpha(x)| \leq \sum_{k=0}^\infty 2^{-k\alpha}\,|1 - e^{2\pi i 2^k h}| = \sum_{0\leq k:\,2^k\leq\frac{1}{h}} 2^{-k\alpha}\cdot(2\pi 2^k h) + \sum_{k:\,2^k>\frac{1}{h}} 2^{-k\alpha}\cdot 2$, since $|1 - e^{2i\theta}| \leq 2$ and $|1 - e^{2i\theta}| = 2|\sin(\theta)| \leq 2\theta$ for all $\theta \geq 0$. If $N \geq 1$ is the smallest integer $n \geq 0$ with $2^n > \frac{1}{h}$, the first dyadic sum equals $2\pi h\sum_{0\leq k\leq N-1} 2^{k(1-\alpha)} = 2\pi h\,\frac{2^{N(1-\alpha)}-1}{2^{1-\alpha}-1} \leq \frac{2^{2-\alpha}\pi h}{2^{1-\alpha}-1}\,2^{(N-1)(1-\alpha)} \leq \frac{2^{2-\alpha}\pi}{2^{1-\alpha}-1}\,h^\alpha$, while the second dyadic sum equals $2\sum_{N\leq k} 2^{-k\alpha} = 2\,\frac{2^{-N\alpha}}{1-2^{-\alpha}} \leq \frac{2}{1-2^{-\alpha}}\,h^\alpha$. Thus $f_\alpha \in h_\alpha$ for $\alpha \in (0,1)$. Note that the considerations for the first dyadic sum break down for $\alpha \geq 1$. We claim that $f_\alpha \notin h_1$ if $\alpha \in [1,\frac{3}{2})$. Indeed, $f_\alpha \in h_1$ means that f is absolutely continuous with an essentially bounded derivative, see Exercise 2.41. But $2(1-\alpha) > -1$ yields $\sum_{k\in\mathbb{Z}} k^2\,|\hat{f}(k)|^2 = \sum_{j\geq 0} j^{2-2\alpha} = \infty$, contradicting Exercise 23.

(ii) If f is smooth, then $|\hat{f}(k)| = \frac{|\widehat{(f^{(m)})}(k)|}{(2\pi|k|)^m} \leq \frac{\|f^{(m)}\|_{L^1[0,1]}}{(2\pi|k|)^m}$ for all $m \geq 1$, in view of (P2) and (P4). Conversely, if $f \in L^1[0,1]$ has rapidly decreasing coefficients, then for any integer $m \geq 1$, applying

Exercise 23 m times, we get that $f^{(m-1)}$ is absolutely continuous. In particular, f is $(m-1)$ times differentiable.

28. Integration by parts in (4.1) for $k = \pm 1$ yields $\hat{f}(\pm 1) = -\frac{e^{\mp 2\pi i}}{2\pi^2}$. Since $\hat{f}(0) = \frac{1}{6}$, we have that $(S_1 f)(x) = \frac{1}{6} - \frac{\cos(2\pi(x-1))}{\pi^2}$. Note that $\sup_{x\in[0,1]} |f(x) - (S_1 f)(x)| \geq |f(0) - (S_1 f)(0)| = \frac{1}{6}$, while $\sup_{x\in[0,1]} |f(x) - \frac{1}{8}| = \frac{1}{8}$ since the function $x \mapsto x(1-x) - \frac{1}{8}$ increases on $[0, \frac{1}{8}]$ from the value $-\frac{1}{8}$ to $\frac{1}{8}$, and decreases on $[\frac{1}{2}, 1]$ from $\frac{1}{8}$ to $-\frac{1}{8}$.

29. (i) Analogous to the proof of Theorem 4.8, since D_n is even, we obtain from (4.11) that $(S_n f)(x_0) = \int_{-1/2}^{1/2} f(x_0 - s) D_n(s)\, ds = \int_{-1/2}^{1/2} f(x_0 + s) D_n(s)\, ds$. Using (4.14), this yields $(S_n f)(x_0) - \theta = \int_{-1/2}^{1/2} \frac{f(x_0-s)+f(x_0+s)-2\theta}{2} D_n(s)\, ds$. Due to (4.10), the expression on the right side is equal to $\int_{-1/2}^{1/2} \frac{f(x_0-s)+f(x_0+s)-2\theta}{2s} [s\cos(2\pi n s) + \frac{s\cos(\pi x)}{\sin(\pi x)} \sin(2\pi n x)]\, ds$. The hypothesis ensures that both functions $s \mapsto \frac{f(x_0-s)+f(x_0+s)-2\theta}{2s}\, s$ and $s \mapsto \frac{f(x_0-s)+f(x_0+s)-2\theta}{2s} \frac{s\cos(\pi x)}{\sin(\pi x)}$ belong to $L^1[-\frac{1}{2}, \frac{1}{2}]$. We conclude by (P5) since $\cos(2\pi n s) = \frac{e^{2\pi i n s} + e^{-2\pi i n s}}{2}$ and $\sin(2\pi n s) = \frac{e^{2\pi i n s} - e^{-2\pi i n s}}{2i}$.

(ii) The considerations made about (2.24) show that the continuous function f_1 is not of bounded variation. However, since $\frac{|f_1(s)+f_1(-s)|}{2} = \left|s \sin\left(\frac{\pi}{s}\right)\right| \leq s$ for $0 < s < \frac{1}{2}$, we can apply Dini's criterion at $x_0 = 0$, with $\theta = 0$. On the other hand, $f_2 \in BV_{\text{per}}[0,1]$ since its restrictions to $[0, \frac{1}{2}]$ and $[\frac{1}{2}, 1]$ are monotone, and Theorem 4.8 applies. However, Dini's criterion is not fulfilled. Indeed, if $\theta \neq 0$, then for $0 < s < e^{-2/|\theta|}$ we have $\frac{1}{\ln(s)} > -\frac{|\theta|}{2}$, so that $|f_2(s) + f_2(-s) - 2\theta| = 2|f_2(s) - \theta| \geq 2|\theta| - 2|f_2(s)| = 2[|\theta| + \frac{1}{\ln(s)}] \geq |\theta|$ and $\int_0^\delta \frac{|\theta|}{s}\, ds = |\theta| \ln(s)\Big|_{s=0}^{s=\delta} = \infty$ for all $\delta \in (0, e^{-2/|\theta|})$. Also, if $\theta = 0$, then for any $\delta \in (0, \frac{1}{2})$ we have $\int_0^\delta \frac{|f_2(s)+f_2(-s)|}{s}\, ds = \int_0^\delta \frac{ds}{-s\ln(s)} = -\ln(-\ln(s))\Big|_{s=0}^{s=\delta} = \infty$.

30. Define the periodic continuous function $g(x) = \int_0^x f(s)\, ds - \hat{f}(0)\, x$. Since g is of bounded variation, being absolutely continuous (see Theorem 2.13), by Theorem 4.8 its Fourier series converges uniformly on $[0,1]$. Since $g'(x) = f(x) - \hat{f}(0)$ a.e., (P4) yields $\hat{g}(k) = \frac{\hat{f}(k)}{2\pi i k}$ for $k \neq 0$. Thus $\sum_{k\neq 0} \frac{\hat{f}(k)}{2\pi i k} e^{2\pi i k x}$ converges uniformly on $[0,1]$. In this context, recall (4.23).

31. The computations made in Section 4.4 show that $(S_n\phi)'(x) = \sum_{k=1}^n 2\cos(2\pi k x) = D_n(x) - 1$. Thus $\sqrt{1 + [(S_n\phi)'(x)]^2} - |(S_n\phi)'(x)| = \frac{1}{\sqrt{1+[(S_n\phi)'(x)]^2} + |(S_n\phi)'(x)|} \leq 1$. Since $\left|\int_0^1 |(S_n\phi)'(x)|\, dx - \|D_n\|_{L^1[0,1]}\right| = \left|\int_0^1 |D_n(x) - 1|\, dx - \int_0^1 |D_n(x)|\, dx\right| \leq \int_0^1 1\, dx = 1$, we can conclude by (4.13).

32. (i) Suppose first that f is also continuous. For $k \neq 0$, set $g(x) = \frac{e^{-2\pi i k x}}{-2\pi i k}$. By the mean-value theorem, given $x_{j-1} < x_j$ in $[0,1]$, there exists $\xi_j \in (x_{j-1}, x_j)$ such that $f(x_j)\,[g(x_j) - g(x_{j-1})] = f(x_j)\,(x_j - x_{j-1})\, e^{-2\pi i k \xi_j} = [f(x_j) - f(\xi_j)]\,(x_j - x_{j-1})\, e^{-2\pi i k \xi_j} + f(\xi_j)\,(x_j - x_{j-1})\, e^{-2\pi i k \xi_j}$. Consequently, using Riemann sums in (4.1), we deduce that, given $\varepsilon > 0$, we have $|\hat{f}(k) - \sum_{j=1}^m f(x_j)\,[g(x_j) - g(x_{j-1})]| \leq \varepsilon$ for all sufficiently small partitions $0 = x_0 < x_1 < \cdots < x_m = 1$ of $[0,1]$. The previous sum can be expressed as $f(1)g(1) - f(x_1)g(0) - \sum_{j=1}^{m-1} [f(x_{j+1}) - f(x_j)]\, g(x_j)$. By periodicity, the first two terms equal $[f(0) - f(x_1)]\, g(0)$ and, due to continuity, this will not exceed ε in absolute value, provided that we choose a sufficiently fine partition. Thus $|\hat{f}(k)| \leq 2\varepsilon + \sum_{j=1}^{m-1} |f(x_{j+1}) - f(x_j)| \cdot |g(x_j)| \leq 2\varepsilon + \frac{V_f[0,1]}{2\pi|k|}$ since $|g(x)| \leq \frac{1}{2\pi|k|}$. In the limit $\varepsilon \to 0$ we obtain $|\hat{f}(k)| \leq \frac{V_f[0,1]}{2\pi|k|}$ for $k \neq 0$.

If f is merely of bounded variation, consider for $n \geq 1$ the continuous functions $f_n(x) = n\int_x^{x+\frac{1}{n}} f(s)\, ds = n\int_0^{1/n} f(x+s)\, ds$. Note that for every finite sequence $0 \leq x_0 < x_1 < \cdots < x_m \leq 1$ we have that $\sum_{j=1}^m |f_n(x_j) - f(x_{j-1})| \leq n\int_0^{1/n} \sum_{j=1}^m |f(x_j + s) - f(x_{j-1} + s)|\, ds \leq V_f[0,1]$ since the integrand is majorised by $V_f[0,1]$. Thus $V_{f_n}[0,1] \leq V_f[0,1]$, and the preamble ensures

$|\widehat{(f_n)}(k)| \le \frac{V_f[0,1]}{2\pi|k|}$ for all $k \neq 0$. On the other hand, a straightforward computation yields $\widehat{(f_n)}(k) = e^{\pi i \frac{k}{n}} \hat{f}(k) \frac{\sin(\frac{\pi k}{n})}{\frac{\pi k}{n}}$, so that $\lim_{n\to\infty} |\widehat{(f_n)}(k)| = |\hat{f}(k)|$, and we can conclude the proof.

(ii) By Theorem 4.8, $(S_n f)(x) \to \frac{f(x+)-f(x-)}{2}$ boundedly on $[0,1]$. Moreover, the right side equals $f(x)$ except perhaps for a countable set of $x \in [0,1]$. By the dominated convergence theorem, $\int_0^1 f(x)g(x)\,dx = \lim_{n\to\infty} \int_0^1 (S_n f)(x) \cdot g(x)\,dx = \lim_{n\to\infty} \sum_{|k|\le n} \hat{f}(k) \int_0^1 g(x)e^{2\pi ikx} dx = \lim_{n\to\infty} \sum_{|k|\le n} \hat{f}(k)\hat{g}(-k)$. The change of index $k \mapsto -k$ yields the statement.

33. Let us first show that $f \in L^1[0,1]$ is positive definite if and only if the defining inequality holds for trigonometric polynomials. The necessity is plain. Conversely, assume that the inequality holds for all trigonometric polynomials and let $u \in C_{\text{per}}[0,1]$. Choose trigonometric polynomials $\{u_n\}_{n\ge1}$ that converge uniformly to u, see Problem 16. The dominated converge theorem yields $\lim_{n\to\infty} \int_0^1 \int_0^1 f(x-y)\overline{u_n(x)}\, u_n(y)\,dxdy = \int_0^1 \int_0^1 f(x-y)\overline{u(x)}\, u(y)\,dxdy$, thus checking the sufficieny.

Assuming that $f \in L^1[0,1]$ is positive definite, the choice $u(x) = e^{2\pi ikx}$ in the defining inequality yields $\hat{f}(k) \ge 0$. On the other hand, if all Fourier coefficients of $f \in L^1[0,1]$ are nonnegative and if $u(x) = \sum_{|k|\le n} c_k\, e^{2\pi ikx}$, then $\int_0^1 \int_0^1 f(x-y)\overline{u(x)}\, u(y)\,dxdy = \sum_{-n\le k,j\le n} \overline{c}_k\, c_j \int_0^1 \int_0^1 f(x-y)e^{-2\pi ik(x-y)}e^{2\pi i(j-k)y}\,dxdy = \sum_{-n\le k,j\le n} \overline{c}_k\, c_j\, \hat{f}(k) \int_0^1 e^{2\pi i(j-k)y}\,dy = \sum_{-n\le k\le n} |c_k|^2\, \hat{f}(k) \ge 0$. The preamble shows now that f is positive definite.

4.6 Notes to Chapter 4

1. In general, nonlinear operations on, say, continuous periodic functions, do not preserve the convergence properties of the Fourier series. For example, there are functions $f \in C_{\text{per}}[0,1]$ having uniformly convergent Fourier series while the Fourier series of f^2 diverges on an uncountable set. Also, one can find $f \in C_{\text{per}}[0,1]$ with a uniformly convergent Fourier series and such that the Fourier series of $|f|$ diverges at some points. For results of this type we refer to Edwards (1967).

2. In addition to the pointwise convergence or the convergence in the mean (of $L^1[0,1]$ or of $L^2[0,1]$) of a Fourier series, the concept of *distributional convergence*, referring to the equality

$$\lim_{n\to\infty} \int_0^1 (S_n f)(x) \cdot g(x)\,dx = \int_0^1 f(x)g(x)\,dx$$

for each infinitely differentiable periodic function g, is a gateway to a transparent correlation with differentiation. The proper framework for these considerations is that of distribution theory – see Chapter 6. It turns out that any trigonometric series $\sum_{k\in\mathbb{Z}} c_k\, e^{2\pi ikx}$ in which $c_k = O(|k|^m)$ for some $m \ge 0$ can be regarded as the Fourier series of a distribution, in terms of which the coefficients c_k are expressible in a way that extends the formulas (1.9), and to which this series is distributionally convergent.

3. The considerations in this chapter offer some insight into the convergence of the Fourier series associated to functions in $L^2[0,1]$ and in $L^1[0,1]$. The general case of $f \in L^p[0,1]$ with $p \ge 1$ is also of interest. A basic result valid for any $p \in [1,\infty)$ is the so-called Hausdorff–Young theorem, which basically says that the operator $f \mapsto \{\hat{f}(k)\}_{k\in\mathbb{Z}}$ is bounded from $L^p[0,1]$ to $l^q(\mathbb{Z})$, where $q = \frac{p}{p-1}$ is the index conjugate to p. The special cases $p = 1$ and $p = 2$ are covered by (P2) and the discussion in Section 4.3, respectively. For $p \in (1,2)$, the result can be proved by interpolating between the cases $p = 1$ and $p = 2$, see Pinsky (2009). As for the case $p > 2$, of central importance is the *Hilbert transform* H, defined on the set $\mathcal{P}$ of trigonometric polynomials by the formula

$$\Big(H\big[\sum_{|k|\le n} c_k\, e^{2\pi ikt}\big]\Big)(x) = -i \sum_{1\le k\le n} c_k e^{2\pi ikx} + i \sum_{-n\le k\le -1} c_k e^{2\pi ikx}\,.$$

The Hilbert transform gives[43] a convenient way of expressing the partial sums $(S_n f)(x)$ in terms of the norm-preserving operators $(T_n^{\pm} f)(x) = e^{\pm 2\pi inx} f(x)$ since

$$(S_n f)(x) = \frac{T_n^+(H[T_n^- f])(x) - T_n^-(H[T_n^+ f])(x)}{2i} + \frac{\hat{f}(n)\, e^{2\pi inx} + \hat{f}(-n)\, e^{-2\pi inx}}{2}$$

for all $x \in [0, 1]$ and $n \geq 1$, whenever $f \in \mathcal{P}$. Clearly $\|Hf\|_{L^2[0,1]} \leq \|f\|_{L^2[0,1]}$ for any $f \in \mathcal{P}$. This allows us to extend H as a bounded linear operator on $L^2[0, 1]$. Moreover, it is relatively easy to show for any integer $k \geq 1$ the existence of a constant $C_{2k} > 0$ such that $\|Hf\|_{L^{2k}[0,1]} \leq C_{2k}\|f\|_{L^{2k}[0,1]}$ for all $f \in \mathcal{P}$. Since $\mathcal{P}$ is dense in $L^{2k}[0, 1]$, H admits an extension as a bounded linear operator on $L^{2k}[0, 1]$ for any integer $k \geq 1$. Interpolation now yields the boundedness of H on all intermediate spaces $L^p[0, 1]$ with $p \in (2, \infty)$. Boundedness for $p \in (1, 2)$ follows by using the duality of norms, namely $\|Hf\|_{L^p[0,1]} = \sup_{g \neq 0,\, g \in \mathcal{P}} \frac{\left|\int_0^1 f(x)\,\overline{g(x)}\,dx\right|}{\|g\|_{L^q[0,1]}}$ with $q = \frac{p}{p-1}$, in combination with Hölder's inequality[44] and the boundedeness result on $L^q[0, 1]$ with $q \in (2, \infty)$; see Pinsky (2009). Consequently, H is a bounded linear operator on $L^p[0, 1]$ whenever[45] $1 < p < \infty$. The existence of a constant $C_p > 0$ such that $\|Hf\|_{L^p[0,1]} \leq C_p\|f\|_{L^{2k}[0,1]}$ for all $f \in \mathcal{P}$ and the density of the trigonometric polynomials in $L^p[0, 1]$ yields $\lim_{n\to\infty} \|S_n f - f\|_{L^p[0,1]} = 0$ for any $f \in L^p[0, 1]$ with $p \in (1, \infty)$. This follows from the above identity that relates $S_n f$ to the conjugate function of $F_n^{\pm} = T_n^{\pm} f$, $H[F_n^{\pm}]$, and from the fact that the validity of (P5) is ensured by the inclusion[46] $L^p[0, 1] \subset L^1[0, 1]$ for $1 < p < \infty$.

Somewhat related to the previous considerations is the fact that if $f \in L^p[0, 1]$ with $1 < p < \infty$, then $S_n f \to f$ a.e. as $n \to \infty$. Carleson (1966) proved this for $p = 2$, and the general case is due to Hunt (1968). These results rely on the fact that convergence is a consequence of so-called weak L^p-inequalities for the distribution function of an associated maximal function, see Hunt (1976). The concept of a maximal function, due to Hardy and Littlewood, is inspired by the result presented in Lemma 4.7: for a locally integrable function $f : \mathbb{R} \to \mathbb{C}$, instead of considering the limit $\lim_{h\downarrow 0}\left(\frac{1}{2h}\int_{x-h}^{x-h} f(t)\,dt\right)$ that equals $f(x)$ a.e., one looks at $f^*(x) = \sup_{h\in(0,1)} \frac{1}{2h}\int_{x-h}^{x-h} |f(t)|\,dt$. It turns out that not only is it much easier to use f^* than f, but an in-depth study of f^* has also the advantage of revealing how various parts of the function f contribute to the convergence process.

4. Many subtle facets of the theory of Fourier series present close connections with complex analysis. While a full account is beyond our scope, we shall provide some background material in Chapter 7.

5. The concept of a Fourier series relies on studying the representation of certain functions as a superposition of the basic building blocks $e_k(x) = e^{2\pi ikx}$ with $k \in \mathbb{Z}$. Note that e_k are the continuous homeomorphisms between the quotient group $\mathbb{R}/2\pi\mathbb{Z}$ and $\mathbb{C} \setminus \{0\}$, which can be identified with the group of linear invertible transformations from $\mathbb{C}$ to $\mathbb{C}$. This group theoretic nature of classical Fourier series can be generalised to a study of continuous homeomorphisms from a topological group[47] G to the space $GL(V)$ of the invertible linear transformations of a finite-dimensional vector space V. A brief excursion into these topics will be provided in Part II.

43 The identity (∗) is obtained for a trigonometric polynomial f by subtracting the identity $e^{-2\pi inx}(H[e^{2\pi int} f(t)])(x) = i\sum_{k\leq -n-1} \hat{f}(k)\, e^{2\pi ikx} - i\sum_{k\geq 1-n} \hat{f}(k)\, e^{2\pi ikx}$ from the identity $e^{2\pi inx}(H[e^{-2\pi int} f(t)])(x) = i\sum_{k\leq n-1} \hat{f}(k)\, e^{2\pi ikx} - i\sum_{k\geq n+1} \hat{f}(k)\, e^{2\pi ikx}$.

44 Stating that $\left|\int_0^1 F(x)G(x)\,dx\right| \leq \left(\int_0^1 |F(x)|^p dx\right)^{\frac{1}{p}}\left(\int_0^1 |G(x)|^q dx\right)^{\frac{1}{q}}$ whenever $F \in L^p[0, 1]$ and $G \in L^q[0, 1]$ with $p > 1$ and $q = \frac{p}{p-1}$, see Hewitt and Stromberg (1965).

45 In contrast to this, $\sup_{f\in\mathcal{P},\, f\neq 0} \frac{\|Hf\|_{L^1[0,1]}}{\|f\|_{L^1[0,1]}} = \infty$. However, if $f \max\{0, \ln|f|\} \in L^1[0, 1]$, then $Hf \in L^1[0, 1]$. For these matters, see Pinsky (2009).

46 This is an immediate consequence of Hölder's inequality.

47 This is a group endowed with a topology such that the group operation and the operation of taking inverses are continuous with respect to the topology.

5

Fourier transforms

For $f \in L^2[-\frac{1}{2}, \frac{1}{2}]$, recall from Chapter 3 the Fourier series representation $f(x) = \sum_{k\in\mathbb{Z}} \hat{f}(k)\, e^{2\pi ikx}$ in $L^2[-\frac{1}{2}, \frac{1}{2}]$, with $\int_{-\frac{1}{2}}^{\frac{1}{2}} |f(x)|^2 dx = \sum_{k\in\mathbb{Z}} |\hat{f}(k)|^2$. This may be thought of as a decomposition of a periodic function (of period 1) into simpler harmonics: we extend $f : \mathbb{R} \to \mathbb{C}$ by setting[1] $f(x+1) = f(x)$ for $x \in \mathbb{R}$. The choice of a unit period was made for convenience: for $f \in L^2[-\frac{T}{2}, \frac{T}{2}]$, extended to a T-periodic function $f : \mathbb{R} \to \mathbb{C}$, the change of variables $F(x) = f(Tx)$ permits us to translate the statements for $F \in L^2[-\frac{1}{2}, \frac{1}{2}]$ into

$$f(x) = \sum_{k\in\mathbb{Z}} \Big(\frac{1}{T} \int_{-\frac{T}{2}}^{\frac{T}{2}} f(y)\, e^{-2\pi ik\frac{y}{T}} dy \Big)\, e^{2\pi ik\frac{x}{T}} \tag{5.1}$$

in $L^2[-\frac{T}{2}, \frac{T}{2}]$, with

$$\int_{-\frac{T}{2}}^{\frac{T}{2}} |f(x)|^2\, dx = \frac{1}{T} \sum_{k\in\mathbb{Z}} \Big| \int_{-\frac{T}{2}}^{\frac{T}{2}} f(y)\, e^{-2\pi ik\frac{y}{T}}\, dy \Big|^2, \tag{5.2}$$

the appropriate harmonics being $e_k(x) = \frac{1}{\sqrt{T}}\, e^{2\pi ik\frac{x}{T}}$ for $k \in \mathbb{Z}$. On the other hand, if $f : \mathbb{R} \to \mathbb{C}$ is not periodic, we can first restrict it to some interval $[-\frac{T}{2}, \frac{T}{2}]$ and thereafter extend the obtained function to a periodic function $f_T : \mathbb{R} \to \mathbb{C}$ (of period T), hoping to recover the function f in the limit $T \to \infty$. Setting $\xi_k = \frac{k}{T}$ for $k \in \mathbb{Z}$ and defining

$$\hat{f}(\xi_k) = \int_{-\frac{T}{2}}^{\frac{T}{2}} f(x)\, e^{-2\pi ix\xi_k}\, dx, \tag{5.3}$$

[1] We may alter, if necessary, the value of $f(\frac{1}{2})$ to $f(-\frac{1}{2})$ since a set of Lebesgue measure zero has no effect at all within the framework of L^2-theory.

if the restriction of f to $[-\frac{T}{2}, \frac{T}{2}]$ belongs to $L^2[-\frac{T}{2}, \frac{T}{2}]$, then (5.1) and (5.2) yield

$$f_T(x) = \sum_{k\in\mathbb{Z}} \hat{f}(\xi_k)\, e^{2\pi i x\xi_k}\, [\xi_k - \xi_{k-1}]\,, \tag{5.4}$$

with the series convergent in $L^2[-\frac{T}{2}, \frac{T}{2}]$, and with

$$\int_{-\frac{T}{2}}^{\frac{T}{2}} |f(x)|^2\, dx = \sum_{k\in\mathbb{Z}} |\hat{f}(\xi_k)|^2\, [\xi_k - \xi_{k-1}]\,. \tag{5.5}$$

The sums on the right-hand side of (5.4) and (5.5) are similar to Riemann sums, with the grid of points $\{\xi_k\}$ getting finer as $T \to \infty$, so passing formally to the limit $T \to \infty$ in (5.3), (5.4) and (5.5), we expect that

$$f(x) = \int_{\mathbb{R}} \hat{f}(\xi)\, e^{2\pi i x\xi}\, d\xi\,, \tag{5.6}$$

with

$$\int_{\mathbb{R}} |f(x)|^2\, dx = \int_{\mathbb{R}} |\hat{f}(\xi)|^2\, d\xi\,, \tag{5.7}$$

where $\hat{f} = \mathcal{F}(f)$ is given by

$$\hat{f}(\xi) = \int_{\mathbb{R}} f(x)\, e^{-2\pi i x\xi}\, dx\,. \tag{5.8}$$

The formulas (5.6), (5.7) and (5.8) are referred to as the Fourier inversion formula, the Plancherel identity and the Fourier transform formula, respectively. We will prove their validity for $f \in L^2(\mathbb{R})$. Note that the Fourier transform (5.8) is well-defined for $f \in L^1(\mathbb{R})$ but[2] $L^2(\mathbb{R}) \not\subset L^1(\mathbb{R})$, so that the meaning of (5.8) for $f \in L^2(\mathbb{R})$ is nontrivial. Even for $f \in L^1(\mathbb{R})$, a direct approach stalls right after defining $\hat{f}$, since we might have $\hat{f} \notin L^1(\mathbb{R})$, see Exercise 10, in which case (5.6) is questionable. However, for the Schwartz class $\mathcal{S}(\mathbb{R})$ of smooth functions that, together with their derivatives, are rapidly decreasing far out, one can rigorously justify the formal approach sketched above. As we shall see, the class $\mathcal{S}(\mathbb{R})$ is tailored to ensure the validity of (5.6), (5.7) and (5.8) in a direct and transparent way. The density of $\mathcal{S}(\mathbb{R})$ in $L^2(\mathbb{R})$ permits us to extend the validity of (5.6), (5.7) and (5.8) to $L^2(\mathbb{R})$. Moreover, the density of $\mathcal{S}(\mathbb{R})$ in $L^1(\mathbb{R})$ is instrumental in deriving some fundamental properties of the Fourier transform acting on functions in $L^1(\mathbb{R})$. The special features of the Fourier transform that are derived in this chapter turn out to be very useful in a wide range of contexts, some of which we will discuss in Part II.

[2] While $L^2[0,T] \subset L^1[0,T]$ for every $T > 0$, note that $x \mapsto \frac{1}{1+|x|}$ belongs to $L^2(\mathbb{R}) \setminus L^1(\mathbb{R})$, while $f(x) = \frac{1}{\sqrt{x}}$ for $x \in (0,1)$, with $f(x) = 0$ otherwise, belongs to $L^1(\mathbb{R}) \setminus L^2(\mathbb{R})$.

5.1 Rapidly decreasing smooth functions

A function $\phi : \mathbb{R} \to \mathbb{C}$ belongs to the Schwartz class $\mathcal{S}(\mathbb{R})$ if ϕ is infinitely many times differentiable and $x^n \phi^{(m)}(x) \to 0$ as $|x| \to \infty$ for all integers $m, n \geq 0$: $\mathcal{S}(\mathbb{R})$ consists of smooth functions whose derivatives (including the function itself) decay at infinity faster than any power: we say that these functions are rapidly decreasing.

For example, an inductive argument shows that $\partial_x^n(e^{-x^2}) = P_n(x)\, e^{-x^2}$ for some polynomial P_n of degree $n \geq 0$, so that $\phi(x) = e^{-x^2}$ belongs to $\mathcal{S}(\mathbb{R})$. On the other hand, $f(x) = \frac{1}{1+x^2}$ and $g(x) = e^{-|x|}$ do not belong to $\mathcal{S}(\mathbb{R})$. Indeed, f is smooth but $x^2 f(x)$ does not decay to 0 for $|x| \to \infty$, and, while g decays at infinity faster than any power, it is not smooth.

One can easily see that $\mathcal{S}(\mathbb{R})$ is a vector space. Moreover, it is an algebra: if $f, g \in \mathcal{S}(\mathbb{R})$, then their product $fg \in \mathcal{S}(\mathbb{R})$. We can also define a notion of convergence: $f_k \to f$ in $\mathcal{S}(\mathbb{R})$ means $\|f_k - f\|_{n,m} = \sup_{x\in\mathbb{R}} \left| x^n\, [f_k^{(m)}(x) - f^{(m)}(x)] \right| \to 0$ as $k \to \infty$, for all integers $n, m \geq 0$. This notion of convergence is compatible with the vector space structure of $\mathcal{S}(\mathbb{R})$: if $f_k \to f$, $g_k \to g$ in $\mathcal{S}(\mathbb{R})$ and $\lambda_k \to \lambda$ in $\mathbb{C}$, then $f_k + g_k \to f + g$ and $\lambda_k f_k \to \lambda f$ in $\mathcal{S}(\mathbb{R})$. However, $\mathcal{S}(\mathbb{R})$ is not a Banach space[3] but only a *Fréchet space*: $\mathcal{S}(\mathbb{R})$ is a complete metric space, with the metric $d(f,g) = \sum_{m,n\geq 0} \frac{\|f-g\|_{n,m}}{2^{n+m}\,(1+\|f-g\|_{n,m})}$, see Exercise 1. For this reason, several properties worked out for Banach spaces in Chapter 3 (e.g. concerning the dual of a Banach space) cannot be taken for granted. This shortcoming is compensated by the fact that the operators of differentiation ∂_x^m are continuous for every integer $m \geq 1$: if $f_k \to f$ in $\mathcal{S}(\mathbb{R})$, then $f_k^{(m)} \to f^{(m)}$ in $\mathcal{S}(\mathbb{R})$. The latter feature is one of the main reasons for introducing the space $\mathcal{S}(\mathbb{R})$ in Fourier analysis. We point out that in distribution theory one introduces the topological vector space of the test functions, $\mathcal{D}(\mathbb{R})$, on which differentiation operators are also continuous, $f : \mathbb{R} \to \mathbb{C}$ being a test function if it is smooth and has compact support[4] (see Chapter 6); with $f_n \to 0$ in $\mathcal{D}(\mathbb{R})$ if concentration and smallness occur simultaneously, that is, if there is a compact set $K \subset \mathbb{R}$ with $\text{supp}(f_n) \subset K$ for all $n \geq 1$ and if $f_n^{(k)} \to 0$ uniformly on K for any integer $k \geq 0$. The reason why we rely on $\mathcal{S}(\mathbb{R})$ rather than $\mathcal{D}(\mathbb{R}) \subset \mathcal{S}(\mathbb{R})$ in studying Fourier transforms is that $\hat{f} \in \mathcal{S}(\mathbb{R})$ whenever $f \in \mathcal{S}(\mathbb{R})$, while the only function $\varphi \in \mathcal{D}(\mathbb{R})$ with $\hat{\varphi} \in \mathcal{D}(\mathbb{R})$ is the zero function $\varphi \equiv 0$; see Exercise 6.14 for the multi-dimensional form of this result. Note that $\mathcal{S}(\mathbb{R})$ is dense in $L^1(\mathbb{R})$ and in $L^2(\mathbb{R})$ since already $\mathcal{D}(\mathbb{R}) \subset \mathcal{S}(\mathbb{R})$ has this property (see Exercise 14).

[3] There is no norm compatible with its topology, see Exercise 6.

[4] That is, $f(x) = 0$ for all $x \in \mathbb{R} \setminus K$ for some compact set $K \subset \mathbb{R}$. The support of f, $\text{supp}(f)$, is the complement of the largest open set on which f vanishes, or, equivalently, the closure of the set of points where f is nonzero .

Since $\mathcal{S}(\mathbb{R}) \subset L^1(\mathbb{R})$, the definition of its Fourier transform $\hat{f}$ of $f \in \mathcal{S}(\mathbb{R})$ by means of (5.8) is not problematic. The following table gathers some of the basic properties of the Fourier transform acting on $\mathcal{S}(\mathbb{R})$: for $h \in \mathbb{R}$ and $\delta \neq 0$, we have

Function in $\mathcal{S}(\mathbb{R})$ (variable x)	Fourier transform $\mathcal{F}$ (variable ξ)
$f(x+h)$	$e^{2\pi ih\xi}\,\hat{f}(\xi)$
$e^{-2\pi ixh}\,f(x)$	$\hat{f}(\xi+h)$
$f(\delta x)$	$\frac{1}{\lvert\delta\rvert}\,\hat{f}\big(\frac{\xi}{\delta}\big)$
$f'(x)$	$2\pi i\xi\,\hat{f}(\xi)$
$-2\pi ix\,f(x)$	$\frac{d}{d\xi}\,\hat{f}(\xi)$

The first two properties listed in the table follow at once from (5.8), while the third is established by a simple change of variables. To prove the fourth property, for a fixed $N > 0$ integration by parts gives

$$\int_{-N}^{N} f'(x)\,e^{-2\pi ix\xi}\,dx = \Big(f(x)\,e^{-2\pi ix\xi}\Big)\Big|_{x=-N}^{x=N} + 2\pi i\xi \int_{-N}^{N} f(x)\,e^{-2\pi ix\xi}\,dx\,.$$

Letting $N \to \infty$ yields the fourth property. To check the fifth property, note that for any sequence $h_n \to 0$ of nonzero real numbers, we have

$$\frac{\hat{f}(\xi+h_n)-\hat{f}(\xi)}{h_n} - \mathcal{F}\Big((-2\pi ix\,f(x)\Big)(\xi) = \int_{\mathbb{R}} f(x)\,e^{-2\pi ix\xi}\,\Big(\frac{e^{-2\pi ixh_n}-1}{h_n} + 2\pi ix\Big)\,dx\,.$$

Since $|e^{-2\pi ixh_n}-1|^2 = 2-2\cos(2\pi xh_n) = 4\sin^2(\pi xh_n) \le 4\pi^2x^2h_n^2$, the fifth property follows by the dominated convergence theorem.

A key feature is that, except for factors of $2\pi i$, the Fourier transform interchanges differentiation and multiplication by x. Note that the function class $\mathcal{S}(\mathbb{R})$ is closed under differentiation and multiplication by polynomials. It is actually invariant under the Fourier transform. Indeed, a repeated application of the fourth and fifth properties listed in the table yields

$$(2\pi i\xi)^p\,\partial_\xi^q\hat{f}(\xi) = \mathcal{F}\Big(\partial_x^p\{(-2\pi ix)^q\,f(x)\}\Big)(\xi) \tag{5.9}$$

for all integers $q, p \ge 0$. In particular, $\hat{f}$ is q-times differentiable. Moreover, since $x \mapsto \partial_x^p\{(-2\pi ix)^q\,f(x)\}$ belongs to $\mathcal{S}(\mathbb{R})$, for every $\xi \in \mathbb{R}$ we have

$$\Big|\mathcal{F}\Big(\partial_x^p\{(-2\pi ix)^q\,f(x)\}\Big)(\xi)\Big| \le \int_{\mathbb{R}} |\partial_x^p\{(-2\pi ix)^q\,f(x)\}|\,dx = M_{q,p} < \infty.$$

Combined with (5.9), this yields that $\xi \mapsto \partial_\xi^q \hat{f}(\xi)$ is rapidly decreasing. Consequently $\hat{f} \in \mathcal{S}(\mathbb{R})$. The next result builds upon this.

Theorem 5.1 *The Fourier transform maps $\mathcal{S}(\mathbb{R})$ onto itself, and (5.6), (5.7) and (5.8) hold.*

Proof Let us start by showing that in the context of $\mathcal{S}(\mathbb{R})$, the formal considerations that led us to (5.6), (5.7) and (5.8) can be justified rigorously.

Assume first that $f \in \mathcal{D}(\mathbb{R})$. Choose $M > 0$ such that $f(x) = 0$ for $|x| > M$. For $T > 2M + 1$, extend $f : [-T/2, T/2] \to \mathbb{C}$ to a continuously differentiable function $f_T : \mathbb{R} \to \mathbb{C}$ of period T. Dirichlet's theorem yields the uniform convergence

$$f(x) = \sum_{k\in\mathbb{Z}} \frac{1}{T}\, e^{2\pi ikx/T} \int_{-T/2}^{T/2} f(y)\, e^{-2\pi iky/T}\, dy, \qquad -\frac{T}{2} \le x \le \frac{T}{2},$$

with Plancherel's identity in the form

$$\int_{-T/2}^{T/2} |f(x)|^2\, dx = \sum_{k\in\mathbb{Z}} \frac{1}{T} \left| \int_{-T/2}^{T/2} f(y)\, e^{-2\pi iky/T}\, dy \right|^2 .$$

Note that the Fourier coefficients $\hat{f}(k) = \int_{-T/2}^{T/2} f(y)\, e^{-2\pi iky/T}\, dy$ are precisely the evaluations of the Fourier transform (5.8) at the points k/T, so that the right-hand sides of the previously displayed relations can be expressed as the Riemann-sums $\sum_{k\in\mathbb{Z}} \frac{1}{T}\, e^{2\pi ikx/T}\, \hat{f}\left(\frac{k}{T}\right)$ and $\sum_{k\in\mathbb{Z}} \frac{1}{T} \left|\hat{f}\left(\frac{k}{T}\right)\right|^2$, respectively. Since $f \in \mathcal{S}(\mathbb{R})$ and for every fixed $x \in \mathbb{R}$, the function $\xi \mapsto \hat{f}(\xi)\, e^{2\pi i\xi x}$ also belongs to $\mathcal{S}(\mathbb{R})$, in the limit $T \to \infty$ the latter sums converge to $\int_{\mathbb{R}} \hat{f}(\xi)\, e^{2\pi i\xi x}\, d\xi$ and $\int_{\mathbb{R}} |\hat{f}(\xi)|^2\, d\xi$. The fact that $f(x) = 0$ for $|x| \ge T/2$ confirms the validity of (5.6) and (5.7) since $\int_{-T/2}^{T/2} |f(x)|^2\, dx = \int_{\mathbb{R}} |f(x)|^2\, dx$.

The proof for $f \in \mathcal{S}(\mathbb{R})$ follows by means of suitable approximations by functions $f_n \in \mathcal{D}(\mathbb{R})$. Choose[5] a function $e : \mathbb{R} \to [0, 1]$ of class $\mathcal{D}(\mathbb{R})$, with $e(x) = 1$ for $|x| \le 1$ and $e(x) = 0$ for $|x| \ge 2$. For an integer $n \ge 1$ we define the smooth cut-off $f_n(x) = f(x)\, e(x/n)$ for $x \in \mathbb{R}$. Note that $f_n \in \mathcal{D}(\mathbb{R})$ coincides with f on $[-n, n]$, vanishes on $(-\infty, -2n] \cup [2n, \infty)$, and is such that $f_n(x) \to f(x)$ with $|f_n(x)| \le |f(x)|$ at every $x \in \mathbb{R}$. Moreover, $\hat{f}_n(\xi) \to \hat{f}(\xi)$ at every $\xi \in \mathbb{R}$ since

$$\begin{aligned} |\hat{f}_n(\xi) - \hat{f}(\xi)| &= \left| \int_{\mathbb{R}} [f_n(x) - f(x)]\, e^{-2\pi ix\xi}\, dx \right| \le \int_{\mathbb{R}} |f_n(x) - f(x)|\, dx \\ &\le 2 \int_{|x|\ge n} |f(x)|\, dx \to 0 \quad \text{for} \quad n \to \infty \end{aligned}$$

by the dominated convergence theorem. Also, the sequence $\{\hat{f}_n\}_{n\ge1}$ is pointwise dominated by a

[5] A procedure for the construction of this type of function consists of defining for $c \in \mathbb{R}$ and $\varepsilon > 0$ the function $\varphi_{c,\varepsilon} \in \mathcal{D}(\mathbb{R})$ by $\varphi_{c,\varepsilon}(x) = \exp\left(-\frac{1}{\varepsilon^2-(x-c)^2}\right)$ if $|x - c| < \varepsilon$ and $\varphi_{c,\varepsilon}(x) = 0$ for $|x - c| \ge \varepsilon$. For real constants $a < b$ and $0 < \varepsilon < \frac{b-a}{2}$, we now define the nondecreasing smooth function $\phi_{a,\varepsilon} : \mathbb{R} \to [0, 1]$ and the nonincreasing smooth function $\psi_{b,\varepsilon} : \mathbb{R} \to [0, 1]$ by $\phi_{a,\varepsilon}(x) = \frac{\int_{-\infty}^x \varphi_{a,\varepsilon}(s)\, ds}{\int_{\mathbb{R}} \varphi_{a,\varepsilon}(s)\, ds}$ and $\psi_{b,\varepsilon}(x) = \frac{\int_x^\infty \varphi_{b,\varepsilon}(s)\, ds}{\int_{\mathbb{R}} \varphi_{b,\varepsilon}(s)\, ds}$ for $x \in \mathbb{R}$. Note that $\phi_{a,\varepsilon}(x) = 0$ for $x \le a - \varepsilon$ and $\phi_{a,\varepsilon}(x) = 1$ for $x \ge a + \varepsilon$, while $\psi_{b,\varepsilon}(x) = 1$ for $x \le b - \varepsilon$ and $\psi_{b,\varepsilon}(x) = 0$ for $x \ge b + \varepsilon$. Then $\theta = \phi_{a,\varepsilon}\psi_{b,\varepsilon} \in \mathcal{D}(\mathbb{R})$, with $\theta(x) = 1$ for $x \in [a + \varepsilon, b - \varepsilon]$ and $\theta(x) = 0$ for $x \le a - \varepsilon$ and $x \ge b + \varepsilon$. Moreover, θ is strictly increasing on $[a - \varepsilon, a + \varepsilon]$ and strictly decreasing on $[b - \varepsilon, b + \varepsilon]$; see Figure 5.1 for $a = -\frac{3}{2}$, $b = \frac{3}{2}$ and $\varepsilon = \frac{1}{2}$.

function $g \in L^1(\mathbb{R}) \cap L^2(\mathbb{R})$. Indeed, since

$$|\hat{f}_n(\xi)| = \Big|\int_{\mathbb{R}} f_n(x)\, e^{2\pi i x\xi}\, dx\Big| \le \int_{\mathbb{R}} |f_n(x|)\, dx \le \int_{\mathbb{R}} |f(x)|\, dx,$$

we may set $g(\xi) = \|f\|_{L^1(\mathbb{R})}$ for $|\xi| \le 1$. To find an appropriate expression to $g(\xi)$ if $|\xi| > 1$, we take advantage of the formula $\hat{f}_n(\xi) = -\frac{1}{4\pi^2\xi^2}\,\widehat{(f_n'')}(\xi)$ for $\xi \neq 0$. Since $f_n''(x) = f''(x)\, e\big(\frac{x}{n}\big) + \frac{2}{n}\, f'(x)\, e'\big(\frac{x}{n}\big) + \frac{1}{n^2}\, f(x)\, e''\big(\frac{x}{n}\big)$, we infer that $\int_{\mathbb{R}} |f_n''(x)|\, dx \le \alpha$ for all $n \ge 1$, where

$$\alpha = \int_{\mathbb{R}} |f''(x)|\, dx + 2 \sup_{s\in\mathbb{R}}\{|e'(s)|\} \int_{\mathbb{R}} |f'(x)|\, dx + \sup_{s\in\mathbb{R}}\{|e''(s)|\} \int_{\mathbb{R}} |f(x)|\, dx.$$

Consequently, for $|\xi| > 1$ and $n \ge 1$ we have

$$|\hat{f}_n(\xi)| = \frac{1}{4\pi^2\xi^2}\Big|\int_{\mathbb{R}} f_n''(x)\, e^{-2\pi i x\xi}\, dx\Big| \le \frac{1}{4\pi^2\xi^2}\int_{\mathbb{R}} |f_n''(x)|\, dx \le \frac{\alpha}{4\pi^2\xi^2},$$

so that we may set $g(\xi) = \frac{\alpha}{4\pi^2\xi^2}$ for $|\xi| > 1$.

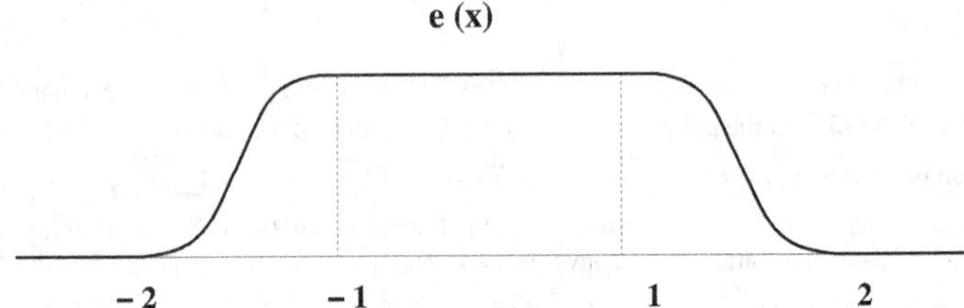

Figure 5.1 The auxiliary test function e.

The $L^1(\mathbb{R})$-domination of $\{\hat{f}_n\}_{n\ge 1}$ by g permits us to apply the dominated convergence theorem to justify (5.6) by a limiting process:

$$f(x) = \lim_{n\to\infty} f_n(x) = \lim_{n\to\infty} \int_{\mathbb{R}} \hat{f}_n(\xi)\, e^{2\pi i \xi x}\, d\xi = \int_{\mathbb{R}} \hat{f}(\xi)\, e^{2\pi i \xi x}\, d\xi,$$

due to the established validity of (5.6) for $f_n \in \mathcal{D}(\mathbb{R})$. On the other hand, the $L^2(\mathbb{R})$-domination allows us to implement the dominated convergence theorem to extend the validity of (5.7) from $\mathcal{D}(\mathbb{R})$ to $\mathcal{S}(\mathbb{R})$ since

$$\|f\|^2_{L^2(\mathbb{R})} = \lim_{n\to\infty} \|f_n\|^2_{L^2(\mathbb{R})} = \lim_{n\to\infty} \|\hat{f}_n\|^2_{L^2(\mathbb{R})} = \|\hat{f}\|^2_{L^2(\mathbb{R})}.$$

To complete the proof, we have to show that the Fourier transform $\mathcal{F}$ maps $\mathcal{S}(\mathbb{R})$ onto $\mathcal{S}(\mathbb{R})$. This follows at once from the invariance of $\mathcal{S}(\mathbb{R})$ under $\mathcal{F}$ by using the identity

$$f(x) = \int_{\mathbb{R}} \hat{f}(-\xi)\, e^{-2\pi i \xi x}\, d\xi \tag{5.10}$$

which displays $f \in \mathcal{S}(\mathbb{R})$ as the Fourier transform of the function $\xi \mapsto \hat{f}(-\xi)$, easily seen to be of class $\mathcal{S}(\mathbb{R})$ since $\hat{f} \in \mathcal{S}(\mathbb{R})$. The validity of (5.10) follows from (5.6) by means of the change of variables $\xi \mapsto (-\xi)$. □

5.2 Fourier transform for square integrable functions

The developments in the previous section permit us to explore the action of the Fourier transform on $L^2(\mathbb{R})$. Given $f \in L^2(\mathbb{R})$, choose $f_n \in \mathcal{S}(\mathbb{R})$ with $\|f_n - f\|_{L^2(\mathbb{R})} \to 0$, see Exercise 14. Using (5.7) for $\mathcal{S}(\mathbb{R})$ we deduce that $\{\hat{f}_n\}_{n\geq 1}$ is a Cauchy sequence in $L^2(\mathbb{R})$:

$$\begin{aligned}\|\hat{f}_n - \hat{f}_m\|_{L^2(\mathbb{R})} &= \|\widehat{f_n - f_m}\|_{L^2(\mathbb{R})} = \|f_n - f_m\|_{L^2(\mathbb{R})} \\ &\leq \|f - f_n\|_{L^2(\mathbb{R})} + \|f - f_m\|_{L^2(\mathbb{R})} \to 0 \quad \text{for} \quad n, m \to \infty .\end{aligned}$$

We identify $\hat{f} \in L^2(\mathbb{R})$ with the limit of the convergent sequence $\{\hat{f}_n\}_{n\geq 1}$ in $L^2(\mathbb{R})$. An inequality analogous to the one above confirms that $\hat{f}$ is not dependent upon the particular approximating functions $f_n \in \mathcal{S}(\mathbb{R})$ employed. Thus $\hat{f}$ is well-defined for $f \in L^2(\mathbb{R})$.

Theorem 5.2 *The Fourier transform $\mathcal{F}$ is a linear length-preserving bijection of $L^2(\mathbb{R})$, with inverse $\mathcal{F}^{-1}(g) = \mathcal{F}(g^\dagger)$ for $g \in L^2(\mathbb{R})$, where $g^\dagger(\xi) = g(-\xi)$ for $\xi \in \mathbb{R}$.*

Proof The linearity of $\mathcal{F}$ follows at once by going through the steps used to define $\mathcal{F}$. We extend the validity of (5.7) from $\mathcal{S}(\mathbb{R})$ to $L^2(\mathbb{R})$ as follows: given some $f \in L^2(\mathbb{R})$, pick $f_n \in \mathcal{S}(\mathbb{R})$ with $\|f_n - f\|_{L^2(\mathbb{R})} \to 0$, see Exercise 14, so that

$$\|f\|_{L^2(\mathbb{R})} = \lim_{n\to\infty} \|f_n\|_{L^2(\mathbb{R})} = \lim_{n\to\infty} \|\hat{f}_n\|_{L^2(\mathbb{R})} = \|\hat{f}\|_{L^2(\mathbb{R})}.$$

This shows that $\mathcal{F}$ is a length-preserving (and, in particular, injective) map of $L^2(\mathbb{R})$ into itself. To prove that $\mathcal{F}$ is onto, we will simply check that the linear operator $\mathcal{F}^{-1}$ defined in the statement is the inverse of $\mathcal{F}$. Given $g \in L^2(\mathbb{R})$, pick $g_n \in \mathcal{S}(\mathbb{R})$ with $\|g_n - g\|_{L^2(\mathbb{R})} \to 0$. Clearly $\|g_n^\dagger - g^\dagger\|_{L^2(\mathbb{R})} \to 0$. By definition, $\mathcal{F}(g^\dagger) = \lim_{n\to\infty} \mathcal{F}(g_n^\dagger)$ in $L^2(\mathbb{R})$. Since $\mathcal{F}$ is continuous on $L^2(\mathbb{R})$, we get $\mathcal{F}(\mathcal{F}(g^\dagger)) = \lim_{n\to\infty} \mathcal{F}(\mathcal{F}(g_n^\dagger))$. We know already that $\mathcal{F}$ is a bijection on $\mathcal{S}(\mathbb{R})$, so that (5.10) applied to $\mathcal{F}^{-1}(g_n)$ yields $\mathcal{F}^{-1}(g_n) = \mathcal{F}(g_n^\dagger)$ for all $n \geq 1$. Thus $\mathcal{F}(\mathcal{F}(g_n^\dagger)) = g_n \to g$, which means that $\mathcal{F}(\mathcal{F}(g^\dagger)) = g$. In particular, $\mathcal{F}$ is onto and hence a bijection. The previous relation then suffices to identify its inverse as in the statement of the theorem. □

Note that $(g^\dagger)^\dagger = g$, so that Theorem 5.2 yields $\mathcal{F}(\mathcal{F}(g)) = g^\dagger$ for all $g \in L^2(\mathbb{R})$. One iteration of this identity leads us to $\mathcal{F}^4 = I$ (the identity) on $L^2(\mathbb{R})$. This feature is expanded upon in Exercise 8.

We conclude our discussion of the Fourier transform on $L^2(\mathbb{R})$ by emphasising that the right-hand sides of (5.8) and (5.6) are not generally valid expressions in this setting, a correct interpretation of these formulas being

$$\hat{f}(\xi) = \lim_{n\to\infty} \int_{-n}^{n} f(x)\, e^{-2\pi i x\xi}\, dx \quad \text{in} \quad L^2(\mathbb{R}), \tag{5.11}$$

and

$$f(x) = \lim_{n\to\infty} \int_{-n}^{n} \hat{f}(\xi)\, e^{2\pi i x\xi}\, d\xi \quad \text{in} \quad L^2(\mathbb{R}). \tag{5.12}$$

Given some $f \in L^2(\mathbb{R})$, for every $n \geq 1$ the integrals in (5.11) and (5.12) are well-defined as the scalar product of two functions in $L^2[-n, n]$. To prove (5.11), note first that if f_n is the product of f and the characteristic function of the interval $[-n, n]$, then $\hat{f}_n(\xi) = \int_{-n}^{n} f(x)\, e^{-2\pi i x\xi}\, dx$ since $f_n \in L^1(\mathbb{R}) \cap L^2(\mathbb{R})$. Now $\|\hat{f} - \hat{f}_n\|^2_{L^2(\mathbb{R})} = \|f - f_n\|^2_{L^2(\mathbb{R})} \to 0$ as $n \to \infty$ by the dominated convergence theorem. The proof of (5.12) is similar. We would like to draw attention to the importance of (5.11). If $f(x) = \frac{1}{1+|x|}$, then $f \in L^2(\mathbb{R})$ so $\hat{f} \in L^2(\mathbb{R})$. However, $\int_{\mathbb{R}} f(x)e^{-2\pi i x\xi} dx$ is not well-defined since $f \notin L^1(\mathbb{R})$. Nevertheless, $\lim_{n\to\infty} \int_{-n}^{n} f(x)e^{-2\pi i x\xi} dx$ exists in $L^2(\mathbb{R})$, so that for all $\xi \in \mathbb{R}$ we can assign a number[6] to this limit, named $\hat{f}(\xi)$.

5.3 Fourier transform for integrable functions

The principal facts about the extension of the Fourier transform from $\mathcal{S}(\mathbb{R})$ to $L^1(\mathbb{R})$ are collected in the following theorem.

Theorem 5.3 *If $f \in L^1(\mathbb{R})$, then (5.8) defines its Fourier transform $\hat{f} : \mathbb{R} \to \mathbb{C}$, enjoying the properties:*

(i) $\hat{f}$ is bounded, with $|\hat{f}(\xi)| \leq \|f\|_{L^1(\mathbb{R})}$ for all $\xi \in \mathbb{R}$;

(ii) $\hat{f}$ is continuous;

(iii) (Riemann–Lebesgue) $\lim_{|\xi|\to\infty} \hat{f}(\xi) = 0$;

(iv) $\widehat{(f * g)} = \hat{f} \cdot \hat{g}$ for all $f, g \in L^1(\mathbb{R})$;

(v) $\hat{f} = 0$ if and only if $f = 0$ a.e.

In particular, Theorem 5.3 shows that the Fourier transform is a $1 : 1$ mapping of $L^1(\mathbb{R})$ into the class of bounded continuous functions which vanish at infinity. The map is not onto, see Exercise 21, the actual image being a very complicated set which is not completely characterised to date.

Proof $\hat{f}(\xi)$ exists as an ordinary Lebesgue integral, with $\left|\int_{\mathbb{R}} f(x)e^{-2\pi i x\xi} dx\right| \leq \int_{\mathbb{R}} |f(x)|\, dx = \|f\|_{L^1(\mathbb{R})}$. The continuity of $\hat{f}$ follows since $|\hat{f}(\xi_1) - \hat{f}(\xi_2)| = \left|\int_{\mathbb{R}} f(x)\,[e^{-2\pi i x\xi_1} - e^{-2\pi i x\xi_2}]\, dx\right| \leq \int_{\mathbb{R}} |f(x)| \left|e^{-2\pi i x\xi_1} - e^{-2\pi i x\xi_2}\right| dx \to 0$ as $|\xi_1 - \xi_2| \to 0$ by dominated convergence. To prove (iii), pick $f_n \in \mathcal{S}(\mathbb{R})$ with $\|f - f_n\|_{L^1(\mathbb{R})} \leq \frac{1}{n}$ for all integers $n \geq 1$; this being possible by density, see Exercise

[6] Since altering a.e. the values of a function in $L^2(\mathbb{R})$ is permissible.

13. Then $\widehat{f_n} \in \mathcal{S}(\mathbb{R})$ and (i) yields $|\widehat{f_n}(\xi) - \hat{f}(\xi)| \le \|f - f_n\|_{L^1(\mathbb{R})} \le \frac{1}{n}$, so that $\limsup_{|\xi|\to\infty} |\hat{f}(\xi)| \le \frac{1}{n} + \limsup_{|\xi|\to\infty} |\widehat{f_n}(\xi)| = \frac{1}{n}$ for any integer $n \ge 1$. This proves (iii). As for (iv), note that by Fubini's theorem $(f * g)(x) = \int_{\mathbb{R}} f(x-y)g(y)\,dy \in L^1(\mathbb{R})$ since $\int_{\mathbb{R}} |(f*g)(x)|\,dx \le \int_{\mathbb{R}}\int_{\mathbb{R}} |f(x-y)g(y)|\,dydx = \big(\int_{\mathbb{R}} |f(\xi)|\,d\xi\big)\big(\int_{\mathbb{R}} |g(y)|\,dy\big) = \|f\|_{L^1(\mathbb{R})} \cdot \|g\|_{L^1(\mathbb{R})}$, using the change of variables $\xi = x - y$ in some intermediate step. The previous estimate also validates the manipulations $\widehat{f * g}(\xi) = \int_{\mathbb{R}} \big(\int_{\mathbb{R}} f(x-y)g(y)\,dy\big)\, e^{-2\pi i x\xi}\,dx = \int_{\mathbb{R}}\int_{\mathbb{R}} f(x-y)\, e^{-2\pi i\xi(x-y)} g(y)\, e^{-2\pi i\xi y}\,dydx = \int_{\mathbb{R}} \hat{f}(\xi)\, g(y)\, e^{-2\pi i\xi y}\,dy = \hat{f}(\xi)\hat{g}(\xi)$ for every $\xi \in \mathbb{R}$. Finally, if $\hat{f} = 0$, then by Fubini's theorem and (5.10), we get $\int_{\mathbb{R}} f(x)g(x)\,dx = \int_{\mathbb{R}} f(x)\big(\int_{\mathbb{R}} \hat{g}(-\xi)\, e^{-2\pi i\xi x}\,d\xi\big)dx = \int_{\mathbb{R}}\int_{\mathbb{R}} f(x)e^{-2\pi i\xi x}dx\,\hat{g}(-\xi)\,d\xi = \int_{\mathbb{R}} \hat{f}(\xi)\,\hat{g}(-\xi)\,d\xi = 0$ for every $g \in \mathcal{S}(\mathbb{R})$. We now show that $\int_a^b f(x)\,dx = 0$ for all real constants $a < b$, which means that the absolutely continuous $F(x) = \int_0^x f(s)\,ds$ vanishes and therefore f, as its a.e. derivative, vanishes a.e. To verify the missing claim, fix a finite interval $[a, b]$ and choose, as in the proof of Theorem 5.1, functions $\psi_n \in \mathcal{D}(\mathbb{R})$ with $0 \le \psi_n(x) \le 1$ for all $x \in \mathbb{R}$, while $\psi_n(x) = 1$ for $x \in [a, b]$ and $\psi_n(x) = 0$ for $x < a - \frac{1}{n}$ and for $x > b + \frac{1}{n}$. By dominated convergence we get $0 = \int_{\mathbb{R}} f(x)\psi_n(x)\,dx \to \int_a^b f(x)\,dx$. The proof is complete. □

In contrast to the case of $L^2(\mathbb{R})$, the inverse Fourier transform cannot be applied directly to "undo" the Fourier transform by means of (5.6) since $\hat{f}$ might not belong to $L^1(\mathbb{R})$ or $L^2(\mathbb{R})$, an example being[7] the function $f \in L^1(\mathbb{R})$ from Exercise 10; see also Exercise 19. The difficulty is circumvented by the *Gauss summability* recipe

$$\lim_{t\downarrow 0} \mathcal{F}^{-1}\big(\hat{f}(\xi)\, e^{-2\pi^2\xi^2 t}\big)(x) = f(x) \;\text{ in }\; L^1(\mathbb{R}) \;\text{ for every }\; f \in L^1(\mathbb{R})\,, \qquad (5.13)$$

with the alternative *Abel summability* recipe described in Exercise 22.

To validate (5.13), note that if for $t > 0$ we set $\theta_t(x) = \frac{1}{\sqrt{2\pi t}} \exp\big(-\frac{x^2}{2t}\big)$ for $x \in \mathbb{R}$, then $\widehat{\theta_t}(\xi) = e^{-2\pi^2\xi^2 t}$, see Exercise 2 and the table from Section 5.1. Exercise 18(ii) yields $\mathcal{F}(\theta_t * f) = \widehat{\theta_t} \cdot \hat{f}$ for $f \in L^1(\mathbb{R})$, so that it suffices to verify that $\lim_{t\downarrow 0} \theta_t * f = f$ in $L^1(\mathbb{R})$. For this, note that $\|\theta_t * f - f\|_{L^1(\mathbb{R})} = \int_{\mathbb{R}} \big| \int_{\mathbb{R}} f(x-y)\theta_t(y)\,dy - f(x)\big|\,dx = \int_{\mathbb{R}} \big| \int_{\mathbb{R}} [f(x-y) - f(x)]\theta_t(y)\,dy\big|\,dx$ since $\int_{\mathbb{R}} \theta_t(y)\,dy = \int_{\mathbb{R}} e^{-\pi s^2}\,ds = 1$. Given $\delta > 0$, we can estimate this by $\int_{\mathbb{R}} \big| \int_{\mathbb{R}} |f(x-y) - f(x)|\,\theta_t(y)\,dy\,dx = \int_{\mathbb{R}} \|\tau_y f - f\|_{L^1(\mathbb{R})}\theta_t(y)\,dy \le \int_{-\delta}^{\delta} \|\tau_y f - f\|_{L^1(\mathbb{R})}\theta_t(y)\,dy + \int_{|y|>\delta} \|\tau_y f - f\|_{L^1(\mathbb{R})}\theta_t(y)\,dy$. The first integral is small for $\delta > 0$ sufficiently small by Exercise 14(ii) since $\int_{\mathbb{R}} \theta_t(y)\,dy = 1$. The second integral is less than $2\|f\|_{L^1(\mathbb{R})} \int_{|y|>\delta} \theta_t(y)\,dy = \frac{4\|f\|_{L^1(\mathbb{R})}}{\sqrt{\pi}} \int_{\frac{\delta}{\sqrt{2t}}}^{\infty} e^{-s^2}\,ds$, which vanishes in the limit $t \downarrow 0$ for any fixed $\delta > 0$ by dominated convergence. This completes the argument.

As a consequence of (5.13), (5.6) holds if $f, \hat{f} \in L^1(\mathbb{R})$ since then

$$\lim_{t\downarrow 0} \int_{\mathbb{R}} e^{-2\pi^2\xi^2 t}\,\hat{f}(\xi)\, e^{2\pi i x\xi}\,d\xi = \int_{\mathbb{R}} \hat{f}(\xi)\, e^{2\pi i x\xi}\,d\xi \;\text{ for every }\; x \in \mathbb{R}\,,$$

by dominated convergence. Since $f_n \to f$ in $L^1(\mathbb{R})$ forces $f_n \to f$ in $L^1[-k, k]$ for every integer $k \ge 1$, Exercise 4.13 and (5.13) yield $f(x) = \int_{\mathbb{R}} \hat{f}(\xi)\, e^{2\pi i x\xi}\,d\xi$ for a.e. $x \in \mathbb{R}$.

[7] Exercise 10 states that $\hat{f} \notin L^1(\mathbb{R})$, while $\hat{f} \notin L^2(\mathbb{R})$ is prevented by $f \notin L^2(\mathbb{R})$, in view of Theorem 5.2.

5.4 Exercises

Exercises 1, 2, 6, 7 and 13 are an integral part of the presented theory and they were confined to this section since the details of their solution are not crucial at first. A number of exercises are routine: 3–5, 11, 12, 14, 15. Finally, Exercises 8–10 represent results that should be known if one would like to go beyond a rudimentary knowledge of the Fourier transform. Exercises 16–23 are somewhat more specialised.

5.4.1 Statements

1. Prove that the metric space $\mathcal{S}(\mathbb{R})$ is complete.

2. If $f(x) = e^{-\pi x^2}$, show that $\hat{f}(\xi) = f(\xi)$.

3. Show that if $f_n \to f$ in $\mathcal{S}(\mathbb{R})$, then $f_n \to f$ in $L^1(\mathbb{R})$ and $f_n \to f$ in $L^2(\mathbb{R})$.

4. Prove the following Riemann-sums approximation result: for $f \in \mathcal{S}(\mathbb{R})$ we have $\lim_{h\downarrow 0} \sum_{k\in\mathbb{Z}} h\, f(hk) = \int_{\mathbb{R}} f(x)\, dx$ and $\lim_{h\downarrow 0} \sum_{k\in\mathbb{Z}} h\, |f(hk)|^2 = \int_{\mathbb{R}} |f(x)|^2\, dx$.

5. Show that $(f, g)_{L^2(\mathbb{R})} = (\hat{f}, \hat{g})_{L^2(\mathbb{R})}$ for all $f,\, g \in L^2(\mathbb{R})$.

6. Show that the vector space $\mathcal{S}(\mathbb{R})$ is not normable, that is, its topology is not induced by a norm.

7. Show that, in contrast to $\mathcal{S}(\mathbb{R})$, $\mathcal{D}(\mathbb{R})$ is not metrizable, that is, its topology is not induced by a metric.

8. By[8] Exercise 3.32(iii), the Hermite functions

$$h_k(x) = \frac{2^{1/4}(-1)^k}{\sqrt{(4\pi)^k k!}}\, e^{\pi x^2} \frac{d^k}{dx^k}\, (e^{-2\pi x^2})\,, \qquad k \geq 0\,,$$

form an orthonormal basis of $L^2(\mathbb{R})$. Note that they all belong to $\mathcal{S}(\mathbb{R})$.

(i) Prove that $\widehat{h_k} = (-i)^k h_k$ and $\mathcal{F}^{-1}(h_k) = i^k h_k$ for all $k \geq 0$.

(ii) Show that for any $f \in L^2(\mathbb{R})$ we have $\hat{f} = \sum_{k=0}^{\infty} \langle f,\, h_k\rangle_{L^2(\mathbb{R})}\, (-1)^k h_k$, so that $L^2(\mathbb{R})$ can be split into the sum $H_0 \oplus H_1 \oplus H_2 \oplus H_3$ of the four eigenspaces of the Fourier transform $H_m = \{f \in L^2(\mathbb{R}) :\ f = \sum_{k=0}^{\infty} \langle f,\, h_{4k+m}\rangle_{L^2(\mathbb{R})}\, h_{4k+m}$ with $m = 0, 1, 2, 3$. Show that these eigenspaces are closed and pairwise orthogonal. (This alternative way to represent the Fourier transform on $L^2(\mathbb{R})$ is due to Wiener. Note that the action of $\mathcal{F}$ upon the subspace H_m is multiplication by $(-i)^m$, that is, a rotation in the plane by 0°, 270°, 180°, 90°, respectively.)

9. Given $a > 0$, show that $\hat{f}(\xi) = \frac{a}{\pi(a^2+\xi^2)}$ if $f(x) = e^{-2\pi a|x|}$.

10. For $f \in L^1(\mathbb{R})$ given by $f(x) = \frac{1}{\sqrt{x}}$ for $x \in (0, 1)$, with $f(x) = 0$ otherwise, show that $\hat{f} \notin L^1(\mathbb{R})$.

11. Show that if $f \in \mathcal{S}(\mathbb{R})$ is positive, then $|\hat{f}(\xi)|^2$ attains its maximum at $\xi = 0$.

12. (i) Compute $\mathcal{F}(\chi_{[-a,a]})$ for $a > 0$ and use the result to estimate $\int_{\mathbb{R}} (\frac{\sin x}{x})^2 dx$.

[8] This statement actually follows from the corresponding property of ψ_k defined in Exercise 3.32(iii), due to $h_k(x) = (2\pi)^{1/4} \psi_k(x\sqrt{2\pi})$ for $k \geq 0$.

(ii) Show that[9] $\frac{\sin x}{x} \notin L^1(\mathbb{R})$ but the improper integral $\lim_{\substack{x\to-\infty \\ y\to\infty}} \int_x^y \frac{\sin s}{s}\, ds$ exists.

13. (i) Show that if $f \in L^1(\mathbb{R})$ and $g \in \mathbb{S}(\mathbb{R})$, then $f * g$ is smooth with $(f * g)^{(n)} = f * g^{(n)}$ for every integer $n \geq 0$. However, we might have $f * g \notin \mathbb{S}(\mathbb{R})$.

(ii) If $f,\, g \in L^1(\mathbb{R})$ are such that $f(x) = g(x) = 0$ for $|x| > M$, show that $(f * g)(x) = 0$ for $|x| > 2M$. Moreover, if $g \in \mathcal{D}(\mathbb{R})$, show that $f * g \in \mathcal{D}(\mathbb{R})$.

14. (i) For $p = 1$ and $p = 2$, prove that $f * g$ is well-defined and belongs to $L^p(\mathbb{R})$ for $f \in L^p(\mathbb{R})$ and $g \in L^1(\mathbb{R})$, with $\|f * g\|_{L^p(\mathbb{R})} \leq \|f\|_{L^p(\mathbb{R})}\|g\|_{L^1(\mathbb{R})}$.

(ii) Prove that the translation operator τ_h, defined by $(\tau_h f)(x) = f(x-h)$, is continuous[10] on $L^2(\mathbb{R})$ and on $L^1(\mathbb{R})$.

(iii) Let[11] $\theta \in \mathcal{D}(\mathbb{R})$ with $\theta \geq 0$, $\theta(0) = 1$ and $\int_{\mathbb{R}} \theta(x)\, dx = 1$. If $\theta_n(x) = n\theta(nx)$ for $n \geq 1$, show that $f * \theta_n \to f$ in $L^p(\mathbb{R})$ for $f \in L^p(\mathbb{R})$ if[12] $p = 1$ or $p = 2$.

(iv) Show that $\mathcal{D}(\mathbb{R})$ is dense in $L^1(\mathbb{R})$ and in $L^2(\mathbb{R})$.

15. Use the Plancherel identity to evaluate $\int_{\mathbb{R}} \frac{1}{(1+x^2)^2}\, dx$.

16. Show that for $f,\, g \in L^1(\mathbb{R})$ we have $\int_{\mathbb{R}} \hat{f}(\xi)\, g(\xi)\, d\xi = \int_{\mathbb{R}} f(\xi)\, \hat{g}(\xi)\, d\xi$.

17. The *Mellin transform* $\mathcal{M}f(\xi) = \int_0^\infty f(y) y^{-2\pi i\xi} \frac{dy}{y}$ is a variant of the Fourier transform that pertains to functions $f : (0,\infty) \to \mathbb{C}$. For $f \in L^1((0,\infty), \frac{dy}{y})$, show that $\mathcal{M}f = \mathcal{F}(f \circ \exp)$. (Thus properties of the Mellin transform can easily be deduced from the corresponding properties of the Fourier transform.)

18. (i) Show that $\widehat{(f * g)} = \hat{f} \cdot \hat{g}$ for $f,\, g \in \mathcal{S}(\mathbb{R})$, and therefore $f * g \in \mathcal{S}(\mathbb{R})$.

(ii) Show that $\widehat{(f * g)} = \hat{f} \cdot \hat{g}$ for $f,\, g \in L^1(\mathbb{R})$.

(iii) Show that $\widehat{(fg)} = \hat{f} * \hat{g}$ for $f,\, g \in L^2(\mathbb{R})$. (In view of Theorem 5.2 and Theorem 5.3, this yields[13] that $F * G \in L^\infty(\mathbb{R})$ for $F, G \in L^2(\mathbb{R})$.)

19. Assume that $f \in L^1(\mathbb{R})$ is such that[14] there exists an increasing sequence $\{x_k\}_{k\in\mathbb{Z}}$ with $\lim_{k\to\pm\infty} x_k = \pm\infty$ and numbers $M_k > 0$ for $k \in \mathbb{Z}$ with f differentiable on (x_k, x_{k+1}) and with $|f'(x)| \leq M_k$ for $x \in (x_k, x_{k+1})$. Show that f has lateral limits $f(x+) = \lim_{y\downarrow x} f(y)$ and $f(x-) = \lim_{y\uparrow x} f(y)$ at every $x \in \mathbb{R}$, and $\lim_{r\to\infty} \int_{-r}^{r} \hat{f}(\xi)\, e^{2\pi i\xi x}\, d\xi = \frac{f(x+)+f(x-)}{2}$. (Thus the expectation that (5.6) can be interpreted for $f \in L^1(\mathbb{R})$ as an improper integral is too optimistic.)

[9] By $\frac{\sin x}{x}$ we mean the continuous function given by $f(x) = \frac{\sin x}{x}$ for $x \neq 0$ and $f(0) = 1$.

[10] That is, $\lim_{h\to 0} \|f(x-h) - f(x)\|_{L^p(\mathbb{R})} = 0$ if $f \in L^p(\mathbb{R})$ for $p \in \{1, 2\}$.

[11] Such functions do exist: in the proof of Theorem 5.1 we constructed $f \in \mathcal{D}(\mathbb{R})$ with $1 = f(0) \geq f(x) \geq 0$ for all $x \in \mathbb{R}$. Set $\theta(x) = f(\alpha x)$, where $\alpha = \int_{\mathbb{R}} f(s)\, ds$.

[12] The statements (i)–(iv) actually hold for any $p \in [1,\infty)$: the presented solution can be easily adapted. On the other hand, only (i) holds (trivially) if $p = \infty$. The failure of (ii)–(iv) for $p = \infty$ can be seen by looking at $f = \chi_{[0,1]}$, in which case $\|\tau_h f - f\|_{L^\infty(\mathbb{R})} = 1$ for any $h \neq 0$, while $f * \theta_n \in \mathcal{D}(\mathbb{R})$ by Exercise 13(ii). Note that a convergence $\phi_n \to f$ in $L^\infty(\mathbb{R})$ with $\phi_n \in \mathcal{D}(\mathbb{R})$ is prevented by the jump discontinuity of f at $x = 1$. Indeed, $\|\phi_n - f\|_{L^\infty(\mathbb{R})} < \frac{1}{3}$ yields $\phi_n(1-\varepsilon) < \frac{1}{3}$ and $\phi_n(1-\varepsilon) > \frac{1}{3}$ for every $\varepsilon \in (0, 1)$.

[13] One can show by other methods that if $F \in L^p(\mathbb{R})$ and $G \in L^q(\mathbb{R})$ for some $1 \leq q, p < \infty$ with $\frac{1}{p} + \frac{1}{q} \geq 1$, then $F * G \in L^r(\mathbb{R})$ with $\frac{1}{r} = \frac{1}{p} + \frac{1}{q} - 1$, see the discussion in Stein (1976).

[14] A typical example being a function that vanishes outside a compact interval K and such that the graph of its restriction to K is a broken line.

20. Construct a sequence of nonnegative functions $\{\phi_k\}_{k\in\mathbb{Z}}$ in $\mathcal{D}(\mathbb{R})$ such that $\sum_{k\in\mathbb{Z}} \phi_k(x) = 1$ with only finitely many values $\phi_k(x) \neq 0$, for every $x \in \mathbb{R}$.

21. For $f \in L^1(\mathbb{R})$ with $\hat{f}$ odd,[15] show the existence of a constant $M > 0$ such that $\left|\int_2^\alpha \frac{\hat{f}(\xi)}{\xi}\, d\xi\right| \leq M$ for all $\alpha \geq 2$. Consequently, if we set $g(\xi) = \frac{x}{\ln(4)}$ for $x \in [0,2]$ and $g(\xi) = \frac{1}{\ln(\xi)}$ for $\xi \geq 2$, and extend it by oddness to $(-\infty, 0)$, there is no $f \in L^1(\mathbb{R})$ with $\hat{f} = g$. (This shows that a bounded continuous function that vanishes at infinity does not have to be the Fourier transform of some $f \in L^1(\mathbb{R})$.)

22. Show that $\lim_{t\downarrow 0} \|\mathcal{F}^{-1}(\hat{f}(\xi)\, e^{-2\pi|\xi|t})(x) - f(x)\|_{L^1(\mathbb{R})} = 0$ if $f \in L^1(\mathbb{R})$.

23. Check that the *Hilbert transform* $\mathcal{H}f = \mathcal{F}^{-1}(\hat{f}(\xi)\,\mathrm{sign}(\xi))$ can be expressed as the principal value integral

$$(\mathcal{H}f)(x) = -\frac{1}{\pi i} \lim_{\varepsilon\downarrow 0} \int_{|x-y|>\varepsilon} \frac{f(y)}{x-y}\, dy \in L^2(\mathbb{R}) \quad \text{for} \quad f \in L^2(\mathbb{R}).$$

Show that $\mathcal{H}$ is norm-preserving on $L^2(\mathbb{R})$. (One of the reasons for the interest presented by $\mathcal{H}$ is the fact that a bounded linear operator on $L^2(\mathbb{R})$ which commutes[16] with translations and dilations and anti-commutes with the map that sends f to $f^\dagger$ has to be of the form $c\mathcal{H}$ for some constant $c \in \mathbb{C}$, see Stein (1976). In addition to this symmetry considerations, $\mathcal{H}$ is important in establishing links with complex analysis, see the discussion in Chapter 7.)

5.4.2 Hints

1. To prove completeness, note that if $\{f_k\}_{k\geq 1}$ is a Cauchy sequence, then for every integer $N \geq 0$, the functions $f_k^{(N)}$ converge uniformly on $\mathbb{R}$ to a continuous function $F_N : \mathbb{R} \to \mathbb{C}$. Use the fundamental theorem of calculus to infer that $F_N = F_0^{(N)}$.

2. Show that $F = \hat{f}$ satisfies $F'(\xi) = -2\pi\xi\, F(\xi)$.

3. Apply the dominated convergence theorem.

4. To $h_n \in (0,1)$ with $h_n \to 0$, associate the functions $f_n : \mathbb{R} \to \mathbb{C}$ given by $f_n(x) = f(nh_k)$ for $x \in [(k-1)h_n, kh_n)$ with $k \in \mathbb{Z}$, so that $\sum_{k\in\mathbb{Z}} h_n\, f(kh_n) = \int_{\mathbb{R}} f_n(x)\, dx$ and $\sum_{k\in\mathbb{Z}} h_n\, |f(kh_n)|^2 = \int_{\mathbb{R}} |f_n(x)|^2\, dx$. Use the dominated convergence theorem to get $\lim_{n\to\infty} \int_{\mathbb{R}} f_n(x)\, dx = \int_{\mathbb{R}} f(x)\, dx$ and $\lim_{n\to\infty} \int_{\mathbb{R}} |f_n(x)|^2\, dx = \int_{\mathbb{R}} |f(x)|^2\, dx$.

5. Recall that $L^2(\mathbb{R})$ is a Hilbert space.

6. If $\rho_N(f) = \max\limits_{0\leq m,n\leq N} \{\|f\|_{m,n}\}$, then for every integer $N \geq 0$ and every $s > 0$, the set $V_s^N = \{f \in \mathcal{S}(\mathbb{R}) : \rho_n(f) < s\}$ is open. If $\mathcal{S}(\mathbb{R})$ were normable, then there would exist N_0 such that for $N \geq N_0$ one can find $\gamma(N) \geq \delta(N) > 0$ such that $V_{\delta(N)}^N \subset B_1 \subset V_{\gamma(N)}^N$, where B_1 is the open unit ball, $B_1 = \{f \in \mathcal{S}(\mathbb{R}) : \|f\| < 1\}$. But $V_{\delta(N)}^N \subset V_{\gamma(N+1)}^{N+1}$ means that a control on $\rho_N(f)$ forces a bound on ρ_{N+1}. This last assertion can be shown to be false by a suitably tailored $f \in \mathcal{S}(\mathbb{R})$.

[15] That is, $\hat{f}(-\xi) = -\hat{f}(\xi)$ for all $\xi \in \mathbb{R}$.

[16] If $S : X \to X$ is a symmetry operation acting on a function space, we say that the operator $T : X \to X$ commutes/anti-commutes with S if $TS = ST$ and $TS = -ST$, respectively. The translations of $L^2(\mathbb{R})$ correspond to $(Sf)(x) = f(x-h)$ for $h \in \mathbb{R}$ and the dilations to $(Sf)(x) = f(ax)$ with $a > 0$.

7. Assume otherwise and show that if $\phi_n \in \mathcal{D}(\mathbb{R})$ with $\phi_n(x) = 1$ for $|x| \leq n$ and $\phi_n(x) = 0$ for $|x| \geq n+1$, then $c_n\phi_n \to 0$ for suitable $c_n > 0$.

8. For (i) note that $\widehat{h_0} = h_0$ and take the Fourier transform of the identity $h_k'(x) = 2\pi x h_k(x) - 2\sqrt{\pi(k+1)}\, h_{k+1}(x)$ to verify that $\{\widehat{h_k}\}_{k\geq 0}$ and $\{(-i)^k h_k\}_{k\geq 0}$ satisfy the same functional recursion formula. For (ii) use Hilbert space theory.

9. Direct computation.

10. By Theorem 5.3, $\hat{f}$ is bounded. Show that $\hat{f} \in L^1(\mathbb{R})$ would yield $\hat{f} \in L^2(\mathbb{R})$.

11. Write down $\hat{f}(\xi) \cdot \overline{\hat{f}(\xi)}$ using (5.8).

12. For (i) rely on Theorem 5.2. For (ii), integrate by parts in $\int_\varepsilon^x \frac{\sin^2(s)}{s^2}\, ds$ and use a substitution to relate it to $\int_{2\varepsilon}^{2x} \frac{\sin(t)}{t}\, dt$.

13. For the first part of (i) use the definition and Exercises 2 and 9 for the second part. The first part of (ii) follows by inspection and the second by relying on (i).

14. For (i) with $p = 2$, show that $\left(\int_{\mathbb{R}} \left(\int_{\mathbb{R}} F(x,y) dy\right)^2 dx\right)^{1/2} \leq \int_{\mathbb{R}} \left(\int_{\mathbb{R}} F^2(x,y)\, dx\right)^{1/2} dy$ for any measurable nonnegative function $F(x,y)$. The case $p = 1$ in (i) follows from Fubini's theorem. In (ii), use approximations by step functions. For (iii), use the inequality stated above and (ii), by splitting the integral expressing $\|f * \theta_n - f\|_{L^p(\mathbb{R})}$ into a small interval centred at the origin and its complement. For (iv), truncate f and use (iii).

15. Use Exercise 9.

16. By inspection, using (5.8).

17. Change variables $y = e^x$.

18. (i)–(ii) follow from Fubini's theorem. For (iii), use the fact that the Fourier transform preserves the $L^2(\mathbb{R})$-inner product

19. Rely on the mean-value theorem for the first part. For the second part, express $\int_{-r}^{r} \hat{f}(\xi)\, e^{2\pi i x\xi} d\xi$ alternatively to be able to use Exercise 12(ii).

20. For a suitable $\phi \in \mathcal{D}(\mathbb{R})$, set $\phi_k(x) = \frac{\phi(x-k)}{\sum_{j\in\mathbb{Z}} \phi(x-j)}$.

21. Show that $\hat{f}(\xi) = -2i \int_0^\infty f(t)\, \sin(2\pi\xi t)\, dt$ and use Fubini's theorem and Exercise 12(ii).

22. Proceed analogously to the proof of (5.13), using Exercise 9.

23. Write the improper integral as a convolution and use Exercise 12.

5.4.3 Solutions

1. Clearly $d(f,g) = \sum_{n,m\geq 0} \frac{\|f-g\|_{n,m}}{2^{n+m}(1+\|f-g\|_{n,m})}$ defines a metric on $\mathcal{S}(\mathbb{R})$: the triangle inequality holds since if a, b, $c > 0$ are such that $a \leq b + c$, then $\frac{a}{1+a} \leq \frac{b}{1+b} + \frac{c}{1+c}$.

Let us now show that the two notions of convergence (metric and by means of the countable family of semi-norms $\|\cdot\|_{n,m}$) coincide. First, assume $d(f_k, f) \to 0$ for $k \to \infty$. For fixed integers $n, m \geq 0$, given $\varepsilon > 0$, choose an integer $k_\varepsilon \geq 1$ with $d(f_k, f) \leq \frac{\varepsilon}{2^{n+m}(1+\varepsilon)}$ for all $k \geq k_\varepsilon$. We get $\frac{\|f_k-f\|_{n,m}}{1+\|f_k-f\|_{n,m}} \leq 2^{n+m}\, d(f_k, f) \leq \frac{\varepsilon}{1+\varepsilon}$ for $k \geq k_\varepsilon$, so $\|f_k - f\|_{n,m} \leq \varepsilon$ for $k \to k_\varepsilon$. Thus $\|f_k - f\|_{n,m} \to 0$ for $k \to \infty$. Conversely, if $\|f_k - f\|_{n,m} \to 0$ for all integers $n, m \geq 0$, given $\varepsilon > 0$, choose the integer $\kappa \geq 1$ with $\frac{1}{2^\kappa} \leq \frac{\varepsilon}{2}$. Since $\sum_{n+m\geq\kappa+1} \frac{1}{2^{n+m}} \leq \sum_{n=0}^\infty \frac{1}{2^n}\left(\sum_{m\geq\kappa+1}^\infty \frac{1}{2^m}\right) = \frac{1}{2^\kappa}\sum_{n=0}^\infty \frac{1}{2^n} = \frac{2}{2^\kappa} \leq \varepsilon$, we have $d(f_k, f) \leq \sum_{n+m\leq\kappa} \|f_k - f\|_{n,m} + \varepsilon$. Consequently $\limsup_{k\to\infty}\{d(f_k, f)\} \leq \varepsilon$. Since $\varepsilon > 0$ was arbitrary, we get $d(f_k, f) \to 0$ for $k \to \infty$.

To prove completeness, note that if $\{f_k\}_{k\geq 1}$ is a Cauchy sequence in $\mathcal{S}(\mathbb{R})$, with respect to the metric d, then $\|f_k - f_j\|_{0,N} \to 0$ for $k, j \to \infty$ ensures that for every fixed integer $N \geq 0$, the functions $\{f_k^{(N)}\}_{k\geq 1}$ converge uniformly on $\mathbb{R}$ to a function $F_N : \mathbb{R} \to \mathbb{C}$. Letting $k \to \infty$ in the relation $f_k^{(N)}(x) - f_k^{(N)}(0) = \int_0^x f_k^{(N+1)}(s)\, ds$, we obtain $F_N(x) - F_N(0) = \int_0^x F_{N+1}(s)\, ds$ for every fixed $x \in \mathbb{R}$. The fundamental theorem of calculus implies that $F_{N+1} = F_N'$, and an iteration of the procedure yields $F_N = F_0^{(N)}$. Finally, note that $\|f_k - f_j\|_{n,m} \leq \varepsilon$ for $n, m \geq \kappa$ means that for any fixed $x \in \mathbb{R}$ we have $\left|x^n\, [f_k^{(m)}(x) - f_j^{(m)}(x)]\right| \leq \varepsilon$ for $n, m \geq \kappa$. Letting $j \to \infty$, we get $\|f_k - F_0\|_{n,m} \leq \varepsilon$ for $k \geq \kappa$. This proves that $F_0 \in S(\mathbb{R})$, with $d(f_k, F_0) \to 0$.

2. Set $F(\xi) = \hat{f}(\xi) = \int_{\mathbb{R}} e^{-\pi x^2} e^{-2\pi i x\xi}\, dx$. Then $F'(\xi) = \int_{\mathbb{R}} (-2\pi i x)\, e^{-\pi x^2}\, e^{-2\pi i x\xi} dx$, which is equal to $i \int_{\mathbb{R}} f'(x)\, e^{-2\pi i x\xi}\, dx$. Since $\hat{f'}(\xi) = 2\pi i\xi\, \hat{f}(\xi)$ for all $f \in S(\mathbb{R})$, we get $F'(\xi) = -2\pi\xi\, F(\xi)$. Thus $G'(\xi) = 0$, where $G(\xi) = e^{\pi\xi^2}\, F(\xi)$. Consequently G is a constant, so that $F(\xi) = F(0)\, e^{-\pi\xi^2}$. The Plancherel identity yields $|F(0)|^2 = 1$. Since $F(0) = \int_{\mathbb{R}} e^{-\pi x^2}\, dx > 0$, we infer that $F(0) = 1$. Thus $F(\xi) = e^{-\pi\xi^2}$.

3. Note that $f_n \to f$ in $S(\mathbb{R})$ ensures the uniform convergence $(1 + x^2)\, f_n(x) \to (1 + x^2)\, f(x)$ on $\mathbb{R}$. Set $M = \sup_{x\in\mathbb{R}} \{(1+x^2)\, |f(x)|\}$ and choose an integer $N \geq 1$ such that $\left|(1+x^2)\, f_n(x) - (1+x^2)\, f(x)\right| \leq 1$ for all $x \in \mathbb{R}$ if $n \geq N$. Then $\left|(1 + x^2)\, f_n(x)\right| \leq 1 + (1 + x^2)\, |f(x)| \leq 1 + M$ for all $x \in \mathbb{R}$ if $n \geq N$, so that $|f_n(x)| \leq \frac{1+M}{1+x^2}$ for all $x \in \mathbb{R}$ if $n \geq N$. This permits us to apply the dominated convergence theorem.

4. Associate to a sequence $h_n \in (0, 1)$ with $h_n \to 0$ the functions f_n given by $f_n(x) = f(kh_n)$ for $x \in [(k - 1)h_n, kh_n)$ with $k \in \mathbb{Z}$, so that $\sum_{k\in\mathbb{Z}} h_n\, f(kh_n) = \int_{\mathbb{R}} f_n(x)\, dx$ and $\sum_{k\in\mathbb{Z}} h_n\, |f(kh_n)|^2 = \int_{\mathbb{R}} |f_n(x)|^2\, dx$. Note that if $m = \sup_{x\in\mathbb{R}} \{|f'(x)|\}$, then $|f(x) - f(y)| = |\int_x^y f'(s)\, ds| \leq m\, |x - y|$ for all $x, y \in \mathbb{R}$. Consequently $|f_n(x) - f(x)| \leq m\, h_n$ for all $x \in \mathbb{R}$, so that $f_n(x) \to f(x)$ uniformly on $\mathbb{R}$. On the other hand, choosing $M > 0$ so that $|f(x)| \leq \frac{M}{1+x^2}$ for all $x \in \mathbb{R}$, since every $x \in \mathbb{R}$ belongs to some interval $[(k - 1)h_n, kh_n)$ for some $k \in \mathbb{Z}$, we have $|f_n(x)| = |f(kh_n)| \leq \frac{M}{1+k^2h_n^2} \leq \frac{3M}{1+x^2}$. We can now apply the dominated convergence theorem.

5. The Fourier transform being length-preserving on $L^2(\mathbb{R})$, the identity follows at once by expressing the scalar product $(f, g)_{L^2(\mathbb{R})}$ in terms of $\|f \pm g\|_{L^2(\mathbb{R})}$ by means of the general polarisation identity (see Chapter 3).

6. Assume $S(\mathbb{R})$ can be normed by $\|\cdot\|$ and let $B_t = \{f \in S(\mathbb{R}) :\ \|f\| < t\}$ be the open ball of radius $t > 0$. Then there is $\varepsilon > 0$ such that $d(f, 0) < \varepsilon$ ensures $f \in B_1$, where d is the metric of $S(\mathbb{R})$, discussed in Exercise 1. For an integer $N \geq 0$, denote $\rho_N(f) = \max_{0\leq m,n\leq N}\{\|f\|_{m,n}\}$. Since

$$\sum_{m+n\geq N+1} \frac{\|f\|_{m,n}}{2^{n+m}(1 + \|f\|_{m,n}\|)} < \sum_{m+n\geq N+1} \frac{1}{2^{n+m}} = \frac{1}{2^{N+1}} \sum_{0\leq j,k} \frac{1}{2^{j+k}}$$
$$\leq \frac{1}{2^{N+1}} \Big(\sum_{0\leq j} \frac{1}{2^j}\Big)\Big(\sum_{0\leq k} \frac{1}{2^k}\Big) = \frac{4}{2^{N+1}}\,,$$

we infer the existence of an integer $N_0 \geq 0$ such that for every integer $N \geq N_0$ there is some $\delta(N) > 0$ ensuring $d(f, 0) < \varepsilon$ and thus $f \in B_1$ whenever $\rho_N(f) < \delta(N)$. On the other hand, for every $s > 0$ and integer $N \geq 0$, the set $V_s^N = \{f \in S(\mathbb{R}) :\ \rho_N(f) < s\}$ is open since if $f \in V_s^N$ is such that $\rho_N(f) = \alpha < s$ and if $g \in S(\mathbb{R})$ is such that $d(f, g) < \mu = \frac{s-\alpha}{4^N(1+s-\alpha)}$, then $g \in V_s^N$. Indeed, $d(f, g) < \mu$ forces $\frac{\|g-f\|_{m,n}}{2^{m+n}(1+\|g-f\|_{m,n})} < \mu$ for all $0 \leq m, n \leq N$, so that $\|g - f\|_{m,n} < \frac{2^{m+n}\mu}{1-2^{m+n}\mu}$ and thus $\rho_N(g - f) < \frac{4^N\mu}{1-4^N\mu}$ since $t \mapsto \frac{t}{1-t}$ is increasing on $[0, 1)$. Therefore $\rho_N(g) \leq \rho_N(g - f) + \rho_N(f) < \frac{4^N\mu}{1-4^N\mu} + \alpha = s$, thus proving the claim. Since $0 \in V_s^N$, we deduce that $B_t \subset V_s^N$ for some $t \in (0, 1)$, so that $B_1 \subset V_{st^{-N}}^N$ since $\|t^{-1}f\| = t^{-1}\|f\|$ and $\rho_N(t^{-1}f) \leq t^{-N}\rho_N(f)$ for every $f \in S(\mathbb{R})$. Setting $\gamma(N) = \max\{st^{-N},\ \delta(N)\}$, we can summarise the previous considerations as follows: were $S(\mathbb{R})$ normable, we would infer the existence of an integer $N_0 \geq 0$ such that for every integer $N \geq N_0$

there exist $\gamma(N) \geq \delta(N) > 0$ with $V^N_{\delta(N)} \subset B_1 \subset V^N_{\gamma(N)}$. In particular, $V^N_{\delta(N)} \subset V^{N+1}_{\gamma(N+1)}$ for some $N \geq 1$, which is a contradiction. To see this, choose $\theta \in \mathcal{D}(\mathbb{R})$ with $\theta(x) \in [0,1]$ for all $x \in \mathbb{R}$, with $\theta(x) = 1$ for $|x| \leq 1$ and $\theta(x) = 0$ for $|x| \geq 2$, and let $\phi(x) = \tau x^{-N-\frac{1}{2}}[1-\theta(x)]\,\theta(x-\xi N)$ for some constants $\tau \in (0,1)$ and $\xi > 5$, to be specified later on. Since $\phi(x) = 0$ for $|x - \xi N| \geq 2$, we can see that $|x^n \phi^{(m)}(x)| \leq \tau\,(N+1)!\,\max_{0\leq j,k\leq N}\{\sup_{|y|\leq 2}|\theta^{(j)}(y)| \cdot \sup_{|t|\leq 2}|\theta^{(k)}(t)|\}$ for all $x \in \mathbb{R}$ and all $0 \leq m,n \leq N$. Consequently, choosing $\tau > 0$ sufficiently small would ensure $\phi \in V^N_{\delta(N)}$ for any $\xi > 5$. The presumed inclusion would further ensure $\phi \in V^{N+1}_{\gamma(N+1)}$ for any $\xi > 5$. But $|(\xi N)^{N+1}\phi(\xi N)| < \gamma(N+1)$ for any $\xi > 5$ is impossible since $(\xi N)^{N+1}\phi(\xi N) = \tau N^{\frac{1}{2}}\xi^{\frac{1}{2}} \to \infty$ for $\xi \to \infty$.

7. If d is a metric compatible with the topology of $\mathcal{D}(\mathbb{R})$, define for every $n \geq 1$ the open set $B_{1/n} = \{f \in \mathcal{D}(\mathbb{R}) : d(0,f) < \frac{1}{n}\}$. The procedure described in the proof of Theorem 5.1 yields the existence of $\phi_n \in \mathcal{D}(\mathbb{R})$ with $\phi_n(x) = 1$ for $|x| \leq n$ and $\phi_n(x) = 0$ for $|x| \geq n+1$. Since $\frac{1}{k}\phi_n \to 0$ in $\mathcal{D}(\mathbb{R})$ as $k \to \infty$, for any $n \geq 1$ we can find some $c_n > 0$ with $c_n\phi_n \in B_{1/n}$. We obtain the contradiction $c_n\phi_n \to 0$ in $\mathcal{D}(\mathbb{R})$ as $n \to \infty$ even though, by construction, $[-n,n] \subset \text{supp}(c_n\phi_n) = \text{supp}(\phi_n)$ for $n \geq 1$.

8. (i) Differentiation of the formula for h_k yields $h_k'(x) = 2\pi x h_k(x) - 2\sqrt{\pi(k+1)}\,h_{k+1}(x)$, the Fourier transform of which leads to $\widehat{h_{k+1}}(\xi) = \frac{i}{2\sqrt{\pi(k+1)}}\,(\widehat{h_k})'(\xi) - \frac{2\pi i\xi}{2\sqrt{\pi(k+1)}}\,\widehat{h_k}(\xi)$. These two identities show that $\{\widehat{h_k}\}_{k\geq 0}$ and $\{(-i)^k h_k\}_{k\geq 0}$ satisfy the same functional recursion formula. Both being initiated by the same function, as $\widehat{h_0} = h_0$ by Exercice 2, we deduce that $\widehat{h_k} = (-i)^k h_k$ for all $k \geq 0$. Starting from this relation, since Theorem 5.2 yields $\mathcal{F}(\mathcal{F}(\mathcal{F}(\phi))) = \mathcal{F}^{-1}(\phi)$ for all $\phi \in L^2(\mathbb{R})$ and $(-i)^{3k} = i^k$, we obtain the second relation.

(ii) Since $\{h_k\}_{k\geq 0}$ is an orthonormal basis of $L^2(\mathbb{R})$, we have $f = \sum_{k=0}^{\infty}\langle f, h_k\rangle_{L^2(\mathbb{R})}\,h_k$ in $L^2(\mathbb{R})$. Theorem 5.2 and (i) yield $\hat{f} = \sum_{k=0}^{\infty}\langle f, h_k\rangle_{L^2(\mathbb{R})}\,(-i)^k h_k$ in $L^2(\mathbb{R})$. The fact that the eigenspaces H_m $(m = 0,1,2,3)$ are orthogonal is clear since $\{h_k\}_{k\geq 0}$ is an orthonormal basis of $L^2(\mathbb{R})$, each eigenspace being closed at it is itself a Hilbert space with orthonormal basis $\{h_{4k+m}\}_{k\geq 0}$.

9. $\hat{f}(\xi) = \int_{\mathbb{R}} e^{-2\pi a|x| - 2\pi i x\xi}\,dx = \int_{-\infty}^0 e^{2\pi x(a-i\xi)}\,dx + \int_0^{\infty} e^{-2\pi x(a+i\xi)}\,dx = \frac{e^{2\pi x(a-i\xi)}}{2\pi(a-i\xi)}\Big|_{x=-\infty}^{x=0} - \frac{e^{-2\pi x(a+i\xi)}}{2\pi(a+i\xi)}\Big|_{x=0}^{x=\infty} = \frac{1}{2\pi(a-i\xi)} + \frac{1}{2\pi(a+i\xi)} = \frac{a}{\pi(a^2+\xi^2)}$.

10. If $\hat{f} \in L^1(\mathbb{R})$, then Theorem 5.3 yields $\hat{f} \in L^2(\mathbb{R})$: $\int_{\mathbb{R}}|\hat{f}(\xi)|^2\,d\xi \leq \sup_{s\in\mathbb{R}}\{|\hat{f}(s)|\}\int_{\mathbb{R}}|\hat{f}(\xi)|\,d\xi \leq \|f\|_{L^1(\mathbb{R})}\|\hat{f}\|_{L^1(\mathbb{R})} < \infty$. But since the Fourier transform is a length-preserving bijection on $L^2(\mathbb{R})$, $\hat{f} \in L^2(\mathbb{R})$ forces $f \in L^2(\mathbb{R})$. However, $\int_{\mathbb{R}}|f(x)|^2\,dx = \int_0^1 \frac{1}{x}\,dx = \infty$.

11. From (5.8) we get $|\hat{f}(\xi)|^2 = \int_{\mathbb{R}} f(x)e^{-2\pi i x\xi}dx \cdot \int_{\mathbb{R}} f(y)e^{2\pi i y\xi}dy = \int_{\mathbb{R}}\int_{\mathbb{R}} f(x)f(y)e^{2\pi i(y-x)\xi}dxdy \leq \int_{\mathbb{R}}\int_{\mathbb{R}} f(x)f(y)dxdy = |\hat{f}(0)|^2$.

12. (i) $\mathcal{F}(\chi_{[-a,a]}) = \int_{-a}^{a} e^{-2\pi i x\xi}\,dx = \frac{e^{-2\pi i x\xi}}{-2\pi i\xi}\Big|_{-a}^{a} = \frac{\sin(2\pi\xi a)}{\pi\xi}$ for $\xi \neq 0$. For $a = \frac{1}{2\pi}$, this, by Theorem 5.2, yields $\int_{\mathbb{R}}(\frac{\sin\xi}{\pi\xi})^2\,d\xi = \int_{\mathbb{R}}\chi^2_{[-\frac{1}{2\pi},\frac{1}{2\pi}]}(x)\,dx = \frac{1}{\pi}$, so that $\int_{\mathbb{R}}(\frac{\sin\xi}{\xi})^2\,d\xi = \pi$.

(ii) Partial integration and the substitution $t = 2s$ yield

$$\int_{\varepsilon}^{x}\frac{\sin^2(s)}{s^2}\,ds = -\left.\frac{\sin^2(s)}{s}\right|_{s=\varepsilon}^{s=x} + \int_{\varepsilon}^{x}\frac{2\sin(s)\cos(s)}{s}\,ds = \frac{\sin^2(\varepsilon)}{\varepsilon} - \frac{\sin^2(x)}{x} + \int_{2\varepsilon}^{2x}\frac{\sin(t)}{t}\,dt.$$

Since $\lim_{\varepsilon\downarrow 0}\frac{\sin^2(\varepsilon)}{\varepsilon} = 0$ and $\left|\frac{\sin^2(x)}{x}\right| \leq \frac{1}{|x|}$ for $x \neq 0$, the improper integral $\int_0^{\infty}\frac{\sin(t)}{t}\,dt$ exists. Since $f(x) = f(-x)$ for all $x \in \mathbb{R}$, the improper integral over $\mathbb{R}$ will also exists and its value is π.

13. (i) Given $h \neq 0$, we have $\frac{(f*g)(x+h)-(f*g)(x)}{h} - (f*g')(x) = \frac{1}{h}\int_{\mathbb{R}} f(x+h-y)g(y)\,dy - \frac{1}{h}\int_{\mathbb{R}} f(x-y)g(y)\,dy - \int_{\mathbb{R}} f(x-y)g'(y)\,dy = \frac{1}{h}\int_{\mathbb{R}} f(x-y)g(y+h)\,dy - \frac{1}{h}\int_{\mathbb{R}} f(x-y)g(y)\,dy - \int_{\mathbb{R}} f(x-y)g'(y)\,dy =$

$\int_{\mathbb{R}} f(x-y)\,[\frac{g(y+h)-g(y)}{h} - g'(y)]\,dy$. By the mean-value theorem $\frac{g(y+h)-g(y)}{h} = g'(\xi_h)$ for some ξ_h between x and $x+h$, so that $\left|\frac{g(y+h)-g(y)}{h} - g'(y)\right| \le 2\sup_{t\in\mathbb{R}}\{|g'(t)|\}$. By dominated convergence we deduce that $f*g$ is differentiable with $(f*g)' = f*g'$. An inductive argument yields the statement for higher derivatives. Exercise 9 with $a = \frac{1}{2\pi}$ and Theorem 5.2 yield $(\mathcal{F}^{-1}(e^{-|\xi|}))(x) = (\mathcal{F}(e^{-|\xi|}))(x) = \frac{2}{1+4\pi^2x^2}$. Consequently, with $f(x) = \frac{2}{1+4\pi^2x^2}$ and $g(x) = e^{-\pi x^2}$, using Exercise 2 and Theorem 5.3, we get $(f*g)(\xi) = \mathcal{F}^{-1}(e^{-\pi\xi^2-|\xi|}) \notin \mathcal{S}(\mathbb{R})$.

(ii) Since $g(y) = 0$ for $|y| > M$, we have that $(f*g)(x) = \int_{\mathbb{R}} f(x-y)g(y)\,dy = \int_{-M}^{M} f(x-y)g(y)\,dy$. This vanishes for $|x| > 2M$ since then $|x-y| \ge |x|-|y| > 2M-M$ for $|y| \le M$. The second statement follows by combining the first part with (i).

14. (i) Since the construction of $(f*g)(x)$ involves only multiplications and integration, the measurability is not an issue. If $p = 1$, Fubini's theorem yields $\int_{\mathbb{R}} |(f*g)(x)|\,dx = \int_{\mathbb{R}} \left|\int_{\mathbb{R}} f(x-y)g(y)\,dy\right| dx \le \int_{\mathbb{R}}\int_{\mathbb{R}} |f(x-y)||g(y)|\,dxdy = \|f\|_{L^1(\mathbb{R})}\|g\|_{L^1(\mathbb{R})}$ for $f, g \in L^1(\mathbb{R})$. If $p = 2$ it suffices to show that $\left(\int_{\mathbb{R}}\left(\int_{\mathbb{R}} F(x,y)dy\right)^2 dx\right)^{\frac{1}{2}} \le \int_{\mathbb{R}}\left(\int_{\mathbb{R}} F^2(x,y)\,dx\right)^{\frac{1}{2}} dy$ for any measurable nonnegative function $F(x,y)$, with the claimed inequality corresponding to $F(x,y) = |f(x-y)g(y)|$. For this, note that if $G(x) = \int_{\mathbb{R}} F(x,y)dy$, then $\|G\|_{L^2(\mathbb{R})} = \left(\int_{\mathbb{R}}\left(\int_{\mathbb{R}} F(x,y)dy\right)^2 dx\right)^{1/2} = \|\int_{\mathbb{R}} F(x,y)dy\|_{L^2(\mathbb{R})}$. Using Fubini's theorem and the Cauchy–Schwarz inequality we obtain $\|\int_{\mathbb{R}} F(x,y)\,dy\|^2_{L^2(\mathbb{R})} = \int_{\mathbb{R}} G(x)\left(\int_{\mathbb{R}} F(x,y)\,dy\right)dx = \int_{\mathbb{R}}\int_{\mathbb{R}} G(x)\,F(x,y)\,dxdy \le \int_{\mathbb{R}}\left(\int_{\mathbb{R}} G^2(x)\,dx\right)^{\frac{1}{2}}\left(\int_{\mathbb{R}} F^2(x,y)\,dx\right)^{\frac{1}{2}} dy = \|G\|_{L^2(\mathbb{R})}\int_{\mathbb{R}}\left(\int_{\mathbb{R}} F^2(x,y)\,dx\right)^{\frac{1}{2}} dy$. If $\|G\|_{L^2(\mathbb{R})} < \infty$, we are done dividing by it. If $\|G\|_{L^2(\mathbb{R})} = \infty$, replace in the above procedure F by $F_n(x,y) = \min\{F(x,y),\, n\}\chi_{[x^2+y^2\le n^2]}$ to ensure that the corresponding G_n satisfies firstly $\|G_n\|_{L^2(\mathbb{R})} < \infty$ and, after repeating the procedure, $\|G_n\|_{L^2(\mathbb{R})} \le \int_{\mathbb{R}}\left(\int_{\mathbb{R}} F_n^2(x,y)\,dx\right)^{\frac{1}{2}} dy$. Monotone convergence shows that the left side converges to $\|G\|_{L^2(\mathbb{R})} = \infty$ and the right side converges to $\int_{\mathbb{R}}\left(\int_{\mathbb{R}} F^2(x,y)\,dx\right)^{\frac{1}{2}} dy$.

(ii) Given $f \in L^p(\mathbb{R})$ with $p = 1$ or $p = 2$, for any $\varepsilon > 0$ we can find an integer $n \ge 1$, constants $c_k \in \mathbb{C}$ and disjoint intervals $[a_k, b_k)$ for $1 \le k \le n$, such that $\|f - g_n\|_{L^p(\mathbb{R})} < \frac{\varepsilon}{3}$ for $g_n(x) = \sum_{k=1}^n c_k\chi_{[a_k,b_k)}(x)$. Note that $\|\tau_h g_n - g_n\|^p_{L^p(\mathbb{R})} = \int_{\mathbb{R}}\left|\sum_{k=1}^n c_k\,[\chi_{[a_k+h,b_k+h)}(x) - \chi_{[a_k,b_k)}(x)]\,dx\right|^p dx \le \int_{\mathbb{R}} n\sum_{k=1}^n |c_k|^p\,|\chi_{[a_k+h,b_k+h)}(x) - \chi_{[a_k,b_k)}(x)|^p dx \le 2nh^p\sum_{k=1}^n |c_k|^p \to 0$ as $h \to 0$. Choosing $\delta > 0$ such that $\|\tau_h g_n - g_n\|_{L^p(\mathbb{R})} < \frac{\varepsilon}{3}$ if $|h| < \delta$, we get $\|\tau_h f - f\|_{L^p(\mathbb{R})} \le \|f - g_n\|_{L^p(\mathbb{R})} + \|\tau_h f - \tau_h g_n\|_{L^p(\mathbb{R})} + \|\tau_h g_n - g_n\|_{L^p(\mathbb{R})} = 2\|f - g_n\|_{L^p(\mathbb{R})} + \|\tau_h g_n - g_n\|_{L^p(\mathbb{R})} < \varepsilon$.

(iii) Using the inequality in the solution to (i) with $F(x,y) = |f(x-y) - f(x)|\,\theta_n(y)$, we obtain $\|f*\theta_n - f\|^2_{L^2(\mathbb{R})} = \int_{\mathbb{R}}\left|\int_{\mathbb{R}} f(x-y)\theta_n(y)\,dy - f(x)\right|^2 dx = \int_{\mathbb{R}}\left|\int_{\mathbb{R}}[f(x-y) - f(x)]\,\theta_n(y)\,dy\right|^2 dx \le \int_{\mathbb{R}}\left(\int_{\mathbb{R}} F(x,y)\,dy\right)^2 dx \le \left(\int_{\mathbb{R}} \|F(x,y)\|_{L^2(dx)}\,dy\right)^2 = \left(\int_{\mathbb{R}} \|f(x-y) - f(x)\|_{L^2(dx)}\,\theta_n(y)\,dy\right)^2$. Now $\int_{\mathbb{R}} \|f(x-y) - f(x)\|_{L^2(dx)}\,\theta_n(y)\,dy = I_\delta + J_\delta$, where $I_\delta = \int_{|y|<\delta} \|f(x-y) - f(x)\|_{L^2(dx)}\,\theta_n(y)\,dy$ and $J_\delta = \int_{|y|\ge\delta} \|f(x-y) - f(x)\|_{L^2(dx)}\,\theta_n(y)\,dy$. By (ii), given $\varepsilon > 0$, there exists $\delta > 0$ such that $\|f(x-y) - f(x)\|_{L^2(dx)} < \frac{\varepsilon}{2}$ if $|y| < \delta$, so that $I_\delta \le \frac{\varepsilon}{2}\int_{\mathbb{R}} \theta_n(y)\,dy = \frac{\varepsilon}{2}\int_{\mathbb{R}} \theta(x)\,dx = \frac{\varepsilon}{2}$. On the other hand, $J_\delta \le 2\|f\|_{L^2(\mathbb{R})}\int_{|y|\ge\delta} \theta_n(y)\,dy = 2\|f\|_{L^2(\mathbb{R})}\int_{|x|\ge n\delta} \theta(x)\,dx \to 0$ as $n \to \infty$ by dominated convergence. This settles the case $p = 2$. The case $p = 1$ follows by implementing the same approach, some of the steps being actually easier.

(iv) If $f \in L^p(\mathbb{R})$, then $f_n \to f$ in $L^p(\mathbb{R})$ by monotone convergence, where $f_n(x) = f(x)$ for $|x| \le n$ and $f_n(x) = 0$ for $|x| > n$. Exercise 13(ii) ensures $f_n * \theta_m \in \mathcal{D}(\mathbb{R})$, with $\lim_{m\to\infty}(f_n * \theta_m) = f_n$ in $L^p(\mathbb{R})$ due to (iii). Combining these two approximation results yields the claim.

15. We use Exercise 9 with $a = 1$ and Theorem 5.2 with $f(x) = e^{-2\pi|x|} \in L^2(\mathbb{R})$ to compute $\int_{\mathbb{R}} \frac{d\xi}{(1+\xi^2)^2} = \pi^2\int_{\mathbb{R}} e^{-4\pi|x|}\,dx = \pi^2\int_0^{\infty} e^{-4\pi x}\,dx + \pi^2\int_{-\infty}^{0} e^{4\pi x}\,dx = \frac{\pi}{2}$.

16. Using (5.8) and Fubini's theorem, we get $\int_{\mathbb{R}} \hat{f}(\xi)\,g(\xi)\,d\xi = \int_{\mathbb{R}}\left(\int_{\mathbb{R}} f(x)e^{-2\pi i\xi x}dx\right)g(\xi)\,d\xi =$

$\int_{\mathbb{R}}\int_{\mathbb{R}} f(x)g(\xi)e^{-2\pi i\xi x}dx\,d\xi$. Interchanging in the double integral x and ξ, the reverse sequence of equalities leads to $\int_{\mathbb{R}} f(\xi)\,\hat{g}(\xi)\,d\xi$.

17. The change of variables $y = e^x$ yields $\mathcal{F}(f \circ \exp)(\xi) = \int_{\mathbb{R}} f(e^x)\,e^{-2\pi ix\xi}dx = \int_0^\infty f(y)\,y^{-2\pi i\xi}\frac{dy}{y}$, with the existence of the integrals ensured since $y^{-2\pi i\xi} = e^{-2\pi i\xi \ln y}$ has absolute value 1.

18. (i) $(\widehat{f * g})(\xi) = \int_{\mathbb{R}}\int_{\mathbb{R}} f(x-y)g(y)\,dy\,e^{-2\pi ix\xi}dx = \int_{\mathbb{R}}\int_{\mathbb{R}} f(x-y)\,e^{-2\pi i(x-y)\xi}\,g(y)\,e^{-2\pi iy\xi}\,dydx = \int_{\mathbb{R}}\int_{\mathbb{R}} f(s)\,e^{-2\pi is\xi}\,g(y)\,e^{-2\pi iy\xi}\,dsdy = \int_{\mathbb{R}} \hat{f}(\xi)\,g(y)\,e^{-2\pi iy\xi}\,dy = \hat{f}(\xi)\cdot\hat{g}(\xi)$ by Fubini's theorem and the intermediate change of variables $(x, y) \mapsto (s, y)$ with $s = x - y$. This proves the formula. The fact that $f * g \in \mathcal{S}(\mathbb{R})$ follows from the formula and Theorem 5.1.

(ii) The same chain of equalities as in (i) yields $(\widehat{f * g}) = \hat{f}\cdot\hat{g}$ if $f, g \in L^1(\mathbb{R})$.

(iii) As a by-product of Theorem 5.2, $\mathcal{F}$ preserves the $L^2(\mathbb{R})$-inner product, see Hilbert space theory (see Chapter 3). Consequently, for $f, g \in L^2(\mathbb{R})$ we have

$$(\widehat{fg})(\xi) = (f(x), \overline{g(x)\,e^{2\pi ix\xi}})_{L^2(dx)} = (\hat{f}(\tau), \mathcal{F}[\overline{g(x)\,e^{2\pi ix\xi}}](\tau))_{L^2(d\tau)}\,.$$

On the other hand, $\mathcal{F}[\overline{g(x)\,e^{2\pi ix\xi}}](\tau) = \mathcal{F}[\overline{g}](\tau-\xi) = \overline{\hat{g}(\xi-\tau)}$ since $\mathcal{F}[\overline{\phi}](\xi) = \int_{\mathbb{R}} \overline{\phi(x)}\,e^{-2\pi ix\xi}dx = \overline{\int_{\mathbb{R}} \phi(x)\,e^{2\pi ix\xi}dx} = \overline{\mathcal{F}^{-1}[\phi](\xi)} = \overline{\mathcal{F}[\phi^\dagger](\xi)}$ for $\phi \in \mathcal{S}(\mathbb{R})$ ensures $\mathcal{F}[\overline{g}] = \overline{\mathcal{F}[g^\dagger]}$ for all $g \in L^2(\mathbb{R})$, in view of the density result from Exercise 14(iv). Thus $(\widehat{fg})(\xi) = (\hat{f}(\tau), \overline{\hat{g}(\xi-\tau)})_{L^2(d\tau)} = \int_{\mathbb{R}} \hat{f}(\tau)\,\hat{g}(\xi-\tau)\,d\tau = \hat{f} * \hat{g}$.

19. Firstly, if $t_n \downarrow x$, then there exists $N \geq 1$ such that all elements t_k that differ from the points x_n belong to at most N intervals of the form (x_j, x_{j+1}), and the mean-value theorem yields $|f(t_n) - f(t_m)| \leq M|t_n - t_m|$ for n and m large enough, where $M > 0$ is the maximum of the corresponding N values of the constants M_j. As a Cauchy sequence, $\{f(t_n)\}_{n\geq1}$ has a limit, called $f(x+)$, that can be seen to be independent of the particular choice of sequence $\{t_n\}_{n\geq1}$ by noticing that the union of two such sequences must converge. On the other hand, by Fubini's theorem and performing the change of variables $s = x - y$, $\int_{-r}^r \hat{f}(\xi)\,e^{2\pi ix\xi}d\xi = \int_{-r}^r \int_{\mathbb{R}} f(y)e^{-2\pi iy\xi}dy\,e^{2\pi ix\xi}d\xi = \int_{\mathbb{R}} f(y)\int_{-r}^r e^{2\pi i(x-y)\xi}d\xi\,dy = \int_{\mathbb{R}} f(y)\,\frac{\sin[2\pi(x-y)r]}{\pi(x-y)}\,dy = \int_{\mathbb{R}} f(x-s)\,\frac{\sin(2\pi sr)}{\pi s}\,ds$ for $x \in \mathbb{R}$, $r > 0$. Since $\lim_{r\to\infty}\int_0^r \frac{\sin(2\pi sr)}{\pi s}\,ds = \lim_{r\to-\infty}\int_r^0 \frac{\sin(2\pi sr)}{\pi s}\,ds = \frac{1}{2}$, see Exercise 12(ii), if we establish that

$$\lim_{r\to\infty}\int_{-r}^0 \frac{\sin(2\pi sr)}{\pi s}\,[f(x-s) - f(x+)]\,ds = \lim_{r\to\infty}\int_0^r \frac{\sin(2\pi sr)}{\pi s}\,[f(x-s) - f(x-)]\,ds = 0\,,$$

then we can conclude that $\lim_{r\to\infty}\int_{-r}^r \hat{f}(\xi)\,e^{2\pi ix\xi}d\xi - \frac{f(x+)+f(x-)}{2} = 0$. The other integral being similar, it suffices to show that $\lim_{r\to\infty}\int_0^r \frac{\sin(2\pi sr)}{\pi s}\,[f(x-s) - f(x-)]\,ds = 0$. If $r \geq T \geq \frac{1}{\pi}$, then $\int_T^r \frac{\sin(2\pi sr)}{\pi s}\,f(x-)\,ds = f(x-)\int_T^r \frac{\sin(2\pi sr)}{\pi s}\,ds \to 0$ for $r \geq T \to \infty$, being the tail of a convergent integral, see Exercise 12(ii), while $\left|\int_T^r \frac{\sin(2\pi sr)}{\pi s}\,f(x-s)\,ds\right| \leq \int_T^r |f(x-s)|\,ds = \int_{-\infty}^{x-T} |f(t)|\,dt \to 0$ for $r \geq T \to \infty$ by dominated convergence. Finally, $\int_0^T \frac{\sin(2\pi sr)}{\pi s}\,[f(x-s) - f(x-)]\,ds = \frac{\widehat{g_T}(r)-\widehat{g_T}(-r)}{-2i}$ with $g_T(s) = \frac{f(x-s)-f(x-)}{\pi s}$ if $0 < s < T$ and $g_T(s) = 0$ otherwise. The mean-value theorem shows that g_T is bounded, so that $g_T \in L^1(\mathbb{R})$. By Riemann–Lebesgue (Theorem 5.3),

$$\lim_{r\to\infty}\int_0^T \frac{\sin(2\pi sr)}{\pi s}\,[f(x-s) - f(x-)]\,ds \to 0$$

for any fixed $T \geq \frac{1}{\pi}$. Combining these limits yields $\lim_{r\to\infty}\int_0^r \frac{\sin(2\pi sr)}{\pi s}\,[f(x-s) - f(x-)]\,ds = 0$.

20. Choose $\phi \in \mathcal{D}(\mathbb{R})$ with $\phi(x) = 1$ for $|x| \leq \frac{3}{2}$, $\phi(x) \geq 0$ for all $x \in \mathbb{R}$ and $\phi(x) = 0$ for $|x| \geq 2$, see the proof of Theorem 5.1. Then $\psi(x) = \sum_{j\in\mathbb{Z}} \phi(x-j) \in [1, 4]$ defines a smooth function since every fixed $x \in \mathbb{R}$ belongs to at least one interval $(k - \frac{3}{2}, k + \frac{3}{2})$ and to precisely four intervals of the form $(k-2, k+2)$ with $k \in \mathbb{Z}$, so that at among the functions $s \mapsto \phi(s-k)$ at least one will be equal to 1 at $s = x$ and at most four will not vanish there. Now set $\phi_k(x) = \frac{\phi(x-k)}{\psi(x)}$ for $k \in \mathbb{Z}$.

21. Note that

$$\hat{f}(\xi) = \int_0^\infty f(x)e^{-2\pi i x\xi}\,dx + \int_{-\infty}^0 f(x)e^{-2\pi i x\xi}\,dx = -2i\int_0^\infty f(t)\,\sin(2\pi\xi t)\,dt,$$

after the change of variables $x = -t$ in the integral over $(-\infty, 0)$. Fubini's theorem yields

$$\Big|\int_2^\alpha \frac{\hat{f}(\xi)}{\xi}\,d\xi\Big| = 2\Big|\int_2^\alpha\int_0^\infty f(t)\,\frac{\sin(2\pi\xi t)}{\xi}\,dtd\xi\Big| = 2\Big|\int_{4\pi t}^{2\pi t\alpha}\int_0^\infty f(t)\,\frac{\sin(y)}{y}\,dtdy\Big|$$

$$\le 2\int_0^\infty |f(t)|\,dt\cdot \sup_{0\le a<b<\infty}\Big\{\Big|\int_a^b \frac{\sin(y)}{y}\,dy\Big|\Big\} < \infty$$

by Exercise 12(ii). The final assertion follows since $\int_2^\alpha \frac{d\xi}{\xi\ln(\xi)} = \ln(\ln(\alpha)) - \ln(\ln(2))$.

22. Setting $\theta_t(x) = \frac{t}{\pi(t^2+x^2)}$ for $t > 0$ and $x \in \mathbb{R}$, by Exercise 9 and Exercise 18(ii), the statement is equivalent to $\lim_{t\downarrow 0}\theta_t * f = f$ in $L^1(\mathbb{R})$. Since $\int_{\mathbb{R}}\theta_t(y)\,dy = 1$, we see that $\|\theta_t * f - f\|_{L^1(\mathbb{R})} = \int_{\mathbb{R}}\big|\int_{\mathbb{R}} f(x-y)\theta_t(y)\,dy - f(x)\big|\,dx = \int_{\mathbb{R}}\big|\int_{\mathbb{R}}[f(x-y) - f(x)]\theta_t(y)\,dy\big|\,dx \le \int_{\mathbb{R}}\int_{\mathbb{R}}|f(x-y) - f(x)|\,\theta_t(y)\,dy\,dx$. Given $\delta > 0$, the upper bound equals $\int_{\mathbb{R}}\|\tau_y f - f\|_{L^1(\mathbb{R})}\theta_t(y)\,dy \le \int_{-\delta}^{\delta}\|\tau_y f - f\|_{L^1(\mathbb{R})}\theta_t(y)\,dy + \int_{|y|>\delta}\|\tau_y f - f\|_{L^1(\mathbb{R})}\theta_t(y)\,dy$. The first integral is small for $\delta > 0$ sufficiently small by Exercise 14(ii) since $\int_{\mathbb{R}}\theta_t(y)\,dy = 1$. The second integral is less than $2\|f\|_{L^1(\mathbb{R})}\int_{|y|>\delta}\theta_t(y)\,dy = \frac{4\|f\|_{L^1(\mathbb{R})}}{\pi}\int_{\delta/t}^\infty \frac{1}{1+s^2}\,ds$, which vanishes in the limit $t\downarrow 0$ for any fixed $\delta > 0$ as the tail of a convergent integral.

23. For $\varepsilon > 0$, define $g_\varepsilon(x) = x^{-1}$ for $|x| > \varepsilon$ and $g_\varepsilon(x) = 0$ for $|x| \le \varepsilon$. Then $g_\varepsilon \in L^2(\mathbb{R})$ and $\int_{|x-y|>\varepsilon}\frac{f(y)}{x-y}\,dy = \int_{|s|>\varepsilon}\frac{f(x-s)}{s}\,ds = (f * g_\varepsilon)(x) = \mathcal{F}^{-1}(\hat{f}\cdot\widehat{g_\varepsilon})$, see Exercise 18(ii). Since $\widehat{g_\varepsilon}(\xi) = \int_{|x|>\varepsilon}\frac{1}{x}\,e^{-2\pi i x\xi}\,dx = \int_{-\infty}^{-\varepsilon}\frac{e^{-2\pi i x\xi}}{x}\,dx + \int_\varepsilon^\infty \frac{e^{-2\pi i x\xi}}{x}\,dx = -2i\int_\varepsilon^\infty \frac{\sin(2\pi y\xi)}{y}\,dy = -2\pi i\,\frac{\xi}{|\xi|}\int_{2\pi\varepsilon|\xi|}^\infty \frac{\sin(s)}{s}\,ds$ for $\xi \neq 0$, by changing variables $y = -x$ in the integral on $(-\infty, -\varepsilon)$ and $s = 2\pi y|\xi|$ in the last step. Exercise 18(ii) yields $\lim_{\varepsilon\downarrow 0}\widehat{g_\varepsilon}(\xi) = -i\frac{\xi}{|\xi|}$ for $\xi \neq 0$. Moreover, from the previous considerations we see that $|\widehat{g_\varepsilon}(\xi)| \le M = 2\pi\sup_{0<\tau<t}\big|\int_\tau^t \frac{\sin(s)}{s}\,ds\big| < \infty$ for all $\xi \neq 0$. Consequently, $\hat{f}\cdot\widehat{g_\varepsilon} \to \hat{f}(\xi)\cdot\mathrm{sign}(\xi)$ in $L^2(\mathbb{R})$ by dominated convergence. Theorem 5.2 yields $\mathcal{F}^{-1}(\hat{f}\cdot\widehat{g_\varepsilon}) \to \mathcal{F}^{-1}(\hat{f}(\xi)\cdot\mathrm{sign}(\xi))$ in $L^2(\mathbb{R})$. Finally, the norm-preserving property follows from the established formula and the fact that $|\hat{f}(\xi)\cdot\mathrm{sign}(\xi)| = |\hat{f}(\xi)|$ for $\xi \neq 0$, in view of Theorem 5.2.

5.5 Note to Chapter 5

It is natural to ask whether there is a reasonable theory of Fourier (direct and inverse) transforms for $L^p(\mathbb{R})$-functions with[17] $p > 1$, $p \neq 2$. Since for this aspect of Fourier analysis there is little to gain by the restriction to one spatial dimension, we will discuss it directly in the multi-dimensional setting – see Chapter 6.

[17] The case of Fourier series is dissimilar, since $L^p[0,1] \subset L^1[0,1]$ for all $p > 1$

6

Multi-dimensional Fourier analysis

The framework of distribution theory makes it possible to deal in a systematic way with the Fourier analysis of functions $f : \mathbb{R}^n \to \mathbb{C}$ for integers $n \geq 1$. In particular, one can treat Fourier series as the special case of the Fourier transform in the setting of periodic distributions, thus avoiding many difficulties that have to be overcome if we extend the classical approach from Chapter 4 to a multi-dimensional setting.[1]

6.1 Fourier transform

We define the Fourier transform of φ in the Schwartz class $\mathcal{S}(\mathbb{R}^n)$ of all C^∞-functions $\varphi : \mathbb{R}^n \to \mathbb{C}$ such that[2] $\sup_{x\in\mathbb{R}^n} |x^\alpha\,(D^\beta\varphi)(x)| < \infty$ for all n-tuples $\alpha, \beta \in \mathbb{N}^n$ by

$$\hat{\varphi}(\xi) = \int_{\mathbb{R}^n} \varphi(x)\, e^{-2\pi i x\cdot\xi}\, dx, \qquad \xi = (\xi_1, \ldots, \xi_n) \in \mathbb{R}^n\,, \quad \varphi \in \mathcal{S}(\mathbb{R}^n) \qquad (6.1)$$

where $x\cdot\xi = \sum_{k=1}^n x_k\xi_k$ is the scalar product of $x,\ \xi \in \mathbb{R}^n$. From (6.1) we see that basic operations of analysis have simple relations with the Fourier transform.

(1) If τ_h denotes the translation[3] by $h \in \mathbb{R}^n$, then for all $\varphi \in \mathcal{S}(\mathbb{R}^n)$

$$\widehat{(\tau_h\varphi)}\,(\xi) = e^{-2\pi i h\cdot\xi}\,\hat{\varphi}(\xi) \quad \text{and} \quad \widehat{(e^{2\pi i x\cdot h}\varphi(x))}\,(\xi) = (\tau_h\hat{\varphi})\,(\xi)\,. \qquad (6.2)$$

[1] Specific issues will be discussed as we progress in this chapter.

[2] Here $x^\alpha = x_1^{\alpha_1} \ldots x_n^{\alpha_n}$ and $D^\beta = \frac{\partial^{\beta_1}}{\partial x_1^{\beta_1}} \cdots \frac{\partial^{\beta_n}}{\partial x_n^{\beta_n}}$ for $x = (x_1, \ldots, x_n) \in \mathbb{R}^n$, and $\alpha = (\alpha_1, \ldots, \alpha_n)$, $\beta = (\beta_1, \ldots, \beta_n)$ with nonnegative integers $\alpha_j,\ \beta_j \geq 0$. For such an n-tuple $\alpha = (\alpha_1, \ldots, \alpha_n)$ we denote $|\alpha| = \alpha_1 + \cdots + \alpha_n$ and $\alpha! = \alpha_1! \ldots \alpha_n!$.

[3] That is, the operator mapping the function $x \mapsto \varphi(x)$ into the function $x \mapsto \varphi(x - h)$.

(2) If A_a denotes the dilation[4] by $a > 0$, then for all $\varphi \in \mathcal{S}(\mathbb{R}^n)$

$$a^n\,\widehat{(A_a\varphi)}(\xi) = \hat{\varphi}(a^{-1}\xi)\,. \tag{6.3}$$

(3) Differentiation and Fourier transformation are related by

$$P(D)\,\hat{\varphi}(\xi) = \widehat{(P(-2\pi ix)\varphi(x))}\,(\xi) \quad \text{and} \quad \widehat{(P(D)\varphi)}(\xi) = P(2\pi i\xi)\,\hat{\varphi}(\xi) \tag{6.4}$$

for all $\varphi \in \mathcal{S}(\mathbb{R}^n)$; here P is a polynomial in the n variables $x_1, \ldots, x_n$ and $P(D)$ is the associated differential operator obtained by replacing x^α by D^α in $P(x)$.

Building on the properties of the Fourier transform acting on $\mathcal{S}(\mathbb{R})$ that were explored in Chapter 5, we can prove the following result.

Theorem 6.1 *The Fourier transform $\mathcal{F}(\phi) = \hat{\phi}$ is bijective on $\mathcal{S}(\mathbb{R}^n)$, with $\mathcal{F}^{-1}(\psi) = \mathcal{F}(\psi^\dagger)$, where $\psi^\dagger(x) = \psi(-x)$.*

Proof First of all, (6.4) ensures that $\hat{\psi} \in \mathcal{S}(\mathbb{R}^n)$ whenever $\psi \in \mathcal{S}(\mathbb{R}^n)$. Given $\psi \in \mathcal{S}(\mathbb{R}^n)$, let us fix $y_k \in \mathbb{R}$ for $k \in \{2, \ldots, n\}$. Since the function $s \mapsto \psi(s, y_2, \ldots, y_n)$ belongs to $\mathcal{S}(\mathbb{R})$, we get $\psi(t, y_2, \ldots, y_n) = \int_{\mathbb{R}}\big(\int_{\mathbb{R}} \psi(s, y_2, \ldots, y_n)e^{2\pi is\tau}\,d\tau\big)e^{-2\pi its}\,ds$ for all $t \in \mathbb{R}$, see Chapter 5. We now repeat this procedure by allowing variations only in the second position of a vector in $\mathbb{R}^n$. Subsequently we vary only the third position. After n steps, we gather by means of Fubini's theorem the obtained information as $\int_{\mathbb{R}^n}\big(\int_{\mathbb{R}^n}\psi^\dagger(y)e^{-2\pi iy\cdot\xi}\,dy\big)e^{-2\pi i\xi\cdot x}\,dx = \int_{\mathbb{R}^n}\big(\int_{\mathbb{R}^n}\psi(y)e^{2\pi iy\cdot\xi}\,dy\big)e^{-2\pi i\xi\cdot x}\,dx = \psi(\xi)$ for all $\xi \in \mathbb{R}^n$, by coupling first the integration variable y_1 with x_1, then y_2 with x_2, and so on. Thus $\mathcal{F}(\mathcal{F}(\psi^\dagger)) = \psi$, which encompasses the entire statement. □

Defining the convolution $\phi * \psi$ of $\phi, \psi \in \mathcal{S}(\mathbb{R}^n)$ by

$$(\phi * \psi)(x) = \int_{\mathbb{R}^n} \phi(x - y)\psi(y)\,dy\,, \qquad x \in \mathbb{R}^n\,,$$

we exhibit another essential feature of the Fourier transform:

$$\widehat{(\phi * \psi)}(\xi) = \hat{\phi}(\xi)\cdot\hat{\psi}(\xi) \quad \text{and} \quad \widehat{(\phi\psi)}(\xi) = (\hat{\phi} * \hat{\psi})\,(\xi) \quad \text{for all} \quad \xi \in \mathbb{R}^n\,. \tag{6.5}$$

Before validating (6.5), note that the inclusion $\mathbb{S}(\mathbb{R}^n) \subset L^1(\mathbb{R}^n)$ and Fubini's theorem ensures that all convolutions are well-defined, with[5] $\phi * \psi \in L^1(\mathbb{R}^n)$. Moreover, the change of variables $y \mapsto x - y$ yields $(\phi * \psi)(x) = \int_{\mathbb{R}^n} \phi(y)\psi(x - y)\,dy$, so that $\phi * \psi = \psi * \phi$. Note that $\int_{\mathbb{R}^n}\big(\int_{\mathbb{R}^n}\phi(x - y)\,e^{-2\pi i(x-y)\cdot\xi}\,dx\big)\psi(y)\,e^{-2\pi iy\cdot\xi}\,dy = \int_{\mathbb{R}^n}\int_{\mathbb{R}^n}\phi(x - y)\psi(y)\,dy\,e^{-2\pi ix\cdot\xi}\,dx$. This proves the first part of (6.5), with the understanding that for the moment we extend the definition of $\mathcal{F}$ to functions that merely belong to $L^1(\mathbb{R}^n)$. Since the product of two functions in $\mathbb{S}(\mathbb{R}^n)$ belongs to $\mathbb{S}(\mathbb{R}^n)$, Theorem 6.1 ensures $\phi * \psi \in \mathbb{S}(\mathbb{R}^n)$. As for the second part of (6.5), since both sides belong to $\mathbb{S}(\mathbb{R}^n)$, Theorem 6.1 enables us to rewrite it as $\phi\cdot\psi = \mathcal{F}^{-1}(\hat{\phi} * \hat{\psi})$. By Theorem 6.1, the right side equals $\mathcal{F}((\hat{\phi} * \hat{\psi})^\dagger)$. Since $(f * g)^\dagger = f^\dagger * g^\dagger$ follows from the definition by the change of variables $y \mapsto -y$, we are left to check that $\phi\cdot\psi = \mathcal{F}((\hat{\phi})^\dagger * (\hat{\psi})^\dagger)$. By the first part of (6.5), the right side equals $\mathcal{F}((\hat{\phi})^\dagger)\cdot\mathcal{F}((\hat{\psi})^\dagger)$, and we unwrap this by Theorem 6.1 as $\mathcal{F}^{-1}(\hat{\phi})\cdot\mathcal{F}^{-1}(\hat{\psi}) = \phi\cdot\psi$.

[4] That is, the operator mapping the function $x \mapsto \varphi(x)$ into the function $x \mapsto \varphi(ax)$.
[5] We shall see that actually $\phi * \psi \in \mathbb{S}(\mathbb{R}^n)$.

Taking advantage of the fact that linear combinations of characteristic functions of rectangles are dense in $L^1(\mathbb{R}^n)$ and in $L^2(\mathbb{R}^n)$, and that these building blocks are products of characteristic functions of intervals $I \subset \mathbb{R}$ that can be approximated by functions in $\mathcal{S}(\mathbb{R})$, we deduce that functions of the type $\sum_{j=1}^m \alpha_j\Big(\prod_{k=1}^n \varphi_k(x_k)\Big)$ with $\alpha_j \in \mathbb{C}$ and $\varphi_k \in \mathcal{S}(\mathbb{R})$ are dense in $L^1(\mathbb{R}^n)$ and in $L^2(\mathbb{R}^n)$. We can now establish the following basic results for the multi-dimensional extension of the Fourier transform.

Theorem 6.2 *The mapping $f \mapsto \hat{f}$ is a bounded linear map from $L^1(\mathbb{R}^n)$ into $L^\infty(\mathbb{R}^n)$, with $\|\hat{f}\|_{L^\infty} \le \|f\|_{L^1}$. Moreover, $\hat{f}$ is uniformly continuous.*

Theorem 6.3 (Riemann–Lebesgue) *If $f \in L^1(\mathbb{R}^n)$ then $\lim_{|\xi|\to\infty} \hat{f}(\xi) = 0$.*

Proof Theorem 6.2 is obvious since the definition (6.1) extends to functions in $L^1(\mathbb{R}^n)$. As for Theorem 6.3, it follows for the characteristic function f of an n-dimensional rectangle $\{x \in \mathbb{R}^n : a_1 \le x_1 < b_1, \dots, a_n \le x_n < b_n\}$ by calculating $\hat{f}$ explicitly as an interated integral. The result will therefore also hold for any finite linear combination of such functions. Given a general $f \in L^1(\mathbb{R}^n)$ and $\varepsilon > 0$, for a linear combination g of this type with $\|f-g\|_{L^1} < \varepsilon$ we have $\hat{f} = \hat{g} + (\hat{f} - \hat{g})$, and $\hat{g}(\xi) \to 0$ as $|\xi| \to \infty$ while $(\hat{f} - \hat{g})$ is uniformly small by Theorem 6.2. □

Theorems 6.2 and 6.3 give necessary conditions for a function $F : \mathbb{R}^n \to \mathbb{C}$ to be the Fourier transform of some $f \in L^1(\mathbb{R}^n)$ but these conditions are far from sufficient. A simple sufficiency criterion remains elusive. Also, while the extension of the Fourier transform from $\mathcal{S}(\mathbb{R}^n)$ to $L^1(\mathbb{R}^n)$ is relatively straightforward, the case $n = 1$ indicates that we cannot expect to make sense of the Lebesgue integral in the definition (6.1) for a general $\varphi \in L^2(\mathbb{R}^n)$. Nevertheless, the multi-dimensional Fourier transform L^2-theory is particularly elegant.

Theorem 6.4 *The Fourier transform $\mathcal{F}$ can be extended from $\mathcal{S}(\mathbb{R}^n)$ to a unitary operator on $L^2(\mathbb{R}^n)$, with inverse $\mathcal{F}^{-1}(f) = \mathcal{F}(f^\dagger)$ for $f \in L^2(\mathbb{R}^n)$.*

Proof Let us first show that $\|\hat{f}\|_{L^2} = \|f\|_{L^2}$ whenever $f \in \mathcal{S}(\mathbb{R}^n)$. For such a function f, define $g \in \mathcal{S}(\mathbb{R}^n)$ by $g(x) = \overline{f(-x)}$ and set $h = f * g \in \mathcal{S}(\mathbb{R}^n)$. Since $\hat{g} = \overline{\hat{f}}$, (6.5) yields $\hat{h} = |\hat{f}|^2$, so that $\|\hat{f}\|_{L^2}^2 = \int_{\mathbb{R}^n} \hat{h}(\xi)\,d\xi = \int_{\mathbb{R}^n} \hat{h}(-\xi)\,d\xi = h(0) = \int_{\mathbb{R}^n} f(x) g(0-x)\,dx = \int_{\mathbb{R}^n} f(x)\,\overline{f(x)}\,dx = \|f\|_{L^2}^2$, using Theorem 6.1 with $\psi = \hat{h}$ in the third equality.

Since $\mathcal{F}$ is a bounded linear operator (in fact, it is an isometry) defined on the dense subspace $\mathcal{S}(\mathbb{R}^n)$ of $L^2(\mathbb{R}^n)$, it admits a unique bounded extension to all of $L^2(\mathbb{R}^n)$. This defines the Fourier transform on $L^2(\mathbb{R}^n)$. The constructive procedure grants $\|\mathcal{F}f\|_{L^2} = \|f\|_{L^2}$ for all $f \in L^2(\mathbb{R}^n)$, so that the range of $\mathcal{F}$ is a closed subspace of the Hilbert space $L^2(\mathbb{R}^n)$. If this subspace were not all of $L^2(\mathbb{R}^n)$, we could find a function $f_0 \in L^2(\mathbb{R}^n)$ such that $f_0 \not\equiv 0$ and $\int_{\mathbb{R}^n} \hat{f}(x)\,\overline{f_0(x)}\,dx = 0$ for all $f \in L^2(\mathbb{R}^n)$. Bringing into play the multiplication identity[6]

$$\int_{\mathbb{R}^n} \hat{\phi}(x)\,\psi(x)\,dx = \int_{\mathbb{R}^n} \phi(x)\,\hat{\psi}(x)\,dx\,, \qquad \phi, \psi \in \mathcal{S}(\mathbb{R}^n)\,, \tag{6.6}$$

[6] The proof is a straightforward application of Fubini's theorem, once we make the two Fourier transforms explicit by means of (6.1).

the density of $S(\mathbb{R}^n)$ in $L^2(\mathbb{R}^n)$ forces $\widehat{f_0}(x) = 0$ for a.e. $x \in \mathbb{R}^n$, contradiciting the fact that $\|f_0\|_{L^2} = \|\widehat{f_0}\|_{L^2} \neq 0$. Consequently $\mathcal{F}$ is a bijection on $L^2(\mathbb{R}^n)$. Finaly, since the formula for its inverse holds on $S(\mathbb{R}^n)$ by Theorem 6.1, density ensures its validity throughout $L^2(\mathbb{R}^n)$. □

With the action of the Fourier transform $\mathcal{F}$ on $L^1(\mathbb{R}^n)$ and $L^2(\mathbb{R}^n)$ clarified, we can easily extend $\mathcal{F}$ to the class of all functions $f = f_1+f_2$ with $f_1 \in L^1(\mathbb{R}^n)$ and $f_2 \in L^2(\mathbb{R}^n)$ by setting[7] $\hat{f} = \widehat{f_1} + \widehat{f_2}$. Since this class contains[8] $L^p(\mathbb{R}^n)$, $1 \le p \le 2$, the Fourier transform is thus defined for all $f \in L^p(\mathbb{R}^n)$ with $1 \le p \le 2$. To extend the Fourier transform much further, in particular, to the spaces[9] $L^p(\mathbb{R}^n)$ with $p > 2$, requires the development of new tools.

6.1.1 The class of tempered distributions

Distributions arise as linear functionals on some space of regular functions (the so-called "testing functions") that is well-behaved with respect to linear operations and differentiation, other desirable operations being multiplication and convolution. For the purposes of Fourier analysis, a well-suited class is the Schwartz class $S(\mathbb{R}^n)$.

Let us first show the existence of a metric that will endow $S(\mathbb{R}^n)$ with the structure of a topological vector space: a linear vector space with a topology that is compatible with the operations of vector addition and scalar multiplication (that is, both operations are continuous). Toward this end we consider a countable family of norms $\{\rho_{\alpha,\beta}\}$, indexed by all the ordered pairs (α,β) of n-tuples of nonnegative integers and defined by

$$\rho_{\alpha,\beta}(\varphi) = \sup_{x\in\mathbb{R}^n} |x^\alpha\, (D^\beta\varphi)(x)|$$

for $\varphi \in S(\mathbb{R}^n)$. Rather than the associated metric $\delta_{\alpha,\beta}(\phi,\psi) = \rho_{\alpha,\beta}(\phi-\psi)$ on $S(\mathbb{R}^n)$, the equivalent metric[10] $d_{\alpha,\beta}(\phi,\psi) = \frac{\delta_{\alpha,\beta}(\phi,\psi)}{1+\delta_{\alpha,\beta}(\phi,\psi)}$ is convenient.[11] Let $d_1, d_2, \ldots$ be an ordering of these metrics . Since $d_n \le 1$ for all $n \ge 1$, we can easily check that $d = \sum_{n\ge 1} \frac{1}{2^n}\, d_n$ is a metric on $S(\mathbb{R}^n)$, defining a topology

7 To see that $\mathcal{F}$ is well-defined, note that if $f = g_1 + g_2$ with $g_1 \in L^1(\mathbb{R}^n)$ and $g_2 \in L^2(\mathbb{R}^n)$, then $g_1 - f_1 = f_2 - g_2 \in L^1(\mathbb{R}^n) \cap L^2(\mathbb{R}^n)$. Since the two definitions of the Fourier transform coincide on $L^1(\mathbb{R}^n) \cap L^2(\mathbb{R}^n)$ with the formula (6.1), we have $\widehat{g_1} - \widehat{f_1} = \widehat{f_2} - \widehat{g_2}$. Thus $\widehat{f_1} + \widehat{f_2} = \widehat{g_1} + \widehat{g_2}$.

8 Write $f \in L^p(\mathbb{R}^n)$ as $f = f_1 + f_2$, where $f_1 = f\,\chi_{[|f|>1]}$ and $f_2 = f\,\chi_{[|f|\le 1]}$. Note that $\int_{\mathbb{R}^n} |f_1(x)|\, dx \le \int_{\mathbb{R}^n} |f(x)|^p\, dx < \infty$, while $\int_{\mathbb{R}^n} |f_2(x)|^2\, dx \le \int_{\mathbb{R}^n} |f(x)|^p\, dx < \infty$.

9 This is in stark contrast with the case of Fourier series, where the inclusion $L^p(\mathbb{S}^n) \subset L^1(\mathbb{S}^n)$ for $p \ge 1$ permits us to define at once the Fourier coefficients of a function $f \in L^p(\mathbb{S}^n)$.

10 Meaning that they define the same topology: a sequence $\{\phi_k\}_{k\ge 1}$ will converge to some limit ϕ in the metric $d_{\alpha,\beta}$ if and only if it converges (to the same limit) in the metric $\delta_{\alpha,\beta}$.

11 Since $s \mapsto \frac{s}{1+s}$ is strictly increasing on $(0,\infty)$, the triangle inequality for $d_{\alpha,\beta}$ is immediate.

suitable for our purposes. Note that $\varphi_k \to \varphi$ in $\mathcal{S}(\mathbb{R}^n)$, with respect to d, if and only if $\varphi_k \to \varphi$ with respect to each d_n (as $k \to \infty$). In particular, both the operation of vector addition and of scalar multiplication are continuous, so that $(\mathcal{S}(\mathbb{R}^n), d)$ is a metric space.[12] Further useful and easily established properties of $\mathcal{S}(\mathbb{R}^n)$ and its topology are the following.

(i) The vector space $\mathcal{S}(\mathbb{R}^n)$ is a separable Fréchet space. A variation of Exercise 5.14 shows that it is dense in any $L^p(\mathbb{R}^n)$ with $1 \le p < \infty$.

(ii) The Fourier transform is a continuous linear map of $\mathcal{S}(\mathbb{R}^n)$ onto itself, with a continuous inverse.

(iii) The product of two functions in $\mathcal{S}(\mathbb{R}^n)$ also belongs to $\mathcal{S}(\mathbb{R}^n)$.

(iv) Although polynomials are not in $\mathcal{S}(\mathbb{R}^n)$, the space $\mathcal{S}(\mathbb{R}^n)$ is closed under multiplication by polynomials and under differentiation, the mapping $\varphi(x) \mapsto x^\alpha\,(D^\beta\varphi)(x)$ being continuous.

(v) $\mathcal{S}(\mathbb{R}^n)$ is closed under translations and multiplications by complex exponentials $x \mapsto e^{ix\cdot\xi}$ with $\xi \in \mathbb{R}^n$, both operations being continuous.

Definition *The collection of all continuous linear functionals $L : \mathcal{S}(\mathbb{R}^n) \to \mathbb{C}$ is called the space of tempered distributions, and is denoted by $\mathcal{S}'(\mathbb{R}^n)$.*

Let us first discuss some important examples of tempered distributions.

(I) Any $f \in L^p(\mathbb{R}^n)$, $1 \le p \le \infty$, defines an element $L_f \in \mathcal{S}'(\mathbb{R}^n)$ by $L_f(\phi) = \int_{\mathbb{R}^n} f(x)\phi(x)\,dx$ for $\phi \in \mathcal{S}(\mathbb{R}^n)$.

Clearly L_f is a linear functional on $\mathcal{S}(\mathbb{R}^n)$. To prove its continuity it suffices to show it at $0 \in \mathcal{S}(\mathbb{R}^n)$. Suppose $\phi_k \to 0$ in $\mathcal{S}(\mathbb{R}^n)$. Choosing $q \ge 1$ so that $\frac{1}{p} + \frac{1}{q} = 1$, Hölder's inequality yields $|L_f(\phi_k)| \le \|f\|_{L^p}\|\phi_k\|_{L^q}$. For $q = \infty$ we can conclude at once that $L_f(\varphi_k) \to 0$ as $k \to \infty$, while for $q \in [1,\infty)$ we can take advantage of the fact that the $L^q(\mathbb{R}^n)$-norm of $\varphi \in \mathcal{S}(\mathbb{R}^n)$ is bounded by a finite linear combination of supremum norms of terms of the form $x^\alpha\varphi(x)$. More precisely, if $a = \sup_{x\in\mathbb{R}^n}|\varphi(x)|$ and $b = \sup_{x\in\mathbb{R}^n}|x|^{2n}|\varphi(x)|$, then, since $\varphi(x) = \varphi(x)\chi_{[|x|\le 1]} + \varphi(x)\chi_{[|x|>1]}$,

$$\Big(\int_{\mathbb{R}^n}|\varphi(x)|^q dx\Big)^{1/q} \le \Big(\int_{|x|\le 1}|\varphi(x)|^q dx\Big)^{1/q} + \Big(\int_{|x|>1}|\varphi(x)|^q dx\Big)^{1/q}$$
$$\le a\Big(\int_{|x|\le 1} 1 dx\Big)^{1/q} + b\Big(\int_{|x|>1}|x|^{-2nq} dx\Big)^{1/q} = a\,k_n + b\,\kappa_n \qquad (6.7)$$

for some constants $k_n, \kappa_n > 0$ that are independent of φ. Summing up, we always deduce that $L_f(\varphi_k) \to 0$ as $k \to \infty$. Thus, $L_f \in \mathcal{S}'(\mathbb{R}^n)$.

(II) If μ is a finite Borel measure on $\mathbb{R}^n$, then the linear functional L_μ defined by $L_\mu(\phi) = \int_{\mathbb{R}^n}\phi(x)\,d\mu(x)$ for $\phi \in \mathcal{S}(\mathbb{R}^n)$, is a tempered distribution.

The proof is similar to the one for (I) since in this case $|L_f(\phi_k)| \le \|\mu\| \cdot \|\phi_k\|_{L^\infty}$.

[12] Attempts to endow it with a Banach space structure are doomed to failure – the dual space $\mathcal{S}'(\mathbb{R}^n)$ is not even metrisable since its topology does not have a countable base of neighbourhoods of 0; see Friedlander (1998).

(III) A *tempered L^p-function*[13] is a Lebesgue measurable function $f : \mathbb{R}^n \to \mathbb{C}$ such that $x \mapsto \frac{f(x)}{(1+|x|^2)^k}$ is in $L^p(\mathbb{R}^n)$ for some $p \in [1, \infty]$ and some integer $k \geq 1$. For each such function, $L_f(\phi) = \int_{\mathbb{R}^n} f(x)\phi(x)\,dx$ defines an element $L_f \in \mathcal{S}'(\mathbb{R}^n)$.

We see this by writing $L_f(\phi) = \int_{\mathbb{R}^n}(1+|x|^2)^k\,\phi(x)\,\frac{f(x)}{(1+|x|^2)^k}\,dx$. Since $\phi(x) \mapsto (1+|x|^2)^k\,\phi(x)$ is a continuous map on $\mathcal{S}(\mathbb{R}^n)$, we can rely on the approach used in example (I).

(IV) A *tempered Borel measure* μ on $\mathbb{R}^n$: a Borel measure such that $\int_{\mathbb{R}^n}(1+|x|^2)^{-k}\,d\mu(x) < \infty$ for some integer $k \geq 1$, defines a tempered distribution.

Writing $L_f(\phi) = \int_{\mathbb{R}^n}(1+|x|^2)^k\,\phi(x)\,\frac{d\mu(x)}{(1+|x|^2)^k}$, since $\phi(x) \mapsto (1+|x|^2)^k\,\phi(x)$ is continuous on $\mathcal{S}(\mathbb{R}^n)$, we can rely on the approach used in example (II).

(V) For any $x_0 \in \mathbb{R}^n$ and any n-tuple β of nonnegative integers, $\phi \mapsto (D^\beta\phi)(x_0)$ defines an element of $\mathcal{S}'(\mathbb{R}^n)$. The special case $\beta = 0$ is the Dirac mass at x_0, δ_{x_0}.

This follows at once from the continuity of the norm $\rho_{0\beta}$ in $\mathcal{S}(\mathbb{R}^n)$.

While the tempered distributions in examples (I)–(IV) can be associated to functions or measures, the case $\beta \neq 0$ in example (V) illustrates the fact that the new concept goes beyond[14] these two classes.

Let us now prove the following characterisation of tempered distributions.

Theorem 6.5 *A linear functional $L : \mathcal{S}(\mathbb{R}^n) \to \mathbb{C}$ belongs to $\mathcal{S}'(\mathbb{R}^n)$ if and only if there exists a constant $C > 0$ and integers $m,\ N \geq 0$ such that*

$$|L(\phi)| \leq C \sum_{|\alpha|\leq m,\ |\beta|\leq N} \rho_{\alpha\beta}(\phi) \quad \textit{for all} \quad \phi \in \mathcal{S}(\mathbb{R}^n).$$

Proof Such an inequality clearly implies the continuity of L. Conversely, suppose L is continuous. Since the topology of $\mathcal{S}(\mathbb{R}^n)$ was defined such that $\phi_k \to \phi$ if and only if $\rho_{\alpha\beta}(\phi_k - \phi) \to 0$ for all (α, β), from the continuity of L at the origin in $\mathcal{S}(\mathbb{R}^n)$ it follows that there exists a set $O_{\varepsilon,m,n} = \{\phi \in \mathcal{S}(\mathbb{R}^n) : \sum_{|\alpha|\leq m,\,|\beta|\leq N}\rho_{\alpha\beta}(\phi) < \varepsilon\}$, where $\varepsilon > 0$ and $m,\ N \geq 0$ are integers, such that $|L(\varphi)| \leq 1$ whenever $\varphi \in O_{\varepsilon,m,N}$. If we set $|||\phi||| = \sum_{|\alpha|\leq m,\,|\beta|\leq n}\rho_{\alpha\beta}(\phi)$ for all $\phi \in \mathcal{S}(\mathbb{R}^n)$, then $\psi = \frac{\delta}{|||\phi|||}\,\phi \in O_{\varepsilon,m,N}$ if $\phi \neq 0$ and $\delta \in (0, \varepsilon)$, and the linearity of L yields $\frac{\delta}{|||\phi|||}|L(\phi)| \leq 1$. This is the desired inequality with $C = \frac{1}{\delta}$. □

Theorem 6.5 suggests the following representation of tempered distributions.

13 When $p = \infty$ such functions are called *functions of polynomial growth*.

14 If, for example, $\phi \mapsto [D^{(1,0,\dots,0)}\phi](0)$ would come from a tempered function or measure, we could find an integer $k \geq 0$ so that $|\partial_{x_1}\phi(0)| \leq M\,\sup_{x\in\mathbb{R}^n}\{(1+|x|^2)^k|\phi(x)|\}$ for all $\phi \in \mathcal{S}(\mathbb{R}^n)$. However, this inequality behaves badly under scaling: if we set $\phi_j(x) = \phi(jx)$ for integers $j \geq 1$, the right side remains uniformly bounded as $j \to \infty$ since with $y = jx$ we get $\sup_{x\in\mathbb{R}^n}\{(1+|x|^2)^k|\phi_j(x)|\} \leq \sup_{y\in\mathbb{R}^n}\{(1+|y|^2)^k|\phi(y)|\}$, while $|\partial_{x_1}\phi_j(0)| = j\,|\partial_{x_1}\phi(0)|$. For $\phi(x) = e^{-|x|^2}$ this type of inequality must therefore fail.

Theorem 6.6 *If $L \in \mathcal{S}'(\mathbb{R}^n)$, then for an appropriate integer $N \geq 0$ and for each n-tuple β with $|\beta| \leq N$, there exist tempered complex Borel measures μ_β on $\mathbb{R}^n$ so that $L(\phi) = \sum_{|\beta|\leq N} \int_{\mathbb{R}^n} D^\beta\phi \, d\mu_\beta$ for all $\phi \in \mathcal{S}(\mathbb{R}^n)$.*

There is no uniqueness corresponding to the above representation. For example, if $n = 1$ and $L = f_0 + \delta_0'$ with $f_0 \in L^1(\mathbb{R})$, then also $L = f' + \delta_0'$ with $f(x) = \int_0^x f_0(s)\,ds$. This lack of uniqueness shows that Theorem 6.6 is not as explicit as it might seem; note also that it is rather difficult to grasp exactly how high-order derivatives are integrated against a tempered measure.

Proof Let $m, N \geq 0$ be the integers associated by means of Theorem 6.5 with $L \in \mathcal{S}'(\mathbb{R}^n)$. The application $\phi \mapsto \prod_{|\beta|\leq N} D^\beta\phi$ maps $\mathcal{S}(\mathbb{R}^n)$ to $X = \prod_{|\beta|\leq N} C_m(\mathbb{R}^n)$, where $C_m(\mathbb{R}^n)$ is the Banach space of continuous functions $f : \mathbb{R}^n \to \mathbb{C}$ with $\sup_{x\in\mathbb{R}^n} |x^\alpha f(x)| < \infty$ for all n-tuples α with $|\alpha| \leq m$, endowed with the norm $\|f\|_m = \sum_{|\alpha|\leq m} \sup_{x\in\mathbb{R}^n} |x^\alpha f(x)|$. This map is injective, but not surjective, its range Y being a proper subspace of the Banach space X. If $\pi : Y \to \mathcal{S}(\mathbb{R}^n)$ is the inverse, then $L \circ \pi : Y \to \mathbb{C}$ defines a linear functional that is bounded, due to the inequality in Theorem 6.5. By the Hahn–Banach theorem (see Chapter 3), this functional admits an extension to some $\theta \in X^*$. But X^* can be identified with a direct product of tempered complex Borel measures[15] μ_β on $\mathbb{R}^n$. In particular, for an arbitrary $\phi \in \mathcal{S}(\mathbb{R}^n)$ we have $\pi^{-1}(\phi) \in Y$ and $L(\phi) = (L\circ\pi)(\pi^{-1}(\phi)) = \sum_{|\beta|\leq N} \int_{\mathbb{R}^n} \pi^{-1}(\phi)\,d\mu_\beta = \sum_{|\beta|\leq N} \int_{\mathbb{R}^n} D^\beta\phi\,d\mu_\beta$. □

Let us now define differentiation on $\mathcal{S}'(\mathbb{R}^n)$. Note that integration by parts yields

$$\int_{\mathbb{R}^n} (D^\beta u)(x)\,\phi(x)\,dx = (-1)^{|\beta|} \int_{\mathbb{R}^n} u(x)\,(D^\beta\phi)(x)\,dx \quad \text{for} \quad u, \phi \in \mathcal{S}(\mathbb{R}^n).$$

Denoting the continuous linear functionals $\phi \mapsto \int_{\mathbb{R}^n} (D^\beta u)(x)\,\phi(x)\,dx$ and $\phi \mapsto \int_{\mathbb{R}^n} u(x)\,\phi(x)\,dx$ on $\mathcal{S}(\mathbb{R}^n)$ by $D^\beta u$ and u, respectively, we can write the equality displayed above in the form

$$\langle D^\beta u, \phi\rangle = (-1)^{|\beta|}\,\langle u, D^\beta\phi\rangle\,. \tag{6.8}$$

Notice that the right side of (6.8) is well-defined whenever $u \in \mathcal{S}'(\mathbb{R}^n)$. Furthermore, $\phi \mapsto \langle u, D^\beta\phi\rangle$ is a continuous linear functional on $\mathcal{S}(\mathbb{R}^n)$, being the composite of two continuous functions. In view of these considerations, it is natural to define the *partial derivative $D^\beta u$ of the tempered distribution u* by means of (6.8). Clearly $D^\beta u \in \mathcal{S}'(\mathbb{R}^n)$. We refer to Exercise 9 for the more intricate issue of integrating tempered distributions.

In a similar way, the *translation operator* on $\mathcal{S}'(\mathbb{R}^n)$ is defined by letting $\tau_h u$, where $u \in \mathcal{S}'(\mathbb{R}^n)$ and $h \in \mathbb{R}^n$, be the continuous linear functional on $\mathcal{S}(\mathbb{R}^n)$ that assigns to $\phi \in \mathcal{S}(\mathbb{R}^n)$ the value $\langle u, \tau_{-h}\phi\rangle$.

The same approach leads us to define the *multiplication of $u \in \mathcal{S}'(\mathbb{R}^n)$ by $\varphi \in$*

[15] This is the content of the Riesz reprersentation theorem, see Reed and Simon (1980a).

$\mathcal{S}(\mathbb{R}^n)$ as the element $\varphi u \in \mathcal{S}'(\mathbb{R}^n)$ given by $\langle \varphi u, \psi \rangle = \langle u, \varphi\psi \rangle$ for $\psi \in \mathcal{S}(\mathbb{R}^n)$. More generally, we can multiply $u \in \mathcal{S}'(\mathbb{R}^n)$ by functions φ in the class $\mathcal{O}(\mathbb{R}^n)$ of all $\varphi \in C^\infty(\mathbb{R}^n)$ such that $D^\beta\varphi$ is of polynomial growth for all n-tuples β, since in this case $\varphi\psi \in \mathcal{S}(\mathbb{R}^n)$ whenever $\psi \in \mathcal{S}(\mathbb{R}^n)$; see also Exercise 11. The functions $x \mapsto (1 + |x|^2)^s$ with $s > 0$, playing an important role in the theory of Sobolev spaces, are examples of functions in $\mathcal{O}(\mathbb{R}^n) \setminus \mathcal{S}(\mathbb{R}^n)$. On the other hand, the product of two tempered distributions cannot be sensibly defined since, for example $f \cdot \delta_0$ cannot be defined for the discontinuous measurable bounded function $f(x) = \frac{x}{|x|}$ (here $x \in \mathbb{R} \setminus \{0\}$), for by rights we should have $\langle f \cdot \delta_0, \phi \rangle = f(0)\phi(0)$ but there is no reasonable way to define $f(0)$; see also Exercise 7.

We can define the *convolution of $u \in \mathcal{S}'(\mathbb{R}^n)$ by $\varphi \in \mathcal{S}(\mathbb{R}^n)$* as follows. First, a direct application of Fubini's theorem shows by means of a simple change of variables that if $u, \phi, \psi \in \mathcal{S}(\mathbb{R}^n)$, then

$$\int_{\mathbb{R}^n} (u * \phi)(x)\, \psi(x)\, dx = \int_{\mathbb{R}^n} u(y)\, (\phi^\dagger * \psi)(y)\, dy\,.$$

If we denote by $u * \phi$ and u the continuous linear functionals on $\mathcal{S}(\mathbb{R}^n)$ given by $\phi \mapsto \int_{\mathbb{R}^n} (u * \phi)(x)\, \psi(x)\, dx$ and $\theta \mapsto \int_{\mathbb{R}^n} u(y)\, \theta(y)\, dy$, respectively, the previously displayed equality can be written in the form

$$\langle u * \phi, \psi \rangle = \langle u, \phi^\dagger * \psi \rangle\,. \tag{6.9}$$

For $u \in \mathcal{S}'(\mathbb{R}^n)$ and $\phi, \psi \in \mathcal{S}(\mathbb{R}^n)$, the right side of (6.9) is well-defined since $\phi^\dagger * \psi \in \mathcal{S}(\mathbb{R}^n)$. Furthermore, $\psi \mapsto \phi^\dagger * \psi$ is continuous from $\mathcal{S}(\mathbb{R}^n)$ to $\mathcal{S}(\mathbb{R}^n)$, being the composition of two continuous maps. Consequently, we can define the convolution $u * \phi \in \mathcal{S}'(\mathbb{R}^n)$ by means of (6.9). The following relations can be easily checked by manipulating (6.8) and (6.9),

$$D^\beta(u * \phi) = (D^\beta u) * \phi = u * (D^\beta \phi) \quad \text{for all} \quad u \in \mathcal{S}'(\mathbb{R}^n), \quad \phi \in \mathcal{S}(\mathbb{R}^n)\,. \tag{6.10}$$

They and the fact that the convolution of two functions has cumulative regularity[16] suggest that perhaps $u * \phi$ is not merely a tempered distribution. This is indeed the case, due to the following regularisation result.

Theorem 6.7 *If $u \in \mathcal{S}'(\mathbb{R}^n)$ and $\phi \in \mathcal{S}(\mathbb{R}^n)$, then $u * \phi$ is the $\mathcal{O}(\mathbb{R}^n)$-function $f(x) = \langle u, \tau_x \phi^\dagger \rangle$.*

Proof We first show that f has the claimed properties. If $e_j \in \mathbb{R}^n$ is the unit vector having the entry 1 on the jth position, $1 \le j \le n$, then $\lim_{s \to 0} \frac{\tau_{x+se_j}\phi^\dagger - \tau_x\phi^\dagger}{s} \to -\tau_x\big(\frac{\partial(\phi^\dagger)}{\partial x_j}\big)$ in $\mathcal{S}(\mathbb{R}^n)$.

16 For j-times and k-times differentiable functions $f, g : \mathbb{R} \to \mathbb{C}$ with $f^{(j)}, g^{(k)} \in L^1(\mathbb{R})$, $f * g$ is $(j+k)$-times differentiable: $(f * g)^{(j+k)}(x) = \int_{\mathbb{R}} f^{(j)}(x-y)\, g^{(k)}(y)\, dy$.

Since $u \in \mathcal{S}'(\mathbb{R}^n)$, we get $\frac{f(x+se_j)-f(x)}{s} = \left\langle u, \frac{\tau_{x+se_j}\phi^\dagger - \tau_x\phi^\dagger}{s}\right\rangle \to \left\langle u, -\tau_x(\frac{\partial(\phi^\dagger)}{\partial x_j})\right\rangle$ as $s \to 0$, so that $f : \mathbb{R}^n \to \mathbb{C}$ has continuous first partial derivatives. Since $\frac{\partial(\phi^\dagger)}{\partial x_j}$ and $\tau_x(\frac{\partial(\phi^\dagger)}{\partial x_j})$ are both in $\mathcal{S}(\mathbb{R}^n)$, we can iterate this approach to infer the existence and continuity of $D^\beta f$ for all n-tuples of nonnegative integers β, with $(D^\beta f)(x) = (-1)^{|\beta|}\langle u, \tau_x D^\beta(\phi^\dagger)\rangle$. Consequently, once we check that f has polynomial growth, since $D^\beta(\phi^\dagger) \in \mathcal{S}(\mathbb{R}^n)$, all derivatives of f would also have this property. To verify that f has polynomial growth, notice that Theorem 6.5 grants the existence of $C > 0$ and of integers $m, N \geq 0$ with $|f(x)| = |\langle u, \tau_x(\phi^\dagger)\rangle| \leq C \sum_{|\alpha|\leq m, |\beta|\leq N} \rho_{\alpha,\beta}(\tau_x(\phi^\dagger))$. But $\rho_{\alpha,\beta}(\tau_x(\phi^\dagger)) = \sup_{y\in\mathbb{R}^n} |y^\alpha (D^\beta(\phi^\dagger))(y+x)| = \sup_{w\in\mathbb{R}^n} |(w-x)^\alpha (D^\beta(\phi^\dagger))(w)|$ is clearly bounded by $M_\phi(1+|x|^2)^m$ for a suitable $M_\phi > 0$.

It remains to identify $u * \phi$ with the function f. For this, it suffices to show that $\langle u * \phi, \psi\rangle = \int_{\mathbb{R}^n} \psi(x)f(x)\,dx$ holds for all $\psi \in \mathcal{S}(\mathbb{R}^n)$. But, $\langle u * \phi, \psi\rangle = \langle u, \phi^\dagger * \psi\rangle = \langle u, \int_{\mathbb{R}^n} \phi^\dagger(x-y)\psi(y)\,dy\rangle = \langle u, \int_{\mathbb{R}^n} (\tau_y\phi^\dagger)(x)\,\psi(y)\,dy\rangle$. The Riemann sums of the integral $\int_{\mathbb{R}^n} (\tau_y\phi^\dagger)(x)\,\psi(y)\,dy$ are easily seen to converge in $\mathcal{S}(\mathbb{R}^n)$, as functions of x, so that the continuity of the linear functional $u \in \mathcal{S}'(\mathbb{R}^n)$ yields $\langle u, \int_{\mathbb{R}^n} (\tau_y\phi^\dagger)(x)\,\psi(y)\,dy\rangle = \int_{\mathbb{R}^n} \langle u, (\tau_y\phi^\dagger)\rangle\,\psi(y)\,dy = \int_{\mathbb{R}^n} f(y)\psi(y)\,dy$, which leads us to the desired equality. □

While a sensible notion of convolution for general elements of $\mathcal{S}'(\mathbb{R}^n)$ is not possible, see Exercise 7, it turns out that if one of the factors $u, v \in \mathcal{S}'(\mathbb{R}^n)$ has compact support, then Fourier analysis is instrumental in defining $u * v \in \mathcal{S}'(\mathbb{R}^n)$, as we shall show in the upcoming section. Here we limit ourselves to a brief discussion of the class $\mathcal{E}'(\mathbb{R}^n)$ of tempered distributions with compact support.

The class $\mathcal{D}(\mathbb{R}^n)$ of *test functions* consists of C^∞-functions $\varphi : \mathbb{R}^n \to \mathbb{C}$ that vanish outside a compact set $K \subset \mathbb{R}^n$. The support $\mathrm{supp}(\phi)$ of $\varphi \in \mathcal{D}(\mathbb{R}^n)$ is the complement of the largest open set on which φ vanishes, being a compact set since by definition it is a bounded and closed subset of $\mathbb{R}^n$. Clearly $\mathcal{D}(\mathbb{R}^n) \subset \mathcal{S}(\mathbb{R}^n)$, and actually $\mathcal{D}(\mathbb{R}^n)$ is dense in $\mathcal{S}(\mathbb{R}^n)$, see Exercise 1. While it is possible to endow $\mathcal{D}(\mathbb{R}^n)$ with the structure of a complete locally convex space by means of the procedure of inductive limits[17] and to consider the distributions $\mathcal{D}'(\mathbb{R}^n)$ that arise as continuous linear functionals[18] on $\mathcal{D}(\mathbb{R}^n)$, our aim is to explore the interplay between distribution theory and Fourier analysis. For this purpose the distribution space $\mathcal{D}'(\mathbb{R}^n)$ is too large: $\mathcal{E}'(\mathbb{R}^n) \subsetneq \mathcal{S}'(\mathbb{R}^n) \subsetneq \mathcal{D}'(\mathbb{R}^n)$ and the Fourier transform theory applies mostly to $\mathcal{S}'(\mathbb{R}^n)$. It is meaningless to say that $u \in \mathcal{S}'(\mathbb{R}^n)$ vanishes at a point $x_0 \in \mathbb{R}^n$, but vanishing in a neighbourhood of x_0 makes sense: if $\Omega \subset \mathbb{R}^n$ is an open set, we say that u vanishes

[17] In this topology, a sequence $\varphi_n \in \mathcal{D}(\mathbb{R}^n)$ is said to converge to zero if the supports of all functions are contained in a fixed compact set and for each n-tuple β, $D^\beta\varphi_n$ converge to zero uniformly as $n \to \infty$; see Reed and Simon (1980a).

[18] The continuity of $w \in \mathcal{D}'(\mathbb{R}^n)$ being defined by the condition that $\lim_{n\to\infty}\langle w, \varphi_n\rangle = 0$ for every sequence $\{\varphi_n\}_{n\geq 1}$ which converges to zero in $\mathcal{D}(\mathbb{R}^n)$. One can express this by means of estimates as follows: for every compact set $K \subset \mathbb{R}^n$, there is a constant $C(K) > 0$ and an integer $N(K) \geq 0$ such that $|\langle w, \varphi\rangle| \leq C(K)\sum_{|\beta|\leq N(K)} \sup_{x\in\mathbb{R}^n} |D^\beta\varphi(x)|$ for all $\varphi \in \mathcal{D}(\mathbb{R}^n)$ with support contained in K. For an authorative full-scale treatment of distribution theory we refer to Hörmander (1983).

in Ω if $\langle u, \varphi\rangle = 0$ for every $\varphi \in \mathcal{D}(\mathbb{R}^n)$ having support in Ω. The *support* of $u \in \mathcal{S}'(\mathbb{R}^n)$, supp$(u)$, is defined as the complement of the largest open set on which u vanishes. For example, $\text{supp}(\delta_0) = \{0\}$, while $\text{supp}(e^{-|x|^2}) = \mathbb{R}$. We define $\mathcal{E}'(\mathbb{R}^n)$ as the class of tempered distributions $u \in \mathcal{S}'(\mathbb{R}^n)$ with compact support. A pleasant feature of any $u \in \mathcal{E}'(\mathbb{R}^n)$ is that $\langle u, \phi\rangle$ can be defined for every C^∞-function $\phi : \mathbb{R}^n \to \mathbb{C}$ as follows: choose[19] a function $\theta \in \mathcal{D}(\mathbb{R}^n)$ with $\theta(x) = 1$ if $|x| \le R$, where $R > 0$ is chosen so that $\text{supp}(u) \subset \{x \in \mathbb{R}^n : |x| < R\}$, and set $\langle u, \phi\rangle = \langle u, \phi\theta\rangle$. This definition does not depend on the choice of θ since a different choice $\tilde\theta$ would yield $\langle u, \phi\tilde\theta\rangle = \langle u, \phi\theta\rangle + \langle u, \phi(\tilde\theta - \theta)\rangle$ and the last term vanishes since $\phi(\tilde\theta - \theta)$ vanishes in a neighbourhood of supp(u).

We conclude this section by presenting some facts that underline the usefulness of differentiation in the sense of distributions.

- Every $u \in \mathcal{S}'(\mathbb{R}^n)$ possesses derivatives $D^\beta u \in \mathcal{S}'(\mathbb{R}^n)$ of all orders. Moreover, $D^\beta : \mathcal{S}'(\mathbb{R}^n) \to \mathcal{S}'(\mathbb{R}^n)$ is a continuous linear map.[20]
- $\frac{\partial}{\partial x_j}\frac{\partial}{\partial x_k} u = \frac{\partial}{\partial x_k}\frac{\partial}{\partial x_j} u$ for all $u \in \mathcal{S}'(\mathbb{R}^n)$ and all $1 \le j, k \le n$, in contrast to the situation encountered for classical functions of several variables.[21]
- Distributions are suitable to deal with jump discontinuities. The simplest example is the Heaviside step function $H(x) = \begin{cases} 1 & \text{for } x > 0, \\ 0 & \text{for } x \le 0, \end{cases}$ with[22] $H' = \delta_0$.
- Distributional derivatives can sometimes extract "finite parts" from divergent integrals.[23] For example, the function $f(x) = \frac{1}{x}\chi_{(0,\infty)}(x)$ is continuous on $\mathbb{R}\backslash\{0\}$ without being locally integrable near the origin, with $\int_{\mathbb{R}} f(x)\phi(x)\,dx$ divergent whenever $\phi(0) \neq 0$. Note that $u(x) = \ln(x)\chi_{(0,\infty)}(x)$ is a tempered L^1-function such that $u' = f$ pointwise on $\mathbb{R} \setminus \{0\}$. If we set $u_\varepsilon(x) = \ln(x)\chi_{(\varepsilon,\infty)}(x)$, then $u_\varepsilon \to u$ in $\mathcal{S}'(\mathbb{R})$ as $\varepsilon \downarrow 0$. It follows that $u'_\varepsilon \to u'$ in $\mathcal{S}'(\mathbb{R})$. But, for $\phi \in \mathcal{S}(\mathbb{R})$, $\langle u'_\varepsilon, \phi\rangle = -\langle u_\varepsilon, \phi'\rangle = -\int_\varepsilon^\infty \ln(x)\phi'(x)\,dx = \phi(\varepsilon)\ln(\varepsilon) + \int_\varepsilon^\infty \frac{\phi(x)}{x}\,dx$. As $\varepsilon \downarrow 0$, this last sum converges to $\langle u', \phi\rangle$ even

19 In Section 6.1 we gave an example of a function $\mathfrak{e} \in \mathcal{D}(\mathbb{R})$ such that $0 \le \mathfrak{e}(s) \le 1$ for all $s \in \mathbb{R}$, $\mathfrak{e}(s) = 1$ for $|s| \le 1$ and $\mathfrak{e}(s) = 0$ for $|s| \ge 2$. Then $\theta(x) = \prod_{k=1}^n \mathfrak{e}(\frac{x_k}{R})$ serves our purposes, with $\theta(x) = 0$ for $|x| \ge 2R\sqrt{n}$ since if $(x_1^2 + \cdots + x_n^2)/R^2 \ge 4n$, then $x_k^2/R^2 \ge 4$ for at least one $k \in \{1, \dots, n\}$.

20 $\mathcal{S}'(\mathbb{R}^n)$ is endowed with the weak* topology: $u_n \in \mathcal{S}'(\mathbb{R}^n)$ converges weak* to $u \in \mathcal{S}'(\mathbb{R}^n)$ if $\langle u_n, \phi\rangle \to \langle u, \phi\rangle$ for all $\phi \in \mathcal{S}(\mathbb{R}^n)$.

21 Where the equality of mixed partial derivatives holds only under certain assumptions, in addition to their mere existence. For example, if the partial derivatives are continuous.

22 Clearly $H \in \mathcal{S}'(\mathbb{R})$ and for $\phi \in \mathcal{S}(\mathbb{R})$ we have $\langle H', \phi\rangle = -\langle H, \phi'\rangle = -\int_0^\infty \phi'(x)\,dx = \phi(0)$.

23 This process is similar to the "renormalisations" used to remove divergences in quantum field theory, see Folland (1999).

though its two constituent terms might not converge individually.[24] Thus $\phi \mapsto \lim_{\varepsilon\downarrow 0}\{\phi(\varepsilon)\ln(\varepsilon) + \int_\varepsilon^\infty \frac{\phi(x)}{x}\,dx\}$ is a tempered distribution, called a "renormalisation" of f, mapping functions ϕ in the subclass $\mathcal{S}_0(\mathbb{R})$ of $\mathcal{S}(\mathbb{R})$, defined by the constraint $\phi(0) = 0$, into $\int_{\mathbb{R}} f(x)\phi(x)\,dx$.

6.1.2 Fourier transform of a tempered distribution

In view of (6.6), the procedure used before to define various operations on $\mathcal{S}'(\mathbb{R}^n)$ motivates us to define the *Fourier transform $\hat{u}$ of the tempered distribution u* to be the continuous linear functional whose value at $\phi \in \mathcal{S}(\mathbb{R}^n)$ is

$$\langle \hat{u}, \phi\rangle = \langle u, \hat{\phi}\rangle\,. \tag{6.11}$$

Theorem 6.8 *The Fourier transform (6.11) is a linear bijection from $\mathcal{S}'(\mathbb{R}^n)$ onto $\mathcal{S}'(\mathbb{R}^n)$, being a weakly continuous[25] extension of $\mathcal{F}$ on $\mathcal{S}(\mathbb{R}^n)$.*

Proof For $f, g \in \mathcal{S}(\mathbb{R}^n)$, Theorem 6.4 and the polarisation identity in the complex Hilbert space $L^2(\mathbb{R}^n)$ yield $\int_{\mathbb{R}^n} f(x)\,\overline{g(x)}\,dx = \int_{\mathbb{R}^n} \hat{f}(\xi)\,\overline{\hat{g}(\xi)}\,d\xi$. Substituting $\overline{\hat{\phi}} = \mathcal{F}^{-1}(\overline{\phi})$ for g, we obtain $\langle \hat{\phi}, f\rangle = \int_{\mathbb{R}^n} \hat{f}(\xi)\,\phi(\xi)\,d\xi = \langle \hat{f}, \phi\rangle$ for all $f, \phi \in \mathcal{S}(\mathbb{R}^n)$. This shows that (6.11) extends the transform $\mathcal{F}$ we defined on $\mathcal{S}(\mathbb{R}^n)$. Moreover, if $\phi_n \to \phi$ in $\mathcal{S}(\mathbb{R}^n)$, then $\widehat{\phi_n} \to \hat{\phi}$ in $\mathcal{S}(\mathbb{R}^n)$, so that (6.11) ensures $\langle \hat{u}, \phi_n\rangle \to \langle \hat{u}, \phi\rangle$ for each $u \in \mathcal{S}'(\mathbb{R}^n)$, proving that $\hat{u}$ is a bounded linear functional on $\mathcal{S}(\mathbb{R}^n)$. Furthermore, if $u_n \to u$ weakly in $\mathcal{S}'(\mathbb{R}^n)$, then $\widehat{u_n} \to \hat{u}$ weakly in $\mathcal{S}'(\mathbb{R}^n)$ since this is a rephrasing of $\langle \widehat{u_n}, \phi\rangle \to \langle \hat{u}, \phi\rangle$ for all $\phi \in \mathcal{S}(\mathbb{R}^n)$, and the latter is granted by (6.11). Clearly $\mathcal{F}$ is linear on $\mathcal{S}'(\mathbb{R}^n)$. Moreover, Theorem 6.1 and (6.11) ensure that $\hat{u} = 0$ forces $u = 0$, proving injectivity, and we can read off (6.11) the inversion formula

$$\langle \mathcal{F}^{-1}u, \phi\rangle = \langle u, \mathcal{F}^{-1}\phi\rangle\,, \qquad u \in \mathcal{S}'(\mathbb{R}^n), \quad \phi \in \mathcal{S}(\mathbb{R}^n)\,, \tag{6.12}$$

which holds since for $v \in \mathcal{S}'(\mathbb{R}^n)$ we have $\langle v, \phi\rangle = \langle v, \mathcal{F}(\mathcal{F}(\phi^\dagger))\rangle = \langle \hat{v}, \mathcal{F}(\phi^\dagger)\rangle = \langle v, \mathcal{F}^{-1}(\phi)\rangle$ in view of Theorem 6.1 and (6.11). This proves that $\mathcal{F}$ is surjective on $\mathcal{S}'(\mathbb{R}^n)$. □

Although certain tempered distributions may be highly singular objects, they can be approximated by smooth functions: by identifying $\mathcal{S}(\mathbb{R}^n)$ and $\mathcal{S}'(\mathbb{R}^n)$ with sequence spaces, one can show that $\mathcal{S}(\mathbb{R}^n)$ is dense in $\mathcal{S}'(\mathbb{R}^n)$, see Reed and Simon (1980a). As a consequence of this, (6.11) is the unique weakly continuous extension of $\mathcal{F}$ from $\mathcal{S}(\mathbb{R}^n)$ to $\mathcal{S}'(\mathbb{R}^n)$.

Let us now list some operational rules for the Fourier transform: for $u \in \mathcal{S}'(\mathbb{R}^n)$ we have

$$\mathcal{F}(D^\alpha u) = (2\pi i)^{|\alpha|}\xi^\alpha\,\hat{u}\,, \qquad \mathcal{F}(x^\alpha u) = \Big(\frac{i}{2\pi}\Big)^{|\alpha|} D^\alpha(\mathcal{F}u)\,, \tag{6.13}$$

[24] A formal passage to the limit yields $\langle u', \phi\rangle = \phi(0)\ln(0) + \int_{\mathbb{R}} f(x)\phi(x)\,dx$, suggesting that u' is obtained from f by subtracting an infinite multiple of δ_0.

[25] Weakly continuous means that if $u_n \to u$ in $\mathcal{S}'(\mathbb{R}^n)$, that is, if $\langle u_n, \phi\rangle \to \langle u, \phi\rangle$ for all $\phi \in \mathcal{S}(\mathbb{R}^n)$, then $\widehat{u_n} \to \hat{u}$ in $\mathcal{S}'(\mathbb{R}^n)$.

any n-tuple α of nonnegative integers, and

$$\mathcal{F}(\tau_h u) = e^{-2\pi i h\cdot\xi}\hat{u}\,, \qquad \tau_h(\hat{u}) = \mathcal{F}(e^{2\pi i h\cdot\xi}u)\,, \tag{6.14}$$

for any $h \in \mathbb{R}^n$. These identities can be proved by simple manipulations, relying on duality relations and the properties of $\mathcal{F}$ on $\mathcal{S}(\mathbb{R}^n)$.

Indeed, for $\phi \in \mathcal{S}(\mathbb{R}^n)$ we have $\langle\mathcal{F}(D^\alpha u), \phi\rangle = \langle D^\alpha u, \hat{\phi}\rangle = (-1)^{|\alpha|}\langle u, D^\alpha\hat{\phi}\rangle = (2\pi i)^{|\alpha|}\langle u, \mathcal{F}(x^\alpha\phi)\rangle$, using (6.4) in the last step. By (6.11), this gives the first part of (6.13)). The second part follows similarly. Similarly, for, say, the first part of (6.14), $\langle\mathcal{F}(\tau_h u), \phi\rangle = \langle\tau_h u, \hat{\phi}\rangle = \langle u, \tau_h(\hat{\phi})\rangle$ and $\tau_h(\hat{\phi})(\xi) = \hat{\phi}(\xi + h) = \int_{\mathbb{R}^n}\phi(x)e^{-2\pi i x\cdot(\xi+h)}dx = \int_{\mathbb{R}^n}(e^{-2\pi i x\cdot h}\phi(x))e^{-2\pi i x\cdot\xi}dx = \mathcal{F}(e^{-2\pi i x\cdot h}\phi(x))$ yields $\langle u, \tau_h(\hat{\phi})\rangle = \langle u, \mathcal{F}(e^{-2\pi i x\cdot h}\phi(x))\rangle = \langle\hat{u}, e^{-2\pi i x\cdot h}\phi(x)\rangle = \langle e^{-2\pi i x\cdot h}\,\hat{u}, \phi\rangle$.

A simple example is the direct/inverse Fourier transform of the Dirac mass

$$\widehat{\delta_0} = 1\,, \qquad \hat{1} = \delta_0\,. \tag{6.15}$$

Indeed, for all $\phi \in \mathcal{S}(\mathbb{R}^n)$ we have $\langle\widehat{\delta_0}, \phi\rangle = \langle\delta_0, \hat{\phi}\rangle = \hat{\phi}(0) = \int_{\mathbb{R}^n}\phi(x)\,dx = \langle 1, \phi\rangle$. Subsequently, $u = \delta_0$ in (6.12) yields $\langle\mathcal{F}^{-1}\delta_0, \phi\rangle = \langle\delta_0, \mathcal{F}^{-1}\phi\rangle = (\mathcal{F}^{-1}\phi)(0) = (\mathcal{F}\phi^\dagger)(0) = \int_{\mathbb{R}^n}\phi(-x)\,dx = \int_{\mathbb{R}^n}\phi(x)\,dx = \langle 1, \phi\rangle$, in view of Theorem 6.1.

The Fourier transform has distinguished features when acting on $\mathcal{E}'(\mathbb{R}^n)$.

Theorem 6.9 *If $u \in \mathcal{E}'(\mathbb{R}^n)$, then $\hat{u} \in \mathcal{O}(\mathbb{R}^n)$ with*[26] $\hat{u}(\xi) = \langle u(x), e^{-2\pi i\xi\cdot x}\rangle$.

Proof Let $E_\xi(x) = e^{2\pi i\xi\cdot x}$ and $f(\xi) = \langle u(x), \theta(x)\,E_{-\xi}(x)\rangle$. Consideration of difference quotients, as in the proof of Theorem 6.7, shows that $D^\beta f$ exists for any n-tuple of nonnegative integers β, with $(D^\beta)f(\xi) = (-2\pi i)^{|\beta|}\langle u, \theta x^\beta E_{-\xi}\rangle$. By Theorem 6.5 there exists $C = C_\beta > 0$ and an integer $N = N(\beta) \geq 0$ such that $|\langle u, \theta x^\beta E_{-\xi}\rangle| \leq C\sum_{|\alpha|\leq N, |\gamma|\leq N} p_{\alpha,\gamma}(\theta x^\beta E_{-\xi})$. Since the right side is bounded by an expression of the form $C'(1 + |\xi|)^N$, we proved that $f \in \mathcal{O}(\mathbb{R}^n)$.

It remains to show that $f = \hat{u}$, or, due to (6.11), $\langle u, \hat{\phi}\rangle = \int_{\mathbb{R}^n} f(\xi)\phi(\xi)\,d\xi$ for every $\phi \in \mathcal{S}(\mathbb{R}^n)$. Both sides being continuous functionals on $\mathcal{S}(\mathbb{R}^n)$, it suffices to verify this equality for ϕ in the dense subset $\mathcal{D}(\mathbb{R}^n)$ of $\mathcal{S}(\mathbb{R}^n)$. In this case, for each integer $m \geq 1$ we can approximate the compact set $K = \mathrm{supp}(\phi)$ by a finite union of cubes of side length $\frac{1}{2^m}$ and volume $\frac{1}{2^{nm}}$, centred at points $\xi_j^m \in K$, $1 \leq j \leq k(m)$. The Riemann sums $S_m = \frac{1}{2^{nm}}\sum_{j=1}^{k(m)} f(\xi_j^m)\phi(\xi_j^m)$ are supported as $m \to \infty$ in a common compact subset of $\mathbb{R}^n$ and converge uniformly to $\int_{\mathbb{R}^n} f(\xi)\phi(\xi)\,d\xi$. Likewise, the corresponding Riemann sums $R_m(x) = 2^{-nm}\sum_{j=1}^{k(m)}\phi(\xi_j^m)E_{-\xi_j^m}(x)$ and their derivatives converge uniformly, for x in any fixed compact set, to $\hat{\phi}(x)$ and its derivatives. Therefore, since u is a continuous functional, $\int_{\mathbb{R}^n} f(\xi)\phi(\xi)\,d\xi = \lim_{m\to\infty} S_m = \lim_{m\to\infty}\sum_{j=1}^{k(m)}\frac{1}{2^{nm}}\langle u(x), \theta(x)E_{-\xi_j^m}(x)\rangle\phi(\xi_j^m) = \lim_{m\to\infty}\langle u(x), \theta(x)\sum_{j=1}^{k(m)}\frac{1}{2^{nm}}E_{-\xi_j^m}(x)\phi(\xi_j^m)\rangle = \lim_{m\to\infty}\langle u, \theta(x)R_m(x)\rangle = \langle u, \theta\hat{\phi}\rangle = \langle u, \hat{\phi}\rangle$, recalling that $\theta(x) = 1$ for $x \in \mathrm{supp}(u)$. □

[26] Choosing $\theta \in \mathcal{D}(\mathbb{R}^n)$ with $\theta(x) = 1$ if $x \in \mathrm{supp}(u)$, we set $\langle u, f\rangle = \langle u, f\theta\rangle$ for any $f \in C^\infty(\mathbb{R}^n)$. This expression is well-defined since for any $\tilde{\theta} \in \mathcal{D}(\mathbb{R}^n)$ with $\tilde{\theta}(x) = 1$ if $x \in \mathrm{supp}(u)$ we have $\langle u, f\tilde{\theta}\rangle = \langle u, f\tilde{\theta}\theta\rangle = \langle u, f\theta\rangle$.

Theorem 6.9 enables us[27] to define

$$u * v = v * u = \mathcal{F}^{-1}(\hat{u}\hat{v}) \in \mathcal{S}'(\mathbb{R}^n) \quad \text{for} \quad u \in \mathcal{E}'(\mathbb{R}^n),\ v \in \mathcal{S}'(\mathbb{R}^n)\,, \tag{6.16}$$

since $fw \in \mathcal{S}'(\mathbb{R}^n)$ for $f \in \mathcal{O}(\mathbb{R}^n)$ and $w \in \mathcal{S}'(\mathbb{R}^n)$, in accordance with the discussion in Section 6.1.1. The motivation for (6.16) is twofold. Firstly, recall (6.5) for $u, \phi \in \mathcal{S}(\mathbb{R}^n)$. Also, note that for $u \in \mathcal{S}'(\mathbb{R}^n)$ and $\phi \in \mathcal{S}(\mathbb{R}^n)$, (6.11) and (6.9) yield $\widehat{u * \phi} = \hat{\phi} \cdot \hat{u}$ since $\langle \widehat{u * \phi}, \psi \rangle = \langle u * \phi, \hat{\psi} \rangle = \langle u, \phi^\dagger * \hat{\psi} \rangle = \langle u, \mathcal{F}(\hat{\phi} \cdot \psi) \rangle = \langle \hat{u}, \hat{\phi}\psi \rangle = \langle \hat{\phi} \cdot \hat{u}, \psi \rangle$ for every $\psi \in \mathcal{S}(\mathbb{R}^n)$, taking into account (6.5) and the fact that $\mathcal{F}(\hat{\phi}) = \phi^\dagger$ by Theorem 6.1. Actually, since $\mathcal{O}(\mathbb{R}^n) \subset \mathcal{S}'(\mathbb{R}^n)$, we have that $\mathcal{O}'(\mathbb{R}^n) = \mathcal{F}^{-1}(\mathcal{O}(\mathbb{R}^n))$ is also a subspace of $\mathcal{S}'(\mathbb{R}^n)$. We can extend the range of validity of (6.16) by defining $u * v = v * u$ for $u \in \mathcal{O}'(\mathbb{R}^n)$ and $v \in \mathcal{S}'(\mathbb{R}^n)$ as $\mathcal{F}^{-1}(\hat{u}\hat{v})$. With the extended definition, convolution is associative: $(u*v)*w = u*(v*w)$ whenever at least two of the three tempered distributions u, v, w are in $\mathcal{O}'(\mathbb{R}^n)$. This follows at once from the definition since $fg \in \mathcal{O}(\mathbb{R}^n)$ if $f, g \in \mathcal{O}(\mathbb{R}^n)$ yields $u*v \in \mathcal{O}'(\mathbb{R}^n)$ if $u, v \in \mathcal{O}'(\mathbb{R}^n)$. Associativity holds even if we restrict ourselves to the case when at least two tempered distributions have compact support, but in this case we must rely on the not that straightforward fact that $u * v \in \mathcal{E}'(\mathbb{R}^n)$ if $u, v \in \mathcal{E}'(\mathbb{R}^n)$, proved in Exercise 15.

Some simple rules for manipulating convolutions are

$$\partial_{x_k}(u * v) = (\partial_{x_k} u) * v = u * (\partial_{x_k} v)\,, \qquad k = 1, \ldots, n, \tag{6.17}$$

$$\tau_n(u * v) = (\tau_h u) * v = u * (\tau_h v)\,, \qquad h \in \mathbb{R}^n\,, \tag{6.18}$$

whenever one of the distributions $u, v \in \mathcal{D}'(\mathbb{R}^n)$ belongs to $\mathcal{E}'(\mathbb{R}^n)$; these are easily derived from (6.16) in combination with (6.13) and (6.14). Note also that (6.16) and (6.15) lead us to the identity

$$u * \delta_0 = \delta_0 * u = u \quad \text{for all} \quad u \in \mathcal{S}'(\mathbb{R}^n)\,, \tag{6.19}$$

in contrast to the convolution of functions in $L^1(\mathbb{R}^n)$ and in $\mathcal{S}(\mathbb{R}^n)$, settings in which Theorem 6.3 prevents the existence of a neutral element in either class. This feature is more than a simple algebraic curiosity, being very useful in the application of Fourier analysis to partial differential equations. To see this, let $L = \sum_{|\beta| \le N} c_\beta D^\beta$ be a differential operator with constant coefficients. A distribution $u_0 \in \mathcal{S}'(\mathbb{R}^n)$ such that $Lu_0 = \delta_0$ is called a *fundamental solution*.

[27] See Exercise 7(iii) for the impossibility of making proper sense of $u * v \in \mathcal{S}'(\mathbb{R}^n)$ for general $u, v \in \mathcal{S}'(\mathbb{R}^n)$. It is, however, possible to go beyond the requirement that one of the distributions has compact support: relying on tensor products, one can replace this by the condition that the restriction of the map $\mu(x, y) = x + y$ to $\text{supp}(u) \times \text{supp}(v)$ is proper, meaning that the inverse image of a compact subset of $\mathbb{R}^n$ is compact. For these matters we refer to Friedlander (1998).

Its importance stems from settings in which the formal calculations

$$L(u_0 * f) = (Lu_0) * f = \delta_0 * f = f$$

can be justified to provide a solution to $Lu = f$:

- due to (6.16), $u = u_0 * f \in \mathcal{S}'(\mathbb{R}^n)$ solves $Lu = f$ for $f \in \mathcal{S}'(\mathbb{R}^n)$ if $u_0 \in \mathcal{E}'(\mathbb{R}^n)$;
- by Theorem 6.8, $u = u_0 * f \in O(\mathbb{R}^n)$ solves $Lu = f$ for $f \in \mathcal{S}(\mathbb{R}^n)$ if $u_0 \in \mathcal{S}'(\mathbb{R}^n)$.

Let us discuss in some detail an important concrete example.

Example Let us show that the tempered $L^1(\mathbb{R}^n)$-function

$$u_0(x) = \begin{cases} xH(x) & \text{if } n = 1\,, \\ \frac{1}{2\pi}\ln|x| & \text{if } n = 2\,, \\ \frac{1}{(2-n)\omega_n}|x|^{2-n} & \text{if } n \geq 3\,, \end{cases} \tag{6.20}$$

is a fundamental solution for the Laplace operator $\Delta = \sum_{k=1}^n \partial_{x_k}^2$; here H is the Heaviside step function, given by $H(x) = 1$ for $x > 0$ and $H(x) = 0$ for $x \leq 0$, and $\omega_n = \frac{2\pi^{n/2}}{\Gamma(n/2)}$ is the surface area of the unit sphere in $\mathbb{R}^n$.

Indeed, if $n = 1$, $u_0 \in \mathcal{S}'(\mathbb{R})$ since $x \mapsto \frac{u_0(x)}{[1+x^2]^2} \in L^1(\mathbb{R})$, and for $\phi \in \mathcal{S}(\mathbb{R})$ we have $\langle \partial_x^2[xH(x)], \phi(x)\rangle = \langle xH(x), \phi''(x)\rangle = \int_0^\infty x\phi''(x)\,dx = \phi(0)$ after two integration by parts.

For $n \geq 2$ the claim is not that straighforward. Note that $u_0 \in \mathcal{S}'(\mathbb{R}^2)$ since $x \mapsto \frac{u_0(x)}{[1+|x|^2]^2} \in L^1(\mathbb{R}^2)$ follows by using polar coordinates $x_1 = r\cos\theta$, $x_2 = r\sin\theta$, with $r \geq 0$ and $\theta \in [0, 2\pi)$, since[28] $\int_{\mathbb{R}^2} \frac{|u_0(x)|}{[1+|x|^2]^2}\,dx = \frac{1}{2\pi}\int_0^{2\pi}\int_0^\infty \frac{r\ln|r|}{(1+r^2)^2}\,dr = \int_0^\infty \frac{r\ln|r|}{(1+r^2)^2}\,dr \leq \int_0^\infty \frac{1}{1+r^2}\,dr = \frac{\pi}{2}$. For $\phi \in \mathcal{S}(\mathbb{R})$ we have $\langle \Delta u_0, \phi\rangle = \langle u_0, \Delta\phi\rangle = \int_{\mathbb{R}^2} u_0(x)\,(\Delta\phi)(x)\,dx$ since $x \mapsto \frac{u_0(x)}{[1+|x|^2]^2} \in L^1(\mathbb{R}^2)$ and $x \mapsto [1+|x|^2]^2(\Delta\phi)(x)$ is bounded. To evaluate the latter expression, we would like to integrate by parts to move all derivatives on u_0. To avoid the singularities of the partial derivatives of u_0 at the origin, we use the dominated convergence theorem to write $\int_{\mathbb{R}^2} \frac{u_0(x)}{[1+|x|^2]^2}[1+|x|^2]^2(\Delta\phi)(x)\,dx = \lim_{\varepsilon\downarrow 0}\int_{\varepsilon<|x|<\varepsilon^{-1}} \frac{u_0(x)}{[1+|x|^2]^2}[1+|x|^2]^2(\Delta\phi)(x)\,dx = \lim_{\varepsilon\downarrow 0}\int_{\varepsilon<|x|<\varepsilon^{-1}} u_0(x)\,(\Delta\phi)(x)\,dx$. Since $u_0(x)$ is smooth with $(\Delta u_0)(x) = 0$ for $x \neq 0$, as confirmed by direct computation, Green's formula $\int_{\varepsilon<|x|<\varepsilon^{-1}} u_0\,\Delta\phi\,dx = -\int_{|x|=\varepsilon^{-1}}(\phi\frac{\partial u_0}{\partial r} - u_0\frac{\partial\phi}{\partial r})\,d\sigma_{\varepsilon^{-1}} + \int_{|x|=\varepsilon}(\phi\frac{\partial u_0}{\partial r} - u_0\frac{\partial\phi}{\partial r})\,d\sigma_\varepsilon$, where $d\sigma_R$ is the measure on the sphere $|x| = R$ of area $2\pi R$, can be applied. Choosing $M > 0$ such that[29] $|\phi(x)| + \left|\frac{\partial\phi}{\partial r}(x)\right| \leq \frac{M}{1+|x|^3}$ for all $x \in \mathbb{R}^2$, we can bound the absolute value of the first integral on the right by $2\pi\varepsilon^{-1}\frac{\varepsilon^{-1}-\varepsilon}{2\pi}\frac{M}{1+\varepsilon^{-3}} \to 0$ as $\varepsilon \downarrow 0$. As for the second integral on the right, we explicitate is as $\frac{1}{2\pi\varepsilon}\int_{|x|=\varepsilon}\phi(x)\,d\sigma_\varepsilon - \frac{\ln(\varepsilon)}{2\pi}\int_{|x|=\varepsilon}\frac{\partial\phi}{\partial r}(x)\,d\sigma_\varepsilon$. Since ϕ is smooth, we have that $\lim_{\varepsilon\downarrow 0}\frac{1}{2\pi\varepsilon}\int_{|x|=\varepsilon}[\phi(x)-\phi(0)]\,d\sigma_\varepsilon = \lim_{\varepsilon\downarrow 0}\frac{1}{2\pi\varepsilon}\int_{|x|=\varepsilon}[\frac{\partial\phi}{\partial r}(x)-\frac{\partial\phi}{\partial r}(0)]\,d\sigma_\varepsilon = 0$ and $\lim_{\varepsilon\downarrow 0}[\varepsilon\ln(\varepsilon)] = 0$

[28] We rely on the inequality $r\ln|r| \leq 1 + r^2$ for $r > 0$, obtained by combining $r\ln(r) \leq 1 + r^2$ for $r \geq 1$ and $r\,|\ln(r)| \leq \frac{1}{e}$ for $r \in (0, 1)$. The latter is a consequence of the montonic behaviour of $r \mapsto -r\ln(r)$ on $(0, 1)$.

[29] Note that $\frac{\partial\phi}{\partial r} = \sum_{k=1}^n \frac{\partial\phi}{\partial x_k}\frac{\partial x_k}{\partial r} = \sum_{k=1}^n \frac{x_k}{r}\frac{\partial\phi}{\partial x_k}$ yields $\left|\frac{\partial\phi}{\partial r}(x)\right| \leq \sup_{1\leq k\leq n}\left|\frac{\partial\phi}{\partial x_k}(x)\right|$ as $r \geq |x_k|$.

shows that the final outcome is $\phi(0)$, as claimed. The same procedure works for $n \geq 3$, the only difference being that we have to replace polar coordinates by spherical coordinates.

Note that the fundamental solution (6.20) is not unique: if $u_0 \in \mathcal{S}'(\mathbb{R}^2)$ is such that $\Delta u_0 = \delta_0$, then $4\pi^2|\xi|^2\,\widehat{u_0}(\xi) = 1$. Due to Theorem 6.9, this means that $u \in \mathcal{S}'(\mathbb{R}^2)$ is a fundamental solution if and only if $|\xi|^2(\hat{u}-\widehat{u_0}) = 0$ in $\mathcal{S}'(\mathbb{R}^2)$. But then $\mathrm{supp}(\hat{u}-\widehat{u_0}) \subset \{0\}$, and thus $(\hat{u}-\widehat{u_0}) \in \mathcal{S}'(\mathbb{R}^2)$ is a finite linear combination of derivatives of δ_0, see Exercise 18(i). Now, the identities (6.13) and (6.15) ensure that u differs from u_0 by a polynomial $P(x)$. A closer look confirms that this necessary condition is also sufficient if the polynomial P is *harmonic*, that is, if $\Delta P = 0$ pointwise. For example, for $n = 2$, all polynomials of degree $N = 1$ are harmonic while the homogeneous polynomials of degree $N \geq 2$ (that is, the degree of each term is N) are the linear combinations of $(x_1 + ix_2)^N$ and $(x_1 - ix_2)^N$, see Stein (1976). This is still not the whole picture since there are many harmonic functions that are not polynomials, e.g. $e^{x_1}\sin(x_2)$. Such functions are excluded from our approach because the fast growth of $e^{x_1}\sin(x_2)$ as $x_1 \to \infty$ prevents it from being a tempered distribution. In fact, we proved that any smooth harmonic function that is not a polynomial must grow too fast at infinity to be a tempered distribution. □

6.2 Fourier series

Doubly periodic functions $f : \mathbb{R}^2 \to \mathbb{C}$ are defined as functions that are 1-periodic in each variable, that is, $f(x+1,y) = f(x,y)$ and $f(x,y) = f(x,y+1)$ for all $(x,y) \in \mathbb{R}^2$. Such functions, encountered in solid-state physics in the theory of crystal lattices, are completely determined by their restriction to a unit square $[-\frac{1}{2}, \frac{1}{2}) \times [-\frac{1}{2}, \frac{1}{2})$, taken for symmetry reasons to be centred at the origin. This leads us naturally to consider functions $f : \mathbb{T}^d \to \mathbb{C}$ defined on the d-dimensional torus $\mathbb{T}^d = [-\frac{1}{2}, \frac{1}{2})^d$, functions that are in $1:1$ correspondence with functions $f : \mathbb{R}^d \to \mathbb{C}$ which are periodic in each coordinate, that is, $f(x_1,\dots,x_k+1,\dots,x_d) = f(x_1,\dots,x_k,\dots,x_d)$ for all $1 \leq k \leq d$ and $(x_1,\dots,x_d) \in \mathbb{R}^d$. The Fourier coefficients of a function $f \in L^1(\mathbb{T}^d)$ are defined by

$$\hat{f}(n) = \int_{\mathbb{T}^d} f(x)\,e^{-2\pi i x\cdot n}dx\,, \quad n \in \mathbb{Z}^d\,, \tag{6.21}$$

the associated Fourier series being $\sum_{n\in\mathbb{Z}^d} \hat{f}(n)\,e^{2\pi i x\cdot n}$. As in the one-dimensional setting of Chapter 4, the meaning of the infinite sum has to be specified.

6.2.1 L^2-convergence

The elegance of Fourier series for functions $f \in L^2(\mathbb{T}^d)$ is a direct consequence of the fact that $\{e^{2\pi i x\cdot n}\}_{n\in\mathbb{Z}^d}$ is an orthonormal basis of $L^2(\mathbb{T}^d)$.

Indeed, since orthonormality is easily checked, $L^2(\mathbb{T}^d)$ being a Hilbert space, it suffices to prove the density of the trigonometric polynomials $\sum_{|n|\le N} c_n\, e^{2\pi i x\cdot n}$ in $L^2(\mathbb{T}^d)$. But by the construction of the multi-dimensional Lebesgue integral, the linear combinations of characteristic functions of rectangles contained in $\mathbb{T}^d$ are dense in $L^2(\mathbb{T}^d)$. If $R = \prod_{k=1}^d [a_k, b_k)$ is such a rectangle, the characteristic function χ_k of each interval $[a_k, b_k)$ is approximated with accuracy ε in $L^2[-\frac{1}{2},\frac{1}{2}]$ by a Fourier partial sum S_k. Then $\chi_R = \prod_{k=1}^d \chi_k$ and $\prod_{k=1}^d \chi_k - \prod_{k=1}^d S_k = (\chi_1 - S_1)\prod_{j=2}^d \chi_j + S_1(\chi_2 - S_2)\prod_{j=3}^d \chi_j + \cdots + \big(\prod_{j=1}^{d-1} S_j\big)(\chi_d - S_d)$, so that $\chi_R - \prod_{k=1}^d S_k = \sum_{j=1}^d (\chi_j - S_j) g_j$, where g_j contains $(j-1)$ factors S_k and $(d-j)$ factors χ_k. By Theorem 4.8, there exists a uniform upper bound $M \ge 1$ for the supremum of all Fourier partial sums of characteristic functions of subintervals of $[-\frac{1}{2},\frac{1}{2}]$. In particular $|g_j(x)| \le M^d$ for all $j = 1,\dots,d$ and all $x \in \mathbb{T}^d$. We get $\|\prod_{k=1}^d \chi_k - \prod_{k=1}^d S_k\|^2_{L^2(\mathbb{T}^d)} = \int_{\mathbb{T}^d} \sum_{1\le j,k\le d} (S_j - \chi_j) g_j\, \overline{(S_k - \chi_k) g_j}\, dx \le M^{2d} \sum_{1\le j,k\le d} \big(\int_{\mathbb{T}^d} |S_j - \chi_j|\cdot|S_K - \chi_k|\, dx\big) \le M^{2d} \sum_{1\le j,k\le d} \|S_j - \chi_j\|_{L^2[-\frac{1}{2},\frac{1}{2}]}\cdot \|S_k - \chi_k\|_{L^2[-\frac{1}{2},\frac{1}{2}]} \le d^2 M^{2d} \varepsilon^2$. Since $\prod_{k=1}^d S_k$ is a trigonometric polynomial, the argumentation is complete.

The applicability of the general considerations on orthonormal bases in Hilbert spaces (Theorem 3.1) being thus ensured, we obtain the following result.

Theorem 6.10 *The Fourier series of $f \in L^2(\mathbb{T}^d)$ converges to f in $L^2(\mathbb{T}^d)$, and Plancherel's identity $\int_{\mathbb{T}^d} |f(x)|^2 dx = \sum_{n\in\mathbb{Z}^d} |\hat{f}(n)|^2$ holds. Conversely, any complex sequence $\{\alpha_n\}_{n\in\mathbb{Z}^d}$ with $\sum_{n\in\mathbb{Z}^d} |\alpha_n|^2 < \infty$ determines a unique $f \in L^2(\mathbb{T}^d)$ with $\hat{f}(n) = \alpha_n$ for every multi-index $n \in \mathbb{Z}^d$.*

6.2.2 Pointwise convergence

If a d-periodic function $f : \mathbb{T}^d \to \mathbb{C}$ is sufficiently smooth, then integration by parts yields $(2\pi i n)^\alpha \hat{f}(n) = \int_{\mathbb{T}^d} (D^\alpha f)(x)\, e^{-2\pi i n\cdot x} dx$ for the multi-indices of integers $n = (n_1,\dots,n_d)$ and $\alpha = (\alpha_1,\dots,\alpha_d)$ with $\alpha_k \ge 0$ for $1 \le k \le d$, where $n^\alpha = n_1^{\alpha_1}\dots n_d^{\alpha_d}$. If $\Delta = \sum_{k=1}^d \partial^2_{x_k}$ is the Laplace operator, we obtain $(1 + 2\pi|n|^2)\hat{f}(n) = \int_{\mathbb{T}^d} [(1-\Delta) f(x)]\, e^{-2\pi i n\cdot x} dx$, as well as $(1 + 2\pi|n|^2)^k \hat{f}(n) = \int_{\mathbb{T}^d} [(1-\Delta)^k f(x)]\, e^{-2\pi i n\cdot x} dx$ for any integer $k \ge 1$. For $f \in C^{2k}(\mathbb{T}^d)$, the continuity of the integrand on the right side yields the estimate

$$|\hat{f}(n)| \le \frac{C}{(1 + 2\pi|n|^2)^k}, \qquad n \in \mathbb{Z}^d, \tag{6.22}$$

for some $C > 0$ that is independent of n. This leads to the following result.

Theorem 6.11 *The Fourier series of a d-periodic function $f \in C^{2k}(\mathbb{T}^d)$, with $2k \ge d + 1$, converges on $\mathbb{R}^d$ absolutely and uniformly to f.*

Proof For $2k \ge d + 1$, the estimate (6.24) ensures that $\sum_{n\in\mathbb{Z}^d} |\hat{f}(n)|$ converges uniformly. Indeed, for any integer $p \geq 1$, the number $\gamma_d(p)$ of multi-indices $n \in \mathbb{Z}^d$ with $|n| = p$ does not exceed[30] $p^{d/2}$, as $0 \leq n_j \leq \sqrt{p}$ for each $j \in \{1, \ldots, d\}$, and $\sum_{p\geq 1} p^{d/2}/p^k \leq \sum_{p\geq 1} p^{-1-\frac{d}{2}} < \infty$. Since uniform convergence implies convergence in $L^2(\mathbb{T}^d)$, Theorem 6.10 identifies the limit as f. □

In dimensions $d > 1$, the problem of pointwise convergence of the Fourier series of a function $f : \mathbb{T}^d \to \mathbb{C}$ that is not sufficiently smooth is quite complicated if absolute convergence fails because there is no natural truncation. Since the integer lattice $\mathbb{Z}^p$ has no natural ordering for $p > 1$, one has to investigate several alternative orderings. For example, if $(S_r f)(x) = \sum_{n_1^2+\cdots+n_d^2 \leq r^2} \hat{f}(n) e^{2\pi i n \cdot x}$ converges as $r \to \infty$, we say that the Fourier series converges *spherically*, while *rectangular convergence* and *square convergence* correspond to the convergence of the partial sums $(S_n f)(x) = \sum_{|k_1|\leq n_1} \cdots \sum_{|k_d|\leq n_d} \hat{f}(k) e^{2\pi i k \cdot x}$ for $|n| = \sum_{k=1}^d n_k \to \infty$ and $\sum_{|k_1|\leq N} \cdots \sum_{|k_d|\leq N} \hat{f}(k) e^{2\pi i k \cdot x}$ for $N \to \infty$, respectively. The presumption that most results from the one-dimensional theory would extend to higher dimensions by induction arguments turns out to be too optimistic: there is little uniform texture in multi-dimensional theory and the previous convergence methods are basically incompatible,[31] so that it has become a common practice to make a choice of convergence method. Without trying to survey the state-of-the-art, let us illustrate the intricacy of the multi-dimensional theory by presenting some important results that capture the spirit of the situation. For example, Carleson (1966) proved that the partial sums of the Fourier series of a function $f \in L^2[0, 1]$ converge almost everywhere. It turns out that the square convergence of the Fourier series of $f \in L^2(\mathbb{T}^d)$ is a simple consequence of the one-dimensonal result, see Fefferman (1971a). On the other hand, in stark contrast to the one-dimensional case, for any $d > 1$ there exists a continuous d-periodic function $f : \mathbb{R}^d \to \mathbb{C}$, having uniformly bounded rectangular partial sums such that $S_n f$ is everywhere divergent, see Fefferman (1971b); Ash and Gluck (1972). Not only does the passage from one to two dimensions present dramatic changes: the spherical partial sums of the Fourier series of the characteristic function of the d-dimensional unit ball $\{(x_1, \ldots, x_d) \in \mathbb{R}^d : x_1^2 + \cdots + x_d^2 \leq 1\}$ diverge at the origin if $d \geq 3$, see Pinsky (1993).

[30] Estimating accurately the number $\gamma_d(p)$ of ways we can represent the integer $p \geq 1$ as a sum of d squares of integers, $p = n_1^2 + \cdots + n_d^2$, different orders being regarded as distinct, is an important and difficult classical problem in number theory. A simple aspect is the fact that $\gamma_2(N) = 0$ if $N \equiv 3 \pmod 4$, since for all integers n_k we either have $n_k^2 \equiv 0 \pmod 4$ or $n_k^2 \equiv 1 \pmod 4$. One can show that asymptotically $\gamma_d(N)$ grows in the limit $N \to \infty$ like $N^{\frac{d}{2}-1}$, see the discussion in Rouse (2012).

[31] Except for the only obvious connection that rectangular convergence implies square convergence, and for the fact that we are not aware of examples of Fourier series which are sphericaly convergent but not square convergent. See Exercise 23 for a trigonometric series that is almost everywhere square convergent and diverges spherically almost everywhere.

Let us now briefly indicate how the multi-dimensional square partial sums can be investigated by iterating the one-dimensional results discussed in Chapter 4. The square Dirichlet kernel D_N^d on $\mathbb{T}^d$ is equal to the product of the one-dimensional Dirichlet kernels, that is,

$$D_N^d(x_1,\dots,x_d) = \prod_{k=1}^{d} D_N(x_k) \quad \text{for} \quad N \geq 0\,,$$

and the Fejér kernel F_N^d on $\mathbb{T}^d$ is defined as the average of the product of the Dirichlet kernels in each variable, that is,

$$F_N^d(x_1,\dots,x_d) = \frac{1}{N^d}\sum_{k_1=0}^{N-1}\cdots\sum_{k_d=0}^{N-1} D_{k_1}(x_1)\dots D_{k_d}(x_d) = \prod_{j=1}^{d}\Big(\frac{1}{N}\sum_{k=0}^{N-1} D_k(x_j)\Big)$$

$$= \prod_{j=1}^{d} F_N(x_j) \quad \text{for} \quad N \geq 1\,.$$

In contrast to the one-dimensional case, it turns out that F_N^d is not equal to[32] the average $\frac{1}{N}\sum_{j=0}^{N-1} D_j^d$ if $d \geq 2$. Nevertheless, D_N^d and F_N^d are both trigonometric polynomials, with

$$(f * D_N^d)(x) = \sum_{|n_1|,\dots,|n_d|\leq N} \hat{f}(n)\, e^{2\pi i n\cdot x}\,, \qquad N \geq 0,$$

(the square partial sums of the multi-dimensional Fourier series) and

$$(f * F_N^d)(x) = \sum_{|n_1|,\dots,|n_d|\leq N-1} \Big(1-\frac{|n_1|}{N}\Big)\dots\Big(1-\frac{|n_d|}{N}\Big)\hat{f}(n)\, e^{2\pi i n\cdot x}\,, \qquad N \geq 1,$$

for $f \in L^1(\mathbb{T}^d)$. The relationship to the one-dimensional Fejér kernel yields at once that F_N^d is an approximate identity, satisfying the multi-dimensional analogue of the relations (A1)-(A3) in Chapter 4. The proof of Theorem 4.4 and that of the first part of Theorem 4.5 can be practically transcribed,[33] yielding the following.

Theorem 6.12 *If $f \in C(\mathbb{T}^d)$, then $f * F_N^d$ converges uniformly to f as $N \to \infty$, while if $f \in L^1(\mathbb{T}^d)$, then $f * F_N^d \to f$ in $L^1(\mathbb{T}^d)$.*

The next result shows that the Fourier coefficients of $f \in L^1(\mathbb{T}^d)$ determine

[32] It also comprises terms of the form $\prod_{j=1}^d D_{k_j}(x_j)$ where $k_1 = \dots = k_d$ fails. For example, $F_2(x_1,x_2) = \frac{1}{4}\Big[\frac{\sin(2\pi x_1)\sin(2\pi x_2)}{\sin(\pi x_1)\sin(\pi x_2)}\Big]^2 \neq \frac{D_0(x_1,x_2)+D_1(x_1,x_2)}{2} = \frac{1}{2}\Big(1+\frac{\sin(3\pi x_1)\sin(3\pi x_2)}{\sin(\pi x_1)\sin(\pi x_2)}\Big)$.

[33] The multi-dimensional counterpart of the second part of Theorem 4.5, namely that $f * F_N^d \to f$ a.e. if $f \in L^1(\mathbb{T}^d)$, also holds true, even if its proof is technically more demanding than that provided in Chapter 4 for $d = 1$; see Grafakos (2008).

uniquely[34] the function. Due to linearity, it suffices to deal with vanishing coefficients.

Corollary (Uniqueness of Fourier coefficients) *If* $f \in L^1(\mathbb{T}^d)$ *is such that* $\hat{f}(n) = 0$ *for all* $n \in \mathbb{Z}^d$, *then* $f = 0$ *a.e.*

Proof Clearly $(f * F_N^d)(x) = 0$ for all $x \in \mathbb{T}^d$ and all $N \geq 1$. Theorem 6.12 yields $\|f\|_{L^1(\mathbb{T}^d)} = 0$, from which we conclude that $f = 0$ a.e. □

The previous considerations show that the issue of converge for the square partial Fourier sums parallels that of the one-dimensional case (see Chapter 4). Given this, it is surprising that localisation in dimension $d \geq 2$ is a complete failure! Recall from Chapter 4 that the behaviour of the partial sums of the Fourier series of $f \in L^1[0, 1]$ at a point depends only on the values of the function near that point. For $d \geq 2$, there are d-periodic continuous functions $f : \mathbb{T}^d \to \mathbb{C}$ that vanish near a point x^* and for which the square Fourier partial sums (and hence *a fortiori* the rectangular partial sums) diverge at the point. Moreover, localisation also fails for the spherical convergence of continuous d-periodic functions in dimension $d \geq 2$, see Ash (1976).

An elegant way around the intricacies of multi-dimensional Fourier series for functions that are not very smooth is to rely on tempered distributions and to demand more regularity than mere integrability or continuity by introducing the scale of Sobolev spaces. This is the approach that we will develop in the remaining part of this chapter.

6.2.3 The tempered distributions approach

In dimension $n = 1$, a trigonometric series $\sum_{k\in\mathbb{Z}} c_k\, e^{2\pi ikx}$, with coefficients c_k that do not necessarily tend to zero for $|k| \to \infty$, is generally not the Fourier series of a function $f \in L^1(\mathbb{R})$, nor is it pointwise convergent. Nevertheless, under rather general conditions, this series converges in $\mathcal{D}'(\mathbb{R})$, defining a *periodic distribution*. The fact that a function $f : \mathbb{R} \to \mathbb{C}$ has period $a \neq 0$ can be expressed as $\tau_a f = f$ in terms of the translation operator and τ_a can be extended to all distributions: we say that $u \in \mathcal{D}'(\mathbb{R})$ is *periodic* with period a if $\tau_a u = u$, i.e., if

$$\langle u, \tau_{-a}\varphi\rangle = \langle u, \varphi\rangle \quad \text{for every} \quad \varphi \in \mathcal{D}(\mathbb{R}).$$

[34] A more refined uniqueness problem is the following: if the multiple trigonometric series $\sum_{n\in\mathbb{Z}^d} c_n\, e^{2\pi in\cdot x}$ converges rectangularly or spherically everywhere to an everywhere finite integrable function f, then it is the Fourier series of f, see Ash (2013) and Bourgain (1996).

In dimensions $n \geq 2$, we say that the distribution $u \in \mathcal{D}'(\mathbb{R}^n)$ is *periodic* if $\tau_h u = u$ for any vector h with integer coordinates. We could consider n linearly independent vectors $h_1, \ldots, h_n \in \mathbb{R}^n$ and impose $\tau_h u = u$ for any vector $h = \sum_{k=1}^n c_k h_k$ with $c_k \in \mathbb{Z}$ (this being the lattice generated by the h_ks). However, if e_k is the unit vector with entry 1 on position k, in the direction of the kth coordinates axis of $\mathbb{R}^n$ ($1 \leq k \leq n$), there will exist a linear bijection $f : \mathbb{R}^n \to \mathbb{R}^n$ such that $f(h_k) = e_k$ for $1 \leq k \leq n$; and then the so-called *pullback* $f^* u \in \mathcal{D}'(\mathbb{R}^n)$, defined[35] by

$$\langle f^* u, \varphi\rangle = \big\langle u(y), \varphi(g(y))\,[\det g'(y)]\big\rangle, \qquad \varphi \in \mathcal{D}(\mathbb{R}^n),$$

where $g = f^{-1}$, is a periodic distribution with respect to the $\mathbb{Z}^n$-lattice in $\mathbb{R}^n$. For this reason we restrict our attention to the standard lattice $\mathbb{Z}^n$.

Example *Dirac's comb* Ш (also called the *shah functional*[36]), defined by $\langle Ш, \varphi\rangle = \sum_{k\in\mathbb{Z}} \varphi(ka)$ for $\varphi \in \mathcal{D}(\mathbb{R})$, is periodic with period a. □

While there are no[37] nonzero periodic distributions in $\mathcal{E}'(\mathbb{R})$, it turns out that every periodic distribution is tempered. Before proving this result, let us look at an instructive special case.

We claim that the periodic distribution $u_0 = \sum_{g\in\mathbb{Z}^n} \tau_g \delta_0$ belongs to $\mathcal{S}'(\mathbb{R}^n)$, and the following identity, called the *Poisson summation formula*, holds:

$$\sum_{g\in\mathbb{Z}^n} \tau_g \delta_0 = \sum_{g\in\mathbb{Z}^n} e^{2\pi i g\cdot x} \quad \text{in} \quad \mathcal{S}'(\mathbb{R}^n). \tag{6.23}$$

Since $\mathcal{F}(e^{2\pi i g\cdot x}) = \tau_g \delta_0$ by (6.14) and (6.15), we see that (6.23) is exactly the statement that Dirac's comb (*alias* the shah functional) Ш is its own Fourier transform.

Indeed, for $\varphi \in \mathcal{S}(\mathbb{R}^n)$ we have

$$|\langle u_0, \varphi\rangle = \Big|\sum_{g\in\mathbb{Z}^n} \varphi(g)\Big| \leq \sum_{g\in\mathbb{Z}^n} |\varphi(g)| \leq \sup_{x\in\mathbb{R}^n}\{(1+|x|^2)^n |\varphi(x)|\} \sum_{g\in\mathbb{Z}^n} (1+|g|^2)^{-n}$$

and $\sum_{g\in\mathbb{Z}^n} (1+|g|^2)^{-n} \leq 1 + n\int_{\mathbb{R}^n} \frac{dx}{(1+|x|^2)^n} \leq 1 + 2n\pi^{n-1}\int_0^\infty \frac{r^{n-1}}{(1+r^2)^n} < \infty$, estimate obtained by changing to spherical coordinates[38] in the last step, after noticing the validity of the inequality

35 The origin of this formula lies in the rule for changing variables in an integral, if $u \in L^1(\mathbb{R}^n)$; see Friedlander (1998) for general coordinate transformations.

36 The regular placing of infinite spikes resembles the shape of the namesake Cyrillic letter.

37 Since $\langle u, \varphi\rangle \neq 0$ for some $\varphi \in \mathcal{D}(\mathbb{R})$ ensures $\langle u, \tau_h\varphi\rangle \neq 0$ for all $h \in \mathbb{Z}^n$.

38 That is, for $x = (x_1, \ldots, x_n) \in \mathbb{R}^n$, set $x_1 = r\cos\varphi_1$, $x_2 = r\sin\varphi_1\cos\varphi_2$, $x_3 = r\sin\varphi_1\sin\varphi_2\cos\varphi_3, \ldots, x_{n-1} = r\sin\varphi_1\sin\varphi_2\ldots\sin\varphi_{n-2}\cos\varphi_{n-1}$, $x_n = r\sin\varphi_1\sin\varphi_2\ldots\sin\varphi_{n-2}\sin\varphi_{n-1}$ with $r = \sqrt{x_1^2+\cdots+x_n^2}$ and $0 \leq \varphi_1, \ldots, \varphi_{n-2} \leq \pi$, $0 \leq \varphi_{n-1} \leq 2\pi$. The Jacobian of this transformation is given by $J(n, r, \varphi_1, \ldots, \varphi_{n-1}) = r^{n-1}(\sin\varphi_1)^{n-2}\ldots(\sin\varphi_{n-3})^2\sin\varphi_{n-2}$.

$(1+|g|^2)^{-n} \le n \int_G \frac{dx}{(1+|x|^2)^n}$ for every $g \in \mathbb{Z}^n \setminus \{0\}$, where G is any of the 2^n unit cubes[39] in $\mathbb{R}^n$ having a vertex at g. Thus $u_0 \in \mathcal{S}'(\mathbb{R}^n)$.

To investigate the second member in (6.23), we develop some technical tools. Let $\psi : \mathbb{R} \to [0,1]$ be a C^∞-function such that $\psi(s) = 1$ for $|s| \le 1/2$ and $\psi(s) = 0$ for $|s| \ge 2/3$. Setting $\Psi(x) = \prod_{k=1}^n \psi(x_k)$ for $x = (x_1, \dots, x_n) \in \mathbb{R}^n$, the sum[40] $\phi(x) = \sum_{g\in\mathbb{Z}^n} \Psi(x-g)$ defines a periodic C^∞-function which never vanishes, and the functions $\phi_g(x) = \Psi(x-g)/\phi(x)$ form a *partition of unity*: $\phi_g \in \mathcal{D}(\mathbb{R}^n)$ for every $g \in \mathbb{Z}^n$ and $\sum_{g\in\mathbb{Z}^n} \phi_g(x) = 1$ for all $x \in \mathbb{R}^n$. Also, note that $\phi_g = \tau_g\phi_0$ for every $g \in \mathbb{Z}^n$, while $\phi_0(x) = 0$ if $|x_k| \ge 2/3$ for some $1 \le k \le n$. Let us now denote by v the second member in (6.23). For $\varphi \in \mathcal{S}(\mathbb{R}^n)$ we have $\langle e^{2\pi i g\cdot x}, \varphi(x)\rangle = \langle 1, e^{2\pi i g\cdot x}\varphi(x)\rangle = \langle \widehat{\delta_0}, e^{2\pi i g\cdot x}\varphi(x)\rangle = \langle \delta_0, \mathcal{F}(e^{2\pi i g\cdot x}\varphi(x))\rangle = \langle \delta_0, \tau_g(\hat{\varphi})\rangle = \hat{\varphi}(-g)$, using (6.15), (6.11) and (6.14) in the second, third and fourth step, respectively. Thus, formally,

$$\langle v, \varphi\rangle = \sum_{g\in\mathbb{Z}^n} \hat{\varphi}(-g), \qquad \varphi \in \mathcal{S}(\mathbb{R}^n). \tag{6.24}$$

But, if $\varphi \in \mathcal{S}(\mathbb{R}^n)$, it follows from (6.13) that

$$\begin{aligned}(1+4\pi^2|g|^2)^n|\hat{\varphi}(g)| &= |\mathcal{F}\{(1-\sum_{k=1}^n \partial_{x_k}^2)^n\varphi(x)\}(g)| = \Big|\int_{\mathbb{R}^n} e^{-2\pi i g\cdot x}(1-\sum_{k=1}^n \partial_{x_k}^2)^n\varphi(x)\,dx\Big| \\ &\le \int_{\mathbb{R}^n} \Big|(1-\sum_{k=1}^n \partial_{x_k}^2)^n\varphi(x)\Big|\,dx = \int_{\mathbb{R}^n} (1+|x|^2)^n\Big|(1-\sum_{k=1}^n \partial_{x_k}^2)^n\varphi(x)\Big|\,(1+|x|^2)^{-n}\,dx \\ &\le C_n\Big(\sum_{k=0}^{2n} \sup_{y\in\mathbb{R}^n}\{(1+|y|^2)^n|\varphi^{(k)}(y)|\}\Big)\int_{\mathbb{R}^n}(1+|x|^2)^{-n}\,dx\end{aligned}$$

for some constant $C_n > 0$, independent of φ. Since $\sum_{g\in\mathbb{Z}^n} \frac{1}{(1+4\pi^2|g|^2)^n} \le \sum_{g\in\mathbb{Z}^n} \frac{1}{(1+|g|^2)^n} < \infty$, we deduce that $v \in \mathcal{S}'(\mathbb{R}^n)$. Moreover, since translation is clearly a continuous map from $\mathcal{S}'(\mathbb{R}^n)$ to $\mathcal{S}'(\mathbb{R}^n)$, it follows that v is a periodic distribution. Consequently, with ϕ as above, we have

$$v = \sum_{g\in\mathbb{Z}^n} \phi_g v = \sum_{g\in\mathbb{Z}^n} (\tau_g\phi_0)v = \sum_{g\in\mathbb{Z}^n} \tau_g(\phi_0 v). \tag{6.25}$$

On the other hand, it follows from $v = \sum_{g\in\mathbb{Z}^n} e^{2\pi i g\cdot x}$ that $(e^{2\pi i x_k} - 1)v = 0$ for $k = 1, \dots, n$. This carries over to $(\phi_0 v)$, so that, in particular $(e^{2\pi i x_k} - 1)\phi_0 v = 0$ for $k = 1, \dots, n$. But $e^{2\pi i x_k} - 1 = 2ie^{\pi i x_k}\sin(\pi x_k) = x_k\,\sigma(x_k)$ for a C^∞-function $\sigma : \mathbb{R} \to \mathbb{C}$ which does not vanish on $(-1,1)$, where the support of the distribution $(\phi_0 v)$ is located. Thus $x_k(\phi_0 v) = 0$ for $k = 1, \dots, n$. Writing $\varphi(x) = \varphi(0) + \sum_{k=1}^n x_k\varphi_k(x)$ with φ_k of class C^∞, see Exercise 8(iii), since the distribution $(\phi_0 v)$ has compact support, we get $\langle \phi_0 v, \varphi\rangle = \varphi(0)\,\langle \phi_0 v, 1\rangle = \alpha\varphi(0)$, which means that $\phi_0 v = \alpha\,\delta_0$ and (6.25) yields $v = \alpha\sum_{g\in\mathbb{Z}^n} \tau_g\delta_0$. Therefore, to validate (6.23), it only remains to show that $\alpha = 1$. For this, note that the characteristic function χ_I of the unit cube $I = (-1/2, 1/2)^n$ belongs to $\mathcal{E}'(\mathbb{R}^n)$. Since $v \in \mathcal{S}'(\mathbb{R}^n)$, the convolution $v * \chi_I$ exists and, using the definition $v = \sum_{g\in\mathbb{Z}^n} e^{2\pi i g\cdot x}$, we can compute $(v * \chi_I)(x) = \sum_{g\in\mathbb{Z}^n} \int_I e^{2\pi i g\cdot(x-y)}\,dy = 1$ since the only nonzero contribution comes from $g = 0$. On the other hand, the representation $v = \alpha\sum_{g\in\mathbb{Z}^n} \tau_g\delta_0$ yields $v * \chi_I = \alpha\sum_{g\in\mathbb{Z}^n} \tau_g\chi_I = 1$ a.e. Hence $\alpha = 1$ and (6.23) is proved.

We now extend the previous considerations to a general periodic distribution.

[39] The fact that each of the n integer components of g is a possible endpoint (left- or right-) of an edge of such an n-dimensional cube shows that their number is 2^n. As for the estimate, by construction $|g| \ge |x|/\sqrt{n}$ for $x \in G$.

[40] For any fixed $x \in \mathbb{R}^n$, at least one and at most finitely many terms are non-zero.

Theorem 6.13 *If $u \in \mathcal{D}'(\mathbb{R}^n)$ is a periodic distribution, then $u \in \mathcal{S}'(\mathbb{R}^n)$ and*

$$u = \sum_{g\in\mathbb{Z}^n} c_g\, e^{2\pi i g\cdot x}, \tag{6.26}$$

where $\{c_g\}_{g\in\mathbb{Z}^n}$ are complex numbers such that

$$|c_g| \le C\,(1+|g|)^N \quad for\ all \quad g\in\mathbb{Z}^n. \tag{6.27}$$

for some $C > 0$ and some integer $N \ge 0$. Conversely, any complex numbers $\{c_g\}_{g\in\mathbb{Z}^n}$ subject to (6.27) define a periodic distribution $\sum_{g\in\mathbb{Z}^n} c_g\, e^{2\pi i g\cdot x}$.

Proof For a periodic distribution $u \in \mathcal{D}'(\mathbb{R}^n)$, if $u_0 \in \mathcal{S}'(\mathbb{R}^n)$ is the first member of (6.23), then $(\phi_0 u) \in \mathcal{E}'(\mathbb{R}^n)$, so that $u_0 * (\phi_0 u) \in \mathcal{S}'(\mathbb{R}^n)$, by (6.16). But $u_0 * (\phi_0 u) = \sum_{g\in\mathbb{Z}^n}(\tau_g\delta_0) * (\phi_0 u) = \sum_{g\in\mathbb{Z}^n}\tau_g(\delta_0 * (\phi_0 u)) = \sum_{g\in\mathbb{Z}^n}\tau_g(\phi_0 u) = \sum_{g\in\mathbb{Z}^n}(\tau_g\phi_0)u = \sum_{g\in\mathbb{Z}^n}\phi_g\, u = u$. Thus $u \in \mathcal{S}'(\mathbb{R}^n)$. Moreover, convoluting the two members of (6.23) by $(\phi_0 u) \in \mathcal{E}'(\mathbb{R}^n)$, since we already established that $u = u_0 * (\phi_0 u)$, from (6.24), (6.16) and (6.14)–(6.15), we get

$$\begin{aligned} u &= \sum_{g\in\mathbb{Z}^n} e^{2\pi i g\cdot x} * (\phi_0 u) = \sum_{g\in\mathbb{Z}^n} \mathcal{F}^{-1}\{\mathcal{F}(e^{2\pi i g\cdot x})\,\widehat{(\phi_0 u)}\} = \sum_{g\in\mathbb{Z}^n} \mathcal{F}^{-1}\{(\tau_g\delta_0)\,\widehat{(\phi_0 u)}\} \\ &= \sum_{g\in\mathbb{Z}^n} \mathcal{F}^{-1}\{\widehat{(\phi_0 u)}(g)\,(\tau_g\delta_0)\} = \sum_{g\in\mathbb{Z}^n} \widehat{(\phi_0 u)}(g)\,\mathcal{F}^{-1}\{(\tau_g\delta_0)\} = \sum_{g\in\mathbb{Z}^n} \widehat{(\phi_0 u)}(g)\, e^{2\pi i g\cdot x}. \end{aligned}$$

This is the desired representation with $c_g = \widehat{(\phi_0 u)}(g)$ for $g \in \mathbb{Z}^n$. The inequalities (6.27) now follow from the considerations made in the proof of Theorem 6.7.

Conversely, note first that for any complex sequence $\{c_g\}_{g\in\mathbb{Z}^n}$, $T = \sum_{g\in\mathbb{Z}^n} c_g\delta_{-g}$ defines an element of $\mathcal{D}'(\mathbb{R}^n)$ since for any $\varphi \in \mathcal{D}(\mathbb{R}^n)$, $\langle T,\varphi\rangle = \sum_{g\in\mathbb{Z}^n} c_g\varphi(-g)$ is, in fact, a finite sum; hence there is no problem with the convergence. We claim that if the sequence $\{c_g\}_{g\in\mathbb{Z}^n}$ is *slowly increasing*, that is, if it satisfies (6.27), then $T \in \mathcal{S}'(\mathbb{R}^n)$. Indeed, let $\varphi_k \to 0$ in $\mathcal{S}(\mathbb{R}^n)$, i.e. $\lim_{k\to\infty}\sup_{x\in\mathbb{R}^n}|x^\alpha(D^\beta\varphi_k)(x)| \to 0$ for all n-tuples of nonnegative integers α and β. Then $M_k = \sup_{x\in\mathbb{R}^n}\{(1+|x|)^{N+2n}|\varphi_k(x)|\} \to 0$ for $k \to \infty$ and

$$|\langle c_g\delta_{-g},\varphi_k\rangle| = |c_g\varphi_k(-g)| \le C(1+|g|)^N\frac{M_k}{(1+|g|)^{N+2n}} \le \frac{C\,M_k}{(1+|g|)^{2n}} \le \frac{C\,M_k}{(1+|g|^2)^n}.$$

We showed in the proof of (6.23) that $\sum_{g\in\mathbb{Z}^n}(1+|g|^2)^{-n} < \infty$. Since $M_k \to 0$ as $k\to\infty$, we deduce that $\langle T,\varphi_k\rangle \to 0$ as $k \to \infty$. Thus (6.27) ensures $T \in \mathcal{S}'(\mathbb{R}^n)$, and therefore $\hat{T} \in \mathcal{S}'(\mathbb{R}^n)$. Now $(\tau_{-g}\delta_0) = \delta_{-g}$ and the fact that $\mathcal{F}\{(\tau_{-g}\delta_0)\} = e^{2\pi i g\cdot x}$, due to (6.14), permit us to compute $\hat{T} = \sum_{g\in\mathbb{Z}^n} c_g\, e^{2\pi i g\cdot x}$. The proof is complete. □

Theorem 6.13 says that there are no issues about the convergence of Fourier series in the sense of distributions: the Fourier series of any periodic distribution u converges to u in $\mathcal{S}'(\mathbb{R}^n)$. In this context, note that the proof of Theorem 6.13, combined with Theorem 6.9, shows that the Fourier coefficients

$$\hat{u}(g) := c_g = \widehat{(\phi_0 u)}(g) = \langle\phi_0 u,\, e^{-2\pi i g\cdot x}\rangle, \qquad g\in\mathbb{Z}^n, \tag{6.28}$$

are determined by the periodic distribution u, using the cut-off function ϕ_0. One can derive a more revealing explicit formula for them:

$$u(x) * (\chi_I(x)\, e^{2\pi i g\cdot x}) = \hat{u}(g)\, e^{2\pi i g\cdot x}, \qquad g\in\mathbb{Z}^n, \tag{6.29}$$

where χ_I is the characteristic function of the unit cube $I = (-1/2, 1/2)^n$. This proves that the specific choice of the cut-off function in (6.28) plays no role.

Let us validate (6.29). For $g, g' \in \mathbb{Z}^n$ we have $e^{2\pi i g'\cdot x} * (\chi_I(x)\, e^{2\pi i g\cdot x}) = \int_I e^{2\pi i g\cdot y} e^{2\pi i g'\cdot(x-y)}dy = e^{2\pi i g'\cdot x}\int_I e^{2\pi i(g-g')\cdot y}dy$ vanishes if $g \neq g'$ and equals to $e^{2\pi i g'\cdot x}$ if $g = g'$. Since $\sum_{g'\in\mathbb{Z}^n} c_g\, e^{2\pi i g'\cdot x}$ converges to u in $\mathcal{S}'(\mathbb{R}^n)$, $\chi_I(x)\, e^{2\pi i g\cdot x} \in \mathcal{E}'(\mathbb{R}^n)$ and it is easy to check[41] that if $U \in \mathcal{E}'(\mathbb{R}^n)$ and $V_k \to V$ in $\mathcal{S}'(\mathbb{R}^n)$, then $V_k * U \to V * U$ in $\mathcal{S}'(\mathbb{R}^n)$, (6.29) follows at once.

Combining Theorem 6.13 with the relations

$$\mathcal{F}^{-1}(e^{2\pi i g\cdot x}) = \tau_{-g}\delta_0 = \delta_{-g}\,, \qquad \mathcal{F}(e^{2\pi i g\cdot x}) = \tau_g\delta_0 = \delta_g\,,$$

which are a consequence of (6.14) and (6.15), we deduce that the periodic distribution $u \in \mathcal{S}'(\mathbb{R}^n)$ given by (6.26) is the Fourier transform of the distribution

$$\mathcal{F}^{-1}(u) = \sum_{g\in\mathbb{Z}^n} c_g\, \delta_{-g}\,, \tag{6.30}$$

while its Fourier transform is

$$\hat{u} = \sum_{g\in\mathbb{Z}^n} c_g\, \delta_g\,. \tag{6.31}$$

The expression on the right side of (6.30) is a measure supported on the lattice $\mathbb{Z}^n$, so that a trigonometric series whose coefficients have a polynomial order of growth should be thought of as the Fourier transform of a measure on the lattice $\mathbb{Z}^n$, the coefficient c_g being the mass at $\xi = g$. The smoother the distribution $\mathcal{F}^{-1}(u)$ is, the more rapidly do the Fourier coefficients $\hat{u}_g$ of u decay at infinity. For example, if $\sum_{g\in\mathbb{Z}^n} |c_g|^2 < \infty$, then $\mathcal{F}^{-1}(u) \in L^2[0,1]$ and $u \in l^2$; this idea is further developed in Section 6.4 (see also Exercise 26).

Let us investigate the possibility of the pointwise validity of (6.26). With this purpose in mind, note that, given a function $f : \mathbb{R} \to \mathbb{C}$, one can construct a periodic function f_p by summing over the integer translates, defining

$$f_p(x) = \sum_{k\in\mathbb{Z}} f(x+k)\,, \tag{6.32}$$

if the symmetric partial sums $\sum_{k=-n}^{n} f(x+k)$ converge.

Theorem 6.14 *If $f \in L^1(\mathbb{R})$, then f_p is finite a.e., satisfies $f_p(x+1) = f(x)$ a.e. and $f_p \in L^1[0,1]$, with Fourier coefficients $\widehat{f_p}(k) = \hat{f}(k)$ for all $k \in \mathbb{Z}$. Moreover, if $\sum_{k\in\mathbb{Z}} |\hat{f}(k)| < \infty$, then the Fourier series of f_p converges and*

$$\sum_{k\in\mathbb{Z}} f(x+k) = f_p(x) = \sum_{k\in\mathbb{Z}} \hat{f}(k)\, e^{2\pi i k x} \quad \text{a.e.} \tag{6.33}$$

In particular, f_p is a.e. equal to a continuous function of period 1.

[41] If $\varphi \in \mathcal{S}(\mathbb{R}^n)$, then $\mathcal{F}^{-1}(\varphi) \in \mathcal{S}(\mathbb{R}^n)$, $\hat{U} \in \mathcal{O}(\mathbb{R}^n)$, so that $\hat{U}\mathcal{F}^{-1}(\varphi) \in \mathcal{S}(\mathbb{R}^n)$ and $\langle V_k * U, \varphi\rangle = \langle \mathcal{F}^{-1}(\hat{V}_k\hat{U}), \varphi\rangle = \langle \hat{V}_k\hat{U}, \mathcal{F}^{-1}(\varphi)\rangle = \langle \hat{V}_k, \hat{U}\mathcal{F}^{-1}(\varphi)\rangle \to \langle \hat{V}, \hat{U}\mathcal{F}^{-1}(\varphi)\rangle = \langle V * U, \varphi\rangle$.

Proof We have $\int_0^1 \big(\sum_{k\in\mathbb{Z}} |f(x+k)|\big)\,dx = \sum_{k=-\infty}^{\infty} \int_k^{k+1} |f(x)|\,dx = \int_{\mathbb{R}} |f(x)|\,dx$, so that f_p is finite a.e. and belongs to $L^1[0,1]$. Similarly, for any integer m, we have $\int_0^1 f_p(x)\,e^{-2\pi imx}dx = \int_0^1 \big(\sum_{k\in\mathbb{Z}} f(x+k)\big)\,e^{-2\pi imx}dx = \sum_{k\in\mathbb{Z}} \int_0^1 f(x+k)\,e^{-2\pi imx}dx = \sum_{k=-\infty}^{\infty} \int_k^{k+1} f(y)\,e^{-2\pi im(y-k)}dy = \sum_{k=-\infty}^{\infty} \int_k^{k+1} f(y)\,e^{-2\pi imy}dy = \int_{\mathbb{R}} f(y)\,e^{-2\pi imy}dy = \hat{f}(m)$. If $\sum_{k\in\mathbb{Z}} |\hat{f}(k)| < \infty$, then Exercise 4.22(i) ensures that $f_p(x) = \sum_{k\in\mathbb{Z}} \hat{f}(k)\,e^{2\pi ikx}$ a.e., thus proving (6.33). □

Example If $f(x) = \frac{\exp(-x^2/(2t))}{\sqrt{2\pi t}}$, where $t > 0$ is a parameter, then $\hat{f}(k) = e^{-2\pi^2k^2t}$, see Section 5.3, and the identity $\sum_{k\in\mathbb{Z}} \frac{1}{\sqrt{2\pi t}} \exp\big(-\frac{(x-k)^2}{2t}\big) = \sum_{k\in\mathbb{Z}} e^{-2\pi^2k^2t}\,e^{2\pi ikx}$ emerges from (6.33). □

The pointwise Poisson summation formula in $\mathbb{R}^n$ is entirely similar to (6.33), the periodisation of $f \in L^1(\mathbb{R}^n)$ being defined by $f_p(x) = \sum_{g\in\mathbb{Z}^n} f(x+g)$, which defines a periodic function $f_p \in L^1([0,1]^n)$ whose Fourier coefficients (for $g \in \mathbb{Z}^n$) are computed as

$$\int_{[0,1]^n} f_p(x)\,e^{-2\pi ig\cdot x}dx = \int_{[0,1]^n} \Big(\sum_{g\in\mathbb{Z}^n} f(x+g)\Big)\,e^{-2\pi ig\cdot x}dx = \sum_{g\in\mathbb{Z}^n} \int_{[0,1]^n} f(x+g)\,e^{-2\pi ig\cdot x}dx$$

$$= \sum_{g\in\mathbb{Z}^n} \int_{g+[0,1]^n} f(y)e^{-2\pi ig\cdot(y-n)}dy = \sum_{g\in\mathbb{Z}^n} \int_{g+[0,1]^n} f(y)e^{-2\pi ig\cdot y}dy = \int_{\mathbb{R}^n} f(y)e^{-2\pi ig\cdot y}dy\,.$$

Thus $\hat{f}_p(g) = \hat{f}(g)$ for every $g \in \mathbb{Z}^n$, suggesting the identity

$$f_p(x) := \sum_{g\in\mathbb{Z}^n} f(x+g) = \sum_{g\in\mathbb{Z}^n} \hat{f}(g)\,e^{2\pi ig\cdot x}\,, \tag{6.34}$$

that, for the moment, we can only interpret as a formal computation that fails without additional conditions on f. For example, it is known that $f \in L^1(\mathbb{R})$ and $\hat{f} \in L^1(\mathbb{R})$ are insufficient to ensure the validity of (6.33): there are examples in which both sides diverge, as well as examples in which both sides converge but to unequal sums – see the discussion in Hogan and Lakey (2004). The insight provided by the distributional form of the Poisson summation (Theorem 6.13) shows that we only need to worry about the pointwise convergence of both sides of (6.34). Note that the existence of positive constants $C, \varepsilon > 0$ such that f and its Fourier transform $\hat{f}$ satisfy the decay estimates[42]

$$|f(x)| \le \frac{C}{(1+|x|)^{n+\varepsilon}}\,, \qquad |\hat{f}(\xi)| \le \frac{C}{(1+|\xi|)^{n+\varepsilon}}\,, \qquad x, \xi \in \mathbb{R}^n\,, \tag{6.35}$$

ensures that both sides of (6.34) converge absolutely and uniformly on $[0,1]^n$. Indeed, $f \in L^1(\mathbb{R}^n)$ as a consequence of its decay. Therefore the series defining $f_p \in L^1([0,1]^n)$ converges absolutely. On the other hand, the considerations made to prove (6.23) actually show that $\sum_{g\in\mathbb{Z}^n}(1+|g|)^{-n-\varepsilon} < \infty$, so that $f_p \in L^1([0,1]^n)$ has the absolutely convergent Fourier series $\sum_{g\in\mathbb{Z}^n} \hat{f}(g)\,e^{-2\pi ig\cdot x}$, and (6.34) follows from Exercise 4.22(i).

42 See Gröchenig (1996) for optimal results in this direction.

6.3 Fourier transform of a measure

While the Fourier transform theory for measures is inherently nonsymmetric, it has nevertheless a discernible structure that we explore in this section.

Let μ be a finite Borel measure on $\mathbb{R}^d$, that is, a nonnegative, countably additive set function defined on the Borel sets of $\mathbb{R}^d$, with $\mu(\mathbb{R}^d) < \infty$. Its Fourier transform, defined by[43]

$$\hat{\mu}(\xi) = \int_{\mathbb{R}^d} e^{-2\pi i \xi\cdot x} d\mu(x), \tag{6.36}$$

has the following properties:

- $\hat{\mu}$ is a uniformly continuous bounded function with $|\hat{\mu}(\xi)| \leq \hat{\mu}(0) = \mu(\mathbb{R}^d)$;
- $\hat{\mu}$ is a *positive-definite function*, meaning that for any integer $n \geq 1$ we have

$$\sum_{j,k=1}^{n} c_j \overline{c_k}\, \hat{\mu}(\xi_j - \xi_k) \geq 0 \tag{6.37}$$

  for every set of complex numbers $\{c_j\}_{1\leq j\leq n}$ and vectors $\{\xi_j\}_{1\leq j\leq n}$ in $\mathbb{R}^d$.

Indeed, from the definition we have $|\hat{\mu}(\xi)| \leq \int_{\mathbb{R}^d} |e^{-2\pi i\xi\cdot x}|\, d\mu(x) = \int_{\mathbb{R}^d} d\mu(x) = \mu(\mathbb{R}^d)$, while $|\hat{\mu}(\xi + h) - \hat{\mu}(\xi)| = \left|\int_{\mathbb{R}^d} e^{-2\pi i\xi\cdot x}(e^{-2\pi i h\cdot x} - 1)\, d\mu(x)\right| \leq \int_{\mathbb{R}^d} \left|e^{-2\pi i h\cdot x} - 1\right| d\mu(x)$, with the last term independent of ξ and vanishing in the limit $h \to 0$ by dominated convergence (since μ is finite). The positive-definite property is checked by direct computation: $\sum_{j,k=1}^{n} c_j\overline{c_k}\, \hat{\mu}(\xi_j - \xi_k) = \sum_{j,k=1}^{n} c_j\overline{c_k} \int_{\mathbb{R}^d} e^{-2\pi i(\xi_j-\xi_k)\cdot x}\, d\mu(x) = \int_{\mathbb{R}^d} \left|\sum_{j=1}^{n} c_j e^{-2\pi i\xi_j\cdot x}\right|^2 d\mu(x) \geq 0$.

Note that positive-definite does not mean positive, e.g. $f(s) = e^{is}$ is positive-definite on $\mathbb{R}$ since in this case $\sum_{j,k=1}^{n} c_j\overline{c_k} f(\xi_j - \xi_k) = |\sum_{j=1}^{n} c_j e^{i\xi_j}|^2 \geq 0$. Let us also point point some basic properties of positive-definite functions:

$$f(0) \geq |f(\xi)| \quad \text{and} \quad f(-\xi) = \overline{f(\xi)} \quad \text{for all} \quad \xi \in \mathbb{R}^d\,, \tag{6.38}$$

which follow directly from (6.37) without a continuity assumption.[44]

Indeed, the choice $n = 2$, $\xi_1 = \xi$, $\xi_2 = 0$ in (6.37) shows that the inequality $(|c_1|^2 + |c_2|^2) f(0) + c_1\overline{c_2}\, f(\xi) + \overline{c_1} c_2\, f(-\xi) \geq 0$ should be valid for all $c_1, c_2 \in \mathbb{C}$ and all $\xi \in \mathbb{R}^d$. Letting $c_1 = 1$ and $c_2 = 0$ yields $f(0) \geq 0$, letting $c_1 = c_2 = 1$ yields $\Im\mathfrak{m}[f(\xi)] = -\Im\mathfrak{m}[f(-\xi)]$, letting $c_1 = 1$ and $c_2 = i$ yields $\Re\mathfrak{e}[f(\xi)] = \Re\mathfrak{e}[f(-\xi)]$, while the choice $c_1 = \overline{c_2} = \sqrt{f(\xi)}$ leads to $f(0) \geq |f(\xi)|$.

The Fourier analysis of measures is particularly well-suited to the study of issues of weak convergence. We will probe some deeper aspects of this fruitful interplay that is especially useful in probability theory (see Part II).

[43] This definition is consistent with the distributional definition (6.11) since for $\varphi \in \mathcal{S}(\mathbb{R}^d)$ we have $\langle\hat{\mu}, \varphi\rangle = \int_{\mathbb{R}^d}\int_{\mathbb{R}^d} e^{-2\pi i\xi\cdot x} d\mu(x)\, \varphi(\xi)\, d\xi = \int_{\mathbb{R}^d} \hat{\varphi}(x)\, d\mu(x) = \langle\mu, \hat{\varphi}\rangle$.

[44] The Fourier transform approach is often the most expedite way to prove that a function is positive-definite, e.g. on $\mathbb{R}$ for $x \mapsto e^{-|x|}$, $x \mapsto e^{-\pi x^2}$, $x \mapsto 1/(1 + x^2)$.

Let us first point out that the space $\mathcal{M}(\mathbb{R}^d)$ of (signed) finite Borel measures on $\mathbb{R}^d$ is the dual of the Banach space $C_0(\mathbb{R}^d)$ of continuous functions on $\mathbb{R}^d$ which vanish at infinity[45] (endowed with the supremum norm), see Folland (1999). The normed vector space $C_c(\mathbb{R}^d)$ of continuous functions with compact support (with the supremum norm) is not complete, being dense in $C_0(\mathbb{R}^d)$. Thus $\mathcal{M}(\mathbb{R}^d)$, the norm being the total variation of a measure, is isomorphic to the dual of $C_c(\mathbb{R}^d)$ and its completion $C_0(\mathbb{R}^d)$. A subtle issue here is the relationship with a fundamental result of Riesz, stating that if $L : C_c(\mathbb{R}^d) \to \mathbb{R}$ is a *positive linear functional* (meaning that $Lf \geq 0$ if $f \geq 0$), then there exists a unique nonnegative regular[46] Borel measure ν on $\mathbb{R}^d$, which is finite on compact sets, such that $Lf = \int_{\mathbb{R}^d} f(x)\, d\nu(x)$ for all $f \in C_c(\mathbb{R}^d)$; see Reed and Simon (1980a). This construction differs from the one leading to $\mathcal{M}(\mathbb{R})$: for example, the Riemann integral gives rise to the Lebesgue measure on $\mathbb{R}$, without being[47] a continuous linear functional on $C_c(\mathbb{R})$.

Let us denote by $\mathcal{M}^+(\mathbb{R}^d)$ the family of finite (nonnegative) Borel measures on $\mathbb{R}^d$. In probability theory, a common terminology for the weak* topology on $\mathcal{M}^+(\mathbb{R}^d)$, in which (see Chapter 3) μ_n converges weak* to μ if $\int_{\mathbb{R}^d} f(x)\, d\mu_n(x) \to \int_{\mathbb{R}^d} f(x)\, d\mu(x)$ for all $f \in C_0(\mathbb{R}^d)$, is the *vague topology*. Of considerable importance in applications, is the notion[48] of *weak convergence of measures*: a sequence $\{\mu_n\}_{n\geq 1}$ converges weakly to a limit measure μ in $\mathcal{M}^+(\mathbb{R}^d)$ if

$$\int_{\mathbb{R}^d} f(x)\, d\mu_n(x) \to \int_{\mathbb{R}^d} f(x)\, d\mu(x) \quad \text{for every bounded continuous function}$$

$f : \mathbb{R}^d \to \mathbb{C}$. Note that a weakly convergent sequence is also vaguely convergent, but the converse might fail since functions in $C_0(\mathbb{R}^d)$ cannot prevent mass from escaping at infinity.[49] It turns out that without loss of mass,[50] vague convergence implies weak convergence:

45 A continuous function $f : \mathbb{R}^d \to \mathbb{C}$ vanishes at infinity if $f(x) \to 0$ for $|x| \to \infty$.

46 Meaning that for every Borel set B in $\mathbb{R}^d$ we have $\nu(B) = \inf\{\nu(U) : U \supset B,\ U \text{ open}\}$ and $\nu(B) = \sup\{\nu(K) : K \subset B,\ K \text{ compact}\}$.

47 Indeed, for $n \geq 1$ the piecewise linear functions given by $f_n(x) = 0$ for $|x| \geq n + 1/n$, $f_n(x) = n(x + n + 1/n)$ for $-n - 1/n \leq x \leq -n$, $f_x(x) = 1$ for $|x| \leq n$, $f_n(x) = -n(x - n - 1/n)$ for $n \leq x \leq n + 1/n$, are such that $\|f_n\|_{C_c(\mathbb{R})} = 1$ and $Lf_n = \int_{-n-1/n}^{n+1/n} f_n(x)\, dx = 2n + 1/n$.

48 Even if this probabilistic terminology conflicts somewhat with that used in functional analysis, we will adhere to it.

49 For example, the measures δ_n converge vaguely to 0 in $\mathcal{M}(\mathbb{R})$ even though their action on constant nonzero functions shows that they do not converge weakly to 0.

50 No mass gain can be experienced in the vague limit: if μ_n converges vaguely to μ in $\mathcal{M}^+(\mathbb{R}^d)$, then $\mu(\mathbb{R}^d) \leq \liminf_{n\to\infty} \mu_n(\mathbb{R}^d)$. Indeed, choosing for each integer $n \geq 1$ a function $f_n \in C_0(\mathbb{R}^d)$ such that $f_n : \mathbb{R}^d \to [0, 1]$ and $f_n(x) = 1$ for $|x| \leq n$, we have $\mu(\mathbb{R}^d) = \sup_{n\geq 1} \int_{\mathbb{R}^d} f_n\, d\mu = \sup_{n\geq 1} \lim_{k\to\infty} \int_{\mathbb{R}^d} f_n\, d\mu_k \leq \liminf_{k\to\infty} \sup_{n\geq 1} \int_{\mathbb{R}^d} f_n\, d\mu_k = \liminf_{k\to\infty} \mu_k(\mathbb{R}^d)$.

Theorem 6.15 *The sequence $\{\mu_n\}_{n\geq 1}$ converges weakly to μ in $\mathcal{M}^+(\mathbb{R}^d)$ if and only if it converges vaguely and $\lim_{n\to\infty}\mu_n(\mathbb{R}^d)=\mu(\mathbb{R}^d)$.*

Proof Necessity is immediate. For sufficiency, it is enough to show that $\lim_{n\to\infty}\int_{\mathbb{R}^d} f(x)\,d\mu_n(x) = \int_{\mathbb{R}^d} f(x)\,d\mu(x)$ for any nonnegative bounded and continuous function $f:\mathbb{R}^d\to[0,\infty)$. Since $\mu(\mathbb{R}^d)<\infty$, as in Exercise 2.17, given $\varepsilon>0$, there is some $N\geq 1$ such that $E=\{x\in\mathbb{R}^d:\ |x|>N\}$ satisfies $\mu(E)\leq\varepsilon$. Let $\varphi:\mathbb{R}^d\to[0,1]$ be in the class $\mathcal{D}(\mathbb{R}^d)$ with $\varphi(x)=1$ for $|x|\leq N$. Then, with $\|f\|=\sup_{x\in\mathbb{R}^d}\{|f(x)|\}$, we have

$$\limsup_{n\to\infty}\Big|\int_{\mathbb{R}^d} f\,d\mu_n-\int_{\mathbb{R}^d}\varphi f\,d\mu_n\Big|\leq\limsup_{n\to\infty}\Big(\|f\|\int_{\mathbb{R}^d}(1-\varphi)\,d\mu_n\Big)$$
$$=\|f\|\limsup_{n\to\infty}\Big(\mu_n(\mathbb{R}^d)-\int_{\mathbb{R}^d}\varphi\,d\mu_n\Big)=\|f\|\Big(\mu(\mathbb{R}^d)-\int_{\mathbb{R}^d}\varphi\,d\mu\Big)\leq\|f\|\,\mu(E)\leq\varepsilon\|f\|\,.$$

Similarly, $\limsup_{n\to\infty}\big|\int_{\mathbb{R}^d} f\,d\mu-\int_{\mathbb{R}^d}\varphi f\,d\mu\big|\leq\varepsilon\|f\|$, so that $\limsup_{n\to\infty}\big|\int_{\mathbb{R}^d} f\,d\mu_n-\int_{\mathbb{R}^d} f\,d\mu\big|\leq 2\varepsilon\|f\|$, which completes the proof. □

While $C_0(\mathbb{R}^d)$ is[51] a separable Banach space, $\mathcal{M}(\mathbb{R}^d)$ is not separable since the measures $\{\delta_x\}_{x\in\mathbb{R}^d}$ form an uncountable set of equidistant points. Consequently, we cannot rely on Theorem 3.20 to claim that any bounded sequence in $\mathcal{M}^+(\mathbb{R}^d)$ admits a vaguely convergent subsequence.[52] Nevertheless, the following result holds.

Theorem 6.16 (Helly's theorem) *Any bounded sequence $\{\mu_n\}_{n\geq 1}$ in $\mathcal{M}^+(\mathbb{R}^d)$ has a vaguely convergent subsequence.*

Proof Let $\{f_n\}_{n\geq 1}$ be dense in $C_0(\mathbb{R}^d)$; for convenience, let $f_n\in C_c(\mathbb{R}^d)$ for every $n\geq 1$ (this being possible as $C_0(\mathbb{R}^d)$ is the closure of $C_c(\mathbb{R}^d)$, with the supremum norm). Extract from the bounded sequence $\{\int_{\mathbb{R}^d} f_1\,d\mu_n\}_{n\geq 1}$ a convergent subsequence $\{\int_{\mathbb{R}^d} f_1\,d\mu_{1,n}\}_{n\geq 1}$. Consider now the bounded sequence $\{\int_{\mathbb{R}^d} f_2\,d\mu_{1,n}\}_{n\geq 1}$ and extract a convergent subsequence $\{\int_{\mathbb{R}^d} f_2\,d\mu_{2,n}\}_{n\geq 1}$ and so on: in step $k+1\geq 3$, extract from the bounded sequence $\{\int_{\mathbb{R}^d} f_{k+1}\,d\mu_{k,n}\}_{n\geq 1}$ a convergent subsequence $\{\int_{\mathbb{R}^d} f_{k+1}\,d\mu_{k+1,n}\}_{n\geq 1}$. By construction, the diagonal subsequence $\{\mu_{k,k}\}_{k\geq 1}$ has the property that $\{\int_{\mathbb{R}^d} f_n\,d\mu_{k,k}\}_{k\geq 1}$ converges for all f_n. Denote $\nu_k=\mu_{k,k}$ for $k\geq 1$.

Given some $f\in C_0(\mathbb{R}^d)$, for $k,\ j\geq 1$ we have

$$\Big|\int_{\mathbb{R}^d} f\,d\nu_k-\int_{\mathbb{R}^d} f\,d\nu_j\Big|\leq\Big|\int_{\mathbb{R}^d}(f-f_n)\,d\nu_k\Big|+\Big|\int_{\mathbb{R}^d}(f-f_n)\,d\nu_j\Big|+\Big|\int_{\mathbb{R}^d} f_n\,d\nu_k-\int_{\mathbb{R}^d} f_n\,d\nu_j\Big|$$
$$\leq\|f-f_n\|_{C_0(\mathbb{R}^d)}\sup_{m\geq 1}\{\mu_m(\mathbb{R}^d)\}+\Big|\int_{\mathbb{R}^d} f_n\,d\nu_k-\int_{\mathbb{R}^d} f_n\,d\nu_j\Big|.$$

[51] The general proof is an adaptation of the case $d=1$ with real values, that we now sketch. Any $f\in C_0(\mathbb{R})$ is uniformly continuous and can therefore be uniformly approximated by a continuous function whose graph consists of finitely many line segments, each of which has endpoints with rational coordinates. There are are only countably many such functions.

[52] While the closed unit ball in the dual X^* of a Banach space X is always compact with respect to the weak*-topology, the latter is not induced by a metric if X^* is not separable (see Chapter 3), so that the characterisation of compactness by the fact that every sequence admits a convergent subsequence is not granted. For example, when $X=l^\infty$, there exists a sequence in the unit ball of X^* without any convergent subsequence – see Brézis (2011).

Therefore $\{\int_{\mathbb{R}^d} f\,d\nu_k\}_{k\geq 1}$ is a Cauchy sequence in $\mathbb{C}$. Its limit $L(f)$ defines a positive linear functional on $C_c(\mathbb{R}^d)$, so that there exists a (nonnegative) regular Borel measure ν on $\mathbb{R}^d$, finite on compact sets, such that $L(f) = \int_{\mathbb{R}^d} f\,d\nu$ for every $f \in C_c(\mathbb{R}^d)$. To see that $\nu \in \mathcal{M}^+(\mathbb{R}^d)$, it suffices to note that if we choose for every integer $n \geq 1$ a function $\varphi_n : \mathbb{R}^d \to [0,1]$ in $C_c(\mathbb{R}^d)$ with $\varphi(x) = 1$ for $|x| \leq n$, then $\nu\{x \in \mathbb{R}^d : |x| \leq n\} \leq \int_{\mathbb{R}^d} \varphi_n d\nu = \lim_{k\to\infty} \int_{\mathbb{R}^d} \varphi_n d\nu_k \leq \sup_{k\geq 1} \nu_k(\mathbb{R}^d)$, so that ν has finite total mass. We now claim that $\{\nu_k\}_{k\geq 1}$ converges vaguely to ν in $\mathcal{M}^+(\mathbb{R}^d)$. Fix $f \in C_0(\mathbb{R}^d)$. To show that $\int_{\mathbb{R}^d} f d\nu_k \to \int_{\mathbb{R}^d} f d\nu$, we rely on the knowledge that $\int_{\mathbb{R}^d} f d\nu_k \to L(f)$, while for every $n \geq 1$, $\int_{\mathbb{R}^d} f_n d\nu_k \to \int_{\mathbb{R}^d} f_n d\nu$ as $k \to \infty$. Moreover, the estimate $|\int_{\mathbb{R}^d} f\,d\mu - \int_{\mathbb{R}^d} f_n\,d\mu| \leq \|f - f_n\|_{C_0(\mathbb{R}^d)} \mu(\mathbb{R}^d)$, valid for every $\mu \in \mathcal{M}(\mathbb{R}^d)$, provides us with some uniformity. More precisely, given $\varepsilon > 0$, choose $k_1 \geq 1$ such that $|L(f) - \int_{\mathbb{R}^d} f\,d\nu_k| \leq \varepsilon/4$ for all $k \geq k_1$, then choose $k_2 \geq k_1$ such that $|\int_{\mathbb{R}^d} f_n\,d\nu_k - \int_{\mathbb{R}^d} f\,d\nu_k| \leq \varepsilon/4$ for all $n, k \geq k_2$, and subsequently choose $k_3 \geq k_2$ such that $|\int_{\mathbb{R}^d} f\,d\nu - \int_{\mathbb{R}^d} f_n\,d\nu| \leq \varepsilon/4$ for all $n \geq k_3$. The set of these inequalities yields $|L(f) - \int_{\mathbb{R}^d} f\,d\nu| \leq |L(f) - \int_{\mathbb{R}^d} f\,d\nu_k| + |\int_{\mathbb{R}^d} f\,d\nu_k - \int_{\mathbb{R}^d} f_n\,d\nu_k| + |\int_{\mathbb{R}^d} f_n\,d\nu_k - \int_{\mathbb{R}^d} f_n\,d\nu| + |\int_{\mathbb{R}^d} f_n\,d\nu - \int_{\mathbb{R}^d} f\,d\nu| \leq 3\varepsilon/4 + |\int_{\mathbb{R}^d} f_n\,d\nu_k - \int_{\mathbb{R}^d} f_n\,d\nu|$ for $n, k \geq k_3$. Letting now $k \to \infty$ yields $|L(f) - \int_{\mathbb{R}^d} f\,d\nu| \leq 3\varepsilon/4$ and since $\varepsilon > 0$ was arbitrary, this means that $L(f) = \int_{\mathbb{R}^d} f\,d\nu$ and the proof is complete. □

Let us now present a useful criterion for weak convergence in $\mathcal{M}^+(\mathbb{R}^d)$, given in in terms of Fourier transforms.

Theorem 6.17 (Lévy's theorem) *Let $\{\mu_n\}_{n\geq 1}$ be a sequence of probabiliy measures*[53] *in $\mathcal{M}^+(\mathbb{R}^d)$ such that $\lim_{n\to\infty} \widehat{\mu_n}(\xi) = \varphi(\xi)$ exists for all $\xi \in \mathbb{R}^d$. If the function $\varphi(\xi)$ is continuous at $\xi = 0$, then $\{\mu_n\}_{n\geq 1}$ converges weakly in $\mathcal{M}^+(\mathbb{R}^d)$ to a probability measure μ whose Fourier transform is φ.*

Proof Since $\sup_{\xi\in\mathbb{R}^d}\{|\widehat{\mu_n}(\xi)|\} = \widehat{\mu_n}(0) = \mu_n(\mathbb{R}^d) = 1$, we have $\varphi(0) = 1 \geq |\varphi(\xi)| \geq 0$ for all $\xi \in \mathbb{R}^d$. Moreover, by Theorem 6.16, the sequence $\{\mu_n\}_{n\geq 1}$ admits a vaguely convergent subsequence $\{\mu_{n_k}\}$, whose limit we denote by $\mu \in \mathcal{M}(\mathbb{R}^d)$. We now claim that $\mu(\mathbb{R}^d) = 1$, so that, by Theorem 6.15, $\{\mu_{n_k}\}$ converges weakly to μ. First, note that $\mu(\mathbb{R}^d) \leq 1$ since, as we pointed out in the preamble of Theorem 6.15, vague limits do not experience mass gain. By definition, we have $1 - \widehat{\mu_{n_k}}(\xi) = \int_{\mathbb{R}^d} (1 - e^{-2\pi i \xi\cdot x})\,d\mu_{n_k}(x)$. Integrating this equation over the cube $C_a = \bigcap_{j=1}^d \{\xi = (\xi_1, \ldots, \xi_d) \in \mathbb{R}^d : |\xi_j| \leq a\}$, where $a > 0$, we obtain $\int_{C_a} [1 - \widehat{\mu_{n_k}}(\xi)]\,d\xi = \int_{\mathbb{R}^d} \big((2a)^d - \prod_{j=1}^d \frac{\sin(2\pi a x_j)}{\pi x_j}\big)\,d\mu_{n_k}(x)$, so that

$$\frac{1}{(2a)^d} \int_{C_a} [1 - \widehat{\mu_{n_k}}(\xi)]\,d\xi = \int_{\mathbb{R}^d} \Big(1 - \prod_{j=1}^d \frac{\sin(2\pi a x_j)}{2\pi a x_j}\Big)\,d\mu_{n_k}(x)$$

$$\geq \int_{\mathbb{R}^d \setminus C_{1/(4a)}} \Big(1 - \prod_{j=1}^d \frac{\sin(2\pi a x_j)}{2\pi a x_j}\Big)\,d\mu_{n_k}(x)$$

as $1 \geq \prod_{j=1}^d \frac{\sin(2\pi a x_j)}{2\pi a x_j}$ if $x \in C_{1/(4a)}$ since $0 \leq \frac{\sin(s)}{s} \leq 1$ for $|s| \leq \pi/2$. On the other hand, the fact that $|\sin(s)| \leq |s|$ for all $s \in \mathbb{R}$ ensures that if $x \in \mathbb{R}^d$ lies outside the cube $C_{1/(4a)}$, then at least one factor in the product $\prod_{j=1}^d \frac{\sin(2\pi a x_j)}{2\pi a x_j}$ has absolute value less than $2/\pi$, and the remaining ones are less than 1 in absolute value, so that

$$\frac{1}{(2a)^d} \int_{C_a} [1 - \widehat{\mu_{n_k}}(\xi)]\,d\xi \geq \Big(1 - \frac{2}{\pi}\Big) \int_{\mathbb{R}^d \setminus C_{1/(4a)}} d\mu_{n_k}(x)\,.$$

[53] That is, $\mu_n(\mathbb{R}^d) = 1$ for all $n \geq 1$.

The dominated convergence theorem yields, after an algebraic manipulation,

$$\limsup_{n_k\to\infty} \int_{\mathbb{R}^d\setminus C_{1/(4a)}} d\mu_{n_k}(x) \le \frac{\pi}{(\pi-2)(2a)^d} \int_{C_a} [1-\varphi(\xi)]\, d\xi \,. \tag{6.39}$$

Take now a function $f : \mathbb{R}^d \to [0,1]$ of class $C_0(\mathbb{R}^d)$ with $f(x) = 1$ for $x \in C_{1/(4a)}$. Since the sequence $\{\mu_{n_k}\}$ converges vaguely to $\mu \in \mathcal{M}(\mathbb{R}^d)$, we have

$$1 - \int_{\mathbb{R}^d\setminus C_{1/(4a)}} d\mu_{n_k}(x) = \mu_{n_k}(C_{1/(4a)}) \le \int_{C_{1/(4a)}} f\, d\mu_{n_k} \le \int_{\mathbb{R}^d} f\, d\mu_{n_k} \to \int_{\mathbb{R}^d} f\, d\mu \le \mu(\mathbb{R}^d)\,.$$

But by the continuity of $\varphi(\xi)$ at $\xi = 0$, the right side of (6.39) can be made arbitrarily small by taking $a > 0$ sufficiently small. Thus $\mu(\mathbb{R}^d) \ge 1$, and actually, in view of the considerations made above, $\mu(\mathbb{R}^d) = 1$, so that $\{\mu_{n_k}\}$ converges weakly to μ. The function $x \mapsto e^{-2\pi i\xi\cdot x}$ being bounded and continuous, we have $\hat{\mu}(\xi) = \int_{\mathbb{R}^d} e^{-2\pi i\xi\cdot x} d\mu(x) = \lim_{n_k\to\infty} \int_{\mathbb{R}^d} e^{-2\pi i\xi\cdot x} d\mu_{n_k}(x) = \varphi(\xi)$. Thus every subsequence of $\{\mu_n\}_{n\ge1}$ has a subsubsequence that converges weakly to the probability measure μ with $\hat{\mu} = \varphi$. This ensures that the original sequence converges weakly to μ. □

Using the theory of weak convergence of measures we can characterise the image of finite Borel measures under the Fourier transform.

Theorem 6.18 (Bochner's theorem) *Any continuous positive-definite function on $\mathbb{R}^d$ is the Fourier transform of a finite Borel measure.*

Proof If $f : \mathbb{R}^d \to \mathbb{C}$ is continuous and positive-definite, then, by (6.38), $|f(\xi)| \le f(0)$ for all $\xi \in \mathbb{R}^d$. Moreover, due to (6.37), for every $\psi \in L^1(\mathbb{R}^d)$ we have $\int_{\mathbb{R}^d}\int_{\mathbb{R}^d} f(\xi-\eta)\,\psi(\xi)\,\overline{\psi(\eta)}\, d\xi d\eta \ge 0$. For $\psi(\xi) = e^{2\pi i x\cdot\xi} e^{-2\varepsilon|\xi|^2}$ with $\varepsilon > 0$, we get

$$\begin{aligned} 0 &\le \int_{\mathbb{R}^d}\int_{\mathbb{R}^d} f(\xi-\eta)\, e^{2\pi i x\cdot(\xi-\eta)}\, e^{-2\varepsilon|\xi|^2}\, e^{-2\varepsilon|\eta|^2}\, d\xi d\eta \\ &= \int_{\mathbb{R}^d}\int_{\mathbb{R}^d} f(u)\, e^{2\pi i x\cdot u}\, e^{-2\varepsilon|u+\eta|^2}\, e^{-2\varepsilon|\eta|^2}\, du d\eta = \Big(\frac{4\pi}{\varepsilon}\Big)^{d/2} \int_{\mathbb{R}^d} f(u)\, e^{2\pi i x\cdot u}\, e^{-\varepsilon|u|^2}\, du \end{aligned}$$

since $\int_{\mathbb{R}^d} e^{-\varepsilon|2\eta+u|^2}\, d\eta = \int_{\mathbb{R}^d} e^{-4\varepsilon|y|^2}\, dy = \prod_{j=1}^d \int_{\mathbb{R}} e^{-4\varepsilon y_j^2} dy_j = \prod_{j=1}^d \big(\frac{2}{\sqrt{\varepsilon}} \int_{\mathbb{R}} e^{-s^2} ds\big)^d = \big(\frac{4\pi}{\varepsilon}\big)^{d/2}$. Therefore

$$m_\varepsilon(x) = \int_{\mathbb{R}^d} e^{2\pi i x\cdot u} e^{-\varepsilon|u|^2} f(u)\, du \ge 0\,, \qquad x \in \mathbb{R}^d\,.$$

We claim that $m_\varepsilon \in L^1(\mathbb{R}^d)$, with $\widehat{m_\varepsilon}(\xi) = e^{-\varepsilon|\xi|^2} f(\xi)$. To see this, use Fubini's theorem and the formula for the Fourier transform of a Gaussian (see Exercise 27) to validate for $\delta > 0$ the following identity

$$\begin{aligned} \int_{\mathbb{R}^d} m_\varepsilon(x)\, e^{-2\pi i\xi\cdot x} e^{-\delta|x|^2} dx &= \int_{\mathbb{R}^d}\int_{\mathbb{R}^d} e^{-2\pi i(\xi-u)\cdot x} e^{-\delta|x|^2} e^{-\varepsilon|u|^2} f(u)\, dx du \\ &= \int_{\mathbb{R}^d}\Big(\int_{\mathbb{R}^d} e^{-2\pi i(\xi-u)\cdot x} e^{-\delta|x|^2} dx\Big) e^{-\varepsilon|u|^2} f(u)\, du = \Big(\frac{\pi}{\delta}\Big)^{\frac{d}{2}} \int_{\mathbb{R}^d} e^{-\pi^2|\xi-u|^2/\delta}\, e^{-\varepsilon|u|^2} f(u)\, du\,. \end{aligned} \tag{6.40}$$

Set $\xi = 0$ above to get

$$\int_{\mathbb{R}^d} m_\varepsilon(x)\, e^{-\delta|x|^2} dx = \Big(\frac{\pi}{\delta}\Big)^{\frac{d}{2}} \int_{\mathbb{R}^d} e^{-\pi^2|u|^2/\delta}\, g_\varepsilon(u)\, du\,, \tag{6.41}$$

with $g_\varepsilon(u) := e^{-\varepsilon|u|^2} f(u)$. Note that g is continuous with $|g_\varepsilon(u)| \le f(0)$, and for $n \ge 1$ we have

$$\Big|(\frac{\pi}{\delta})^{\frac{d}{2}} \int_{\mathbb{R}^d} e^{-\pi^2|u|^2/\delta} g_\varepsilon(u)\, du - g_\varepsilon(0)\Big| = \Big|(\frac{\pi}{\delta})^{\frac{d}{2}} \int_{\mathbb{R}^d} e^{-\pi^2|u|^2/\delta} [g_\varepsilon(u) - g_\varepsilon(0)]\, du\Big|$$

$$\le (\frac{\pi}{\delta})^{\frac{d}{2}} \int_{|u|\ge\frac{1}{n}} e^{-\pi^2|u|^2/\delta} |g_\varepsilon(u) - g_\varepsilon(0)|\, du + (\frac{\pi}{\delta})^{\frac{d}{2}} \int_{|u|<\frac{1}{n}} e^{-\pi^2|u|^2/\delta} |g_\varepsilon(u) - g_\varepsilon(0)|\, du$$

$$\le 2f(0)\,(\frac{\pi}{\delta})^{\frac{d}{2}} \int_{|u|\ge\frac{1}{n}} e^{-\pi^2|u|^2/\delta}\, du + \sup_{|u|\le\frac{1}{n}} \{|g_\varepsilon(u) - g_\varepsilon(0)|\}(\frac{\pi}{\delta})^{\frac{d}{2}} \int_{\mathbb{R}^d} e^{-\pi^2|u|^2/\delta}\, du$$

$$= 2f(0)\,\pi^{\frac{d}{2}} \int_{|y|\ge\frac{1}{n\sqrt{\delta}}} e^{-\pi^2|y|^2}\, dy + \sup_{|u|\le\frac{1}{n}} \{|g_\varepsilon(u) - g_\varepsilon(0)|\}\,.$$

Let now $\delta \downarrow 0$ to infer by dominated convergence the upper bound $\sup_{|u|\le\frac{1}{n}} \{|g_\varepsilon(u) - g_\varepsilon(0)|\}$, valid for any $n \ge 1$. Taking into account the continuity of g_ε, we see that this bound vanishes for $n \uparrow \infty$. Consequently, Fatou's lemma (see Chapter 2) in combination with (6.41) shows that

$$\int_{\mathbb{R}^d} m_\varepsilon(x)\, dx \le \liminf_{\delta\downarrow 0} \Big\{(\frac{\pi}{\delta})^{\frac{d}{2}} \int_{\mathbb{R}^d} e^{-\pi^2|u|^2/\delta} g_\varepsilon(u)\, du\Big\} = g_\varepsilon(0) = f(0)\,.$$

Thus $m_\varepsilon \in L^1(\mathbb{R}^d)$ with $\|m_\varepsilon\|_{L^1(\mathbb{R}^d)} \le f(0)$. We now compute the Fourier transform of m_ε with $\varepsilon > 0$ fixed, taking advantage of the identity (6.40):

$$\widehat{m_\varepsilon}(\xi) = \int_{\mathbb{R}^d} m_\varepsilon(x)\, e^{-2\pi i\xi\cdot x}\, dx = \lim_{\delta\downarrow 0} \int_{\mathbb{R}^d} m_\varepsilon(x)\, e^{-2\pi i\xi\cdot x}\, e^{-\delta|x|^2}\, dx$$

$$= \lim_{\delta\downarrow 0} \Big(\frac{\pi}{\delta}\Big)^{\frac{d}{2}} \int_{\mathbb{R}^d} e^{-\pi^2|\xi-u|^2/\delta}\, e^{-\varepsilon|u|^2} f(u)\, du = e^{-\varepsilon|\xi|^2} f(\xi)\,,$$

where we have used the continuity and boundedness of f in the last step, analogous to the considerations made after the relation (6.41). The bounded nonnegative sequence $\{\alpha_n\}_{n\ge1}$, where $\alpha_n = \|m_{1/n}\|_{L^1(\mathbb{R}^d)} \le f(0)$ for $n \ge 1$, admits a convergent subsequence $\{\alpha_{n_k}\}$. Note that unless $f \equiv 0$ (a trivial case), we have $\alpha_{n_k} > 0$, due to (6.41), and, moreover, the limit α of this sequence cannot be 0. Indeed, the left side of (6.41) is less than $\|m_\varepsilon\|_{L^1(\mathbb{R}^d)}$ and in the limit $\varepsilon \downarrow 0$, by dominated convergence (recall that f is bounded), the right side converges to $(\frac{\pi}{\delta})^{\frac{d}{2}} \int_{\mathbb{R}^d} e^{-\pi^2|\xi-u|^2/\delta} f(u)\, du$, which, using the considerations made above after relation (6.41), converges to $f(0)$ for $\delta \downarrow 0$; this would mean that $f(0) = 0$ and thus $f \equiv 0$. We conclude that $\mu_{n_k} = \frac{1}{\alpha_{n_k}} m_{1/n_k}(x)\, dx$ define probability measures on $\mathbb{R}^d$, with $\widehat{\mu_{n_k}}(\xi) = \frac{1}{\alpha_{n_k}} e^{-\frac{1}{n_k}|\xi|^2} f(\xi)$ converging to $\frac{1}{\alpha} f(\xi)$ at every $\xi \in \mathbb{R}^d$. Since $\widehat{\mu_{n_k}}(0) = 1$ for every n_k, we get $\alpha = f(0)$. By Theorem 6.17, $\frac{1}{\alpha} f(\xi)$ is the Fourier transform of a probability measure on $\mathbb{R}^d$, so that f is the Fourier transform of a Borel measure with total mass $f(0)$. □

6.4 The Fourier transform on $L^p(\mathbb{R})$-spaces

In Section 6.1 we shwed that the Fourier transform is a bounded linear operator from $L^1(\mathbb{R}^n)$ to $L^\infty(\mathbb{R}^n)$, and from $L^2(\mathbb{R}^n)$ onto $L^2(\mathbb{R}^n)$. These properties permit us to define the Fourier transform of a function in $L^p(\mathbb{R}^n)$ with $1 < p < 2$ since any such function f is of the form $f = f_1 + f_2$ with $f_1 = f\chi\{x \in \mathbb{R}^n : |f(x)| < 1\} \in L^1(\mathbb{R}^n)$ and $f_2 = f\chi\{x \in \mathbb{R}^n : |f(x)| \ge 1\} \in L^2(\mathbb{R}^n)$. A more precise result is the following.

Theorem 6.19 (Hausdorff–Young theorem) *If $f \in L^p(\mathbb{R}^n)$ with $1 < p < 2$, then $\hat{f} \in L^q(\mathbb{R}^n)$ where $\frac{1}{p} + \frac{1}{q} = 1$, and $\|\hat{f}\|_{L^q(\mathbb{R}^n)} \le \|f\|_{L^p(\mathbb{R}^n)}$.*

Proof We rely on the *Riesz–Thorin interpolation theorem*,[54] the proof of which can be found in Stein and Weiss (1971). Since, according to Theorem 6.2 and Theorem 6.4, we have $\|\hat{f}\|_{L^\infty(\mathbb{R}^n)} \le \|f\|_{L^1(\mathbb{R}^n)}$ and $\|\hat{f}\|_{L^2(\mathbb{R}^n)} \le \|f\|_{L^2(\mathbb{R}^n)}$ for all $f \in L^1(\mathbb{R}^n) \cap L^2(\mathbb{R}^n)$, we deduce that $\|\hat{f}\|_{L^q(\mathbb{R}^n)} \le \|f\|_{L^p(\mathbb{R}^n)}$ for all $f \in L^1(\mathbb{R}^n) \cap L^2(\mathbb{R}^n)$. This is a dense subset of $f \in L^p(\mathbb{R}^n)$, since by dominated convergence any $f \in L^p(\mathbb{R}^n)$ is the $L^p(\mathbb{R}^n)$-limit of the functions $f_m = f\chi\{x \in \mathbb{R}^n : |f(x)| \le m\} \in L^1(\mathbb{R}^n) \cap L^2(\mathbb{R}^n)$, so that the linear map $f \mapsto \hat{f}$ extends uniquely to a bounded linear map $T : L^p(\mathbb{R}^n) \to L^q(\mathbb{R}^n)$ with $\|T\| \le 1$. The fact that $f \mapsto \hat{f}$ is continuous from $L^p(\mathbb{R}^n)$ to $\mathcal{S}'(\mathbb{R}^n)$, see Section 6.1.1, permits us to identify T with the Fourier transform. □

It is of interest to note that the unit bound for the operator norm in Theorem 6.19 is not sharp: one can show that for $1 < p < 2$ we have, with $q = \frac{p}{p-1}$,

$$\|\hat{f}\|_{L^q(\mathbb{R}^n)} \le \Big(\frac{p^{1/p}}{q^{1/q}}\Big)^{\frac{n}{2}} \|f\|_{L^p(\mathbb{R}^n)} \quad \text{for all} \quad f \in L^p(\mathbb{R}^n). \tag{6.42}$$

Equality is achieved in (6.42) if and only if $f(x) = a\, e^{-(Mx,x)+\sum_{k=1}^n b_k x_k}$ with M an $n \times n$ real, symmetric, positive-definite matrix[55] and arbitary $a, b_1, \dots, b_n \in \mathbb{C}$; see Lieb (1990). This is in contrast with the limiting case $p = 2$: although in this case the sharp bound is also 1, this is attained by all $f \in L^2(\mathbb{R}^n)$, see Theorem 6.4.

Theorem 6.19 does not extend to the $L^p(\mathbb{R}^n)$-spaces with $p > 2$ due to insufficient decrease at infinity: there exists a bounded function $f \in C^\infty(\mathbb{R}^n) \cap L^p(\mathbb{R}^n)$ for all $p > 2$, with all its derivatives bounded, such that $\hat{f} \in \mathcal{S}'(\mathbb{R}^n)$ is not even a measure in any open subset of $\mathbb{R}^n$, see Hörmander (1995).

6.5 Sobolev spaces

We saw that the Fourier transform converts differentiation into multiplication on $\mathcal{S}'(\mathbb{R}^n)$, being at the same time an isometry on $L^2(\mathbb{R}^n)$. These features suggest the use of L^2-norms as a way to measure the regularity of tempered distributions. The Sobolev spaces $H^s(\mathbb{R}^n)$ are Fourier transforms of weighted $L^2(\mathbb{R}^n)$-spaces, having a Hilbert space structure and with the added benefit of

[54] Stating that if $p_1, p_2, q_1, q_2 \in [1, \infty]$, if $M_1, M_2 > 0$ and if T is a linear map from $L^{p_1}(\mathbb{R}^n) \cap L^{p_2}(\mathbb{R}^n)$ to $L^{q_1}(\mathbb{R}^n) \cap L^{q_2}(\mathbb{R}^n)$ such that $\|Tf\|_{L^{q_1}(\mathbb{R}^n)} \le M_1 \|f\|_{L^{p_1}(\mathbb{R}^n)}$ and $\|Tf\|_{L^{q_2}(\mathbb{R}^n)} \le M_2 \|f\|_{L^{p_2}(\mathbb{R}^n)}$ for all $f \in L^{p_1}(\mathbb{R}^n) \cap L^{p_2}(\mathbb{R}^n)$, and if $\frac{1}{p} = \frac{t}{p_1} + \frac{1-t}{p_2}$ and $\frac{1}{q} = \frac{t}{q_1} + \frac{1-t}{q_2}$ for some $t \in (0, 1)$, then $\|Tf\|_{L^q(\mathbb{R}^n)} \le M_1^t M_2^{1-t} \|f\|_{L^p(\mathbb{R}^n)}$ for all $f \in L^{p_1}(\mathbb{R}^n) \cap L^{p_2}(\mathbb{R}^n)$.

[55] That is, the real matrix M is equal to its transpose and $(Mx, x) > 0$ for $x \in \mathbb{R}^n \setminus \{0\}$.

an innate ability of interacting nicely with (distributional) derivatives: while the partial derivatives of an $L^2(\mathbb{R}^n)$-function could leave the space and even cease to be functions, we will see that the partial derivatives of a function in the Sobolev space $H^s(\mathbb{R}^n)$ end up in another Sobolev space, namely $H^{s-1}(\mathbb{R}^n)$.

Definition *For $s \in \mathbb{R}$, the Sobolev space $H^s(\mathbb{R}^n)$ is the vector space of all $u \in \mathcal{S}'(\mathbb{R}^n)$ such that $\xi \mapsto (1+|\xi|^2)^{s/2}\hat{u}(\xi) \in L^2(\mathbb{R}^n)$. For $u \in H^s(\mathbb{R}^n)$ we set $\|u\|_s = \int_{\mathbb{R}^n} |\hat{u}(\xi)|^2(1+|\xi|^2)^s\, d\xi$.*

Example Theorem 6.4 shows that $H^0(\mathbb{R}^n) = L^2(\mathbb{R}^n)$. □

Since for any $s \in \mathbb{R}$ the function $\xi \mapsto (1+|\xi|^2)^{s/2}$ belongs to $O(\mathbb{R}^n)$, the map

$$\Lambda_s u := \mathcal{F}^{-1}\Big((1+|\xi|^2)^{s/2}\hat{u}(\xi)\Big) \tag{6.43}$$

is a continuous linear operator on $\mathcal{S}'(\mathbb{R}^n)$, actually an isomorphism with $(\Lambda_s)^{-1} = \Lambda_{-s}$. Note that $H^s(\mathbb{R}^n) = \{u \in \mathcal{S}'(\mathbb{R}^n) : \Lambda_s u \in L^2(\mathbb{R}^n)\}$ with $\|u\|_s = \|\Lambda_s u\|_{L^2(\mathbb{R}^n)}$ a norm on $H^s(\mathbb{R}^n)$, induced by the inner product $(u,v)_s = \int_{\mathbb{R}^n} (\Lambda_s u)\,\overline{\Lambda_s v}\, d\xi$. This representation has several direct but very important consequences:

(i) $H^s(\mathbb{R}^n)$ is a Hilbert space for all $s \in \mathbb{R}$;
(ii) $\mathcal{S}(\mathbb{R}^n)$ is dense in $H^s(\mathbb{R}^n)$ for all $s \in \mathbb{R}$;
(iii) Λ_{t-s} is a unitary isomorphism from $H^t(\mathbb{R}^n)$ onto $H^s(\mathbb{R}^n)$ for all $s, t \in \mathbb{R}$;
(iv) ∂_x^α is a bounded linear map from $H^s(\mathbb{R}^n)$ to $H^{s-|\alpha|}(\mathbb{R}^n)$ for all $s \in \mathbb{R}$ and all n-tuples α of nonnegative integers.

Indeed, (i) holds since Λ_s is a unitary isomorphism (with inverse Λ_{-s}) from $H^s(\mathbb{R}^n)$ to $L^2(\mathbb{R}^n)$, while (ii) follows from (i) since $\mathcal{S}(\mathbb{R}^n)$ is dense in $L^2(\mathbb{R}^n)$ and is invariant under Λ_{-s}. Property (iii) is also clear since $\Lambda_t : H^t(\mathbb{R}^n) \to L^2(\mathbb{R}^n)$ and $\Lambda_{-s} : L^2(\mathbb{R}^n) \to H^s(\mathbb{R}^n)$ are both unitary isomorphisms and for $u \in \mathcal{S}(\mathbb{R}^n)$ it is easy to check that $\Lambda_t\Lambda_{-s}u = \Lambda_{t-s}u$. Thus the density of $\mathcal{S}(\mathbb{R}^n)$, see (ii), proves (iii). For (iv), note that for $u \in H^s(\mathbb{R}^n) \subset \mathcal{S}'(\mathbb{R}^n)$, due to (6.13), we have $\Lambda_{s-|\alpha|}(\partial_x^\alpha u) = (2\pi i)^{|\alpha|}\mathcal{F}^{-1}((1+|\xi|^2)^{(s-|\alpha|)/2}\xi^\alpha\hat{u}(\xi)) \in L^2(\mathbb{R}^n)$ since $|\xi^\alpha| \le (1+|\xi|^2)^{|\alpha|/2}$.

Note that although the Fourier transform is a unitary isomorphism of $L^2(\mathbb{R}^n)$, it fails to map $H^s(\mathbb{R}^n)$ into $H^s(\mathbb{R}^n)$ for every $s \in \mathbb{R}$; see Exercise 34.

By the definition and by (6.13), $u \in H^{s+1}(\mathbb{R}^n)$ if and only if $u \in H^s(\mathbb{R}^n)$ and $\partial_{x_k}u \in H^s(\mathbb{R}^n)$ for $k = 1, \ldots, n$, with

$$\|u\|_{s+1}^2 = \|u\|_s^2 + \frac{1}{4\pi^2}\sum_{k=1}^n \|\partial_{x_k}u\|_s^2 . \tag{6.44}$$

Iterating this observation shows that for any integer $m \ge 1$, the space $H^m(\mathbb{R}^n)$ consists of all $u \in L^2(\mathbb{R}^n)$ whose distributional derivatives $\partial_x^\alpha u \in L^2(\mathbb{R}^n)$ for any n-tuple α of nonnegative integers with $|\alpha| \le m$; distributional derivatives that can be identified with a locally integrable functions are called *weak derivatives*.

We can also work out the characterisation for negative integers: $u \in H^{-m}(\mathbb{R}^n)$, with $m \geq 1$ an integer, if and only if it has the representation

$$u = v_0 + \sum_{0<|\alpha|\leq m} \partial_x^\alpha v_\alpha$$

where $v_0,\ v_\alpha \in L^2(\mathbb{R}^n)$ and

$$c_m \sum_{|\alpha|\leq m} \|v_\alpha\|_0^2 \leq \|u\|_{-m}^2 \leq C_m \sum_{|\alpha|\leq m} \|v_\alpha\|_0^2 \tag{6.45}$$

for some universal constants $c_m,\ C_m > 0$; roughly speaking, $H^{-m}(\mathbb{R}^n)$ consists of the tempered distributions which are sums of derivatives of order at most m of functions in $L^2(\mathbb{R}^n)$.

Indeed, take $v_\alpha \in L^2(\mathbb{R}^n)$ such that $\widehat{v_\alpha}(\xi) = (\frac{\xi}{2\pi i})^\alpha \frac{\hat{u}(\xi)}{\sum_{|\alpha|\leq m}\xi^{2\alpha}}$, so that the representation follows by applying the inverse Fourier transform in $S'(\mathbb{R}^n)$ to the relation $\hat{u}(\xi) = \sum_{|\alpha|\leq m}(2\pi i\xi)^\alpha\, \widehat{v_\alpha}(\xi)$. On the other hand, since $\sum_{|\alpha|\leq m}\|v_\alpha\|_0^2 = \sum_{|\alpha|\leq m}(\frac{1}{2\pi})^{2|\alpha|}\int_{\mathbb{R}^n}\frac{\xi^{2\alpha}|\hat{u}(\xi)|^2}{\left(\sum_{|\alpha|\leq m}\xi^{2\alpha}\right)^2}\,d\xi$, the estimate (6.45) follows if we set $c_m = \inf_{\xi\in\mathbb{R}^n}\left\{\frac{\sum_{|\alpha|\leq m}\xi^{2\alpha}}{(1+|\xi|^2)^m}\right\}$ and $c_M = (2\pi)^{2m}\sup_{\xi\in\mathbb{R}^n}\left\{\frac{\sum_{|\alpha|\leq m}\xi^{2\alpha}}{(1+|\xi|^2)^m}\right\}$.

When $s \in \mathbb{R}$ is not an integer, membership to $H^s(\mathbb{R}^n)$ amounts to some kind of Hölder condition in an $L^2(\mathbb{R}^n)$-sense: since the integer scale is well-described, it suffices to deal with the case $0 < s < 1$, in which case $u \in L^2(\mathbb{R}^n)$ belongs to $H^s(\mathbb{R}^n)$ if and only if

$$\int_{\mathbb{R}^n}\int_{\mathbb{R}^n}\frac{|u(x+y)-u(x)|^2}{|y|^{n+2s}}\,dydx < \infty\,. \tag{6.46}$$

Indeed, for $0 < s < 1$ the inequality[56]

$$(1+|\xi|^2)^s \leq 1+|\xi|^{2s} \leq 2(1+|\xi|^2)^s\,, \qquad \xi\in\mathbb{R}^n\,, \tag{6.47}$$

yields

$$\|u\|_s^2 \leq \int_{\mathbb{R}^n}|\hat{u}(\xi)|^2\,(1+|\xi|^{2s})\,ds \leq 2\|u\|_s^2\,. \tag{6.48}$$

On the other hand, note that the Fourier transform of the $L^2(\mathbb{R}^n)$-function $x \mapsto u(x+y) - u(x)$ is $\xi \mapsto (e^{2\pi i\xi\cdot y} - 1)\hat{u}(\xi)$ and[57]

$$\int_{\mathbb{R}^n}|e^{2\pi i\xi\cdot y} - 1|^2|y|^{-n-2s}\,dy = I_s|\xi|^{2s}\,, \qquad \xi\in\mathbb{R}^n\setminus\{0\}\,, \tag{6.49}$$

[56] The upper bound of which holds since $1 + |\xi|^{2s} \leq 2$ if $|\xi| \leq 1$, while for $|\xi| > 1$ we have $1 + |\xi|^{2s} \leq 2|\xi|^{2s} \leq 2(1 + |\xi|^2)^s$. The lower bound follows by adding the inequalities $1 \geq (1 + |\xi|^2)^{s-1}$ and $|\xi|^{2s} \geq |\xi|^2(1 + |\xi|^2)^{s-1}$.

[57] Indeed, for $\xi \neq 0$, there is an orthogonal transformation R of $\mathbb{R}^n$ mapping ξ into $(|\xi|, 0, \ldots, 0) \in \mathbb{R}^n$. Since $\det(R) = 1$ and $R^* = R^{-1}$ ensures $R\xi\cdot w = \xi\cdot R^{-1}w$ for all $w \in \mathbb{R}^n$, the change of variables $w = Ry$ shows that the integral on the left is a function of $|\xi|$ only, while replacing ξ by ξ/t and changing variables $y \to yt$ for $t > 0$ shows that it is homogeneous in $|\xi|$, of degree $2s$.

for some constant $I_s > 0$. Using (6.49) and Theorem 6.4, for $u \in L^2(\mathbb{R}^n)$ we get

$$\begin{aligned}
&\int_{\mathbb{R}^n}\int_{\mathbb{R}^n} \Big|\big(\mathcal{F}^{-1}[(e^{2\pi i\xi\cdot y}-1)\hat{u}(\xi)]\big)(x)\Big|^2 |y|^{-n-2s}\,dydx\\
&= \int_{\mathbb{R}^n}\Big(\int_{\mathbb{R}^n} \Big|\big(\mathcal{F}^{-1}[(e^{2\pi i\xi\cdot y}-1)\hat{u}(\xi)]\big)(x)\Big|^2 dx\Big)|y|^{-n-2s}\,dy\\
&= \int_{\mathbb{R}^n}\Big(\int_{\mathbb{R}^n} \big|(e^{2\pi i\xi\cdot y}-1)\hat{u}(\xi)\big|^2 d\xi\Big)|y|^{-n-2s}\,dy\\
&= \int_{\mathbb{R}^n}\Big(\int_{\mathbb{R}^n} |(e^{2\pi i\xi\cdot y}-1)|^2\,|y|^{-n-2s}\,dy\Big)|\hat{u}(\xi)|^2\,d\xi = I_s\int_{\mathbb{R}^n}|\xi|^{2s}\,|\hat{u}(\xi)|^2\,d\xi\,,
\end{aligned}$$

so that

$$\int_{\mathbb{R}^n}(1+|\xi|^{2s})\,|\hat{u}(\xi)|^2\,d\xi = \int_{\mathbb{R}^n}|\hat{u}(\xi)|^2\,d\xi + \frac{1}{I_s}\int_{\mathbb{R}^n}\int_{\mathbb{R}^n}|u(x+y)-u(x)|^2|y|^{-n-2s}\,dydx\,.$$

Now (6.48) yields (6.46).

For simplicity, let us denote $\mathcal{F}^{-1}(u) = \check{u}$ for $u \in \mathcal{S}'(\mathbb{R}^n)$. In our further considerations we will rely on some operational rules for manipulating Sobolev functions and Fourier transforms thereof:

$$\overline{(\hat{u})} = \mathcal{F}^{-1}(\overline{u})\,,\qquad \overline{(\check{u})} = \widehat{(\overline{u})}\quad\text{and}\quad \widehat{(u^\dagger)} = (\hat{u})^\dagger = \check{u}\,,\qquad u \in \mathcal{S}'(\mathbb{R}^n)\,. \tag{6.50}$$

Here $\overline{v}$, $u^\dagger \in \mathcal{S}'(\mathbb{R}^n)$ are defined for v, $u \in \mathcal{S}'(\mathbb{R}^n)$ by setting $\langle\overline{v},\varphi\rangle = \overline{\langle v,\overline{\varphi}\rangle}$ and $\langle u^\dagger,\varphi\rangle = \langle u,\varphi^\dagger\rangle$, respectively, for $\varphi \in \mathcal{S}(\mathbb{R}^n)$. Moreover, if $u \in H^s(\mathbb{R}^n)$, then $u^\dagger$, $\overline{u} \in H^s(\mathbb{R}^n)$ and

$$\|\overline{u}\|_s = \|u^\dagger\|_s = \|u\|_s\,. \tag{6.51}$$

The validity of (6.50) follows by duality, using (6.11), Theorem 6.1, the change of variables $x \mapsto (-x)$, and the fact that the explicit defining formula shows that the same relations hold in $\mathcal{S}(\mathbb{R}^n)$. As for (6.51), if $u \in H^s(\mathbb{R}^n)$, then $(1+|\xi|^2)^{\frac{s}{2}}\,\hat{u}(\xi) \in L^2(\mathbb{R}^n)$ yields $(1+|\xi|^2)^{\frac{s}{2}}\,\hat{u}(-\xi) \in L^2(\mathbb{R}^n)$ by the change of variables $\xi \mapsto -\xi$, so that, in view of (6.50), $(1+|\xi|^2)^{\frac{s}{2}}\,\widehat{(u^\dagger)}(\xi) = (1+|\xi|^2)^{\frac{s}{2}}\,\overline{\mathcal{F}(\overline{u})(\xi)} \in L^2(\mathbb{R}^n)$ which shows that $u^\dagger$, $\overline{u} \in H^s(\mathbb{R}^n)$ and (6.51) holds.

The Sobolev spaces $H^{-s}(\mathbb{R}^n)$ with negative exponents can be viewed as dual spaces of the Sobolev spaces $H^s(\mathbb{R}^n)$ with positive exponents, and *vice-versa*.

Theorem 6.20 *For any $s \in \mathbb{R}$, the pairing between $\mathcal{S}'(\mathbb{R}^n)$ and $\mathcal{S}(\mathbb{R}^n)$ induces a unitary isomorphism from $H^{-s}(\mathbb{R}^n)$ to the dual $(H^s(\mathbb{R}^n))^*$ of the Hilbert space $H^s(\mathbb{R}^n)$: if $u \in H^{-s}(\mathbb{R}^n)$, then the functional $\varphi \mapsto \langle\overline{u},\varphi\rangle$ on $\mathcal{S}(\mathbb{R}^n)$ extends continuously to $H^s(\mathbb{R}^n)$, with operator norm $\|u\|_{-s}$, and every element of $(H^s(\mathbb{R}^n))^*$ arises in this fashion.*

Proof If $u \in H^{-s}(\mathbb{R}^n)$ and $\varphi \in \mathcal{S}(\mathbb{R}^n)$, then

$$\begin{aligned}
\langle\overline{u},\varphi\rangle &= \langle\mathcal{F}^{-1}(\overline{u}),\hat{\varphi}\rangle = \langle(1+|\xi|^2)^{-\frac{s}{2}}(\mathcal{F}^{-1}[\overline{u}])(\xi),\,(1+|\xi|^2)^{\frac{s}{2}}\hat{\varphi}(\xi)\rangle\\
&= \int_{\mathbb{R}^n}\Big((1+|\xi|^2)^{-\frac{s}{2}}(\mathcal{F}^{-1}[\overline{u}])(\xi)\Big)\Big((1+|\xi|^2)^{\frac{s}{2}}\hat{\varphi}(\xi)\Big)\,d\xi
\end{aligned}$$

since $\xi \mapsto (1+|\xi|^2)^{-\frac{s}{2}}(\mathcal{F}^{-1}[\overline{u}])(\xi)$ and $\xi \mapsto (1+|\xi|^2)^{\frac{s}{2}}\hat{\varphi}(\xi)$ are both $L^2(\mathbb{R}^n)$-functions. Since $\overline{\mathcal{F}^{-1}[\overline{u}]} = \hat{u}$ by (6.50), we get

$$\langle \overline{u}, \varphi\rangle = \int_{\mathbb{R}^n} \overline{(1+|\xi|^2)^{-\frac{s}{2}}\hat{u}(\xi)}\,(1+|\xi|^2)^{\frac{s}{2}}\hat{\varphi}(\xi)\,d\xi$$

so that $|\langle \overline{u}, \varphi\rangle| \le \|u\|_{-s}\|\varphi\|_{-s}$. Thus the functional $\varphi \mapsto \langle \overline{u}, \varphi\rangle$ extends to a continuous functional on $H^s(\mathbb{R}^n)$, with norm at most $\|u\|_{-s}$. The norm actually equals $\|u\|_{-s}$ since the outcome of the functional acting on $f = \Lambda_{-2s}\hat{u} \in H^s(\mathbb{R}^n)$ is $\int_{\mathbb{R}^n} \overline{(1+|\xi|^2)^{-s}|\hat{u}(\xi)|^2\,d\xi} = \|u\|_{-s}^2 = \|u\|_{-s}\|f\|_s$.

Finally, if $U \in (H^s(\mathbb{R}^n))^*$, then $U \circ \Lambda_{-s}$ is a bounded linear functional on $L^2(\mathbb{R}^n)$, so, by Theorem 3.13, there exists $h \in L^2(\mathbb{R}^n)$ such that $U(\Lambda_{-s}g) = \int_{\mathbb{R}^n} g(\xi)\,\overline{h(\xi)}\,d\xi$ for all $g \in L^2(\mathbb{R}^n)$. Using Theorem 6.4 and (6.50), for $f \in S(\mathbb{R}^n)$ and with $w = \Lambda_s(h^\dagger) \in H^{-s}(\mathbb{R}^n)$, we get

$$\begin{aligned} U(f) &= \int_{\mathbb{R}^n} (\Lambda_s f)(\xi)\,\overline{h(\xi)}d\xi = \int_{\mathbb{R}^n} \big(\mathcal{F}[\Lambda_s f]\big)(\xi)\,\overline{\hat{h}(\xi)}d\xi = \int_{\mathbb{R}^n} \big((1+|\xi|^2)^{s/2}\hat{f}(\xi)\big)\overline{(\mathcal{F}^{-1}[h^\dagger])(\xi)}d\xi \\ &= \int_{\mathbb{R}^n} \big((1+|\xi|^2)^{s/2}\hat{f}(\xi)\big)\overline{(\mathcal{F}[\Lambda_{-s}w])(\xi)}\,d\xi = \int_{\mathbb{R}^n} \big((1+|\xi|^2)^{s/2}\hat{f}(\xi)\big)\overline{(1+|\xi|^2)^{-s/2}\hat{w}(\xi)}\,d\xi \\ &= \langle\overline{(1+|\xi|^2)^{-s/2}\hat{w}(\xi)},\,(1+|\xi|^2)^{s/2}\hat{f}(\xi)\rangle = \langle\overline{\hat{w}},\,\hat{f}\rangle = \langle\mathcal{F}^{-1}[\overline{w}],\,\hat{f}\rangle = \langle\overline{w},\,f\rangle \end{aligned}$$

in view of (6.11). □

In view of the characterisation of the elements of $H^s(\mathbb{R}^n)$ for integers $s \ge 1$ as functions with weak derivatives (up to order s) in $L^2(\mathbb{R}^n)$, it is natural to wonder how this relates to ordinary differentiability. This question has an elegant answer.

Theorem 6.21 (Sobolev embedding theorem) *If $s > k + n/2$ for some integer $k \ge 0$, then all functions $f \in H^s(\mathbb{R}^n)$ are continuously differentiable of order k, and for $|\alpha| \le k$ their partial derivatives $\partial_x^\alpha f$ vanish at infinity.*

Proof If $f \in H^s(\mathbb{R}^n)$ with $s > k + n/2$ and $|\alpha| \le k$, then $\partial_x^\alpha f \in L^2(\mathbb{R}^n)$ and (6.13) yields

$$\begin{aligned} (2\pi)^{-|\alpha|}\int_{\mathbb{R}^n} \big|(\mathcal{F}[\partial_x^\alpha f])(\xi)\big|\,d\xi &= \int_{\mathbb{R}^n} \big|\xi^\alpha \hat{f}(\xi)\big|\,d\xi \le \int_{\mathbb{R}^n} (1+|\xi|^2)^{k/2}|\hat{f}(\xi)|\,d\xi \\ &\le \Big(\int_{\mathbb{R}^n} (1+|\xi|^2)^s|\hat{f}(\xi)|^2\,d\xi\Big)^{1/2}\Big(\int_{\mathbb{R}^n} (1+|\xi|^2)^{k-s}\,d\xi\Big)^{1/2} = \|f\|_s\Big(\int_{\mathbb{R}^n} (1+|\xi|^2)^{k-s}\,d\xi\Big)^{1/2}. \end{aligned}$$

Passing to spherical coordinates shows that the second factor above is finite precisely when $2(k-s)+n-1 < -1$, that is, when $s > k + n/2$. Thus $\mathcal{F}[\partial_x^\alpha f] \in L^1(\mathbb{R}^n)$ and the conclusion follows by Theorem 6.2 and Theorem 6.3. □

The spaces $H^s(\mathbb{R}^n)$ are $S(\mathbb{R}^n)$-modules, that is, $\varphi \in S(\mathbb{R}^n)$ and $f \in H^s(\mathbb{R}^n)$ imply $\varphi f \in H^s(\mathbb{R}^n)$. This is a consequence of the following much more precise result.

Theorem 6.22 *Suppose $\varphi \in L^2(\mathbb{R}^n)$ satisfies $\int_{\mathbb{R}^n}(1+|\xi|^2)^p|\hat{\varphi}(\xi)|^2 d\xi = C < \infty$ for some $p > 0$; in particular, φ is a.e. equal to a continuous functions that vanishes at infinity. Then the multiplication operator $M_\varphi : f \mapsto \varphi f$ is bounded on $H^s(\mathbb{R}^n)$ for any $s \in \mathbb{R}$ with $|s| \le p$.*

Proof Let us first note that the continuity and vanishing of φ at infinity follow from Theorem 6.2 and Theorem 6.3 since $\mathcal{F}((\hat{\varphi})^\dagger) = \varphi$, and $\hat{\varphi} \in L^1(\mathbb{R}^n)$ ensures $(\hat{\varphi})^\dagger \in L^1(\mathbb{R}^n)$ by the change of variables $\xi \mapsto -\xi$. To check the validity of the relation $\mathcal{F}((\hat{\varphi})^\dagger) = \varphi$, note that since $\hat{\varphi} \in L^2(\mathbb{R}^n)$, we may choose $\psi_n \in \mathcal{S}(\mathbb{R}^n)$ converging to $\hat{\varphi}$ in $L^2(\mathbb{R}^n)$. Then $\psi_n^\dagger \to (\hat{\varphi})^\dagger$ in $L^2(\mathbb{R}^n)$ and Theorem 6.4 yields $\mathcal{F}(\psi_n^\dagger) \to \mathcal{F}((\hat{\varphi})^\dagger)$ in $L^2(\mathbb{R}^n)$. But, by Theorem 6.1, $\mathcal{F}(\psi_n^\dagger) = \mathcal{F}^{-1}(\psi_n) \to \varphi$ in $L^2(\mathbb{R}^n)$, which proves the desired relation.

Fix s with $|s| \leq p$. Since $\Lambda_s : H^s(\mathbb{R}^n) \to L^2(\mathbb{R}^n)$ and $\Lambda_{-s} : L^2(\mathbb{R}^n) \to H^s(\mathbb{R}^n)$ are unitary, the claim is equivalent to $\Lambda_s M_\varphi \Lambda_{-s}$ being a bounded operator on $L^2(\mathbb{R}^n)$. For $f \in L^2(\mathbb{R}^n)$ we have

$$\begin{aligned}\big(\mathcal{F}[\Lambda_s M_\varphi \Lambda_{-s} f]\big)(\xi) &= (1+|\xi|^2)^{s/2}\big(\mathcal{F}[\varphi(\Lambda_{-s} f)]\big)(\xi) = (1+|\xi|^2)^{s/2}\big(\hat{\varphi} * (\widehat{\Lambda_{-s} f})\big)(\xi)\\ &= (1+|\xi|^2)^{s/2}\int_{\mathbb{R}^n} \hat{\varphi}(\xi-\eta)\,(1+|\eta|^2)^{-s/2}\,\hat{f}(\eta)\,d\eta = \int_{\mathbb{R}^n} K(\xi,\eta)\,\hat{f}(\eta)\,d\eta\end{aligned}$$

where $K(\xi,\eta) = (1+|\xi|^2)^{s/2}(1+|\eta|^2)^{-s/2}\hat{\varphi}(\xi-\eta)$. The inequality[58]

$$(1+|\xi|^2)^{s/2}(1+|\eta|^2)^{-s/2} \leq 2^{|s|/2}(1+|\xi-\eta|^2)^{|s|/2}, \qquad \xi,\eta \in \mathbb{R}^n, \tag{6.52}$$

yields $|K(\xi,\eta)| \leq 2^{|s|/2}(1+|\xi-\eta|^2)^{|s|/2}|\hat{\varphi}(\xi-\eta)|$, so that

$$\begin{aligned}&\int_{\mathbb{R}^n}\Big|\int_{\mathbb{R}^n} K(\xi,\eta)\,\hat{f}(\eta)\,d\eta\Big|^2 d\xi \leq 2^{|s|}\int_{\mathbb{R}^n}\Big(\int_{\mathbb{R}^n}(1+|\xi-\eta|^2)^{|s|/2}|\hat{\varphi}(\xi-\eta)|\,|\hat{f}(\eta)|\,d\eta\Big)^2 d\xi\\ &\leq 2^{|s|}\int_{\mathbb{R}^n}\Big(\int_{\mathbb{R}^n}(1+|\xi-\eta|^2)^{|s|/2}|\hat{\varphi}(\xi-\eta)|d\eta\Big)\Big(\int_{\mathbb{R}^n}(1+|\xi-\eta|^2)^{|s|/2}|\hat{\varphi}(\xi-\eta)|\,|\hat{f}(\eta)|^2\,d\eta\Big)d\xi\\ &\leq 2^{|s|}C\int_{\mathbb{R}^n}\Big(\int_{\mathbb{R}^n}(1+|\xi-\eta|^2)^{|s|/2}|\hat{\varphi}(\xi-\eta)|\,|\hat{f}(\eta)|^2\,d\eta\Big)d\xi\\ &\leq 2^{|s|}C\int_{\mathbb{R}^n}\Big(\int_{\mathbb{R}^n}(1+|\zeta|^2)^{|s|/2}|\hat{\varphi}(\zeta)|\,|\hat{f}(\eta)|^2\,d\eta\Big)d\zeta \leq 2^{|s|}C^2\int_{\mathbb{R}^n}|\hat{f}(\eta)|^2\,d\eta = 2^{|s|}C^2\|f\|_{L^2(\mathbb{R}^n)}\end{aligned}$$

using the Cauchy–Schwarz inequality in the second step and the change of variables $\zeta = \xi - \eta$ to get to the last line. The claimed boundedness is thus validated. □

Sobolev spaces possess the remarkable property that localisation and boundedness enable the trade of higher regularity for compactness – that is, sequences bounded in a high exponent space, and confined to a compact domain, have convergent subsequences in lower exponent spaces.

Theorem 6.23 (Rellich's compactness theorem) *Let $\{f_k\}_{k\geq 1}$ be a sequence in $H^s(\mathbb{R}^n)$ such that $\sup_{k\geq 1}\|f_k\|_s < \infty$ and there exists some $M > 0$ with $f_k(x) = 0$ if $|x| \geq M$, for all $k \geq 1$. Then there exists a subsequence $\{f_{k_j}\}$ which converges in $H^t(\mathbb{R}^n)$ for all $t < s$.*

Proof By Theorem 6.9, $\widehat{f_k} \in \mathcal{O}(\mathbb{R}^n)$ for $k \geq 1$. Pick some $\varphi \in \mathcal{D}(\mathbb{R}^n)$ such that $\varphi(x) = 1$ for $|x| \leq M$. Then $f_k = \varphi f_k$ and the discussion after (6.16), in combination with (6.50), yields

$$\hat{\varphi} * \widehat{f_k} = \mathcal{F}^{-1}(\varphi^\dagger f_k^\dagger) = \mathcal{F}^{-1}(f_k^\dagger) = \widehat{f_k}\,. \tag{6.53}$$

We now take advantage of the fact that

$$(u * \theta)(x) = \langle u, \tau_x(\theta^\dagger)\rangle \in \mathcal{O}(\mathbb{R}^n) \quad \text{if} \quad u \in \mathcal{S}'(\mathbb{R}^n),\ \theta \in \mathcal{S}(\mathbb{R}^n). \tag{6.54}$$

[58] To verify it, note that $|\xi| \leq |\xi-\eta| + |\eta|$ yields $|\xi|^2 \leq 2(|\xi-\eta|^2 + |\eta|^2)$, so that $1+|\xi|^2 \leq 2(1+|\xi-\eta|^2)(1+|\eta|^2)$. If $s \geq 0$, it suffices to raise both sides to the power $s/2$, while if $s < 0$, we interchange ξ and η and replace s by $-s$ to obtain (6.52) from the first case.

To validate (6.54), use Theorem 6.1, (6.11) and (6.50) to infer that for $\psi \in \mathcal{S}(\mathbb{R}^n)$ we have

$$\begin{aligned}\langle u * \theta, \psi\rangle &= \langle \mathcal{F}^{-1}(\hat{u}\hat{\theta}), \psi\rangle = \langle \hat{u}\hat{\theta}, \check{\psi}\rangle = \langle \hat{u}, \hat{\theta}\check{\psi}\rangle \\ &= \langle u, \mathcal{F}(\hat{\theta}\check{\psi})\rangle = \langle u, \theta^\dagger * \psi\rangle = \Big\langle u, \int_{\mathbb{R}^n} \theta(y-x)\psi(y)\, dy\Big\rangle .\end{aligned}$$

The integral can be approximated by finite Riemann sums $\sum_{j=1}^N \theta(y_j - x)\psi(y_j)\, \Delta y_j$ in the topology of $\mathcal{S}(\mathbb{R}^n)$, so that

$$\begin{aligned}\langle u * \theta, \psi\rangle &= \lim_{N\to\infty} \sum_{j=1}^N \langle u, \theta(y_j - x)\rangle\, \psi(y_j)\, \Delta y_j = \lim_{N\to\infty} \sum_{j=1}^N \langle u, \tau_{y_j}\theta^\dagger\rangle\, \psi(y_j)\, \Delta y_j \\ &= \int_{\mathbb{R}^n} \langle u, \tau_y(\theta^\dagger)\rangle\, \psi(y)\, dy\end{aligned}$$

since $\langle u, \tau_y(\theta^\dagger)\rangle\, \psi(y) \in \mathcal{S}(\mathbb{R}^n)$ as $\langle u, \tau_x(\theta^\dagger)\rangle \in O(\mathbb{R}^n)$ and $\psi \in \mathcal{S}(\mathbb{R}^n)$. This confirms (6.54).

Setting $u = \widehat{f_k}$ and $\theta = \hat{\varphi}$ in (6.54) enables us to identify the expressions in (6.53) pointwise with an absolutely convergent integral, so that

$$\begin{aligned}\widehat{f_k}(\xi) &= \big\langle \widehat{f_k}, \tau_\xi[(\hat{\varphi})^\dagger]\big\rangle = \big\langle (1+|\eta|^2)^{s/2}\widehat{f_k}(\eta), (1+|\eta|^2)^{-s/2}\hat{\varphi}(\xi-\eta)\big\rangle \\ &= \int_{\mathbb{R}^n} \big((1+|\eta|^2)^{s/2}\widehat{f_k}(\eta)\big)(1+|\eta|^2)^{-s/2}\hat{\varphi}(\xi-\eta)\, d\eta\end{aligned}$$

since $\eta \mapsto (1+|\eta|^2)^{s/2}\widehat{f_k}(\eta)$ belongs to $L^2(\mathbb{R}^n)$. Using (6.52), we get

$$\begin{aligned}(1+|\xi|^2)^{s/2}|\widehat{f_k}(\xi)| &\le \int_{\mathbb{R}^n} (1+|\eta|^2)^{s/2}|\widehat{f_k}(\eta)|\, (1+|\xi|^2)^{s/2}(1+|\eta|^2)^{-s/2}|\hat{\varphi}(\xi-\eta)|\, d\eta \\ &\le 2^{|s|/2} \int_{\mathbb{R}^n} (1+|\eta|^2)^{s/2}|\widehat{f_k}(\eta)|\, (1+|\xi-\eta|^2)^{|s|/2}|\hat{\varphi}(\xi-\eta)|\, d\eta \le C\|f_k\|_s\end{aligned}$$

for some constant $C > 0$, independent of $\xi \in \mathbb{R}^n$ and of $k \ge 1$, since $\hat{\varphi} \in \mathcal{S}(\mathbb{R}^n)$. Likewise, since $\partial_\xi^\alpha(\hat{\varphi} * \widehat{f_k}) = (\partial_\xi^\alpha\hat{\varphi}) * \widehat{f_k}$, we see that $(1+|\xi|^2)^{s/2}|\partial_\xi^\alpha \widehat{f_k}(\xi)|$ is bounded by a constant that is independent of $\xi \in \mathbb{R}^n$, of the n-tuple of nonnegative integers α, and of the integer $k \ge 1$. In particular, the smooth functions f_k and their first partial derivatives are uniformly bounded on the compact sets $K_m = \{\xi \in \mathbb{R}^n : |\xi| \le m\}$ for all integers $m \ge 1$. The mean-value theorem and the Arzelà–Ascoli theorem (see Chapter 3), combined with a Cantor diagonalisation process, show the existence of a subsequence $\{f_{k_j}\}$ which converges uniformly on all sets K_m with $m \ge 1$.

We claim that $\{f_{k_j}\}$ is a Cauchy sequence in every $H^t(\mathbb{R}^n)$-space, with $t < s$. Indeed, given $R > 0$, we can write $\|f_{k_j} - f_{k_i}\|_t^2 = \int_{\mathbb{R}^n} (1+|\xi|^2)^t|\widehat{f_{k_j}}(\xi) - \widehat{f_{k_i}}(\xi)|^2 d\xi$ as the sum of the integrals over the regions $|\xi| \le R$ and $|\xi| > R$. Since $(1+|\xi|^2)^{t/2} \le (1+R^2)^{|t|}$ if $|\xi| \le R$ while $(1+|\xi|^2)^t \le (1+R^2)^{t-s}(1+|\xi|^2)^s$ if $|\xi| > R$, we get

$$\|f_{k_j} - f_{k_i}\|_t^2 \le (1+R^2)^{|t|}\Big(\int_{|\xi|\le R} 1\, d\xi\Big) \sup_{|\xi|\le R}\{|\widehat{f_{k_j}}(\xi) - \widehat{f_{k_i}}(\xi)|^2\} + (1+R^2)^{t-s}\|f_{k_j} - f_{k_i}\|_s^2 .$$

Given $\varepsilon > 0$, the fact that $t < s$ enables us to choose $R > 0$ sufficiently large to make the second terms less that $\varepsilon/2$; and having fixed such an R, we can make the first term less that $\varepsilon/2$ by choosing k_i and k_j sufficiently large. Since $H^t(\mathbb{R}^n)$ is a Hilbert space, the proof is complete. □

6.6 Periodic Sobolev spaces

In analogy to the $H^s(\mathbb{R}^n)$-spaces one can introduce the *periodic Sobolev spaces*

$$H^s(\mathbb{T}^n) = \{u = \sum_{g\in\mathbb{Z}^n} c_g\, e^{2\pi i g\cdot x} \in \mathcal{S}'(\mathbb{R}^n) : \sum_{g\in\mathbb{Z}^n} (1+|g|^2)^s |c_g|^2 < \infty\} \qquad (6.55)$$

for $s \in \mathbb{R}$. Theorem 6.13 shows that any periodic distribution belongs to some $H^s(\mathbb{T}^n)$-space. This is in contrast to the case of the $H^s(\mathbb{R}^n)$-spaces, where the constant function $1 \in \mathcal{S}'(\mathbb{R}^n)$ fails to belongs to any Sobolev space. Analogs of the results in this section can be proved for $H^s(\mathbb{T}^n)$ by adapting the distributional approach to the periodic setting, as outlined in Folland (1999). Alternatively, one can dispense with this heavy technical machinery and rely on simpler considerations that use the association with sequences of complex numbers, see Iorio and Iorio (2001); see e.g. Exercise 37.

6.7 Exercises

The exercises range from illustrations and extensions of the ideas and techniques presented in the text (Exercises 1–7, 10, 11, 20, 22, 27–30, 32, 34–36), to problems that go beyond the acquisition of skills, representing theoretical results (Exercises 8, 9, 13–19, 21, 23–26, 37). We have also selected some challenging and quite involved topics (Exercises 12, 31, 33).

6.7.1 Statements

1. Show that $\mathcal{D}(\mathbb{R}^n)$ is dense in $\mathcal{S}(\mathbb{R}^n)$.

2. As an indication of how rich the class $\mathcal{D}(\mathbb{R})$ is, show that for any given real sequence $\{a_k\}_{k\geq 0}$ there is some $\varphi \in \mathcal{D}(\mathbb{R})$ with $\varphi^{(k)}(0) = a_k$ for $k \geq 0$.

3. Show that if $u : \mathbb{R} \to \mathbb{R}$ is a tempered L^1-function that is pointwise differentiable on $\mathbb{R} \setminus \{a\}$, with the pointwise derivative $v = u'$ such that $|v(x)| \leq M$ a.e. on $\mathbb{R} \setminus [a-\varepsilon, a+\varepsilon]$ and[59] $v \in L^1[a-\varepsilon, a+\varepsilon]$ for some $M, \varepsilon > 0$, then the limits $u^+(a) = \lim_{x\downarrow a} u(x)$, $u^-(a) = \lim_{x\uparrow a} u(x)$ exist and $u' = v + [u^+(a) - u^-(a)]\,\delta_a$ in $\mathcal{S}'(\mathbb{R})$.

4. Show that there exist $\phi, \psi \in \mathcal{S}(\mathbb{R}^n)$ with $\phi * \psi \equiv 0$ although $\phi \not\equiv 0$ and $\psi \not\equiv 0$. On the other hand, $\phi * \phi \equiv 0$ with $\phi \in \mathcal{S}(\mathbb{R}^n)$ is only possible if $\phi \equiv 0$.

5. Show that $\phi \mapsto \int_{\mathbb{R}} \phi(x)\, e^x \cos(e^x)\, dx$ defines a tempered distribution even if $x \mapsto e^x \cos(e^x)$ fails to have polynomial growth.

[59] Note that this does not require the differentiability of u at a since in Lebesgue integrals the value of the function at a particular point is of no relevance. In particular, whether $v \in L^1[a-\varepsilon, a+\varepsilon]$ or not can be solely determined by the behaviour of $v(x)$ for $x \neq a$ near a.

6. Show that $\phi \mapsto \lim_{\varepsilon\downarrow 0}\int_{|x|>\varepsilon}\frac{\phi(x)}{x}\,dx$ defines a tempered distribution, called *principal value* and denoted by pv, while $\phi \mapsto \lim_{\varepsilon\downarrow 0}\int_{|x|>\varepsilon}\frac{\phi(x)}{|x|}\,dx$ is not an element of $\mathcal{S}'(\mathbb{R})$.

7. (i) Compute $x\cdot \text{pv}$ and $x\cdot\delta_0$ in $\mathcal{S}'(\mathbb{R})$.

(ii) Show that one cannot define the product of δ_0 and pv in $\mathcal{S}'(\mathbb{R})$, in the sense that such a product cannot be commutative[60] and associative with respect to the multiplication[61] by the smooth function of polynomial growth $\varphi(x) = x$.

(iii) Show that it is not possible to define an associative convolution of tempered distributions.[62]

8. (Taylor's formula with integral remainder) (i) Show that for $\phi \in C^\infty(\mathbb{R})$ we have

$$\phi(t) = \sum_{k=0}^{m-1}\frac{t^k}{k!}\,\phi^{(k)}(0) + \frac{1}{(m-1)!}\int_0^t (t-s)^{m-1}\phi^{(m)}(s)\,ds\,, \qquad t\in\mathbb{R},\quad m\geq 1\,.$$

(ii) Show that if $\phi \in \mathcal{S}(\mathbb{R})$ is such that $\phi^{(j)}(0) = 0$ for $j = 0,\dots,m-1$ with $m\geq 1$ an integer, then $\phi(t) = t^m\psi(t)$ for some $\psi\in\mathcal{S}(\mathbb{R})$ with $\sup_{t\in\mathbb{R}}|t^p\psi^{(k)}(t)| \leq \frac{1}{(m-1)!}\sup_{t\in\mathbb{R}}|t^p\phi^{(k+m)}(t)|$ for all integers $k,\ p\geq 0$. Moreover, $\psi\in\mathcal{D}(\mathbb{R})$ if $\phi\in\mathcal{D}(\mathbb{R})$.

(iii) Show that $f\in C^\infty(\mathbb{R}^n)$ can be written for any integer $m\geq 1$ in the form

$$f(x) = \sum_{|\alpha|\leq m-1}\frac{x^\alpha}{\alpha!}\,(D^\alpha f)(0) + m\sum_{|\alpha|=m}\frac{x^\alpha}{\alpha!}\int_0^1 (1-t)^{m-1}(D^\alpha f)(xt)\,dt\,, \qquad x\in\mathbb{R}^n\,.$$

9. [Primitives of tempered distributions] (i) Show that $\phi\in\mathcal{S}(\mathbb{R})$ is a derivative of some $\psi\in\mathcal{S}(\mathbb{R})$ if and only $\int_{\mathbb{R}}\phi(x)\,dx = 0$.

(ii) Given $v\in\mathcal{S}'(\mathbb{R})$, a solution $u\in\mathcal{S}'(\mathbb{R})$ of $u' = v$ is called a primitive of v. Show that any $v\in\mathcal{S}'(\mathbb{R})$ admits a primitive, and all primitives differ by a constant.

10. (i) We say that $u\in\mathcal{S}'(\mathbb{R})$ is even/odd if $\langle u,\,\phi^\dagger\rangle = \pm\langle u,\,\phi\rangle$ for every $\phi\in\mathcal{S}(\mathbb{R})$. Show that if $u\in\mathcal{S}'(\mathbb{R})$ is odd/even, then $\mathcal{F}^{-1}(u) = \pm\mathcal{F}(u)$.

(ii) Compute $\mathcal{F}(\text{pv})$ and $\mathcal{F}(H)$ in $\mathcal{S}'(\mathbb{R})$, H being the Heaviside step function.

11. Show that a measurable function $f:\mathbb{R}^n\to\mathbb{C}$ defines a continuous map $\phi\mapsto f\phi$ from $\mathcal{S}(\mathbb{R}^n)$ into $\mathcal{S}(\mathbb{R}^n)$ if and only if $f\in\mathcal{O}(\mathbb{R}^n)$.

12. The nonlinear partial differential equation

$$u_t - u_{txx} + 3uu_x = 2u_xu_{xx} + uu_{xxx}\,,$$

in dimensionless space-time variables (x,t), models small-amplitude axial-radial deformation waves in thin cylindrical hyperelastic rods, see the discussion in Constantin and Strauss (2000). Show that the concept of distributional derivative is useful in finding solitary wave solutions: deformations $u(x,t) = f(x-ct)$ having a shape f that is practically confined to a finite region (in the sense that f and its derivatives decay fast far out) and that is shifted in time, at constant speed $c>0$.

13. Show that there exist $u\in\mathcal{E}'(\mathbb{R}^n)$ and $\phi\in\mathcal{S}(\mathbb{R}^n)$ with $\langle u,\,\phi\rangle\neq 0$ despite $\phi(x)=0$ for all $x\in\operatorname{supp}(u)$, while $\langle u,\,\phi\rangle = 0$ if $\operatorname{supp}(u)\cap\operatorname{supp}(\phi)=\emptyset$.

[60] Meaning $u\cdot v = v\cdot u$.
[61] Meaning that $\varphi\cdot(u\cdot v) = (\varphi\cdot u)\cdot v$.
[62] Meaning that $u*(v*w) = (u*v)*w$.

14. Show that if $\phi \in \mathcal{D}(\mathbb{R}^n)$ then $\hat{\phi}$ cannot vanish on any open nonempty subset of $\mathbb{R}^n$ unless $\phi \equiv 0$. (Thus the Fourier transform does not map $\mathcal{D}(\mathbb{R}^n)$ into itself.)

15. Show that $u * v \in \mathcal{E}'(\mathbb{R}^n)$ if $u, v \in \mathcal{E}'(\mathbb{R}^n)$.

16. Show that both sets $O(\mathbb{R}^n) \setminus O'(\mathbb{R}^n)$ and $O'(\mathbb{R}^n) \setminus O(\mathbb{R}^n)$ are nonempty.

17. (i) Let $u : (-\infty, 0) \cup (0, \infty) \to \mathbb{C}$ be locally integrable[63] on $\mathbb{R} \setminus \{0\}$. Show that if there exists $C > 0$ and an integer $n \geq 1$ such that $|u(x)| \leq \frac{C}{|x|^n}$ for $x \in (-1, 0) \cup (0, 1)$, then there is a tempered distribution $U \in \mathcal{S}'(\mathbb{R})$ with $\langle U, \phi \rangle = \int_{\mathbb{R}} u(x)\phi(x)\,dx$ for every $\phi \in \mathcal{S}(\mathbb{R})$ with $\mathrm{supp}(\phi) \subset \mathbb{R} \setminus \{0\}$.

(ii) Show that if $u(x) = \frac{1}{x^2}$ for $x \neq 0$, then

$$\langle U, \phi \rangle = \lim_{\varepsilon \downarrow 0} \Big[\int_{|x|>\varepsilon} \frac{\phi(x)}{x^2}\,dx - \frac{2}{\varepsilon}\,\phi(0) \Big] + 2\phi(0)\,, \qquad \phi \in \mathcal{S}(\mathbb{R})\,,$$

is a solution in $\mathcal{S}'(\mathbb{R})$ to (i).

(iii) Show that the singularity of infinite order at the origin of the function $u(x) = e^{\frac{1}{x^2}}$ for $x \neq 0$ prevents the existence of some $U \in \mathcal{S}'(\mathbb{R})$ with $\langle U, \phi \rangle = \int_{\mathbb{R}} u(x)\phi(x)\,dx$ for every $\phi \in \mathcal{S}(\mathbb{R})$ with $\mathrm{supp}(\phi) \subset \mathbb{R} \setminus \{0\}$. (These examples illustrate the power/limitations of distribution theory with regard to isolated singularities.)

18. (i) Prove that if $u \in \mathcal{S}'(\mathbb{R})$ has $\mathrm{supp}(u) = \{0\}$, then there exists an integer $n \geq 0$ such that $u = \sum_{k=0}^{n} c_k\, D^k \delta_0$ for some constants $c_k \in \mathbb{C}$.

(ii) How much arbitraryness is there in the renormalisation of $f(x) = \frac{1}{x}\chi_{(0,\infty)}(x)$ described at the end of Section 6.1.1?

19. (Division in $\mathcal{S}'(\mathbb{R})$) (i) Given $v \in \mathcal{E}'(\mathbb{R})$ and an integer $m \geq 1$, show the existence of some $u \in \mathcal{S}'(\mathbb{R})$ with $x^m u = v$ in $\mathcal{S}'(\mathbb{R})$.

(ii) If $f \in C^\infty(\mathbb{R})$ with $f(x) > 0$ for $x \neq 0$ is such that the equation $fu = 1$ has a solution $u \in \mathcal{S}'(\mathbb{R})$, show that f cannot have a zero of infinite order[64] at $x = 0$.

20. (i) Show that $\cos(2\pi n x) \to 0$ in $\mathcal{S}'(\mathbb{R})$, while the sequence converges pointwise only for $x \in \mathbb{Z}$ and fails to converge in $L^p[0, 1]$ for any $p \geq 1$.

(ii) For $n \geq 1$, set $f_n = n^2 \chi_{(-\frac{1}{n}, \frac{1}{n})}$. Show that $f_n \to 0$ a.e. but f_n does not converge in $\mathcal{S}'(\mathbb{R})$.

21. (i) Show that the Fourier transform is rotation-invariant: if R is a rotation,[65] then $\widehat{f \circ R} = \hat{f} \circ R$ for all $f \in L^1(\mathbb{R}^n)$.

(ii) Prove that a differential operator $L = \sum_{|\beta| \leq N} c_\beta(x)\, D^\beta$ with continuous coefficients c_β is invariant under translations and rotations[66] if and only if there is a polynomial P in one variable such that $L = P(\Delta)$, where $\Delta = \sum_{k=1}^{n} \partial_{x_k}^2$ is the Laplace operator. (Thus

63 Meaning that $\int_K |u(x)|\,dx < \infty$ for each compact set $K \subset \mathbb{R} \setminus \{0\}$.

64 A prototype is $f(0) = 0$ and $f(x) = \exp\big(-\frac{1}{x^2}\big)$ for $x \neq 0$.

65 Defined as a map $x \to Rx$ on $\mathbb{R}^n$ with the real $n \times n$ matrix A orthogonal, that is, such that $AA^* = I_n$, where A^* is the transpose of A. In $\mathbb{R}^2 \simeq \mathbb{C}$ all rotations are of one of two forms: the proper rotations $Rz = ze^{i\theta}$ (for some $\theta \in \mathbb{R}$) with $\det(R) = 1$, and the improper rotations $Rz = \overline{z}\, e^{i\theta}$ (for some $\theta \in \mathbb{R}$) with $\det(R) = -1$. In $\mathbb{R}^3$, any rotation consists of a rotation by a fixed angle about a fixed axis (axis–angle representation); see Stein and Shakarchi (2003).

66 This means that for every $f \in C^\infty(\mathbb{R}^n)$ we have that $L(f \circ T) = (Lf) \circ T$ for all rotations $x \mapsto Rx$ and all translations $x \mapsto x^* + x$ of $\mathbb{R}^n$.

the Laplace operator is essentially the only differential operator that is invariant under translations and rotations.)

(iii) Show that if $f : \mathbb{R}^3 \to \mathbb{C}$ is a radial function[67] then $\hat{f}$ is also radial, with

$$\hat{f}(\xi) = \frac{2}{|\xi|}\int_0^\infty f_0(r)\, r \sin(2\pi r|\xi|)\, dr\,.$$

(In dimensions $n = 2$ and $n \geq 4$, a more complicated formula holds, involving Bessel functions; see Stein and Weiss (1971).)

22. Show that in the (x,t)-plane $E(x,t) = \begin{cases} 1/2 & \text{if } \; t-|x|>0\,, \\ 0 & \text{if } \; t-|x|<0\,, \end{cases}$ is a fundamental solution of the wave operator $\Box = \partial_t^2 - \partial_x^2$.

23. Show that the series $\sum_{k=1}^\infty 2^k e^{2\pi 4^k ix} \sin^{2^k}(2\pi y)$ is square convergent at almost every point and is spherically divergent almost everywhere.

24. (Riemann–Lebesgue) Prove that $\lim_{|n|\to\infty} |\hat{f}(n)| = 0$ for any $f \in L^1(\mathbb{T}^d)$.

25. (Weierstrass approximation) Prove that every $f \in C(\mathbb{T}^d)$ is a uniform limit of trigonometric polynomials.

26. Show that if the Fourier transform (6.31) of the periodic distribution u is a measure of finite total mass, that is, $\sum_{g\in\mathbb{Z}^n} |c_g| < \infty$, then u can be identified with a continuous function. (While the continuity of the periodic function u does not guarantee that its Fourier transform is a measure of finite total mass, the hypothesis that $u \in h_\alpha$ (Hölder class) with $\alpha > 1/2$ will, see Exercise 4.27(i).)

27. For $a > 0$, show that $\mathcal{F}[\exp(-\pi a|x|^2)](\xi) = a^{-\frac{n}{2}} \exp(-\pi|\xi|^2/a)$ in $\mathbb{R}^n$.

28. For $a > 0$, show that

$$\mathcal{F}[\exp(-2\pi a|x|](\xi) = \frac{ac_n}{(a^2+|\xi|^2)^{\frac{n+1}{2}}} \quad \text{in} \quad \mathbb{R}^n\,,$$

with $c_n = \pi^{-\frac{n+1}{2}}\Gamma(\frac{n+1}{2})$. (The Γ-function, defined by $\Gamma(t) = \int_0^\infty s^{t-1}e^{-s}ds$ for $t > 0$, satisfies $\Gamma(m) = (m-1)!$ for all integers $m \geq 1$.)

29. For $0 < \alpha < n$, compute the Fourier transform of $|x|^{-\alpha}$ in $\mathbb{R}^n$.

30. (i) Are there discontinuous positive-definite functions on $\mathbb{R}^d$?

(ii) Show that $|f(\xi) - f(\eta)| \leq 4f(0)\,|f(0) - f(\xi-\eta)|$ for all $\xi, \eta \in \mathbb{R}^d$, whenever $f : \mathbb{R}^d \to \mathbb{C}$ is a positive-definite function. (In particular, continuity at $\xi = 0$ implies uniform continuity on $\mathbb{R}^d$.)

31. Show that if $f \in L^1(\mathbb{R}^n)$ with $n \geq 2$ is a function with the property that there exist functions $f_0, f_1, \dots, f_n : \mathbb{R} \to \mathbb{C}$ so that for all $x = (x_1, \dots, x_n) \in \mathbb{R}^n$,

$$f(x_1,\dots,x_n) = \prod_{k=1}^n f_k(x_k) = f_0(\sqrt{x_1^2+\cdots+x_n^2})\,,$$

then $f(x) = \alpha e^{-\beta|x|^2}$ for some $\alpha, \beta \in \mathbb{C}$ with $\Re e(\beta) < 0$.

[67] That is, $f(Rx) = f(x)$ for all rotations R or, equivalently (since any two orthonormal bases correspond to each other by means of a rotation), $f(x) = f_0(|x|)$ with $f_0 : [0,\infty) \to \mathbb{C}$.

32. Using complex analysis, find the Fourier transform of the general Gaussian $f(x) = e^{-\beta|x|^2+\sum_{k=1}^n \gamma_k x_k}$ in $\mathbb{R}^n$, for arbitrary $\beta, \gamma_1, \dots, \gamma_n \in \mathbb{C}$ with $\Re e(\beta) > 0$.

33. Show that if $p, q \in [1, \infty]$ are such that $\hat{f} \in L^q(\mathbb{R}^n)$ whenever $f \in L^p(\mathbb{R}^n)$, then $1 \leq p \leq q \leq 2$ and $\frac{1}{p} + \frac{1}{q} = 1$.

34. Show that $\delta_0 \in H^s(\mathbb{R}^n)$ if and only if $s < -n/2$, while its Fourier transform $1 \notin H^t(\mathbb{R}^n)$ for any $t \in \mathbb{R}$.

35. Show that $f(x) = e^{-|x|-|x|^3}$ belongs to $H^s(\mathbb{R})$ if and only if $s < 3/2$.

36. Let $\theta \in \mathcal{D}(\mathbb{R}^n)$ be such that $\theta(x) = 1$ for $|x| \leq 1$ and set $f(x) = \theta(x)\,|x|^\lambda$, where[68] $\lambda > -n$. Show that $f \in H^s(\mathbb{R}^n)$ if and only if $\lambda > s - n/2$, whereas f is k-times continuously differentiable on $\mathbb{R}^n$ if and only if $\lambda > k$. (Since for some $M > 1$ we have $f(x) = 0$ for $|x| \geq M$, this example elucidates Theorem 6.21.)

37. The Fourier coefficients of the periodic distribution $u = \sum_{k\in\mathbb{Z}} c_k e^{2\pi i k x} \in \mathcal{S}'(\mathbb{R})$ are defined by $\hat{u}(k) = c_k$ for $k \in \mathbb{Z}$.

(i) Show that for any $s \in \mathbb{R}$, $H^s(\mathbb{T})$ is a Hilbert space with respect to the inner product

$$(f, g)_s = \sum_{k\in\mathbb{Z}} (1 + |k|^2)^s \hat{f}(k)\,\overline{\hat{g}(k)},$$

with the dual space $(H^s(\mathbb{T}))^*$ isometrically isomorphic to $H^{-s}(\mathbb{T})$ by means of the duality pairing $\langle u, v\rangle_s = \sum_{k\in\mathbb{Z}} \hat{u}(k)\,\hat{v}(k)$ for $u \in H^{-s}(\mathbb{T})$ and $v \in H^s(\mathbb{T})$. Moreover, the trigonometric polynomials are dense in $H^s(\mathbb{T})$.

(ii) If $m \geq 1$ is an integer, show that $f \in H^m(\mathbb{T})$ if and only if the distributional derivatives $f^{(j)} \in \mathcal{S}'(\mathbb{R})$ belong to $L^2(\mathbb{T})$ for $0 \leq j \leq m$, and there exist constants $c_m, C_m > 0$ such that $c_m\|f\|_m^2 \leq \sum_{j=0}^m \|f^{(j)}\|_0^2 \leq C_m\|f\|_m^2$ for all $f \in H^m(\mathbb{T})$.

(iii) Show that any $f \in H^s(\mathbb{T})$ equals a.e. a continuous function if $s > 1/2$.

(iv) Show that $H^s(\mathbb{T})$ is a Banach algebra for $s > 1/2$.

(v) Prove that if $q > p$, then any bounded sequence in $H^q(\mathbb{T})$ has a subsequence that converges in $H^p(\mathbb{T})$.

6.7.2 Hints

1. Choose some $\theta \in \mathcal{D}(\mathbb{R}^n)$ such that $\theta(x) = 1$ for $|x| \leq 1$. For a given $\phi \in \mathcal{S}(\mathbb{R}^n)$, let $\phi_k(x) = \theta(\frac{x}{k})\,\phi(x)$ for $k \geq 1$. Then $\phi_k \in \mathcal{D}(\mathbb{R}^n)$ and $\phi_k \to \phi$ in $\mathcal{S}(\mathbb{R}^n)$.

2. Take $\theta \in \mathcal{D}(\mathbb{R})$ with $\mathrm{supp}(\theta) \subset [-1, 1]$ and $\theta(x) = 1$ if $|x| \leq a$ for some $a \in (0, 1)$. Show that $\sum_{k\geq 0} \frac{a_k}{k!}\, x^k \theta(\frac{x}{\varepsilon_k})$ serves the purpose for suitable choice of $\varepsilon_k \in (0, 1)$. (One way to construct functions $\varphi \in \mathcal{D}(\mathbb{R})$ is to take power series $\sum_{k\geq 0} c_k x^k$ that converge on some $(-R, R)$ with $R > 0$, and to multiply them with some $\phi_0 \in \mathcal{D}(\mathbb{R})$ such that $\mathrm{supp}(\phi_0) \subset (-R, R)$ and $\phi_0(x) = 1$ near $x = 0$. However, this approach does not work if, say, $a_k = k!k^k$ for $k \geq 1$, since in this case $c_k k! = a_k$ yields a zero radius of convergence as $\frac{1}{R} = \limsup_{k\to\infty} |c_k|^{1/k}$.)

3. For the first part, note that $u(x) - u(y) = \int_x^y u'(s)\,ds$ for $x > y > a$. For the second part, use the definition of distributional derivatives and observe that $v\phi \in L^1(\mathbb{R})$ whenever $\phi \in \mathcal{S}(\mathbb{R})$.

4. Take Fourier transforms.

[68] Using spherical coordinates we see that this is the necessary and sufficient condition for f to be locally integrable, since f is compactly supported and infinitely differentiable on $\mathbb{R}^n \setminus \{0\}$.

5. Note that $(\sin(e^x))' = e^x \cos(e^x)$ and pay attention to the points $x_n = \ln(2n\pi)$ with $n \geq 1$.

6. For the first part, rely on $\int_{|x|\geq\varepsilon} \frac{\phi(x)}{x}\,dx = \int_{|x|\geq 1} \frac{\phi(x)}{x}\,dx + \int_{\varepsilon}^{1} \frac{\phi(x)-\phi(0)}{x}\,dx + \int_{-1}^{-\varepsilon} \frac{\phi(x)-\phi(0)}{x}\,dx$. For the second part, choose $\phi(x) = e^{-x^2}$.

7. Items (i)-(ii) are easy. For (iii), choose $u = 1$, $v = \delta_0'$, $w = H$ (the Heaviside step function).

8. In (i) use integration by parts. For (ii), the change of variable of integration $s = t\tau$ in (i) produces the candidate $\psi(t) = \frac{1}{(m-1)!} \int_0^1 (1-\tau)^{m-1} \phi^{(m)}(t\tau)\,d\tau$. For (iii), evaluate (i) at $t = 1$ for the function $t \mapsto f(xt)$.

9. For (i), let $\psi(t) = \int_{-\infty}^{t} \phi(s)\,ds$. For (ii), choose $\theta \in \mathcal{D}(\mathbb{R})$ with $\int_{\mathbb{R}} \theta(t)\,dt = 1$ and let $\omega : \mathcal{S}(\mathbb{R}) \to \mathcal{S}(\mathbb{R})$ be the linear map $\phi \mapsto \omega_\phi$, where $\omega_\phi(t) = \int_t^\infty \big(\phi(s) - \theta(s) \int_{\mathbb{R}} \phi(\tau)d\tau\big)\,ds$. Then that all primitives $u \in \mathcal{S}'(\mathbb{R})$ of v are of the form $\langle u, \phi\rangle = \langle v, \omega_\phi\rangle + \langle c, \phi\rangle$ for some constant c.

10. Rely on the definitions for (i). For (ii), use (i), Exercise 7(i), Exercise 9(ii) and the properties of the Fourier transform.

11. Note that $f(x)\,e^{-|x|^2} \in \mathcal{S}(\mathbb{R}^n)$ implies $f \in C^\infty(\mathbb{R}^n)$. To show that f has polynomial growth, assume the opposite and construct $\phi \in \mathcal{S}(\mathbb{R}^n)$, obtained as a weighted infinite sum of translates of a suitabe test function, such that $f\phi$ is unbounded.

12. Formal differentiation suggests $u(x,t) = f(x - ct)$ with $f(s) = c\,e^{-|s|}$. Interpret it as a solution in $\mathcal{S}'(\mathbb{R}^2)$ of the equation written in the form $\partial_t(1 - \partial_x^2)u + (1 - \partial_x^2)uu_x + \partial_x(u^2 + \frac{1}{2}\,u_x^2) = 0$.

13. Choose $u = \partial_{x_1}\delta_0$ and $\phi(x) = x_1\,e^{-|x|^2}$ for the first part. For the second part, construct $\theta \in \mathcal{D}(\mathbb{R})$ with $\theta u = u$ and $\theta\phi = 0$.

14. Assume without loss of generality that $\hat{\phi}$ vanishes near the origin. Use (6.14) and the power series expansion of $e^{-2\pi i\xi\cdot x}$ to obtain $\hat{\phi}(\xi) = \sum_\alpha \frac{\xi^\alpha}{\alpha!}\,(D^\alpha\hat{\phi})(0)$.

15. Show that $\mathrm{supp}(u * v) \subset \mathrm{supp}(u) + \mathrm{supp}(v) = \{x + y \in \mathbb{R}^n : x \in \mathrm{supp}(u),\ y \in \mathrm{supp}(v)\}$ using Exercise 13 and the properties of the Fourier transform.

16. Consider the elements δ_0 and 1.

17. For (i) rely on Exercise 8(iii) and Theorem 6.6. In (ii) use the dominated convergence theorem. For (iii), choose $\theta \in \mathcal{D}(\mathbb{R})$ with $\theta \geq 0$, $\mathrm{supp}(\theta) \subset [1,2]$ and such that $\theta(x) = 1$ if $x \in [a,b]$, for some $a, b \in (1,2)$, and look at the action on $\theta_n(x) = e^{-n}\,\theta(nx)$ as $n \to \infty$.

18. For (i), use Theorem 6.6 to show that there is a integer $n \geq 0$ such that $\langle u, \phi\rangle = 0$ if $\phi \in \mathcal{S}(\mathbb{R})$ satisfies $\phi(0) = \phi'(0) = \cdots = \phi^{(n)}(0) = 0$, and subsequently rely on Exercise 8. For (ii), use (i).

19. For (i), use Exercise 18(i) in combination with Exercise 8(iii) and start with $v = 0$. For (ii), assume the existence of a solution $u \in \mathcal{S}'(\mathbb{R})$ and investigate $\langle u, f\theta_\varepsilon\rangle$, where $\theta_\varepsilon(x) = \frac{1}{\varepsilon}\,\theta(\frac{x}{\varepsilon})$ for some $\theta \in \mathcal{D}(\mathbb{R})$ with $\theta \geq 0$, $\theta(x) = 1$ for $|x| \leq 1$ and $\mathrm{supp}(\theta) \subset [-2,2]$.

20. (i) To prove pointwise divergence for $x \in \mathbb{R} \setminus \mathbb{Q}$, estimate $\frac{1}{N}\sum_{n=0}^{N-1} \cos(2\pi nx)$ as $N \to \infty$. To rule out convergence in $L^p[0,1]$, identify $f = 0$ as the only possible limit and show that $\int_0^1 |\cos(2\pi nx)|\,dx > \frac{1}{\pi}$. In (ii), choose $\phi \in \mathcal{D}(\mathbb{R})$ with $\phi(x) = 1$ for $|x| \leq 1$.

21. For (i) rely on the definitions. In (ii), show that translation-invariance forces all coefficients c_β to be constant. Write now $Q(\xi) = \sum_{|\beta|\leq N}(2\pi i\xi)^\beta$ as $\sum_{0\leq j\leq N} Q_j(\xi)$ with each polynomial Q_j homogeneous of degree j and use (i) to show that each $Q_j(\xi)$ depends only on $|\xi|^2$. For (iii) rely on (i) and on spherical coordinates.

22. Check that $\langle \Box E, \phi\rangle = \phi(0,0)$ for every $\phi \in \mathcal{S}(\mathbb{R}^2)$.

23. Show that convergence holds for $|\sin(2\pi y)| < 1$ and estimate the difference between the spherical partial sums for $N = 4^k$ and $N = 4^k - 1$.

24. Write $f = f_1 + f_2$ with $f_2 \in L^2(\mathbb{T}^d)$ and $f \in L^1(\mathbb{T}^d)$ such that $\|f_1\|_{L^1(\mathbb{T}^d)} < \varepsilon$.

25. Use the multi-dimensional Fejér kernels.

26. Rely on (6.26).

27. Note that this reduces to the one-dimensional case dealt with in Exercise 5.2.

28. Write $e^{-2\pi a|x|}$ as an "average" of Gaussians $e^{-t|x|^2}$, by proving the validity of the identity $e^{-2\pi a|x|} = \int_0^\infty \frac{e^{-s}}{\sqrt{\pi s}}\, e^{-\pi^2 a^2|x|^2/s} ds$.

29. Show first the validity of the identity $\int_{\mathbb{R}^n} \varphi(x-y)\, \frac{\exp(-|y|^2/4t)}{(4\pi t)^{n/2}}\, dy = \int_{\mathbb{R}^n} e^{-4\pi^2 t|\xi|^2} e^{2\pi i x\cdot\xi} \hat{\varphi}(\xi)\, d\xi$ for every $\varphi \in S(\mathbb{R}^n)$ and every $t > 0$. Subsequently, multiply both sides by t^k and integrate.

30. (i) Look for f vanishing everywhere but at $\xi = 0$. (ii) With $\xi_1 = \xi \in \mathbb{R}^d$, $\xi_2 = \eta \in \mathbb{R}^d$ and $\xi_3 = 0 \in \mathbb{R}^d$, use the fact that the determinant of the positive-definite Hermitian matrix $\{f(\xi_j - \xi_k)\}_{1\le j,k\le 3}$ is nonnegative.

31. Assuming $f \not\equiv 0$, show that $f_k \in L^1(\mathbb{R})$ for $1 \le k \le n$, with $\prod_{k=1}^n \widehat{f_k}(\xi_k) = g(\xi_1^2 + \cdots + \xi_n^2)$, where $g : [0,\infty) \to \mathbb{C}$ is some bounded and continuous function satisfying the functional equation $g(t+s) = g(t)g(s)$ for all $s, t \ge 0$.

32. Use analytic continuation and Exercise 27.

33. Use the closed graph theorem to show the existence of some $C > 0$ with $\|\hat{f}\|_{L^q(\mathbb{R}^n)} \le C\|f\|_{L^p(\mathbb{R}^n)}$ for all $f \in L^p(\mathbb{R}^n)$. To infer from this that $q = \frac{p}{p-1}$, look at the effect of replacing $f(x)$ by $f(x/t)$, with $t > 0$, on the norms of f and $\hat{f}$. To show that $q \ge p$, consider the functions $f_m(x) = e^{-[\sin(1/m) - i\cos(1/m)]\,|x|^2}$ for $m \ge n$, using Exercise 32.

34. Take the Fourier transform.

35. Compute the second distributional derivative.

36. Note that the (classical) derivative $\partial_x^\alpha f$ is smooth except at 0 and there exists a constant $C_{\alpha,\lambda} > 0$ such that $|(\partial_x^\alpha f)(x)| \le C_{\alpha,\lambda}|x|^{\lambda-|\alpha|}$ for $x \ne 0$. Show that it coincides with the distributional derivative of f, if the latter belongs to $L^2(\mathbb{R}^n)$.

37. For (i) adapt the solution of Exercise 3.11 and use Theorem 3.13 for the duality. Use (6.13) for (ii) and for (iii) rely on Exercise 4.22. For (iv), use the existence, for $s > 0$, of constants $m_s, M_s > 0$ such that $m_s(a^s + b^s) \le (a+b)^s \le M_s(a^s + b^s)$ for all $a, b \ge 0$, combined with the fact that if $\alpha \in l^1(\mathbb{Z})$ and $\beta \in l^2(\mathbb{Z})$, then $\sum_{k\in\mathbb{Z}}(\sum_{j\in\mathbb{Z}} |\alpha(j)|\,|\beta(k-j)|)^2 \le \|\alpha\|_{l^1}^2 \|\beta\|_{l^2}^2$. For (v), adapt the proof of Theorem 3.31(i) to the embedding from $H^q(\mathbb{T})$ to $H^p(\mathbb{T})$.

6.7.3 Solutions

1. Choose $\theta \in \mathcal{D}(\mathbb{R}^n)$ with $\theta(x) = 1$ for $|x| \le 1$. Let $\phi \in S(\mathbb{R}^n)$. For $k \ge 1$, let $\phi_k \in \mathcal{D}(\mathbb{R}^n)$ be given by $\phi_k(x) = \theta(\frac{x}{k})\,\phi(x)$ for $k \ge 1$. Then there are constants $C_N > 0$, $N = 1, 2, \ldots$ such that $\sum_{|\alpha|,|\beta|\le N} d_{\alpha,\beta}(\phi, \phi_k) \le C_N \sum_{|\alpha|\,|\beta|\le N} \sup_{|x|\ge k}\{|x^\alpha D^\beta \phi(x)|\}$. Since $|x^\alpha D^\beta \phi(x)| \le \frac{1}{|x|} \sum_{|\gamma|=|\alpha|+1} d_{\gamma,\beta}(0,\phi)$ for $|x| \ge 1$, the right side above converges to 0 as $k \to \infty$.

2. Choose $\theta \in \mathcal{D}(\mathbb{R})$ with $\text{supp}(\theta) \subset [-1,1]$ and such that $\theta(x) = 1$ if $|x| \le a$, for some $a \in (0,1)$. For any integer $k \ge 0$ choose $M_k > 0$ such that $\sup_{x\in\mathbb{R}} |\theta^{(j)}(x)| \le M_k$ for all $j \in \{0, \ldots, k\}$. We now claim that for a suitable choice of $\varepsilon_k \in (0,1)$, defining $\varphi_n \in \mathcal{D}(\mathbb{R})$ by $\varphi_0(x) = a_0\theta(x)$ and $\varphi_k(x) = \frac{a_k}{k!}\, x^k\, \theta(\frac{x}{\varepsilon_k})$ for $k \ge 1$ ensures

$$\sup_{x\in\mathbb{R}} |\varphi_k^{(j)}(x)| \le \frac{1}{2^j}, \qquad 0 \le j \le k-1. \tag{$*$}$$

Once $(*)$ is granted, $\varphi(x) = \sum_{k\ge 0} \varphi_k(x)$ will define $\varphi \in \mathcal{D}(\mathbb{R})$ with $\text{supp}(\varphi) \subset [-1,1]$ since $\sum_{k\ge 0} \varphi_k^{(j)}(x)$ converges uniformly on $\mathbb{R}$ for every fixed $j \ge 0$. Moreover, $\varphi(0) = a_0$ and, since $\theta^{(j)}(0) = 0$ for all $j \ge 1$, we also have that $\varphi^{(k)}(0) = a_k$ for all $k \ge 1$. It remains to validate $(*)$. Fix $k \ge 1$. Since $\text{supp}(\varphi_k) \subset [-\varepsilon_k, \varepsilon_k]$, for $0 \le j \le k-1$ we get that $|\varphi_k^{(j)}(x)| =$

$\left|\frac{a_k}{k!}\sum_{i=0}^{j} C_j^i\,(x^k)^{(j-i)}\frac{1}{\varepsilon_k^i}\,\theta^{(i)}(\frac{x}{\varepsilon_k})\right| \le \frac{|a_k|}{k!}\sum_{i=0}^{j} C_j^i\,k!\,\varepsilon_k^{k-j+i}\,\frac{1}{\varepsilon_k^i}\,M_k \le |a_k|\sum_{i=0}^{j} C_j^i\,\varepsilon_k M_k \le |a_k|\,2^k\,\varepsilon_k M_k$. Consequently, it suffices to choose $\varepsilon_k < \frac{1}{M_k(1+|a_k|)2^{2k}}$ for $k \ge 1$.

3. If $x_n \in (a, a+\varepsilon)$ for all $n \ge 1$ and $\lim_{n\to\infty} x_n = a$, then $|u(x_n) - u(x_m)| \le \left|\int_{x_n}^{x_m} v(s)\,ds\right| \le \int_a^{a+\varepsilon} |v(s)|\chi_{[x_n,x_m]}(s)\,ds \to 0$ as $n, m \to \infty$ by dominated convergence (see Chapter 2), so that $\{u(x_n)\}_{n\ge1}$ is a Cauchy sequence. Its limit $u^+(a) \in \mathbb{R}$ is easily seen to be independent of the sequence $\{x_n\}_{n\ge1}$. Similary one shows the existence of $u^-(a) \in \mathbb{R}$. Since $u \in S'(\mathbb{R})$ and $v\phi \in L^1(\mathbb{R})$ for $\phi \in S(\mathbb{R})$, we have $u'(\phi) = -u(\phi') = -\int_{\mathbb{R}} u(x)\,\phi'(x)\,dx = -\lim_{\delta\to0}\{\int_{-\infty}^{a-\delta} u(x)\,\phi'(x)\,dx + \int_{a+\delta}^{\infty} u(x)\,\phi'(x)\,dx\} = -\lim_{\delta\to0}\{u(a-\delta)\phi(a-\delta) - \int_{-\infty}^{a-\delta} v(x)\,\phi(x)\,dx + u(a+\delta)\phi(a+\delta) - \int_{a+\delta}^{\infty} v(x)\,\phi(x)\,dx\} = [u^+(a) - u^-(a)]\,\phi(a) + \lim_{\delta\to0}\{\int_{-\infty}^{a-\delta} v(x)\,\phi(x)\,dx + \int_{a+\delta}^{\infty} v(x)\,\phi(x)\,dx\} = [u^+(a) - u^-(a)]\,\phi(a) + \int_{\mathbb{R}} v(x)\,\phi(x)\,dx$.

4. Let $f, g \in \mathcal{D}(\mathbb{R}^n)$ be nonzero elements with $\mathrm{supp}(f) \cap \mathrm{supp}(g) = \emptyset$. Then $\phi = \mathcal{F}^{-1}(f)$ and $\psi = \mathcal{F}^{-1}(g)$ are in $S(\mathbb{R}^n)$ with $\phi * \psi = 0$ while $\phi \not\equiv 0$ and $\psi \not\equiv 0$, since $\mathcal{F}$ is a bijection on $S(\mathbb{R}^n)$ and $\mathcal{F}(\phi * \psi) = \hat\phi \cdot \hat\psi = f \cdot g = 0$. On the other hand, if $\phi \in S(\mathbb{R}^n)$ is such that $\phi * \phi = 0$, application of the Fourier transform yields $\hat\phi \cdot \hat\phi = 0$, so that $\hat\phi = 0$ and $\phi = 0$.

5. Since $|\phi'(x)| \le \frac{\sup_{s\in\mathbb{R}}\{(1+s^2)|\phi'(s)|\}}{1+x^2}$, Theorem 6.5 shows that $\phi \mapsto -\int_{\mathbb{R}} \sin(e^x)\,\phi'(x)\,dx$ is in $S'(\mathbb{R})$. Indeed, $\int_{\mathbb{R}} |\phi'(x)|\,dx \le \sup_{s\in\mathbb{R}}\{(1+s^2)|\phi'(s)|\}\int_{\mathbb{R}} \frac{1}{1+x^2}\,dx = \pi \sup_{s\in\mathbb{R}}\{(1+s^2)|\phi'(s)|\}$. For $\phi \in S(\mathbb{R})$ integration by parts yields $-\int_{\mathbb{R}} \sin(e^x)\,\phi'(x)\,dx = \int_{\mathbb{R}} e^x \cos(e^x)\,\phi(x)\,dx$. Consequently $\phi \mapsto \int_{\mathbb{R}} e^x \cos(e^x)\,\phi(x)\,dx$ is a tempered distribution. As for the growth issue, at $x_n = \ln(2\pi n)$ with the integer $n \ge 1$, we have $\lim_{n\to\infty} \frac{e^{x_n}\cos(e^{x_n})}{[1+x_n^2]^k} = \lim_{n\to\infty} \frac{2n\pi}{[1+\ln^2(2n\pi)]^k} = \infty$ for any integer $k \ge 1$.

6. While $\frac{1}{x} \notin L^1(\mathbb{R})$, for $\phi \in S(\mathbb{R})$ we have $\int_{|x|\ge\varepsilon} \frac{\phi(x)}{x}\,dx = \int_{|x|\ge1} \frac{\phi(x)}{x}\,dx + \int_{\varepsilon}^{1} \frac{\phi(x)}{x}\,dx + \int_{-1}^{-\varepsilon} \frac{\phi(x)}{x}\,dx = \int_{|x|\ge1} \frac{\phi(x)}{x}\,dx + \int_{\varepsilon}^{1} \frac{\phi(x)-\phi(0)}{x}\,dx + \int_{-1}^{-\varepsilon} \frac{\phi(x)-\phi(0)}{x}\,dx$ since $\phi(0)\int_{\varepsilon}^{1} \frac{1}{x}\,dx = -\phi(0)\ln\varepsilon = -\phi(0)\int_{-1}^{-\varepsilon} \frac{1}{x}\,dx$. By the mean-value theorem we have $\phi(x) - \phi(0) = x\,\phi'(\xi_x)$ for some ξ_x between 0 and x, so that $\left|\int_{\varepsilon}^{1} \frac{\phi(x)-\phi(0)}{x}\,dx + \int_{-1}^{-\varepsilon} \frac{\phi(x)-\phi(0)}{x}\,dx\right| \le 2\sup_{|s|\le1} |\phi'(s)| \le 2\rho_{0,1}(\phi)$. On the other hand, $\left|\int_{|x|\ge1} \frac{\phi(x)}{x}\,dx\right| \le \int_{|x|\ge1} \frac{(1+x^2)\,|\phi(x)|}{1+x^2}\,dx \le \sup_{s\in\mathbb{R}}\{(1+s^2)\,|\phi(s)|\}\int_{\mathbb{R}} \frac{dx}{1+x^2} \le \pi[\rho_{0,0}(\phi) + \rho_{2,0}(\phi)]$. Consequently the linear functional $\mathrm{pv} : S(\mathbb{R}) \to \mathbb{C}$ satisfies $|\langle \mathrm{pv}, \phi\rangle| \le \pi[\rho_{0,1}(\phi) + \rho_{0,0}(\phi) + \rho_{2,0}(\phi)]$ for all $\phi \in S(\mathbb{R})$. By Theorem 6.5, $\mathrm{pv} \in S'(\mathbb{R})$. In the second functional we cannot take advantage of cancellations: one has $\int_{|x|\ge\varepsilon} \frac{e^{-x^2}}{|x|}\,dx = 2\int_{\varepsilon}^{\infty} \frac{e^{-x^2}}{x}\,dx \ge \frac{2}{e}\int_{\varepsilon}^{1} \frac{dx}{x} = \frac{-2\ln\varepsilon}{e} \to \infty$ as $\varepsilon \downarrow 0$.

7. (i) For $\phi \in S(\mathbb{R})$ we have $\langle x \cdot \mathrm{pv}, \phi\rangle = \langle \mathrm{pv}, x\phi(x)\rangle = \lim_{\varepsilon\downarrow0}\int_{|x|\ge\varepsilon} \phi(x)\,dx = \int_{\mathbb{R}} \phi(x)\,dx = \langle 1, \phi\rangle$ and $\langle x \cdot \delta_0, \phi\rangle = \langle \delta_0, x\phi(x)\rangle = 0$, so that $x \cdot \mathrm{pv} = 1$ and $x \cdot \delta_0 = 0$.

(ii) Assuming it were possible, we would have $x \cdot (\mathrm{pv} \cdot \delta_0) = (x \cdot \mathrm{pv}) \cdot \delta_0 = 1 \cdot \delta_0 = \delta_0$, while $x \cdot (\delta_0 \cdot \mathrm{pv}) = (x \cdot \delta_0) \cdot \mathrm{pv} = 0 \cdot \mathrm{pv} = 0$.

(iii) Assuming it were possible, since $\delta_0 * u = u * \delta_0 = u$ for all $u \in S'(\mathbb{R})$, we would get $1 * (\delta_0' * H) = 1 * (\delta_0 * H)' = 1 * H' = 1 * \delta_0 = 1$ while $(1 * \delta_0') * H = (1 * \delta_0)' * H = 1' * H = 0 * H = 0$.

8. (i) follows by integrating $\frac{1}{(m-1)!}\int_0^t (t-s)^{m-1}\varphi^{(m)}(s)ds$ m-times by parts. For (ii), the change of variable of integration $s = t\tau$ in (i) yields $\phi(t) = t^m\psi(t)$, where $\psi(t) = \frac{1}{(m-1)!}\int_0^1 (1-\tau)^{m-1}\phi^{(m)}(t\tau)\,d\tau$. Clearly $\psi \in C^\infty(\mathbb{R})$. The desired estimates hold since $0 \le \tau^k(1-\tau)^{m-1} \le 1$ for $\tau \in [0,1]$. The fact that $\psi \in \mathcal{D}(\mathbb{R})$ if $\phi \in \mathcal{D}(\mathbb{R})$ follows from $\phi(t) = t^m\psi(t)$. We obtain (iii) by evaluating (i) at $t = 1$ for the function $t \mapsto f(xt)$ since $\frac{1}{k!}\,\partial_t^k f(xt) = \frac{1}{k!}\big(\sum_{j=1}^n x_j \frac{\partial}{\partial y_j}\big)^k f(y)\Big|_{y=xt} = \sum_{|\alpha|=k} \frac{x^\alpha}{\alpha!}\,(D^\alpha f)(xt)$.

9. (i) Necessity is clear. Conversely, if $\phi \in S(\mathbb{R})$ is such that $\int_{\mathbb{R}} \phi(s)\,ds = 0$, then (6.7) shows that $\psi(t) = \int_{-\infty}^t \phi(s)\,ds$ defines a $C^\infty(\mathbb{R})$-function with $\psi' = \phi$. Moreover, for integers $p \ge 0$ and $q \ge 1$ we have $\sup_{t\in\mathbb{R}} |t^p\psi^{(q)}(t)| = \sup_{t\in\mathbb{R}} |t^p\phi^{(q-1)}(t)| < \infty$. For integers $p \ge 0$ and $t \le -1$ we have

$|t^p\psi(t)| \le (1+t^2)^p|\psi(t)| \le (1+t^2)^p \int_{-\infty}^t |\phi(s)|\,ds \le \int_{\mathbb{R}} (1+s^2)^p|\phi(s)|\,ds < \infty$ due to (6.7), since $s \mapsto (1+s^2)^p\phi(s)$ belongs to $\mathcal{S}(\mathbb{R})$. A similar estimate will hold for $t \ge 1$ since $\psi(t) = -\int_t^\infty \phi(s)\,ds$. Finally, $\sup_{|t|\le 1} |t^p\psi(t)| \le \sup_{|t|\le 1} |\psi(t)| \le \int_{\mathbb{R}} |\phi(s)|\,ds < \infty$ due to (6.7), so that $\psi \in \mathcal{S}(\mathbb{R})$.

(ii) Choose $\theta \in \mathcal{D}(\mathbb{R})$ with $\int_{\mathbb{R}} \theta(t)\,dt = 1$ and define on $\mathcal{S}(\mathbb{R})$ the linear map $\phi \mapsto \omega_\phi$ by $\omega_\phi(t) = \int_t^\infty (\phi(s) - \theta(s)\int_{\mathbb{R}} \phi(\tau)d\tau)\,ds$. By (i), $\omega_\phi \in \mathcal{S}(\mathbb{R})$. Moreover, an approach analogous to that in (i) shows that for every integer $N \ge 0$ there exists a constant $\kappa_N(\theta) > 0$ such that $\sum_{p,q\le N} \rho_{p,q}(\omega_\phi) \le \kappa_N(\theta) \sum_{j,k\le 2N+2} \rho_{j,k}(\phi)$ for all $\phi \in \mathcal{S}(\mathbb{R})$. We now claim that for any constant $c \in \mathbb{C}$, $\langle u, \phi\rangle = \langle v, \omega_\phi\rangle + \langle c, \phi\rangle$ defines a primitive $u \in \mathcal{S}'(\mathbb{R})$ of $v \in \mathcal{S}'(\mathbb{R})$, and that this formula[69] exhausts all primitives. Indeed, the previous estimates show that $u \in \mathcal{S}'(\mathbb{R})$ by means of Theorem 6.5 and clearly $\langle u', \phi\rangle = -\langle u, \phi'\rangle = \langle v, \phi\rangle$ for every $\phi \in \mathcal{S}(\mathbb{R})$, since $\omega_{\phi'}(t) = -\phi(t)$ and $\langle c, \phi'\rangle = 0$. Also, if $u_1, u_2 \in \mathcal{S}'(\mathbb{R})$ are such that $u_1' = u_2' = v$, then, setting the constant $c = \langle u_1 - u_2, \theta\rangle$, every $\phi \in \mathcal{S}(\mathbb{R})$ determines $\psi \in \mathcal{S}(\mathbb{R})$ with $\psi' = \phi$ by means of $\psi(t) = \phi(t) - \theta(t)\int_{\mathbb{R}} \phi(\tau)d\tau$, and $\langle u_1 - u_2, \phi\rangle = \langle u_1 - u_2, \psi'\rangle + \int_{\mathbb{R}} \phi(\tau)d\tau\,\langle u_1 - u_2, \theta\rangle = \langle u_1 - u_2, \psi'\rangle + \langle c, \phi\rangle = -\langle u_1' - u_2', \psi\rangle + \langle c, \phi\rangle$ yields $u_1 - u_2 = c$.

10. (i) By definition, $\langle \mathcal{F}^{-1}u, \phi\rangle = \langle u, \mathcal{F}^{-1}\phi\rangle$ for $u \in \mathcal{S}'(\mathbb{R})$, $\phi \in \mathcal{S}(\mathbb{R})$. Since $(\mathcal{F}^{-1}\phi)(\xi) = \int_{\mathbb{R}^n} \phi(x)\,e^{2\pi i x\cdot\xi}dx = \int_{\mathbb{R}^n} \phi(x)\,e^{-2\pi i x\cdot(-\xi)}dx = (\mathcal{F}\phi)(-\xi)$, if u is even/odd, then $\langle \mathcal{F}^{-1}u, \phi\rangle = \langle u, \mathcal{F}^{-1}\phi\rangle = \langle u, (\mathcal{F}\phi)^\dagger\rangle = \pm\langle u, \mathcal{F}\phi\rangle = \pm\langle \mathcal{F}u, \phi\rangle$, that is, $\mathcal{F}^{-1}u = \pm\mathcal{F}u$.

(ii) Exercise 7(i), (6.13) and (6.15) yield $\frac{i}{2\pi}\frac{d}{d\xi}\mathcal{F}(\mathrm{pv}) = \mathcal{F}(x\cdot\mathrm{pv}) = \mathcal{F}(1) = \delta_0$, so that $\frac{d}{d\xi}\mathcal{F}(\mathrm{pv}) = -2\pi i\,\delta_0$. Since we know that $(-2\pi i)\frac{d}{d\xi}H(\xi) = -2\pi i\,\delta_0$, we infer from Exercise 9(ii) that $\mathcal{F}(\mathrm{pv}) = -2\pi i\,H + c$ for some constant $c \in \mathbb{C}$. Now note that pv is odd since $\langle \mathrm{pv}, \phi^\dagger\rangle = -\langle \mathrm{pv}, \phi\rangle$ holds for all $\phi \in \mathcal{S}(\mathbb{R})$, due to the relation $\int_{|x|>\varepsilon} \frac{\phi(-x)}{x}\,dx = \int_{|y|>\varepsilon} \frac{\phi(y)}{y}\,dy$ that follows by the change of variables $y = -x$. Using (i), we deduce that $\langle -2\pi iH + c, \phi^\dagger\rangle = -\langle -2\pi iH + c, \phi\rangle$ for all $\phi \in \mathcal{S}(\mathbb{R})$, or $-2\pi i\int_0^\infty \phi(-x)\,dx + c\int_{\mathbb{R}} \phi(-x)\,dx = 2\pi i\int_0^\infty \phi(x)\,dx - c\int_{\mathbb{R}} \phi(x)\,dx$. Changing variables $y = -x$, we get $\int_0^\infty \phi(-x)\,dx = \int_{-\infty}^0 \phi(y)\,dy$ and $\int_{\mathbb{R}} \phi(-x)\,dx = \int_{\mathbb{R}} \phi(y)\,dy$, so that $(2\pi i - 2c)\int_{\mathbb{R}} \phi(y)\,dy = 0$ for all $\phi \in \mathcal{S}(\mathbb{R})$. Thus $c = \pi i$ and $\mathcal{F}(\mathrm{pv}) = -2\pi i\,H + \pi i = \begin{cases} -\pi i & \text{if } \xi > 0, \\ \pi i & \text{if } \xi < 0. \end{cases}$
Applying $\mathcal{F}$ to the previous relation yields $\mathcal{F}(\mathcal{F}(\mathrm{pv})) = -2\pi i\mathcal{F}(H) + \pi i\mathcal{F}(1) = -2\pi i\mathcal{F}(H) + \pi i\,\delta_0$ by (6.15). Consequently $\langle \mathcal{F}(H), \phi\rangle = \frac{1}{2}\langle \delta_0, \phi\rangle - \frac{1}{2\pi i}\langle \mathcal{F}(\mathcal{F}(\mathrm{pv})), \phi\rangle$ for all $\phi \in \mathcal{S}(\mathbb{R})$. Since $\langle \mathcal{F}(\mathcal{F}(\mathrm{pv})), \phi\rangle = \langle \mathcal{F}(\mathrm{pv}), \hat{\phi}\rangle = \langle \mathrm{pv}, \mathcal{F}(\hat{\phi})\rangle = \langle \mathrm{pv}, \phi^\dagger\rangle = -\langle \mathrm{pv}, \phi\rangle$ by (6.11), Theorem 6.1 and the fact that pv is odd, we get $\langle \mathcal{F}(H), \phi\rangle = \frac{1}{2}\langle \delta_0, \phi\rangle + \frac{1}{2\pi i}\langle \mathrm{pv}, \phi\rangle$. Thus $\mathcal{F}(H) = \frac{1}{2}\,\delta_0 + \frac{1}{2\pi i}\,\mathrm{pv}$.

11. Continuity holds clearly if $f \in \mathcal{O}(\mathbb{R}^n)$. Conversely, note that $f\phi \in \mathcal{S}(\mathbb{R}^n)$ for $\phi(x) = e^{-|x|^2}$ implies $f \in C^\infty(\mathbb{R}^n)$. To show that f must have polynomial growth, assume the contrary. Then there is a sequence $\{x_k\}_{k\ge 1}$ in $\mathbb{R}^n$ such that $|f(x_k)| \ge k\,(1 + |x_k|^2)^k$ for every $k \ge 1$. The sequence $\{x_k\}_{k\ge 1}$ is not bounded, as otherwise it would have an accumulation point at which f could not be finite. Passing, if needed, to a subsequence, we may assume that $|x_{k+1} - x_k| > 4\sqrt{n}$ for all $k \ge 1$. Let now $\theta \in \mathcal{D}(\mathbb{R}^n)$ be such that $\theta(x) = 1$ for $|x| \le 1$ and $\theta(x) = 0$ for $|x| \ge 2\sqrt{n}$. Then $\phi(x) = \sum_{k\ge 1}(1 + |x|^2)^{-k}\theta(x - x_k) \in \mathcal{S}(\mathbb{R}^n)$ with $|f(x_k)\phi(x_k)| \ge k$ for all $k \ge 1$. This being prevented by $f\phi \in \mathcal{S}(\mathbb{R}^n)$ proves that f must have polynomial growth. We now show that $D^\beta f$ has this property by induction on $|\beta|$. Since there are integers c_α with $\partial_{x_j}D^\beta(f\phi) = (\partial_{x_j}D^\beta f)\phi + \sum_{|\alpha|\le|\beta|} c_\alpha(D^\alpha f)(D^{\beta-\alpha}\phi)$ for all $\phi \in \mathcal{S}(\mathbb{R}^n)$, the polynomial growth of $D^\alpha f$ for all $|\alpha| \le |\beta|$ ensures that $d_{\gamma,0}((D^\alpha f)(D^{\beta-\alpha}\phi)) < \infty$ for all n-tuples of nonnegative integers γ, so that $f\phi \in \mathcal{S}(\mathbb{R}^n)$ forces $d_{\gamma,0}((\partial_{x_j}D^\beta f)\phi) < \infty$. Implementing the approach used for f, we infer that $(\partial_{x_j}D^\beta f)$ has polynomial growth.

12. Formally plug-in $u(x,t) = f(x-ct)$ to get the ordinary differential equation $-cf' + cf''' + 3ff' = 2f'f'' + ff'''$ in the variable $s = x - ct$. Integrating on $(-\infty, s]$ leads to $-cf + cf'' + \frac{3}{2}f^2 = \frac{1}{2}(f')^2 +$

69 To guess it, note that, given $v \in \mathcal{S}'(\mathbb{R})$, if $u' = v$ has a solution $u \in \mathcal{S}'(\mathbb{R})$, then for every $\phi \in \mathcal{S}(\mathbb{R})$ we should have $\langle v, \omega_\phi\rangle = \langle u', \omega_\phi\rangle = -\langle u, \partial_t\omega_\phi\rangle = \langle u, \phi - \theta\int_{\mathbb{R}} \phi(\tau)d\tau\rangle$ $= \langle u, \phi\rangle - \langle u, \theta\rangle\int_{\mathbb{R}} \phi(\tau)d\tau$, so that $\langle u, \phi\rangle = \langle v, \omega_\phi\rangle + \langle c, \phi\rangle$ with $c = \langle u, \theta\rangle$.

ff''. Multiply this by f' and integrate on $(-\infty, s]$ to obtain $-\frac{c}{2}f^2 + \frac{c}{2}(f')^2 + \frac{1}{3}f^3 = \frac{1}{2}f(f')^2$, or $[(f')^2 - f^2](c-f) = 0$. Assuming f to be continuous, on the open set where $f \neq c$ we have $f' = \pm f$, so that $f(s) = a\,e^{\pm s}$ on each constituent open interval. Coupled with the decay assumption, we arrive at the candidate $f_c(s) = c\,e^{-|s|}$. However, the 'peakon' $u_c(x,t) = c\,e^{-|x-ct|}$ fails to be differentiable, and therefore cannot be a classical solution. The best we can hope for is that it satisfies the equation in some weak sense. As a bounded continuous function, $u_c \in \mathcal{S}'(\mathbb{R}^2)$ has distributional derivatives of any order but the nonlinear terms in the equation require special attention. Note that $\partial_x u_c = c\,e^{-|x-ct|}\mathrm{sgn}(ct-x) \in L^\infty(\mathbb{R}^2)$ since for any $\phi \in \mathcal{S}(\mathbb{R}^2)$, $\langle \partial_x u_c, \phi\rangle = -\langle u_c, \partial_x\phi\rangle = -c\int_{\mathbb{R}}\Big(\int_{-\infty}^{ct} e^{x-ct}\phi_x(x,t)\,dx + \int_{ct}^{\infty} e^{ct-x}\phi_x(x,t)\,dx\Big)dt = c\int_{\mathbb{R}}\int_{\mathbb{R}} e^{-|x-ct|}\mathrm{sgn}(ct-x)\,\phi(x,t)\,dxdt$ after integration by parts. Consequently $\partial_x u_c$, $(\partial_x u_c)^2$, $u_c(\partial_x u_c) \in \mathcal{S}'(\mathbb{R}^2)$ and we can attempt the weak formulation $\partial_t(1-\partial_x^2)u + (1-\partial_x^2)uu_x + \partial_x(u^2 + \frac{1}{2}u_x^2) = 0$ in $\mathcal{S}'(\mathbb{R}^2)$. For functions u that are three times differentiable, the pointwise interpretation of the expression on the left is precisely the given equation. To show that $u_c \in \mathcal{S}'(\mathbb{R}^2)$ is a solution, for $\phi \in \mathcal{S}(\mathbb{R}^2)$ we compute $\langle \partial_t(1-\partial_x^2)u_c, \phi\rangle = -\langle u_c, \partial_t(1-\partial_x^2)\phi\rangle = -c\int_{\mathbb{R}}\Big(\int_{-\infty}^{ct} e^{x-ct}(1-\partial_x^2)\phi_t\,dx + \int_{ct}^{\infty} e^{ct-x}(1-\partial_x^2)\phi_t\,dx\Big)dt = -2c\int_{\mathbb{R}}\phi_t(ct,t)\,dt$ after two integration by parts, $\langle (1-\partial_x^2)(u_c\,\partial_x u_c), \phi\rangle = \langle u_c\,\partial_x u_c, (1-\partial_x^2)\phi\rangle = c^2\int_{\mathbb{R}}\Big(\int_{-\infty}^{ct} e^{2x-2ct}(1-\partial_x^2)\phi\,dx - \int_{ct}^{\infty} e^{2ct-2x}(1-\partial_x^2)\phi\,dx\Big)dt = -3c^2\int_{\mathbb{R}}\int_{\mathbb{R}} e^{-2|x-ct|}\mathrm{sgn}(ct-x)\,\phi(x,t)\,dxdt - 2c^2\int_{\mathbb{R}}\phi_x(ct,t)\,dt$ after two integration by parts, $\langle \partial_x(u_c^2 + \frac{1}{2}(\partial_x u_c)^2), \phi\rangle = -\langle u_c^2 + \frac{1}{2}(\partial_x u_c)^2, \phi_x\rangle = -\langle \frac{3}{2}u_c^2, \phi_x\rangle = -\frac{3}{2}c^2\int_{\mathbb{R}}\Big(\int_{-\infty}^{ct} e^{2x-2ct}\phi_x\,dx + \int_{ct}^{\infty} e^{2ct-2x}\phi_x\,dx\Big)dt = 3c^2\int_{\mathbb{R}}\int_{\mathbb{R}} e^{-2|x-ct|}\mathrm{sgn}(ct-x)\,\phi(x,t)\,dxdt$. We conclude since $\int_{\mathbb{R}}[\phi_t(ct,t) + c\phi_x(ct,t)]\,dt = \int_{\mathbb{R}}(\frac{\partial}{\partial t}\phi(ct,t))\,dt = 0$ for $\phi \in \mathcal{S}(\mathbb{R}^2)$.

13. $u = \partial_{x_1}\delta_0$ and $\phi(x) = x_1\,e^{-|x|^2}$ are such that $\mathrm{supp}(u) = \{0\}$ and $\mathrm{supp}(\phi) = \mathbb{R}$, despite $\phi(0) = 0$, with $\langle u, \phi\rangle = -\langle \delta_0, \partial_{x_1}\phi\rangle = 1$. Assume now that $\mathrm{supp}(u) \cap \mathrm{supp}(\phi) = \emptyset$, with $\mathrm{supp}(\phi)$ not necessarily compact as perhaps $\phi \notin \mathcal{D}(\mathbb{R})$. Every $x \in \mathrm{supp}(u)$ has an open neighbourhood N_x such that $N_x \cap \mathrm{supp}(\phi) = \emptyset$. Indeed, $\mathrm{supp}(\phi)$ being closed ensures $\inf_{y\in\mathrm{supp}(\phi)}|x-y| = \delta_x > 0$ and we can set $N_x = \{y \in \mathbb{R}^n : |x-y| < \frac{\delta_x}{4\sqrt{n}}\}$. Since $\mathrm{supp}(u) \subset \bigcup_{x\in\mathrm{supp}(u)} N_x$ is compact, we can find a finite number of these open sets $\{N_j\}_{j=1,\dots,m}$ that cover it. If we denote by N_j^2 the open balls concentric with N_j and having a diameter $2\sqrt{n}$-times larger, then open sets $O_1 = \bigcup_{j=1}^m N_j$ and $O_2 = \bigcup_{j=1}^m N_j^2$ are such that $\mathrm{supp}(u) \subset O_1 \subset O_2$ and $O_2 \cap \mathrm{supp}(\phi) = \emptyset$. We now construct $\theta \in \mathcal{D}(\mathbb{R}^n)$ with $0 \le \theta(x) \le 1$ for all $x \in \mathbb{R}^n$, $\mathrm{supp}(\theta) \subset O_2$ and $\theta(x) = 1$ for $x \in O_1$. For this, choose for any $j \in \{1,\dots,m\}$ a function $\theta_j \in \mathcal{D}(\mathbb{R}^n)$ such that $0 \le \theta_j(x) \le 1$ for all $x \in \mathbb{R}^n$, $\theta_j(x) = 1$ for $x \in N_j$ and $\theta_j(x) = 0$ for $x \notin O_2$. Such a function can be obtained by translating and scaling the function in $\mathcal{D}(\mathbb{R}^n)$ provided towards the end of Section 6.1.1: if $0 \le \varphi(x) \le 1$ for all $x \in \mathbb{R}^n$, $\varphi(x) = 1$ for $|x| \le 1$ and $\varphi(x) = 0$ for $|x| > 2\sqrt{n}$, then $\psi(x) = \varphi(\frac{4\sqrt{n}(x-x^*)}{\delta_x})$ is such that $0 \le \psi(x) \le 1$ for all $x \in \mathbb{R}^n$, $\psi(x) = 1$ for $|x - x^*| \le \frac{\delta_x}{4\sqrt{n}}$ and $\psi(x) = 0$ for $|x - x^*| > \frac{\delta_x}{2}$. All requirements are met by $\theta = 1 - \prod_{j=1}^m(1-\theta_j)$. We conclude as follows. By construction, $\theta u = u$ and $\theta\phi = 0$, so that $\langle u, \phi\rangle = \langle \theta u, \phi\rangle = \langle u, \theta\phi\rangle = \langle u, 0\rangle = 0$.

14. If $\hat\phi$ vanishes on a neighbourhood of ξ^*, replacing ϕ by $x \mapsto \phi(x)e^{-2\pi i\xi^*\cdot x}$, we may assume that $\xi^* = 0$, in view of (6.14). Since ϕ has compact support and $e^{-2\pi i\xi\cdot x} = \sum_{k\ge 0}\frac{(-2\pi i\xi\cdot x)^k}{k!}$, we have that $\hat\phi(\xi) = \int_{\mathbb{R}^n}\phi(x)\,e^{-2\pi i\xi\cdot x}dx = \sum_\alpha \frac{\xi^\alpha}{\alpha!}\int_{\mathbb{R}^n}(-2\pi i x)^\alpha\phi(x)\,dx$, in view of the identity $(x_1 + \dots x_n)^k = \sum_{|\alpha|=k}\frac{k!}{\alpha!}x^\alpha$. Since $\int_{\mathbb{R}^n}(-2\pi i x)^\alpha\phi(x)\,dx = (D^\alpha\hat\phi)(0)$ by (6.13), we obtain $\hat\phi(\xi) = \sum_\alpha\frac{\xi^\alpha}{\alpha!}(D^\alpha\hat\phi)(0)$. By assumption $(D^\alpha\hat\phi)(0) = 0$ for all α, so that $\hat\phi = 0$ and $\phi = 0$.

15. The set $K = \mathrm{supp}(u) + \mathrm{supp}(v) = \{x + y : x \in \mathrm{supp}(u), y \in \mathrm{supp}(v)\}$ is compact since every sequence in K clearly admits a subsequence converging to some limit in K. Proving $\mathrm{supp}(u*v) \subset K$ amounts to showing that $\langle u * v, \phi\rangle = 0$ for every $\phi \in \mathcal{D}(\mathbb{R}^n)$ with $\mathrm{supp}(\phi) \subset \mathbb{R}^n \setminus K$. For such a function ϕ we have $\langle u * v, \phi\rangle = \langle \mathcal{F}^{-1}(\hat u\hat v), \phi\rangle = \langle \hat u\hat v, \mathcal{F}^{-1}(\phi)\rangle = \langle \hat u, \hat v\mathcal{F}^{-1}(\phi)\rangle = \langle \hat u, \hat v\mathcal{F}(\phi^\dagger)\rangle = \langle \hat u, \mathcal{F}(v*\phi^\dagger)\rangle = \langle u, \mathcal{F}(\mathcal{F}(v*\phi^\dagger))\rangle$, in view of (6.16), (6.12), the definition of multiplication of $w \in \mathcal{S}'(\mathbb{R}^n)$ by a function in $\mathcal{O}(\mathbb{R}^n)$, Theorem 6.9, and Theorem 6.1. Since $\mathcal{F}(v * \phi^\dagger) = \hat v\,\widehat{(\phi^\dagger)} \in \mathcal{S}(\mathbb{R}^n)$ ensures $v*\phi^\dagger \in \mathcal{S}(\mathbb{R}^n)$, Theorem 6.1 yields $\mathcal{F}(\mathcal{F}(v*\phi^\dagger)) = (v*\phi^\dagger)^\dagger$, so that $\langle u*v, \phi\rangle = \langle u, (v*\phi^\dagger)^\dagger\rangle$. Recall from Theorem 6.7 that $v * \phi^\dagger$ is the function $\varphi(x) = \langle v, \tau_x\phi\rangle$. We claim that $\varphi \in \mathcal{D}(\mathbb{R}^n)$

with $\text{supp}(\varphi) \subset K_1 = \{w - y : y \in \text{supp}(\phi), w \in \text{supp}(v)\}$. Indeed, clearly $\tau_x\phi \in \mathcal{D}(\mathbb{R}^n)$ with $\text{supp}(\tau_x\phi) = \{x + y : y \in \text{supp}(\phi)\}$ and for $x \notin K_1$ we have that $\text{supp}(v) \cap \text{supp}(\tau_x\phi^\dagger) = \emptyset$, so that $\varphi(x) = 0$ by Exercise 13. Consequently $\varphi^\dagger \in \mathcal{D}(\mathbb{R}^n)$ with $\text{supp}(\varphi^\dagger) \subset K_2 = \{-\xi : \xi \in K_1\}$. We see that $\text{supp}(\varphi^\dagger) \cap \text{supp}(u) = \emptyset$ since $K_2 \cap \text{supp}(u) = \emptyset$ due to $\text{supp}(\phi) \subset \mathbb{R}^n \setminus K$. Exercise 13 permits us to conclude that $\langle u * v, \phi\rangle = 0$.

16. By (6.15), δ_0 belongs to the first set and 1 to the second.

17. (i) The most natural way to proceed is to avoid the possible singularity at $x = 0$. Given $\phi \in \mathcal{S}(\mathbb{R})$, By Exercise 8(ii) there exists $\psi \in \mathcal{S}(\mathbb{R})$ with $\phi(x) = \sum_{k=0}^{n-1} \frac{x^k}{k!}\phi^{(k)}(0) + x^n\psi(x)$, so that we may set $\langle U, \phi\rangle = \int_{|x|\ge 1} u(x)\phi(x)\,dx + \int_{|x|<1} u(x)x^n\psi(x)\,dx$. Since by hypothesis $x \mapsto u(x)x^n$ is locally integrable on $\mathbb{R}$ and $\sup_{|x|\le 1}|\psi(x)| \le \sup_{|x|\le 1}|\phi^{(n)}(x)|$ due to Exercise 8(iii), Theorem 6.6 ensures $U \in \mathcal{S}'(\mathbb{R})$. Note that if $\text{supp}(\phi) \subset \mathbb{R}\setminus\{0\}$, then $x^n\psi(x) = \phi(x)$, so that $\langle U, \phi\rangle = \langle u, \phi\rangle$.

(ii) The procedure used in (i) leads us to define $\langle U, \phi\rangle = \int_{|x|\ge 1} \frac{\phi(x)}{x^2}\,dx + \int_{|x|<1}\psi(x)\,dx$ for $\phi \in \mathcal{S}(\mathbb{R})$, where $\psi \in \mathcal{S}(\mathbb{R})$ is given by $\psi(x) = \frac{\phi(x)-\phi(0)-x\phi'(0)}{x^2}$ for $x \neq 0$. By dominated convergence $\int_{|x|<1}\psi(x)\,dx = \lim_{\varepsilon\downarrow 0}\int_{\varepsilon<|x|<1}\psi(x)\,dx$, while $\int_{-1}^{-\varepsilon}\frac{\phi(0)}{x^2}\,dx + \int_\varepsilon^1 \frac{\phi(0)}{x^2}\,dx = (\frac{2}{\varepsilon} - 2)$ and $\int_{-1}^{-\varepsilon}\frac{\phi'(0)}{x}\,dx + \int_\varepsilon^1 \frac{\phi'(0)}{x}\,dx = 0$, so that $\langle U, \phi\rangle = \lim_{\varepsilon\downarrow 0}\big[\int_{|x|>\varepsilon}\frac{\phi(x)}{x^2} - \frac{2}{\varepsilon}\phi(0)\big] + 2\phi(0)$.

(iii) Repeated application of l'Hospital's rule shows that the origin is a singularity of infinite order since $\lim_{x\downarrow 0} x^n e^{\frac{1}{x^2}} = \infty$ for every integer $n \ge 0$. Choose now $\theta \in \mathcal{D}(\mathbb{R})$ with $\theta \ge 0$, $\text{supp}(\theta) \subset [1, 2]$ and such that $\theta(x) = 1$ if $x \in [a, b]$, for some $a, b \in (1, 2)$. If $\theta_n(x) = e^{-n}\,\theta(nx)$ for $n \ge 1$, then $\theta_n \ge 0$, $\text{supp}(\theta_n) \subset [\frac{1}{n}, \frac{2}{n}]$ and $\theta_n \to 0$ in $\mathcal{S}(\mathbb{R})$ since for all integers $q, p \ge 0$ we have that $\sup_{x\in\mathbb{R}}|x^q\theta_n^{(p)}(x)| \le (\frac{2}{n})^q e^{-n}n^p \sup_{1\le y\le 2}|\theta^{(p)}(y)| \to 0$ as $n \to \infty$. The existence of $U \in \mathcal{S}'(\mathbb{R})$ with $\langle U, \phi\rangle = \int_\mathbb{R} u(x)\phi(x)\,dx$ for all $\phi \in \mathcal{S}(\mathbb{R})$ with $\text{supp}(\phi) \subset \mathbb{R}\setminus\{0\}$ would yield $\langle U, \theta_n\rangle \to 0$. However, $\langle U, \theta_n\rangle = \int_\mathbb{R} e^{\frac{1}{x^2}}\theta_n(x)\,dx \ge e^{-n}\int_{a/n}^{b/n} e^{\frac{1}{x^2}}\,dx \ge e^{-n}\,\frac{b-a}{n}\,e^{\frac{n^2}{b^2}} \to \infty$ for $n \to \infty$.

18. (i) By Theorem 6.6, there exists $C > 0$ and an integer $n \ge 0$ with $|\langle u, \phi\rangle| \le C\sum_{0\le q,p\le n}\rho_{q,p}(\phi)$ for all $\phi \in \mathcal{S}(\mathbb{R})$. Choose $\psi \in \mathcal{D}(\mathbb{R})$ with $\psi(x) = 1$ for $|x| \le 1$ and $\psi(x) = 0$ for $|x| \ge 2$. Then for every $\varepsilon > 0$ we have that $\psi_\varepsilon(x) = \psi(\frac{x}{\varepsilon})$ belongs to $\mathcal{D}(\mathbb{R})$, with $\psi_\varepsilon(x) = 1$ for $|x| \le \varepsilon$ and $\text{supp}(\psi_\varepsilon) \subset [-2\varepsilon, 2\varepsilon]$. Consequently $\langle u, \phi\rangle = \langle u, \psi\phi\rangle$ for all $\phi \in \mathcal{S}(\mathbb{R})$. If $\phi(0) = \phi'(0) = \cdots = \phi^{(n)}(0) = 0$, Exercise 8(ii) yields $\phi(x) = x^{n+1}\varphi(x)$ for some $\varphi \in \mathcal{S}(\mathbb{R})$, and for every $0 \le q, p \le n$ we find that $\rho_{q,p}(\phi) \le M_{q,p}\varepsilon^{n+1-p+q}$ for some constant $M_{q,p} > 0$ that depends on $\sup_{|x|\le 2}|\psi^{(k)}(x)|$ and $\sup_{|x|\le 2}|\phi^{(k)}(x)|$ with $0 \le k \le n+1$ but is independent of $\varepsilon \in (0, 1)$. Letting $\varepsilon \to 0$, we get that $\langle u, \phi\rangle = 0$ for all $\phi \in \mathcal{S}(\mathbb{R})$ with $\phi(0) = \phi'(0) = \cdots = \phi^{(n)}(0) = 0$. By Exercise 8(iii), we can write any $\phi \in \mathcal{S}(\mathbb{R})$ as $\phi(x) = \sum_{k=0}^n \frac{x^k}{k!}\phi^{(k)}(0) + x^{n+1}\varphi(x)$ with $\varphi \in \mathcal{S}(\mathbb{R})$. Then $\langle u, \phi\rangle = \langle u, \psi\phi\rangle = \sum_{k=0}^n \frac{\phi^{(k)}(0)}{k!}\langle u, x^k\psi(x)\rangle$, due to the previously established property. Since $\langle D^k\delta_0, \phi\rangle = (-1)^k\langle\delta_0, \phi^{(k)}\rangle = (-1)^k\phi^{(k)}(0)$, we get $\langle u, \phi\rangle = \langle\sum_{k=0}^n c_k\,D^k\delta_0, \phi\rangle$, with $c_k = \frac{(-1)^k}{k!}\langle u, x^k\psi(x)\rangle$ for $k = 0, \dots, n$. Thus $u = \sum_{k=0}^n c_k\,D^k\delta_0$.

(ii) If $U_1, U_2 \in \mathcal{S}'(\mathbb{R})$ are two renormalisations, then $U_1 - U_2$ has support $\{0\}$, so that $U_1 - U_2 = \sum_{k=0}^n c_k\,D^k\delta_0$ for some integer $n \ge 0$ and some constants $c_k \in \mathbb{C}$. Clearly, adding to the renormalisation described at the end of Section 6.1.1 a distribution of type $\sum_{k=0}^n c_k\,D^k\delta_0$ yields another renormalisation. Consequently, there are infinitely many renormalisations.

19. (i) Let us first discuss the case $v = 0$. In this setting we claim that a nontrivial solution $u \in \mathcal{S}'(\mathbb{R})$ to $x^n u = 0$ in $\mathcal{S}'(\mathbb{R})$ must have $\text{supp}(u) = \{0\}$. Indeed, if $\phi \in \mathcal{D}(\mathbb{R})$ has $\text{supp}(\phi) \subset \mathbb{R}\setminus\{0\}$, then $\varphi(x) = x^{-n}\phi(x)$ belongs to $\mathcal{D}(\mathbb{R})$ and $0 = \langle x^n u, \varphi\rangle = \langle u, x^n\varphi\rangle = \langle u, \phi\rangle$. Exercise 18(i) yields $u = \sum_{k=0}^m c_k\,D^k\delta_0$ for some integer $m \ge 0$ and some constants $c_k \in \mathbb{C}$. For $\phi \in \mathcal{S}(\mathbb{R})$ we compute $\langle x^n\,D^k\delta_0, \phi\rangle = (-1)^k\langle\delta_0, D^k(x^n\phi)\rangle = \begin{cases} 0 \text{ if } k < n, \\ (-1)^k n!\,\phi^{(k-n)}(0) \text{ if } k \ge n, \end{cases}$ and this shows that the general solution of $x^n u = 0$ in $\mathcal{S}'(\mathbb{R})$ is $u = \sum_{k=0}^{n-1} c_k\,D^k\delta_0$ for some constants $c_k \in \mathbb{C}$, $0 \le k \le n-1$. Now, given $v \in \mathcal{E}'(\mathbb{R})$, if $u_1, u_2 \in \mathcal{S}'(\mathbb{R})$ solve $x^n u = v$ in $\mathcal{S}'(\mathbb{R})$, then $x^n(u_1 - u_2) = 0$

in $\mathcal{S}'(\mathbb{R})$, so that the general solution to $x^n u = v$ in $\mathcal{S}'(\mathbb{R})$ differs by an expression of the form $\sum_{k=0}^{n-1} c_k D^k\delta_0$, with $c_k \in \mathbb{C}$ constants for $0 \le k \le n-1$, from a particular solution. To find a particular solution, if $\mathrm{supp}(v) \subset [-M, M]$, choose $\theta \in \mathcal{D}(\mathbb{R})$ with $\theta(x) = 1$ for $|x| \le M$. By Exercise 8(iii), any $\phi \in \mathcal{S}(\mathbb{R})$ is of the form $\phi(x) = \sum_{k=0}^{n-1} \frac{x^k}{k!}\, \phi^{(k)}(0) + x^n\, \psi(x)$ with $\psi \in \mathcal{S}(\mathbb{R})$. Consequently, for a solution $u \in \mathcal{S}'(\mathbb{R})$ we have $\langle u, \phi\rangle = \langle \theta u, \phi\rangle = \langle u, \theta\phi\rangle = \sum_{k=0}^{n-1} \frac{\phi^{(k)}(0)}{k!}\langle u, x^k\theta\rangle + \langle u, x^n\psi\theta\rangle = \sum_{k=0}^{n-1} \frac{\phi^{(k)}(0)}{k!}\langle u, x^k\theta\rangle + \langle x^n u, \psi\theta\rangle = \sum_{k=0}^{n-1} \frac{\phi^{(k)}(0)}{k!}\langle u, x^k\theta\rangle + \langle v, \psi\theta\rangle = \langle \sum_{k=0}^{n-1} c_k D^k\delta_0, \phi\rangle + \langle v, \psi\theta\rangle$, where $c_k = \frac{(-1)^k\phi^{(k)}(0)}{k!}\langle u, x^k\theta\rangle$ for $0 \le k \le n-1$. The explicit formula for ψ from Exercise 8(iii) shows by means of Theorem 6.6 that $\langle u, \phi\rangle = \langle v, \theta\psi\rangle$ defines an element of $\mathcal{S}'(\mathbb{R})$. This is a particular solution since $\langle x^n u, \phi\rangle = \langle u, x^n\phi\rangle = \langle v, \phi\theta\rangle$ by definition[70] and $\langle v, \phi\theta\rangle = \langle v, \phi\rangle$.

(ii) Choose $\theta \in \mathcal{D}(\mathbb{R})$ with $\theta \ge 0$, $\theta(x) = 1$ if $|x| \le 1$ and $\mathrm{supp}(\theta) \subset [-2, 2]$. For $\varepsilon > 0$ define $\theta_\varepsilon \in \mathcal{D}(\mathbb{R})$ by $\theta_\varepsilon(x) = \frac{1}{\varepsilon}\,\theta(\frac{x}{\varepsilon})$. If $u \in \mathcal{S}'(\mathbb{R})$ were a solution of $fu = 1$ in $\mathcal{S}'(\mathbb{R})$, by Theorem 6.6 we could find a constant $C > 0$ and an integer $n \ge 0$ such that $|\langle u, \phi\rangle| \le C \sum_{0 \le q, p \le n} \sup_{x\in\mathbb{R}} |x^q\phi^{(p)}(x)|$ for all $\phi \in \mathcal{S}(\mathbb{R})$. On the other hand, due to Exercise 8(ii), the fact that $f^{(k)}(0) = 0$ for all $k \ge 0$ yields the existence of a constant $M > 0$, independent of $\varepsilon \in (0, 1)$, such that $\sup_{|x|\le 2\varepsilon} |f^{(k)}(x)| \le \varepsilon M$ for all $0 \le k \le n$. Since $\mathrm{supp}(f\theta_\varepsilon) \subset [-2\varepsilon, 2\varepsilon]$, we get $\lim_{\varepsilon\downarrow 0}\langle u, f\theta_\varepsilon\rangle = 0$. However, this is in contradiction with $\langle u, f\theta_\varepsilon\rangle = \langle fu, \theta_\varepsilon\rangle = \langle 1, \theta_\varepsilon\rangle = \int_{\mathbb{R}} \theta_\varepsilon(x)\,dx = \int_{\mathbb{R}} \theta(y)\,dy \ge 1$, using the change of variables $y = \frac{x}{\varepsilon}$ and the fact that $\theta(x) \ge 0$ for all $x \in \mathbb{R}$, with $\theta(x) = 1$ for $|x| \le 1$.

20. (i) For $\phi \in \mathcal{S}(\mathbb{R})$, integration by parts yields $|\langle \cos(2\pi nx), \phi\rangle| = \frac{1}{2\pi n}\left|\int_{\mathbb{R}} [\sin(2\pi nx)]'\phi(x)\,dx\right| = \frac{1}{2\pi n}\left|\int_{\mathbb{R}} \sin(2\pi nx)\phi'(x)\,dx\right| \le \frac{1}{2\pi n}\int_{\mathbb{R}} |\phi'(x)|\,dx \to 0$ as $n \to \infty$. On the other hand, the numerical sequence $\{\cos(2\pi nx)\}_{n\ge 1}$ clearly converges (to 1) if $x \in \mathbb{Z}$, while if $x = \frac{q}{p}$ for q, p relatively prime integers with $p > 1$, then $\cos(2\pi p^k x) = 1$ and $\cos(2\pi(p^k + 1)x) = \cos(2\pi\frac{q}{p}) \ne 1$ shows that $\{\cos(2\pi nx)\}_{n\ge 1}$ fails to converge. For $x \in \mathbb{R} \setminus \mathbb{Q}$ we have $\left|\frac{1}{N}\sum_{n=0}^{N-1} \cos(2\pi nx)\right| \le \frac{1}{N}\left|\sum_{n=0}^{N-1} e^{2\pi inx}\right| = \frac{|1-e^{2\pi iNx}|}{N|1-e^{2\pi ix}|} \le \frac{2}{N|1-e^{2\pi ix}|} \to 0$ as $N \to \infty$. If $\lim_{n\to\infty}\cos(2\pi nx) = l$, then the left side converges to l, see the considerations preceding Theorem 4.4. But the only option $l = 0$ leads to the contradiction $\cos(4\pi nx) = 2\cos^2(2\pi nx) - 1 \to -1$. Finally, set $f_n(x) = \cos(2\pi nx)$ for $n \ge 1$. If $f_n \to f$ in $L^p[0, 1]$ for some $p \ge 1$, Hölder's inequality yields $f_n \to f$ in $L^1[0, 1]$. Property (P2) from Section 4.1 forces $\hat{f}(k) = \lim_{n\to\infty}\widehat{f_n}(k) = 0$ for every $k \in \mathbb{Z}$ since $\widehat{f_n}(k) = 0$ for $n \ne |k|$. By the uniqueness of Fourier coefficients (see Chapter 4) we infer $f = 0$ a.e. But $f_n \to 0$ in $L^1[0, 1]$ leads to $\lim_{n\to\infty}\int_0^1 |\cos(2\pi nx)|dx = 0$ and we actually have that $\int_0^1 |\cos(2\pi nx)|dx \ge -\sum_{k=0}^{n-1}\int_{\frac{4k+1}{4n}}^{\frac{4k+3}{4n}} \cos(2\pi nx)dx = \sum_{k=0}^{n-1}\frac{1}{\pi n} = \frac{1}{\pi}$.

(ii) For every $x \ne 0$ there is some integer $n_0 \ge 1$ with $|x| > \frac{1}{n_0}$, so that $f_n(x) = 0$ for all $n \ge n_0$ and $\lim_{n\to\infty} f_n(x) = 0$. Choose now some $\phi \in \mathcal{D}(\mathbb{R})$ with $\phi(x) = 1$ for $|x| \le 1$. Then $\langle f_n, \phi\rangle = n^2\int_{-1/n}^{1/n}\phi(x)\,dx = 2n \to \infty$. This rules out convergence in $\mathcal{S}'(\mathbb{R})$.

21. (i) Since $\det(R) = 1$ and $R^{-1} = R^*$ yields $(R\xi)\cdot y = \xi\cdot(R^{-1}y)$ for all $\xi, y \in \mathbb{R}^n$, for $f \in L^1(\mathbb{R}^n)$ we get by means of the change of variables $y = Rx$ that $\widehat{f\circ R}(\xi) = \int_{\mathbb{R}^n} f(Rx)\,e^{-2\pi i\xi\cdot x}dx = \int_{\mathbb{R}^n} f(y)\,e^{-2\pi i\xi\cdot(R^{-1}y)}dy = \int_{\mathbb{R}^n} f(y)\,e^{-2\pi i(R\xi)\cdot y}dy = \hat{f}(R\xi)$ for all $\xi \in \mathbb{R}^n$. Thus $\widehat{f\circ R} = \hat{f}\circ R$.

(ii) For the translation $Tx = x^* + x$ of $\mathbb{R}^n$ we have $[(Lf)\circ T](x) = \sum_{|\beta|\le N} c_\beta(x^* + x)\,(D^\beta f)(x^* + x)$ and $[L(f\circ T)](x) = \sum_{|\beta|\le N} c_\beta(x)\,(D^\beta f)(x^* + x)$ for all $f \in C^\infty(\mathbb{R}^n)$. Induction of $|\beta|$ shows that all c_β should be constants[71]. For $f \in \mathcal{S}(\mathbb{R}^n)$ we get $\mathcal{F}[L(f\circ R)](\xi) = \sum_{|\beta|\le N} c_\beta(2\pi i\xi)^\beta\,\widehat{f\circ R}(\xi) = \sum_{|\beta|\le N} c_\beta(2\pi i\xi)^\beta\,\hat{f}(R\xi)$ and $\mathcal{F}[(Lf)\circ R](\xi) = [\mathcal{F}(Lf)\circ R](\xi) = \sum_{|\beta|\le N} c_\beta(2\pi iR\xi)^\beta\,\hat{f}(R\xi)$ due to (i).

[70] The first n terms of the Taylor expansion of $x^n\phi(x)$ around $x = 0$ vanish.

[71] Indeed, for $\beta = 0$ choose $f \equiv 1$. If the property holds up to j and $|\beta| = j + 1$, choose $f(x) = x^\beta$ to infer that c_β is constant.

Since $e^{-|x|^2} \in \mathcal{S}(\mathbb{R}^n)$, by Theorem 6.1 there exists[72] $f \in \mathcal{S}(\mathbb{R}^n)$ with $\hat{f}(R\xi) = e^{-|R\xi|^2} \neq 0$ for all $\xi \in \mathbb{R}^n$. Consequently, if $Q(\xi) = \sum_{|\beta|\le N} c_\beta (2\pi i\xi)^\beta$, then $Q(\xi) = Q(R\xi)$ for all rotations R. Since for every $\xi \neq 0$ in $\mathbb{R}^n$ we can find a rotation[73] R with $\xi = |\xi|\, R\xi_0$, where $\xi_0 = (0,\dots,0,1) \in \mathbb{R}^n$, we get $Q(\xi) = Q(|\xi|\, R\xi_0) = Q(|\xi|\,\xi_0)$. Thus $Q(\xi)$ depends only on $|\xi|$. Since Q_0 is a constant and $Q_j(\xi) = \lim_{s\downarrow 0}\big(s^{-j}\sum_{k=j}^{N} Q_k(s\xi)\big)$, induction on j shows that every $Q_j(\xi)$ depends only on $|\xi|$. By homogeneity, $Q_j(\xi) = \alpha_j|\xi|^j$ for some constant $\alpha_j \in \mathbb{C}$. Since $|\xi|^j$ is a polynomial in n variables only when j is even, we deduce that $Q(\xi) = \sum_{0\le 2k\le N} \alpha_{2k}|\xi|^{2k}$. Consequently $L = \sum_{0\le 2k\le N} \frac{-\alpha_{2k}}{4\pi^2}\,\Delta^k$.

(iii) Part (i) implies that $\hat{f}$ is radial. Changing variables to spherical coordinates $x_1 = r\cos\varphi_1$, $x_2 = r\sin\varphi_1\cos\varphi_2$, $x_3 = r\sin\varphi_1\sin\varphi_2$ with $r = \sqrt{x_1^2+x_2^2+x_3^2}$, $\varphi_1 \in [0,\pi]$, $\varphi_2 \in [0,2\pi]$, having Jacobian determinant $r^2\sin\varphi_1$, we compute $\hat{f}(\xi) = \hat{f}(|\xi|,0,0) = \int_{\mathbb{R}^3} e^{-2\pi i x_1|\xi|} f_0(|x|)\,dx = \int_0^\infty\int_0^{2\pi}\int_0^\pi e^{-2\pi i|\xi| r\cos\varphi_1} f(r) r^2 \sin\varphi_1\, d\varphi_1 d\varphi_2 dr = \frac{2}{|\xi|}\int_0^\infty f(r) r\sin(2\pi|\xi|r)\,dr$.

22. $E \in \mathcal{S}'(\mathbb{R}^2)$ since $(x,t) \mapsto \frac{E(x,t)}{(1+x^2+t^2)^2}$ is in $L^1(\mathbb{R}^2)$ as $\int_{\mathbb{R}}\int_{\mathbb{R}} \frac{|E(x,t)|}{(1+x^2+t^2)^2}\,dxdt \le \int_{\mathbb{R}}\int_{\mathbb{R}} \frac{dxdt}{(1+x^2+t^2)^2} \le \big(\int_{\mathbb{R}} \frac{dx}{1+x^2}\big)\big(\int_{\mathbb{R}} \frac{dt}{1+t^2}\big) < \infty$. For $\phi \in \mathcal{S}(\mathbb{R}^2)$ we have $\langle \Box E, \phi\rangle = \langle E, \Box\phi\rangle = \frac{1}{2}\int_{\mathbb{R}}\int_{|x|}^\infty \phi_{tt}\,dtdx - \frac{1}{2}\int_0^\infty\int_{-t}^t \phi_{xx}\,dxdt = -\frac{1}{2}\int_{\mathbb{R}} \phi_t(x,|x|)\,dx - \frac{1}{2}\int_0^\infty \phi_x(t,t)\,dt + \frac{1}{2}\int_0^\infty \phi_x(-t,t)\,dt = -\frac{1}{2}\int_0^\infty \phi_t(x,x)\,dx - \frac{1}{2}\int_{-\infty}^0 \phi_t(x,-x)\,dx - \frac{1}{2}\int_0^\infty \phi_x(t,t)\,dt + \frac{1}{2}\int_0^\infty \phi_x(-t,t)\,dt = -\frac{1}{2}\int_0^\infty \phi_t(x,x)\,dx - \frac{1}{2}\int_0^\infty \phi_t(-y,y)\,dy - \frac{1}{2}\int_0^\infty \phi_x(t,t)\,dt + \frac{1}{2}\int_0^\infty \phi_x(-t,t)\,dt = -\frac{1}{2}\int_0^\infty [\phi_t(s,s)+\phi_x(s,s)]\,ds - \frac{1}{2}\int_0^\infty [\phi_t(-s,s)-\phi_x(-s,s)]\,ds$, after the change of variables $y = -x$ in the next-to-last step. Since $\frac{d}{ds}\phi(\pm s,s) = \pm\phi_x(\pm s,s) + \phi_t(\pm s,s)$, we get that $\langle \Box E, \phi\rangle = \phi(0,0)$.

23. Since $\sin(s) = \frac{e^{is}-e^{-is}}{2i}$, for each fixed $k \ge 1$ the series $\sum_{q,p\in\mathbb{Z}} c_{qp} e^{2\pi i(qx+py)}$ contributes terms with the $2^{k+1}+1$ indices $\{(4^k, j-2^k)\}_{j=0,\dots,2^{k+1}}$. Given $N \ge 1$, for $|\sin(2\pi y)| < 1$ we get that $\big|\sum_{N_1\le|q|,|p|\le N_2} c_{qp} e^{2\pi i(qx+py)}\big| \le \sum_{N_1\le 4^k\le N_2} 2^k|\sin(2\pi y)|^{2^k}(2^{k+1}+1) \to 0$ as $N_2 \ge N_1 \to \infty$ since $\sum_{j\ge 1} j(2j+1)\,|\sin(2\pi y)|^j < \infty$ by the ratio test for power series. On the other hand, if $\sin(2\pi y) \neq 0$, then the spherical partial sums diverge since the difference between the spherical partial sums for $N = 4^k$ and $N = 4^k - 1$ equals $2^k e^{2\pi 4^k ix}\sin^{2^k}(2\pi y)$.

24. Consider the measurable sets $M_k = \{x \in \mathbb{T}^d : |f(x)| \le k\}$ for integers $k \ge 1$. The dominated convergence theorem ensures $\lim_{k\to\infty}\int_{\mathbb{T}^d} |f(x)|\chi_{\mathbb{T}^d\setminus M_k}(x)\,dx = 0$. Given $\varepsilon > 0$, choose $k \ge 1$ such that $f_1 = f\chi_{\mathbb{T}^d\setminus M_k}$ satisfies $\|f_1\|_{L^1(\mathbb{T}^d)} < \varepsilon$. Clearly $f_2 = f - f_1 = f\chi_{M_k} \in L^2(\mathbb{T}^d) \subset L^1(\mathbb{T}^d)$ since $\int_{\mathbb{T}^d} |f_2(x)|^2 dx \le k\int_{\mathbb{T}^d} |f(x)|\,dx < \infty$. For any $n \in \mathbb{Z}^d$ we have that $\hat{f}(n) = \widehat{f_1}(n) + \widehat{f_2}(n)$, and $|\widehat{f_1}(n)| = \big|\int_{\mathbb{T}^d} |f_1(x)|\,dx < \varepsilon$, while $\sum_{n\in\mathbb{Z}^d} |\widehat{f_2}(n)|^2 = \int_{\mathbb{T}^d} |f_2(x)|^2 dx < \infty$ yields $\lim_{|n|\to\infty} |\widehat{f_2}(n)| = 0$. Consequently $\limsup_{|n|\to\infty} |\hat{f}(n)| \le \varepsilon$. Since $\varepsilon > 0$ was arbitrary, we must have $\lim_{|n|\to\infty} |\hat{f}(n)| = 0$.

25. By Theorem 6.12, $f * F_N^d \to f$ uniformly, and each $f * F_N^d$ is a trigonometric polynomial.

26. Formula (6.26) shows that the series converges absolutely.

27. We have

$$\mathcal{F}[\exp(-\pi a|x|^2)](\xi) = \int_{\mathbb{R}^n} e^{-2\pi i\xi\cdot x} e^{-\pi a|x|^2}\,dx = \prod_{k=1}^n \int_{\mathbb{R}} e^{-2\pi i\xi_k x_k} e^{-\pi a x_k^2}\,dx_k$$

$$= \prod_{k=1}^n \Big(\frac{1}{\sqrt{a}}\int_{\mathbb{R}} e^{-2\pi i s\frac{\xi_k}{\sqrt{a}}} e^{-\pi s^2}\,ds\Big) = \prod_{k=1}^n \Big(\frac{1}{\sqrt{a}}\, e^{-\pi\xi_k^2/a} ds\Big) = a^{-\frac{n}{2}}\exp(-\pi|\xi|^2/a),$$

[72] See also Exercise 22.

[73] The "rolling ball" analogy permits us to visualise the possibility of using a rotation to map the unit vector ξ_0, that points towards the "north pole" of the unit ball in $\mathbb{R}^n$, in the direction of an arbitrary vector $\xi \neq 0$. This can be achieved by the rotation matrix R whose columns are the coordinates of a new orthonormal basis of $\mathbb{R}^n$ with $\frac{\xi}{|\xi|}$ in the "north pole" position.

after the change of variables $s = x_k\sqrt{a}$ in step three, and using Exercise 5.2 in the fourth step.

28. Let us first show the validity of the identity $e^{-2\pi|t|} = \int_0^\infty \frac{e^{-s}}{\sqrt{\pi s}}\, e^{-\pi^2 t^2/s} ds$ for all $t \in \mathbb{R}$. For this, note that $\int_0^\infty e^{-s} e^{-s\tau^2} ds = \frac{1}{1+\tau^2}$ for $\tau \in \mathbb{R}$ while, in view of Exercise 5.9 and (5.12), $e^{-2\pi|t|} = \int_{\mathbb{R}} \frac{1}{\pi(1+\tau^2)} e^{2\pi i t\tau} d\tau$. We may substitute in the last relation for $\frac{1}{1+\tau^2}$ to get

$$
\begin{aligned}
e^{-2\pi|t|} &= \frac{1}{\pi}\int_{\mathbb{R}}\int_0^\infty e^{-s} e^{-s\tau^2} e^{2\pi i t\tau} ds d\tau \\
&= \frac{1}{\pi}\int_0^\infty \int_{\mathbb{R}} e^{-s\tau^2} e^{2\pi i t\tau} d\tau\, e^{-s}\, ds = \int_0^\infty \frac{e^{-s}}{\sqrt{\pi s}}\, e^{-\pi^2 t^2/s} ds,
\end{aligned}
$$

using Exercise 6.27 in the last step. Thus

$$
\begin{aligned}
\mathcal{F}[\exp(-2\pi a|x|)](\xi) &= \int_{\mathbb{R}^n} e^{-2\pi i\xi\cdot x} e^{-2\pi a|x|} dx = \int_{\mathbb{R}^n} e^{-2\pi i\xi\cdot x}\Big(\int_0^\infty \frac{e^{-s}}{\sqrt{\pi s}}\, e^{-\pi^2 a^2|x|^2/s} ds\Big) dx \\
&= \int_0^\infty \frac{e^{-s}}{\sqrt{\pi s}}\, \mathcal{F}[\exp(-\pi^2 a^2|x|^2/s)](\xi)\, ds = \frac{1}{a^n \pi^{(n+1)/2}} \int_0^\infty e^{-s} s^{(n-1)/2} e^{-|\xi|^2 s/a^2}\, ds,
\end{aligned}
$$

using Exercise 6.27 in the last step. The change of variables $u = s(1 + \frac{|\xi|^2}{a^2})$ yields the desired formula.

29. Let us first show that for $t > 0$ we have

$$
\int_{\mathbb{R}^n} \varphi(x-y)\, \frac{\exp(-|y|^2/4t)}{(4\pi t)^{n/2}}\, dy = \int_{\mathbb{R}^n} e^{-4\pi^2 t|\xi|^2} e^{2\pi i x\cdot\xi} \hat{\varphi}(\xi)\, d\xi\,, \qquad \varphi \in \mathcal{S}(\mathbb{R}^n)\,. \qquad (*)
$$

Indeed, by Theorem 6.1, (6.16) and Exercise 6.27, the right side equals $\mathcal{F}^{-1}[e^{-4\pi^2 t|\xi|^2}\hat{\varphi}(\xi)](x) = (\mathcal{F}^{-1}[e^{-4\pi^2 t|\xi|^2}] * \varphi)(x) = ([\frac{\exp(-|y|^2/4t)}{(4\pi t)^{n/2}}] * \varphi)(x)$. Multiply now each side of $(*)$ by t^k with[74] $-1 < k < \frac{n}{2} - 1$, integrate on $(0, \infty)$ and apply Fubini to obtain

$$
\int_{\mathbb{R}^n}\Big(\int_0^\infty t^k\, \frac{\exp(-|y|^2/4t)}{(4\pi t)^{n/2}}\, dt\Big)\varphi(x-y)\, dy = \int_{\mathbb{R}^n} \hat{\varphi}(\xi)\, e^{2\pi i x\cdot\xi}\Big(\int_0^\infty e^{-4\pi^2 t|\xi|^2} t^k\, dt\Big) d\xi
$$

for every $\varphi \in \mathcal{S}(\mathbb{R}^n)$. For $y \neq 0$, the change of variable $s = \frac{|y|^2}{4t}$ transforms the inner integral on the left to $\frac{1}{4^{k+1}\pi^{n/2}|y|^{n-2k-2}} \int_0^\infty e^{-s} s^{\frac{n}{2}-k-2} ds = \frac{\Gamma(\frac{n}{2}-k-1)}{4^{k+1}\pi^{n/2}}\, \frac{1}{|y|^{n-2k-2}}$, while for $\xi \neq 0$ the change of variable $u = -4\pi^2 t|\xi|^2$ transforms the inner integral on the right to $\frac{1}{(4\pi^2|\xi|^2)^{k+1}} \int_0^\infty e^{-u} u^k\, du = \frac{\Gamma(k+1)}{(4\pi^2|\xi|^2)^{k+1}}$. Consequently

$$
\frac{\Gamma(\frac{n}{2}-k-1)}{\Gamma(k+1)\pi^{\frac{n}{2}-2k-2}} \int_{\mathbb{R}^n} \frac{\varphi(x-y)}{|y|^{n-2k-2}}\, dy = \int_{\mathbb{R}^n} \hat{\varphi}(\xi)\, \frac{e^{2\pi i x\cdot\xi}}{|\xi|^{2k+2}}\, d\xi\,, \qquad \varphi \in \mathcal{S}(\mathbb{R}^n)\,.
$$

Taking above $x = 0$, from (6.11) we see that the right side is the Fourier transform of the tempered distribution defined by the locally integrable function $\xi \mapsto |\xi|^{-2k-2}$, acting on φ, while, after changing variables $y \mapsto -y$, the left side equals, up to a multiplicative constant factor, the action of the tempered distribution defined by the locally integrable function $y \mapsto |y|^{2k+2-n}$, acting on φ. The identification $\alpha = 2k + 2 \in (0, n)$ permits us to conclude that the Fourier transform of $|\xi|^{-\alpha}$ is $C_{n,\alpha}|x|^{\alpha-n}$, where $C_{n,\alpha} = \pi^{2k+2-\frac{n}{2}}\,\Gamma(\frac{n}{2} - k - 1)/\Gamma(k + 1)$.

30. (i) It is easy to see that f given by $f(0) = 1$ and $f(\xi) = 0$ for $\xi \neq 0$ is positive-definite.

(ii) For all $\xi_1, \xi_2, \xi_3 \in \mathbb{R}^d$, the complex 3×3 matrix $\Xi = \{f(\xi_j - \xi_k)\}_{1\leq j,k\leq 3}$ is Hermitian positive-definite matrix, meaning that its transpose Ξ^t equals the matrix obtained by taking the complex

[74] Note that $k < \frac{n}{2} - 1$ ensures the integrability of the left side, while $k > -1$ guarantees the integrability of the right side, both in the sense of $L^1(\mathbb{R}^n)$.

conjugates of the entries of Ξ and $(\Xi Z, Z) \geq 0$ for all $Z \in \mathbb{C}^3$. Thus the operator $\Xi : \mathbb{C}^3 \to \mathbb{C}^3$ is self-adjoint and diagonalisable, with real and nonnegative eigenvalues $\lambda_1, \lambda_2, \lambda_3$ (see Chapter 3); in particular, in the orthonormal basis in which Ξ is diagonal $\det(\Xi) = \lambda_1\lambda_2\lambda_3 \geq 0$. In particular, if $\xi_1 = \xi, \xi_2 = \eta$ and $\xi_3 = 0$, then the matrix

$$\begin{pmatrix} f(0) & f(\xi-\eta) & f(\xi) \\ \overline{f(\xi-\eta)} & f(0) & f(\eta) \\ \overline{f(\xi)} & \overline{f(\eta)} & f(0) \end{pmatrix}$$

has a nonnegative determinant. This gives

$$\begin{aligned} 0 &\leq f^3(0) + f(\eta)\overline{f(\xi-\eta)}\,\overline{f(\xi)} + \overline{f(\eta)f(\xi-\eta)}f(\xi) - f(0)\left[|f(\eta)|^2 + |f(\xi)|^2 + |f(\xi-\eta)|^2\right] \\ &= f^3(0) - f(0)\left\{|f(\xi)|^2 + |f(\eta)|^2\right\} - 2\Re e\left\{f(\eta)\overline{f(\xi)}\left[f(0) - f(\xi-\eta)\right]\right\} \\ &\leq f^3(0) - f(0)\left\{|f(\xi) - f(\eta)|^2 + |f(\xi-\eta)|^2\right\} + 2f^2(0)\,|f(0) - f(\xi-\eta)| \end{aligned}$$

using the fact that $|f(\xi)|, |f(\eta)| \leq f(0)$, see (6.38). Since $f(0) \geq 0$, re-arranging gives

$$\begin{aligned} |f(0) - f(\xi-\eta)|^2 &\leq f^2(0) - |f(\xi-\eta)|^2 + 2f(0)\,|f(0) - f(\xi-\eta)| \\ &= (f(0) + |f(\xi-\eta)|)\,(f(0) - |f(\xi-\eta)|) + 2f(0)\,|f(0) - f(\xi-\eta)| \\ &\leq 4f(0)\,|f(0) - f(\xi-\eta)| \end{aligned}$$

since $|f(\xi-\eta)| \leq f(0)$, in view of (6.38).

31. Assuming $f \not\equiv 0$, Fubini's theorem yields $f_k \in L^1(\mathbb{R})$ for $1 \leq k \leq n$, and, passing to spherical coordinates, we see that $\int_0^\infty |f_0(r)|\, r^{n-1}\, dr < \infty$. Taking the Fourier transform we obtain

$$\hat{f}(\xi_1, \ldots, \xi_n) = \prod_{k=1}^{n} \widehat{f_k}(\xi_k) = g(\xi_1^2 + \cdots + \xi_n^2), \qquad \xi \in \mathbb{R}^n, \tag{$*$}$$

since

$$\hat{f}(\xi) = \int_{\mathbb{R}^n} f(x)\, e^{-2\pi i x\cdot\xi}\, dx = |\xi|^{-n} \int_{\mathbb{R}^n} f_0(y|\xi|)\, e^{-2\pi i y\cdot\eta}\, dy$$

with $\eta = \xi/|\xi|$ for $\xi \neq 0$, due to the change of variable $y = |\xi|x$. Due to Theorem 6.2, the function $g : [0, \infty) \to \mathbb{C}$ in $(*)$ is continuous and bounded. Moreover, since $\widehat{f_k}(0) = 0$ for some $1 \leq k \leq n$ forces $g \equiv 0$ in view of $(*)$, we must have $\widehat{f_k}(0) \neq 0$ for all $1 \leq k \leq n$. The choice $\xi_j = 0$ for $j \neq k$ yields $\widehat{f_k}(\xi) = c_k g(\xi^2)$ for some constant $c_k \in \mathbb{C}$, and setting $\xi_1 = \cdots = \xi_n = 0$ in $(*)$ identifies the constant $c_1 \ldots c_n = g^{1-n}(0)$. Thus $(*)$ is equivalent to $g^{1-n}(0)g(\xi_1^2)\ldots g(\xi_n^2) = g(\xi_1^2 + \cdots + \xi_n^2)$ for all $\xi \in \mathbb{R}^n$, and the continuous and bounded function $G : [0, \infty) \to \mathbb{C}$, given by $G(s) = g(s)/g(0)$, satisfies the functional equation $G(t + s) = G(t)G(s)$. Since $G(0) = 1$, there exists some $\delta > 0$ with $|G(s) - 1| < 1/2$ for $0 \leq s \leq 2\delta$. Setting $G(s) = R(s)e^{i\theta(s)}$ where $R : [0, 2\delta] \to (0, \infty)$ and $\theta : [0, 2\delta] \to (-\pi/2, \pi/2)$ are continuous with $R(0) = 1$ and $\theta(0) = 0$, we obtain that $R(t + s) = R(t)R(s)$ and $\theta(t + s) = \theta(t) + \theta(s)$ for all $t, s \in [0, \delta]$. Consequently the two continuous functions $\ln R(t)$ and $\theta(t)$ both satisfy the functional equation $F(t + s) = F(t) + F(s)$ for all $t, s \in [0, 2\delta]$ with $t + s \leq 2\delta$. But then $\int_0^{t+s} F(u)\, du - \int_0^t F(u)\, du - \int_0^s F(u)\, du = \int_t^{t+s} F(u)\, du - \int_0^s F(u)\, du = \int_0^s F(u + t)\, du - \int_0^s F(u)\, du = sF(t)$ for all $t, s \in [0, \delta]$. Since the left side is invariant under the interchange of t and s, we must have $sF(t) = tF(s)$ for all $t, s \in [0, \delta]$, so that $F(t)/t$ is constant on $(0, \delta]$, and, by continuity, on $[0, \delta]$. This shows that on $[0, \delta]$ we have $R(t) = e^{b_1 t}$ and $\theta(t) = b_2 t$ for some constants $b_1, b_2 \in \mathbb{R}$, that is, $G(t) = e^{(b_1 + ib_2)t}$ for $t \in [0, \delta]$. The functional equation satisfied by G permits us to extend this to all intervals $[n\delta, (n + 1)\delta]$ with $n \geq 1$ an integer, since $G(t + n\delta) = G(t)G^n(\delta) = e^{(b_1+ib_2)(t+n\delta)}$ for $t \in [0, \delta]$. Thus $G(t) = e^{(b_1+ib_2)t}$ and $g(t) = g(0)e^{(b_1+ib_2)t}$ for all $t \geq 0$. Theorem 6.3 and $(*)$ force $b_1 < 0$.

It remains to show the admissibility of any g of the form $g(t) = ae^{(b_1+ib_2)t}$ with $a \in \mathbb{C} \setminus \{0\}$ and $b_1, b_2 \in \mathbb{R}$ with $b_1 < 0$. With $\hat{f}(\xi) = a\, e^{(b_1+ib_2)|\xi|^2} \in S(\mathbb{R}^n)$, Theorem 6.1 yields $f(x) = a \int_{\mathbb{R}^n} e^{(b_1+ib_2)|\xi|^2} e^{-2\pi i x\cdot\xi}\, d\xi$. We obtain that $(\partial_{x_k} f)(x_k) = -2\pi i a \int_{\mathbb{R}^n} \xi_k e^{(b_1+ib_2)|\xi|^2} e^{-2\pi i x\cdot\xi}\, d\xi =$

$\frac{4\pi^2 ax_k}{2(b_1+ib_2)}\int_{\mathbb{R}^n} e^{(b_1+ib_2)|\xi|^2} e^{-2\pi ix\cdot\xi} d\xi$ after integration by parts. Thus $(\partial_{x_k} f)(x) = \frac{2\pi}{b_1+ib_2} f(x)$ for $1 \le k \le n$, or $\partial_{x_k}(e^{-\frac{\pi^2}{b_1+ib_2}|x|^2} f(x)) = 0$, which means that $f(x) = \alpha e^{\frac{\pi^2}{b_1+ib_2}|x|^2}$ for some constant $\alpha \in \mathbb{C}$. Set $\beta = \frac{\pi^2}{b_1+ib_2} = \frac{\pi^2(b_1-ib_2)}{b_1^2+b_2^2}$ to obtain the desired representation.

32. By Exercise 27 we have $\int_{\mathbb{R}} e^{-\pi at^2} e^{-2\pi ist} dt = \frac{\sqrt{\pi}}{\sqrt{a}} e^{-\pi s^2/a}$ for all $a > 0$. The integral is well-defined for arbitrary $a, s \in \mathbb{C}$ with $\Re e(a) > 0$, so by analytic continuation the formula remains valid, with the square root chosen in the right half-plane. The algebraic identity $\beta\sum_{k=1}^n (x_k - a_k)^2 + 2\pi i\sum_{k=1}^n (x_k-a_k)(\xi_k-b_k) - \beta\sum_{k=1}^n a_k^2 + 2\pi i\sum_{k=1}^n a_k(\xi_k-b_k) = \beta\sum_{k=1}^n x_k^2 - \sum_{k=1}^n \gamma_k x_k + 2\pi i\sum_{k=1}^n x_k\xi_k$ with $a_k = \frac{\Re e(\gamma_k)}{2\Re e(\beta)} \in \mathbb{R}$ and $b_k = \frac{\Im m(\gamma_k)}{2\pi} - \frac{\Im m(\beta)\Re e(\gamma_k)}{2\pi\Re e(\beta)} \in \mathbb{R}$ for $1 \le k \le n$ now yields

$$\mathcal{F}[e^{-\beta|x|^2+\sum_{k=1}^n \gamma_k x_k}](\xi) = \int_{\mathbb{R}^n} e^{-\beta|x|^2+\sum_{k=1}^n \gamma_k x_k} e^{-2\pi ix\cdot\xi} dx$$

$$= e^{\beta|\xi|^2-2\pi ia\cdot(\xi-b)} \int_{\mathbb{R}^n} e^{-\beta\sum_{k=1}^n (x_k-a_k)^2 - 2\pi i\sum_{k=1}^n (x_k-a_k)(\xi_k-b_k)} dx$$

$$= e^{\beta|\xi|^2-2\pi ia\cdot(\xi-b)} \prod_{k=1}^n \int_{\mathbb{R}} e^{-\beta(x_k-a_k)^2-2\pi i(x_k-a_k)(\xi_k-b_k)} dx_k$$

$$= e^{\beta|\xi|^2-2\pi ia\cdot(\xi-b)} \prod_{k=1}^n \int_{\mathbb{R}} e^{-\beta t^2-2\pi i(\xi_k-b_k)t} dt = e^{\beta|\xi|^2-2\pi ia\cdot(\xi-b)} \prod_{k=1}^n \Big(\frac{\sqrt{\pi}}{\sqrt{\beta}} e^{-\pi^2(\xi_k-b_k)^2/\beta}\Big)$$

$$= \Big(\frac{\sqrt{\pi}}{\sqrt{\beta}}\Big)^n e^{-\pi^2|\xi-b|^2/\beta-2\pi ia\cdot(\xi-b)+\beta|a|^2} = \Big(\frac{\sqrt{\pi}}{\sqrt{\beta}}\Big)^n e^{-\pi^2\langle\xi-\gamma/2,x-\gamma/2\rangle/\beta}$$

since $\beta a + \pi ib = \gamma/2$ for $a = (a_1, \dots, a_n)$, $b = (b_1, \dots, b_n)$, $\gamma = (\gamma_1, \dots, \gamma_n)$. Note that in the final expression $\langle z, z\rangle$ stands for $\sum_{k=1}^n z_k^2$ even if $z \in \mathbb{C}^n \setminus \mathbb{R}^n$.

33. Since the map $L^p(\mathbb{R}^n) \ni f \mapsto S'(\mathbb{R}^n)$ is continuous, the hypothesis ensures that the graph of $f \mapsto \hat{f}$ from $L^p(\mathbb{R}^n)$ to $L^q(\mathbb{R}^n)$ is closed. The closed graph theorem (see Chapter 3) yields the existence of some constant $C > 0$ such that

$$\|\hat{f}\|_{L^q(\mathbb{R}^n)} \le C\|f\|_{L^p(\mathbb{R}^n)} \quad \text{for all} \quad f \in L^p(\mathbb{R}^n). \tag{$*$}$$

Given $t > 0$ and $0 \not\equiv f \in S(\mathbb{R}^n)$, the explicit integral formula for $\hat{f}$ shows that if we replace $f(x)$ by $f(x/t)$, then $\hat{f}(\xi)$ is replaced by $t^n\hat{f}(t\xi)$, so that $(*)$ yields $t^{n(1-1/q)}\|\hat{f}\|_{L^q(\mathbb{R}^n)} \le Ct^{n/p}\|f\|_{L^p(\mathbb{R}^n)}$ for all $t > 0$. Reasons of homogeneity (letting $t \downarrow 0$ and $t \uparrow \infty$) force $\frac{1}{p} + \frac{1}{q} = 1$. Let us now define $f_m(x) = e^{-b_m|x|^2} \in S(\mathbb{R}^n)$ for integers $m \ge 1$, where $b_m = \sin\big(\frac{1}{m}\big) + i\cos\big(\frac{1}{m}\big)$ with square root $\sqrt{b_m} = \cos\big(\frac{\pi}{4} - \frac{1}{2m}\big) + i\sin\big(\frac{\pi}{4} - \frac{1}{2m}\big)$ chosen in the right half-plane. By Exercise 32, $\widehat{f_m}(\xi) = \big(\frac{\sqrt{\pi}}{\sqrt{b_m}}\big)^n e^{-\pi^2|\xi|^2/b_m}$. Computing $\|f_m\|_{L^p(\mathbb{R}^n)}^p = \int_{\mathbb{R}^n} e^{-p\Re e(b_m)|x|^2} dx = \big(\frac{\pi}{p\Re e(b_m)}\big)^{\frac{n}{2}}$ and $\|\widehat{f_m}\|_{L^q(\mathbb{R}^n)}^q = \pi^{\frac{nq}{2}}\int_{\mathbb{R}^n} e^{-q\pi^2\Re e(b_m)|\xi|^2} d\xi = \pi^{\frac{nq}{2}}\big(\frac{1}{q\pi\Re e(b_m)}\big)^{\frac{n}{2}}$, we see that $(*)$ requires

$$C \ge \pi^{\frac{n}{2}}\Big(\frac{1}{q\pi\,\Re e(b_m)}\Big)^{\frac{n}{2q}}\Big(\frac{\pi}{p\,\Re e(b_m)}\Big)^{-\frac{n}{2p}} = \pi^{\frac{n}{2}(1-\frac{1}{p}-\frac{1}{q})}\big(p^{\frac{1}{p}}/q^{\frac{1}{q}}\big)^{\frac{n}{2}}\big(\Re e(b_m)\big)^{\frac{n}{2}(\frac{1}{p}-\frac{1}{q})}$$

for all integers $m \ge 1$. Since $\Re e(b_m) \downarrow 0$ for $m \uparrow \infty$, we find that $q \ge p$.

34. We have $\delta_0 \in S'(\mathbb{R}^n)$ with $\widehat{\delta_0} = 1$ and, passing to spherical coordinates, $\int_{\mathbb{R}^n}(1 + |\xi|^2)^s d\xi < \infty$ if and only if $\int_0^\infty (1 + r^2)^s r^{n-1} ds < \infty$, which is equivalent to $n + 2s > 0$. On the other hand, the Fourier transform of $1 \in S'(\mathbb{R}^n)$ is the measure δ_0.

35. Two integrations by parts show that $u(x) = e^{-|x|-|x|^3} \in L^2(\mathbb{R})$ is such that for all $\varphi \in S(\mathbb{R})$, $\langle u'', \varphi\rangle = \langle u, \varphi''\rangle = \int_0^\infty e^{-x-x^3}\varphi''(x)\,dx + \int_{-\infty}^0 e^{x+x^3}\varphi''(x)\,dx = \int_0^\infty (1+3x^2)e^{-x-x^3}\varphi'(x)\,dx - \int_{-\infty}^0 (1+3x^2)e^{x+x^3}\varphi'(x)\,dx = -2\varphi(0) + \int_0^\infty \{(1+3x^2)^2-6x\}e^{-x-x^3}\varphi(x)\,dx - \int_{-\infty}^0 \{(1+3x^2)^2-6x\}e^{x+x^3}\varphi'(x)\,dx,$

i.e. $\langle u'',\varphi\rangle = \langle -2\delta_0+f,\varphi\rangle$ with $f(x) = \{(1+3x^2)^2-6x\}e^{-|x|-|x|^3}\,\mathrm{sgn}(x) \in L^2(\mathbb{R})$. Thus $u'' = -2\delta_0+f$ in $S'(\mathbb{R})$ and Exercise 34 yields $u'' \in H^s(\mathbb{R})$ if and only if $s < -1/2$, so that $u \in H^s(\mathbb{R})$ if and only if $s < 3/2$.

36. The (classical) derivative $\partial_x^\alpha f$ is of class C^∞ on $\mathbb{R}^n \setminus \{0\}$, being of the form $(\partial_x^\alpha f)(x) = P_\alpha(x)\,|x|^{\lambda-2|\alpha|}$ for $0 < |x| < 1$, where $P_\alpha(x)$ is a polynomial of degree $|\alpha|$, homogeneous of degree $|\alpha|$, so that there exists a constant $C_{\alpha,\lambda} > 0$ with $|(\partial_x^\alpha f)(x)| \le |x|^{\lambda-|\alpha|}$ for $0 < |x| < 1$. Setting $f_m(x) = \theta(x)\,(|x|^2 + \frac{1}{m^2})^{\lambda/2}$ for $m \ge 1$, we have $f_m \in \mathcal{D}(\mathbb{R}^n)$, $\lim_{m\to\infty} f_m(x) = f(x)$ for $x \ne 0$, and there exists $M > 0$ such that $|f_m(x)| \le M\,|\theta(x)|\,(|f(x)| + 1)$ for all $x \ne 0$ and all $m \ge 1$. Using spherical coordinates we see that $x \mapsto M\,|\theta(x)|\,(|f(x)| + 1)$ belongs to $L^1(\mathbb{R}^n)$, so that $\langle \partial_x^\alpha f_m, \varphi\rangle = (-1)^{|m|}\langle f_m, \partial_x^\alpha\varphi\rangle = (-1)^{|m|}\int_{\mathbb{R}^n} f_m(x)\,(\partial_x^\alpha\varphi)(x)\,dx \to (-1)^{|m|}\int_{\mathbb{R}^n} f(x)\,(\partial_x^\alpha\varphi)(x)\,dx = \langle u_\alpha,\varphi\rangle$ for every $\varphi \in S(\mathbb{R}^n)$, where $u_\alpha \in S'(\mathbb{R}^n)$ is the distributional derivative of $f \in S'(\mathbb{R}^n)$. Moreover, since $(|x|^2 + \frac{1}{m^2})^{-1} \le |x|^{-1}$ for $x \ne 0$, there exist constants $M_\alpha > 0$ such that $|(\partial_x^\alpha f_m)(x)| \le M_\alpha\,|\theta(x)|\,(|(\partial_x^\alpha f)(x)| + 1)^2$ for all $x \ne 0$ and all $m \ge 1$. If $\partial_x^\alpha f \in L^2(\mathbb{R}^n)$, by dominated convergence, this estimate together with the fact that $\lim_{m\to\infty}(\partial_x^\alpha f_m)(x) = (\partial_x^\alpha f)(x)$ for $x \ne 0$ yield $\langle \partial_x^\alpha f_m, \varphi\rangle = \int_{\mathbb{R}^n}(\partial_x^\alpha f_m)(x)\,\varphi(x)\,dx \to \int_{\mathbb{R}^n}(\partial_x^\alpha f)(x)\,\varphi(x)\,dx$ for every $\varphi \in S(\mathbb{R}^n)$, so that $u_\alpha = \partial_x^\alpha f \in L^2(\mathbb{R}^n)$. Using spherical coordinates, the explicit expression for $\partial_x^\alpha f$ reveals that $\partial_x^\alpha f \in L^2(\mathbb{R}^n)$ if and only if $\lambda - |\alpha| > -n/2$, so $f \in H^k(\mathbb{R}^n)$ if and only if the integer $k \ge 1$ is such that $\lambda > k - n/2$, whereas f is k-times continuously differentiable if and only if $\lambda > k$.

37. (i) It is easy to see that $(\cdot,\cdot)_s$ is an inner product on $H^s(\mathbb{T})$. To show that $H^p(\mathbb{T})$ is complete, consider a Cauchy sequence $\{f_n\}_{n\ge1}$. Since $\mathbb{C}$ is complete, if $\sum_{j\in\mathbb{Z}}(1 + k^2)^s|\widehat{f_n}(k) - \widehat{f_m}(k)|^2 < \varepsilon$ for $m, n \ge n_0(\varepsilon)$, there is a sequence $\{a(k)\}_{k\in\mathbb{Z}}$ with $\widehat{f_n}(k) \to a(k)$ as $n \to \infty$ for every fixed $k \in \mathbb{Z}$, so that passing to the limit in the previous inequality, we get $\sum_{k=-N}^N(1+k^2)^s|a(k)-\widehat{f_m}(k)|^2 \le \varepsilon$ for $m \ge n_0(\varepsilon)$ and every fixed integer $N \ge 1$. We now claim that $f(x) = \sum_{k\in\mathbb{Z}} a(k)e^{2\pi ikx}$ belongs to $H^s(\mathbb{T})$, and $\lim_{n\to\infty}\|f_n - f\|_s = 0$. Indeed, $\sum_{k\in\mathbb{Z}}(1 + k^2)^s|a(k)|^2 = \sum_{k\in\mathbb{Z}}(1 + k^2)^s|a(k) - \widehat{f_n}(k) + \widehat{f_n}(k)|^2 \le 2\sum_{k\in\mathbb{Z}}(1 + k^2)^s|a(k) - \widehat{f_n}(k)|^2 + 2\sum_{k\in\mathbb{Z}}(1 + k^2)^s|\widehat{f_n}(k)|^2 \le 2\varepsilon + \|f_n\|_s^2 < \infty$, while $\|f_m - f\|_s^2 \le \varepsilon$ for $m \ge n_0(\varepsilon)$. To prove the density of trigonometric polynomials, for $f \in H^s(\mathbb{T})$ set $F_n(x) = \sum_{k=-n}^n \hat{f}(k)\,e^{2\pi ikx}$ for $n \ge 1$, with $\|f - F_n\|_s^2 = \sum_{|k|\ge n+1}(1 + k^2)^s|\hat{f}(k)|^2 \to 0$ as $n \to \infty$.

To identify the dual of $H^s(\mathbb{T})$, note that if $f \in H^{-s}(\mathbb{T})$, then $\langle f,g\rangle_s$ defines a continuous linear functional on $H^s(\mathbb{T})$. Conversely, if $L : H^s(\mathbb{T}) \to \mathbb{C}$ is an element of $(H^s(\mathbb{T}))^*$, then by Theorem 3.13 there exists a unique $\varphi \in H^s(\mathbb{T})$ such that $Lg = (g,\varphi)_s = \sum_{k\in\mathbb{Z}}\hat{g}(k)\overline{\hat{\varphi}(k)} = \sum_{k\in\mathbb{Z}}\hat{g}(k)\overline{(1 + k^2)^s\hat{\varphi}(k)}$ for all $g \in H^s(\mathbb{T})$. Since $\{(1 + k^2)^{s/2}\hat{\varphi}(k)\}_{k\in\mathbb{Z}} \in l^2(\mathbb{Z})$, we see that the periodic distribution $f \in S'(\mathbb{R})$ with Fourier coefficients $\hat{f}(k) = (1 + k^2)^{s/2}\overline{(1 + k^2)^{s/2}\hat{\varphi}(k)}$ belongs to $H^{-s}(\mathbb{T})$ and $\langle f,g\rangle_s = \sum_{k\in\mathbb{Z}}\hat{f}(k)\hat{g}(k) = Lg$ for all $g \in H^s(\mathbb{T})$.

(ii) Let $f \in H^m(\mathbb{T})$ for some integer $m \ge 0$. Then $\sum_{k\in\mathbb{Z}}(1 + k^2)^m|\hat{f}(k)|^2 < \infty$ and thus $\sum_{k\in\mathbb{Z}}(2\pi k)^{2j}|\hat{f}(k)|^2 < \infty$ for all $0 \le j \le m$, meaning in view of (6.13) that the distribution $\partial_x^j f$ belongs to $L^2(\mathbb{T})$. Moreover, by (6.13), $\sum_{j=0}^m\|\partial_x^j f\|_0^2 = \sum_{k\in\mathbb{Z}}\big(\sum_{j=0}^m(2\pi k)^{2j}\big)|\hat{f}(k)|^2 \le (2\pi)^{2m}\sum_{k\in\mathbb{Z}}(1+k^2)^m|\hat{f}(k)|^2 = (2\pi)^{2m}\|f\|_s^2$. Conversely, if $\partial_x^j f \in L^2(\mathbb{T})$ for all $0 \le j \le m$, then (6.13) shows that $\{(2\pi ik)^j\hat{f}(k)\}_{k\in\mathbb{Z}}$ belongs to $l^2(\mathbb{Z})$. Thus

$$\|f\|_m^2 = \sum_{k\in\mathbb{Z}}(1 + k^2)^m|\hat{f}(k)|^2 = \sum_{k\in\mathbb{Z}}\Big(\sum_{j=0}^m\binom{j}{k}k^{2j}\Big)|\hat{f}(k)|^2 = \sum_{j=0}^m\binom{j}{k}\Big(\sum_{k\in\mathbb{Z}}k^{2j}|\hat{f}(k)|^2\Big)$$

$$= \sum_{j=0}^m\binom{j}{k}(2\pi)^{-2j}\|\partial_x^j f\|_{L^2[0,1]}^2 \ge (2\pi)^{-2m}\sum_{j=0}^m\|\partial_x^j f\|_{L^2[0,1]}^2 = (2\pi)^{-2m}\sum_{j=0}^m\|\partial_x^j f\|_0^2\,.$$

(iii) If $f \in H^s(\mathbb{T})$ with $s > 1/2$, then the Cauchy–Schwarz inequality yields $\big(\sum_{k\in\mathbb{Z}}|\hat{f}(k)|\big)^2 = \big(\sum_{k\in\mathbb{Z}}\frac{(1+k^2)^{s/2}|\hat{f}(k)|}{(1+k^2)^{s/2}}\big)^2 \le \big(\sum_{k\in\mathbb{Z}}(1 + k^2)^s|\hat{f}(k)|^2\big)\big(\sum_{k\in\mathbb{Z}}(1 + k^2)^{-s}\big) = \|f\|_s^2\sum_{k\in\mathbb{Z}}(1 + k^2)^{-s} < \infty$ and we conclude by Exercise 4.22.

(iv) Let us first establish for $s > 0$ the existence of constants m_s, $M_s > 0$ such that

$$m_s(a^s + b^s) \le (a+b)^s \le M_s(a^s + b^s) \quad \text{for all} \quad a,\, b \ge 0\,. \tag{$*$}$$

Indeed, in the nontrivial case $a > 0$, after dividing by a, $(*)$ amounts to $m_s(1 + r^s) \le (1 + r)^s \le M_s(1 + r^s)$ for all $r \ge 0$, which holds with $m_s = \inf_{r\ge 0}\{\frac{(1+r)^s}{(1+r^s)}\} > 0$ and $M_s = \sup_{r\ge 0}\{\frac{(1+r)^s}{(1+r^s)}\} \ge 1$. We will also rely on the fact that if $\alpha \in l^1(\mathbb{Z})$ and $\beta \in l^2(\mathbb{Z})$, then

$$\sum_{k\in\mathbb{Z}}\Big(\sum_{j\in\mathbb{Z}}|\alpha(j)|\,|\beta(k-j)|\Big)^2 \le \sum_{k\in\mathbb{Z}}\Big(\sum_{j\in\mathbb{Z}}|\alpha(j)|\Big)\Big(\sum_{j\in\mathbb{Z}}|\alpha(j)|\,|\beta(k-j)|^2\Big) = \|\alpha\|_{l^1}\sum_{k\in\mathbb{Z}}\sum_{j\in\mathbb{Z}}|\alpha(j)|\,|\beta(k-j)|^2$$

$$\le \|\alpha\|_{l^1}\Big(\sum_{j\in\mathbb{Z}}|\alpha(j)|\Big)\Big(\sum_{k\in\mathbb{Z}}|\beta(k)|^2\Big) = \|\alpha\|_{l^1}^2\,\|\beta\|_{l^2}^2\,. \tag{$**$}$$

If $f, g \in H^s(\mathbb{T})$, then f and g are a.e. equal to continuous functions, see (iii), and therefore their Fourier series converge uniformly and absolutely to these continuous functions, so that $\widehat{(fg)}(k) = \int_0^1 f(x)g(x)e^{-2\pi ikx}\,dx = \int_0^1\big\{\sum_{j\in\mathbb{Z}}\hat{f}(j)e^{2\pi ijx}\big\}g(x)e^{-2\pi ikx}\,dx = \sum_{j\in\mathbb{Z}}\hat{f}(j)\int_0^1 g(x)e^{-2\pi i(k-j)x}\,dx = \sum_{j\in\mathbb{Z}}\hat{f}(j)\hat{g}(k-j)$. Using $(*)$, we get $(1+k^2)^{s/2}\big|\sum_{j\in\mathbb{Z}}\hat{f}(j)\hat{g}(k-j)\big| \le (1+|k|)^s\big|\sum_{j\in\mathbb{Z}}\hat{f}(j)\hat{g}(k-j)\big| \le M_s(1+|k|^s)\big|\sum_{j\in\mathbb{Z}}\hat{f}(j)\hat{g}(k-j)\big| \le M_s\big|\sum_{j\in\mathbb{Z}}\hat{f}(j)\hat{g}(k-j)\,(1+M_s|k-j|^s+M_s|j|^s)\big| \le M_s^2\sum_{j\in\mathbb{Z}}\{|\hat{f}(j)|\,|\hat{g}(k-j)| + |j|^s|\hat{f}(j)|\,|\hat{g}(k-j)| + |\hat{f}(j)|\,|\hat{g}(k-j)|\,|k-j|^s\}$. Since both sequences $\{|j|^s\hat{f}(j)\}_{j\in\mathbb{Z}}$ and $\{|m|^s\hat{g}(m)\}_{m\in\mathbb{Z}}$ belong to $l^2(\mathbb{Z})$, while $\{\hat{f}(j)\}_{j\in\mathbb{Z}}$, $\{\hat{g}(m)\}_{m\in\mathbb{Z}} \in l^1(\mathbb{Z})\cap l^2(\mathbb{Z})$, see (iii), we deduce from $(**)$ that $fg \in H^s(\mathbb{T})$.

(v) Clearly $H^q(\mathbb{T}) \subset H^p(\mathbb{T})$. If I is the embedding map from $H^q(\mathbb{T})$ to $H^p(\mathbb{T})$, define for integers $n \ge 1$ the finite rank operators $I_n : H^q(\mathbb{T}) \to H^p(\mathbb{T})$ by $(I_nf)(x) = \sum_{k=-n}^n \hat{f}(k)\,e^{2\pi ikx}$ for $f \in H^q(\mathbb{T})$. Then $\|(I-I_n)f\|_p^2 = \sum_{|k|\ge n+1}(1+k^2)^p|\hat{f}(k)|^2 \le (1+n^2)^{p-q}\sum_{|k|\ge n+1}(1+k^2)^q|\hat{f}(k)|^2 \le (1+n^2)^{p-q}\|f\|_q^2 \to 0$ as $n\to\infty$. Going through the argumentation made in the proof of Theorem 3.31(i) yields the claim.

6.8 Notes to Chapter 6

1. It is possible to develop the entire theory of periodic distributions directly, rather than relying on the general theory of distributions $\mathcal{D}'(\mathbb{R}^n)$. We illustrate this alternative approach in the simpler one-dimensional case. Let $\mathcal{P}(\mathbb{R})$ be the complex vector space of all C^∞-functions $f : \mathbb{R}\to\mathbb{C}$ which are periodic (with unit period). Equipped with the sequence of norms

$$\|f\|_n = \sum_{0\le k\le n}\ \sup_{x\in[0,1]}\{|f^{(k)}(x)|\}\,, \qquad n \ge 0\,,$$

$\mathcal{P}(\mathbb{R})$ is a Fréchet space. A *periodic distribution* u is a continuous linear functional on $\mathcal{P}(\mathbb{R})$, and its action on $f \in \mathcal{P}(\mathbb{R})$ is denoted by $[u, f]$, to distinguish it from the notation $\langle u, f\rangle$ used in $\mathcal{D}'(\mathbb{R})$, which can also make sense but be different. The continuity amounts to the condition that for some integer $N \ge 0$ there exists a constant $C_N > 0$ such that $|[u, f]| \le C_N\|f\|_N$ for all $f \in \mathcal{P}(\mathbb{R})$. The space of all these periodic distribution is denoted by $\mathcal{P}'(\mathbb{R})$. The technical advantage of this procedure is the fact that $\mathcal{P}(\mathbb{R})$ is a Fréchet space, in contrast to $\mathcal{D}(\mathbb{R})$, and all separately continuous bilinear forms on the product of Fréchet spaces are also jointly continuous[75]. Nevertheless,

[75] This is a consequence of the Baire category theorem, and the proof runs along similar lines to the approach used in the solution to Exercise 3.50(i). Note that if X is a non-normable locally convex space, then the evaluation map $X^*\times X\to\mathbb{C}$, $(f, x)\mapsto f(x)$, is jointly discontinuous even though it is separately continuous; see Köthe (1979).

$\mathcal{P}'(\mathbb{R}) \subset \mathcal{D}'(\mathbb{R})$: to define $\langle u, \varphi \rangle$ for $u \in \mathcal{P}'(\mathbb{R})$ and $\varphi \in \mathcal{D}(\mathbb{R})$, consider the map $\sigma : \mathcal{D}(\mathbb{R}) \to \mathcal{P}(\mathbb{R})$, given by $(\sigma\varphi)(x) = \sum_{k\in\mathbb{Z}} \varphi(x-k)$, and set $\langle u, \varphi \rangle = [u, \sigma\varphi]$.

2. Linear maps K between function spaces are sometimes given by kernels $k(x, y)$, such that

$$(Kf)(x) = \int_{\mathbb{R}^m} k(x,y)f(y)\,dy \quad \text{for} \quad x \in \mathbb{R}^n\,;$$

see Chapter 3 for examples. If $k \in \mathcal{D}'(\mathbb{R}^{n+m})$, then $(\phi, \psi) \to \langle k, \phi(x)\psi(y)\rangle$ is a bilinear functional on $\mathcal{D}(\mathbb{R}^n)\times\mathcal{D}(\mathbb{R}^m)$. For fixed ψ, it becomes a linear functional on $\mathcal{D}(\mathbb{R}^n)$, and for fixed ϕ it becomes a linear functional on $\mathcal{D}(\mathbb{R}^m)$, both of these being distributions. One of the substantial results of distribution theory, Schwartz's kernel theorem, says that the converse is true: any linear map that is sequentially continuous from $\mathcal{D}(\mathbb{R}^n)$ to $\mathcal{D}'(\mathbb{R}^m)$ is generated by a distribution kernel (see the discussion in Friedlander (1998)). This result puts differential operators on the same footing as integral operators. For example, the differential operator $\sum_{|\alpha|\le m} a_\alpha(x)\,\partial_x^\alpha$ with smooth coefficients a_α, on $\mathbb{R}^n$, has the kernel $k(x,y) = \sum_{|\alpha|\le m} a_\alpha(x)\,\partial_x^\alpha \delta(x-y) \in \mathcal{D}'(\mathbb{R}^{2n})$.

3. Distribution theory is a natural framework for the study of linear partial differential equations with smooth coefficients; see Gårding (1997) for an insightful account of the historical background. One of the early successes of distribution theory (proved by using the Fourier transform) was the fact that every linear differential operator L with constant complex coefficients has a (distributional) fundamental solution — this is the celebrated Malgrange-Ehrenpreis theorem, see Malgrange (1955) and Ehrenpreis (1954). Therefore, all partial differential equations with constant coefficients have a local solution: for any $x_0 \in \mathbb{R}^n$ there exists an open neighbourhood O of x_0 such that for any $f \in \mathcal{D}(O)$ the equation $Lu = f$ has a solution $u \in \mathcal{D}'(O)$. While this aspect will be discussed in more detail in Part II, let us point out that the situation is radically different for operators with variable coefficients: for example, the operator $L = x\partial_y^2 + \partial_x$ is not locally solvable at any point $(0, y_0) \in \mathbb{R}^2$, while the operator $L = x\partial_y^2 - \partial_x$ is locally solvable everywhere; see Bramanti (2014).

4. For $1 \le p < \infty$ and for an integer $k \ge 0$, the Sobolev spaces $W_p^k(\mathbb{R}^n)$ consist of the functions $f \in L^p(\mathbb{R}^n)$ with distributional derivatives $\partial_x^\alpha f \in L^p(\mathbb{R}^n)$ for all n-tuples α of nonnegative integers with $|\alpha| \le k$. Endowed with the norm $\|f\|_{W_p^k} = \left(\sum_{|\alpha|\le k} \|\partial_x^\alpha f\|_{L^p(\mathbb{R}^n)}^p\right)^{1/p}$, these are Banach spaces in which $\mathcal{D}(\mathbb{R}^n)$ is dense; see Brézis (2011). Moreover, $W_2^k(\mathbb{R}^n) = H^k(\mathbb{R}^n)$, so that the spaces $H^s(\mathbb{R}^n)$ extend naturally the spaces $W_2^k(\mathbb{R}^n)$. It is possible to also extend $W_p^k(\mathbb{R}^n)$ for $1 < p < \infty$, allowing for real exponents. In contrast to the case $p = 2$, where the construction pursued in Section 6.5 relied on the fact that $\mathcal{F}$ and $\mathcal{F}^{-1}$ are unitary operators on $L^2(\mathbb{R}^n)$, for $p \in (1,2) \cup (2,\infty)$ the extension relies on the deep *Michlin-Hörmander multiplier theorem*, stating that there exists a constant $c > 0$ such that, for any continuous function $m : \mathbb{R}^n \to \mathbb{C}$ with bounded and continuous partial derivatives $\partial_x^\alpha m$ up to order $l > n/2$, and any $f \in L^p(\mathbb{R}^n)$, we have

$$\|\mathcal{F}^{-1}(m\hat{f})\|_{L^p(\mathbb{R}^n)} \le c\,\|f\|_{L^p(\mathbb{R}^n)} \sup_{x\in\mathbb{R}^n} \sup_{|\alpha|\le l} \{|(\partial_x^\alpha m)(x)|\}\,.$$

Due to the difficulties encountered with the range of the Fourier transform on $L^p(\mathbb{R}^n)$-spaces, discussed in Section 6.4, the operator $f \mapsto \mathcal{F}^{-1}(m\hat{f})$ is initially defined for $f \in S(\mathbb{R}^n)$, and subsequently shown to admit a bounded extension on $L^p(\mathbb{R}^n)$. This result is suggestive of setting $H_p^k(\mathbb{R}^n) = \{f \in S'(\mathbb{R}^n) : \mathcal{F}^{-1}(m\hat{f}) \in L^p(\mathbb{R}^n)\}$ for $m(x) = (1 + |x|^2)^{s/2}$; see the discussion in Haroske and Triebel (2008). Note that one of the most important features of Sobolev spaces is their innate ability to trade differentiability for integrability; for example, Theorem 6.21 shows that if $f \in H^{3/4}(\mathbb{R})$, then $f \in L^{2016}(\mathbb{R}) \subset L^2(\mathbb{R}) \cap L^\infty(\mathbb{R})$. However, this trade is only one-way: one cannot sacrifice integrability to gain differentiability, as integrable functions need not have a well-defined weak derivative. Refinements of Sobolev spaces (e.g. Besov spaces, Triebel-Lizorkin spaces and their weighted versions) are often employed in settings in which one imposes a non-homogeneous behaviour on the dependent variables.

7
A glance at some advanced topics

Fourier analysis is a wide ranging, inexhaustible area of mathematics. Its sheer breadth precludes an exhaustive list of advanced topics. In this chapter we discuss aspects of two such topics: the very strong links to complex analysis and the relevance of Fourier analysis to the theory of pseudodifferential operators that arise in many areas of analysis and mathematical physics.

7.1 Complex analysis techniques

There is a fruitful interplay between complex function theory and Fourier analysis. We illustrate it by discussing two topics: the first concerns Fourier series and the second Fourier transforms. Before delving into these topics we present an outline of complex function theory, intended to convey the principal facts; additional material can be found in Burckel (1979).

7.1.1 Basic facts about analytic functions

Complex analysis is a fascinating area of mathematics, astonishingly elegant and powerful. Several special factors ensure that the single requirement of complex differentiability implies that functions belonging to this class (comprising a plethora of intricate functions arising in applications) have a strong inner bond which entails remarkable structural properties. The fact that the complex numbers are a field, the infinite degree of freedom associated with variations in a plane (instead of only on a line, for functions of one real variable), balanced by restrictive features of the topology of the plane (compared to dimensions higher than two), illustrate the fine interplay of algebra, analysis and geometry that is a hallmark of this area of mathematics of great aesthetic appeal. Moreover, through the computational tools of contour integrals, conformal mapping, and power series, complex analysis has wide applicability in physics and engineering.

Review of elementary facts

The complex numbers, defined formally as the field extension $\mathbb{C} = \mathbb{R}[i]$ of $\mathbb{R}$, are represented in the Euclidean plane by[1] $z = (x, y) = x + iy$, where the "imaginary" number $i = (0, 1)$ satisfies $i^2 = -1$ for the multiplication rule $(x, y) \cdot (x', y') = (xx' - yy', xy' + x'y)$. The latter is consistent with the Euclidean inner product in $\mathbb{R}^2$ and justifies the identification of $\mathbb{C}$ and $\mathbb{R}^2$. Complex conjugation is motivated by the fact that a quadratic polynomial with real coefficients having the root $z = x + iy$ must also admit the root $\bar{z} = x - iy$. Calling the Euclidean distance $\sqrt{x^2 + y^2}$ of $z = x + iy$ to $0 \in \mathbb{C}$ the *modulus* of z, note that $1/z = \bar{z}/|z|^2$ for $z \neq 0$. Recall also the polar description of a complex number: $z = [r, \theta] = r(\cos\theta + i \sin\theta)$ with $r = |z|$ and $\theta = \arg z \in \mathbb{R}/2\pi\mathbb{Z}$. In the polar representation it is convenient not to restrict the angle θ to $[0, 2\pi)$ – the ensuing loss of uniqueness (or multi-valuedness) of $\arg z$ requires caution but turns out to be the ultimate source of powerful results such as Cauchy's integral formula. At a more elementary level, one reason for allowing this is to accommodate the multiplication formula $[r_1, \theta_1]\,[r_2, \theta_2] = [r_1 r_2, \theta_1 + \theta_2]$ as a consequence of $\arg(z_1 z_2) = \arg(z_1) + \arg(z_2)$ and $|z_1 z_2| = |z_1|\,|z_2|$ (this is also related to the addition formulas for sin and cos).

Complex differentiation

Let $U \subset \mathbb{C}$ be an open set. A function $f : U \to \mathbb{C}$ is called *analytic* if the *complex derivative*

$$f'(z_0) = \lim_{z \to z_0} \frac{f(z) - f(z_0)}{z - z_0} \tag{7.1}$$

exists (in $\mathbb{C}$) independently of the mode of approach of z to z_0, for every $z_0 \in U$. Clearly (7.1) ensures continuity throughout U. To appreciate the strength of the requirement (7.1), note that although the function $f(z) = \bar{z}$ is smooth (infinitely differentiable) as a function from $\mathbb{R}^2$ to $\mathbb{R}^2$, it does not possess a complex derivative at $z_0 = 0 \in \mathbb{C}$ since the limits along the coordinate axes differ, being equal to ± 1, respectively. To make this point systematically, let us write $f = u + iv$, where u and v are real-valued functions of the real variables

[1] The real number x is called the *real part* of z, $\Re z$, while the real number y is called the *imaginary part* of z, $\Im z$.

x and y; thus we regard f as a function from an open set of $\mathbb{R}^2$ into $\mathbb{R}^2$. With f viewed in these real-variable terms, within the class of C^1-functions,[2] by straightforward calculation one can check that, if we define[3]

$$\frac{df}{dz} = \frac{1}{2}\Big(\frac{df}{dx} + \frac{1}{i}\,\frac{df}{dy}\Big), \qquad \frac{df}{d\bar{z}} = \frac{1}{2}\Big(\frac{df}{dx} - \frac{1}{i}\,\frac{df}{dy}\Big),$$

then

$$\begin{aligned} Df_{(x_0,y_0)}\begin{pmatrix} x-x_0 \\ y-y_0 \end{pmatrix} &= \begin{pmatrix} u_x(x_0,y_0) & u_y(x_0,y_0) \\ v_x(x_0,y_0) & v_y(x_0,y_0) \end{pmatrix}\begin{pmatrix} x-x_0 \\ y-y_0 \end{pmatrix} \\ &= \frac{df}{dz}(z_0)\,(z-z_0) + \frac{df}{d\bar{z}}(z_0)\,\overline{(z-z_0)}\,. \end{aligned}$$

Interpreting (7.1) as $f(z) - f(z_0) \approx f'(z_0)\,(z - z_0)$, since the Jacobian matrix of f, viewed in real-variable terms, is the matrix of the linear transformation that best approximates $f(x,y) - f(x_0,y_0)$ at (x_0,y_0), we see that the existence of the complex derivative at z_0 is equivalent to the requirement that $\frac{df}{d\bar{z}}(z_0) = 0$, i.e. the validity of the *Cauchy–Riemann equations*

$$u_x = v_y\,, \qquad u_y = -v_x\,. \tag{7.2}$$

The requirement of continuous differentiability can be relaxed. For example, if f is continuous on U and, as a distribution, satisfies (7.2), then f is analytic – see Gray and Morris (1978). This is a particular instance of the regularity properties of hypo-elliptic systems of partial differential equations, of which the Cauchy–Riemann system (7.2) is a paradigm example. However, the mere existence of the partial derivatives of u and v, subject to (7.2), does not suffice. For example, it is readily seen that the partial derivatives of the real and imaginary parts of the function given by $f(z) = \exp(-z^{-4})$ if $z \neq 0$ and $f(0) = 0$ exist everywhere and satisfy (7.2); nevertheless f fails to have a complex derivative at the origin – even continuity fails there, since $f(z) \to \infty$ as $z \to 0$ with $\arg z = \pi/4$. Moreover, the weakening to the validity of (7.2) a.e. on U (in the sense of the planar Lebesgue measure) for f continuous, that is, with the partial derivatives of u and v existing and satisfying (7.2) a.e. in U, might fail – see Urysohn (1923) for a counter-example.

The Cauchy–Riemann equations (7.2) lead also to a geometric interpretation of (7.1). Indeed, denoting $a = u_x(x_0,y_0) \in \mathbb{R}$, $b = v_x(x_0,y_0) \in \mathbb{R}$, (7.2) mean that the (real) Jacobian of f at (x_0,y_0) has the form $\left(\begin{smallmatrix} a & -b \\ b & a \end{smallmatrix}\right)$, and we can write this matrix as $\left(\begin{smallmatrix} \sqrt{a^2+b^2} & 0 \\ 0 & \sqrt{a^2+b^2} \end{smallmatrix}\right)\left(\begin{smallmatrix} \cos\theta & -\sin\theta \\ \sin\theta & \cos\theta \end{smallmatrix}\right)$, where $|f'(z_0)| = \sqrt{a^2+b^2}$ and $\theta = \arg f'(z_0)$, since

$$f'(z_0) = u_x(x_0,y_0) + i\,v_x(x_0,y_0)\,. \tag{7.3}$$

Geometrically, the above matrices have simple meanings: $\left(\begin{smallmatrix} \cos\theta & -\sin\theta \\ \sin\theta & \cos\theta \end{smallmatrix}\right)$ represents a (counter-clockwise) rotation around the origin by the angle θ, while

[2] A property that is inherited by u and v from f and also transferred from these components to f, unlike the mere condition of differentiability.

[3] Here $\dfrac{df}{dx} = u_x + iv_x$ and $\dfrac{df}{dy} = u_y + iv_y$, where $u_x = \dfrac{\partial u}{\partial x}$, $v_x = \dfrac{\partial v}{\partial x}$, $u_y = \dfrac{\partial u}{\partial y}$ and $v_y = \dfrac{\partial v}{\partial y}$.

$\begin{pmatrix} \sqrt{a^2+b^2} & 0 \\ 0 & \sqrt{a^2+b^2} \end{pmatrix}$ represents the multiplication of all vectors of $\mathbb{R}^2$ by $\sqrt{a^2+b^2}$; alternatively, this can be seen by writing $f'(z_0)(z-z_0)$ in polar form. Consequently, the effect of an analytic mapping f on infinitesimal vectors emanating from z_0 (mapped by $f'(z_0)$ into vectors emanating from $f(z_0)$, in accordance with the above interpretation of the effect of the real Jacobian matrix) is an expansion and a rotation: the length of $f'(z_0)$ is the amplification factor and the argument of $f'(z_0)$ is the angle of rotation. In real calculus one can visualise the derivative of a function $f : \mathbb{R} \to \mathbb{R}$ as the slope of the graph $y = f(x)$. In the spirit of the previous geometric considerations, thinking of $f'(x)$ as a point on the real axis of $\mathbb{R}$, the local effect of f on an infinitesimal vector at x is to expand it by $|f'(x)|$ and to rotate it by $\arg f'(x)$: the orientation is preserved if $f'(x) > 0$ (when $\arg f'(x) = 0$) and reversed if $f'(x) < 0$ (when $\arg f'(x) = \pi$). With this geometric interpretation in mind, the complex derivative is a natural generalisation. This points out to an important feature of analytic maps: if two curves meet in z_0 at some arbitrary angle φ, then, their images under an analytic map f with $f'(z_0) \neq 0$ will also meet at angle φ. We say that analytic maps are *conformal* (angle-preserving) at points where $f'(z_0) \neq 0$.

Using the definition (7.1), one can easily check the following basic properties of analytic functions: if $f, g : U \to \mathbb{C}$ are analytic, then their sum and product are analytic, with $(f+g)' = f'+g'$ and $(fg)' = f'g+fg'$; if $g(z_0) \neq 0$, then f/g is analytic at z_0 and $(f/g)' = (f'g-fg')/g^2$. Moreover, if $f : U \to V$ and $g : V \to \mathbb{C}$ are analytic on some domains $U, V \subset \mathbb{C}$, then $g \circ f$ is analytic and the chain rule $(g \circ f)'(z) = g'(f(z))f'(z)$ holds for all $z \in U$.

Some sources of analytic functions are the following.

- *Polynomials.* Note that any polynomial in two real variables with complex coefficients, $p(x, y)$, can be re-written as a polynomial in the two independent variables z and $\bar{z}$ since $x = (z+\bar{z})/2$ and $y = (z-\bar{z})/(2i)$; analyticity amounts to independence of the $\bar{z}$-variable.
- *Power series.* As we shall see, analytic functions are given locally by power series of the form $\sum_{n\geq 0} a_n z^n$ for suitable coefficients $a_n \in \mathbb{C}$ (for example, $e^z = \sum_{n\geq 0} z^n/n!$). While analytic functions are a generalisation of analytic polynomials, they are more flexible objects, being e.g. closed under uniform limits, exponentiation, and many other operations as well.
- *Differential equations.* Analytic functions arise when one solves differential equations. For example, the constant-coefficient differential equations $f'' \pm f = 0$ give rise to the transcendental functions $\sin(z)$, $\cos(z)$ and e^z.
- *Integration.* While the indefinite integral $\int (x^2+ax+b)^{1/2}\,dx$ can be expressed in terms of trigonometric functions, $\int (x^3+ax+b)^{1/2}\,dx$ leads to elliptic

functions and higher-degree polynomials lead to hyperelliptic surfaces of higher genus.

- *Geometric function theory.* Some analytic functions can be specified by simply-connected domains $U \subsetneq \mathbb{C}$ since (by the Riemann mapping theorem – see below) there is an essentially unique analytic homeomorphism $f : \mathbb{D} \to U$; here $\mathbb{D} = \{z \in \mathbb{C} : |z| < 1\}$ is the unit disc.

Complex integration

Complex path integration is not merely a re-working of standard real variable theory, it enjoys remarkable properties that are not found in its real sibling.

Let us first introduce some useful terminology. A (planar) *curve* is a continuous function[4] from a compact interval $[a, b]$ in $\mathbb{R}$ into $\mathbb{C}$, a one-to-one curve is called an *arc* and a closed curve γ (for which $\gamma(a) = \gamma(b)$) is called a *loop*, a loop being simple if γ is one-to-one on $[a, b)$. A *Jordan curve* is a simple closed curve. The *Jordan curve theorem* asserts the intuitively obvious fact[5] that every Jordan curve $J \subset \mathbb{C}$ divides the plane into two disjoint, open, connected components:[6] an "interior" region (the unique bounded component of $\mathbb{C} \setminus J$) and an "exterior" region (the unique unbounded component of $\mathbb{C} \setminus J$), so that any curve connecting a point of one region to a point of the other intersects J somewhere. We say that a Jordan curve is *positively directed* if its interior always lies to the left when one travels along it in the indicated direction. For example, the positive orientation of the unit circle $\mathbb{S} = \{z \in \mathbb{C} : |z| = 1\}$ is counter-clockwise.

From an utilitarian point of view there is no need to integrate over curves more general than piecewise C^1: a curve $\gamma : [a, b] \to \mathbb{C}$ is called *piecewise continuously differentiable* if there exists a partition $a = t_0 < t_1 < \cdots < t_n = b$ such that γ is differentiable in (t_{k-1}, t_k), has left- and right-hand derivatives at t_k and t_{k-1} respectively and the derivative function $\gamma'(t)$ is continuous on $[t_{k-1}, t_k]$, for $k = 1, \ldots, n$. The simplest example of a piecewise C^1-curve that fails to

4 Two such maps represent the same directed curve if they sweep out the same geometrical figure, in the same direction of travel.

5 At least for closed curves whose simple visualisation identifies their interior unambiguously; to dispel the idea that everything here is self-evident, note that there are Jordan curves with nonzero area. For a proof and for related material we refer to Burckel (1979).

6 Recall that a topological space is connected if it cannot be represented as the union of two or more open sets. Remarkably, a planar open connected set is path-connected as any two points can be joined by a piecewise C^1-arc; to see this, fix a point and note that the set of points which can be joined to it by a piecewise C^1-arc is simultaneously relatively open and relatively closed, and must therefore be the whole set. This is not so in general: for the closed and connected planar set $\{(x, y) : x = 0 \text{ or } y = \sin(1/x)\}$ there is no path that reaches the y-axis from beyond.

be (overall) continuously differentiable is a broken polygonal line, defined by $\gamma(t) = a_{k-1}+(t-k+1)(a_k-a_{k-1})$, where $a_k \in \mathbb{R}$ are its distinct vertices (corners).

Let U be the interior of a piecewise C^1 Jordan curve γ and let $f : \overline{U} = U \cup \gamma \to \mathbb{C}$ be a continuous function. The path integral of f about γ is defined by

$$\int_\gamma f(z)\,dz = \int_a^b f(\gamma(t))\,\gamma'(t)\,dt\,.$$

This integral[7] is well-approximated by the Riemann sums $\sum_{k=1}^n f(z_k)\,(z_{k+1}-z_k)$, where $z_1,\dots,z_n$ are points along γ such that the labelling respects the direction of travel (here $z_{n+1} = z_1$) and the spacing is sufficiently fine; see Fig. 7.1 ($n = 6$). Thus we see that the number $\int_\gamma f(z)\,dz$ depends only upon the directed curve γ itself, and not upon the particular map employed to parametrise it: we can talk about the integral along a curve when the curve is described geometrically, without reference to a specific parametrisation. However, the direction of γ is important: if it is reversed, the integral changes sign. Throughout this section we always assume loops to be positively-directed, unless something is said to the contrary.

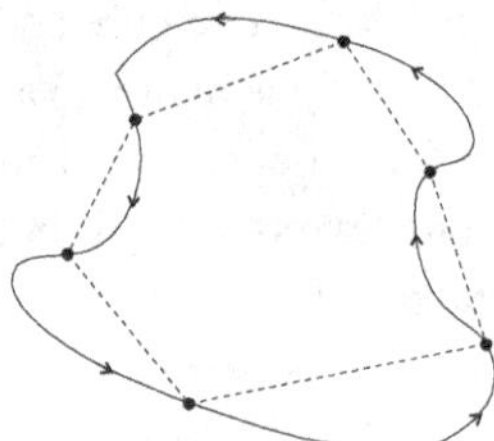

Figure 7.1 Choosing points close together along the curve permits us to approximate the path integral by Riemann sums.

It is easy and instructive to check the following computations of path integrals about the circle $|\gamma| = \{z \in \mathbb{C} : |z - z_0| = R\}$ (directed counter-clockwise) for arbitrary $z_0 \in \mathbb{C}$ and $R > 0$:

$$\int_\gamma (z-z_0)^n\,dz = \begin{cases} 0 & \text{if the integer} \quad n \neq -1\,, \\ 2\pi i & \text{if} \quad n = -1\,. \end{cases} \tag{7.4}$$

For this, it is convenient to consider the parametrisation $\gamma(t) = z_0 + R\,e^{2\pi i t}$ with $t \in [0,1]$. Note that the same outcome is obtained if γ is the boundary of the square $U = \{z_0+(x+iy) : |x| < R, |y| < R\}$ centred at z_0, e.g., for $\gamma : [0,1] \to \mathbb{C}$, $\gamma(t) = z_0+(8t-1)R-Ri$ if $0 \le t \le 1/4$, $\gamma(t) = z_0+R+(8t-3)Ri$ if $1/4 \le t \le 1/2$, $\gamma(t) = z_0+(5-8t)R+Ri$ if $1/2 \le t \le 3/4$, $\gamma(t) = z_0+-R+(7-8t)Ri$ if $3/4 \le t \le 1$.

[7] Not be confused with the integral $\int_\gamma f(z)\,|dz| = \lim_{n\to\infty} \sum_{k=1}^n f(z_k)\,|z_{k+1} - z_k|$ with respect to arc-length (integral which does not depend on the orientation of γ).

The first remarkable result of complex analysis exhibits a deep connection between complex differentiation and path integrals.

Theorem 7.1 (Cauchy's theorem) *Let $U \subset \mathbb{C}$ be a domain*[8] *and let $\gamma \subset U$ be a piecewise C^1, simple, positively directed, closed curve such that its interior lies wholly in U. Then for any analytic function $f : U \to \mathbb{C}$, we have $\int_\gamma f(z)\,dz = 0$.*

Proof The standard proof of this result – see e.g. Stein and Shakarchi (2003) – requires only the existence of the complex derivative at each point of U, and not its continuity: the approach consists of approximating γ with polygonal arcs, cutting the interior into small triangles, noticing that on these triangles $f(z) \approx az + b$ for suitable constants $a, b \in \mathbb{C}$, and using the fact that the path integrals of $(az + b)$ about each triangle vanishes. We present an alternative proof that, while relying on the additional hypothesis of C^1-continuity of f (viewed in real-variables terms), reveals readily the heart of the matter. Using the divergence theorem for the vector fields $F_1 = -(v, u)$ and $F_2 = (u, -v)$, if $f = u + iv$ and if D is the interior of $\gamma(t) = x(t) + iy(t)$ for $t \in [0, 1]$, we express the integral as

$$\int_\gamma f(z)\,dz = \int_0^1 f(\gamma(t))\,\gamma'(t)\,dt = \int_0^1 (ux' - vy')\,dt + i\int_0^1 (vx' + uy')\,dt$$
$$= \iint_D \operatorname{div} F_1\,dxdy + i\iint_D \operatorname{div} F_2\,dxdy = -\iint_D (v_x + u_y)\,dxdy + i\iint_D (u_x - v_y)\,dxdy = 0$$

by the Cauchy–Riemann equations (7.2). □

The existence of primitives gives a first manifestation of the importance of Cauchy's theorem. In real analysis, primitives (or anti-derivatives) of functions are of great utility. Since every continuous function $f : [0, 1] \to \mathbb{R}$ has a primitive (see Chapter 2), it is natural to look to integration to produce primitives for continuous functions $f : U \to \mathbb{C}$ defined on planar regions $U \subset \mathbb{C}$. A *primitive* of f is an analytic function $F : U \to \mathbb{C}$ such that $F'(z) = f(z)$ for all $z \in U$. A simple application of the chain rule and the fundamental theorem of calculus on each of the subintervals where γ is C^1 shows that if γ is a piecewise C^1-curve starting at $z_1 \in \mathbb{C}$ and ending at $z_2 \in \mathbb{C}$, then $\int_\gamma f(z)\,dz = F(z_2) - F(z_1)$ for every continuous function $f : U \to \mathbb{C}$ having a primitive F. In particular, if γ is a piecewise-C^1 loop, then $\int_\gamma f(z)\,dz = 0$ since the endpoints coincide. This shows that the conclusion of Cauchy's theorem is a necessary condition for the existence of a primitive. The converse of this and of Cauchy's theorem is also true.

Theorem 7.2 (Morera's theorem) *Let $U \subset \mathbb{C}$ be a domain and let $f : U \to \mathbb{C}$ be continuous. If $\int_\gamma f(z)\,dz = 0$ for every piecewise-C^1 closed curve consisting of a finite number of straight line segments and such that its interior lies wholly in U, then f is analytic in U.*

[8] Recall that a domain is an open and connected set.

Proof Since analyticity is a local property, we restrict our attention to an open ball $|z - z_0| < \varepsilon$ that lies wholly in U. Define locally $F(z) = \int_\gamma f(w)\,dw$, where the path integral is taken along any piecewise C^1-curve γ from z_0 to z in U, composed of a finite number of straight line segments. The definition of F is unambiguous since if we choose an alternative path, then the integral of f about the loop (or loops, as finite self-intersections are possible) obtained by joining the first path, with the original orientation, to the new path, with the reverse orientation, vanishes by hypothesis. Then, for $|z - z_0| < \varepsilon/2$, we have $F(z) - F(z_0) = \int_\zeta f(w)\,dw$, where ζ is the straight line segment from z_0 to z. Since f is continuous at z_0, we can write $f(w) = f(z_0) + \psi(w)$ with $\psi(w) \to 0$ as $w \to z_0$. Therefore $F(z) - F(z_0) = \int_\zeta f(z_0)\,dw + \int_\zeta \psi(w)\,dw = f(z_0) \int_\zeta 1\,dw + \int_\zeta \psi(w)\,dw$. Since the constant function 1 has w as a primitive, we obtain $F(z) - F(z_0) = (z - z_0)\,f(z_0) + \int_\zeta \psi(w)\,dw$. Expressing from this $[F(z) - F(z_0)]/(z - z_0)$ and using the estimate $\left|\int_\zeta \psi(w)\,dw\right| \le |z - z_0| \sup_{w\in\zeta}\{|\psi(w)|\}$ yields that F has the complex derivative $f(z_0)$ at z_0. □

Representation formulas are very useful since they permit us to recover certain functions on some sets from their behaviour on suitable (considerable smaller) subsets. For analytic functions this type of procedure is provided by the following result.

Theorem 7.3 (Cauchy's integral formula) *Let $U \subset \mathbb{C}$ be a domain and let $\gamma \subset U$ be a piecewise C^1, simple, positively directed, closed curve with interior D lying wholly in U. If $f : U \to \mathbb{C}$ is analytic, then*

$$f(z_0) = \frac{1}{2\pi i} \int_\gamma \frac{f(z)}{z - z_0}\,dz\,, \quad z_0 \in D\,. \tag{7.5}$$

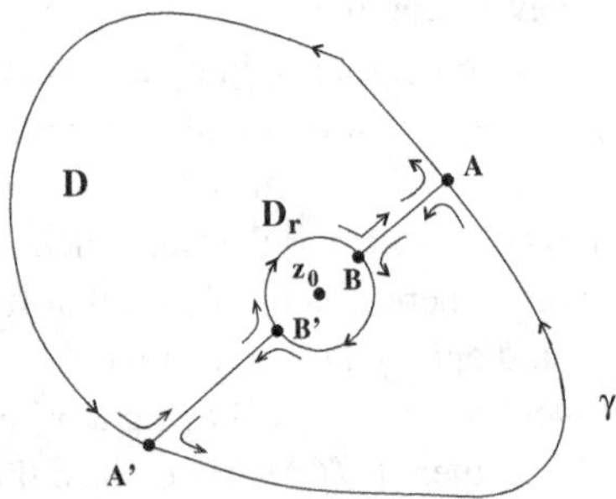

Figure 7.2 Two cuts permit us to relate the path integrals along γ and a small circle around z_0, thus obtaining Cauchy's integral formula.

Proof Fix some $z_0 \in D$, excise from D a small disc D_r of radius $r > 0$, centred at z_0, and connect a point B on the small circle ζ (the boundary of this disc) to a point $A \in \gamma$ by a straight line segment η whose interior points lie entirely within D; see[9] Figure 7.2. Let B' be the other intersection of the half-line from A to B with ζ (possibly $B' = B$), and let A' be the first time this half-line intersects γ, so that the interior of the straight line segment η' between B' and A' lies

[9] Since both γ and ζ are disjoint compact subsets of $\mathbb{C}$, $\inf_{z\in\gamma,\, w\in\zeta}\{|z - w|\} > 0$ is attained, providing us with a permissible choice of the pair of points (A, B).

entirely within D. Now integrate $(z-z_0)^{-1}f(z)$ counter-clockwise about γ from the point A to A', then in along η' to B', then clockwise about ζ from B' to B, and finally along η from B at A. Also, integrate $(z-z_0)^{-1}f(z)$ along η from A to B, then clockwise about ζ from the point B to B', then in along η' to A', then finally counter-clockwise about γ from A' to A. Since the integrand is analytic in a domain containing the closure of the interior $D\backslash(D_0\cup\zeta)$ of these simple, closed, piecewise C^1-curves, by Cauchy's theorem the two path integrals must vanish. Since the two opposing integrals along η and η' cancel by a change of sign, we get $0=\int_\gamma \frac{f(z)}{z-z_0}\,dz+\int_\zeta \frac{f(z)}{z-z_0}\,dz$. The parametrisation $\{z_0+re^{2\pi it} : t\in[0,1]\}$ for ζ permits us to write $\int_\zeta \frac{f(z)}{z-z_0}\,dz = 2\pi i\int_0^1 f(z_0+re^{2\pi it})\,dt$. This expression should be independent of r, being equal to $\int_\gamma \frac{f(z)}{z-z_0}\,dz$, so that we can evaluate it by letting $r\downarrow 0$: $\int_\zeta \frac{f(z)}{z-z_0}\,dz = 2\pi i\lim_{r\downarrow 0}\int_0^1 f(z_0+Re^{2\pi it})\,dt = 2\pi i f(z_0)$, in view of the continuity of f. This completes the proof. □

Making the substitution $z=z_0+re^{2\pi it}$ with $0\le t\le 1$ in Cauchy's integral formula, we see that analytic functions have the *mean value property*:

$$f(z_0)=\int_0^1 f(z_0+re^{2\pi it})\,dt$$

whenever f is analytic in a domain containing the closed disc $|z-z_0|\le r$. This result pertains to the size of an analytic function:

Theorem 7.4 (Maximum modulus principle) *Let $f : U\to\mathbb{C}$ be analytic in some domain $U\subset\mathbb{C}$. If $|f(z_0)|=\sup_{z\in U}|f(z)|$ for some $z_0\in U$, then f is constant.*

Proof Let $z_0\in U$ be such that $|f(z_0)|=M$, where $M=\sup_{z\in U}|f(z)|$. The mean-value property of f and the continuity of $|f|$ imply that $|f(z)|=|f(z_0)|$ on every circle $|z-z_0|=r$ such that U contains the closed disc $|z-z_0|\le r$. Consequently, $|f(z)|=M$ in a neighbourhood of z_0. Therefore, the set $\mathcal{M}=\{z\in U : |f(z)|=M\}$ is open and closed in U. Since U is connected and $\mathcal{M}\neq\emptyset$, we must have $\mathcal{M}=U$. If $M=0$ we are done, and for $M>0$, let $u=\Re e(f)$, $v=\Im m(f)$, and use the fact that $u^2+v^2=M^2$ in combination with (7.2) to infer, by differentiating with respect to the x- and y-variables, that $uu_x+vv_x=0$ and $-uv_x+vu_x=0$. Multiplying the first relation by u, the second one by v and adding up yields $u_x(u^2+v^2)=0$. Thus $u_x=0$ and, similarly, $v_x=0$. Now (7.2) ensures that both functions u and v are constant. □

A more striking consequence of Cauchy's formula is that *if f is analytic in U, then it is infinitely differentiable, and its higher-order complex derivatives, given by*

$$f^{(n)}(z_0)=\frac{n!}{2\pi i}\int_\gamma \frac{f(z)}{(z-z_0)^{n+1}}\,dz\,,\quad z_0\in D\,,\quad n\ge 1\,, \tag{7.6}$$

are also analytic in U; here $0!=1$. This statement follows at once if we notice that the integrand in (7.5) depends on z_0 only through the function $1/(z-z_0)$, which is infinitely differentiable; differentiating under the integral – a convolution of $f|_\gamma$ and $1/z$ – we get (7.6). If $d(z_0,\gamma)=\inf_{z\in\gamma}\{|z-z_0|\}=R>0$,

$\sup_{z\in\gamma}\{|f(z)|\} = M$ and the length of γ is L, then (7.6) yields the bound

$$|a_n| \leq \frac{ML}{2\pi R^{n+1}} \quad \text{for} \quad a_n = \frac{f^{(n)}(z_0)}{n!}\,, \; n \geq 0\,.$$

In particular, $\limsup_{n\geq 1} |a_n|^{1/n} \leq 1/R$ and the power series $\sum_{n\geq 0} a_n(z-z_0)^n$ has radius of convergence[10] at least R. This suggests that analytic functions have a power series representation. Note that power series are analytic in the interior of their disc of convergence,[11] so that for every piecewise-C^1, simple, directed curve γ lying inside this disc, the rapid convergence of the power series permits us to infer the continuity of the sum and the interchange of limit and integral: $\int_\gamma \sum_{n\geq 0} c_n(z-z_0)^n\, dz = \sum_{n\geq 0} \int_\gamma c_n(z-z_0)^n\, dz = 0$ and we conclude by Morera's theorem. A more striking point is the converse.

Theorem 7.5 *If $f : U \to \mathbb{C}$ is analytic in a domain U containing the closure of the disc $D_0 = \{z \in \mathbb{C} : |z - z_0| < R\}$, then f is expressible in D_0 as the power series $\sum_{n\geq 0} a_n(z-z_0)^n$ with coefficients*

$$a_n = \frac{1}{2\pi i}\int_\gamma \frac{f(w)}{(w-z_0)^{n+1}}\, dw\,, \qquad n \geq 0\,,$$

where γ denotes the boundary $\{w \in \mathbb{C} : |w - z_0| = R\}$ of D_0.

Proof Cauchy's formula yields

$$f(z) = \frac{1}{2\pi i}\int_\gamma \frac{f(w)}{w-z}\, dw = \frac{1}{2\pi i}\int_\gamma \frac{f(w)}{w-z_0}\Big(1 - \frac{z-z_0}{w-z_0}\Big)^{-1} dw\,, \qquad z \in D_0\,.$$

The inequality $\left|\frac{z-z_0}{w-z_0}\right| = \frac{|z-z_0|}{R} < 1$, for $w \in \gamma$ and $z \in D_0$, validates the geometric series expansion $\big(1 - \frac{z-z_0}{w-z_0}\big)^{-1} = \sum_{n\geq 0} \big(\frac{z-z_0}{w-z_0}\big)^n$, the convergence being uniform on γ. Therefore, we can interchange sum and integral, finishing the proof. □

Example The *complex exponential* e^z and the *trigonometric functions* $\cos(z)$ and $\sin(z)$ are defined by power series that converge in the whole complex plane:

$$e^z = \sum_{n=0}^{\infty} \frac{z^n}{n!}\,, \qquad \cos(z) = \sum_{n=0}^{\infty} (-1)^n \frac{z^{2n}}{(2n)!}\,, \qquad \sin(z) = \sum_{n=0}^{\infty} (-1)^n \frac{z^{2n+1}}{(2n+1)!}\,,$$

10 A complex power series centred at $z_0 \in \mathbb{C}$ is an infinite sum of the form $\sum_{n\geq 0} c_n(z-z_0)^n$, with complex coefficients $c_0, c_1, \ldots$. The sum converges absolutely inside the circle $|z - z_0| = C$ and diverges outside it; here $C = 1/\limsup_{n\to\infty} \sqrt[n]{|c_n|}$, with $C = 0$ and $C = \infty$ possible occurrences. Indeed, if $|z - z_0| < C$, then $|c_n|\,|z-z_0|^n \leq a^n$ for some constant $a \in (0,1)$ and all sufficiently large integers n, which ensures geometrically fast convergence, while if $|z - z_0| > C$, then $|c_n|\,|z-z_0|^n \geq 1$ infinitely often, preventing convergence. On the boundary of the disc of convergence, $|z - z_0| = C$, one can have either convergence or divergence: $\sum_{n\geq 1} z^n/n^2$ converges at every point of the unit circle, $\sum_{n\geq 1} nz^n$ diverges at every point of the unit circle, while $\sum_{n\geq 1} z^n/n$ converges at every point of the unit circle except $z = 1$. To check the last statement, note that $(1-z)\sum_{n\geq 1} z^n/n = z - \sum_{n\geq 2} z^n/n(n-1)$ for $|z| < 1$.

11 This being an elaboration of the fact that a power series is complex differentiable in its disc of convergence, its derivative being the power series (with the same radius of convergence) obtained by term-wise differentiation.

and they agree[12] with the usual functions of a real variable for $z \in \mathbb{R}$. A simple calculation exhibits a connection between these functions: $\cos(z) = (e^{iz} + e^{-iz})/2$ and $\sin(z) = (e^{iz} - e^{-iz})/(2i)$. In particular, $e^{i\pi} = -1$. Note that the equality on the real axis of expressions representing power series with infinite radius of convergence forces equality throughout the complex plane since the coefficients are determined by the Taylor expansions of the restrictions to the real axis. Thus[13] $\cos^2(z) + \sin^2(z) = 1$ on $\mathbb{C}$. Also, $e^{z_1+z_2} = e^{z_1}e^{z_2}$ for all $z_1, z_2 \in \mathbb{C}$, and therefore

$$e^{x+iy} = e^x\left(\cos(y) + i\sin(y)\right), \qquad x, y \in \mathbb{R},$$

since we already know that $e^{iy} = \cos(y) + i\sin(y)$; this makes the connection with the polar form. A straightforward calculation based on the above identities yields $\cos(x+iy) = \cos(x)\cosh(y) - i\sin(x)\sinh(y)$ and $\sin(x+iy) = \sin(x)\cosh(y) + i\cos(x)\sinh(y)$ for $x, y \in \mathbb{R}$, so that the apparent boundedness of $\sin(z)$ and $\cos(z)$ fails as we move away from the real axis. □

Certain fallacies in real analysis become theorems in complex analysis.

- The existence of one (complex) derivative guarantees the existence of derivatives of any order; see (7.6).
- Another striking result is *Montel's compactness theorem*: a bounded sequence $\{f_n\}_{n\geq 1}$ of analytic functions in a domain $U \subset \mathbb{C}$ is pre-compact (it has a subsequence converging uniformly on compact subsets of U).

  To prove this, note that if $f : U \to \mathbb{C}$ is analytic with $|f(z)| \leq M$ for all $z \in U$, then $|f'(z_0)| \leq M/d(z_0, \partial U)$, where $d(z_0, \partial U) = \inf_{z \notin U} |z - z_0| > 0$ for $z_0 \in U$. Indeed, for any $0 < R < d(z_0, \partial U)$, the disc $|z - z_0| \leq R$ lies in U, and the stated inequality follows from (7.6) with $n = 1$, letting $R \uparrow d(z_0, \partial U)$. For any compact set $K \subset U$, since $\inf_{z_0 \in K} d(z_0, \partial U) > 0$, we obtain that the restrictions of f_n to K are equi-continuous and we can apply the Arzelà–Ascoli theorem (see Chapter 3).
- If $f_n : U \to \mathbb{C}$ are analytic and converge uniformly to some $f : U \to \mathbb{C}$, then f is analytic and f_n' converge to f' uniformly on compact subsets[14] of U.

  The analyticity of f follows from Morera's theorem. Subsequently, we observe that every compact set $K \subset U$ can be covered by finitely many open discs centred at points in K and whose closure lies within U. The boundary arcs of these discs form a piecewise C^1-map γ that, by construction, contains K in its interior. The uniform convergence of f_n' to f' on K follows now by means of (7.6) with $n = 1$.

An important consequence of Theorem 7.4 is that *an analytic function has at least one singularity on its circle of convergence*. Indeed, if the function can be extended analytically from the disc centred at z_0, of radius $R > 0$ (the radius of convergence of the series), to the concentric disc of radius $R_1 > R$, then the radius of convergence is at least R_1, so that there must be some obstruction to making such an extension.

Example The Fibonacci sequence $\{a_n\}_{n\geq 0}$ is defined recursively by $a_0 = a_1 = 1$ and $a_n = a_{n-1} + a_{n-2}$ for $n \geq 2$. Setting $f(z) = \sum_{n\geq 0} a_n z^n$, we get $f(z) = (z + z^2)f(z) + 1$ by identifying the power series coefficients on each side. Thus $f(z) = 1/(1 - z - z^2)$. Since this function has singularities

12 The fact that the coefficients are those of the Taylor expansions of the ordinary functions in elementary calculus is not accidental.

13 Checking this directly from the power series expansions is tedious.

14 Note than $f_n(z) = z^n/n$ converge to zero uniformly in $|z| < 1$ but $f_n'(z) = z^{n-1}$ converge to zero only locally uniformly. This proves that the result is sharp.

at $z = (-1 \pm \sqrt{5})/2$, the radius of convergence must be $(-1+\sqrt{5})/2$, so that $\limsup_{n\geq 1} |a_n|^{1/n} = (1+\sqrt{5})/2$ (the golden ratio). □

The previous example is symptomatic of the following general result: *a power series represents a rational function*[15] *if and only if its coefficients satisfy a recurrence relation.* This result leads us to discuss singularities. We call $z_0 \in \mathbb{C}$ an *isolated singularity* of the function f if f is analytic in $U \setminus \{z_0\}$ for some domain U that contains z_0. To acquire a bilateral series expansion for f around such a point we rely on the Cauchy formula for the annulus.

Theorem 7.6 (Laurent series expansion) *Let $f : U \to \mathbb{C}$ be analytic in a domain $U \subset \mathbb{C}$ containing the closure of the annulus $A(z_0, r, R) = \{z \in \mathbb{C} : r < |z - z_0| < R\}$ with $0 < r < R$. Then $f(z) = \sum_{k=-\infty}^{\infty} a_k (z - z_0)^k$ in $A(z_0, r, R)$, where*

$$a_k = \frac{1}{2\pi i}\int_{|w-z_0|=R} \frac{f(w)}{(w-z_0)^{k+1}}\,dw\,, \qquad k \geq 0\,,$$

$$a_k = \frac{1}{2\pi i}\int_{|w-z_0|=r} \frac{f(w)}{(w-z_0)^{k+1}}\,dw\,, \qquad k < 0\,,$$

with absolute convergence in the interior and uniform convergence on compact subsets of $A(z_0, r, R)$.

Proof The straight line segment $\eta = \{z_0 + tr + (1-t)R : 0 \leq t \leq 1\}$ cuts the annulus radially. Proceeding as in the proof of Cauchy's integral formula, by integrating $f(w)/(w-z)$, for a fixed point $z \in A(z_0, r, R)$, over the composite curve consisting of the circle $|w - z_0| = R$ (counterclockwise from $z_0 + R$ and back), then in along the cut, clockwise around the circle $|w - z_0| = r$, and finally out again along the cut to the starting point yields

$$f(z) = \frac{1}{2\pi i}\int_{|w-z_0|=R} \frac{f(w)}{w-z}\,dw - \frac{1}{2\pi i}\int_{|w-z_0|=r} \frac{f(w)}{w-z}\,, \qquad z \in A(z_0, r, R)\,.$$

Note that $\frac{1}{w-z} = \frac{1}{w-z_0}\left(1 - \frac{z-z_0}{w-z_0}\right)^{-1}$ with $\left|\frac{z-z_0}{w-z_0}\right| < 1$ if $|w - z_0| = R$ and $z \in A(z_0, r, R)$, while $\frac{1}{w-z} = -\frac{1}{z-z_0}\left(1-\frac{w-z_0}{z-z_0}\right)^{-1}$ with $\left|\frac{w-z_0}{z-z_0}\right| < 1$ if $|w-z_0| = r$ and $z \in A(z_0, r, R)$. We can therefore expand into geometric series: $\left(1-\frac{z-z_0}{w-z_0}\right)^{-1} = \sum_{n\geq 0}\left(\frac{z-z_0}{w-z_0}\right)^n$ for $|w-z_0| = R$ and $(1-\frac{w-z_0}{z-z_0})^{-1} = \sum_{n\geq 0}(\frac{w-z_0}{z-z_0})^n$ for $|w - z_0| = r$, respectively, the convergence being uniform. The interchange of sum and integral being thus granted, we obtain the claim. □

In Theorem 7.6, set $M = \sup_{r\leq|w-z_0|\leq R}\{|f(w)|\}$. Then we can estimate the coefficients of the Laurent series by $|a_n| \leq MR^{-n}$ and $|a_{-n}| \leq Mr^n$ for $n > 0$, so that $\limsup_{n\to\infty} \sqrt[n]{|a_n|} \leq 1/R$ for the power series $\sum_{n=0}^{\infty} a_n(z - z_0)^n$, while $\limsup_{n\to\infty} \sqrt[n]{|a_{-n}|} \leq r$ for the power series $\sum_{n=1}^{\infty} a_{-n}\left(\frac{1}{z-z_0}\right)^n$. Consequently, the function f *can be expressed as the sum of a function analytic on $|z - z_0| < R$ and a function analytic on $|z - z_0| > r$.* As a special case, if $f : U \setminus \{z_0\} \to \mathbb{C}$ is analytic on a domain U that contains z_0, then near the isolated singularity

15 A fraction such that both the numerator and the denominator are polynomials.

z_0 we have a Laurent series expansion $f(z) = \sum_{k=-\infty}^{\infty} a_k(z-z_0)^k$. We define $\mathrm{ord}(f, z_0) = k_0$ if $a_{k_0} \neq 0$ but $a_k = 0$ for all $k < k_0$, allowing the values $k_0 = \pm\infty$. If $k_0 \geq 0$, the apparent singularity at z_0 is removable – the function f has an analytic extension to U (since the terms with negative index in the power series vanish); in particular, if $k_0 > 0$, then f has a *zero* of order/multiplicity k_0 at z_0. If $k_0 < 0$ is finite, we say that f has a *pole* or order $|k_0|$ at z_0; in this case f has a *finite Laurent series*

$$f(z) = \frac{a_{-|k_0|}}{(z-z_0)^{|k_0|}} + \cdots + \frac{a_{-1}}{z-z_0} + \sum_{n=0}^{\infty} a_n(z-z_0)^n \tag{7.7}$$

near z_0. Note that, in either of these cases, we can write

$$f(z) = (z-z_0)^{k_0} g(z)\,, \tag{7.8}$$

where g is analytic on U and $g(z_0) \neq 0$. If $\mathrm{ord}(f, z_0) = -\infty$ we say that f has an *essential singularity* at z_0. For example, the function $e^{1/z}$ has an essential singularity at $z = 0$ since its Laurent series is $\sum_{n=0}^{\infty} \frac{1}{n!\,z^n}$. Analytic functions have a wild behaviour near an essential singularity.

Theorem 7.7 (Casorati–Weierstrass theorem) *Let $f : \{z \in \mathbb{C} :\ 0 < |z - z_0| < R\} \to \mathbb{C}$ be analytic for some $z_0 \in \mathbb{C}$ and some $R > 0$. Then z_0 is an essential singularity of f if and only if the image of $\{z \in \mathbb{C} :\ 0 < |z - z_0| < r\}$ under f is dense in $\mathbb{C}$ for every $r \in (0, R)$.*

Proof One implication is easy since if z_0 is removable, then $\lim_{z\to z_0} f(z)$ exists and is finite, while if z_0 is a pole, then it is clear from (7.8) that $\lim_{z\to z_0} |f(z)| = \infty$. Assume now that z_0 is an essential singularity and the claim is false. Then there exist $w \in \mathbb{C}$, $\varepsilon > 0$ and $r \in (0, R)$ such that $|f(z) - w| > \varepsilon$ for all $z \in \mathbb{C} \setminus \{z_0\}$ with $|z - z_0| < r$. The function $\phi(z) = 1/[f(z) - w]$ is then analytic and bounded in $\{z \in \mathbb{C} :\ 0 < |z - z_0| < r\}$. Hence z_0 is a removable singularity of ϕ. In view of (7.8), we may therefore write $\phi(z) = (z - z_0)^k g(z)$ for some integer k and some function g that is analytic in the interior of the circle $|z - z_0| = r$, with $g(z_0) \neq 0$. It then follows that $1/\phi(z)$ has at most a pole at z_0, which means that $z \mapsto f(z) - w$ has at most a pole, contradicting the hypothesis that the Laurent series of f at z_0 has infinitely many terms of negative order. □

Remark (i) A strengthening of Theorem 7.7, due to Picard, ensures that f not only comes arbitrarily close to every complex number, but for every $r \in (0, R)$, the image of $\{z \in \mathbb{C} :\ 0 < |z - z_0| < r\}$ under f contains every complex number with at most one exception; see Burckel (1979). Since the function $\exp(1/z)$ omits the value 0, it is necessary to allow for this one omission.

(ii) The proof of Theorem 7.7 shows that z_0 is a removable singularity of f if and only if f is bounded near z_0, while poles are characterised by the behaviour $\lim_{z\to z_0} |f(z)| = \infty$. □

The power series expansion is an important tool for the local study of analytic functions. We now provide some powerful illustrations of its impact.

Theorem 7.8 (Liouville's theorem) *If $f : \mathbb{C} \to \mathbb{C}$ is analytic and has polynomial growth at infinity, that is, for some $C > 0$ and some integer $n \geq 0$ we have*

$|f(z)| \le C(1+|z|^n)$ for all $z \in \mathbb{C}$, then f is a polynomial of degree at most n. In particular, if f is bounded, then it is a constant.

Proof If $f(z) = \sum_{k\ge 0} a_k z^k$, the hypothesis and the formulas in Theorem 7.5 yield $|a_k| \le C(1+R^n)R^{-k}$ for all $k \ge 0$ and all $R > 0$. Letting $R \to \infty$ yields $a_k = 0$ for $k \ge n+1$. □

Theorem 7.9 (The fundamental theorem of algebra) *The only (complex) root-free polynomials $p(z) = \sum_{k=0}^{n} a_k z^k$ with complex coefficients are the constant polynomials: $a_0 \neq 0$ and $a_k = 0$ for $k \ge 1$.*

Proof If p is nonvanishing in $\mathbb{C}$, then $1/p$ is analytic and bounded on $\mathbb{C}$. □

Combining the previous two results with Picard's theorem, we see that the only analytic functions $f : \mathbb{C} \to \mathbb{C}$ that are one-to-one are of the form $f(z) = az + b$ with $a \neq 0$; indeed, $f(1/z)$ cannot have an essential singularity at $z = 0$, and a pole there means that f is a polynomial.

A critical role is played by the *residue* of a function f at the centre z_0 of an annular region of analyticity for f, defined by $\text{Res}(f, z_0) = a_{-1}$, using the notation of Theorem 7.6. The designation stems from the fact that, by (7.4),

$$\frac{1}{2\pi i}\int_{|z-z_0|=\rho} f(z)\,dz = a_{-1} \quad \text{for every} \quad \rho \in (r, R),$$

so that this is the sole surviving term when the Laurent series of f is integrated term-by-term. Actually, the integral of f along any piecewise C^1 simple loop γ in the annulus $A(z_0, r, R)$ equals[16] a_{-1}. Thus, in the language of differential geometry, the residue is intrinsically an invariant of the 1-form $f(z)\,dz$ (being invariant under change of coordinates), and not of the function $f(z)$. For the calculation of the residue of f at a pole z_0 of order $n \ge 1$, note that

$$\text{Res}(f, z_0) = \frac{\phi^{(n-1)}(z_0)}{(n-1)!} \quad \text{with} \quad \phi(z) = (z-z_0)^n f(z), \tag{7.9}$$

due to (7.7). However, this formula fails at an essential singularity since in this case the function ϕ in (7.9) does not have a complex derivative at z_0.

Example The rational function $f = P/Q$, where P and Q polynomials in z, has a "simple" (i.e. order 1) pole at z_0 if Q has a simple root at that point and $P(z_0) \neq 0$. In this case

$$\text{Res}(f, z_0) = \lim_{z\to z_0}\{(z-z_0)f(z)\} = \lim_{z\to z_0} \frac{P(z)}{\frac{Q(z)-Q(z_0)}{z-z_0}} = \frac{P(z_0)}{Q'(z_0)}\,.$$

16 Its value equals to that of the integral about the circle $\nu = \{z \in \mathbb{C} : |z - z_0| = \rho^*\}$, where $R > \rho^* > \sup_{z\in\gamma}\{|z - z_0|\}$. To see this, make two cuts, similar to the approach pursued in proving Cauchy's integral formula: find $A \in \gamma$ and $B \in \nu$ attaining $\inf_{z\in\gamma, w\in\nu} |z - w|$, define B' as the other intersection of the half-line from B to A (note that the ν circumscribes γ) with the circle ν, and let $A' \in \gamma$ be the closest point to B' on the straight line segment from B' to B; then proceed as in the proof of Theorem 7.3.

Thus the residue of $z^2/(z^2-1)$ at $z_0 = 1$ is 1. On the other hand, to compute the residue of $e^{1/z}$ at the essential singularity $z_0 = 0$, we cannot rely on (7.9). Instead, the power series expansion of the exponential function yields the Laurent series $e^{1/z} = \sum_{n=0}^{\infty} \frac{1}{n!z^n}$, so that $\mathrm{Res}(f, 0) = 1$. □

In the process of evaluating various types of integral it is useful to consider functions that have several singularities. This leads to the following result.

Theorem 7.10 (The residue theorem) *Let $f : U \to \mathbb{C}$ be analytic in the domain $U \subset \mathbb{C}$, apart from a finite set of isolated singularities, and let $\gamma \subset U$ be a piecewise C^1, simple, directed, closed curve, with interior D lying wholly in U. If γ avoids the singularities of f, then*[17]

$$\int_\gamma f(z)\, dz = 2\pi i \sum_{z_0 \in D} Res(f, z_0)\,.$$

Proof The integral over γ may be broken up into integrals over "smaller curves", each of which surrounds just one of the isolated singularities; see Figure 7.3. Each such integral reduces, just like in the approach performed while discussing the invariance of the residue, to an integral around a circle centred at the singularity. The desired formula follows at once. □

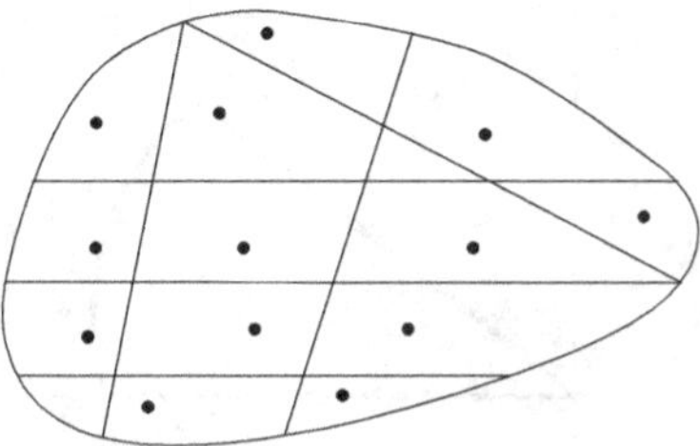

Figure 7.3 Reduction to "smaller" curves.

The residue theorem can be used to systematically evaluate various real integrals that are not readily found by relying on real variable methods. Broadly speaking, the approach consists of regarding these integrals as integrals along the real axis of a function that is analytic with the exception of a number of isolated singularities in some suitable domain, and subsequently deforming the path of integration at the expense of computing some residues. We now illustrate the evaluation of some definite real integrals by residue calculus.

- *Rational functions of* $\sin(\theta)$ *and* $\cos(\theta)$. For any rational function $R(x, y)$ we can evaluate $\int_0^{2\pi} R(\cos(\theta), \sin(\theta))\, d\theta$ by setting $z = e^{i\theta}$ and convert it to an integral of an analytic function over the unit circle, using the fact that $\cos(\theta) = (z + 1/z)/2$, $\sin(\theta) = (z - 1/z)/(2i)$, and $dz = iz\, d\theta$ if $z = e^{i\theta}$.

[17] The sum is finite since the residue at a point where f has a complex derivative must vanish.

Example For $a \in (0,1)$ we see that $\int_0^{2\pi} \frac{d\theta}{1-2a\cos(\theta)+a^2} = \int_{|z|=1} \frac{i\,dz}{az^2-(a^2+1)z+a} = \int_{|z|=1} \frac{i\,dz}{a(z-1/a)(z-a)} = \frac{2\pi}{1-a^2}$ since $z = a$ is the only singularity of the integrand inside $|z| = 1$, and by (7.9) the residue at this simple pole equals $i/(a^2-1)$. □

- *Rational functions on* $\mathbb{R}$. One can compute[18] $\int_{\mathbb{R}} R(x)\,dx$ via residues whenever the rational function $R(x) = \frac{P(x)}{Q(x)}$ is such that $\int_{\mathbb{R}} R(x)\,dx < \infty$.

Example To compute $\int_{\mathbb{R}} \frac{dx}{1+x^{2n}}$ for integers $n \geq 1$, consider the function $f(z) = 1/(1+z^{2n})$ having n simple poles at $\zeta_k = e^{(2k+1)\pi i/(2n)}$ with $k = 0, \dots, 2n-1$. Integrating f over the piecewise C^1-curve γ_R consisting of the straight line segment from 0 to $R > 1$, followed by the arc of circle $\gamma_R^+ = \{z = Re^{it} : 0 \leq t \leq \pi/n\}$ and finally by the straight line segment from $Re^{\pi i/n}$ to 0 (see Figure 7.4), we obtain by the residue theorem that

$$\int_0^R \frac{dx}{1+x^{2n}} + \int_0^{\pi/n} \frac{iRe^{it}}{1+R^{2n}e^{2\pi int}}\,dt - \int_0^R \frac{e^{\pi i/n}}{1+s^{2n}}\,ds = 2\pi i \operatorname{Res}(f,\zeta_1).$$

Since $z^{2n}+1 = z^{2n} - \zeta_1^{2n} = (z^n+\zeta_1^n)(z^n-\zeta_1^n) = (z^n+\zeta_1^n)(z-\zeta_1)\sum_{k=0}^{n-1} z^k \zeta_1^{n-1-k}$ yields $\operatorname{Res}(f,\zeta_1) = 1/(2n\zeta_1^{2n-1})$, in view of (7.9), and the middle integral vanishes for $R \to \infty$, we get $(1-\zeta_1^2)\int_0^\infty \frac{dx}{1+x^{2n}} = \frac{\pi i}{n\zeta_1^{2n-1}}$. Using the fact that $\zeta_1^{2n} = -1$ gives $\int_0^\infty \frac{dx}{1+x^{2n}} = \frac{\pi i}{n(\zeta_1 - 1/\zeta_1)} = \frac{\pi/(2n)}{\sin[\pi/(2n)]}$. The integrand being even, the requested value is twice this number. □

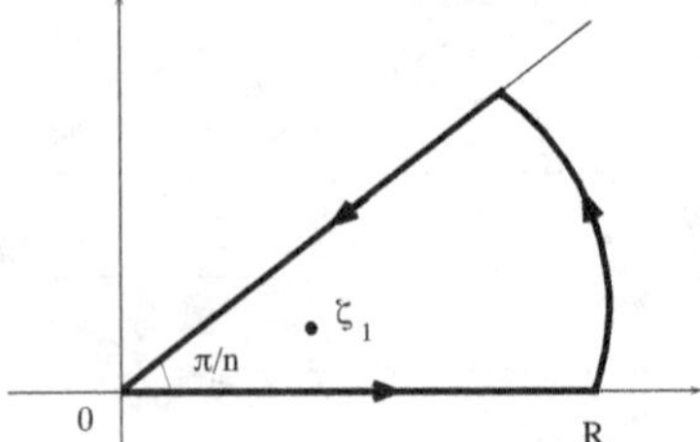

Figure 7.4 The boundary of the pie-slice used in computing $\int_0^\infty \frac{dx}{1+x^{2n}}$.

- *Contour integration.* In certain cases, extending a nonrational real function f to the complex plane and taking advantage of the decay behaviour, one can choose contours γ_R such that $\int_{\mathbb{R}} f(x)\,dx = \lim_{R\to\infty} \int_{\gamma_R} f(z)\,dz$, and, by computing the residues of f at its singularities, the latter integrals can be easily obtained.

Example The function $x \mapsto \frac{1}{\cosh(\pi x)}$ is its own Fourier transform: $\int_{\mathbb{R}} \frac{e^{-2\pi ix\xi}}{\cosh(\pi x)}\,dx = \frac{1}{\cosh(\pi\xi)}$. To see this, fix $\xi \in \mathbb{R}$ and define the function $f(z) = \frac{e^{-2\pi iz\xi}}{\cosh(\pi z)}$, whose singularities occur when $e^{2\pi z} = -1$, that is, at the discrete set of points $\zeta_k = (2k+1)i/2$ with $k \in \mathbb{Z}$. Consider the rectangle with piecewise C^1-boundary γ_R drawn in Figure 7.5, with a fixed height but whose width is adjustable. The only singularities of f inside such a rectangle are at the points $\alpha = i/2$ and $\beta = 3i/2$. Recognising in the expression $(z-\alpha)f(z) = 2e^{-2\pi iz\xi}e^{\pi z}\frac{z-\alpha}{e^{2\pi z}-e^{2\pi\alpha}}$ the reciprocal of the difference quotient for the function $e^{2\pi z}$ at $z = \alpha$, we see that $\lim_{z\to\alpha}(z-\alpha)f(z) = e^{\pi\xi}/(\pi i)$, so that f has a simple pole at α, with residue $e^{\pi\xi}/(\pi i)$. Similarly we show that f has a simple

[18] Instead of relying on partial fraction decompositions and trigonometric substitutions.

pole at β, with residue $-e^{3\pi\xi}/(\pi i)$. Using the fact that $\cosh(\pi z)$ is periodic with period $(2i)$, we can write the integral of f over γ_R as

$$(1-e^{4\pi\xi})\int_{-R}^{R}\frac{e^{-2\pi ix\xi}}{\cosh(\pi x)}\,dx + i\int_0^2 \frac{e^{-2\pi i\xi(R+iy)}}{\cosh(\pi(R+iy))}\,dy - i\int_0^2 \frac{e^{-2\pi i\xi(-R+iy)}}{\cosh(\pi(-R+iy))}\,dy\,.$$

By the residue theorem, the above expression equals $2(e^{\pi\xi}-e^{3\pi\xi})$ for every $R>0$. Moreover, since the absolute value of the second integrand is bounded from above by

$$\frac{2e^{2\pi\xi y}}{|e^{\pi(R+iy)}+e^{-\pi(R+iy)}|} \le \frac{2e^{2\pi\xi y}}{|e^{\pi(R+iy)}|-|e^{-\pi(R+iy)}|} = \frac{2e^{2\pi\xi y}}{e^{\pi R}-e^{-\pi R}}\,,$$

and the same bound holds for the last integrand, in the limit $R\to\infty$ the last two integrals have no contribution and $\int_{\mathbb{R}}\frac{e^{-2\pi ix\xi}}{\cosh(\pi x)}\,dx = \frac{2(e^{\pi\xi}-e^{3\pi\xi})}{1-e^{4\pi\xi}} = \frac{1}{\cosh(\pi\xi)}$ emerges. □

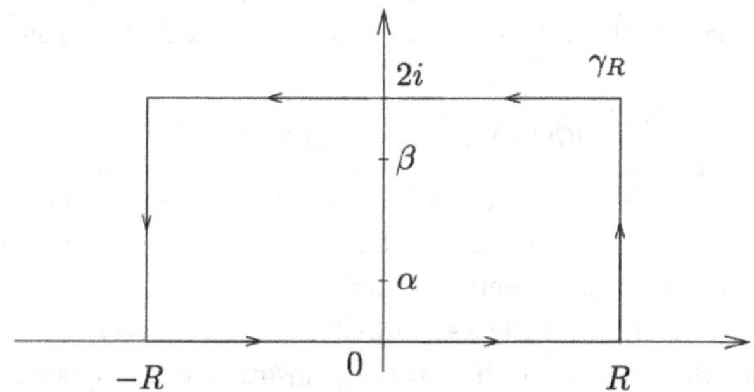

Figure 7.5 The contour used to compute the Fourier transform of $x \mapsto 1/\cosh(\pi x)$.

Geometric function theory

Let us now discuss some aspects of a geometric, qualitative nature rather than an analytical, quantitative one. The central issue is the local geometric behaviour of an analytic function.

Let $f: U\to\mathbb{C}$ be analytic in some domain $U\subset\mathbb{C}$. If f vanishes at $z_0\in U$ but is not identically zero, then we can factor out the leading term and write $f(z)=(z-z_0)^n g(z)$ for some integer $n\ge 1$ and a function $g(z)$ that is analytic in an open set containing z_0, with $g(z_0)\ne 0$; see (7.8). This result is sometimes alluded to as the *Weierstrass preparation theorem*: it shows that analytic functions behave locally like polynomials times units (invertible functions) – the difference between a general analytic map and a polynomial being that the latter has a global factorisation, while the former generally requires a different factorisation of this type in the neighbourhood of each point. A useful consequence is the following theorem.

Theorem 7.11 *The zeros of any nonconstant analytic function $f : U \to \mathbb{C}$ are isolated if $U \subset \mathbb{C}$ is a domain.*[19]

Proof Let $U_0 \subset U$ be the (relatively) closed set of limit points of the zeros of f. If $U_0 \neq \emptyset$, choose some $z_0 \in U$. Since, by continuity, $f(z_0) = 0$, we infer that $f \equiv 0$ in a neighbourhood of z_0 since otherwise the factorisation (7.8) ensures that f has no zeros near z_0, except z_0 itself. Thus U_0 is open. Since U is connected, this forces $U_0 = U$, that is, $f \equiv 0$ in U. □

Theorem 7.11 shows that analytic functions are very "rigid", being determined by a rather small amount of information: *if $U \subset \mathbb{C}$ is a domain and the analytic functions $f, g : U \to \mathbb{C}$ coincide on a countable set $U_0 \subset U$ that has an accumulation point in*[20] *U, then $f = g$ in U* (this is the **identity theorem**).

Viewing analytic functions $f = u + iv$ as mappings (u, v) from domains in $\mathbb{R}^2$ to $\mathbb{R}^2$, due to (7.2 and (7.3), the determinant of the Jacobian at $z_0 = x_0 + iy_0$ equals

$$u_x^2(x_0, y_0) + v_x^2(x_0, y_0) = |f'(z_0)|^2 .$$

Therefore, by the inverse function theorem, if $f : U \to \mathbb{C}$ is analytic in some domain $U \subset \mathbb{C}$ and $f'(z_0) \neq 0$ at $z_0 \in U$, then there are open neighbourhoods V and W of z_0 and $f(z_0)$, respectively, such that f is a bijection from V onto W, with an analytic inverse $g : W \to V$ satisfying $g'(f(z)) = 1/f'(z)$ for $z \in V$. To understand what happens if $f'(z_0) = 0$ it is instructive to first consider the map $z \to z^n$ near $z_0 = 0$, for some integer $n \geq 2$: the origin is again an interior point of the image since for $r > 0$ sufficiently small, any point in $\{z \in \mathbb{C} : |z| < r^n\}$ is the image of some point in $\{z \in \mathbb{C} : |z| < r\}$. These considerations, combined with the factorisation (7.8), show that nonconstant analytic maps are open (that is, they map open sets into open sets).

Theorem 7.12 *Let $f : U \to \mathbb{C}$ be nonconstant and analytic in a domain $U \subset \mathbb{C}$. If $n \geq 1$ is the order of the zero of $z \mapsto f(z) - f(z_0)$ at $z_0 \in U$, then there is an open neighbourhood $V \subset U$ of z_0, a number $r > 0$, and an analytic bijection $\phi : V \to \{w \in \mathbb{C} : |w| < r\}$ with $\phi'(z) \neq 0$ and $f(z) = f(z_0) + [\phi(z)]^n$ for all $z \in V$. In particular, f is an n-to-one mapping from $V \setminus \{z_0\}$ to $\{w \in \mathbb{C} : 0 < |w - f(z_0)| < r^n\}$ and $f(z_0)$ is an interior point of $f(U)$.*

Proof Let $V_0 \subset U$ be an open ball centred at z_0, chosen to ensure that $f(z) - f(z_0) = (z - z_0)^n g(z)$ with $g : V_0 \to \mathbb{C}$ analytic and such that $g(z) \neq 0$ for $z \in V_0$; see (7.8). The approach used in the proof of Morera's theorem shows that functions that are analytic in V_0 have a primitive. Let $H_0 : V_0 \to \mathbb{C}$ be analytic and such that $H_0' = g'/g$. Then $(ge^{-H_0})' = 0$ in V_0 and hence

[19] Note that f could be constant on each connected component of U if U is merely open.

[20] This follows by applying Theorem 7.11 to $(f - g)$. Note that an accumulation of zeros on the boundary of U would be possible. For example, as a composition of the two analytic functions $z \to 1/z$ from the punctured plane $U = \mathbb{C} \setminus \{0\}$ to $\mathbb{C}$ and $z \to \sin(z)$ on $\mathbb{C}$, $z \mapsto \sin(1/z)$ is analytic on U and vanishes at $z_n = 1/(n\pi) \to 0$.

$g(z)e^{-H_0(z)} = c$ (constant) in V_0. Choosing $a \in \mathbb{C}$ such that $e^a = c$ and setting $H = H_0 + a$, we obtain that $g(z) = e^{H(z)}$ in V_0. Note that the analytic function $h : V_0 \to \mathbb{C}$ given by $h = e^{H/n}$ is such that $h^n = g$ throughout V_0. If we take $\phi(z) = (z - z_0)h(z)$, then $\phi(z_0) = 0$ and $\phi'(z_0) \neq 0$, and the implicit function theorem provides us with an open neighbourhood $V \subset V_0$ of z_0 in which the desired representation of f holds. □

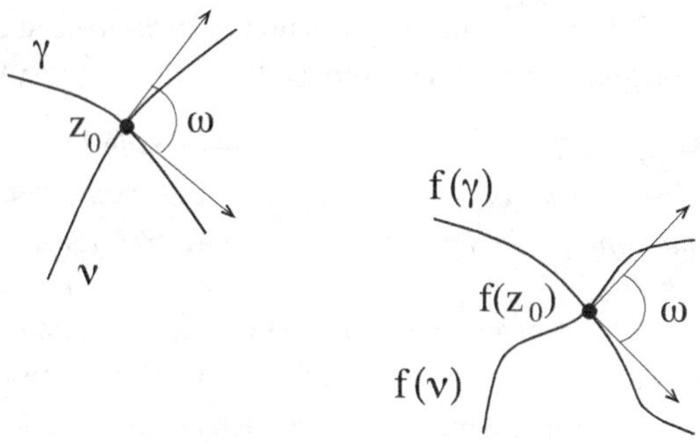

Figure 7.6 Conformal maps preserve angles.

Let us now discuss the relationship between conformal and analytic maps. Recall that the angle between two nonzero complex numbers w and z (taken in that order) is the oriented angle (in the interval $(-\pi, \pi]$) formed between the vectors in $\mathbb{R}^2$ corresponding to these two points, being uniquely determined by the two quantities $\frac{(z,w)}{|z|\,|w|}$ and $\frac{(z,-iw)}{|z|\,|w|}$, where $(\cdot,\cdot)$ stands for the inner product in $\mathbb{R}^2$, so that $(w, z) = \Re\mathrm{e}(z\overline{w})$. We say that two C^1-curves $\gamma, \nu : (0, 1) \to \mathbb{C}$ intersect at $z_0 = \gamma(t_0) = \nu(t_0)$, for some $t_0 \in (0, 1)$, at the angle ω if this is the angle between the unit tangent vectors[21] at z_0, $\gamma'(t_0)$ and $\nu'(t_0)$. A C^1 (real) function $f : U \to \mathbb{C}$ in a domain $U \subset \mathbb{C}$ that contains both curves maps these curves into the C^1-curves $f(\gamma)$ and $f(\nu)$ which intersect at $f(z_0)$; see Figure 7.6. Note that the tangents to these curves at $f(z_0)$ are given by $Df_{z_0}(\gamma'(t_0))$ and $Df_{z_0}(\nu'(t_0))$. If f is analytic in U, the discussion related to (7.3) shows that if $f'(z_0) \neq 0$, then the angle formed between the curves γ and ν at z_0 equals the angle formed between the curves $f(\gamma)$ and $f(\nu)$ at $f(z_0)$. Conversely, let us assume that the C^1 (real) function $f : U \to \mathbb{C}$ in a domain $U \subset \mathbb{R}^2$ preserves angles (i.e., is *conformal*). Fix some $z_0 \in U$ and denote $\frac{\partial f}{\partial z}(z_0) = a \in \mathbb{C}$, $\frac{\partial f}{\partial \overline{z}}(z_0) = b \in \mathbb{C}$. Since $Df_{z_0}z = az + b\overline{z}$, by the considerations preceding (7.2), we see that conformal means $\arg\{(az_1 + b\overline{z_1})\,\overline{(az_2 + b\overline{z_2}}\} - \arg(z_1\overline{z_2}) \in 2\pi\mathbb{Z}$ for all $z_1, z_2 \in \mathbb{C}$ with $|z_1| = |z_2| = 1$; here we use the multi-valued convention for the argument. Using the fact that $\arg(zw) = \arg(z) + \arg(w)$ for $z, w \in \mathbb{C}$, we can write the previous relation as $\arg(a + b\overline{z_1}^2) - \arg(a + b\overline{z_2}^2) \in 2\pi\mathbb{Z}$ for

21 We assume that near t_0 both curves are parametrised by arc-length; see Chapter 2.

all $z_1, z_2 \in \mathbb{C}$ with $|z_1| = |z_2| = 1$. This forces $b = 0$ since otherwise, with $a/b = Re^{i\theta}$ for some[22] $R > 0$ and some $\theta \in (-\pi, \pi]$, with $z_1 = e^{-i\theta/2}$ and $z_2 = e^{-i(\theta+\varphi)/2}$, the identity becomes $\arg(R + e^{i\varphi}) \in 2\pi\mathbb{Z}$ for all $\varphi \in (-\pi, \pi]$, which clearly fails. Consequently $b = 0$, meaning that f has to be analytic in U. Moreover, if f is analytic in U and has a zero of order $n \geq 2$ at some z_0 in U, then Theorem 7.12 shows that f blows up angles n times at z_0 since $z \mapsto z^n$ does this at the origin. Thus we have proved the following result.

Theorem 7.13 *If $f : U \to \mathbb{C}$ is analytic in the domain $U \subset \mathbb{C}$ and $f'(z) \neq 0$ for all $z \in U$, then f is a conformal mapping. Conversely, if $f : U \to \mathbb{C}$ is a conformal C^1-mapping, then f is analytic and $f'(z) \neq 0$ for all $z \in U$.*

The special structural properties of analytic functions open up possibilities that are not available for merely smooth functions. For example, by Theorem 7.8, a loop in a domain in which the nonconstant function f is analytic can only enclose a finite number of zeros[23] of f. Moreover, one can count their number:

Theorem 7.14 (The argument principle) *Let $f : U \to \mathbb{C}$ be analytic in the domain $U \subset \mathbb{C}$, and let $\gamma \subset U$ be a piecewise C^1, simple, directed, closed curve, with interior D lying wholly in U. If f has no zeros on γ, then the number of zeros of f in D, counted with multiplicity, is given by*

$$\mathcal{N}(f, 0) = \sum_{z_0 \in D} ord(f, z_0) = \frac{1}{2\pi i} \int_\gamma \frac{f'(z)}{f(z)}\, dz\,.$$

Proof By definition, $\mathrm{ord}(f, z_0) = 0$ if $f(z_0) \neq 0$, and in this case f'/f is analytic in an open disc centred at z_0. On the other hand, if $f(z_0) = 0$, then the factorisation (7.8) yields

$$\frac{f'(z)}{f(z)} = \frac{k_0}{z - z_0} + \frac{g'(z)}{g(z)} \quad \text{near} \quad z = z_0\,,$$

where $k_0 = \mathrm{ord}(f, z_0) \geq 1$ is the multiplicity of the zero. Consequently, f'/f has a simple pole at z_0, with residue k_0. We conclude by applying the residue theorem. □

Winding number, homology and homotopy

So far, we have mainly considered path integrals along simple, piecewise C^1 loops but occasionaly we have encountered some more general constructions.[24]

[22] Note that $R = 0$ means $a = 0$ and the identity is clearly violated.

[23] In contrast to this, recall from Chapter 5 the existence of compactly supported smooth functions.

[24] For example, in defining the residue of an analytic function at the isolated singularity z_0 we used the fact that $\int_{|z-z_0|=r_2} f(z)\,dz - \int_{|z-z_0|=r_1} f(z)\,dz = 0$ if $0 < r_1 < r_2 < R$, where $R > 0$ is such that f is analytic in the punctured disc $\{z \in \mathbb{C} : 0 < |z - z_0| < R\}$.

To address the issue of integration along curves that form the boundary of more complicated domains than interiors of simple loops or annular regions we must briefly address some aspects of algebraic topology, more specifically, the notions of *winding number* and *homology* for planar curves. Note that the passage from simple closed curves to self-intersecting curves is quite involved since, while the intuitive idea of decomposing the latter into a finite number of simple loops is applicable for an *analytic closed curve*,[25] defined as the image $f(\gamma)$ of a circle γ under a map f that is analytic in a neighbourhood of γ, intricacies arise even for C^∞-regular curves.

Example Consider the simple C^∞-closed curve $\gamma : [0,1] \to \mathbb{C}$ given by $\gamma(0) = 1$, $\gamma(t) = e^{2\pi i t}\{1 + \exp\left(\frac{1}{t(t-1/2)}\right)\sin\left(\frac{1}{t(t-1/2)}\right)\}$ for $0 < t < \frac{1}{2}$, and $\gamma(t) = e^{2\pi i t}$ for $\frac{1}{2} \le t \le 1$. The image of γ under $f(z) = z^2$ is a smooth loop with infinitely many self-intersections since for every $n \ge 1$, the value $e^{4\pi i t_n}$ is attained at $t = t_n = \frac{1}{4} + \sqrt{\frac{1}{16} - \frac{2}{(n+5)\pi}} \in (0, \frac{1}{2})$ and at $t = t_n + \frac{1}{2}$; note that t_n was chosen so that $\sin\left(\frac{1}{t_n(t_n-1/2)}\right) = 0$.

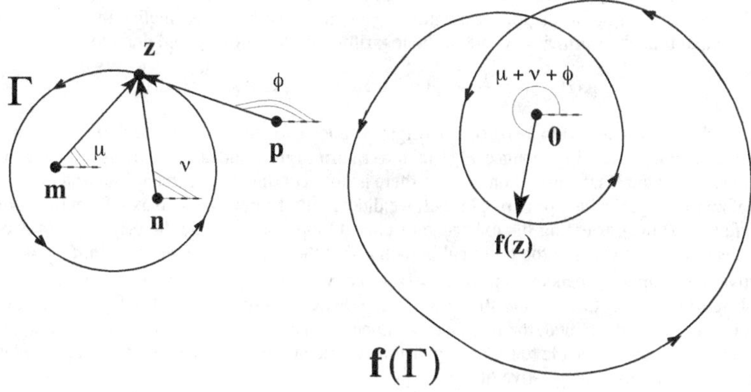

Figure 7.7 The image of the circle Γ under the cubic map $f(z) = (z-m)(z-n)(z-p)$, with m, n inside Γ and p outside Γ, is the self-intersecting smooth curve $f(\Gamma)$: m, n and p are mapped into 0, the argument of $f(z)$ equals to the sum of the arguments of $z-m$, $z-n$, $z-p$, and as z goes round Γ once, each of the vectors $z-m$ and $z-n$ execute a complete revolution, while the direction of $z-p$ merely oscillates. The case when Γ is the unit circle $|z| = 1$ and $m = 0$, $n = 1/2$, $p = 2$, shows that $f'(z) \neq 0$ on Γ does not prevent self-intersections of $f(\Gamma)$.

For a piecewise C^1 parametrisation $\gamma : [0,1] \to \mathbb{C}$, the integral expression on the right side in the formula of Theorem 7.14 becomes $\frac{1}{2\pi i}\int_0^1 \frac{f'(\gamma(t))\gamma'(t)}{f(\gamma(t))}\,dt$, which is of the form $\frac{1}{2\pi i}\int_{f(\gamma)} \frac{dz}{z-\alpha}$, with the path $f(\gamma)$ not passing through α. Note that if γ is a piecewise C^1, simple, directed, closed curve, with interior D lying wholly in a domain of analyticity of f, then its image $f(\gamma)$ will be

25 In this case $f(\gamma)$ has at most finitely many self-intersections and the complement of $f(\gamma)$ has only a finite number of components; see Miller and Olin (1984).

a piecewise C^1, directed, closed curve but self-intersections might occur; see Figure 7.7. It is therefore reasonable to define for any piecewise C^1, closed, directed curve $\Gamma : [0,1] \to \mathbb{C}$ its *winding number* with respect to a point α (not on the path) to be

$$W(\Gamma, \alpha) = \frac{1}{2\pi i} \int_\Gamma \frac{dz}{z - \alpha}.$$

This integral can be written in the form $\frac{1}{2\pi i} \int_0^1 \frac{\nu'(t)}{\nu(t)}\, dt$ for $\nu(t) = \Gamma(t) - \alpha$. Formally, it equals $\log(\nu(1)) - \log(\nu(0))$, but the situation is more intricate since we have to deal with the multi-valuedness of the complex logarithm function $\log(z)$, introduced as the "inverse" of e^z.

For $z \neq 0$ we define $\log(z)$ to be any complex number such that $e^{\log(z)} = z$; thus

$$\log(z) = \ln |z| + i \arg(z) \quad \text{for} \quad 0 \neq z \in \mathbb{C}.$$

$\log(z)$ is multi-valued since, by the adopted convention, $\arg(z)$ takes infinitely many values, differing from each other by integer multiples of 2π; in particular, $\log(\sqrt{3} + i) = \ln 2 + i\pi/6 + 2k\pi i$ with $k \in \mathbb{Z}$. Restricting $\arg(z)$ to its principal value $\mathrm{Arg}(z)$, given by the representative in $(-\pi, \pi]$, yields the principal branch or principal value of the logarithm, written $\mathrm{Log}(z)$, and defined by

$$\mathrm{Log}(z) = \ln |z| + i\,\mathrm{Arg}(z) \quad \text{for} \quad z \in \mathbb{C} \setminus \{x \in \mathbb{R} :\ x \leq 0\};$$

in particular, $\mathrm{Log}(\sqrt{3}+i) = \ln 2 + i\pi/6$. The origin is called a *logarithmic branch point* for the multi-valued function $F(z) = \log(z)$ since $F(z)$ fails to return to its value at the start as z travels along any loop that encircles the origin once since there is no[26] continuous argument function on $\mathbb{C} \setminus \{0\}$. To define the single-valued principal branch we did cut out the negative real axis from the domain of definition (thus preventing the existence of closed loops that encircle the origin). A drawback with restricting ourselves to the principal branch is that the familiar rule $\ln(ab) = \ln(a) + \ln(b)$ for positive real numbers breaks down: for $a = (i-1)/\sqrt{2}$ with $a^2 = -i$, we have $\mathrm{Log}(a) = 3\pi i/4$ and $\mathrm{Log}(a^2) = \pi i/2$. Concerning the power-series expansion of the complex logarithm, since a power series is single-valued, the most we can hope for is a local representation. The origin being a branch point, it is advisable to try an expansion centred at $z = 1$. The power series representation of the real logarithm is suggestive of the *Ansatz*

$$\mathrm{Log}(1+z) = z - \frac{z^2}{2} + \frac{z^3}{3} - \frac{z^4}{4} + \frac{z^5}{5} - \frac{z^6}{6} + \ldots$$

The radius of convergence of the power series on the right side is $R = 1$ because $\limsup_{n\to\infty} n^{1/n} = 1$. Since multiplication by $(1+z)$ yields the power series $z + \sum_{n\geq 1} \frac{z^{2n}}{2n(2n-1)} - \sum_{n\geq 1} \frac{z^{2n+1}}{2n(2n+1)}$, which, as $\sum_{n\geq 1} \frac{1}{2n(2n-1)} < \infty$, converges absolutely for all $|z| \leq 1$, the power series displayed above converges everywhere on the unit circle $|z| = 1$, except at $z = -1$. It is now convenient to work with $\log(z)$, since it accommodates the rule $\log(ab) = \log(a) + \log(b)$ for $a, b \in \mathbb{C} \setminus \{0\}$, so that its power-series representation near $z_0 \neq 0$ is obtained by writing $\log(z) = \log(z/z_0) + \log(z_0)$ and taking advantage of the power-series available for $|z/z_0 - 1| < 1$.

For $w \in \mathbb{C}$ we now define for $z \in \mathbb{C} \setminus \{0\}$ the complex power z^w as $e^{w \log(z)}$. This function inherits multi-valuedness from the complex logarithm. Its principal branch is $\exp(w\,[\ln |z| + i\mathrm{Arg}(z)])$. Note that this convention leads to the surprising outcome $\sqrt[3]{-1} = e^{\mathrm{Log}(-1)/3} = e^{i\pi/3} = \frac{1+i\sqrt{3}}{2} \neq -1$.

26 Indeed, if $a : \mathbb{C} \setminus \{0\} \to \mathbb{R}$ were such a function, then $b : \mathbb{R} \to \mathbb{R}$, given by $b(t) = \frac{a(e^{it}) - t}{2\pi}$, would have to be continuous and $\mathbb{Z}$-valued, hence constant. But $b(t) = k \in \mathbb{Z}$ leads to the contradiction $2\pi k = a(1) = a(e^{2\pi i}) = 2\pi(k+1)$.

In this context, the following result about polar representations is very useful: *if $v : [0, 1] \to \mathbb{C}$ is a piecewise C^1, closed curve with the origin not on its path, then*

$$v(t) = R(t)\, e^{2\pi i\theta(t)}\,, \qquad t \in [0, 1]\,, \tag{7.10}$$

for some functions $R : [0, 1] \to (0, \infty)$ and $\theta : [0, 1] \to \mathbb{R}$ which enjoy at each point $t \in [0, 1]$ whatever smoothness properties v does, and such that $R(1) = R(0)$ and $\theta(1) - \theta(0) \in \mathbb{Z}$; in particular, the winding number of a closed path is an integer. As its name suggests, the winding number is the number of times the loop Γ winds around the point α: $W(\Gamma, \alpha)$ is simply the net number of revolutions $[\theta(1) - \theta(0)]$ of the direction of $(z - \alpha)$ as z traces out Γ once in its given sense.

To validate (7.10), let $R(t) = |v(t)|$ and define the absolutely continuous function $\theta : [0, 1] \to \mathbb{C}$ by $\theta(t) = \frac{1}{2\pi i}\int_0^t \frac{R(s)v'(s) - R'(s)v(s)}{R(s)v(s)}\, ds$. These functions have the desired regularity. The results in Chapter 2 show that $\frac{v(t)}{R(t)}\, e^{-2\pi i\theta(t)}$ is constant on $[0, 1]$ since for all $t \in (0, 1)$, except for at most finitely many points, $\frac{d}{dt}\big(\frac{v(t)}{R(t)}\, e^{-2\pi i\theta(t)}\big) = 0$. Now $v(1) = v(0)$ and $R(1) = R(0)$ ensure $\theta(1) - \theta(0) \in \mathbb{Z}$.

From the definition of the winding number we see at once that, given a piecewise C^1, closed, directed curve Γ, the function $\alpha \mapsto W(\Gamma, \alpha)$, defined for $\alpha \in \mathbb{C}$ not on the path, is continuous. Since this function takes its values in the integers, it must be constant on any open connected set D not intersecting Γ. Any loop partitions the complex plane into a number of connected components, precisely one of which is unbounded.[27] The (constant) integer value of the winding number in this unbounded component must be 0 since for $|\alpha|$ sufficiently large the integrand in the definition of $W(\Gamma, \alpha)$ can be made arbitrarily small, and estimating the integral we obtain a number of very small absolute value. We claim that if the point α crosses Γ transversally[28] from one component of $\mathbb{C} \setminus \Gamma$ to an adjacent component, avoiding in this process the points of self-intersection of Γ, then $W(\Gamma, \alpha)$ increases or decreases by one, depending on the orientation: if α is moving from left to right (right to left) as it crosses Γ, the winding number $W(\Gamma, \alpha)$ increases (decreases).

To see why, consider an initial position $\alpha \in D$ and a terminal position $\alpha' \in D'$, belonging to different components D and D' of $\mathbb{C} \setminus \Gamma$, sufficiently close to $z_0 \in \Gamma$ so that the intersection of Γ with an open disc $\mathbb{D}_0$, centred at $z_0 \in \Gamma$ and of radius $\varepsilon > 0$, is an arc v connecting $z_1 \in \Gamma$ to $z_2 \in \Gamma$ along γ and separating the points $\alpha, \alpha' \in \mathbb{D}_0$; here Γ is oriented from z_1 to z_2, see Figure 7.8. By Cauchy's theorem, in the evaluation of $W(\Gamma, \alpha)$ we may replace v by the arc along the circular boundary of $\mathbb{D}_0$ joining z_1 to z_2 within D. Similarly, in the computation of $W(\Gamma, \alpha')$, we may replace v by the arc along the circular bounday of $\mathbb{D}_0$ joining z_2 to z_1 within D'.

[27] Since Γ lies within a circle centred at the origin (of sufficiently large radius), the outside of this circle belongs the unique unbounded connected component of the open set $\mathbb{C} \setminus \Gamma$.

[28] By hypothesis, Γ admits a tangent at all points, except perhaps at finitely many. We say that Γ intersects a C^1-curve transversally at $z_0 \in \Gamma$ if the two tangent lines at z_0 are not identical; see Guillemin and Pollack (1974).

Consequently $W(\Gamma,\alpha) - W(\Gamma,\alpha') = \pm\frac{1}{2\pi i}\int_{|z-z_0|=\varepsilon}\frac{dz}{z-z_0}$, the orientation of the circular path being counter-clockwise if the crossing occurs from right to left. The claim now follows from (7.4).

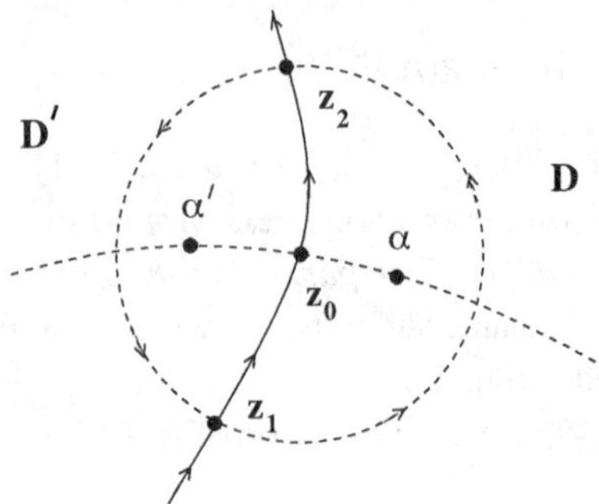

Figure 7.8 Each transversal crossing of the loop triggers an alteration of the winding number by one.

The above considerations provide an easy way to find the winding number for complicated loops: start counting from the unbounded component (where the winding number is zero), and move from a component to the adjacent one using the crossing rule, adding or substracting one at each crossing of the loop; see Figure 7.9.

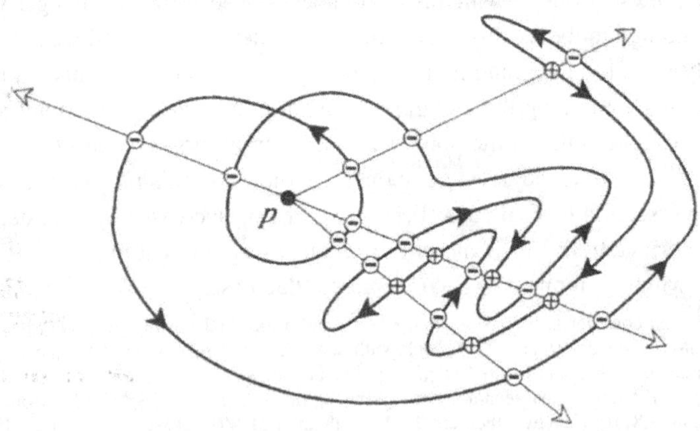

Figure 7.9 Each intersection point is marked with ⊕ or ⊖ according as the winding number increases or decreases as the loop is crossed along the ray emanating from p.

The Jordan curve theorem ensures that a simple loop divides the complex plane into just two components, its inside and its outside: $W(\Gamma,\alpha) = 1$ for

α inside, while $W(\Gamma, \alpha) = 0$ for α outside. However, a loop that is not simple will partition the plane into a larger number of components[29] and it is no longer obvious which points are to be considered inside, and which outside. The winding number allows us make this distinction: the outside consists of those components for which the winding number vanishes,[30] while the remaining components constitute the inside.

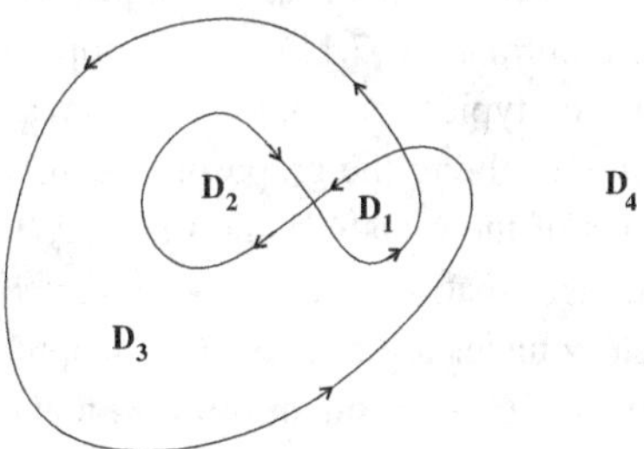

Figure 7.10 The values of the winding number for the depicted loop, dividing the complement of the complex plane into four components, are: 0 for D_4, 1 for D_3, 0 for D_2, 2 for D_1. Thus $D_1 \cup D_3$ is the inside and $D_2 \cup D_4$ is the outside.

Just as the winding number only changes when a continuously moving point crosses a fixed loop, the winding number of an evolving loop can only change (by ± 1) if it crosses a fixed point. Thus if a loop γ can be continuously deformed into another loop ν without ever crossing a point α, then $W(\gamma, \alpha) = W(\nu, \alpha)$. To further develop this idea, it helps to look at loops in a new way. Imagine a closed rubber band ν lying on a planar surface, e.g. in the shape of the piecewise C^1 loop depicted in Figure 7.10. Choosing the origin in the domain D_3, we may now deform ν into the shape γ of the circle $|z| = R_0$, with $R_0 > 0$ sufficiently small so that the disc $|z| \leq R_0$ does not intersect ν, by pulling it radially onto γ without crossing the origin. For this, we rely on (7.10) to define the one-parameter family of maps $h_s : [0,1] \to \mathbb{C}$ (for $0 \leq s \leq 1$) by

$$h_s(t) = \begin{cases} R(t)\, e^{2\pi i\theta(t)} + 2s(R_0\, e^{2\pi i\theta(t)} - R(t)\, e^{2\pi i\theta(t)}), & 0 \leq s \leq \frac{1}{2}, \\ R_0\, e^{2\pi i\{\theta(t)+(2s-1)[kt-\theta(t)]\}}, & \frac{1}{2} \leq s \leq 1, \end{cases} \tag{7.11}$$

where $k = \theta(1) - \theta(0) \in \mathbb{Z}$. Note that the loops ν and γ are the images of h_0 and h_1, respectively, and, as t varies from 0 to 1, $h_0(t)$ moves once round ν and $h_1(t)$ performs k net rotations round γ. Moreover, as s varies from 0 to

29 In this context, for finitely many components, Euler's formula from graph theory states that $v - e + c = 2$, where v is the number of vertices (self-intersections), e is the number of edges (arcs between two vertices) and c is the number of components; see Giblin (1981). In particular, $v = 1$, $e = 2$ and $c = 3$ for the self-intersecting loop Γ in Figure 7.7.

30 Since, intuitively, the winding number of a curve with respect to a point is the net number of times the curve surrounds the point.

1, the image of h_s gradually (and reversibly) changes from ν into γ. A helpful visualisation of this process consists of thinking of the loop ν as a planar elastic band in extension, surrounding a solid vertical cylinder with a basis circle of radius R_0, that will contract to γ once it is released. The contraction can be realised by means of the maps H_s with $0 \le s \le \frac{1}{2}$, defined on the punctured plane $\mathbb{C} \setminus \{0\}$ by setting $H_s(w) = w + 2s(\hat{w} - w)$ for $w \in \mathbb{C} \setminus \{0\}$, where $\hat{w} = R_0\, e^{2\pi i\theta(t)}$ is the intersection of γ with the segment joining $w = R(t)\, e^{2\pi i\theta(t)}$ to the origin. The two-stage deformation (7.11), of the loop ν with winding number k around the origin to the typical circular loop γ, is achieved following this contraction by smoothly evolving the graph of $t \mapsto \theta(t)$ into the straight-line graph $t \mapsto kt$ by means of the transformation $t \mapsto \theta(t) + (2s-1)[kt - \theta(t)]$, for $\frac{1}{2} < s \le 1$. The transformation $R_0\, e^{2\pi it} \to R(t)\, e^{2\pi i\theta(t)}$ traces the loop ν as the image of the circle γ under a piecewise C^1 mapping. Consider the effect of an analytic function $f : U \to \mathbb{C}$ on an infinitesimal circle $\gamma_0 \subset U$ centred at some point $z_0 \in U$. Using the power series expansion, see also (7.8), in a small open disc $\mathbb{D}_0$ centred at z_0, we can write $f(z) - f(z_0) = (z - z_0)^k g(z)$ for some integer $k \ge 1$ and some analytic function $g : \mathbb{D}_0 \to \mathbb{C}$ with $g(z) \ne 0$ for all $z \in \mathbb{D}_0$. If $\gamma_0 \subset \mathbb{D}_0$ and if $f(z_0)$ is not on the loop $\Gamma_0 = f(\gamma_0)$, we have $W(\Gamma_0, f(z_0)) = \frac{1}{2\pi i}\int_{\Gamma_0} \frac{dw}{w - f(z_0)} = \frac{1}{2\pi i}\int_{\gamma_0} \frac{f'(z)\,dz}{f(z) - f(z_0)} = \frac{1}{2\pi i}\int_{\gamma_0} \frac{k\,dz}{z - z_0} + \frac{1}{2\pi i}\int_{\gamma_0} \frac{g'(z)\,dz}{g(z)} =$ k in view of (7.4), since the second integral vanishes by Cauchy's formula, g'/g being analytic in $\mathbb{D}_0$. Thus, as z revolves round z_0 once, $f(z)$ completes k revolutions round $f(z_0)$. This shows that the multiplicity $k \ge 1$ of the zero z_0 of the analytic function $[f(z) - f(z_0)]$ is a geometric fingerprint: the local effect of f near z_0 is to map simple loops around z_0 into loops winding k times around $f(z_0)$. In particular, the winding number of images of simple loops under analytic maps is always positive.[31]

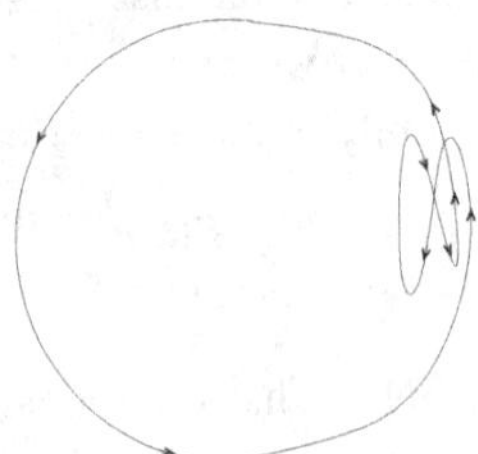

Figure 7.11 The deformation of the loop depicted in Figure 7.10, with the origin located in D_3, by means of h_s defined in (7.11), is nearly circular for values of s close to 1. Note that self-intersections persist throughout the deformation (7.11) since if $\nu(t_1) = \nu(t_2)$ for some $0 \le t_1 < t_2 < 1$, then $h_s(t_1) = h_s(t_1)$ for all $s \in [0, 1]$.

[31] In contrast to this, a general continuous mapping might be orientation-reversing, e.g. the complex conjugation $z \mapsto \bar{z}$ reverses the orientation of circles centred at the origin.

The development of a more general theory of integration requires the concept of a *cycle* in a domain $U \subset \mathbb{C}$, defined as a (formal) integral linear combination of closed loops. That is, given the loops ν_j and the integers n_j with $1 \le j \le k$, the cycle $\nu = n_1\nu_1 + \cdots + n_k\nu_k$ is a linear combination, with the integer coefficients reflecting orientation and multiplicity, and we set

$$\int_\nu f(z)\,dz = n_1 \int_{\nu_1} f(z)\,dz + \cdots + n_k \int_{\nu_k} f(z)\,dz$$

for a continuous function $f : U \to \mathbb{C}$, while the winding number of the cycle ν with respect to the point $z_0 \in \mathbb{C} \setminus (\nu_1 \cup \cdots \cup \nu_k)$ is defined by $W(\nu, z_0) = \sum_{j=1}^{k} n_j W(\nu_j, z_0)$. We say that a cycle ν in U is *homologous to zero* if $W(\nu, z_0) = 0$ for all $z_0 \in \mathbb{C}\setminus U$. Since a loop ν_j represents a collection of complex numbers, it might be tempting to take $n_j\nu_j$ to mean the loop obtained by multiplying these numbers by n_j. This is not meant: we speak of "formal" linear combinations. Thus, with the obvious formal addition and subtraction, the cycles in U form an abelian group – the free abelian group generated by all loops in U. This prompts the use of algebraic methods. Two cycles ν and ν' in U are *homologous* if $\nu - \nu'$ is homologous to zero, and U is called *simply connected* if all cycles in U are homologous to zero. U being simply connected means that a closed curve in U cannot wind around any point which does not belong to the domain – homology is a rigorous mathematical method for defining and categorising holes in a shape.

Examples $\mathbb{C}$ is simply connected, since in this case $\mathbb{C} \setminus U = \emptyset$ (the condition is vacuous). The punctured plane $U = \mathbb{C} \setminus \{0\}$ is not simply connected since $W(\nu_0, 0) = 1$ by (7.4), where ν_0 is the unit circle centred at the origin. In this case, every cycle ν in U is homologous to a multiple of ν_0 since $\nu - W(\nu, 0)\,\nu_0$ is homologous to zero. □

The cycles in U that are homologous to zero are clearly a subgroup of the group of all cycles and the quotient group $\mathcal{H}_1(U)$ is called the *first homology group*[32] of U. In these algebraic terms, U is simply connected if and only if $\mathcal{H}_1(U) = \{0\}$.

Examples Note that $\mathcal{H}_1(\mathbb{C}\setminus\{0\}) = \mathbb{Z}$, the isomorphism being given by the winding number $W(\nu, 0)$. Since $\mathcal{H}_1(\mathbb{C}) = \{0\}$, the topological distinction between the complex plane and the punctured plane $\mathbb{C} \setminus \{0\}$ is detected by the first homology group. □

The following global Cauchy theorem holds.

Theorem 7.15 *Let γ be a null-homologous cycle in the domain $U \subset \mathbb{C}$. If $f : U \to \mathbb{C}$ is analytic, then $\int_\gamma f(z)\,dz = 0$ and $W(\gamma, z_0)\, f(z_0) = \frac{1}{2\pi i}\int_\gamma \frac{f(z)}{z - z_0}\,dz$ for $z_0 \in U \setminus \gamma$.*

[32] There is is a more general higher-dimensional theory, but here we consider only one-dimensional submanifolds (curves).

Proof Note that the second relation follows from the first, using the definition of $W(\gamma, z)$ and the fact that the map $\tilde{f} : U \to \mathbb{C}$ defined by $\tilde{f}(z_0) = f'(z_0)$ and $\tilde{f}(z) = (f(z) - f(z_0))/(z - z_0)$ for $z \neq z_0$, is analytic in U, due to (7.8). To prove that $\int_\gamma f(z)\,dz = 0$, let K be the complement of the exterior of γ, that is, the union of γ and of the support of $W(\gamma, \cdot)$ (i.e., the subset of $\mathbb{C} \setminus \gamma$ where the winding number of γ does not vanish). As a bounded and closed subset of $\mathbb{C}$, K is compact.

Let us first construct a C^∞-function $\varphi : \mathbb{C} \to [0, 1]$ such that $\varphi(z) = 0$ if $z \in \mathbb{C} \setminus U$ and for every $z_0 \in K$ there exists some $r_0 > 0$ with the property that the closed disc $|z - z_0| \leq r_0$ lies in U and $\varphi(z) = 1$ if $|z - z_0| \leq r_0$ (that is, $\varphi \equiv 1$ in a neighbourhood of K). Indeed, if $z_0 \in K$ then $z_0 \in U$ since $W(\gamma, \cdot)$ vanishes on $\mathbb{C} \setminus U$. Thus there exists some $R_0 > 0$ such that $\{z : |z - z_0| < 2R_0\} \subset U$. Let $\varphi_0 : \mathbb{C} \to [0, 1]$ be of class C^∞ and such that $\varphi_0(z) = 1$ if $|z - z_0| < R_0$, while $\varphi_0(z) = 0$ if $|z - z_0| > 2R_0$ (see Section 6.1.1). Since K is compact, there is a finite subcover of its open cover $\bigcup_{z_0 \in K} \{z : |z - z_0| < R_0\}$. The (finite) sum of the functions associated by the above procedure produces a C^∞-function $\phi : \mathbb{C} \to [0, \infty)$ such that $\phi(z) = 0$ if $z \in \mathbb{C} \setminus U$ and $\phi(z) \geq 1$ if $z \in K$. Take now a C^∞-function $\psi : \mathbb{R} \to [0, 1]$ such that $\psi(t) = 1$ for $|t| \leq 1/4$ and $\psi(t) = 0$ for $|t| \geq 1/2$. Then the function $\varphi(z) = 1 - \psi(\phi(z))$ has all the desired properties; in particular, $\varphi \equiv 1$ on the pre-image of $(1/2, \infty)$ under ϕ, and this is an open set containing K.

Let $R > 0$ be such that the exterior domain $|z| > R/2$ lies entirely within $\mathbb{C} \setminus U$; in particular, $|z| < R$ if $z \in U$. In complex notation Green's formula becomes

$$\int_{|z|=r_2} F\,dz + G\,d\bar{z} - \int_{|z|=r_1} F\,dz + G\,d\bar{z} = 2i \iint_{r_1<|z|<r_2} \Big(\frac{\partial F}{\partial \bar{z}} - \frac{\partial G}{\partial z}\Big)\,dxdy$$

for all $F, G : \{r_1 \leq |z| \leq r_2\} \to \mathbb{C}$ of class C^1. Fix $z_0 \in K$ and apply the above formula to[33] $F(z) = \varphi(z) f(z)/(z - z_0)$ and $G \equiv 0$ in an annulus $\varepsilon < |z - z_0| < R + |z_0|$ with $\varepsilon < r_0$ (where $r_0 > 0$ is such that $|z - z_0| \leq r_0$ lies in U and $\varphi(z) = 1$ for $|z - z_0| \leq r_0$) to obtain

$$-\int_{|z-z_0|=\varepsilon} \frac{f(z)}{z - z_0}\,dz = 2i \iint_{\varepsilon<|z-z_0|<R+|z_0|} \frac{\partial \varphi}{\partial \bar{z}} \frac{f(z)}{z - z_0}\,dxdy\,.$$

By Theorem 7.3, the left side equals $-2\pi i f(z_0)$. Since $\varphi(z) = 1$ for $|z - z_0| \leq r_0$, we may write the previous equality as

$$f(z_0) = -\frac{1}{\pi} \iint_U \frac{\partial \varphi}{\partial \bar{z}} \frac{f(z)}{z - z_0}\,dxdy \quad \text{for all} \quad z_0 \in K\,.$$

Integrating along γ, Fubini's theorem (see Chapter 2) yields

$$\frac{1}{2\pi i} \int_\gamma f(z_0)\,dz_0 = \frac{1}{\pi} \iint_U \frac{\partial \varphi}{\partial \bar{z}} f(z)\,W(\gamma, z)\,dxdy = 0$$

since $W(\gamma, \cdot)$ is supported on K and $\partial\varphi/\partial\bar{z} = 0$ there. □

Remark An interesting consequence of Theorem 7.15 is that *any analytic function f in a simply connected domain U has a primitive.*[34] □

A far-reaching generalisation of the global Cauchy theorem is the following result.

Theorem 7.16 *Let $U \subset \mathbb{C}$ be a domain, let $S \subset U$ be a set without accumulation points in U and let $f : U \setminus S \to \mathbb{C}$ be analytic (so that the points in S are*

[33] It is understood that $F(z) = 0$ for $z \notin U$.

[34] This follows by simply fixing a point $z_0 \in U$ and defining the value $F(z)$ of the primitive at $z \in U$ by the integral of f along any piecewise C^1-curve in U joining z_0 to z. The null-homology of closed curves in U ensures that this definition is independent of the curve, and $F'(z) = f(z)$ is a local feature, already verified in the proof of Morera's theorem.

isolated singularities of f). If γ is a null-homologous cycle in U and $\gamma \cap S = \emptyset$, then

$$\int_\gamma f(z)\,dz = 2\pi i \sum_{s \in S} W(\gamma, s)\, Res(f, s),$$

and the sum has only finitely many non-zero terms.

Proof The closure I of the set $\{z_0 \in U \backslash \gamma : W(\gamma, z_0) \neq 0\}$ is compact (being closed and bounded). Consequently $I \cap S$ is finite, which proves the second claim, since non-zero terms in the sum can only occur when $W(\gamma, s) \neq 0$. For each of the finitely many $s \in I \cap S$, let γ_s be a circle centred at s, of sufficiently small radius to ensure that the closed disc it bounds lies entirely within U and its intersection with S is just the point s. Then $\Gamma = \gamma - \sum_{s \in I \cap S} W(\gamma, s)\gamma_s$ is homologous to zero in $U \setminus S$ and Theorem 7.15 yields $0 = \int_\Gamma f(z)\,dz = \int_\gamma f(z)\,dz - \sum_{s \in I \cap S} W(\gamma, s) \int_{\gamma_s} f(z)\,dz$. The claim follows since $\int_{\gamma_s} f(z)\,dz = 2\pi i\,\mathrm{Res}(f, s)$ by the residue theorem. □

Homotopy is a mathematical formulation of the intuitive idea of a continuous transition between two geometrical configurations. In many circumstances, this is an efficient way to show that a closed curve is null-homologous. Given a domain $U \subset \mathbb{C}$, we say that two piecewise C^1-curves $\gamma_0, \gamma_1 : [0, 1] \to U$ with common endpoints $\gamma_0(0) = \gamma_1(0) = \alpha$ and $\gamma_0(1) = \gamma_1(1) = \beta$ are *homotopic* in U if there exists a continuous map $\gamma : [0, 1] \times [0, 1] \to U$ such that for each $0 \leq s \leq 1$ the curve $\gamma_s : [0, 1] \to U$ is piecewise C^1 with $\gamma_s(0) = \alpha$ and $\gamma_s(1) = \beta$; the intuition being that we continuously deform γ_0 into γ_1 without ever leaving U and keeping the endpoints fixed (see Figure 7.12).

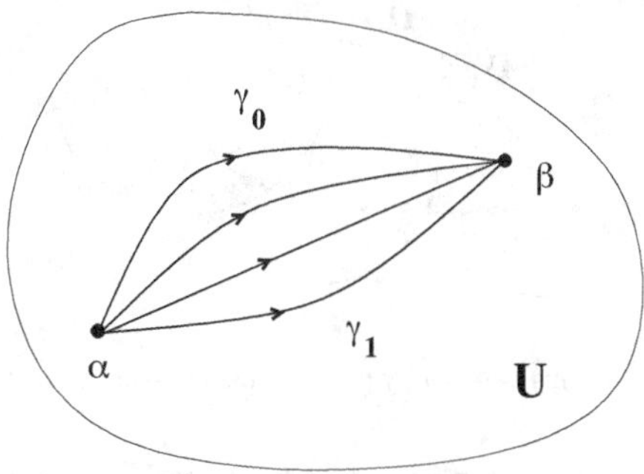

Figure 7.12 Homotopy of curves.

Example In the definition of homotopic curves it is permissible that the curves are closed. For example, the homotopy $\gamma : [0, 1] \times [0, 1] \to \mathbb{C}$ given by $\gamma(s, t) = s + (1 - s)e^{2\pi i t}$ shrinks the unit circle $\gamma_0(t) = e^{2\pi i t}$ to the point $\gamma_1(t) = 1$. Such loops are called *null-homotopic*. □

Theorem 7.17 (Homotopy invariance of the integral) *If γ_0 and γ_1 are two homotopic piecewise-C^1 paths in the domain $U \subset \mathbb{C}$, then $\int_{\gamma_0} f(z)\,dz = \int_{\gamma_1} f(z)\,dz$ for any analytic map $f : U \to \mathbb{C}$.*

Proof Let $f : U \to \mathbb{C}$ be analytic and let $\gamma : [0,1] \times [0,1] \to U$ be a homotopy between γ_0 and γ_1. The key to the proof lies in showing that the integrals over paths having the same endpoints and being close to each other must be equal. Since γ is continuous, its image will be compact and therefore at a positive distance $2\varepsilon > 0$ from the closed set $\mathbb{C} \setminus U$; we assume, without loss of generality, that $U \neq \mathbb{C}$. Moreover, the uniform continuity of γ yields the existence of an integer $n \geq 2$ such that $|\gamma(s,t) - \gamma(s',t')| < \varepsilon$ if $|s - s'| + |t - t'| \leq 1/n$. Fix $s_1, s_2 \in [0,1]$ with $|s_1 - s_2| < 1/n$ and choose discs $D_0, D_1, \ldots, D_k$ of radius ε, and consecutive points $z_0, \ldots, z_{k+1}$ on γ_{s_1} and $w_0, \ldots, w_{k+1}$ on γ_{s_2} such that $z_0 = w_0 = \alpha$, $z_{k+1} = w_{k+1} = \beta$, the union of these discs covers both curves, and $z_j, z_{j+1}, w_j, w_{j+1} \in D_j$ for $j = 0, \ldots, k$; see Figure 7.13. On each disc D_j there is[35] a primitive F_j of f. We have

$$\begin{aligned}\int_{\gamma_{s_1}} f(z)\,dz - \int_{\gamma_{s_2}} f(z)\,dz &= \sum_{j=0}^{k}[F_j(z_{j+1}) - F_j(z_j)] - \sum_{j=0}^{k}[F_j(w_{j+1}) - F_j(w_j)] \\ &= \sum_{j=0}^{k}([F_j(z_{j+1}) - F_j(w_{j+1})] - [F_j(z_j) - F_j(w_j)]) \\ &= ([F_k(z_{k+1}) - F_k(w_{k+1})] - [F_0(z_0) - F_0(w_0)]) = 0\end{aligned}$$

because $z_0 = w_0$, $z_{k+1} = w_{k+1}$ and in the open and connected set $D_j \cap D_{j+1}$ the primitives F_j and F_{j+1} must differ by a constant, say c_j, so that $F_{j+1}(z_{j+1}) - F_{j+1}(w_{j+1}) = F_j(z_{j+1}) - F_j(w_{j+1})$ for $0 \leq j \leq k$. Repeating the argument n times we can go from γ_0 to γ_1. □

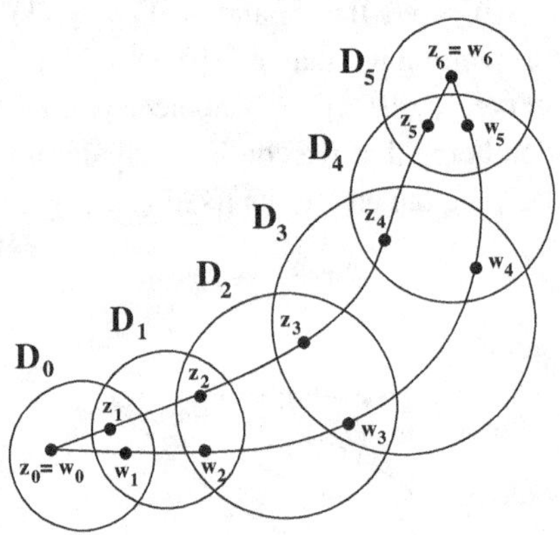

Figure 7.13 Covering two nearby paths having the same endpoints with discs.

Remark (i) The result that the integrals over nearby paths with the same endpoints are equal (established in the proof of Theorem 7.17) allows us to define $\int_\gamma f(z)\,dz$ for analytic f when γ is just continuous, and not necessarily piecewise-C^1: we simply set the integral equal to $\int_{\gamma^*} f(z)\,dz$, where γ^* is a smooth path sufficiently close[36] to the path γ. We can thus extend the notion of

35 The local existence of primitives was shown in the proof of Morera's theorem. The existence of a global primitive is a more delicate issue, as illustrated by the case of the logarithm, which arises as a primitive of $1/z$.

36 Obtained by convoluting with a smooth function; see Chapter 6.

homotopy to just continuous paths by simply dropping the condition that each curve is piecewise C^1. Then Theorem 7.17 also holds for continuous paths.

(ii) A consequence of Theorem 7.17 is that *homotopic closed curves are homologous*. The converse is not true in general: the curve γ depicted in Fig. 7.14 is null-homologous in $\mathbb{C} \setminus \{-1, 1\}$ since $W(\gamma, \pm 1) = 0$ but, at least intuitively,[37] it is clear that it is not null-homotopic in U. □

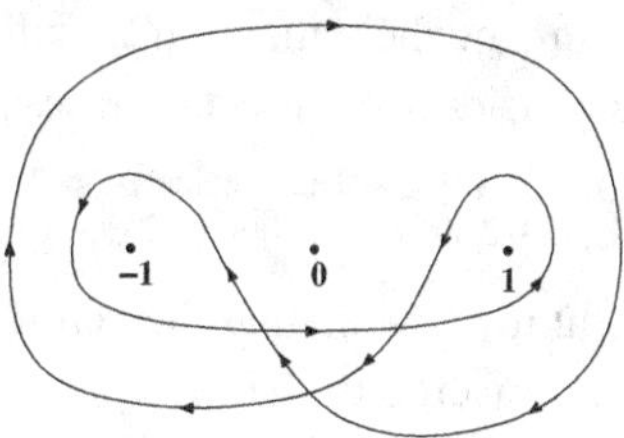

Figure 7.14 A null-homologous curve that is not null-homotopic in $\mathbb{C} \setminus \{-1, 1\}$.

An important application of Theorem 7.16 is the following continuity result, ensuring that an analytic function can be perturbed slightly without changing the number of its zeros.

Theorem 7.18 (Rouché's theorem) *Let $U \subset \mathbb{C}$ be an open set and let γ be a closed null-homologous path in U and having an interior.*[38] *If $f, g : U \to \mathbb{C}$ are analytic on U, and $|f(z) - g(z)| < |f(z)|$ for z on γ, then f and g have the same number of zeros in the interior of γ.*

Proof By hypothesis, f and g have no zero on γ. If $F = g/f$, since $|F(z) - 1| < 1$ for z on γ, we deduce that the closed path $F(\gamma)$ is contained in the disc $|z - 1| < 1$. Since the origin lies outside this disc, we have $W(F(\gamma), 0) = 0$. Thus

$$0 = W(F(\gamma), 0) = \int_{F(\gamma)} \frac{dw}{w} = \int_\gamma \frac{F'(z)}{F(z)}\, dz = \int_\gamma \frac{g'(z)}{g(z)}\, dz - \int_\gamma \frac{f'(z)}{f(z)}\, dz\,.$$

As a consequence of the factorisation (7.8), the function f'/f is analytic in the interior of γ, with exception of singularities in the forms of simple poles with residue $n \geq 1$ at a zero of f of order $n \geq 1$. Invoking Theorem 7.16 we infer the statement. □

[37] A rigorous proof is harder. Given a domain $U \subset \mathbb{C}$, the class of (continuous) loops $\gamma \subset U$ with basepoint $p = \gamma(0) = \gamma(1) \in U$ forms a group with respect to the product operation $(\gamma_1 \cdot \gamma_2)(t) = \gamma_1(t)$ if $0 \leq t < 1/2$ and $(\gamma_1 \cdot \gamma_2)(t) = \gamma_1(t)$ if $1/2 \leq t < 1/2$, denoted $\pi_1(U, p)$ and called the *fundamental group* of U at p; see Munkres (2000). One can show (using van Kampen's theorem) that there is a surjective homeomorphism from $\pi_1(\mathbb{C} \setminus \{-1, 1\}, 0)$ onto $\mathbb{Z}$ that is not one-to-one, whose kernel consists of all null-homologous loops that are not null-homotopic (e.g. the curve depicted in Figure 7.14); see Schlag (2014).

[38] There exists closed curves (continuous images of the unit circle $|z| = 1$) that fill the unit square $[0, 1] \times [0, 1]$; see Salem and Zygmund (1945). These *Peano curves* γ_0 have no interior but are null-homologous in $U = (-1, 2) \times (-1, 2)$, being null-homotopic by means of the homotopy $\gamma(s, t) = \gamma_0(0) + (1 - s)[\gamma_0(t) - \gamma_0(0)]$.

Needham (2000) provides the following amusing interpretation of Rouché's theorem. Imagine walking a dog round and round a tree in a park, both you and the dog finally returning to your starting points, with the dog being on one of those adjustable leashes. Let us say that the tree is located at the origin of $\mathbb{C}$, your path is the image path traced by $f(z)$ as z completes a simple loop γ, while the dog's position is $g(z)$; think of γ as the well-trodden path, while f and g are deviations from it. The requirement that the leash does not stretch to the tree is $|f(z) - g(z)| < |f(z)|$ for $z \in \gamma$, and in these circumstances the dog is forced to walk round the tree the same number of times that you do (since by the argument principle the number of zeros of f inside γ equals the winding number of the curve $f(\gamma)$ around the origin).

Remark The inequality in Rouché's theorem is sufficient for f to have the same number of roots as g inside γ but it is not necessary: take $g = 2f$. □

Example All five roots of $p(z) = z^5 + 15z + 1$ are inside $|z| < 2$ since $|z|^5 = 32 > |15z + 1|$ for $|z| = 2$, and only one is inside $|z| < 3/2$ since $|z^5 + 1| \leq (3/2)^5 + 1 < |15z|$ for $|z| = 3/2$. □

Rouché's theorem is useful in showing that the zeros of the analytic functions in a converging sequence converge to the zeros of the limiting function, in a sense made precise by the following result.

Theorem 7.19 (Hurwitz's theorem) *Let $U \subset \mathbb{C}$ be an domain set and let $f_n : U \to \mathbb{C}$ be analytic functions which converge, uniformly on compact subsets of U, to an analytic function $f : U \to \mathbb{C}$. If f has a zero of order $n_0 \geq 1$ at $z_0 \in U$, then there exists $r > 0$ such that for all n large enough, f_n has exactly n_0 zeros in the disc $|z - z_0| \leq r$, counting multiplicities, and these zeros converge to z_0 as $n \to \infty$.*

Proof Choose $r > 0$ sufficently small so that the closed disc $|z - z_0| \leq r$ is contained in U and $f(z) \neq 0$ for $0 < |z - z_0| \leq r$. If $\delta = \inf_{|z-z_0|=r} |f(z)| > 0$, then for some $N \geq 1$ we have $|f(z) - f_n(z)| \leq \delta/2$ on $|z - z_0| = r$, for $n \geq N$. Thus, for all $n \geq N$, $|f(z) - f_n(z)| \leq \delta/2 < \delta \leq |f(z)|$ on $|z - z_0| = r$ and, by Rouché's theorem, f_n and f have the same number of zeros (counting multiplicity) inside the circle (counting multiplicities). □

An interesting corollary of Hurwitz's theorem is that if a zero-free sequence of analytic functions $f_n : U \to \mathbb{C}$ converges uniformly on compact subsets of the domain U to some $f : U \to \mathbb{C}$, then $f \equiv 0$ in U if $f(z)$ is ever zero in U.

Example To appreciate the essential complex setting in Hurwitz's theorem, note that the zero-free real functions $f_n(x) = x^2 + 1/n$ (with $n \geq 1$) converge uniformly on $\mathbb{R}$ to $f(x) = x^2$. □

A consequence of the interplay between analyticity and topology is the following result.

Theorem 7.20 (Darboux's theorem) *Let $f : U \to \mathbb{C}$ be analytic in the domain $U \subset \mathbb{C}$, and let $\gamma \subset U$ be a piecewise C^1, simple, directed loop, with interior D lying wholly in U. If f is one-to-one on γ, then f is one-to-one throughout $\overline{D}$ and $f(D)$ is the inside of the Jordan curve $f(\gamma)$.*

Proof Clearly $f(\gamma)$ is a piecewise C^1, simple, directed loop (and thus a Jordan curve). Since if $w \notin f(\gamma)$, then $(f - w)$ is zero-free on γ and $\frac{1}{2\pi i} \int_\gamma \frac{f'(z)}{f(z)-w}\, dz = \frac{1}{2\pi i} \int_{f(\gamma)} \frac{dz}{z-w}$, the argument principle and the definition of the winding number $W(f(\gamma), w)$ ensure that $(f - w)$ has a unique zero in D for w inside $f(\gamma)$, and no zero for w outside $f(\gamma)$. □

In the context of Darboux's theorem, let us call two domains U and U' in $\mathbb{C}$ *conformally equivalent* if there is an analytic bijection $f : U \to U'$. A remarkable result (with no analogue in several complex variables), whose proof is beyond our scope (see e.g. Burckel (1979)), identifies the class of open domains $U \subset \mathbb{C}$ which are conformally equivalent (see Figure 7.15) to the unit disc $\mathbb{D} = \{z \in \mathbb{C} : |z| < 1\}$:

Theorem 7.21 (Riemann mapping theorem) *A domain*[39] $U \subsetneq \mathbb{C}$ *is conformally equivalent to* $\mathbb{D}$ *if and only if it is simply connected. Moreover, the conformal bijection* $\phi : \mathbb{D} \to U$ *is uniquely specified by the values assigned to* $\phi(0)$ *and* $\arg \phi'(0)$.

While it is unreasonable to expect an explicit formula in the general case, for some simply connected domains U it is possible to write down the conformal bijection $\phi : \mathbb{D} \to U$ explicitly, e.g. the conformal bijection between the upper half-plane $\mathbb{H} = \{z \in \mathbb{C} : \Im m(z) > 0\}$ and the interior of any polygon are known as the Schwarz–Christoffel maps – see Stein and Shakarchi (2003). The map $z \mapsto (i - z)/(i + z)$ is a conformal bijection of $\mathbb{H}$ onto $\mathbb{D}$. Indeed, the function is clearly analytic in $\mathbb{H}$ and takes values in $\mathbb{D}$ since any $z \in \mathbb{H}$ is closer to i than to $-i$. A direct computation of its imaginary part confirms that the formal inverse, $w \mapsto i(1 - w)/(1 + w)$, maps $\mathbb{D}$ into $\mathbb{H}$, and proves the claim. Note that since $x \in \mathbb{R}$ is equidistant from i and $-i$, the real axis is mapped into the unit circle $|z| = 1$.

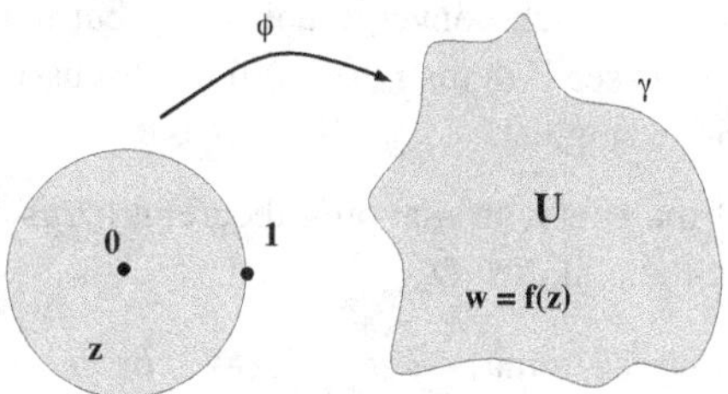

Figure 7.15 A simply connected domain U bounded by a Jordan curve γ is conformally equivalent to the unit disc $|z| = 1$.

An important issue related to the Riemann mapping theorem is whether the bijection extends to the boundary in a nice way. The Schwarz–Christoffel maps admit an extension to a bijection between the closed domains that is analytic except at the pre-images of the corners of the polygon. However, for general domains pathologies may occur and the conformal bijection might not have a

[39] Note that Liouville's theorem shows that $\mathbb{C}$ is not conformally equivalent to $\mathbb{D}$. For multiply-connected domains the situation is quite different: for example, an annulus $\{z \in \mathbb{C} : r_1 < |z| < R_1\}$ with $0 < r_1 < R_1$ is conformally equivalent to another annulus $\{z \in \mathbb{C} : r_2 < |z| < R_2\}$, with $0 < r_2 < R_2$, if and only if $R_1/r_1 = R_2/r_2$; see Burckel (1979).

continuous extension to the boundary: the *Carathéodory theorem* states that a continuous extension of the conformal bijection to the boundary of $\mathbb{D}$ exists if and only if the boundary of U is a continuous path (the continuous image of a segment). Most domains that are relevant in applications of conformal maps are bounded by a finite number of regular arcs, and the regularity properties of the arcs are basically inherited by the function; see the discussion in Pommerenke (2002). Even for Jordan domains U (domains bounded by a closed Jordan curve γ), questions involving length and area arise. While in $\mathbb{D}$ the conformal map f transforms sets of positive area to sets of positive area, distortion at the boundary is possible. If the boundary γ of the Jordan domain U is a rectifiable curve,[40] then ϕ and ϕ^{-1} preserve sets of zero length – see the discussion in Koosis (1998) – but if the Jordan curve γ fails to be rectifiable, all hell breaks loose: a subset of the curve γ having positive area may correspond under ϕ to a subset of the circle $|z| = 1$ having zero Lebesgue measure, and compression of $|z| = 1$ may also happen (in the sense that a subset of positive measure may be mapped onto a set of zero length of γ); see the discussion in Zalcman (1974). In terms of the power series representation of the conformal bijection $\phi : \mathbb{D} \to U$, one can show that for domains U bounded by a Jordan curve γ the power series of $\phi(z)$ converges uniformly on $|z| = 1$ (see Zalcman (1974)) and, if γ is rectifiable, it converges absolutely up to $|z| = 1$; see Koosis (1998). The sharpness of these results is highlighted by the construction of examples of power series which converge uniformly but not absolutely on the closed unit disc $|z| \leq 1$; see Zalcman (1974) for a discussion of a particularly appealing geometric example.

In the context of the Riemann mapping theorem, let us determine the conformal self-maps of the unit disc $\mathbb{D}$.

Theorem 7.22 (Schwarz lemma) *Let $f : \mathbb{D} \to \mathbb{C}$ be analytic. If $f(0) = 0$ and $|f(z)| \leq 1$ for all $z \in \mathbb{D}$, then*

$$|f(z)| \leq |z|\,, \qquad z \in \mathbb{D}\,; \tag{7.12}$$

in particular, $|f'(0)| \leq 1$. If equality holds in (7.12) at some $z \neq 0$, or if $|f'(0)| = 1$, then there exists $\lambda \in \mathbb{C}$ with $|\lambda| = 1$ such that $f(z) = \lambda z$ for all $z \in \mathbb{D}$.

Proof Using (7.8), write $f(z) = zg(z)$ with $g : \mathbb{D} \to \mathbb{C}$ analytic. If $|z| = r \in (0, 1)$, then $|g(z)| = |g(z)|/|z| \leq 1/r$, so that, by the maximum principle, $|g(z)| \leq 1/r$ if $|z| \leq r$. Letting $r \uparrow 1$ yields (7.12), and $|f'(0)| \leq 1$ follows by taking $z \to 0$ in (7.12). If $|f(z_0)| = |z_0|$ for some $z_0 \in \mathbb{D} \setminus \{0\}$, then $|g(z_0)| = 1 \geq |g(z)|$ for all $z \in \mathbb{C}$, so that the strict maximum principle ensures that g is constant, say $g(z) = \lambda$ for all $z \in \mathbb{C}$, and thus $f(z) = \lambda z$ for all $z \in \mathbb{D}$. If $|f'(0)| = 1$, then $|g(0)| = 1 \geq |g(z)|$ for all $z \in \mathbb{C}$ and the maximum principle again forces g to be a constant. □

[40] That is, the continuous function $\gamma : [0, 1] \to \mathbb{C}$ is of finite variation (see Chapter 2).

An estimate analogous to (7.12) holds for any disc: if F is analytic for $|z-z_0| < R$, with $|F(z)| \leq M$ and $F(z_0) = 0$, then $|F(z)| \leq \frac{M}{R}|z - z_0|$ for $|z - z_0| < R$. Indeed, if $M > 0$, then the change of variable $z \mapsto Rz + z_0$ maps the unit disc $\mathbb{D}$ onto the disc $|z - z_0| < R$, and we can apply (7.12) to $f(z) = F(rz + z_0)/M$.

For a fixed angle $\varphi \in [0, 2\pi)$, the rotation $z \mapsto e^{i\varphi}z$ is a conformal self-map of $\mathbb{D}$ that fixes the origin. Actually, these are the only conformal self-maps of $\mathbb{D}$ that leave 0 fixed: if f is such a map, apply Theorem 7.22 to f and to its inverse f^{-1} to infer that $|f(z)| = |z|$ for all $z \in \mathbb{D}$, so that the analytic function $z \mapsto f(z)/z$ has constant modulus and therefore must be constant, that is, $f(z) = \lambda z$ for some $\lambda \in \mathbb{D}$ with $|\lambda| = 1$.

Theorem 7.23 *The conformal self-maps of $\mathbb{D}$ are of the form $f(z) = e^{i\varphi}\,\frac{z-a}{1-\bar{a}z}$, where $a \in \mathbb{D}$ and $\varphi \in [0, 2\pi)$.*

Proof Given $a \in \mathbb{D}$, clearly $g(z) = \frac{z-a}{1-\bar{a}z}$ is analytic in $\mathbb{D}$, and it is easy to check that its inverse (wherever it is defined) is given by $g^{-1}(w) = \frac{w+a}{1+\bar{a}w}$. Since $g(a) = 0$ and $|e^{i\theta}-a| = |e^{-i\theta}-\bar{a}| = |1-\bar{a}e^{i\theta}|$ for $\theta \in [0, 2\pi)$, we see that $|g(z)| = 1$ for $|z| = 1$, so that, by the maximum principle, g maps $\mathbb{D}$ to itself, and so does $f(z) = e^{i\varphi}\,g(z)$. Conversely, if f is a conformal self-map of $\mathbb{D}$, set $a = f^{-1}(0)$ and consider the conformal self-map $f \circ g^{-1}$ of $\mathbb{D}$ with $(f \circ g^{-1})(0) = 0$. The considerations preceding Theorem 7.23 show the existence of some $\varphi \in [0, 2\pi)$ such that $(f \circ g^{-1})(w) = e^{i\varphi}w$ for all $w \in \mathbb{D}$. Writing $w = g(z)$, we get $f(z) = e^{i\varphi}g(z)$, as claimed. □

Note that in Theorem 7.23 the parameters a and $\varphi \in [0, 2\pi)$ are uniquely determined by the conformal map $F = f^{-1}$: $a = F(0)$ and $\varphi \in [0, 2\pi)$ specifies the argument of $F'(0) = \frac{1}{f'(a)} = (1 - |a|^2)e^{-i\varphi}$. It is instructive to compare this with the statement of the Riemann mapping theorem.

The conformal self-maps of $\mathbb{D}$ identified in Theorem 7.23 are special cases of the *Möbius transformations* (or *fractional linear transformations*) $z \mapsto \frac{az+b}{cz+d}$, where a, b, c, $d \in \mathbb{C}$ are constants subject to $ad - bc \neq 0$ (this condition prevents the transformation from being constant and thus uninteresting). In discussing Möbius transformations it is useful to introduce the *extended complex plane*, consisting of $\mathbb{C}$ together with the "point at infinity", ∞. One way to visualise $\mathbb{C}_\infty = \mathbb{C} \cup \{\infty\}$ is through stereographic projection, as follows (see Figure 7.16). Consider the unit two-dimensional sphere $\mathbb{S}^2 = \{(x, y, z) \in \mathbb{R}^3 : x^2 + y^2 + z^2 = 1\}$ and identify the complex plane $\mathbb{C}$ with its equatorial plane $z = x + iy \leftrightarrow (x, y, 0)$. If $P = (X, Y, Z) \in \mathbb{S}^2$ is any point other than the north pole $N = (0, 0, 1)$, we define its stereographic projection to be the point $z = x + iy \leftrightarrow (x, y, 0)$ where the straight line through N and P meets the plane $Z = 0$, the explicit relevant formulas being $x = X/(1 - Z)$, $y = Y/(1 - Z)$, and $X = 2x/(1+x^2+y^2)$, $Y = 2y/(1+x^2+y^2)$, $Z = (x^2+y^2-1)/(1+x^2+y^2)$. Geometrically or algebraically one can check that circles of longitude on the sphere $\mathbb{S}^2$ correspond to straight lines through the origin (in the plane $\mathbb{C}$), while circles of latitude on the sphere $\mathbb{S}^2$ correspond to circles centred at the origin (in the plane $\mathbb{C}$). As the radii of the latter circles tend to infinity, the circles of latitude on $\mathbb{S}^2$ approach the north pole, which justifies the association between the north pole and the point at infinity. Moreover, some algebraic manipulations (see Gamelin (2001) for details) confirm that under stereographic projection, circles on the sphere $\mathbb{S}^2$ correspond to circles and straight lines in the plane; in particular, circles through the north pole correspond to straight lines in the plane and for this reason we regard a straight line in the plane as a circle through ∞.

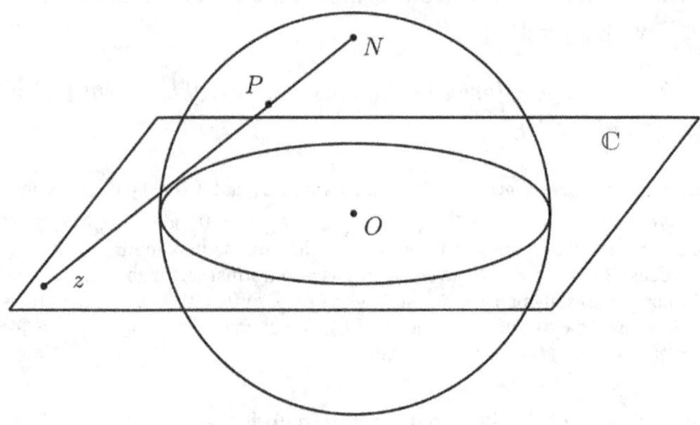

Figure 7.16 The stereographic projection.

We consider Möbius transformations in the extended plane, and the formulas involving ∞ are to be interpreted as the corresponding limits, since in this wider context we can keep track of the behaviour of a Möbius transformation $z \to w = \frac{az+b}{cz+d}$ both near the singularity $z = -d/c$ (if $c \neq 0$) and as $z \to \infty$, and a Möbius transformation is a continuous one-to-one map of the extended z-plane onto the extended w-plane, with a continuous inverse,[41] satisfying the following properties.

- If $c = 0$ then the Möbius transformation is conformal in the entire complex plane, while for $c \neq 0$, it is conformal on $\mathbb{C} \setminus \{-d/c\}$.
- The Möbius transformations form a group – if we identify the transformation $f(z) = \frac{az+b}{cz+d}$ with the matrix $\left(\begin{smallmatrix} a & b \\ c & d \end{smallmatrix}\right)$, then the composition of two transformations corresponds to the product of their matrices, and the inverse trans-

[41] In this context, the distance between two points on the sphere is the length of the chord connecting the points, and taking this as a new distance between the stereographic images of these points, we obtain a metric for the extended complex plane. Since the sphere is compact, the extended complex plane $\mathbb{C}_\infty$ is compact as well, being a one-point compactification of $\mathbb{C}$.

formation corresponds to the inverse matrix. The simplest transformations correspond to matrices of the form $\left(\begin{smallmatrix}1 & \alpha\\ 0 & 1\end{smallmatrix}\right)$, $\left(\begin{smallmatrix}\beta & 0\\ 0 & 1\end{smallmatrix}\right)$ with $\beta \neq 0$, $\left(\begin{smallmatrix}0 & 1\\ 1 & 0\end{smallmatrix}\right)$; the first of these, $z \mapsto z + \alpha$, is called a *translation*, the second, $z \mapsto \beta z$, is a *rotation* if $|\beta| = 1$ and a *homothety* if $\beta > 0$ (so that, writing $\beta = |\beta| \cdot \beta/|\beta|$, the general case is the result of a homothety followed by a rotation), while the third, $z \mapsto 1/z$, is called an *inversion*. If $c \neq 0$, the decomposition $\frac{az+b}{cz+d} = \frac{bc-ad}{c^2(z+d/c)} + \frac{a}{c}$ shows that the general Möbius transformation is composed by a translation, an inversion, a rotation, and a homothety followed by another translation. If $c = 0$, the inversion falls out and we do not need the last translation.

- The *cross ratio* of four distinct points in the extended complex plane $\mathbb{C}_\infty$, given by $[z_1, z_2, z_3, z_4] = \frac{(z_1-z_2)(z_3-z_4)}{(z_1-z_4)(z_3-z_2)}$ (and interpreted by taking limits in the obvious way if one of the points is ∞) is invariant under Möbius transformations; see Boas (1987) for a proof. Therefore the Möbius transformation $z \mapsto w$ defined by $[w, w_2, w_3, w_4] = [z, z_2, z_3, z_4]$ carries z_2, z_3, z_4 (in that order) to w_2, w_3, w_4. Moreover, one can show (see Ahlfors (1979)) that the images of three different points determine uniquely a Möbius transformation and that the cross ratio $[z_1, z_2, z_3, z_4]$ is real if and only if the four points lie on a circle or on a straight line. Consequently, every Möbius transformation of a straight line or circle is again a straight line or a circle (but circles may be transformed into lines and vice-versa, the image of a line L being a circle if and only if $c \neq 0$ and the pole $z = -d/c$ is not on the line L).

A form of Theorem 7.22 that is invariant under conformal self-maps of $\mathbb{D}$ establishes an interesting link with hyperbolic geometry.

Theorem 7.24 (Pick's lemma) *If $f : \mathbb{D} \to \mathbb{D}$ is analytic, then*

$$|f'(z)| \leq \frac{1 - |f(z)|^2}{1 - |z|^2}, \qquad z \in \mathbb{D}, \tag{7.13}$$

the inequality being strict unless f is a conformal self-map of $\mathbb{D}$.

Proof Fix $z_0 \in \mathbb{D}$ and set $w_0 = f(z_0) \in \mathbb{D}$. If $g(z) = \frac{z+z_0}{1+\overline{z_0}\,z}$ and $h(w) = \frac{w-w_0}{1-\overline{w_0}\,w}$ are conformal self-maps of $\mathbb{D}$ mapping 0 to z_0 and w_0 to 0, respectively, then the analytic function $h \circ f \circ g : \mathbb{D} \to \mathbb{D}$ satisfies $(h \circ f \circ g)(0) = 0$. Theorem 7.22 and the chain rule yield now $|h'(w_0)f'(z_0)g'(0)| \leq 1$, so that $g'(0) = 1 - |z_0|^2$ and $h'(w_0) = 1/(1 - |w_0|^2)$ lead us to (7.13). Note that if f is a conformal self-map of $\mathbb{D}$, then so is $h \circ f \circ g$, and $(h \circ f \circ g)(0) = 0$ then Theorem 7.23 ensures $|h'(w_0)f'(z_0)g'(0)| = 1$, so that equality holds in (7.13). Conversely, equality in (7.13) means $|h'(w_0)f'(z_0)g'(0)| = 1$, and Theorem 7.22 yields then that $h \circ f \circ g$ is multiplication by some $\lambda \in \mathbb{D}$ with $|\lambda| = 1$. Composing by g^{-1} on the right and by g^{-1} on the left, we see that f is a conformal self-map of $\mathbb{D}$. □

If f is a conformal self-map of $\mathbb{D}$, Theorem 7.24 yields $|f'(z)| = \frac{1-|w|^2}{1-|z|^2}$ for

$w = f(z)$, or $\frac{|dw|}{1-|w|^2} = \frac{|dz|}{1-|z|^2}$ in the notation of differential geometry, and therefore

$$\int_{f\circ\gamma} \frac{|dw|}{1-|w|^2} = \int_\gamma \frac{|dz|}{1-|z|^2} \tag{7.14}$$

holds for any piecewise C^1-curve γ in $\mathbb{D}$. Defining for such a curve its *length in the hyperbolic metric*[42] by $2\int_\gamma \frac{|dz|}{1-|z|^2}$, the identity (7.14) shows that hyperbolic lengths are invariant under conformal self-maps of $\mathbb{D}$. We now define the *hyperbolic distance* $\rho(z_1, z_2)$ between $z_1, z_2 \in \mathbb{D}$ as the infimum of the hyperbolic lengths of all piecewise C^1-curves in $\mathbb{D}$ from z_1 to z_2. Since, by (7.14), conformal self-maps of $\mathbb{D}$ preserve the hyperbolic lengths of such curves, we see that $\rho(f(z_1), f(z_2)) = \rho(z_1, z_2)$ for every conformal self-map f of $\mathbb{D}$.

Theorem 7.25 *Any two distinct points $z_1, z_2 \in \mathbb{D}$ are joined by a unique shortest curve in $\mathbb{D}$ in the hyperbolic metric, namely, a diameter of $\mathbb{D}$ or the arc of the circle passing through z_1 and z_2, orthogonal to the unit circle $|z| = 1$.*

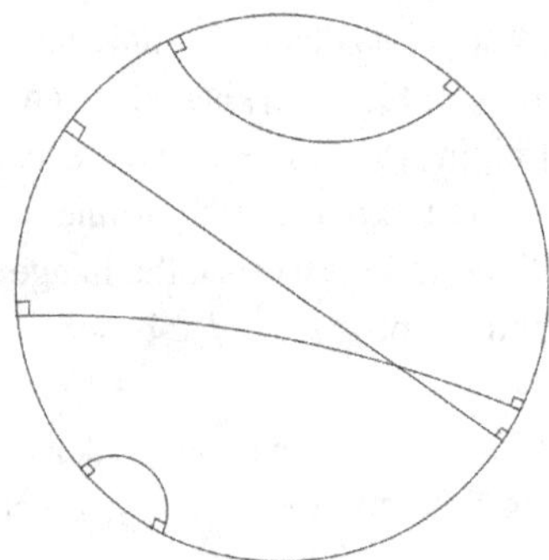

Figure 7.17 Some hyperbolic geodesics: they play the role that straight lines play in classical planar geometry, satisyfing all axioms of Euclidean geometry except the parallel axiom (that for any given point not on a given line, there is exactly one line through the point that does not meet the given line).

Proof Let f be a conformal self-map of $\mathbb{D}$ such that $f(z_1) = 0$. By multipliying f with a suitable unimodular $\lambda \in \mathbb{C}$ (that is, by performing a rotation) we can arrange that $f(z_2) = r \in (0, 1)$. Taking Theorem 7.23 into account, we can explicitate $f(z) = e^{i\varphi} \frac{z-z_1}{1-\overline{z_1}\, z}$, where $\varphi = \mathrm{Arg}(\frac{1-\overline{z_1}\, z_2}{z_2-z_1})$. Let us now show that the straight line segment from 0 to r is the unique piecewise C^1-path of shortest hyperbolic length. For this, let $\gamma(t) = x(t) + iy(t)$, $0 \le t \le 1$, be a piecewise C^1-path in $\mathbb{D}$, with $\gamma(0) = 0$ and $\gamma(1) = r$. Note that $\gamma_0(t) = x(t)$, $0 \le t \le 1$, defines a piecewise C^1-path along the real segment $(-1, 1) \subset \mathbb{D}$, with $\gamma_0(0) = 0$, $\gamma_0(1) = r$, and $\int_{\gamma_0} \frac{|dz|}{1-|z|^2} = \int_0^1 \frac{|x'(t)|}{1-x^2(t)} \le \int_0^1 \frac{|x'(t)|}{1-x^2(t)-y^2(t)} \le \int_\gamma \frac{|dz|}{1-|z|^2}$. If $y(t) \neq 0$ at some $t \in [0, 1]$, then the first inequality is strict and the path γ_0 along the

[42] The factor 2, often omitted, adjusts the hyperbolic metric so that it has curvature -1.

real axis is shorter than γ. Furthermore, if $x'(t_0) < 0$ at some $t_0 \in (0, 1)$, then $x(1) = r > x(0) = 0$ ensures the existence of some $\varepsilon \in (0, t_0)$ and of some $\delta \in (t_0, 1)$ such that $x(\varepsilon) = x(\delta)$, and we can reduce the integral $\int_0^1 \frac{|x'(t)|}{1-x^2(t)}$ by deleting the interval (ε, δ). Consequently the integral is minimal when $t \mapsto \gamma(t)$ is real and nondecreasing, in which case the path is the straight line segment from 0 to r, of hyperbolic length $2\int_0^r \frac{dt}{1-t^2} = \int_0^r \left(\frac{1}{1-t} + \frac{1}{1+t}\right) dt = \ln\left(\frac{1+r}{1-r}\right)$. The statement follows since the conformal self-map $f^{-1}(w) = \frac{e^{-i\varphi}w+z_1}{1+\overline{z_1}\,e^{-i\varphi}w}$ of $\mathbb{D}$ maps circles and lines orthogonal to the unit circle onto circles and lines orthogonal to the unit circle, leaving $|z| = 1$ invariant – to clarify when straight line segments occur as geodesics, note that one can easily verify that for $z_1 \neq 0$ the pole $w = -e^{i\varphi}/\overline{z_1}$ of f^{-1} belongs to the real axis if and only if $\overline{z_1}\,z_2 \in \mathbb{R}$ (that is, z_1 and z_2 are collinear with the origin), while for $z_1 = 0$ clearly f^{-1} maps the straight line segment $(-1, 1)$ into another diameter of $\mathbb{D}$. □

The above proof of Theorem 7.25 permits us to draw two conclusions:

(1) if v is a hyperbolic geodesic through $z_1 \in \mathbb{D}$, then there exists a conformal self-map f of $\mathbb{D}$ that maps v to the real axis and z_1 to 0;

(2) the hyperbolic distance from $z_1 \in \mathbb{D}$ to 0 is $\ln\left(\frac{1+|z_1|}{1-|z_1|}\right)$ and becomes unbounded as z_1 approaches the boundary of $\mathbb{D}$.

Moreover, we can now give a geometric reinterpretation of Theorem 7.24.

Theorem 7.26 *Every analytic map $f : \mathbb{D} \to \mathbb{D}$ is a contraction with respect to the hyperbolic distance ρ, meaning that $\rho(f(z_1), f(z_2)) \leq \rho(z_1, z_2)$ for all $z_1, z_2 \in \mathbb{D}$, with strict inequality for all $z_1 \neq z_2$ unless f is a conformal self-map of $\mathbb{D}$, in which case there is equality for all $z_1, z_2 \in \mathbb{D}$.*

Proof If γ is the geodesic from z_1 to z_2, then $f \circ \gamma$ is a smooth curve from $f(z_1)$ to $f(z_2)$. Theorem 7.24 and the definition of the hyperbolic distance ρ yield $\rho(f(z_1), f(z_2)) \leq 2\int_{f\circ\gamma} \frac{|dw|}{1-|w|^2} = 2\int_\gamma \frac{|f'(z)|\,|dz|}{1-|f(z)|^2} \leq 2\int_\gamma \frac{|dz|}{1-|z|^2} = \rho(z_1, z_2)$. If f is not a conformal self-map of $\mathbb{D}$, then the strict inequality in (7.13) leads to a strict inequality if $z_1 \neq z_2$. □

7.1.2 Fourier series convergence by change of variables

We saw in Chapter 4 that the Fourier series of a continuous periodic function $f : \mathbb{R} \to \mathbb{R}$ may diverge on a dense set. A nice application of complex analysis to Fourier series shows that the this divergence phenomenon can be removed by a suitable change of variables; see Zalcman (1974) for the interesting history of this result. For this purpose it is convenient to regard a continuous function $f : \mathbb{R} \to \mathbb{R}$ which is periodic of period 1 as a continuous real-valued map $f(e^{2\pi it})$ on the unit circle $\gamma = \{z \in \mathbb{C} : |z| = 1\}$.

Theorem 7.27 *Let $f : \gamma \to \mathbb{R}$ be a continuous function on the unit circle γ. There is a continuous bijection $\phi : \gamma \to \gamma$ such that the Fourier series of $f \circ \phi$ converges uniformly on γ.*

Proof Adding a constant to f, we may assume that $\int_0^1 f(x)\,dx = 0$. Furthermore, translating the variable, if necessary, we may assume that $f(0) = f(1) = 0$. Then, by the mean value theorem, there is at least one $x_0 \in (0,1)$ with $f(x_0) = 0$. Let $M_1 = \sup_{x\in[0,x_0]} |f(x)|$ be attained at $x_1 \in (0,x_0)$ and $M_2 = \sup_{x\in[x_0,1]} |f(x)|$ be attained at $x_2 \in (x_0,1)$, and consider the continuous periodic function of bounded variation $\omega : [0,1] \to \mathbb{R}$ given by

$$\omega(x) = \begin{cases} \Big(1 + \max\limits_{0\le s\le x} |f(s)|\Big) \sin(\pi x/x_0)\,, & 0 \le x \le x_1\,, \\ (1+M_1)\sin(\pi x/x_0)\,, & x_1 \le x \le x_0\,, \\ -\Big(1 + \max\limits_{x_0\le s\le x} |f(s)|\Big) \sin(\pi[x-x_0]/[1-x_0])\,, & x_0 \le x \le x_2\,, \\ -(1+M_2)\sin(\pi[x-x_0]/[1-x_0])\,, & x_2 \le x \le 1\,. \end{cases}$$

Then $\tilde{f} = f + \omega$ is periodic, continuous, and of bounded variation, with $\tilde{f}(0) = \tilde{f}(x_0) = \tilde{f}(1) = 0$ and $\tilde{f}(x) > 0$ on $(0,x_0)$, $\tilde{f}(x) < 0$ on $(x_0,1)$. Since the Fourier series of a continuous and periodic function of bounded variation converges uniformly (see Chapter 4), it suffices to prove the theorem for $\tilde{f}$. Define now the continuous periodic function $g : [0,1] \to \mathbb{R}$ by $g(x) = x/x_0$ for $x \in [0,x_0]$ and $g(x) = (1-x)/(1-x_0)$ for $x \in (x_0,1]$. Clearly g is strictly increasing on $(0,x_0)$ and strictly decreasing on $(x_0,1)$. Then the image of $[0,1]$ under the map $H(x) = g(x) + i f(x)$ is a simple closed Jordan curve ν in the plane. Let $\Phi(z) = \sum_{n\ge 0} a_n z^n$ be a conformal bijection of $\mathbb{D}$ onto the interior U of ν such that $\Phi(-1) = H(0)$. By the discussion of the Riemann mapping theorem in Section 7.1.1, Φ extends to a bijection from $\overline{\mathbb{D}} = \mathbb{D} \cap \gamma$ to $U \cap \nu$ and the series $\sum_{n\ge 0} a_n z^n$ converges uniformly on γ. The desired continuous bijection ϕ of γ is obtained by setting $\phi(e^{2\pi i t}) = H^{-1}(\Phi(e^{2\pi i t}))$ for $t \in [0,1]$. Indeed, this is clealy a continuous bijection, and $\tilde{f}(\phi(e^{2\pi i t})) = \tilde{f} \circ H^{-1}(\Phi(e^{2\pi i t})) = \Im\mathrm{m}\,[\Phi(e^{2\pi i t})]$, which has a uniformly convergent Fourier series since $\Phi(e^{2\pi i t})$ does. □

7.1.3 Paley–Wiener theorems

In Chapter 5 we saw that there is a correspondence between the decay at infinity for a function f and the smoothness of its Fourier transform $\hat{f}$. Analyticity may be thought of as the ultimate in smoothness, and compact support as the ultimate in decay at infinity. It is rather easy to see that compact support of f permits the extension[43] of $\hat{f}$ from $\mathbb{R}$ to an analytic function on $\mathbb{C}$, with a specific growth rate at infinity (see below). It is remarkable that one can also go in the reverse direction. However, since this direction is technically more involved we will first develop some useful tools by discussing a variant of the maximum principle applicable to analytic functions with a moderate growth rate in unbounded sectors.

[43] Note that if $f : \mathbb{R} \to \mathbb{R}$ is analytic, in the sense that at every point it admits a convergent power series expansion in an open interval centred at that point, then (using the available real power series expansions) f is the restriction to $\mathbb{R}$ of some analytic function $F : U \to \mathbb{C}$ in some domain $U \subset \mathbb{C}$ containing $\mathbb{R}$. However, this domain need not contain some strip $\{z \in \mathbb{C} : |\Im\mathrm{m}(z)| < \varepsilon\}$ for some $\varepsilon > 0$. For example, the poles $z_n = n \pm i/n$ (with $n \ge 1$) of the analytic function $F(z) = \sum_{n\ge 1} 2^{-n}[(z-n)^2 + 1/n^2]^{-1}$ prevent this type of extension even if its restriction f to $\mathbb{R}$ belongs to $L^1(\mathbb{R}) \cap L^\infty(\mathbb{R})$.

Phragmen–Lindelöf's theorem

Let $f : \overline{U} \to \mathbb{C}$ be continuous on the closure of a domain $U \subset \mathbb{C}$ in which f is analytic. Then $|f|$ is bounded on $\overline{U}$. Moreover, if $|f(z)| \le M$ for $z \in \partial U = \overline{U} \setminus U$, then, by the maximum modulus theorem, $|f(z)| \le U$ throughout $\overline{U}$. This version of the maximum modulus principle – a bound on the boundary holds also in the interior – fails in general for unbounded domains, for example of sectorial type[44]

$$U_\alpha = \{z \in \mathbb{C} : \; |\mathrm{Arg}(z)| < \alpha\pi\}$$

with angular opening $2\alpha\pi$; here $\alpha \in (0, 1)$.

Example For any $c > 0$, the function $f(z) = e^{cz}$ is analytic in the right half-plane $\Delta = \{z \in \mathbb{C} : \Re e(z) > 0\}$ and continuous on $\overline{\Delta}$. While $|f(z)| = 1$ on $\partial\Delta$ (since $|e^{icy}| = 1$ for $y \in \mathbb{R}$), clearly $e^{cx} \to \infty$ for $x \to \infty$. □

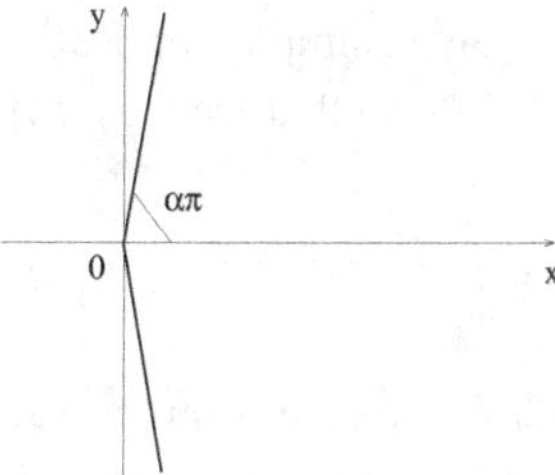

Figure 7.18 A sector U_α with angular opening $2\alpha\pi$, symmetrically disposed about the positive half-line.

One possible way out is to impose some growth restrictions on $|f(z)|$ as $|z| \to \infty$ in U_α. The obvious condition is boundedness of f in $\overline{U_\alpha}$ but maybe we can expect to improve upon this. Indeed, a glance at the proof of Liouville's theorem (Theorem 7.8) shows that if $f : \mathbb{C} \to \mathbb{C}$ is analytic and $|f(z)| \le M(1 + |z|^\varepsilon)$ throughout $\mathbb{C}$, for some $\varepsilon \in (0, 1)$, then f is a constant. Perhaps the *a priori* boundedness requirement on the analytic function in $\overline{U_\alpha}$ could be relaxed.

Example It is instructive to look again at the functions $f(z) = e^{cz}$ in the right half-plane Δ (as before, $c > 0$). Since $|e^{cz}| = e^{cR\cos(\theta)}$ for $z = Re^{i\theta}$ with $|\theta| < \pi/2$, we see that the modulus grows indefinitely on any half-line (in Δ) which starts at the origin. This suggests that perhaps growth at this rate along one half-line influences the growth along other half-lines, so that in this case interior unboundedness is traceable at the boundary if the opening were less than π. □

Let $U \subset \mathbb{C}$ be an unbounded domain. An analytic function $f : U \to \mathbb{C}$ is said to be of *exponential type* $c \ge 0$ if for each $\varepsilon > 0$ there is some $M_\varepsilon > 0$

[44] Performing a rotation and a translation in the complex plane shows that the location of the sector is inconsequential. However, as we shall see, the size of the angular opening matters.

such that $|f(z)| \le M_\varepsilon e^{(c+\varepsilon)|z|}$ for all $z \in U$. For example, in $\mathbb{C}$, polynomials are of minimal type 0, $z \mapsto e^{cz}$ has type c, while $z \mapsto e^{e^z}$ is not of exponential type.

Theorem 7.28 (Phragmen–Lindelöf's theorem) *Consider a sector U_α with*[45] *$\alpha < 1/2$. Let $f : U_\alpha \to \mathbb{C}$ be analytic of exponential type c in U_α and continuous on $\overline{U_\alpha}$. Then $\sup_{z \in \partial U_\alpha} |f(z)| = M < \infty$ ensures $|f(z)| \le M$ for $z \in U_\alpha$.*

Proof Pick $\beta \in (1,2)$ with $\alpha\beta < 1$. For any $a > 0$, $f_1(z) = \exp(-az^\beta) f(z)$ is continuous on $\overline{U_\alpha}$, analytic in U_α, and $|f_1(z)| \le M/2$ for $z \in U_\alpha$ with $|z| \ge R$ for some $R > 0$ sufficiently large. Since $|f_1(z)| \le M$ on the rays $\operatorname{Arg}(z) = \pm\alpha\pi$ that form the boundary ∂U_α of U_α, using the maximum modulus principle for f_1 in the bounded domain $\{z \in U_\alpha : |z| < R\}$, we get $\sup_{z \in \overline{U_\alpha}} |f_1(z)| \le M$. Thus $|f(z)| \le M\,|\exp(az^\beta)|$ for all $z \in \overline{U_\alpha}$. Letting $a \to 0$ yields the claim. □

One-dimensional Paley–Wiener theorems

Let $f \in L^2(\mathbb{R})$ be supported in the interval $[-A, A]$ for some $A > 0$, that is, be such that $f(x) = 0$ for $|x| > A$. A simple application of the Cauchy–Schwarz inequality shows that $f \in L^1(\mathbb{R})$, so that we may write the Fourier transform $\hat{f}$ as

$$\hat{f}(\xi) = \int_{\mathbb{R}} f(x) e^{-2\pi i \xi x} dx, \qquad \xi \in \mathbb{R}. \tag{7.15}$$

This function $\hat{f} : \mathbb{R} \to \mathbb{C}$ admits an extension to an analytic function $\hat{f} : \mathbb{C} \to \mathbb{C}$ of exponential type $2\pi A$. Indeed, simply replace the real variable ξ with the complex variable $\zeta = \xi + i\eta$ (with $\eta \in \mathbb{R}$) and substitute this in the definition (7.15):

$$\hat{f}(\xi + i\eta) = \int_{\mathbb{R}} [f(x) e^{2\pi x\eta}]\, e^{-2\pi i \xi x} dx$$

is the Fourier transform of the $L^1(\mathbb{R})$-function $x \mapsto f(x)e^{2\pi x\eta}$. The fact that $\hat{f}(\xi + i\eta)$ is analytic in $\zeta = \xi + i\eta$ is easily verified by computing its complex derivative using the dominated convergence theorem, and the fact that it is of exponential type is confirmed by the estimate

$$|\hat{f}(\zeta)| = \left| \int_{-A}^{A} [f(x) e^{2\pi x\eta}]\, e^{-2\pi i \xi x} dx \right| \le e^{2\pi A|\eta|} \int_{-A}^{A} |f(x)|\, dx\,.$$

Remarkably, the converse also holds true.

Theorem 7.29 (Paley–Wiener theorem) *Let $F : \mathbb{C} \to \mathbb{C}$ be analytic of exponential type $2\pi A$ (for some $A > 0$) and such that its restriction to the real axis, $F\big|_{\mathbb{R}}$, belongs to $L^2(\mathbb{R})$. Then $F\big|_{\mathbb{R}} = \hat{f}$ for some $f \in L^2(\mathbb{R})$ with $f(x) = 0$ for $|x| > A$.*

[45] The case $f(z) = e^z$ shows that this constraint is needed.

Proof Since $F|_{\mathbb{R}} \in L^2(\mathbb{R})$, see Theorem 5.2, there is some $f \in L^2(\mathbb{R})$ with $\hat{f} = F|_{\mathbb{R}}$. It is to be proved that $f(x) = 0$ for a.e. $|x| > A$, since we may alter f a.e. within $L^2(\mathbb{R})$.

Let us introduce the analytic function $h : \mathbb{C} \to \mathbb{C}$ of exponential type at most $2\pi A$, given by

$$h(z) = \int_{-1/2}^{1/2} F(z+y)\,dy\,, \qquad z \in \mathbb{C}\,.$$

The growth of h on $\mathbb{R}$ is subject to the constraints

$$|h(x)|^2 \le \int_{-1/2}^{1/2} |F(x+y)|^2 dy \le \int_{\mathbb{R}} |F(x)|^2 dx = \|F|_{\mathbb{R}}\|^2_{L^2(\mathbb{R})} < \infty\,,$$

$$\|h\|^2_{L^2(\mathbb{R})} \le \int_{-1/2}^{1/2} \int_{\mathbb{R}} |F(x+y)|^2 dy dx = \|F|_{\mathbb{R}}\|^2_{L^2(\mathbb{R})} < \infty\,,$$

due to the Cauchy–Schwarz inequality and Fubini's theorem. For $B > A$, the function $z \mapsto e^{2\pi i B z} h(z)$ is analytic of exponential type in $\mathbb{C}$, is bounded on $\mathbb{R}$ and bounded[46] on the positive imaginary half-axis $i\mathbb{R}_+ = \{iy : y > 0\}$. Applying the Phragmen–Lindelöf theorem in each the two quarters of the upper half-plane delimited by $\mathbb{R}$ and by $i\mathbb{R}_+$ supplies us with the bound

$$|h(Re^{i\theta})| \le M\, e^{2\pi BR \sin\theta} \qquad \text{for} \quad R > 0, \quad 0 \le \theta \le \pi\,, \tag{7.16}$$

where $M = \sup_{z \in \mathbb{R} \cup i\mathbb{R}_+} |h(z)|$. For a fixed $a > 0$ and $x > B$, consider now the function $z \mapsto e^{2\pi i x z} h(z)/(1 - iaz)$, analytic in $\mathbb{C} \setminus \{-i/a\}$ and with a simple pole at $z = -i/a$. Cauchy's theorem for the simple, closed, piecewise C^1-path γ_R depicted in Figure 7.19 yields

$$\int_{-R}^{R} \frac{e^{2\pi i x\xi} h(\xi)}{1 - ia\xi}\, d\xi = -i \int_0^{\pi} \frac{e^{2\pi i x R e^{i\theta}}}{1 - iaRe^{i\theta}}\, h(Re^{i\theta})\, Re^{i\theta}\, d\theta\,.$$

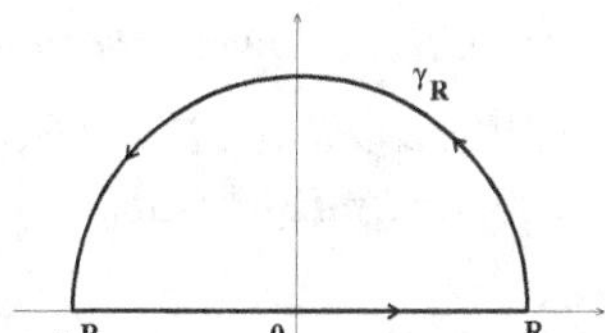

Figure 7.19 For $R > 1/a$, the path γ_R consists of the straight line segment from $-R$ to R, followed by the positively directed semicircle $|z| = R$ with $\Im m(z) \ge 0$.

Using (7.16), the modulus of the right side can be estimated for $R > 1/a$ as follows:

$$R \int_0^{\pi} \frac{e^{-2\pi x R \sin\theta}\, |h(Re^{i\theta})|}{aR - 1}\, d\theta \le \frac{MR}{aR - 1} \int_0^{\pi} e^{-2\pi(x-B)R\sin\theta} d\theta \to 0 \quad \text{for} \quad R \to \infty$$

by dominated convergence. Consequently, due to (5.12), the inverse Fourier transform of the $L^2(\mathbb{R})$-function $\xi \mapsto h(\xi)/(1 - ia\xi)$ vanishes for $x > B$. For the sequence $a_n = 1/n \to 0$ we have that $1/(1 - ia_n\xi) - 1 = i\xi/(n - i\xi)$ is of modulus less than 1 and converges pointwise to zero on $\mathbb{R}$. Consequently $h(\xi)/(1 - ia_n\xi) \to h(\xi)$ in $L^2(\mathbb{R})$. Passing to the inverse Fourier transform $\mathcal{F}^{-1}$, we obtain that $\mathcal{F}^{-1}(h(\xi)/(1 - ia_n\xi)) \to (\mathcal{F}^{-1}h)(x)$ in $L^2(\mathbb{R})$. In particular, this convergence holds

[46] Since for $\varepsilon = \pi(B - A)$, the hypothesis grants the existence of some $M_\varepsilon > 0$ with $|e^{2\pi i B(iy)} h(iy)| = e^{-2\pi By}|h(iy)| \le e^{-2\pi By} M_\varepsilon e^{(2\pi A + \varepsilon)y} = M_\varepsilon e^{\pi(A-B)y}$ for $y > 0$.

in $L^2(B,\infty)$, and we infer that $\int_B^\infty |(\mathcal{F}^{-1}h)(x)|^2 dx = 0$. Thus $\mathcal{F}^{-1}h$ vanishes a.e. on (B,∞). Using (5.11) and Fubini's theorem, for $z \in \mathbb{R}$ we get

$$h(z) = \int_{-1/2}^{1/2} \Big(\lim_{n\to\infty} \int_{-n}^{n} f(\xi) e^{-2\pi i (z+y)\xi} d\xi \Big) dy = \lim_{n\to\infty} \int_{-1/2}^{1/2} \int_{-n}^{n} f(\xi) e^{-2\pi i (z+y)\xi} d\xi dy$$

$$= \lim_{n\to\infty} \int_{-n}^{n} \int_{-1/2}^{1/2} f(\xi) e^{-2\pi i (z+y)\xi} d\xi dy = \lim_{n\to\infty} \int_{-n}^{n} f(\xi) e^{-2\pi i z\xi} \Big(\int_{-1/2}^{1/2} e^{-2\pi i y\xi} dy \Big) d\xi$$

$$= \lim_{n\to\infty} \int_{-n}^{n} f(\xi) e^{-2\pi i z\xi} \frac{\sin(\pi\xi)}{\pi\xi} d\xi = \mathcal{F}(fg),$$

where $g \in L^2(\mathbb{R})$ is given by $g(\xi) = \frac{\sin(\pi\xi)}{\pi\xi}$ for $\xi \neq 0$, while $g(0) = 1$. Since $fg \in L^1(\mathbb{R})$, we infer that $\mathcal{F}^{-1}h = fg$ (see Chapter 5). Since $(\mathcal{F}^{-1}h)(x) = 0$ and $g(x) \neq 0$ for a.e. $x > B$, we conclude that $f = 0$ a.e. on (B,∞). Similarly, $f = 0$ a.e. on $(-\infty, B)$. The proof is complete. □

One natural way to extend the above theorem is to replace "compact support" with some type of decay at infinity. For the proof of this kind of Paley–Wiener type theorems we refer to the discussion in Katznelson (1968).

Theorem 7.30 *For $f \in L^2(\mathbb{R})$, the following two conditions are equivalent:*

(i) f is the restriction to the real axis of a function F that is analytic in the strip $\{z \in \mathbb{C} : |\Im m(z)| < a\}$ and satisfying $\sup_{\{y\in\mathbb{R}:\, |y|<a\}} \Big\{ \int_{\mathbb{R}} |F(x+iy)|^2\, dx \Big\} < \infty$;

(ii) $e^{2\pi a|\xi|} \hat{f}(\xi) \in L^2(\mathbb{R})$.

Theorem 7.31 *For $f \in L^2(\mathbb{R})$, the following two conditions are equivalent:*

(i) $\hat{f}$ is the boundary value[47] of a function F that is analytic in the upper half-plane $\Delta = \{z \in \mathbb{C} : \Im m(z) > 0\}$ and satisfies $\sup_{\{\eta>0\}} \Big\{ \int_{\mathbb{R}} |F(\xi + i\eta)|^2\, d\xi \Big\} < \infty$;

(ii) $f(x) = 0$ for a.e. $x < 0$.

Higher-dimensional Paley–Wiener theorems

All three Paley–Wiener theorems from the previous subsection extend in a natural way to $n \geq 2$ dimensions, with the understanding that by an analytic function of several variables we mean[48] that the function is continuous and analytic in each of the complex variables separately. However, there are some subtle points, e.g. the higher-dimensional correspondent of $[0,\infty)$ is the notion of a cone $\Gamma_{a,\theta}$ about $a \in \mathbb{R}^n$ (with $|a| = 1$) of opening angle θ: $\Gamma_{a,\theta} = \{\xi \in \mathbb{R}^n : \sum_{k=1}^n \xi_k a_k \geq |\xi| \cos\theta\}$, and the "boundary value" has to be understood in the sense of tempered distributions – see Reed and Simon (1980b).

47 In the sense that $\lim_{\eta\downarrow 0} \int_{\mathbb{R}} |F(\xi + i\eta) - \hat{f}(\xi)|^2\, d\xi = 0$.
48 It turns out that there are several equivalent definitions.

7.1.4 Hardy spaces

The theory of Hardy spaces links the theory of analytic functions in the disc $\mathbb{D} = \{z \in \mathbb{C} : |z| < 1\}$ to the Fourier analysis of functions on the boundary $|z| = 1$. Part of the inspiration is a celebrated theorem of Fatou that a bounded analytic function f on $\mathbb{D}$ has radial (indeed nontangential[49]) boundary limits $\lim_{r\uparrow 1} f(re^{i\theta})$ for almost every $\theta \in [0, 2\pi]$. Hardy pioneered the quest of spaces of analytic functions for which such results could be obtained. A number of far-reaching interesting results were provided throughout the twentieth century. This section gives a short introduction to this deep and powerful theory.

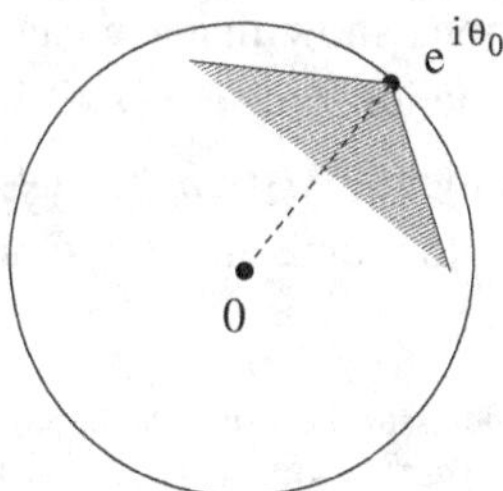

Figure 7.20 A sector with angular opening strictly less than π, symmetric with respect to the ray from the origin through $e^{i\theta_0}$.

Given $f \in L^2[0, 2\pi]$, let us expand it into its Fourier series

$$f(t) = \sum_{k\in\mathbb{Z}} \hat{f}(k)\, e^{ikt} \quad \text{in} \quad L^2[0, 2\pi]\,, \tag{7.17}$$

with coefficients $\hat{f}(k) = \frac{1}{2\pi}\int_0^{2\pi} f(s)\, e^{-iks}\, ds$ for $k \in \mathbb{Z}$. Split (7.17) into two pieces

$$f^+(t) = \sum_{k\geq 0} \hat{f}(k)\, e^{ikt}\,, \qquad f^-(t) = \sum_{k<0} \hat{f}(k)\, e^{ikt}\,. \tag{7.18}$$

Now put $z = e^{it}$ and think of it as a point on the boundary of the unit disc $\mathbb{D} = \{z \in \mathbb{C} : |z| < 1\}$, parametrised by the angle $0 \leq t < 2\pi$. Then f^+ has an analytic extension into $\mathbb{D}$,

$$h(z) = \sum_{k\geq 0} \hat{f}(k)\, z^k\,, \qquad z \in \mathbb{D}\,. \tag{7.19}$$

Indeed, for all $k \geq 0$ we have $|\hat{f}(k)| \leq \frac{1}{2\pi} \|f\|_{L^1[0,2\pi]} \leq \frac{1}{\sqrt{2\pi}} \|f\|_{L^2[0,2\pi]} := M$ by the Cauchy–Schwarz inequality in $L^2[0, 2\pi]$. Since $\lim_{k\to\infty} M^{1/k} = 1$, these

49 In the sense that $re^{i\theta}$ tends to $e^{i\theta_0}$ from within sectors that are symmetric about the radius from 0 out to $e^{i\theta_0}$, of total opening $2\alpha < \pi$ (see Figure 7.20). From the point of view of boundary regularity, the nontangential approach is much more powerful than the radial approach.

estimates ensure that the radius of convergence of the series in (7.19) is at least 1. Moreover, with the notation $h_r(t) = h(re^{it})$ for $0 < r < 1$, the orthogonality of the family $\{e^{ikt}\}_{k\geq 0}$ in $L^2[0, 2\pi]$ yields that

$$\|h_r\|^2_{L^2[0,2\pi]} = \int_0^{2\pi} |h_r(t)|^2 dt = 2\pi \sum_{k\geq 0} |\hat{f}(k)|^2 r^{2k} \tag{7.20}$$

increases with r, and

$$\sup_{0<r<1} \{\|h_r\|^2_{L^2[0,2\pi]}\} = 2\pi \sum_{k\geq 0} |\hat{f}(k)|^2 = \|f^+\|^2_{L^2[0,2\pi]} \leq \|f\|^2_{L^2[0,2\pi]},$$

taking into account the Plancherel identity and Bessel's inequality (see Section 3.2). These considerations show that we can associate with the Fourier series (7.17) an analytic function $h : \mathbb{D} \to \mathbb{C}$ with restricted growth near the boundary: the integral means $M_2(r, h) = \left(\frac{1}{2\pi} \int_0^{2\pi} |h(re^{it})|^2 \, dt\right)^{1/2}$ remain bounded as $r \uparrow 1$. Note that for some analytic functions $h : \mathbb{D} \to \mathbb{C}$ we have that $M_2(r, h) \to \infty$ as $r \uparrow 1$.

Example Analytic continuation addresses the issue whether for an analytic function $h : \mathbb{D} \to \mathbb{C}$, given, say, by a power series $\sum_{n\geq 0} a_n z^n$, there exists a domain $U \subset \mathbb{C}$, larger than $\mathbb{D}$, and an analytic function $H : U \to \mathbb{C}$ such that $H(z) = h(z)$ for all $z \in \mathbb{D}$. In general, the answer cannot be easily discerned. Indeed, $\sum_{n\geq 0} z^n$ converges in $\mathbb{D}$, having sum $h(z) = 1/(1 - z)$ which is analytic in $\mathbb{C} \setminus \{1\}$, so that h has an analytic continuation to $\mathbb{C} \setminus \{1\}$ which should be regarded as the natural domain of definition of h. On the other hand, the circle of convergence of a *lacunary power series*[50] $\sum_{n\geq 0} a_n z^{\lambda_n}$ is a natural boundary for the analytic function given by its sum; the main feature of this type of series being that the sparsity of the exponents appearing in the series induces a certain homogeneity of behaviour. For a proof of this result (known as the *Hadamard gap theorem*) we refer to Segal (2008). Here we only illustrate it by showing that $|z| = 1$ is a natural boundary of the power series $h(z) = \sum_{n\geq 1} z^{n!}$, with radius of convergence 1. For $r \in (0, 1)$ and $\theta = 2\pi p/q$ with $p, q \geq 2$ relatively prime positive integers, we write $h(re^{i\theta}) = h_1(re^{i\theta}) + h_2(re^{i\theta})$ with $h_1(z) = \sum_{n\geq 1}^{q-1} z^n$ and $h_2(z) = \sum_{n\geq q} z^{n!}$. Since h_1 is a polynomial, $h_1(re^{i\theta})$ tends to some finite limit as $r \uparrow 1$. On the other hand, the fact that $n!$ is divisible by q for $n \geq q$ yields $(e^{i\theta})^{n!} = 1$ for $n \geq q$, so that $h_2(re^{i\theta}) \uparrow \infty$ as $r \uparrow 1$. Consequently $h(re^{i\theta}) \uparrow \infty$ as $r \uparrow 1$ and so $e^{i\theta}$ is a singularity of any analytic extension H of h (otherwise the limit would exist and be finite). The claim follows since the points $\{e^{2\pi i p/q}\}$ are dense in $|z| = 1$. □

Spaces of analytic functions

For $0 < p \leq \infty$, the *Hardy spaces* $\mathcal{H}^p$ $(0 < p \leq \infty)$ are classes of analytic functions $h : \mathbb{D} \to \mathbb{C}$ whose elements satisfy certain growth conditions toward the boundary $|z| = 1$: setting

$$M_p(r, h) = \Big(\frac{1}{2\pi} \int_0^{2\pi} |h(re^{it})|^p \, dt\Big)^{1/p} \text{ for } 0 < p < \infty, \qquad M_\infty(r, h) = \max_{|z|=r} |h(z)|,$$

we say that $h \in H^p$ if $M_p(r, h)$ remains bounded as $r \uparrow 1$. In particular, $\mathcal{H}^\infty$ is the class of bounded analytic functions in $\mathbb{D}$, while relations similar to (7.20)

50 The defining property of which is that there exists $\lambda > 1$ such that $\lambda_{n+1} > \lambda\lambda_n$ for all $n \geq 0$.

show that H^2 is the class of power series $\sum_{n\geq 0} a_n z^n$ with $\sum_{n\geq 0} |a_n|^2 < \infty$. For $p \geq 1$ the Hardy space $\mathcal{H}^p$ is a normed linear space, with norm

$$||h||_{\mathcal{H}^p} = \sup_{0<r<1} \{M_p(r,h)\}\,.$$

Note that (7.20) ensures that $M_2(r,h)$ increases with r, while from the maximum modulus principle it follows that $M_\infty(r,h)$ also increases with r. Even if the proof is more involved for other values of p, $M_p(r,h)$ increases with r for any $p > 0$; see Duren (2000). It is easy to check that $\mathcal{H}^\infty \subset \mathcal{H}^q \subset \mathcal{H}^p$ for all $0 < p < q < \infty$.

The Hardy functions were defined by their mean growth toward the boundary and the following theorem (for its proof, we refer to Duren (2000)) establishes the existence and characterises the radial limits of functions in $\mathcal{H}^p$:

Theorem 7.32 *Let $h \in \mathcal{H}^p$ $(0 < p \leq \infty)$. Then the radial limit $\tilde{h}(e^{i\theta}) = \lim_{r\uparrow 1} h(re^{i\theta})$ exists for a.e. $\theta \in [-\pi,\pi)$. Moreover, for $0 < p < \infty$ we have*

$$\lim_{r\uparrow 1} \int_0^{2\pi} |h(re^{i\theta})|^p d\theta = \int_0^{2\pi} |\tilde{h}(e^{i\theta})|^p d\theta\,, \qquad \lim_{r\uparrow 1} \int_0^{2\pi} |h(re^{i\theta}) - \tilde{h}(e^{i\theta})|^p d\theta = 0\,.$$

By abuse of notation we also denote by h the a.e. radial limit[51] of $h \in \mathcal{H}^p$. As a consequence of Theorem 7.32, for $p \geq 1$ we can consider $\mathcal{H}^p$ as a closed subspace of $L^p[0,2\pi]$ and therefore $\mathcal{H}^p$ with $1 \leq p \leq \infty$ is a Banach space by itself.[52] Moreover, since $\mathcal{H}^2$ is a closed space of the Hilbert space $L^2[0,2\pi]$, it is also Hilbert space with the inner product

$$(f,g) = \lim_{r\uparrow 1} \frac{1}{2\pi} \int_0^{2\pi} f(re^{i\theta})\,\overline{g(re^{i\theta})}\,d\theta = \frac{1}{2\pi} \int_0^{2\pi} f(e^{i\theta})\,\overline{g(e^{i\theta})}\,d\theta\,, \qquad f,g \in \mathcal{H}^2\,.$$

An important observation is that, see (7.19), the coefficients of the power series representation of a function $h \in \mathcal{H}^2$ coincide with the Fourier coefficients with nonnegative index of its boundary function $h \in L^2[0,2\pi]$. The following result expresses this connection more precisely.

Theorem 7.33 *Let $h(z) = \sum_{n\geq 0} a_n z^n$ belong to $\mathcal{H}^1$, and let $\{\hat{h}(k)\}_{k\in\mathbb{Z}}$ be the Fourier coefficients of its boundary function $h(e^{it})$. Then $\hat{h}(n) = a_n$ for $n \geq 0$, and $\hat{h}(k) = 0$ for $k < 0$. Furthermore, for $1 \leq p \leq \infty$ the class $\mathcal{H}^p$ is exactly the class of $L^p[0,2\pi]$-functions whose Fourier coefficients vanish for all negative indices.*

[51] Actually, the nontangential limit exists a.e.; see Duren (2000).

[52] If $0 < p < 1$ then the space $\mathcal{H}^p$ is not normable but is a complete metric space with the metric $d(f,g) = \left(\frac{1}{2\pi}\int_0^{2\pi} |\tilde{h}(e^{i\theta})|^p d\theta\right)^{1/p}$; see the discussion in Duren (2000).

Proof We present the proof of the first part, referring to Duren (2000) for the second part. For $h \in \mathcal{H}^1$, the Fourier coefficients of its boundary function $h \in L^1[0, 2\pi]$ are given by

$$\hat{h}(k) = \frac{1}{2\pi}\int_0^{2\pi} h(e^{it})\, e^{-ikt} dt = \lim_{r \uparrow 1} \frac{1}{2\pi}\int_0^{2\pi} h(re^{it})\, e^{-ikt} dt,$$

in view of Theorem 7.32. But for $0 < r < 1$ we have

$$\frac{1}{2\pi}\int_0^{2\pi} h(re^{it})\, e^{-ikt} dt = \frac{r^{k-1}}{2\pi i}\int_{|z|=r} h(z) z^{-k-1} dz = \begin{cases} 0 & \text{if} \quad k \le -1, \\ a_n\, r^{n-1} & \text{if} \quad k = n \ge 0, \end{cases}$$

taking advantage of Theorem 7.5. The first claim is obtained by letting $r \uparrow 1$. □

We conclude our discussion by presenting a procedure that allows one to operate within the setting of $\mathcal{H}^p$-spaces with zero-free functions which, due to the absence of zeros, can be raised to arbitrary powers and therefore permit us to move from one $\mathcal{H}^p$-space to a more convenient one, say $\mathcal{H}^1$. We saw in Section 7.1 that the zeros of an analytic function $h : \mathbb{D} \to \mathbb{C}$ cannot cluster inside $\mathbb{D}$ unless the function vanishes identically. In general, all we can say about the zeros of an analytic function $h \not\equiv 0$ is that they are isolated and their limit points lie on $|z| = 1$: given any such sequence $\{z_k\}_{k \ge 1}$ in $\mathbb{D}$ and any sequence $\{n_k\}_{k \ge 1}$ of positive integers, one can show the existence of an analytic function $h : \mathbb{D} \to \mathbb{C}$ that has zeros of orders n_k at z_k and no others; this is Weierstrass' theorem (see Burckel (1979)). However, if $h \in \mathcal{H}^p$ for some $0 < p \le \infty$, the inherent growth restriction toward the boundary of $\mathbb{D}$ forces the zeros to converge rapidly toward the limit points on $|z| = 1$. More precisely, the zeros $\{z_k\}_{k \ge 1}$ of such a function must satisfy $\sum_{k \ge 1}(1 - |z_k|) < \infty$; for a proof we refer to Duren (2000). Nothing more can be said about the zeros. In fact, given any sequence $\{z_k\}_{k \ge 1}$ in $\mathbb{D}$ such that $\sum_{k \ge 1}(1 - |z_k|) < \infty$, there exists $h \in \mathcal{H}^\infty$ whose zeros are precisely $\{z_k\}_{k \ge 1}$. While we omit the details of the proof (for this, see Duren (2000)), let us provide the idea behind it. Since the functions $b_k(z) = (z - z_k)/(1 - \overline{z_k}\, z)$ represent analytical bijections of $\mathbb{D}$ onto itself, with a simple zero at $z = z_k$, it is natural to consider an infinite product of such factors. Indeed, if $m \ge 0$ is the number of occurences of 0 in the sequence, the infinite product (viewed as a limit of the finite products)

$$B(z) = z^m \prod_{k \ge 1:\ z_k \neq 0} \frac{|z_k|}{z_k}\, \frac{z - z_k}{1 - \overline{z_k}\, z}, \tag{7.21}$$

converges uniformly in each disc $|z| \le R < 1$, each z_k is a zero of B with multiplicity equal to the number of times it occurs in the sequence, and B has no other zeros in $\mathbb{D}$. Moreover, $|B(z)| < 1$ in $\mathbb{D}$ and $|B(e^{i\theta})| = 1$ a.e. on $[-\pi, \pi]$. A function of the form (7.21) is called a *Blaschke product*, and we permit the set $\{z_k\}$ to be finite, or even empty, in which cases $B(z)$ is a polynomial in z. Given a function $f \in \mathcal{H}^p$ with $p > 0$, we can form the Blaschke product $B \in \mathcal{H}^\infty$ with the zeros of f and try to divide out the zeros of f by considering the quotient $F = f/B$. The resulting analytic function $F : \mathbb{D} \to \mathbb{C}$ actually belongs to $\mathcal{H}^p$ and $\|F\|_{\mathcal{H}^p} = \|f\|_{\mathcal{H}^p}$, and $f = FB$ is called the *canonical factorisation* of f; see the discussion in Katznelson (1968). This key result is technically very useful in the analysis of Hardy spaces.

Example Every function $f \in \mathcal{H}^1$ can be written as a product $f = f_1 f_2$ with $f_1, f_2 \in \mathcal{H}^2$: let $f = BF$ be the canonical factorisation of f and set $f_1 = F^{1/2}$, $f_2 = BF^{1/2}$. □

Spaces of harmonic functions

Let $u(z)$ be real-valued and *harmonic* in $|z| < R$ for some $R > 0$, meaning that u is twice continuously differentiable and satisfies $u_{xx} + u_{yy} = 0$; we write throughout $z = x + iy$. Then the real-valued function

$$v(x,y) = -\int_0^x u_y(s,0)\,ds + \int_0^y u_x(x,l)\,dl\,, \qquad |z| < R\,, \tag{7.22}$$

called the (normalised) *harmonic conjugate* of u, has the property that $v(0) = 0$ and $F(z) = u(z) + iv(z)$ is analytic in $|z| < R$, since the Cauchy–Riemann equations hold there. In particular, we find that u and v are smooth (infinitely differentiable) in $|z| < R$. Conversely, the Cauchy–Riemann equations show that the real and imaginary parts of an analytic function are always harmonic. Some other properties of harmonic functions are inherited from complex analysis.

- If u vanishes in an open subset U of the disc $|z| < R$, then u vanishes identically.[53]

  However, note that, in contrast to the zeros of analytic functions, the zeros of harmonic functions are not isolated. One can show that if $u(z_0) = 0$ for some $z_0 \in \mathbb{C}$ with $|z_0| < R$, then, near z_0, the level set $u^{-1}(0)$ consists of $n \geq 1$ analytic curves which, for $n \geq 2$, intersect at z_0 at equal angles $2\pi/n$. Indeed, if $n \geq 1$ is the order of the zero of $z \mapsto F(z) - F(z_0)$ at z_0, then, see Theorem 7.12, there is an open ball V_0, centred at z_0 and contained within the disc $|z| < R$, such that $F(z) - F(z_0) = [\phi(z)]^n$ in V_0, for some analytic bijection ϕ from V_0 to some open ball $|z| < r$ with $\phi(z_0) = 0$. Note that $u(z) = 0$ for $z \in V_0 \setminus \{z_0\}$ is equivalent to $\phi(z) \in \bigcup_{k=0}^{n-1}\{se^{2\pi ik/n} : s \in \mathbb{R},\ s \neq 0\}$ (the union of $2n$ rays out of the origin).

- u has the *mean value property*: $u(z_0) = \frac{1}{2\pi}\int_0^{2\pi} u(z_0 + re^{i\theta})\,d\theta$ for $|z_0| + r < R$.

  This follows by taking the real parts in the mean value property for the analytic function F.

- **The maximum principle** *A nonconstant function u that is harmonic in $B_R = \{z \in \mathbb{C} : |z| < R\}$ and continuous on the closure $\overline{U}$ of a domain $U \subset B_R$ can attain* $\max_{z\in\overline{U}}\{u(z)\}$ *only on the boundary ∂U of U.*

  Indeed, if the maximum M of u on $\overline{U}$ is attained at some $z_0 \in U$, then the continuity of u in U and the mean value property ensure that $u(z) = M$ on all circles $|z - z_0| = r$ contained in U. Hence $f = M$ on an open neighbourhood of z_0. Therefore, the nonempty set $\{z \in U : f(z) = M\}$ is open and closed in U. Since U is connected, we must have $f = M$ throughout U.

[53] This follows from Theorem 7.11 since v also vanishes in U, so that $F \equiv 0$ in U.

In general, a harmonic function and its conjugate behave alike (with respect to smoothness and growth properties), but there are some rather surprising exceptions. We will attempt to shed some light on both aspects, which are of relevance to trigonometric series. In our discussion we rely on some representation formulas.

If $F(z) = \sum_{n\geq 0} a_n z^n$ is the power series development of F for $|z| < R$, writing $z = re^{i\theta}$ with $-\pi \leq \theta < \pi$ and $0 \leq r < R$, since $\Re\{a_n e^{in\theta}\} = [a_n\, e^{in\theta} + \overline{a_n}\, e^{-in\theta}]/2$, we get

$$u(re^{i\theta}) = \Re\{F(z)\} = \sum_{k\in\mathbb{Z}} A_k r^{|k|} e^{ik\theta}$$

with $A_k = \overline{a_{|k|}}/2$ if $k < 0$, $A_0 = \Re\{a_0\}$, $A_k = a_k/2$ if $k > 0$. The above formula for u can be put in closed form. Indeed, if $R > 1$, evaluating the derived formula on $|z| = 1$ yields

$$u(re^{i\theta}) = \frac{1}{2\pi}\int_{-\pi}^{\pi} u(e^{it}) \sum_{k\in\mathbb{Z}} r^{|k|} e^{ik(\theta-t)}\, dt\,, \qquad 0 \leq r < 1\,,$$

since $\int_{-\pi}^{\pi} e^{iqt} e^{ik(\theta-t)}\, dt = 2\pi e^{ik\theta}\delta_{qk}$ for all $k,\, q \in \mathbb{Z}$. Summing the two geometric series (for $k \geq 0$ and $k < 0$) we get

$$\sum_{k\in\mathbb{Z}} r^{|k|} e^{ik\phi} = \frac{1}{1 - re^{i\phi}} + \frac{re^{-i\phi}}{1 - re^{-i\phi}} = \frac{1 - r^2}{1 - 2r\cos\phi + r^2}\,, \qquad 0 \leq r < 1\,.$$

The function

$$P_r(\phi) = \frac{1 - r^2}{1 - 2r\cos\phi + r^2}\,, \qquad 0 \leq r < 1\,, \quad -\pi \leq \phi < \pi\,, \tag{7.23}$$

is called the *Poisson kernel* for $|z| < 1$. We have therefore derived *Poisson's representation*: if u is harmonic for $|z| < R$ with $R > 1$, then

$$u(re^{i\theta}) = \frac{1}{2\pi}\int_{-\pi}^{\pi} \frac{(1 - r^2)\, u(e^{it})}{1 - 2r\cos(\theta - t) + r^2}\, dt\,, \qquad 0 \leq r < 1\,. \tag{7.24}$$

On the other hand, we have

$$v(re^{i\theta}) = \Im\mathfrak{m}\{F(z)\} = -\sum_{k\in\mathbb{Z}} i\,\mathrm{sgn}(k)\, A_k r^{|k|} e^{ik\theta}\,,$$

with $\mathrm{sgn}(0) = 0$. Since by direct summation of two geometric series (for $k > 0$ and $k < 0$) we find the *conjugate Poisson kernel*

$$Q_r(\phi) = \frac{2r\sin\phi}{1 - 2r\cos\phi + r^2} = -\sum_{k\in\mathbb{Z}} i\,\mathrm{sgn}(k)\, r^{|k|} e^{ik\theta}\,, \qquad 0 \leq r < 1\,, \tag{7.25}$$

the following formula emerges:

$$v(re^{i\theta}) = \frac{1}{2\pi}\int_{-\pi}^{\pi} \frac{2r\sin(\theta - t)\,u(e^{it})}{1 - 2r\cos(\theta - t) + r^2}\,dt\,, \qquad 0 \le r < 1\,. \tag{7.26}$$

The Poisson kernel $P_r(\phi)$ has the following properties:

(a) $P_r(\phi) > 0$ for $0 < r < 1$ and $-\pi \le \phi < \pi$;

(b) for each $0 < r < 1$, $\int_{-\pi}^{\pi} P_r(\phi)\,d\phi = 2\pi$;

(c) $P_r(\phi) \to 0$ uniformly for $\delta \le |\theta| \le \pi$ as $r \uparrow 1$, for any $\delta > 0$;

(d) $P_r(\phi)$ is harmonic in $|z| < 1$ with normalised harmonic conjugate $Q_r(\phi)$.

Items (a) and (c) are immediate from the formula (7.23), while (b) follows from the series representation for $P_r(\phi)$. As for (d), note that $P_r(\phi) = 1 + \sum_{n\ge1}(z^n + \overline{z}^n)$ and $Q_r(\phi) = i\sum_{n\ge1}(\overline{z}^n - z^n)$ with $z = re^{i\theta}$, so that $P_r(\phi) + iQ_r(\phi) = 1 + 2\sum_{n\ge1} z^n = (1+z)/(1-z)$ for $|z| < 1$.

It is remarkable that some version of the Poisson representation (7.24) holds under more general conditions than the one used in the previous considerations, of restricted growth type near the boundary. For harmonic functions $u : \mathbb{D} \to \mathbb{R}$, define for $0 < r < 1$ and $0 < p \le \infty$ the integral means $M_p(r,u)$ analogous to the classes $\mathcal{H}^p$ of analytic functions in $\mathbb{D}$; we say that u is in the *Hardy class* $\mathfrak{h}^p$ if $M_p(r,u)$ is bounded as $r \uparrow 1$. Since[54] $a^p \le (a+b)^p \le 2^{p-1}(a^p + b^p)$ for $a, b \ge 0$ and $0 < p < \infty$, we see that an analytic function F belongs to $\mathcal{H}^p$ if and only if the harmonic functions given by its real and imaginary parts are both in $\mathfrak{h}^p$. Despite these strong links between the analytic Hardy spaces $\mathcal{H}^p$ and the harmonic Hardy spaces $\mathfrak{h}^p$, there are some subtle differences[55]. To explore the similarities, it is instructive to consider some $u \in \mathfrak{h}^2$. If v is the (normalised) harmonic conjugate, defined for $|z| < 1$ by (7.22), and $u(z) + iv(z) = a_0 + \sum_{n\ge1}(a_n + ib_n)z^n$ with $a_n, b_n \in \mathbb{R}$, then

$$\begin{aligned} u(re^{i\theta}) &= a_0 + \sum_{n\ge1} r^n[a_n\cos(n\theta) - b_n\sin(n\theta)]\,,\\ v(re^{i\theta}) &= \sum_{n\ge1} r^n[a_n\sin(n\theta) + b_n\cos(n\theta)]\,, \end{aligned} \qquad \text{for } 0 < r < 1,\ -\pi \le \theta < \pi,$$

the orthonormality of the sequence $\{e^{ikt}\}_{k\in\mathbb{Z}}$ in $L^2[-\pi,\pi]$ yields

$$[M_2(r,u)]^2 = a_0^2 + \frac{1}{2}\sum_{n\ge1} r^{2n}[a_n^2 + b_n^2]\,, \qquad [M_2(r,v)]^2 = \frac{1}{2}\sum_{n\ge1} r^{2n}[a_n^2 + b_n^2]\,.$$

Hence $M_2(r,v) \le M_2(r,u)$, with equality if and only if $a_0 = 0$. It turns out that this feature is replicated for any $1 < p < \infty$ (for a proof, see Duren (2000)).

[54] This inequality follows from the fact that the map $s \mapsto s^p + (1-s)^p$ is decreasing on $(0,1/2)$ and increasing on $(1/2,1)$, so that $[a/(a+b)]^p + [b/(a+b)]^p \ge 2^{1-p}$ for $a, b > 0$.

[55] For example, see the considerations in Section 4.1 of Duren (2000), the function $r \mapsto M_p(r,u)$ is nondecreasing for $u \in \mathfrak{h}^p$ with $p \ge 1$, while, in contrast to the case of $\mathcal{H}^p$-spaces, for $u > 0$ in $\mathfrak{h}^p$ with $p \in (0,1)$, the function $r \mapsto M_p(r,u)$ is strictly decreasing.

Theorem 7.34 *For $1 < p < \infty$ there exists a constant $C_p > 0$ such that $M_p(r, v) \leq C_p M_p(r, u)$ for all $u \in \mathfrak{h}^p$. In particular, if $u \in \mathfrak{h}^p$, then its harmonic conjugate v is also of class $\mathfrak{h}^p$.*

Theorem 7.34 breaks down for $p = 1$ and for $p = \infty$. In the latter case, since $\mathfrak{h}^\infty \subset \mathfrak{h}^p$ for any $p > 0$, the result provides nevertheless useful information about the harmonic conjugate. As for the case $p = 1$, it turns out that the harmonic conjugate of $u \in \mathfrak{h}^1$ belongs to $\mathfrak{h}^q$ for all $q \in (0, 1)$; see Duren (2000).

Examples (i) The listed properties of the Poisson kernel $P_r(\phi)$ show that it belongs to $\mathfrak{h}^1$, but its analytic completion $(1+z)/(1-z) \notin \mathcal{H}^1$ since $(1+e^{i\theta})/(1-e^{i\theta}) = i\cot(\theta/2) \notin L^1[-\pi, \pi]$.

(ii) The map $f(z) = (1+z)/(1-z)$ is analytic in $\mathbb{D}$, with real part $P_r(\phi) > 0$, so that its values lie in the right half-plane and therefore $F(z) = i\log(f(z))$ is analytic in $\mathbb{D}$ and has bounded real part $u \in \mathfrak{h}^\infty$. Since $\log((1+r)/(1-r)) \uparrow \infty$ for $r \uparrow 1$, we see that the imaginary part of F is unbounded in $\mathbb{D}$, so that the harmonic conjugate of u does not belong to $\mathfrak{h}^\infty$. However, Theorem 7.34 shows that $F \in \mathcal{H}^p$ for all $0 < p < \infty$, a fact not easy to check by direct calculation. □

A consequence of Theorem 7.32 and Theorem 7.34 is the fact that every $u \in \mathfrak{h}^p$ with $1 \leq p \leq \infty$ has radial (actually, nontangential) limits a.e. toward the boundary $|z| = 1$ of $\mathbb{D}$, and its harmonic conjugate v shares this property. The case $p < 1$ stands apart:[56] there are examples of harmonic functions u belonging to all classes $\mathfrak{h}^p$ with $p \in (0, 1)$, yet the harmonic conjugate v has a radial limit almost nowhere, so that $v \notin \mathfrak{h}^q$ for any $q \in (0, 1)$; see Duren (2000). As anticipated above, we will now prove a representation formula simiar to (7.24) for $u \in \mathfrak{h}^p$ with $p > 1$, which permits the recovery of $u \in \mathfrak{h}^p$ from its boundary values on $|z| = 1$.

Theorem 7.35 *The Poisson representation formula (7.24) holds for $u \in \mathfrak{h}^p$ with $1 < p \leq \infty$, where $u(e^{i\theta})$ is the (nontangential) boundary value of u.*

Proof For a fixed $r \in (0, 1)$, we have by (7.24) that

$$u(\rho re^{i\theta}) = \frac{1}{2\pi}\int_{-\pi}^{\pi} P_r(\theta - t)\, u(\rho e^{it})\, dt\,, \qquad 0 \leq \rho < 1\,, \quad -\pi \leq \theta < \pi\,.$$

Theorem 7.34 ensures that the harmonic conjugate v of u belongs to $\mathfrak{h}^1$, so that $F = u + iv \in \mathcal{H}^1$ and Theorem 7.32 yields $\int_{-\pi}^{\pi} |u(\rho e^{it}) - u(e^{it})|\, dt \to 0$ as $\rho \uparrow 1$. The properties of the convolution – see relation (4.6) in Chapter 4 – show now that, as $\rho \uparrow 1$, the right side of the displayed relation converges to $\frac{1}{2\pi}\int_{-\pi}^{\pi} P_r(\theta - t)\, u(e^{it})\, dt$ in $L^1[-\pi, \pi]$. But, since u is smooth in $\mathbb{D}$, the left side of the displayed relation converges to $u(re^{i\theta})$ in $L^1[-\pi, \pi]$. Consequently, (7.24) holds for a.e. $\theta \in [-\pi, \pi]$. The right side being continuous (due to the continuity of P_r), (7.24) must hold for all $\theta \in [-\pi, \pi]$. Since $r \in (0, 1)$ was arbitrary, the proof is complete. □

The proof of Theorem 7.35 is not suitable for $p = 1$ since we might have $F \notin \mathcal{H}^1$. In the case of the Poisson kernel $u(re^{i\theta}) = P_r(\theta)$ with $0 \leq r < 1$ and

56 Note also that $u(z) = \sin\left(\frac{1}{1-|z|}\right)$ is smooth and bounded for $|z| < 1$ but fails to have a radial limit at all points of the boundary $|z| = 1$.

$-\pi \le \theta < \pi$, see (7.23), we have $\lim_{r\uparrow 1} u(re^{i\theta}) = 0$ save for $\theta = 0$, so that (7.24) fails. Instead, we have the following result.

Theorem 7.36 *If $u \in \mathfrak{h}^1$, there is a finite signed measure μ on $[-\pi, \pi]$ with*

$$u(re^{i\theta}) = \frac{1}{2\pi}\int_{-\pi}^{\pi} P_r(\theta - t)\, d\mu(t)\,, \qquad 0 \le r < 1\,, \quad -\pi \le \theta < \pi\,.$$

Proof While $L^1[-\pi,\pi]$ is not the dual of any Banach space (see Chapter 3), the space $\mathcal{M}[-\pi,\pi]$ of finite signed measures μ on $[-\pi,\pi]$, with the norm $\|\mu\|$ given by the total variation of μ, is the dual of the Banach space $C[-\pi,\pi]$ of continuous real-valued functions on $[-\pi,\pi]$ with the supremum norm. To any real-valued $f \in L^1[-\pi,\pi]$ we can associate a measure $\mu_f \in \mathcal{M}[-\pi,\pi]$ with $\|\mu_f\| = \|f\|_{L^1[-\pi,\pi]}$ by setting $\int_{-\pi}^{\pi} g(t)\, dt = \int_{-\pi}^{\pi} g(t) f(t)\, dt$ for real-valued $g \in C[-\pi,\pi]$. Since $u \in \mathfrak{h}^1$, there exists a constant $C > 0$ such that the functions $u_n(\theta) = u((1-1/n)e^{i\theta})$ satisfy $\|u_n\|_{L^1[-\pi,\pi]} \le C$ for all $n \ge 1$. By Theorem 3.20 from Chapter 3, there is a subsequence $\{u_{n_k}\}$ and some $\mu \in \mathcal{M}[-\pi,\pi]$ with $\lim_{n_k\to\infty} \int_{-\pi}^{\pi} g(\theta) u_{n_k}(\theta)\, d\theta = \int_{-\pi}^{\pi} g(\theta)\, d\mu(\theta)$ for all real-valued $g \in C[-\pi,\pi]$. But $u_{n_k}(z)$ is harmonic for $|z| < 1/(1-1/n)$, so that, by (7.24),

$$u_{n_k}(re^{i\theta}) = \frac{1}{2\pi}\int_{-\pi}^{\pi} P_r(\theta - t) u_{n_k}(e^{it})\, dt\,, \qquad 0 \le r < 1\,, \quad -\pi \le \theta < \pi\,.$$

Since $P_r \in C[-\pi,\pi]$ for any fixed $r < 1$, the right side converges to $\frac{1}{2\pi}\int_{-\pi}^{\pi} P_r(\theta - t)\, d\mu(t)$ for $n_k \to \infty$. Since $u_{n_k}(re^{i\theta}) = u((1-1/n_k)re^{i\theta})$ and u is smooth in $\mathbb{D}$, the right side converges to $u(re^{i\theta})$, thus proving the claim. □

Example For the Poisson kernel $u(re^{i\theta}) = P_r(\theta)$ with $0 \le r < 1$ and $-\pi \le \theta < \pi$, Theorem 7.36 holds for the measure $\mu = \delta_0$ (the unit point mass at 0). In contrast to this, the approach used in the proof of Theorem 7.35 shows that $\mu(t) = u(e^{it})\, dt$ if $(u + iv) \in \mathcal{H}^1$. One can show (see Katznelson (1968)) that, given $u \in \mathfrak{h}^1$, if the boundary values on $|z| = 1$ of the harmonic conjugate v are in $L^1[-\pi,\pi]$, then $(u + iv) \in \mathcal{H}^1$. □

The preceding results permit us to elucidate the issue of *conjugate Fourier series* mentioned in the introduction to Chapter 4; for convenience, we use here the basic periodicity interval $[-\pi,\pi]$ instead of $[0,1]$ and we restrict our attention to real-valued integrable functions. Recall that a formal trigonometric series

$$\sum_{k\in\mathbb{Z}} A_k e^{ik\theta}\,, \tag{7.27}$$

is called a Fourier series if there exists some real-valued $\varphi \in L^1[-\pi,\pi]$ such that

$$A_k = \frac{1}{2\pi}\int_{-\pi}^{\pi} \varphi(t)\, e^{-ikt}\, dt \quad \text{for all} \quad k \in \mathbb{Z}\,. \tag{7.28}$$

The formal series

$$\sum_{k\in\mathbb{Z}} -i\,\mathrm{sgn}(k)\, A_k e^{ik\theta}\,, \tag{7.29}$$

is called the *conjugate trigonometric series* of (7.27). Since φ is real-valued, setting

$$a_n = \frac{1}{\pi}\int_{-\pi}^{\pi} \varphi(t)\cos(nt)\,dt\,, \quad b_n = \frac{1}{\pi}\int_{-\pi}^{\pi} \varphi(t)\sin(nt)\,dt\,, \qquad n \geq 0\,,$$

we find that $A_0 = a_0/2$ while $A_n = (a_n + ib_n)/2$ and $A_{-n} = (a_n - ib_n)/2$ for $n \geq 1$. Consequently, (7.27) and (7.29) become

$$\frac{a_0}{2} + \sum_{n\geq 1} [a_n \cos(n\theta) - b_n \sin(n\theta)] \quad \text{and} \quad \sum_{n\geq 1} [b_n \cos(n\theta) + a_n \sin(n\theta)]\,, \tag{7.30}$$

respectively. The nomenclature *conjugate* is easily justified. If (7.27) is the Fourier series (7.28) of some real-valued $\varphi \in L^1[-\pi,\pi]$, consider the Poisson integral

$$u(z) = \frac{1}{2\pi}\int_{-\pi}^{\pi} P_r(\theta - t)\varphi(t)\,dt \quad \text{for} \quad z = re^{i\theta} \quad \text{with} \quad r > 1\,.$$

Then u is smooth and for the Laplace operator $\Delta = \partial_x^2 + \partial_y^2$ we have $\Delta u = \frac{1}{2\pi}\int_{-\pi}^{\pi} \Delta P_r(\theta - t)\varphi(t)\,dt = 0$ since the Poisson kernel $u_0(z) = P_r(\theta)$ is harmonic in $\mathbb{D}$. Moreover, by the properties of the convolution, see relation (4.6) in Chapter 4, $\|u(re^{i\theta})\|_{L^1[-\pi,\pi]} \leq \|P_r\|_{L^1[-\pi,\pi]}\|\varphi\|_{L^1[-\pi,\pi]} = \|\varphi\|_{L^1[-\pi,\pi]}$, so that $u \in \mathfrak{h}^1$. Since, as discussed above, the analytic completion of the harmonic function $u_0(z) = P_r(\theta)$ is $z \mapsto (1+z)/(1-z)$, the analytic completion of u is easily obtained as

$$F(z) = u(z) + iv(z) = \frac{1}{2\pi}\int_{-\pi}^{\pi} \frac{1 + ze^{-it}}{1 - ze^{-it}}\,\varphi(t)\,dt \quad \text{for} \quad |z| < 1\,. \tag{7.31}$$

The expansion

$$\frac{1 + ze^{-it}}{1 - ze^{-it}} = 1 + \frac{2ze^{-it}}{1 - ze^{-it}} = 1 + 2\sum_{n\geq 1} e^{-int} z^n \quad \text{for} \quad |z| < 1\,,$$

and the relations (7.28) now give $F(z) = A_0 + 2\sum_{n\geq 1} A_n z^n$. Setting $z = re^{i\theta}$ and taking real and imaginary parts, for $0 \leq r < 1$ and $-\pi \leq \theta < \pi$, we find

$$\begin{cases} u(z) = \dfrac{a_0}{2} + \displaystyle\sum_{n\geq 1} [a_n \cos(n\theta) - b_n \sin(n\theta)]\, r^n\,, \\ v(z) = \displaystyle\sum_{n\geq 1} [b_n \cos(n\theta) + a_n \sin(n\theta)]\, r^n\,, \end{cases} \quad |z| < 1\,. \tag{7.32}$$

Since $u \in \mathfrak{h}^1$, see the comments after Theorem 7.34, $v \in \mathfrak{h}^q$ for any $0 < q < 1$. Consequently $F \in \mathcal{H}^q$ so that, by Theorem 7.32, $u(z)$ and $v(z)$ have a radial limit a.e., with $\varphi(\theta) = \lim_{r\uparrow 1} u(re^{i\theta})$ for a.e. $\theta \in [-\pi,\pi]$.

To prove the above claim, let us first show that if $\phi \in L^1[-\pi,\pi]$, then $\lim_{r\uparrow 1}(P_r * \phi) = 2\pi\phi$ in $L^1[-\pi,\pi]$. Indeed, for $\phi \in L^1[-\pi,\pi]$ we have $\|P_r * \phi\|_{L^1[-\pi,\pi]} \le \|P_r\|_{L^1[-\pi,\pi]}\|\phi\|_{L^1[-\pi,\pi]} = 2\pi\|\phi\|_{L^1[-\pi,\pi]}$. Given $\varepsilon > 0$, choose (see Chapter 5) some continuous $\Phi : [-\pi,\pi] \to \mathbb{C}$ such that $\Phi(-\pi) = \Phi(\pi)$ and $\|\phi - \Phi\|_{L^1[-\pi,\pi]} < \varepsilon/(5\pi)$. Then $\|P_r * \phi - 2\pi\phi\|_{L^1[-\pi,\pi]} = \|P_r * (\phi - \Phi) + 2\pi(\Phi - \phi) + (P_r * \Phi - 2\pi\Phi)\|_{L^1[-\pi,\pi]} \le 4\pi\|\phi - \Phi\|_{L^1[-\pi,\pi]} + \|P_r * \Phi - 2\pi\Phi\|_{L^1[-\pi,\pi]} < \varepsilon$ if $r < 1$ is near 1 since Exercise 12 in Chapter 4 shows that $P_r * \Phi \to 2\pi\Phi$ uniformly on $[-\pi,\pi]$ as $r \uparrow 1$. This proves the preliminary claim. For the initial claim, note that $2\pi\, u(re^{i\theta}) = (P_r * \varphi)(\theta) \to 2\pi\, \varphi(\theta)$ in $L^1[-\pi,\pi]$ forces $\varphi(\theta) = u(e^{i\theta})$ for a.e. $\theta \in [-\pi,\pi]$ since $\lim_{n\to\infty} u((1-1/n)e^{i\theta}) = u(e^{i\theta})$ a.e. on $[-\pi,\pi]$ and, by Fatou's lemma (see Chapter 2), $\int_{-\pi}^{\pi} \liminf_{n\to\infty} |u((1-1/n)e^{i\theta}) - \varphi(\theta)|\, d\theta = 0$.

The function

$$\tilde{\varphi}(\theta) = \lim_{r\uparrow 1} v(re^{i\theta})$$

is called the *conjugate function* of $\varphi(\theta)$. Do we have $\tilde{\varphi} \in L^1[-\pi,\pi]$? If so, then, see the comments in the example after Theorem 7.36, $F = u + iv \in \mathcal{H}^1$ and Theorem 7.32 yields $v(re^{i\theta}) \to \tilde{\varphi}(\theta)$ in $L^1[-\pi,\pi]$. From (7.32) we get $\frac{1}{\pi}\int_{-\pi}^{\pi} \tilde{\varphi}(\theta)\cos(n\theta)\, d\theta = b_n = \lim_{r\uparrow 1} \frac{1}{\pi}\int_{-\pi}^{\pi} v(re^{i\theta})\cos(n\theta)\, d\theta$ etc., so that (7.29) is the Fourier series of $\tilde{\varphi} \in L^1[-\pi,\pi]$. Conversely, if (7.29) is the Fourier series of some $f \in L^1[-\pi,\pi]$, then we go through the above stages, starting with $\tilde{v} = \frac{1}{2\pi} P_r * f \in \mathfrak{h}^1$ instead of u, considering its analytic completion $\tilde{v} + i\tilde{u}$, which leads to the conclusion that $\lim_{r\uparrow 1} \tilde{v}(re^{i\theta}) = f(\theta)$ for a.e. $\theta \in [-\pi,\pi]$, and also to a formula similar to (7.32). By inspection of the coefficients of this formula and of (7.32) we see that $\tilde{v}(z) = v(z)$ and $\tilde{u}(z) = a_0/2 - u(z)$ for $|z| < 1$. Since $\lim_{r\uparrow 1} v(re^{i\theta}) = \tilde{\varphi}(\theta)$ for a.e. $\theta \in [-\pi,\pi]$, we infer that $f = \tilde{\varphi}$. These considerations show that $\tilde{\varphi} \in L^1[-\pi,\pi]$ if and only if (7.29) is the Fourier series of some integrable function.[57]

Example After shifting from $[0,1]$ to $[-\pi,\pi]$, Exercise 21 in Chapter 4 shows that $\sum_{n\ge 2} \frac{\cos(nt)}{\ln(n)}$ is the Fourier series of some $\varphi \in L^1[-\pi,\pi]$, while the conjugate series $\sum_{n\ge 2} \frac{\sin(nt)}{\ln(n)}$ converges at every $t \in [-\pi,\pi]$ without being the Fourier series of some integrable function. Thus the analytic completion $F(z) = \sum_{n\ge 2} \frac{z^n}{\ln(n)} \notin \mathcal{H}^1$ and $\sum_{n\ge 2} \frac{\sin(nt)}{\ln(n)} \notin L^1[-\pi,\pi]$, a conclusion that cannot be deduced from the mere fact that the series is not a Fourier series. □

Hardy spaces for the upper half-plane

In Section 7.1, after the statement of the Riemann mapping theorem, we showed that $z \mapsto (i-z)/(i+z)$ is a conformal map from the upper half-plane $\mathbb{H}$ onto $\mathbb{D}$. We therefore expect a rich theory dealing with the boundary behaviour (toward the real line $\mathbb{R}$) of functions that are analytic in $\mathbb{H}$.

For $p \in (0,\infty)$, the *Hardy space* $\mathcal{H}^p(\mathbb{H})$ is the family of all analytic functions $F : \mathbb{H} \to \mathbb{C}$ such that $\sup_{\{y>0\}} \int_{\mathbb{R}} |F(x+iy)|^p dx < \infty$, while $\mathcal{H}^\infty(\mathbb{H})$ is the set of

[57] At the heart of the matter is the fact that if the boundary value f of $F \in \mathcal{H}^p$ (with $p < 1$) belongs to $L^1[-\pi,\pi]$, then $F \in \mathcal{H}^1$ and F is the Poisson integral of f, $F = \frac{1}{2\pi} P_r * f$; see Katznelson (1968). The Poisson kernel links the theory of trigonometric series with the theory of analytic functions in $\mathbb{D}$.

functions analytic and bounded in $\mathbb{H}$. Many of the features highlighted for the real and complex Hardy spaces of the disc $\mathbb{D}$ transfer at once to $\mathbb{H}$ by way of the conformal map. We illustrate this by the following result.

Theorem 7.37 *Let $f : \mathbb{H} \to \mathbb{R}$ be harmonic and bounded. Then $\lim_{y\downarrow 0} f(x + iy) = f_0(x)$ exists for a.e. $x \in \mathbb{R}$ and $f(x+iy) = \frac{1}{\pi}\int_{\mathbb{R}} \frac{y}{(x-s)^2+y^2} f_0(s)\,ds$ for $y > 0$.*

Proof Let $u(w) = f(z)$ with the correspondence $w = (i - z)/(i + z)$ between $z \in \mathbb{H}$ and $w \in \mathbb{D}$. Then $u \in \mathfrak{h}^\infty \subset \mathfrak{h}^2$ so that, by Theorem 7.34 and Theorem 7.35, for some $u_0 \in L^2[-\pi,\pi]$ we have

$$u(w) = \frac{1}{2\pi}\int_{-\pi}^{\pi} \frac{1-r^2}{1-2r\cos(\theta-t)+r^2}\, u_0(t)\,dt \quad \text{for} \quad w = re^{i\theta} \quad \text{with} \quad r < 1. \tag{7.33}$$

Let $t \in [-\pi,\pi]$ correspond to $s \in \mathbb{R}$ in the conformal bijection from $\mathbb{H}$ to $\mathbb{D}$, i.e., $\zeta = e^{it} = \frac{i-s}{i+s}$ with $dt = \frac{d\zeta}{i\zeta}$. Then $\zeta = \frac{2i}{i+s} - 1$ yields $d\zeta = -\frac{2i\,ds}{(i+s)^2}$, so that $\frac{d\zeta}{i\zeta} = \frac{2ds}{1+s^2}$. Thus, fow $w = re^{i\theta} = \frac{i-z}{i+z}$ with $z = x + iy \in \mathbb{H}$, we have

$$\begin{aligned}\frac{1-r^2}{1-2r\cos(\theta-t)+r^2}\,dt &= \frac{1-|re^{i\theta}|^2}{|re^{i\theta}-e^{it}|^2}\,dt = \frac{1-|w|^2}{|w-\zeta|^2}\frac{d\zeta}{i\zeta} = \frac{1-\left|\frac{i-z}{i+z}\right|^2}{\left|\frac{i-z}{i+z}-\frac{i-s}{i+s}\right|^2}\frac{2\,ds}{1+s^2}\\ &= \frac{|i+z|^2-|i-z|^2}{|(i+s)(i-z)-(i-s)(i+z)|^2}\,2\,ds = \frac{4y}{|2i(s-z)|^2}\,2ds = \frac{2y\,ds}{|z-s|^2}.\end{aligned}$$

Setting $f_0(s) = u_0(t)$ in (7.33) yields the claim. □

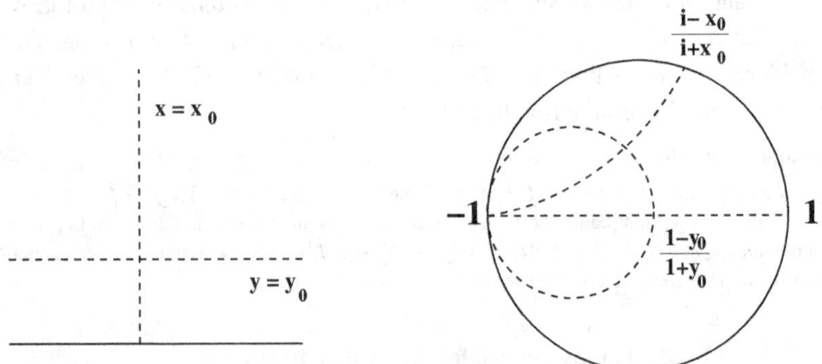

Figure 7.21 The conformal image under $z \mapsto (i-z)/(i+z)$ of a straight line $\{x + iy_0 : x \in \mathbb{R}\}$ with $y_0 > 0$ is a circle tangent to $|z| = 1$ at $z = -1$, centred at $-y_0/(1+y_0)$ and of radius $1/(1+y_0)$, while a half-line $l_{x_0} = \{x_0 + iy : y > 0\}$ is mapped for $x_0 \neq 0$ into an arc of the circle centred at $(\frac{i}{x_0} - 1)$ and of radius $\frac{1}{|x_0|}$, while for $x_0 = 0$ the image is the straight line segment $(-1,1) \subset \mathbb{R}$. In particular, the radial limits in $\mathbb{D}$ generally do not correspond to approaching $\mathbb{R}$ vertically down from the upper half-plane $\mathbb{H}$.

Theorem 7.37 might give the impression that $\mathcal{H}^p(\mathbb{H})$ is just obtained from $\mathcal{H}^p$ by conformal mapping, associating $w = (i-z)/(i+z) \in \mathbb{D}$ with $z \in \mathbb{H}$ and setting $f(w) = F(z)$. However, this is misleading (see Figure 7.21). For example, for $p \geq 1$, while $f \in \mathcal{H}^p$ if $F \in \mathcal{H}^p(\mathbb{H})$, the converse implication is false for $p < \infty$, as indicated by the presence of a factor (depending on p) which also comes in: $f \in \mathcal{H}^p$ only ensures $\frac{F(z)}{(z+i)^{2/p}} \in \mathcal{H}^p(\mathbb{H})$; see Koosis (1998). For this reason, even if the main features of the Hardy spaces of the disc have an analogy in the upper half-plane, it is generally quite cumbersome to keep track of the additional factor in using the conformal mapping $z \mapsto (i-z)/(i+z)$ to translate features of the $\mathcal{H}^p$-spaces into those of the $\mathcal{H}^p(\mathbb{H})$-spaces. Instead, generally it is advisable to work directly in $\mathbb{H}$, and it even turns out that the computational details are simpler. We refer to Koosis (1998) for a systematic approach.

7.2 Pseudodifferential operators

The theory of pseudodifferential operators, rooted in Fourier analysis, is a glorious achievement of mathematics in the last half-century. We endeavour to discuss some basic aspects of singular integrals and of variable coefficient operators as stepping-stones to a comprehensive theory of pseudodifferential operators.

There is an important representation of the conjugate $\tilde{\varphi}$ of $\varphi \in L^1[-\pi,\pi]$, discussed in Section 7.1 in the context of the connection between the Hardy space $\mathfrak{h}^1$ and trigonometric series. If we take the imaginary part of (7.31), we find

$$v(re^{i\theta}) = \frac{1}{2\pi}\int_{-\pi}^{\pi} Q_r(\theta - t)\varphi(t)\,dt\,, \qquad 0 \le r < 1,\ -\pi \le \theta < \pi\,,$$

where $Q_r(\phi)$ is the conjugate Poisson kernel (7.25). Since $\tilde{\varphi}(\theta)$ was defined for a.e. $\theta \in [-\pi,\pi]$ as $\lim_{r\uparrow 1} v(re^{i\theta})$, in view of the explicit limit

$$\lim_{r\uparrow 1} Q_r(\phi) = \frac{\sin(\phi)}{1-\cos(\phi)} = \cot\Big(\frac{\phi}{2}\Big) := Q_1(\phi)\,, \qquad -\pi \le \phi < \pi\,,$$

it is tempting to reverse the order of the operations and write formally

$$\tilde{\varphi}(\theta) = \frac{1}{2\pi}\int_{-\pi}^{\pi} \cot\Big(\frac{\theta - t}{2}\Big)\varphi(t)\,dt = \frac{1}{2\pi}(Q_1 * \varphi)(\theta)\,, \qquad -\pi \le \theta < \pi\,. \quad (7.34)$$

The difficulty, however, is that $Q_1 \notin L^1[-\pi,\pi]$ because of the singularity at 0, so that the convolution (7.34) is, as yet, undefined. The proper interpretation is as the symmetric limit

$$\tilde{\varphi}(\theta) = \lim_{\varepsilon\downarrow 0}\frac{1}{2\pi}\Big\{\int_{-\pi}^{\theta-\varepsilon} + \int_{\theta+\varepsilon}^{\pi}\Big\}\cot\Big(\frac{\theta - t}{2}\Big)\varphi(t)\,dt\,, \qquad \text{for a.e.} \quad -\pi < \theta < \pi\,,$$

called the *principal value*,[58] and written alternatively, after changing variables,[59]

$$\tilde{\varphi}(\theta) = \lim_{\varepsilon \downarrow 0} \frac{1}{\pi} \int_{\varepsilon}^{\pi} \frac{\varphi(\theta - s) - \varphi(\theta + s)}{2\tan(s/2)}\, ds\,. \tag{7.35}$$

The discussion in Section 7.1 shows that the existence of $\tilde{\varphi}$, given $\varphi \in L^1[-\pi, \pi]$, is a delicate result. For example, in contrast to the case when φ is Hölder continuous, for a merely continuous function $\varphi : [-\pi, \pi] \to \mathbb{R}$, the a.e. existence of the limit in (7.35) is not due to the smallness of $|\varphi(\theta - s) - \varphi(\theta + s)|$, but comes from an interference phenomenon that cancels positive and negative contributions.[60] Moreover, the conjugation $\varphi \mapsto \tilde{\varphi}$ is not a local operation; that is, it is not true[61] that if $\varphi_1, \varphi_2 \in L^1[-\pi, \pi]$ are such that $\varphi_1(t) = \varphi_2(t)$ a.e. in some interval I, then $\tilde{\varphi}_1(t) = \tilde{\varphi}_2(t)$ a.e. on I.

The map $H : \varphi \to \tilde{\varphi}$, given by (7.35), is called the *Hilbert transform* and is the quintessential example of a *singular integral*. One can show that $H : L^p[-\pi, \pi] \to L^p[-\pi, \pi]$ is a bounded linear operator for $1 < p < \infty$, and that, for $\alpha \in (0, 1)$, it maps the Hölder class[62] h_α into h_α; see Koosis (1998). In contrast to this, we saw in the context of the hardy space $\mathfrak{h}^1$ that H does not map $L^1[-\pi, \pi]$ to $L^1[-\pi, \pi]$, and an example discussed above shows that H does not map $L^\infty[-\pi, \pi]$ to $L^\infty[-\pi, \pi]$. The Hilbert transform H is the key to the convergence questions of Fourier series in $L^p[-\pi, \pi]$ (with $1 < p < \infty$). To see this, notice first that for any integer $n \geq 0$, since e^{int} is the boundary value on $|z| = 1$ of the analytic function z^n, we have $H[\cos(nt)] = \sin(nt)$ and $H[\sin(nt)] = -\cos(nt)$, so that

$$H[e^{\pm int}] = \mp i e^{\pm int}\,, \qquad n \geq 0, \quad t \in [-\pi, \pi]\,.$$

This shows that the in terms of the orthonormal basis $\{e^{ikt}\}_{k \in \mathbb{Z}}$ of $L^2[-\pi, \pi]$, the operator H has a particularly simple representation as a *multiplication operator*

[58] The proper interpretation being in the sense of tempered distributions (see Chapter 6).

[59] To see this, for, say, $\theta \in (0, \pi)$, perform for $\varepsilon \in (0, \theta)$ the change of variables $t = \theta - s$ in the integral on $[-\pi, \theta - \varepsilon]$ and the change of variables $t = \theta + s$ in the integral on $[\theta + \varepsilon, \pi]$, thus obtaining the expression $\frac{1}{2\pi} \int_{\varepsilon}^{\theta + \pi} \frac{\varphi(\theta - s)}{\tan(s/2)}\, ds - \frac{1}{2\pi} \int_{\varepsilon}^{\pi - \theta} \frac{\varphi(\theta + s)}{\tan(s/2)}\, ds$, with the understanding that we extended φ to an a.e. defined 2π-periodic function on $\mathbb{R}$ by setting $\varphi(s + 2k\pi) = \varphi(s)$ for $s \in (-\pi, \pi)$ and $k \in \mathbb{Z}$. This is the desired expression, see (7.35), up to the additive remainder $\frac{1}{2\pi} \int_{\pi}^{\theta + \pi} \frac{\varphi(\theta - s)}{\tan(s/2)}\, ds + \frac{1}{2\pi} \int_{\pi - \theta}^{\pi} \frac{\varphi(\theta + s)}{\tan(s/2)}\, ds$. But this sum vanishes: write the second integral as $\frac{1}{2\pi} \int_{\pi - \theta}^{\pi} \frac{\varphi(\theta + s - 2\pi)}{\tan(s/2)}\, ds$ and change the variable $\tau = 2\pi - s$.

[60] We refer to Koosis (1998) for the construction of a continuous function $f : \mathbb{R} \to \mathbb{R}$, periodic of period 2π, such that $\int_0^\pi \frac{|\varphi(\theta - s) - \varphi(\theta + s)|}{s}\, ds = \infty$ for every $\theta \in \mathbb{R}$.

[61] For example, if $\varphi(t)$ vanishes for $t \in (a, b)$, then $\tilde{\varphi}(t)$ is real-analytic on (a, b); see Katznelson (1968). In particular, $\tilde{\varphi}(t) = \frac{1}{\pi} \ln\left(\frac{1 - \cos(t)}{1 - \sin(t)}\right) \notin L^\infty[-\pi, \pi]$ for $\varphi = \chi[0, \pi/2]$; see Koosis (1998).

[62] Of period 2π in this context, rather than of period 1, as discussed in Chapter 3 and Chapter 4.

with *multipliers*

$$\lambda(k) = -\mathrm{sgn}(k) \quad \text{for} \quad k \in \mathbb{Z}.$$

Moreover, the Fourier partial sums $(S_N f)(t) = \sum_{k=-N}^{N} \hat{f}(k)\, e^{ikt}$ of $f \in L^1[-\pi,\pi]$ can be expressed in terms of H. For this, write for any trigonometric polynomial $p_n(t) = \sum_{k=-n}^{n} c_k e^{ikt}$ with $n \geq N \geq 1$,

$$e^{iNt} H[e^{-iNs} p_n(s)](t) = -i \sum_{k=N+1}^{n} c_k e^{ikt} + i \sum_{k=-n}^{N-1} c_k e^{ikt},$$

$$e^{-iNt} H[e^{iNs} p_n(s)](t) = -i \sum_{k=-N+1}^{n} c_k e^{ikt} + i \sum_{k=-n}^{-N-1} c_k e^{ikt}.$$

When we substract, we get

$$e^{iNt} H[e^{-iNs} p_n(s)](t) - e^{-iNt} H[e^{iNs} p_n(s)](t) = 2i \sum_{k=-N}^{N} c_k e^{ikt} - ic_N e^{iNt} - ic_{-N} e^{iNt}.$$

This identity allows us to represent the partial sums $S_N f$ in terms of the norm-preserving operators $f(t) \mapsto e^{\pm iNt} f(t)$ and the operator H:

$$(S_N f)(t) = \frac{1}{2}\, \hat{f}(N)\, e^{iNt} + \frac{1}{2}\, \hat{f}(-N)\, e^{-iNt} + \frac{e^{iNt} H[e^{-iNs} p_n(s)](t) - e^{-iNt} H[e^{iNs} p_n(s)](t)}{2i}. \tag{7.36}$$

By the Riemann–Lebesgue lemma (see Chapter 4), the first two terms in (7.36) tend to zero when $N \to \infty$, and the Hölder inequality yields

$$|\hat{f}(\pm N)| = \Big| \frac{1}{2\pi} \int_{-\pi}^{\pi} f(t) e^{\mp iNt} dt \Big| \leq \frac{1}{2\pi} \int_{-\pi}^{\pi} |f(t)|\, dt \leq \Big(\int_{-\pi}^{\pi} |f(t)|^p\, dt \Big)^{\frac{1}{p}} (2\pi)^{-\frac{1}{p}}.$$

In particular, the representation (7.36) shows that H cannot be bounded[63] on $L^1[-\pi,\pi]$. On the other hand, we have the following.

Theorem 7.38 $\lim_{N\to\infty} \|S_N f - f\|_{L^p[-\pi,\pi]} = 0$ *if* $f \in L^p[-\pi,\pi]$ *with* $p \in (1,\infty)$.

Proof Since $H : L^p[-\pi,\pi] \to L^p[-\pi,\pi]$ is a bounded linear operator, the representation (7.36) and the subsequently displayed inequality ensure the existence of a constant $C_p > 0$ such that $\|S_N f - f\|_{L^p[-\pi,\pi]} \leq C_p \|f\|_{L^p[-\pi,\pi]}$ for all $f \in L^p[-\pi,\pi]$. Since for trigonometric polynomials f we have that $S_N f = f$ for all $N \geq 1$ large enough, from the density of the trigonometric polynomials in $L^p[-\pi,\pi]$ we obtain the desired convergence on the entire space $L^p[-\pi,\pi]$. □

[63] Indeed, the validity of the estimate $\|Hf\|_{L^1[-\pi,\pi]} \leq C\|f\|_{L^1[-\pi,\pi]}$ for all $f \in L^1[-\pi,\pi]$, would force $\|S_N f\|_{L^1[-\pi,\pi]} \leq (C+1)\|f\|_{L^1[-\pi,\pi]}$ which is false, see Exercise 24 of Chapter 4.

Similarly, on the real line, the Hilbert transform H is defined formally as the singular integral

$$(Hf)(x) = \frac{1}{\pi} \lim_{\varepsilon\to 0,\, M\to\infty} \int_{\varepsilon<|y|<M} \frac{f(x-y)}{y}\, dy\,.$$

One can prove that $H : L^p(\mathbb{R}) \to L^p(\mathbb{R})$ is a bounded linear operator for $1 < p < \infty$, while the fact that $(Hf)(x) = \frac{1}{\pi} \ln\left(\frac{|x-a|}{|x-b|}\right)$ for $x \neq a, b$ does not belong to $L^1(\mathbb{R})$ or to $L^\infty(\mathbb{R})$ although $f = \chi^{[a,b]} \in L^1(\mathbb{R}) \cap L^\infty(\mathbb{R})$ shows that H does not map $L^1(\mathbb{R})$ to $L^1(\mathbb{R})$ or $L^\infty(\mathbb{R})$ to $L^\infty(\mathbb{R})$; see Pinsky (2009). In particular,

$$(Hf)(x) = -i \int_{\mathbb{R}} \operatorname{sgn}(\xi)\, \hat{f}(\xi)\, e^{2\pi i x\xi}\, d\xi\,, \qquad f \in L^2(\mathbb{R})\,. \tag{7.37}$$

The Hilbert transform is the inspiration and a paradigm for far-reaching generalisations that play an important role in mathematical physics. Using the Fourier transform, we can define on the Euclidean space $\mathbb{R}^d$ *Fourier multiplier operators* T_m for some function m, under some minimal measurability and growth assumptions on m, by the formula

$$(T_m f)(x) = \int_{\mathbb{R}^d} \hat{f}(\xi)\, m(\xi)\, e^{2\pi i x\cdot\xi}\, d\xi\,. \tag{7.38}$$

Examples include constant coefficient differential operators, for which $m(\xi)$ is a polynomial, e.g. the Laplace operator $\partial_x^2 + \partial_y^2$ is a Fourier multiplier with symbol $m(\xi_1, \xi_2) = -4\pi^2(\xi_1^2 + \xi_2^2)$. In one dimension $d = 1$, the Hilbert transform H is a Fourier multiplier with symbol $m(\xi) = -i\operatorname{sgn}(\xi)$. As it is the case with the Hilbert transform H, Fourier multipliers T_m can also be naturally identified with convolution operators by the identity

$$(T_m f)(x) = \int_{\mathbb{R}^d} K(x-y) f(y)\, dy\,,$$

where the (possibly distributional) kernel K is simply the inverse Fourier transform of m. Convolution operators of this type fall into many classes, depending on the singular behaviour of the kernel K, or, equivalently, on the asymptotic regularity and decay of the symbol m. Important and significant efforts concern finding conditions for a singular integral operator to be bounded on $L^p(\mathbb{R}^d)$-spaces.

The Fourier multipliers operators are part of a larger class of operators, known as *pseudodifferential operators*. They can be introduced in several ways, one being as follows. The Fourier multipliers T_m were defined in (7.38) while the Fourier inversion formula shows that *spatial multipliers* $f(x) \mapsto \sigma(x) f(x)$, where $\sigma : \mathbb{R}^d \to \mathbb{C}$ is a smooth function, can be written as

$$\sigma(x)f(x) = \int_{\mathbb{R}^d} \sigma(x)\, \hat{f}(\xi)\, e^{2\pi i x\cdot\xi}\, d\xi\,. \tag{7.39}$$

These two classes of operators can be generalised by defining the *pseudodifferential operator* of the form

$$\Psi_\sigma f(x) = \int_{\mathbb{R}^d} \sigma(x,\xi)\, \hat{f}(\xi)\, e^{2\pi i x\cdot\xi}\, d\xi\,, \tag{7.40}$$

where $\sigma(x,\xi)$, the *symbol*, belongs to a suitable class of functions and is subject to some suitable smoothness and decay assumptions. Thus, if σ is purely a function of ξ, then Ψ_σ is a Fourier multiplier, while if σ is purely a function of x, then Ψ_σ is a spatial multiplier. There are many different symbol classes in use, tailored to the specific context in which they arise. For example, the *standard symbols of order k* consist of smooth symbols $\sigma(x,\xi)$ that obey estimates of the form: for any pair of multi-indices α, β, there is a constant $C_{\alpha,\beta} > 0$ such that

$$\Big|\prod_{j=1}^{d} \partial_{\xi_j}^{\alpha_j} \prod_{j=1}^{d} \partial_{x_j}^{\beta_j} \sigma(x,\xi)\Big| \le C_{\alpha,\beta}\,(1+|\xi|)^{k-|\alpha|}\,, \tag{7.41}$$

where $|\alpha| = \sum_{j=1}^{d} \alpha_j$. The index k pinpoints how the corresponding pseudodifferential operator acts on certain function spaces; an operator of order $k > 0$ "differentiates" to order k, while an operator of order $k < 0$ "integrates" to order $(-k)$. More precisely, the pseudodifferential operator Ψ_σ associated by means of (7.40) to a standard symbol σ of class k maps the Sobolev space $H^s(\mathbb{R}^d)$ continuously into the Sobolev space $H^{s-k}(\mathbb{R}^d)$ for any index s. The class of pseudodifferential operators with standard symbols of order $n \ge 1$ includes the variable coefficient linear partial differential operators $\sum_{|\alpha|\le n} a_\alpha(x)\,\partial_x^\alpha$, with the coefficients $a_\alpha(x)$ smooth functions.

There are other classes of linear operators which are suggested by the study of the convergence of Fourier series or Fourier integrals: *oscillatory integrals* of the type

$$T_\lambda f(x) = \int_{\mathbb{R}} a(x,y)\, f(y)\, e^{2\pi i\lambda\phi(x,y)}\, dy\,, \tag{7.42}$$

where $a : \mathbb{R}\times\mathbb{R} \to \mathbb{C}$ is a suitable cutoff function, $\phi : \mathbb{R}\times\mathbb{R} \to \mathbb{R}$ is a smooth phase function, and λ is a real parameter. A classical subject, of great relevance to applications in mathematical physics, concerns the control of the asymptotic behaviour of the oscillatory integral in the limit $\lambda \to \infty$, for specific cutoff and phase functions (see Part II).

We shall not attempt a survey of the calculus of pseudodifferential operators since this would take us too far afield. We refer to Hörmander (1985) for a masterful exposition of the subject; other good sources are Alinhac and Gérard (2007) and Taylor (1981).

Afterword

This being a textbook, it would seem appropriate to offer to the aspiring mathematician/prospective graduate student, from the author's perspective, some subjective points of view that could be worth pondering at the beginning of the journey towards advanced mathematics.

There is a strong parallel between mountain climbing and mathematics: both are very hard work, and both give their enthusiasts lots of pleasure in what they achieve, along with a view of the world that most people don't get. Both activities expose the usual fallacy of only looking at the immediate, obvious risks, and not taking a long-term view. Also, just like after completing the struggle to find a route to the summit, other possible routes may be discerned for the descent or for subsequent ascents; so in mathematics, once an approach is found, other mathematicians can usually find an alternative that is often much better and/or shorter. Indeed, one can push the comparison further. Learning about an unknown but well-established subject is to some extent similar to taking a trip to a vast unknown area that is considered to be a tourist attraction. There are guidebooks, maps and route signs. It is advisable to start by taking the main road and having a look at some of the recommended attractions. Once one is familiar with the basic layout, a favourable first impression might entice one to look more closely. Often old, practically forgotten trails have much to offer. In following the lure to explore uncharted territory by venturing off the trail one should, however, be wary of inadequate preparation. In the mathematical context, the more you know about the background and the more techniques you master, the vaster are the opportunities that you can try out. In particular, a proof is the outcome of the interaction between creative imagination and critical reasoning; further, rigorous formal proofs become really important in the advent of a crisis, e.g. a counter-intuitive behaviour, or when a paradox of some kind arises. For this reason, one ought to master the basic tools that are available, seeking also to get acquainted with recent technical advances. One should try direct and indirect approaches, as well as different techniques – keep in mind the humorous analogue of the likely outcome of looking at night for lost keys under a street lamp just because the light there is better. On the other hand, it is instructive and inspiring to study old-fashioned material since often modern techniques offer possibilities that were not available earlier. Not investing the time and effort to acquire modern techniques is putting oneself at a great disadvantage, even if one is very adept at using basic tools. Ignorance of technical details leads to a limited understanding, which then diminishes the chances of extending the results to a different scenario. It is also instructive to find the limits of applicability of a specific type of consideration, thus preventing unjustified extensions and the apparent expansion of its validity. However, in this process one should not automatically equate technical sophistication with deep insight. The apparent opportunities to use fancy terminology and intricate technical considerations to oversell relatively minor advances are actually counterproductive in the long-term since unrealistic expectations will not be met and cannot mask the inconsequential nature of the whole enterprise. The intellectual attractiveness of a mathematical argument can sometimes be a glittering deception, reaching ultimate heights in splendid isolation. On the other hand, to some extent, it is advisable to practice mathematics as a pursuit of curiosity in its purest form, remaining wary about the willingness to elaborate upon any idea, however absurd. The interplay of ideas is very important, as progress depends on the abstract helping the concrete and on the concrete feeding the abstract. The training to master certain techniques is an important part, but should not represent an end in itself. Technique should be learned in a context that entices and offers a perspective. Most mathematical structures were designed within a specific context and derive some meaning from that context. A meaningful context often represents a bridge to another area, offers a synthesis of diverse mathematical points of view, or represents some good starting point for uncharted territory. Mathematics is much more than some sort of tool for science and technology. Assigning to it the role of merely a toolbox reduces a possibly creative engagement with the subject to an almost automatic undertaking, an algorithm-like process that is often better left to machines. The accurate yet mindless manipulation of symbols impedes aesthetic sensitivity and the inspiration for new ideas.

Appendix
Historical notes

For a historical perspective we provide some brief information about the pioneers of the areas of investigation we have touched upon.

Abel, Niels Henrik (1802–1829): Norwegian mathematician with pioneering contributions to quintic equations and mathematical analysis

Alaoglu, Leonidas (1914–1981): Canadian mathematician of Greek origin, with contributions to functional analysis

Arzelà, Cesare (1847–1912): Italian mathematician with contributions to the theory of functions

Ascoli, Giulio (1843–1896): Italian mathematician with contributions to the theory of functions of a real variable and to Fourier series

d'Alembert, Jean-Baptiste le Rond (1717–1783): French mathematician and philosopher (he played an important role in the intellectual ferment leading to the French Revolution)

Archimedes (c. 287 B.C. – 212 B.C.): Greek mathematician, physicist, engineer, and astronomer (active in Syracuse, Sicily), regarded as one of the leading scientists in classical antiquity

Baire, René-Louis (1874–1932): French mathematician with important contributions to mathematical analysis

Banach, Stefan (1892–1945): Polish mathematician; one of the founders of functional analysis, with major contributions to integration theory and set theory

Bergman, Stefan (1895–1977): Polish-born American mathematician with important contributions to complex analysis

Bernoulli, Daniel (1700–1782): Swiss mathematician and physicist, famous for his applications of mathematics to fluid mechanics and for his pioneering work in probability and statistics

Bernstein, Sergei Natanovich (1880–1968): Russian mathematician known for contributions to constructive function theory, a field studying the connection between smoothness properties of a function and its approximations by polynomials

Bessel, Friedrich Wilhelm (1784–1846): German mathematician and astronomer

Blaschke, Wilhelm Johann Eugen (1885–1962): Austrian mathematician with important contributions to complex analysis and geometry

Bloch, André (1893–1948): French mathematician known for his contributions to complex analysis

Bochner, Salomon (1899–1982): American mathematician known for his wide-ranging work in mathematical analysis

du Bois-Reymond, Paul David (1831–1889): German mathematician with notable contributions to the calculus of variations and Fourier series

Borel, Félix Édouard Justine Émile (1871–1956): French mathematician; one of the pioneers of measure theory and probability theory

Cantelli, Francesco Paolo (1875–1966): Italian mathematician with important contributions to probability theory and mathematical finance

Cantor, Georg (1845–1918): German mathematician born in Saint Petersburg, best known as the inventor of set theory

Carathéodory, Constantin (1873–1950): German mathematician of Greek origin with significant contributions to complex analysis, the calculus of variations and measure theory

Casorati, Felice (1835–1890): Italian mathematician with contributions to complex analysis and difference equations

Cavalieri, Bonaventura Francesco (1598–1647): Italian mathematician famous for his geometric principle that the volumes of two objects are equal if the areas of all their corresponding cross-sections are equal, thus anticipating to some extent integration theory

Cauchy, Baron Augustin-Louis (1789–1857): French mathematician, an early pioneer of rigorous analysis; his writings cover the entire range of mathematics and mathematical physics

Chebyshev, Pafnuty Lvovich (1821–1894): Russian mathematician known for his work in the field of probability, statistics and number theory

Christoffel, Elwin Bruno (1829–1900): German mathematician and physicist with contributions to complex analysis and differential geometry

Daniell, Percy John (1889–1946): British mathematician born in Chile, with notable contributions to abstract integration theory

Darboux, Jean-Gaston (1842–1917): French mathematician with several important contributions to geometry and mathematical analysis

De Morgan, Augustus (1806–1871): British mathematician and logician; he formulated De Morgan's laws and is credited with making the idea of mathematical induction rigorous

Dini, Ulisse (1845–1912): Italian mathematician with contributions to real analysis

Dirac, Paul Adrien Maurice (1902–1984): English theoretical physicist who made fundamental contributions to quantum mechanics

Dirichlet, Johann Peter Gustav Lejeune (1805–1859): German mathematician with important contributions to number theory and mathematical analysis

Eberlein, William Frederick (1917–1986): American mathematician with contribution to mathematical analysis and mathematical physics

Egorov, Dmitri Fyodorovich (1869–1931): Russian mathematician known for significant contributions to differential geometry and mathematical analysis

Euclid (third century B.C.): Greek mathematician (active in Alexandria, Egypt) who wrote one of the most influential works in the history of mathematics, the *Elements*, serving as the main textbook for teaching mathematics (especially geometry) from the time of its publication until the late nineteenth century

Euler, Leonhard (1707–1783): Swiss mathematician and physicist (who spent most of his professional life in St. Petersburg and in Berlin), considered to be one of the greatest mathematicians of all time

Fatou, Pierre (1878–1929): French mathematician and astronomer, known for major contributions to several branches of analysis

Fejér, Leopold (1880–1959): Hungarian mathematician with notable contributions to Fourier series, founder of a highly successful Hungarian school of analysis

Fibonacci, Leonardo (1170–1250): Italian mathematician who popularised Hindu–Arabic numerals in Europe and introduced the sequence of Fibonacci numbers

Fourier, Jean Baptiste Joseph (1768–1830): French mathematician and physicist, best known for initiating the application of trigonometric series to the problem of heat transfer

Fredholm, Erik Ivar (1866–1927): Swedish mathematician with major contributions to integral equations

Fréchet, Maurice René (1878–1973): French mathematician with major contributions to toplogy and analysis

Fubini, Guido (1879–1943): Italian mathematician known for his work on integration theory

Gauss, Carl Friedrich (1777–1855): German mathematician with significant contributions to number theory, algebra, statistics, analysis, differential geometry, mechanics

Gram, Jørgen Pedersen (1850-1916): Danish actuary and mathematician

Gibbs, Josiah Willard (1839–1903): American theorerical physicist with important contributions to thermodynamics and statistical mechanics, who also developed vector calculus

Haar, Alfréd (1885–1933): Hungarian mathematician with important contributions to mathematical analysis and topological groups

Hahn, Hans (1879–1934): Austrian mathematician with notable contributions to mathematical analysis

Hamel, Georg Karl Wilhelm (1877–1954): German mathematician with important contributions to axiomatic theory

Hamilton, Sir William Rowan (1805–1865): Anglo-Irish physicist, astronomer and mathematician with outstanding contributions to classical mechanics and algebra

Hardy, Godfrey Harold (1877–1947): English mathematician known for his achievements in mathematical analysis and number theory

Hausdorff, Felix (1868–1942): German mathematician with significant contributions to topology, set theory, function theory, and measure theory

Heaviside, Oliver (1850–1925): English engineer, mathematician, and physicist with vast contributions to mathematical physics, who also (independently) co-formulated vector calculus

Helly, Eduard (1884–1943): Austrian mathematician known for his work in function spaces and discrete geometry

Hermite, Charles (1822–1901): French mathematician with important contributions to algebra and analysis

Hilbert, David (1862–1943): German mathematician, one of the most influential and universal mathematicians of the nineteenth and early twentieth centuries

Hölder, Otto (1899–1937): German mathematician with important contributions to mathematical analysis and algebra

Hörmander, Lars Valter (1931–2012): Swedish mathematician, the foremost contributor to the modern theory of linear partial differential equations

Hurwitz, Adolf (1859–1919): German mathematician with contributions to complex analysis, number theory and dynamical systems

Jordan, Marie Ennemond Camille (1838–1922): French mathematician, known for his foundational work in group theory, topology and linear algebra

Kähler, Erich (1906–2000): German mathematician with wide-ranging geometrical interests

Kolmogorov, Andrey Nikolaevich (1903–1987): Russian mathematician with significant contributions to probability theory, topology, logic, harmonic analysis, turbulence, classical mechanics and computer science

Lagrange, Joseph-Louis (1736–1813): Italian-born French mathematician and astronomer (active in Prussia and in France), with fundamental contributions to analysis, number theory and classical and celestial mechanics

Laguerre, Edmond Nicolas (1834–1886): French mathematician with contributions to orthogonal polynomials

Lambert, Johann Heinrich (1727–1777): Swiss mathematician

de Laplace, Pierre-Simon (1749–1827): French mathematician whose work was pivotal to the development of mathematical analysis, mathematical astronomy and statistics

Laurent, Pierre Alphonse (1813–1854): French mathematician with contributions to power series

Lebesgue, Henri Léon (1875–1941): French mathematician, famous for his theory of integration

Leibniz, Gottfried Wilhelm (1646–1716): German mathematician and philosopher who developed the infinitesimal calculus independently of Newton; his notation has been widely used ever since it was published

Lévy, Paul Pierre (1886–1971): French mathematician with significant contributions to probabilty theory

Lindelöf, Ernst (1870–1946): Finnish mathematician with simportant contributions to topology, complex analysis and differential equations

Liouville, Joseph (1809–1882): French mathematician with significant contributions to complex analysis, number theory, differential equations and mechanics

Lipschitz, Rudolf Otto Sigismund (1832–1903): German mathematician with many important contributions to differential equations, algebraic number theory and mechanics

von Lindemann, Carl Louis Ferdinand (1852–1939): German mathematician known for providing the first proof that π is transcendental

Luzin, Nikolai Nikolaevich (1883–1950): Russian mathematician famous for his work in descriptive set theory and mathematical analysis

Markov, Andrey Andreyevich (1856–1922): Russian mathematician known for his work on stochastic processes

Mellin, Hjalmar (1854–1933): Finnish mathematician with contributions to the theory of functions

Michelson, Albert Abraham (1852–1913): American physicist famous for his experimental work, especially on the measurement of the speed of light and for providing in 1887 the first strong evidence against the then prevalent aether theory, thus initiating a direction of research that eventually led to the physical theory of special relativity

Mikhlin, Solomon Grigorevich (1908–1990): Russian mathematician who worked in the fields of linear elasticity, singular integrals and numerical analysis

Montel, Paul Antoine Aristide (1876–1975): French mathematician with important contributions to complex analysis

Morera, Giacinto (1856–1909): Italian engineer and mathematician, with contributions to complex analysis and elasticity

Möbius, August Ferdinand (1790–1868): German mathematician and theoretical astronomer with important contributions to projective geometry

von Neumann, John (1903–1957): Hungarian-American mathematician with foundational contributions to operator theory, game theory and computing

Newton, Sir Isaac (1642–1727): English physicist, mathematician, and theologian, considered by many to be the greatest and most influential scientist who ever lived; his monograph *Philosophiae Naturalis Principia Mathematica*, published in 1687, lays the foundations for most of classical mechanics

Nikodym, Otto Marcin (1887–1974): Polish mathematician with notable contributions to measure theory

Paley, Raymond Edward Alan Christopher (1907–1933): English mathematician with contributions to Fourier analysis

Parseval, Marc-Antoine (1755–1836): French mathematician with contributions to Fourier analysis

Pauli, Wolfgang Ernst (1900–1958): Austrian theoretical physicist, one of the pioneers of quantum physics

Peano, Giuseppe (1858–1932): Italian mathematician; founder of mathematical logic and set theory, and with key contributions to differential equations

Phragmén, Lars Edvard (1863–1937): Swedish mathematician known for his contributions to complex analysis

Picard, Émile (1856–1941): French mathematician with important contributions to function theory and differential equations

Pick, Georg Alexander (1859–1942): Austrian mathematician known for his work in complex analysis and discrete geometry

Plancherel, Michel (1885–1967): Swiss mathematician with contributions to Fourier analysis

Poisson, Siméon Denis (1781–1840): French mathematician and physicist, with lasting contributions to mechanics, probability theory and mathematical analysis

Radon, Johann Karl August (1887–1956): Austrian mathematician known for a number of lasting contributions to measure theory and integral geometry

Rellich, Franz (1906–1955): Austrian mathematician with important contributions to mathematical physics and the theory of partial differential equations

Riemann, Georg Friedrich Bernhard (1826–1866): German mathematician with lasting contributions to analysis, number theory and differential geometry

Riesz, Frigyes (1880–1956): Hungarian mathematician with fundamental contributions to functional analysis

Riesz, Marcel (1886–1969): Hungarian mathematician (brother of F. Riesz) with fundamental contributions to harmonic analysis

Rouché, Eugène (1832–1910): French mathematician with contributions to complex analysis and linear algebra

Schauder, Juliusz Pawel (1899–1943): Polish mathematician known for his pioneering work in functional analysis and partial differential equations

Schmidt, Erhard (1876–1959): German mathematician with important contributions to functional analysis

Schmulyan, Vitold Lvovich (1914–1944): Russian mathematician known for his work in functional analysis

Schwarz, Karl Hermann Amandus (1843–1921): German mathematician, known for his work in complex analysis

Schwartz, Laurent-Moïse (1915–2002): French mathematician who pioneered the theory of distributions

Sobolev, Sergei Lvovich (1908–1989): Russian mathematician with important contributions to mathematical analysis and partial differential equations

Souslin, Mikhail Yakovlevich (1894–1919): Russian mathematician with major contributions to general topology and descriptive set theory

Steinhaus, Wladyslaw Hugo Dionizy (1887–1972): one of the architects of the school of mathematics which flowered in Poland between the two world wars; his most notable contribution is the proof of the uniform boundedness principle, given along with S. Banach

Tauber, Alfred (1866–1942): Austrian mathematician with important results on series

Taylor, Brooke (1685–1731): English mathematician, best known for the Taylor series

Thorin, Olof (1912–2004): Swedish mathematician, best known for his work on interpolation of operators.

Tonelli, Leonida (1885–1946): Italian mathematician most noted for his contributions to integration theory

Vitali, Giuseppe (1875–1932): Italian mathematician with significant contributions to several branches of mathematical analysis

Volterra, Vito (1860–1940): Italian mathematician with important contributions to mathematical analysis and mathematical biology

Weierstrass, Karl Theodor Wilhelm (1815–1897): German mathematician, considered to be the founder of modern analysis (despite his late start – he was appointed to a university position in 1856, after a long activity as a high-school teacher); his emphasis on rigour allowed him to make significant advances

Wiener, Norbert (1894–1964): American mathematician with important contributions to stochastic processes and Fourier analysis, regarded as the originator of cybernetics

Young, William Henry (1863–1942): English mathematician with contributions to mathematical analysis

Zorn, Max August (1906–1993): German mathematician with notable contributions to algebra

References

Ahlfors, L. V. 1979. *Complex Analysis*. McGraw-Hill, New York.

Albiac, F., and Kalton, N. J. 2006. *Topics in Banach Space Theory*. Springer, New York.

Alinhac, S., and Gérard, P. 2007. *Pseudo-differential Operators and the Nash-Moser Theorem*. American Mathematical Society, Providence, RI.

Aliprantis, C. D., and Burkinshaw, O. 1999. *Problems in Real Analysis. A Workbook with Solutions*. Academic Press, Inc., San Diego, CA.

Ambrosio, L., DaPrato, G., and Mennucci, A. 2011. *Introduction to Measure Theory and Integration*. Edizioni della Scuola Normale Superiore di Pisa, Pisa.

Androulakis, G., Beanland, K., Dilworth, S. J., and Sanacory, F. 2006. Embedding l^∞ into the space of bounded operators on certain Banach spaces. *Bull. London Math. Soc.*, **38**, 979–990.

Ash, J. M. 1976. Multiple trigonometric series. Pages 76–96 of: Ash, J. M. (ed), *Studies in Harmonic Analysis*. Mathematical Association of America, Washington, DC.

Ash, J. M., and Gluck, L. 1972. A divergent multiple Fourier series of power series type. *Studia Math.*, **44**, 477–491.

Ash, M. J. 2013. A survey of multidimensional generalizations of Cantor's uniqueness theorem for trigonometric series. Pages 49–61 of: Bilyk, D., Carli, L. De, Petukhov, A., Stokolos, A. M., and Wick, B. D. (eds), *Recent Advances in Harmonic Analysis and Applications*. Springer, New York.

Benedetto, J. J. 1997. *Harmonic Analysis and Applications*. CRC Press, Boca Raton, FL.

Boas, R. P. 1987. *Invitation to Complex Analysis*. Random House, New York.

Bôcher, M. 1906. Introduction to the theory of Fourier's series. *Ann. of Math.*, **7**, 81–152.

Borzellino, J. E., and Sherman, M. 2012. When is a trigonometric polynomial not a trigonometric polynomial? *Amer. Math. Monthly*, **119**(5), 422–425.

Bourgain, J. 1996. Spherical summation and uniqueness of multiple trigonometric series. *Internat. Math. Res. Notices*, **3**, 93–107.

Bramanti, M. 2014. *An Invitation to Hypoelliptic Operators and Hörmander's Vector Fields*. Springer, Cham, Switzerland.

Brézis, H. 2011. *Functional Analysis, Sobolev Spaces and Partial Differential Equations*. Springer, New York.

Bruckner, A. M., and Leonard, J. T. 1966. Derivatives. *Amer. Math. Monthly*, **73**, 24–56.

Bruckner, A. M., Bruckner, J. B., and Thomson, B. S. 1997. *Real Analysis*. Prentice-Hall, Upper Saddle River.

Burckel, R. B. 1979. *An Introduction to Classical Complex Analysis*. Birkhäuser Verlag, Basel.

Bustamante, J., and Jiménez, M. A. 2000. Chebyshev and Hölder approximation. *Aportaciones Mat. Comun.*, **27**, 23–31.

Carleson, L. 1966. On convergence and growth of partial sums of Fourier series. *Acta Math.*, **116**, 135–157.

Chernoff, P. R. 1980. Convergence of Fourier series. *Amer. Math. Monthly*, **87**, 399–400.

Ciesielski, Z. 1960. On the isomorphisms of the spaces H_α and m. *Bull. Acad. Pol. Sci. Sér. Sci. Math. Astronom. Phys.*, **8**, 217–222.

Constantin, A., and Strauss, W. A. 2000. Stability of a class of solitary waves in compressible elastic rods. *Phys. Lett. A*, **270**, 140–148.

Conway, J. B. 1990. *A Course in Functional Analysis*. Springer-Verlag, New York.

Crone, L. 1971. A characterization of matrix operators on l^2. *Math. Z.*, **123**, 315–317.

Day, M. M. 1940. The spaces L^p with $0 < p < 1$. *Bull. Amer. Math. Soc.*, **46**, 816–823.

Day, M. M. 1941. Reflexive Banach spaces not isomorphic to uniformly convex spaces. *Bull. Amer. Math. Soc.*, **47**, 313–317.

Dirichlet, P. G. L. 1829. Sur la convergence des séries trigonométriques qui servent à représenter une fonction arbitraire entre des limites données. *J. Reine Angew. Math.*, **4**, 157–169.

du Bois-Reymond, P. 1874. Über die sprungweise Wertänderungen analytischer Funktionen. *Math. Ann.*, 241–261.

du Bois-Reymond, P. 1876. Untersuchungen über die Convergenz und Divergenz der Fourierschen Darstellungsformeln. *Abh. Math.-Phys. Cl. K. Bay. Akad. Wiss.*, **12**, 1–13.

Dunham, W. 2005. Touring the Calculus gallery. *Amer. Math. Monthly*, **112**, 1–19.

Duren, P. 2000. *Theory of H^p Spaces*. Dover, New York.

Dvoretzky, A., and Rogers, C. 1950. Absolute and unconditional convergence in normed linear spaces. *Proc. Nat. Acad. Sci. USA*, **36**(192–197).

Dym, H., and McKean, H. P. 1972. *Fourier Series and Integrals*. Academic Press, New York–London.

Edwards, R. E. 1967. *Fourier Series: a Modern Introduction*. Vol. I. & II. Holt, Rinehart and Winston, Inc., New York.

Ehrenpreis, L. 1954. Solution of some problems of division. I. *Amer. J. Math.*, **76**, 883–903.

Elekes, M., and Keleti, T. 2006. Is Lebesgue measure the only σ-finite invariant Borel measure? *J. Math. Anal. Appl.*, **321**, 445–451.

Evans, L. C. 1990. *Weak Convergence Methods for Nonlinear Partial Differential Equations*. CBMS Series, Amer. Math. Soc., Providence, RI.

Evans, L. C., and Gariepy, R. F. 1992. *Measure Theory and Fine Properties of Functions*. CRC Press, Boca Raton, FL.

Federer, H. 1969. *Geometric Measure Theory*. Springer, Heidelberg.

Fefferman, C. 1971a. On the convergence of Fourier series. *Bull. Amer. Math. Soc.*, **77**, 744–745.

Fefferman, C. 1971b. On the divergence of Fourier series. *Bull. Amer. Math. Soc.*, **77**, 191–195.

Fejér, L. 1904. Untersuchungen über Fouriersche Reihen. *Math. Ann.*, **58**, 51–69.

Feldman, M. B. 1981. A proof of Lusin's theorem. *Amer. Math. Monthly*, **88**, 191–192.

Folland, G. B. 1999. *Real Analysis: Modern Techniques and their Applications*. Wiley-Interscience, New York.

Friedlander, F. G. 1998. *Introduction to the Theory of Distributions*. Cambridge University Press, Cambridge.

Friedman, A. 1982. *Foundations of Modern Analysis*. Dover Publications, Inc., New York.

Gamelin, T. W. 2001. *Complex Analysis*. Springer, New York.

Gårding, L. 1997. *Some Points of Analysis and their History*. American Mathematical Society, Providence, RI.

Gelbaum, B. R., and Olmsted, J. M. H. 2003. *Counterexamples in Analysis*. Dover Publications, Inc., Mineola, NY.

Gibbs, J. W. 1898. Letter to the Editor. *Nature*, **59**, 606.

Giblin, P. J. 1981. *Graphs, Surfaces and Homology. An Introduction to Algebraic Topology*. Chapman and Hall, London and New York.

Goffman, C. 1977. A bounded derivative which is not Riemann integrable. *Amer. Math. Monthly*, **84**, 205–206.

Gohberg, I., Goldberg, S., and Kaashoek, M. A. 2003. *Basic Classes of Linear Operators*. Birkhäuser Verlag, Basel.

Grafakos, L. 2008. *Classical Fourier Analysis*. Springer, New York.

Gray, J. D., and Morris, S. A. 1978. When is a function that satisfies the Cauchy–Riemann equations analytic? *Amer. Math. Monthly*, **85**, 246–256.

Gröchenig, K. 1996. An uncertainty principle related to the Poisson summation formula. *Studia Math.*, **121**, 87–104.

Guillemin, V., and Pollack, A. 1974. *Differential Topology*. Prentice Hall, Inc., Englewood Cliffs, NJ.

Hamadouche, D. 2000. Invariance principles in Hölder spaces. *Portugaliae Math.*, **57**, 127–151.

Hamilton, R. S. 1982. The inverse function theorem of Nash and Moser. *Bull. Amer. Math. Soc.*, **7**, 65–222.

Haroske, D. D., and Triebel, H. 2008. *Distributions, Sobolev Spaces, Elliptic Equations*. Europ. Math. Soc., Züric.

Henstock, R. 1991. *The General Theory of Integration*. Oxford University Press, Oxford.

Hewitt, E., and Hewitt, R. E. 1979. The Gibbs-Wilbraham phenomenon. An episode in Fourier Analysis. *Arch. Hist. Exact Sci.*, **21**, 129–160.

Hewitt, E., and Stromberg, K. 1965. *Real and Abstract Analysis*. Springer-Verlag, New York.

Hogan, J. A., and Lakey, J. D. 2004. *Time-frequency and Time-scale Methods*. Springer, Berlin.

Hörmander, L. 1983. *The Analysis of Linear Partial Differential Operators*. Vol. I. Springer-Verlag, Berlin.

Hörmander, L. 1985. *The Analysis of Linear Partial Differential Operators. III. Pseudodifferential Operators.* Springer Verlag, Berlin.

Hörmander, L. 1995. *Lectures on Harmonic Analysis.* Lund University, Lund, Sweden.

Hunt, R. A. 1968. On the convergence of Fourier series. Pages 235–255 of: Haimo, D. T. (ed), *Orthogonal Expansions and their Continuous Analogues.* Southern Illinois University Press, Carbondale.

Hunt, R. A. 1976. Developments related to the a.e. convergence of Fourier series. Pages 20–37 of: Ash, J. M. (ed), *Studies in Harmonic Analysis.* Mathematical Association of America, Washington, DC.

Ingelstam, L. 1963. Hilbert algebras with identity. *Bull. Amer. Math. Soc.*, **69**, 794–796.

Iorio, R. J., and Iorio, V. M. 2001. *Fourier Analysis and Partial Differential Equations.* Cambridge University Press, Cambridge.

James, R. C. 1951. A non-reflexive Banach space isometric with its second conjugate space. *Proc. Nat. Acad. Sci. USA*, **37**, 174–177.

James, R. C. 1982. Bases in Banach spaces. *Amer. Math. Monthly*, **89**, 625–640.

Jordan, C. 1881. Sur la série de Fourier. *C. R. Acad. Sci. Paris*, **92**, 228–230.

Kahane, J.-P. 2000. Baire's category theorem and trigonometric series. *J. d'Analyse Math.*, **80**, 143–181.

Kahane, J. P., and Katznelson, Y. 1966. Sur les ensembles de divergence des séries trigonométriques. *Studia Math.*, **26**, 305–306.

Katznelson, Y. 1968. *An Introduction to Harmonic Analysis.* J. Wiley & Sons, Inc., New York.

Kolmogorov, A. 1926. Une série de Fourier–Lebesgue divergente partout. *C. R. Acad. Sci. Paris*, **183**, 1327–1328.

Koosis, P. 1998. *Introduction to H_p Spaces.* Cambridge University Press, Cambridge.

Köthe, G. 1969. *Topological Vector Spaces.* Vol. I. Springer-Verlag, Berlin.

Köthe, G. 1979. *Topological Vector Spaces.* Vol. II. Springer-Verlag, Berlin.

Krantz, S. G. 1999. *A Panorama of Harmonic Analysis.* Mathematical Association of America, Washington, DC.

Kupka, J. 1986. Measure theory: the heart of the matter. *Math. Intell.*, **8**, 47–56.

Lax, P. D. 2002. *Functional Analysis.* Wiley-Interscience, New York.

Leoni, G. 2009. *A First Course in Sobolev Spaces.* American Mathematical Society, Providence, RI.

Lévy-Leblond, J.-M. 1997. If Fourier had known Argand ... A geometrical point of view on Fourier transforms. *Math. Intell.*, **19**, 63–71.

Lieb, E. H. 1990. Gaussian kernels have only Gaussian maximizers. *Invent. Math.*, **102**, 179–208.

Lindenstrauss, J., and Tzafriri, L. 1977. *Classical Banach Spaces.* Springer, Berlin-Heidelberg-New York.

Loomis, L. H. 1953. *An Introduction to Abstract Harmonic Analysis.* D. Van Nostrand Company, Inc., Toronto-New York-London.

Lunardi, A. 2013. *Analytic Semigroups and Optimal Regularity in Parabolic Problems.* Birkhäuser Verlag, Basel.

Malgrange, B. 1955. Existence et approximation des solutions des équations aux dérivées partielles et des équations de convolution. *Ann. Inst. Fourier*, **6**, 271–355.

Mazur, S., and Ulam, S. 1932. Sur les transformations isométriques d'espaces vectoriels normés. *C. R. Acad. Sci. Paris*, **194**, 946–948.

Michelson, A. A. 1898. Letter to the Editor. *Nature*, **59**, 544–545.

Miller, T. L., and Olin, R. F. 1984. Analytic curves. *Amer. Math. Monthly*, **91**, 127–130.

Mukherjea, A., and Pothoven, K. 1986. *Real and Functional Analysis*. Plenum Press, New York & London.

Munkres, J. R. 2000. *Topology*. Prentice Hall Inc., NJ.

Natanson, I. P. 1955. *Theory of Functions of a Real Variable*. F. Ungar Publishing Co., New York.

Needham, T. 2000. *Visual Complex Analysis*. Oxford University Press, Oxford.

Niven, I. 1947. A simple proof that π is irrational. *Bull. Amer. Math. Soc.*, **53**, 509.

Peano, G. 1890. Sur une courbe, qui remplit toute une aire plane. *Math. Ann.*, **36**, 157–160.

Pinsky, M. A. 1993. Fourier inversion for piecewise smooth functions in several variables. *Proc. Amer. Math. Soc.*, **118**, 903–910.

Pinsky, M. A. 2009. *Introduction to Fourier Analysis and Wavelets*. American Mathematical Society, Providence, RI.

Piranian, G., Titus, C. J., and Young, G. S. 1952. Conformal mappings and Peano curves. *Michigan Math. J.*, **1**, 69–72.

Pommerenke, C. 2002. Conformal maps at the boundary. Pages 39–74 of: Kuhnau, R. (ed), *Handbook of Complex Analysis: Geometric Function Theory*. North-Holland, Amsterdam.

Reed, M., and Simon, B. 1980a. *Methods of Mathematical Physics. I: Functional Analysis*. Academic Press, Inc., San Diego, CA.

Reed, M., and Simon, B. 1980b. *Methods of Mathematical Physics. II: Fourier Analysis, Self-adjointness*. Academic Press, Inc., San Diego, CA.

Riesz, F., and Sz.-Nagy, B. 1955. *Functional Analysis*. F. Ungar Publ. Co., New York.

Rouse, J. 2012. Explicit bounds for sums of squares. *Math. Res. Lett.*, **19**, 359–376.

Rudin, W. 1974. *Real and Complex Analysis*. McGraw-Hill Book Co., New York.

Salem, R., and Zygmund, A. 1945. Lacunary power series and Peano curves. *Duke Math. J.*, **12**, 569–578.

Schlag, W. 2014. *A Course in Complex Analysis and Riemann Surfaces*. American Mathematical Society, Providence RI.

Segal, S. L. 2008. *Nine Introductions in Complex Analysis*. Elsevier, Amsterdam.

Siegmund-Schultze, R. 2008. Henri Lebesgue. Page 796 of: Gowers, T., Barrow-Green, J., and Leader, I. (eds), *Princeton Companion to Mathematics*. Princeton University Press.

Srivastava, S. M. 1998. *A Course on Borel Sets*. Springer-Verlag, New York.

Stein, E. M. 1976. Harmonic analysis on $\mathbb{R}^n$. Pages 97–135 of: Ash, J. M. (ed), *Studies in Harmonic Analysis*. Mathematical Association of America, Washington, DC.

Stein, E. M., and Shakarchi, R. 2003. *Fourier Analysis. An Introduction*. Princeton University Press, Princeton, NJ.

Stein, E. M., and Weiss, G. 1971. *Introduction to Fourier Analysis on Euclidean Spaces*. Princeton University Press, Princeton, NJ.

Strauss, W. A. 2008. *Partial Differential Equations. An Introduction*. John Wiley & Sons, Ltd., Chichester.

Suslin, M. Y. 1917. Sur une définition des ensembles measurables B sans nombres transfinis. *C. R. Acad. Sci. Paris*, **164**, 88–91.

Taylor, M. 1981. *Pseudo-differential Operators*. Princeton University Press, Princeton, NJ.

Urysohn, P. 1923. Sur une fonction analytique partout continue. *Fund. Math.*, **4**, 144–150.

van Douwen, E. K. 1989. Fubini's theorem for null sets. *Amer. Math. Monthly*, **96**, 718–721.

van Neerven, J. M. A. M. 1997. The norm of a complex Banach lattice. *Positivity*, **1**, 381–390.

Wagon, S. 1985. *The Banach-Tarski Paradox*. Cambridge University Press, Cambridge.

Weiner, J. L., and Wilkens, G. R. 2005. Quaternions and rotations in $\mathbb{E}^4$. *Amer. Math. Monthly*, **112**, 69–76.

Wheeler, G. F., and Crummett, W. P. 1987. The vibrating string controversy. *Amer. J. Phys.*, **55**, 33–37.

Wilbraham, H. 1848. On a certain periodic function. *Cambridge & Dublin Math. J.*, **3**, 198–201.

Yosida, K. 1995. *Functional Analysis*. Springer-Verlag, Berlin.

Zalcman, L. 1974. Real proofs of complex theorems (and vice versa). *Amer. Math. Monthly*, **81**, 115–137.

Zhu, K. 2007. *Operator Theory in Function Spaces*. American Mathematical Society, Providence, RI.

Zygmund, A. 1959. *Trigonometrical Series*. Cambridge University Press, Cambridge.

Index